Solid Mechanics and Its Applications

Volume 197

Series Editor

G. M. L. Gladwell
Department of Civil Engineering, University of Waterloo, Waterloo, Canada

For further volumes:
http://www.springer.com/series/6557

Aims and Scope of the Series

The fundamental questions arising in mechanics are: *Why*?, *How*?, and *How much*? The aim of this series is to provide lucid accounts written by authoritative researchers giving vision and insight in answering these questions on the subject of mechanics as it relates to solids.

The scope of the series covers the entire spectrum of solid mechanics. Thus it includes the foundation of mechanics; variational formulations; computational mechanics; statics, kinematics and dynamics of rigid and elastic bodies: vibrations of solids and structures; dynamical systems and chaos; the theories of elasticity, plasticity and viscoelasticity; composite materials; rods, beams, shells and membranes; structural control and stability; soils, rocks and geomechanics; fracture; tribology; experimental mechanics; biomechanics and machine design.

The median level of presentation is the first year graduate student. Some texts are monographs defining the current state of the field; others are accessible to final year undergraduates; but essentially the emphasis is on readability and clarity.

M. Reza Eslami · Richard B. Hetnarski
Józef Ignaczak · Naotake Noda
Naobumi Sumi · Yoshinobu Tanigawa

Theory of Elasticity and Thermal Stresses

Explanations, Problems and Solutions

Springer

M. Reza Eslami
Department of Mechanical Engineering
Amirkabir University of Technology
Tehran
Iran

Richard B. Hetnarski
Department of Mechanical Engineering
Rochester Institute of Technology
Rochester, NY
USA

Józef Ignaczak
Institute of Fundamental Technological Research
Polish Academy of Sciences
Warsaw
Poland

Naotake Noda
Department of Mechanical Engineering
Shizuoka University
Hamamatsu
Japan

Naobumi Sumi
Faculty of Education
Shizuoka University
Suruga-ku Shizuoka
Japan

Yoshinobu Tanigawa
Department of Mechanical Engineering
Osaka Prefecture University
Sakai
Japan

ISSN 0925-0042
ISBN 978-94-007-9931-8 ISBN 978-94-007-6356-2 (eBook)
DOI 10.1007/978-94-007-6356-2
Springer Dordrecht Heidelberg New York London

Printed on acid-free paper

Springer is part of Springer Science+Business Media (www.springer.com)

The authors dedicate this book
to their wives:

Minoo Eslami

Leokadia Elizabeth Hetnarski

Krystyna Ignaczak

Tosiko Noda

Eiko Sumi

Seiko Tanigawa

They made painful sacrifices while the authors toiled for long years at the preparation of these four books, and of other publications not mentioned here.

They deserve words of highest appreciation and admiration – and love

Preface

Fecimus, quod potuimus,
faciant meliora potentes.

All the authors of this book have done research in and taught numerous courses on the Theory of Elasticity, Thermoelasticity and Thermal Stresses. Coincident with that teaching, we have jointly created three textbooks in these critical fields of Mechanics:

1. Richard B. Hetnarski and Józef Ignaczak, *The Mathematical Theory of Elasticity*.
2. Naotake Noda, Richard B. Hetnarski and Yoshinobu Tanigawa, *Thermal Stresses*.
3. Richard B. Hetnarski and M. Reza Eslami, *Thermal Stresses—Advanced Theory and Applications*.

These publications are a result of our dedication to teaching engineering students on these subjects of Mechanics. Publication details of our three textbooks will be found at the end of this *Preface*.

The new book that we now present here is the crowning achievement of our activities in these fields. It comprises the problems contained in the three listed books, together with detailed solutions and explanations. Thus, Part I is related to the book *The Mathematical Theory of Elasticity*, Part II covers the problems in the book *Thermal Stresses*, and Part III covers problems in the book *Thermal Stresses—Advanced Theory and Applications*.

The three parts are augmented by Part IV, *Numerical Methods*, that covers three important topics: the Method of Characteristics, the Finite Element Method for Coupled Thermoelasticity, and the Boundary Element Method for Coupled Thermoelasticity. A full chapter in Part IV is devoted to the Method of Characteristics. The need for numerical methods in the solution of dynamic problems is dictated by the well-known difficulty of obtaining exact solutions. The Method of Characteristics serves to reduce the hyperbolic partial differential equations of dynamic problems to a family of ordinary differential equations, each of which is valid along a different family of characteristic lines. These equations are more suitable for numerical analysis because their use makes it possible to obtain the solutions through a step-by-step integration procedure. The method has the advantage of giving a simple description of the wave fronts, and can readily find numerical solutions to problems with any type of input functions.

Part IV contains a chapter that treats the Finite Element Method for Coupled Thermoelasticity. The method of finite elements described in that chapter is based on the Galerkin method and presents classical formulation for problems of coupled thermoelasticity. The formulation may be modified to be applicable to the uncoupled thermoelasticity problems simply by removing the coupling term from the energy equation. The method is also made applicable to problems of generalized thermoelasticity, by taking into account the terms containing the relaxation times associated with the Lord-Shulman, the Green-Lindsay, or the Green-Naghdi models.

Part IV also dedicates a chapter to a description of the Boundary Element Method for Coupled Thermoelasticity. The formulation of the Laplace transform boundary element method is based on the generalized thermoelasticity theory of the Lord-Shulman model. The unique feature of this formulation is that a single heat excitation principle solution is used to derive the boundary element formulation.

We consider this new book to be an indispensable companion to all who study any of the initial three books. In it, we present not only the problems contained in these books, together with their careful and often extensive solutions, but also explanations in the form of introductions that appear at the beginning of chapters in Parts I, II and III. Therefore, this book links the three listed books into one consistent entity of four publications.

Note that in Part I, the chapter numbers correspond to chapters in the book *The Mathematical Theory of Elasticity*, except that they are shifted by one, i.e., Chap. 1 in this book corresponds to Chap. 2 in *MTE*, Chap. 2 in this book corresponds to Chap. 3 in *MTE*, etc.

Note also that the notations in Parts I, II, and III are respectively the same as in the three listed books; since not all notations are the same in each of the three books, some notations in different parts of this book differ from each other.

The quality and style of figures differed in the three initial books, thus they differ in the corresponding parts of the new book. Of necessity, we note an overlapping of the material covered in various parts of the new book. Such occurrences are marked by cross-references at the beginning of some chapters.

We took the opportunity to list all discovered errors that exist in the second editions of the books *The Mathematical Theory of Elasticity* and *Thermal Stresses*, and in the book *Thermal Stresses—Advanced Theory and Applications*.

References to the literature are placed in footnotes. At the end of the book, we provide a brief list of important books on the theory and applications, and also the books that are devoted to solving of problems. More extensive lists of references to the literature appear in our three original books.

We express our thanks to Jonathan W. Plant, Executive Editor for Mechanical, Aerospace & Nuclear Engineering at Taylor & Francis/CRC Press, who granted us permission to use the text of the problems and associated figures contained in *The Mathematical Theory of Elasticity*, second edition, as well as in *Thermal Stresses*, second edition. Without such permission, this book could not have been published.

A positive attitude toward publication of the book by Professor G. M. L. Gladwell, the Series Editor, is highly appreciated.

We are indebted to Nathalie Jacobs, Senior Publishing Editor/Engineering at Springer, for undertaking the publication of the book and for her assistance in the execution of this project.

The authors' names are placed on the title page and below in alphabetical order.

April 2013

M. Reza Eslami
Richard B. Hetnarski
Józef Ignaczak
Naotake Noda
Naobumi Sumi
Yoshinobu Tanigawa

Publication information on the three books referred to in *Preface*:

1. Richard B. Hetnarski and Józef Ignaczak, *The Mathematical Theory of Elasticity*. 2nd ed., CRC Press, Boca Raton, 2011.
2. Naotake Noda, Richard B. Hetnarski and Yoshinobu Tanigawa, *Thermal Stresses* 2nd ed., Taylor & Francis, New York, 2003.
3. Richard B. Hetnarski and M. Reza Eslami, *Thermal Stresses—Advanced Theory and Applications*, Springer, 2009.

Contents

Part I
The Mathematical Theory of Elasticity

Chapter 1
Mathematical Preliminaries

In this chapter the basic definitions of vector and tensor algebra, elements of tensor differential and integral calculus, and concept of a convolutional product for two time-dependent tensor fields are recalled. These concepts are then used to solve particular problems related to the Mathematical Preliminaries.

1.1 Some Formulas in Tensor Algebra

A *vector* will be understood as an element of a *vector space* V. The inner product of two vectors $\mathbf{u}$ and $\mathbf{v}$ from V will be denoted by $\mathbf{u} \cdot \mathbf{v}$. If *Cartesian coordinates* are introduced in such a way that the set of vectors $\{\mathbf{e}_i\} = \{\mathbf{e}_1, \mathbf{e}_2, \mathbf{e}_3\}$ with an origin $\mathbf{0}$ stands for an *orthonormal basis*, and if $\mathbf{u}$ is a vector and $\mathbf{x}$ is a point of E^3, then Cartesian coordinates of $\mathbf{u}$ and $\mathbf{x}$ are given by

$$u_i = \mathbf{u} \cdot \mathbf{e}_i, \quad x_i = \mathbf{x} \cdot \mathbf{e}_i \tag{1.1}$$

Apart from the *direct (vector or tensor) notation* we use *indicial notation* in which subscripts range from 1 to 3 and summation convention over repeated subscripts is observed. For example,

$$\mathbf{u} \cdot \mathbf{v} = \sum_{i=1}^{3} u_i v_i = u_i v_i \tag{1.2}$$

From the definition of an orthonormal basis $\{\mathbf{e}_i\}$ it follows that

$$\mathbf{e}_i \cdot \mathbf{e}_j = \delta_{ij} \quad (\mathrm{i, j} = 1, 2, 3) \tag{1.3}$$

where δ_{ij} is called the *Kronecker symbol* defined by

M. Reza Eslami et al., *Theory of Elasticity and Thermal Stresses*, Solid Mechanics and Its Applications 197, DOI: 10.1007/978-94-007-6356-2_1,

$$\delta_{ij} = \begin{cases} 1 & \text{if} \quad \mathrm{i} = \mathrm{j} \\ 0 & \text{if} \quad \mathrm{i} \neq \mathrm{j} \end{cases} \tag{1.4}$$

We introduce the *permutation symbol* ε_{ijk}, also called the *alternating symbol*, defined by

$$\varepsilon_{ijk} = \begin{cases} 1 & \text{if (i, j, k) is an even permutation of}(1,\ 2,\ 3) \\ -1 & \text{if (i, j, k) is an odd permutation of}(1,\ 2,\ 3) \\ 0 & \text{otherwise, that is, if two subscripts are repeated} \end{cases} \tag{1.5}$$

The permutation symbol will be used for the definition of the *vector product* $\mathbf{u} \times \mathbf{v}$ of two vectors $\mathbf{u}$ and $\mathbf{v}$

$$(\mathbf{u} \times \mathbf{v})_i = \varepsilon_{ijk} u_j v_k \tag{1.6}$$

We may observe that the following identity holds true

$$\varepsilon_{mis}\varepsilon_{jks} = \delta_{mj}\delta_{ik} - \delta_{mk}\delta_{ij} \tag{1.7}$$

An alternative definition of the permutation symbol, given in terms of the vectors $\mathbf{e}_i$, is

$$\varepsilon_{ijk} = \mathbf{e}_i \cdot (\mathbf{e}_j \times \mathbf{e}_k) \tag{1.8}$$

Using this definition of ε_{ijk}, a generalized form of Eq. (1.7) is obtained

$$\varepsilon_{ijk}\varepsilon_{pqr} = \begin{vmatrix} \delta_{ip} & \delta_{iq} & \delta_{ir} \\ \delta_{jp} & \delta_{jq} & \delta_{jr} \\ \delta_{kp} & \delta_{kq} & \delta_{kr} \end{vmatrix} \tag{1.9}$$

Letting $\mathrm{k} = \mathrm{r}$ in this identity we obtain Eq. (1.7).

The permutation symbol ε_{ijk} can be also used to calculate a 3×3 determinant

$$\varepsilon_{ijk} a_i b_j c_k = \begin{vmatrix} a_1 & a_2 & a_3 \\ b_1 & b_2 & b_3 \\ c_1 & c_2 & c_3 \end{vmatrix} \tag{1.10}$$

A second-order tensor is defined as a linear transformation from V to V, that is, a tensor $\mathbf{T}$ is a linear mapping that associates with each vector $\mathbf{v}$ a vector $\mathbf{u}$ by

$$\mathbf{u} = \mathbf{T}\mathbf{v} \tag{1.11}$$

The components of $\mathbf{T}$ are denoted by T_{ij}

$$T_{ij} = \mathbf{e}_i \cdot \mathbf{T}\mathbf{e}_j \tag{1.12}$$

so the relation (1.11) in index notation takes the form

$$u_i = T_{ij} v_j \tag{1.13}$$

The Kronecker symbol δ_{ij} represents an *identity tensor* that in direct notation is written as $\mathbf{1}$.

A *product of two tensors* $\mathbf{A}$ and $\mathbf{B}$ is defined by

$$(\mathbf{AB})\mathbf{v} = \mathbf{A}(\mathbf{Bv}) \quad \text{for every vector } \mathbf{v} \tag{1.14}$$

Thus, in Cartesian coordinates

$$(\mathbf{AB})_{ij} = A_{ik} B_{kj} \tag{1.15}$$

The *transpose* of $\mathbf{T}$, denoted by $\mathbf{T}^{\mathrm{T}}$, is defined as a unique tensor satisfying the property

$$\mathbf{Tu} \cdot \mathbf{v} = \mathbf{u} \cdot \mathbf{T}^{\mathrm{T}}\mathbf{v} \quad \text{for every } \mathbf{u} \text{ and } \mathbf{v} \tag{1.16}$$

From this definition it follows that

$$T_{ik} = T^{T}_{ki} \tag{1.17}$$

If $\mathbf{T} = \mathbf{T}^{\mathrm{T}}$, then the tensor $\mathbf{T}$ is *symmetric*. Also, if $\mathbf{T} = -\mathbf{T}^{\mathrm{T}}$ then the tensor $\mathbf{T}$ is *skew* or *asymmetric*. Therefore, $\mathbf{T}$ is symmetric if $T_{ij} = T_{ji}$, and skew if $T_{ij} = -T_{ji}$. Every tensor $\mathbf{T}$ can be expressed by a sum of a symmetric tensor sym $\mathbf{T}$ and skew tensor skw $\mathbf{T}$, that is,

$$\mathbf{T} = \text{sym } \mathbf{T} + \text{skw } \mathbf{T} \tag{1.18}$$

where

$$\text{sym } \mathbf{T} = \frac{1}{2}(\mathbf{T} + \mathbf{T}^{\mathrm{T}}) \tag{1.19}$$

and

$$\text{skw } \mathbf{T} = \frac{1}{2}(\mathbf{T} - \mathbf{T}^{\mathrm{T}}) \tag{1.20}$$

In index notation

$$T_{ij} = T_{(ij)} + T_{[ij]} \tag{1.21}$$

where

$$T_{(ij)} = \frac{1}{2}(T_{ij} + T_{ji}) \tag{1.22}$$

and

$$T_{[ij]} = \frac{1}{2}(T_{ij} - T_{ji}) \tag{1.23}$$

If a tensor **P** is skew, then there exists a vector ω, called the *axial vector* corresponding to **P**, such that

$$\mathbf{Pu} = \omega \times \mathbf{u} \quad \text{for every vector } \mathbf{u} \tag{1.24}$$

and it follows from (1.24) that

$$P_{ij} = -\varepsilon_{ijk}\omega_k \tag{1.25}$$

and

$$\omega_i = -\varepsilon_{ijk}P_{jk} \tag{1.26}$$

For any tensor **T** the *trace* of **T** is denoted by tr (**T**). In index notation

$$\text{tr } (\mathbf{T}) = T_{ii} \tag{1.27}$$

The determinant of **T** is denoted by det (**T**) and it is

$$\det (\mathbf{T}) = \begin{vmatrix} T_{11} & T_{12} & T_{13} \\ T_{21} & T_{22} & T_{23} \\ T_{31} & T_{32} & T_{33} \end{vmatrix} \tag{1.28}$$

1.2 Alternative Definitions of a Vector and of a Tensor Using an Orthogonal Tensor

We say that **Q** is orthogonal if and only if

$$\mathbf{Q}^{\mathrm{T}}\mathbf{Q} = \mathbf{Q}\mathbf{Q}^{\mathrm{T}} = \mathbf{1} \tag{1.29}$$

For an orthonormal basis $\mathbf{e}_i$ and orthogonal **Q**, the vectors

$$\mathbf{e}_i' = \mathbf{Q}\mathbf{e}_i \tag{1.30}$$

form an orthonormal basis.

Also, for two orthonormal bases $\mathbf{e}_i$ and $\mathbf{e}_i'$, there exists a unique orthogonal tensor **Q** such that (1.30) holds true.

For a vector **w** with components in $\mathbf{e}_i$ denoted by w_i, and with components in $\mathbf{e}_i'$ denoted by w_i', we have

$$w_i' = Q_{ji}w_j \tag{1.31}$$

$$w_j = Q_{ji}w_i' \tag{1.32}$$

Similarly, for a tensor **T** we get

$$T'_{ij} = Q_{ki} Q_{lj} T_{kl} \tag{1.33}$$

$$T_{kl} = Q_{ki} Q_{lj} T'_{ij} \tag{1.34}$$

where

$$Q_{ji} = \mathbf{e}'_i \cdot \mathbf{e}_j \tag{1.35}$$

The set of three quantities w_j, or nine quantities T_{kl}, referred to $\mathbf{e}_i$, which transform to another set w'_i or T'_{ij}, referred to $\mathbf{e}'_i$, according to (1.31)–(1.32), or (1.33)–(1.34), is defined as a vector, or a tensor, respectively.

1.3 Further Definitions

The *tensor product of two vectors* **a** and **b** denoted by $\mathbf{a} \otimes \mathbf{b}$ is defined by

$$(\mathbf{a} \otimes \mathbf{b})\mathbf{u} = (\mathbf{b} \cdot \mathbf{u})\mathbf{a} \quad \text{for any vector } \mathbf{u} \tag{1.36}$$

In components

$$(\mathbf{a} \otimes \mathbf{b})_{ij} = a_i b_j \tag{1.37}$$

Clearly

$$\text{tr}\,(\mathbf{a} \otimes \mathbf{b}) = \mathbf{a} \cdot \mathbf{b} \tag{1.38}$$

Similarly, the *inner product of two tensors* **A** and **B** is defined by

$$\mathbf{A} \cdot \mathbf{B} = \text{tr}\,(\mathbf{A}^T \mathbf{B}) = A_{ij} B_{ij} \tag{1.39}$$

The *magnitude of* **A** is defined by

$$|\mathbf{A}| = (\mathbf{A} \cdot \mathbf{A})^{1/2} \tag{1.40}$$

Also, for any tensor **T** the following relation holds

$$\mathbf{T} = T_{ij} \mathbf{e}_i \otimes \mathbf{e}_j \tag{1.41}$$

and the nine tensors $\mathbf{e}_i \otimes \mathbf{e}_j$ are orthonormal in the sense

$$(\mathbf{e}_i \otimes \mathbf{e}_j) \cdot (\mathbf{e}_k \otimes \mathbf{e}_l) = \delta_{ki} \delta_{jl} \tag{1.42}$$

Equation (1.41) constitutes the decomposition formula for **T** in terms of the nine orthonormal tensors $\mathbf{e}_i \otimes \mathbf{e}_j$.

Also note that if $\mathbf{A}$ is a symmetric tensor and $\mathbf{B}$ is a skew tensor then [see MTE-2e, Example 2.1.5]

$$\mathbf{A} \cdot \mathbf{B} = 0 \tag{1.43}$$

and for an orthonormal basis $\mathbf{e}_i$ and an orthogonal tensor $\mathbf{Q}$ the vectors $\mathbf{e}'_i = \mathbf{Q}\mathbf{e}_i$ form an orthonormal basis [see MTE-2e, Example 2.1.6], and, for two orthonormal bases $\mathbf{e}_i$ and $\mathbf{e}'_i$, there exists a unique orthogonal tensor $\mathbf{Q}$ such that $\mathbf{e}'_i = \mathbf{Q}\mathbf{e}_i$.

1.4 Eigenproblem for a Second Order Tensor

We call λ an eigenvalue corresponding to an eigenvector $\mathbf{u}$ of a tensor $\mathbf{T}$ if

$$\mathbf{T}\mathbf{u} = \lambda \mathbf{u} \tag{1.44}$$

For a symmetric tensor $\mathbf{T}$ there exists an orthonormal basis $\{\mathbf{n}_i\}$ defined by three eigenvectors of $\mathbf{T}$ corresponding to three eigenvalues λ_i of $\mathbf{T}$ such that

$$\mathbf{T} = \sum_{i=1}^{3} \lambda_i \, \mathbf{n}_i \otimes \mathbf{n}_i \tag{1.45}$$

Here

$$\mathbf{T}\mathbf{n}_i = \lambda_i \, \mathbf{n}_i \quad \text{(no sum on i)} \tag{1.46}$$

The inverse of $\mathbf{T}$, denoted by $\mathbf{T}^{-1}$, is defined by

$$\mathbf{T}\mathbf{T}^{-1} = \mathbf{T}^{-1}\mathbf{T} = \mathbf{1} \tag{1.47}$$

The tensor $\mathbf{T}^{-1}$ is closely related to that of an orthogonal tensor. A tensor $\mathbf{A}$ is said to be an orthogonal if $\mathbf{A}$ is invertible, that is if $\mathbf{A}^{-1}$ exists and $\mathbf{A}^{-1} = \mathbf{A}^{\mathrm{T}}$. Thus, $\mathbf{A}$ is an orthogonal tensor if and only if

$$\mathbf{A}^{\mathrm{T}}\mathbf{A} = \mathbf{A}\mathbf{A}^{\mathrm{T}} = \mathbf{1} \tag{1.48}$$

For any invertible tensors $\mathbf{A}$ and $\mathbf{B}$

$$(\mathbf{A}\mathbf{B})^{-1} = \mathbf{B}^{-1}\mathbf{A}^{-1} \tag{1.49}$$

$$(\mathbf{A}^{-1})^{-1} = \mathbf{A} \tag{1.50}$$

$$(\mathbf{A}^{\mathrm{T}})^{-1} = (\mathbf{A}^{-1})^{\mathrm{T}} \tag{1.51}$$

Also, a tensor $\mathbf{A}$ is invertible if and only if its matrix $[\mathbf{A}]$ is invertible with $[\mathbf{A}]^{-1} = [\mathbf{A}^{-1}]$.

1.5 A Fourth-Order Tensor $\mathbf{C}$

A fourth-order tensor $\mathbf{C}$ is defined as a linear transformation that assigns to a second-order tensor $\mathbf{U}$ another second-order tensor $\mathbf{V}$

$$\mathbf{V} = \mathbf{C}\,[\mathbf{U}] \tag{1.52}$$

or, in components,

$$V_{ij} = C_{ijkl}\, U_{kl} \tag{1.53}$$

The components C_{ijkl} are defined in terms of the basis $\{\mathbf{e}_i\}$ by

$$C_{ijkl} = (\mathbf{e}_i \otimes \mathbf{e}_j) \cdot \mathbf{C}\,[(\mathbf{e}_k \otimes \mathbf{e}_l)] \tag{1.54}$$

Let $\{\mathbf{e}_i'\}$ be another orthonormal basis, such that

$$\mathbf{e}_i' = \mathbf{Q}\mathbf{e}_i \tag{1.55}$$

and let C'_{ijkl} be components of $\mathbf{C}$ with respect to $\{\mathbf{e}_i'\}$. A fourth-order tensor $\mathbf{C}$ may be also defined as the set of 81 quantities C_{mnpq} that transform to C'_{ijkl} according to the formula

$$C'_{ijkl} = Q_{mi}\, Q_{nj}\, Q_{pk}\, Q_{ql}\, C_{mnpq} \tag{1.56}$$

The *transpose* $\mathbf{C}^{\mathrm{T}}$ of $\mathbf{C}$ is defined as a unique fourth-order tensor that satisfies the relation

$$\mathbf{A} \cdot \mathbf{C}\,[\mathbf{B}] = \mathbf{C}^{\mathrm{T}}\,[\mathbf{A}] \cdot \mathbf{B} \text{ for all second-order tensors } \mathbf{A} \text{ and } \mathbf{B} \tag{1.57}$$

In components

$$C^T_{ijkl} = C_{klij} \tag{1.58}$$

Also,

$$|\mathbf{C}| = \sup_{|\mathbf{A}|=1} \{|\mathbf{C}\,[\mathbf{A}]\,|\} \tag{1.59}$$

is defined as the magnitude $|\mathbf{C}|$ of $\mathbf{C}$. Clearly,

$$|\mathbf{C}\,[\mathbf{A}]| \;\leq |\mathbf{C}|\;\; |\mathbf{A}| \text{ for every } \mathbf{A} \tag{1.60}$$

1.6 Tensor Fields

For a scalar function $\mathrm{f} = \mathrm{f}(\mathbf{x}),\ \mathbf{x} \in \mathrm{R} \subset E^3$, the *gradient* of f at $\mathbf{x}$ is defined by

$$\mathbf{u} = \nabla \mathrm{f} \tag{1.61}$$

or, in components,

$$u_i = \mathrm{f}_{,i} = \mathbf{e}_i \cdot \nabla \mathrm{f} \tag{1.62}$$

The *del operator* ∇ is defined by

$$\nabla = \mathbf{e}_k \frac{\partial}{\partial x_k} \tag{1.63}$$

For a vector function $\mathbf{v} = \mathbf{v}(\mathbf{x})$, the *gradient* of $\mathbf{v}$ is defined by

$$\mathbf{V} = \nabla \mathbf{v} \tag{1.64}$$

or, in components,

$$V_{ij} = (\nabla \mathbf{v})_{ij} = v_{i,j} \tag{1.65}$$

The *divergence* of $\mathbf{v}$, div $\mathbf{v}$, and the *curl* of $\mathbf{v}$, curl $\mathbf{v}$, are defined, respectively, by

$$\operatorname{div} \mathbf{v}(\mathbf{x}) = \operatorname{tr}(\nabla \mathbf{v}) = v_{i,i} \tag{1.66}$$

and

$$(\operatorname{curl} \mathbf{v}) \times \mathbf{a} = (\nabla \mathbf{v} - \nabla \mathbf{v}^{\mathrm{T}})\,\mathbf{a} \quad \text{for every } \mathbf{a} \tag{1.67}$$

or

$$(\operatorname{curl} \mathbf{v})_i = \varepsilon_{ijk} v_{k,j} \tag{1.68}$$

The *symmetric gradient* of $\mathbf{v}$, denoted by $\hat{\nabla}\mathbf{v}$, is defined by

$$\hat{\nabla}\mathbf{v} = \operatorname{sym}\,(\nabla \mathbf{v}) = \frac{1}{2}(\nabla \mathbf{v} + \nabla \mathbf{v}^{\mathrm{T}}) \tag{1.69}$$

Similarly, if $\mathbf{T}$ is a tensor field, the *divergence* of $\mathbf{T}$ and the *curl* of $\mathbf{T}$ are defined, respectively, in components, by

$$(\operatorname{div} \mathbf{T})_i = T_{ij,j} \tag{1.70}$$

and

$$(\operatorname{curl} \mathbf{T})_{ij} = \varepsilon_{ipq} T_{jq,p} \tag{1.71}$$

The Laplacian of a scalar field f, of a vector field $\mathbf{v}$, and of a tensor field $\mathbf{T}$, are defined, respectively, by

$$\Delta f = f_{,kk} \tag{1.72}$$

$$(\Delta \mathbf{v})_i = v_{i,kk} \tag{1.73}$$

$$(\Delta \mathbf{T})_{ij} = T_{ij,kk} \tag{1.74}$$

Also, instead of Δ we often use ∇^2

$$\Delta = \nabla^2 \tag{1.75}$$

1.7 Integral Theorems

The divergence theorem for a tensor field **T**.

Let **T** be a tensor field on $R \subset E^3$, let **n** be a unit outer normal vector to ∂R. Then

$$\int_{\partial R} \mathbf{T}\mathbf{n}\, da = \int_{R} \operatorname{div} \mathbf{T}\, dv \tag{1.76}$$

Stokes' Theorem
Let **u** and **T** denote a vector and tensor fields, respectively, on R, and let C be a closed curve in R. Then

$$\oint_{C} \mathbf{u}\cdot\mathbf{s}\, dt = \int_{S} (\operatorname{curl} \mathbf{u})\cdot\mathbf{n}\, da \tag{1.77}$$

$$\oint_{C} \mathbf{T}\mathbf{s}\, dt = \int_{S} (\operatorname{curl} \mathbf{T})^{T}\mathbf{n}\, da \tag{1.78}$$

where S is a surface contained in R and bounded by C, **n** is the unit vector normal to S, and **s** is a unit vector tangent to C.

1.8 Irrotational and Solenoidal Fields

A vector field **u** on R is said to be *irrotational* in R if

$$\operatorname{curl} \mathbf{u} = \mathbf{0} \text{ on R} \tag{1.79}$$

A vector field **u** on R is said to be *solenoidal* in R if

$$\int_{S} \mathbf{u}\cdot\mathbf{n}\, da = 0 \quad \text{for every closed regular surface S in R} \tag{1.80}$$

Theorem on irrotational fields
Let R be a simply connected region of E^3

(a) If $\mathbf{u}$ is a vector field on R and curl $\mathbf{u} = \mathbf{0}$, then there exists a scalar field f such that $\mathbf{u} = \nabla f$.
(b) If $\mathbf{T}$ is a tensor field on R and curl T $= 0$, then there exists a vector field $\mathbf{v}$ such that $\mathbf{T} = \nabla \mathbf{v}$.

Theorem on solenoidal fields

(a) If $\mathbf{u}$ is a vector field on R and $\int_S \mathbf{u} \cdot \mathbf{n}\, da = 0$ for every closed surface $S \subset R$, then there exists a vector field $\mathbf{w}$ such that $\mathbf{u} = \text{curl}\, \mathbf{w}$.
(b) If $\mathbf{T}$ is a tensor field on R and $\int_S \mathbf{T}^T \mathbf{n}\, da = \mathbf{0}$ for every closed surface $S \subset R$, then there exists a tensor field $\mathbf{W}$ such that $\mathbf{T} = \text{curl}\, \mathbf{W}$.

Helmholtz's Theorem
If $\mathbf{u}$ is a vector field on R then there exist a scalar field f and a vector field $\mathbf{v}$ such that

$$\mathbf{u} = \nabla f + \text{curl}\, \mathbf{v} \tag{1.81}$$

and

$$\text{div}\, \mathbf{v} = 0 \tag{1.82}$$

1.9 Time-Dependent Fields

Let f and g be scalar fields on R × T where R is a region of E^3, and $T = [0, \infty)$ is the time interval. The convolution $f * g$ of f and g is defined by

$$[f * g](\mathbf{x}, t) = \int_0^t f(\mathbf{x}, t-\tau) g(\mathbf{x}, \tau)\, d\tau \tag{1.83}$$

We list properties of convolution that are useful in applications.

Let f, g, and h be scalar fields on $R \times T$, continuous in time. Then

(a) $f * g = g * f$
(b) $(f * g) * h = f * (g * h) = f * g * h$
(c) $f * (g + h) = f * g + f * h$
(d) $f * g = 0 \;\; \Rightarrow \;\; f = 0$ or $g = 0$
(e) $L\{f * g\} = L\{f\}\, L\{g\}$ where L is the Laplace transform with respect to t, that is, for any function $h = h(\mathbf{x}, t)$

$$L\{h\} = \int_0^\infty e^{-pt}\, h(\mathbf{x}, t)\, dt \tag{1.84}$$

The convolution product of two scalar functions can be applied to mixed fields. Thus, if $\mathbf{u}$ is a vector field and f is a scalar field then

$$[f * \mathbf{u}](\mathbf{x}, t) = \int_0^t f(\mathbf{x}, t-\tau)\mathbf{u}(\mathbf{x}, \tau)\, d\tau \tag{1.85}$$

If $\mathbf{A}$ and $\mathbf{B}$ are time-dependent tensor fields, then

$$[\mathbf{A} * \mathbf{B}](\mathbf{x}, t) = \int_0^t \mathbf{A}(\mathbf{x}, t-\tau) \cdot \mathbf{B}(\mathbf{x}, \tau)\, d\tau \tag{1.86}$$

If $\mathbf{A}$ is a tensor field and $\mathbf{u}$ is a vector field, then

$$[\mathbf{A} * \mathbf{u}](\mathbf{x}, t) = \int_0^t \mathbf{A}(\mathbf{x}, t-\tau)\mathbf{u}(\mathbf{x}, \tau)\, d\tau \tag{1.87}$$

In components

$$[f * \mathbf{u}]_i = [f * u_i] \tag{1.88}$$
$$[\mathbf{A} * \mathbf{B}] = [A_{ij} * B_{ij}] \tag{1.89}$$
$$[\mathbf{A} * \mathbf{u}]_i = [A_{ij} * u_j] \tag{1.90}$$

1.10 Problems and Solutions Related to the Mathematical Preliminaries

Problem 1.1. Use the properties of the alternator ε_{ijk} introduced in Sect. 1.1 to show that

$$(\mathbf{a} \times \mathbf{b}) \times \mathbf{c} = (\mathbf{a} \cdot \mathbf{c})\mathbf{b} - (\mathbf{b} \cdot \mathbf{c})\mathbf{a} \tag{1.91}$$

$$(\mathbf{a} \times \mathbf{b}) \times (\mathbf{c} \times \mathbf{d}) = [\mathbf{a} \cdot (\mathbf{c} \times \mathbf{d})]\mathbf{b} - [\mathbf{b} \cdot (\mathbf{c} \times \mathbf{d})]\mathbf{a} \tag{1.92}$$

where $\mathbf{a}$, $\mathbf{b}$, $\mathbf{c}$, and $\mathbf{d}$ are arbitrary vectors.

Solution. To show (1.91) we write the LHS of (1.91) in components and obtain

$$[(\mathbf{a} \times \mathbf{b}) \times \mathbf{c}]_i = \varepsilon_{ijk}(\mathbf{a} \times \mathbf{b})_j c_k = \varepsilon_{ijk}\, \varepsilon_{jpq}\, a_p\, b_q\, c_k = \varepsilon_{jki}\, \varepsilon_{jpq}\, a_p\, b_q\, c_k \tag{1.93}$$

Now, by using the $\varepsilon - \delta$ identify [see Eq. (1.7)] we obtain

$$\varepsilon_{jki}\,\varepsilon_{jpq} = \delta_{kp}\,\delta_{iq} - \delta_{kq}\,\delta_{ip} \tag{1.94}$$

Therefore substituting (1.94) into (1.93) we receive

$$[(\mathbf{a} \times \mathbf{b}) \times \mathbf{c}]_i = a_k\,c_k\,b_i - b_k\,c_k\,a_i \tag{1.95}$$

or

$$[(\mathbf{a} \times \mathbf{b}) \times \mathbf{c}]_i = (\mathbf{a} \cdot \mathbf{c})b_i - (\mathbf{b} \cdot \mathbf{c})a_i \tag{1.96}$$

Equation (1.96) is equivalent to (1.91), and this proves (1.91). To show (1.92) we replace the vector $\mathbf{c}$ in (1.91) by $\mathbf{c} \times \mathbf{d}$, and arrive at (1.92).

Problem 1.2. Show that for any vector $\mathbf{u}$ and a unit vector $\mathbf{n}$ the following decomposition formula holds true

$$\mathbf{u} = \mathbf{u}^{\perp} + \mathbf{u}^{||} \tag{1.97}$$

where

$$\mathbf{u}^{\perp} = (\mathbf{u} \cdot \mathbf{n})\,\mathbf{n} \quad \text{and} \quad \mathbf{u}^{||} = \mathbf{n} \times (\mathbf{u} \times \mathbf{n}) \tag{1.98}$$

Also, show that

$$\mathbf{u}^{\perp} \cdot \mathbf{u}^{||} = 0, \quad \mathbf{u} \cdot \mathbf{n} = \mathbf{u}^{\perp} \cdot \mathbf{n}, \quad \mathbf{u}^{||} \cdot \mathbf{n} = 0 \tag{1.99}$$

Note. If $\mathbf{u} = \mathbf{u}(\mathbf{x})$ is a vector field defined on a surface S in E^3, $\mathbf{n} = \mathbf{n}(\mathbf{x})$ is a unit outward normal vector field on S, and P is a plane tangent to S at $\mathbf{x}$, then $\mathbf{u}^{\perp}$ and $\mathbf{u}^{||}$ represent the *normal* and *tangent* parts of $\mathbf{u}$, respectively, with respect to P.

Solution. The relation (1.91) in components reads

$$u_i = (\mathbf{u}^{\perp})_i + (\mathbf{u}^{||})_i \tag{1.100}$$

where

$$(\mathbf{u}^{\perp})_i = (u_k\,n_k)\,n_i, \quad (\mathbf{u}^{||})_i = \varepsilon_{ijk}\,n_i\,\varepsilon_{kpq} u_p n_q \tag{1.101}$$

Since by the $\varepsilon - \delta$ identify

$$\varepsilon_{ijk}\,\varepsilon_{kpq} = \varepsilon_{kij}\,\varepsilon_{kpq} = \delta_{ip}\delta_{jq} - \delta_{iq}\delta_{pj} \tag{1.102}$$

therefore

$$\begin{aligned}(\mathbf{u}^{||})_i &= \left(\delta_{ip}\,\delta_{jq} - \delta_{iq}\,\delta_{pj}\right) n_j\,n_q\,u_p \\ &= n_q\,n_q\,u_i - n_p\,u_p\,n_i\end{aligned} \tag{1.103}$$

Now, if we substitute $(\mathbf{u}^{\perp})_i$ and $(\mathbf{u}^{||})_i$ from (1.101) and (1.103), respectively, into the RHS of (1.100), and take into account that $n_q\, n_q = 1$, we arrive at the LHS of (1.100) which proves (1.100).

The relations (1.93) hold true, since by (1.101) and (1.103)

$$\mathbf{u}^{\perp} \cdot \mathbf{u}^{||} = (\mathbf{u}^{\perp})_i\, (\mathbf{u}^{||})_i = (u_k\, n_k) n_i\, (u_i - n_i\, u_p\, n_p) = (u_k\, n_k)^2 - (u_k\, n_k)^2 = 0 \tag{1.104}$$

$$\mathbf{u}^{\perp} \cdot \mathbf{n} = [(u_k\, n_k) n_i] n_i = u_k\, n_k = \mathbf{u} \cdot \mathbf{n} \tag{1.105}$$

and

$$\mathbf{u}^{||} \cdot \mathbf{n} = (\mathbf{u}^{||})_i n_i = (u_i - n_i\, u_p\, n_p) n_i = 0 \tag{1.106}$$

Problem 1.3. Show that an alternative form of Eqs. (1.107) and (1.108) in Problem 1.2 reads

$$\mathbf{u} = \mathbf{u}^{\perp} + \mathbf{u}^{||} \tag{1.107}$$

where

$$\mathbf{u}^{\perp} = (\mathbf{n} \otimes \mathbf{u})\, \mathbf{n} \quad \text{and} \quad \mathbf{u}^{||} = (\mathbf{1} - \mathbf{n} \otimes \mathbf{n})\, \mathbf{u} \tag{1.108}$$

In Eq. (1.108) the symbol $\otimes$ represents the tensor product of two vectors, and $\mathbf{1}$ is a unit second-order tensor [see Eq. (1.36)].

Solution. The tensor product of vectors $\mathbf{a}$ and $\mathbf{b}$ is defined as a second order tensor $\mathbf{P}$ with the components

$$P_{ij} = a_i\, b_j \tag{1.109}$$

or in direct notation

$$\mathbf{P} = \mathbf{a} \otimes \mathbf{b} \tag{1.110}$$

Therefore, Eq. (1.108) in components read

$$(\mathbf{u}^{\perp})_i = n_i\, u_j\, n_j \quad \text{and} \quad (\mathbf{u}^{||})_i = (\delta_{ij} - n_i\, n_j) u_j \tag{1.111}$$

Substituting (1.111) into the RHS of (1.107) written in components, we arrive at the LHS of (1.107) written in components. This proves (1.107).

Problem 1.4. Let $\mathbf{T} = \mathbf{T}(\mathbf{x})$ be a symmetric tensor field defined on a surface S in E^3, $\mathbf{n} = \mathbf{n}(\mathbf{x})$ a unit outward normal vector field on S, and P a plane tangent to S at $\mathbf{x}$. Show that

$$\mathbf{T} = \mathbf{T}^{\perp} + \mathbf{T}^{||} \tag{1.112}$$

where

$$\mathbf{T}^{\perp} = 2\,\mathrm{sym}\,(\mathbf{n} \otimes \mathbf{Tn}) - (\mathbf{n} \cdot \mathbf{Tn})\mathbf{n} \otimes \mathbf{n} \tag{1.113}$$

and

$$\mathbf{T}^{||} = (\mathbf{1} - \mathbf{n} \otimes \mathbf{n})\,\mathbf{T}\,(\mathbf{1} - \mathbf{n} \otimes \mathbf{n}) \tag{1.114}$$

Also, show that

$$\mathbf{T}^{\perp} \cdot \mathbf{T}^{||} = 0, \quad \mathbf{Tn} = \mathbf{T}^{\perp}\mathbf{n}, \quad \mathbf{T}^{||}\,\mathbf{n} = 0 \tag{1.115}$$

Note. The tensors $\mathbf{T}^{\perp}$ and $\mathbf{T}^{||}$ represent the *normal* and *tangential* parts of $\mathbf{T}$, respectively, with respect to the plane P.

Solution. Equations (1.113) and (1.114), respectively, in components, take the form

$$(\mathbf{T}^{\perp})_{ij} = n_i\, T_{jk}\, n_k + n_j\, T_{ik}\, n_k - n_k\, T_{kp}\, n_p\, n_i\, n_j \tag{1.116}$$

and

$$(\mathbf{T}^{||})_{ij} = (\delta_{ip} - n_i\, n_p) T_{pq} (\delta_{qj} - n_q\, n_j) = T_{ij} - n_j\, T_{iq}\, n_q - n_i\, T_{pj}\, n_p + n_i\, n_j\, n_p\, T_{pq} \tag{1.117}$$

Since $\mathbf{T}$ is a symmetric tensor, therefore,

$$n_i\, T_{jk}\, n_k = n_i\, T_{pj}\, n_p \tag{1.118}$$

Writing (1.112) in components, and substituting (1.116) and (1.117) into the RHS of (1.112) we arrive at the LHS of (1.112) which proves (1.112).

To prove $(1.115)_1$ note that

$$\mathbf{T}^{\perp} \cdot \mathbf{T}^{||} = (\mathbf{T}^{\perp})_{ij}\, (\mathbf{T}^{||})_{ij} \tag{1.119}$$

If we note that

$$(\delta_{ip} - n_i\, n_p) n_i = 0 \quad \text{and} \quad \left(\delta_{qj} - n_q\, n_j\right) n_j = 0 \tag{1.120}$$

then substituting (1.116) and (1.117) into (1.119) and taking into account (1.120) we obtain $(1.115)_1$. To show $(1.115)_2$ we write the RHS of $(1.115)_2$ in components to obtain

$$\begin{aligned}(\mathbf{T}^{\perp}\mathbf{n})_i &= (n_i\, T_{jk}\, n_k + n_j\, T_{ik}\, n_k - n_k\, T_{kp}\, n_p\, n_i\, n_j)\, n_j \\ &= T_{ik}\, n_k + n_i\, n_j\, n_k\, T_{jk} - n_i\, n_p\, n_k\, T_{kp} \\ &= T_{ik}\, n_k.\end{aligned} \tag{1.121}$$

Hence the RHS of $(1.115)_2$ = the LHS of $(1.115)_2$ and this proves $(1.115)_2$.

Finally, writing the LHS of $(1.115)_3$ in components we obtain

$$(\mathbf{T}^{||}\mathbf{n})_i = (\mathbf{T}^{||})_{ij}\, n_j = (\delta_{ip} - n_i\, n_p)\, T_{pq}(\delta_{qj} - n_q\, n_j)n_j \tag{1.122}$$

If Eq. (1.120)$_2$ is substituted into (1.122) we obtain (1.115)$_3$. This completes solution to Problem 1.4.

Problem 1.5. Show that if S is a plane $x_3 = 0$ with the unit outward normal vector $\mathbf{n} = (0, 0, -1)$ then the decomposition formula (1.107) in Problem 1.4 reads

$$\begin{bmatrix} T_{11} & T_{12} & T_{13} \\ T_{21} & T_{22} & T_{23} \\ T_{31} & T_{32} & T_{33} \end{bmatrix} = \begin{bmatrix} 0 & 0 & T_{13} \\ 0 & 0 & T_{23} \\ T_{31} & T_{32} & T_{33} \end{bmatrix} + \begin{bmatrix} T_{11} & T_{12} & 0 \\ T_{21} & T_{22} & 0 \\ 0 & 0 & 0 \end{bmatrix}$$

Solution. Substituting $n_1 = 0,\ n_2 = 0,\ n_3 = -1$, into Eqs. (1.116) and (1.117), respectively, in the solution of Problem 1.4, we obtain

$$\mathbf{T}^{\perp} = \begin{bmatrix} 0 & 0 & T_{13} \\ 0 & 0 & T_{23} \\ T_{31} & T_{32} & T_{33} \end{bmatrix}$$

and

$$\mathbf{T}^{||} = \begin{bmatrix} T_{11} & T_{12} & 0 \\ T_{21} & T_{22} & 0 \\ 0 & 0 & 0 \end{bmatrix}$$

Hence

$$\mathbf{T} = \mathbf{T}^{\perp} + \mathbf{T}^{||}$$

which proves the decomposition formula of Problem 1.5.

Problem 1.6. Let $\mathbf{T}$ be a second-order tensor with components T_{ij}, and let $\mathbf{T} \neq \mathbf{0}$. Show that

$$\det \mathbf{T} = \varepsilon_{ijk} T_{i1}\, T_{j2}\, T_{k3} \tag{1.123}$$

$$\varepsilon_{pqr}(\det \mathbf{T}) = \varepsilon_{ijk} T_{ip}\, T_{jq}\, T_{kr} \tag{1.124}$$

$$\varepsilon_{ijk}\varepsilon_{pqr}(\det \mathbf{T}) = \begin{vmatrix} T_{ip} & T_{iq} & T_{ir} \\ T_{jp} & T_{jq} & T_{jr} \\ T_{kp} & T_{kq} & T_{kr} \end{vmatrix} \tag{1.125}$$

Solution. To show (1.123) we use the result (1.123) of Eq. (1.10)

$$\varepsilon_{ijk}\, a_i\, b_j\, c_k = \begin{vmatrix} a_1 & a_2 & a_3 \\ b_1 & b_2 & b_3 \\ c_1 & c_2 & c_3 \end{vmatrix} \tag{1.126}$$

where **a**, **b**, and **c** are arbitrary vectors. By letting

$$a_i = T_{i1}, \quad b_i = T_{i2}, \quad c_i = T_{i3} \tag{1.127}$$

in (1.126) we obtain

$$\varepsilon_{ijk}\, T_{i1}\, T_{j2}\, T_{k3} = \begin{vmatrix} T_{11} & T_{21} & T_{31} \\ T_{12} & T_{22} & T_{32} \\ T_{13} & T_{23} & T_{33} \end{vmatrix} \tag{1.128}$$

Since

$$\det(\mathbf{T}) = \det(\mathbf{T}^{\mathrm{T}}) \tag{1.129}$$

Equation (1.128) is equivalent to (1.123), and this proves (1.123).

To show (1.124) we let

$$a_i = T_{ip}, \quad b_i = T_{iq}, \quad c_i = T_{ir} \tag{1.130}$$

in (1.126), where p, q, and r are fixed numbers from the set $\{1, 2, 3\}$, and obtain

$$\varepsilon_{ijk}\, T_{ip}\, T_{jq}\, T_{kr} = \begin{vmatrix} T_{1p} & T_{2p} & T_{3p} \\ T_{1q} & T_{2q} & T_{3q} \\ T_{1r} & T_{2r} & T_{3r} \end{vmatrix} \tag{1.131}$$

Next, multiplying (1.123) by ε_{pqr}, we get

$$\varepsilon_{pqr} \det(\mathbf{T}) = \varepsilon_{pqr}\, \varepsilon_{ijk}\, T_{i1}\, T_{j2}\, T_{k3} \tag{1.132}$$

Since by Eq. (1.9)

$$\varepsilon_{pqr}\, \varepsilon_{ijk} = \begin{vmatrix} \delta_{pi} & \delta_{pj} & \delta_{pk} \\ \delta_{qi} & \delta_{qj} & \delta_{qk} \\ \delta_{ri} & \delta_{rj} & \delta_{rk} \end{vmatrix} \tag{1.133}$$

therefore, substituting (1.133) into (1.132) and multiplying T_{i1}, T_{j2}, and T_{k3}, respectively, by the first, second, and third column of the determinant on the RHS of (1.133), we obtain

$$\varepsilon_{pqr} \det(\mathbf{T}) = \begin{vmatrix} T_{p1} & T_{p2} & T_{p3} \\ T_{q1} & T_{q2} & T_{q3} \\ T_{r1} & T_{r2} & T_{r3} \end{vmatrix} \tag{1.134}$$

Since $\det(\mathbf{T}) = \det(\mathbf{T}^{\mathrm{T}})$, the RHS of (1.131) is identical to the RHS of (1.134), and this proves (1.124).

Finally, to show (1.125) we multiply (1.124) by ε_{ijk} and obtain

$$\varepsilon_{ijk}\, \varepsilon_{pqr} (\det \mathbf{T}) = \varepsilon_{ijk}\, \varepsilon_{abc}\, T_{ap}\, T_{bq}\, T_{cr} \tag{1.135}$$

or by Eq. (1.9)

$$\varepsilon_{ijk}\,\varepsilon_{pqr}(\det\,\mathbf{T}) = \begin{vmatrix} \delta_{ia} & \delta_{ib} & \delta_{ic} \\ \delta_{ja} & \delta_{jb} & \delta_{jc} \\ \delta_{ka} & \delta_{kb} & \delta_{kc} \end{vmatrix}\; T_{ap}\,T_{bq}\,T_{cr} \tag{1.136}$$

Now, multiplying the first, second, and third column of the determinant on the RHS of (1.136) by $T_{ap},\ T_{bq}$, and T_{cr}, respectively, we get

$$\varepsilon_{ijk}\,\varepsilon_{pqr}(\det\,\mathbf{T}) = \begin{vmatrix} T_{ip} & T_{iq} & T_{ir} \\ T_{jp} & T_{jq} & T_{jr} \\ T_{kp} & T_{kq} & T_{kr} \end{vmatrix} \tag{1.137}$$

This proves (1.125), and a solution to Problem 1.6 is complete.

Problem 1.7. Let $\mathbf{T}$ be a second-order tensor with components T_{ij} such that det $\mathbf{T} \neq 0$, and let $\hat{\mathbf{T}}$ be the tensor with components

$$\hat{T}_{ij} = \frac{1}{2}\varepsilon_{ipq}\varepsilon_{jrs}T_{pr}T_{qs} \tag{1.138}$$

Show that

$$\mathbf{T}\hat{\mathbf{T}}^{\mathrm{T}} = \hat{\mathbf{T}}^{\mathrm{T}}\mathbf{T} = (\det\,\mathbf{T})\mathbf{1} \tag{1.139}$$

$$\mathbf{T}^{-1} = (\det\,\mathbf{T})^{-1}\hat{\mathbf{T}}^{\mathrm{T}} \tag{1.140}$$

Note. The matrix $[\hat{T}_{ij}]$ is called the *cofactor* of the matrix $[T_{ij}]$, while $[\hat{T}_{ij}^T]$ is called the adjoint of $[T_{ij}]$.

Solution. The relation (1.139) in components takes the form

$$T_{ik}\,\widehat{T}_{kj}^{T} = \widehat{T}_{ik}^{T}\,T_{kj} = (\det\,\mathbf{T})\,\delta_{ij} \tag{1.141}$$

Using (1.192) we obtain

$$T_{ik}\,\widehat{T}_{kj}^{T} = T_{ik}\,\widehat{T}_{jk} = \frac{1}{2}\varepsilon_{jab}\,\varepsilon_{kcd}\,T_{ac}\,T_{bd}\,T_{ik} \tag{1.142}$$

Since, in view of (1.139) in Problem (1.6)

$$\varepsilon_{pqr}(\det\,\mathbf{T}^{\mathrm{T}}) = \varepsilon_{ijk}\,T_{pi}\,T_{qj}\,T_{rk} \tag{1.143}$$

and

$$\det(\mathbf{T}^{\mathrm{T}}) = \det(\mathbf{T}) \tag{1.144}$$

therefore, Eq. (1.142) can be written in the form

$$T_{ik}\,\widehat{T}_{kj}^{T} = \frac{1}{2}\varepsilon_{jab}\,\varepsilon_{iab}(\det\,\mathbf{T}) \tag{1.145}$$

Now, using the $\varepsilon - \delta$ identity (1.7)

$$\varepsilon_{mis}\, \varepsilon_{jks} = \delta_{mj}\, \delta_{ik} - \delta_{mk}\, \delta_{ij} \tag{1.146}$$

which is equivalent to

$$\varepsilon_{jkb}\, \varepsilon_{imb} = \delta_{mk}\, \delta_{ij} - \delta_{mj}\, \delta_{ik} \tag{1.147}$$

and letting $k = m = a$ in (1.147) we obtain

$$\varepsilon_{jab}\, \varepsilon_{iab} = 2\, \delta_{ij} \tag{1.148}$$

Thus, because of (1.145) and (1.148) we obtain

$$T_{ik}\, \widehat{T}^T_{kj} = (\det \mathbf{T})\, \delta_{ij} \tag{1.149}$$

which proves the second part of (1.141).

To prove that

$$T_{ik}\, \widehat{T}^T_{kj} = \widehat{T}^T_{ik}\, T_{kj} \tag{1.150}$$

we note that

$$\begin{aligned} \widehat{T}^T_{ik}\, T_{kj} = \widehat{T}_{ki}\, T_{kj} &= \frac{1}{2}\, \varepsilon_{kab}\, \varepsilon_{icd}\, T_{ac}\, T_{bd}\, T_{kj} \\ &= \frac{1}{2}\, \varepsilon_{icd}\, \varepsilon_{jcd} (\det\ \mathbf{T}) = (\det\ \mathbf{T}) \delta_{ij} \end{aligned} \tag{1.151}$$

and this completes the proof of (1.139).

To show (1.140) we note that

$$\mathbf{T}\mathbf{T}^{-1} = \mathbf{T}^{-1}\, \mathbf{T} = \mathbf{1} \tag{1.152}$$

and by virtue of (1.139)

$$\mathbf{T}\, \hat{\mathbf{T}}^{\mathrm{T}} = (\det\ \mathbf{T})\, \mathbf{1} \tag{1.153}$$

Multiplying (1.153) by $\mathbf{T}^{-1}$, taking into account (1.152) as well as the relations

$$\mathbf{A\, B\, C} = (\mathbf{A\, B})\, \mathbf{C} = \mathbf{A}\, (\mathbf{B\, C}) \tag{1.154}$$

where $\mathbf{A}$, $\mathbf{B}$, and $\mathbf{C}$ are arbitrary matrices, we obtain

$$\widehat{\mathbf{T}}^{\mathrm{T}} = (\det\ \mathbf{T})\mathbf{T}^{-1} \tag{1.155}$$

and this proves (1.140).

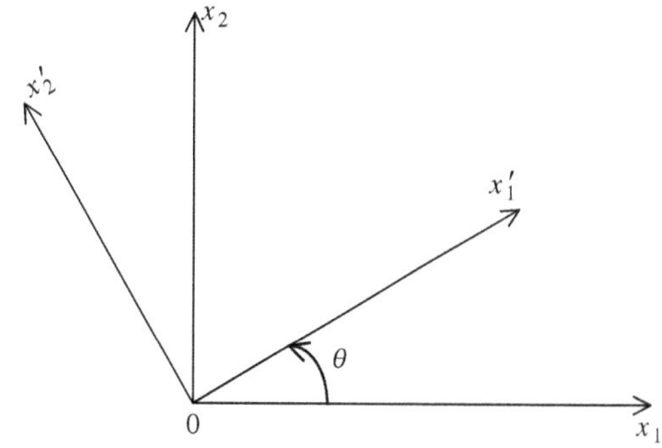

Fig. 1.1 Coordinate axes

Problem 1.8. The x_i' system is obtained by rotating the x_i system about the x_3 axis through an angle $0 < \theta < \pi/2$, as shown in Fig. 1.1. Let $\mathbf{T}$ be a symmetric second-order tensor referred to the x_i system. Show that

$$\mathbf{T}' = (\mathbf{T}')^{\mathrm{T}} \tag{1.156}$$

$$\begin{aligned} T_{11}' &= T_{11}\cos^2\theta + T_{12}\sin 2\theta + T_{22}\sin^2\theta \\ T_{12}' &= \frac{1}{2}(T_{22} - T_{11})\sin 2\theta + T_{12}\cos 2\theta \\ T_{22}' &= T_{11}\sin^2\theta - T_{12}\sin 2\theta + T_{22}\cos^2\theta \end{aligned} \tag{1.157}$$

and

$$\begin{aligned} T_{13}' &= T_{13}\cos\theta + T_{23}\sin\theta \\ T_{23}' &= -T_{13}\sin\theta + T_{23}\cos\theta \\ T_{33}' &= T_{33} \end{aligned} \tag{1.158}$$

Also, show that an alternative form of the transformation formulas (1.157) and (1.158) reads

$$\begin{aligned} &T_{11}' + T_{22}' = T_{11} + T_{22} \\ &T_{22}' - T_{11}' + 2\,\mathrm{i}\,T_{12}' = \exp\,(2\mathrm{i}\theta)\;(T_{22} - T_{11} + 2\,\mathrm{i}\,T_{12}) \\ &T_{13}' - \mathrm{i}\,T_{23}' = \exp\,(\mathrm{i}\,\theta)\;(T_{13} - T_{23}) \\ &T_{33}' = T_{33} \end{aligned} \tag{1.159}$$

where $\mathrm{i} = \sqrt{-1}$.

Hence, if the coordinates $(x_1', \ x_2', \ x_3')$ are identified with the cylindrical coordinates (r, θ, x_3), we find

$$
\begin{aligned}
&T_{rr} + T_{\theta\theta} = T_{11} + T_{22} \\
&T_{\theta\theta} - T_{rr} + 2\mathrm{i}T_{r\theta} = \exp(2\mathrm{i}\theta)\,(T_{22} - T_{11} + 2\mathrm{i}T_{12}) \\
&T_{r3} - \mathrm{i}T_{\theta 3} = \exp(\mathrm{i}\theta)\,(T_{13} - \mathrm{i}T_{23}) \\
&T_{33} = T_{33}
\end{aligned} \tag{1.160}
$$

[Do not sum over r and θ in Eq. (1.160)].
Hint. Use the formula $\mathbf{T}' = \mathbf{Q}^{\mathrm{T}}\mathbf{T}\mathbf{Q}$ where $\mathbf{Q}^{\mathrm{T}}$ is the matrix

$$
\mathbf{Q}^{\mathrm{T}} = \begin{bmatrix} \cos\theta & \sin\theta & 0 \\ -\sin\theta & \cos\theta & 0 \\ 0 & 0 & 1 \end{bmatrix}
$$

Solution. The relations (1.156)–(1.158) follow from the formula

$$
\mathbf{T}' = \mathbf{Q}^{\mathrm{T}}\,\mathbf{T}\mathbf{Q} \tag{1.161}
$$

where $\mathbf{Q}^{\mathrm{T}}$ is the matrix

$$
\mathbf{Q}^{\mathrm{T}} = \begin{bmatrix} \cos\theta & \sin\theta & 0 \\ -\sin\theta & \cos\theta & 0 \\ 0 & 0 & 1 \end{bmatrix} \tag{1.162}
$$

To show that (1.156) and (1.158) are equivalent to (1.159) use the identities

$$
\exp(\mathrm{i}k\theta) = \cos k\theta + \mathrm{i}\sin k\theta, \quad k = 1, 2 \tag{1.163}
$$

$$
\cos 2\theta = \cos^2\theta - \sin^2\theta \tag{1.164}
$$

Finally, using the correspondence

$$
x_1' = r, \quad x_2' = \theta, \quad x_3' = x_3 \tag{1.165}
$$

and

$$
\begin{aligned}
&T_{11}' = T_{rr}, \quad T_{22}' = T_{\theta\theta}, \quad T_{12}' = T_{r\theta} \\
&T_{13}' = T_{r3}, \quad T_{23}' = T_{\theta 3}, \quad T_{33}' = T_{33}
\end{aligned} \tag{1.166}
$$

we transform (1.159) into (1.160), and this completes a solution to Problem 1.8.

Problem 1.9. A tensor $\mathbf{T}$ is said to be positive definite if $\mathbf{u} \cdot \mathbf{T}\mathbf{u} > 0$ for every $\mathbf{u} \neq \mathbf{0}$. Show that if $\mathbf{T}$ is invertible, then $\mathbf{T}\mathbf{T}^{\mathrm{T}}$ and $\mathbf{T}^{\mathrm{T}}\mathbf{T}$ are positive definite.

Solution. We are to show that $\mathbf{T}\mathbf{T}^{\mathrm{T}}$ and $\mathbf{T}^{\mathrm{T}}\,\mathbf{T}$ satisfy the inequalities

$$
\mathbf{u} \cdot (\mathbf{T}\mathbf{T}^{\mathrm{T}})\mathbf{u} > 0 \quad \text{for every } \mathbf{u} \neq \mathbf{0} \tag{1.167}
$$

and

$$\mathbf{u} \cdot (\mathbf{T}^{\mathrm{T}}\ \mathbf{T})\mathbf{u} > 0 \quad \text{for every } \mathbf{u} \neq \mathbf{0} \tag{1.168}$$

To prove (1.167) and (1.168) note that (1.167) and (1.168) are equivalent to

$$(\mathbf{T}^{\mathrm{T}}\mathbf{u}) \cdot (\mathbf{T}^{\mathrm{T}}\mathbf{u}) > 0 \quad \text{for every } \mathbf{u} \neq \mathbf{0} \tag{1.169}$$

and

$$(\mathbf{T}\mathbf{u}) \cdot (\mathbf{T}\mathbf{u}) > 0 \quad \text{for every } \mathbf{u} \neq \mathbf{0} \tag{1.170}$$

The equivalency is implied by the identities

$$\mathbf{u} \cdot (\mathbf{T}\mathbf{T}^{\mathrm{T}})\ \mathbf{u} = (\mathbf{T}^{\mathrm{T}}\mathbf{u}) \cdot (\mathbf{T}^{\mathrm{T}}\mathbf{u}) \tag{1.171}$$

and

$$\mathbf{u} \cdot (\mathbf{T}^{\mathrm{T}}\ \mathbf{T})\ \mathbf{u} = (\mathbf{T}\mathbf{u}) \cdot (\mathbf{T}\mathbf{u}) \tag{1.172}$$

Now, since $\mathbf{T}$ is invertible, $\mathbf{T}^{\mathrm{T}}$ is invertible. Hence

$$\mathbf{T}\mathbf{u} \neq \mathbf{0} \quad \text{for every } \mathbf{u} \neq \mathbf{0} \tag{1.173}$$

and

$$\mathbf{T}^{\mathrm{T}}\mathbf{u} \neq \mathbf{0} \quad \text{for every } \mathbf{u} \neq \mathbf{0} \tag{1.174}$$

As a result, the inequalities (1.169), (1.170), (1.173), and (1.174) imply that $\mathbf{T}\,\mathbf{T}^{\mathrm{T}}$ and $\mathbf{T}^{\mathrm{T}}\,\mathbf{T}$ are positive definite, and this completes solution of Problem 1.9.

Problem 1.10. Show that eigenvalues and eigenvectors for the matrix

$$\mathbf{T} = \begin{bmatrix} 1 & 0 & 1 \\ 0 & 2 & 0 \\ 1 & 0 & 3 \end{bmatrix} \tag{1.175}$$

are given by

$$\lambda_1 = 2 - \sqrt{2}, \quad \lambda_2 = 2, \quad \lambda_3 = 2 + \sqrt{2} \tag{1.176}$$

and

$$\begin{aligned} \mathrm{n}_1^{(1)} &= \pm\frac{1}{\sqrt{2}}\frac{1}{1-\sqrt{2}}\frac{1}{\sqrt{2+\sqrt{2}}} \\ \mathrm{n}_2^{(1)} &= 0 \end{aligned} \tag{1.177}$$

$$\begin{aligned} \mathrm{n}_3^{(1)} &= \pm\frac{1}{\sqrt{2}}\frac{1}{\sqrt{2+\sqrt{2}}} \\ \mathrm{n}_1^{(2)} &= 0, \quad \mathrm{n}_2^{(2)} = \pm 1, \quad \mathrm{n}_3^{(2)} = 0 \end{aligned} \tag{1.178}$$

$$n_1^{(3)} = \pm\frac{1}{\sqrt{2}}\frac{1}{1+\sqrt{2}}\frac{1}{\sqrt{2-\sqrt{2}}}$$
$$n_2^{(3)} = 0 \tag{1.179}$$
$$n_3^{(3)} = \pm\frac{1}{\sqrt{2}}\frac{1}{\sqrt{2-\sqrt{2}}}$$

In Eqs. (1.176)–(1.179) λ_i is an eigenvalue corresponding to the eigenvector $\mathbf{n}^{(i)}\,(i = 1, 2, 3)$.

Solution. By using steps leading to a solution of an eigenproblem for the tensor **T** given by (1.175), we find that an eigenvalue λ corresponding to an eigenvector **n** is a solution of the algebraic equation

$$\det\,(\mathbf{T} - \lambda\,\mathbf{1}) = 0 \tag{1.180}$$

while a unit eigenvector **n** corresponding to λ satisfies the equations

$$(\mathbf{T} - \lambda\,\mathbf{1})\,\mathbf{n} = \mathbf{0} \tag{1.181}$$

$$|\mathbf{n}| = 1 \tag{1.182}$$

Also, it is easy to check that λ_1, λ_2, and λ_3 given by (1.176) satisfy (1.180); and $\mathbf{n}^{(1)}, \mathbf{n}^{(2)}$, and $\mathbf{n}^{(3)}$ given by (1.194), (1.195), and (1.196), respectively, satisfy Eqs. (1.181) and (1.182). Hence λ_i and $\mathbf{n}^{(i)}(i = 1, 2, 3)$ respectively, given by (1.176) and (1.177)–(1.179) are the eigenvalues and eigenvectors for the matrix **T** given by Eq. (1.176). This completes a solution to Problem 1.10.

Problem 1.11. Let **T** be the tensor represented by the matrix (1.183) in Problem 1.10, and let $\{\mathbf{e}_i^*\}$ be the orthonormal basis obtained from Eqs. (1.185)–(1.187) in Problem 1.10 in which the upper signs are postulated. Define the tensor $\mathbf{Q}^{\mathrm{T}}$ in terms of components by

$$Q_{ij}^{T} = \mathbf{e}_i^* \cdot \mathbf{e}_j \tag{1.183}$$

Show that

$$\mathbf{T}^* = \mathbf{Q}^{\mathrm{T}}\mathbf{T}\mathbf{Q} \tag{1.184}$$

is a tensor represented by a diagonal matrix. Also, compute the components T_{11}^*, T_{22}^*, and T_{33}^*, and show that

$$\mathrm{tr}\;\mathbf{T}^* = \mathrm{tr}\;\mathbf{T} = 6 \tag{1.185}$$

Solution. The orthonormal basis $\{\mathbf{e}_i^*\}$ obtained from Eqs. (1.185)–(1.187) in Problem 1.10 is defined by

$$\mathbf{e}_1^* = \left[\frac{1}{\sqrt{2}}\,\frac{1}{1-\sqrt{2}}\,\frac{1}{\sqrt{2+\sqrt{2}}},\ 0,\ \frac{1}{\sqrt{2}}\,\frac{1}{\sqrt{2+\sqrt{2}}}\right]$$
$$\mathbf{e}_2^* = [0,\ 1,\ 0] \tag{1.186}$$
$$\mathbf{e}_3^* = \left[\frac{1}{\sqrt{2}}\,\frac{1}{1+\sqrt{2}}\,\frac{1}{\sqrt{2-\sqrt{2}}},\ 0,\ \frac{1}{\sqrt{2}}\,\frac{1}{\sqrt{2-\sqrt{2}}}\right]$$

Since

$$\mathbf{e}_1 = [1, 0, 0]\,, \quad \mathbf{e}_2 = [0, 1, 0], \quad \mathbf{e}_3 = [0, 0, 1] \tag{1.187}$$

Eq. (1.183) implies that

$$\begin{aligned}
Q_{11}^T &= \frac{1}{\sqrt{2}}\,\frac{1}{1-\sqrt{2}}\,\frac{1}{\sqrt{2+\sqrt{2}}}, \quad Q_{12}^T = 0\\
Q_{13}^T &= \frac{1}{\sqrt{2}}\,\frac{1}{\sqrt{2+\sqrt{2}}}, \quad Q_{21}^T = 0\\
Q_{22}^T &= 1, \quad Q_{23}^T = 0,\\
Q_{31}^T &= \frac{1}{\sqrt{2}}\,\frac{1}{1+\sqrt{2}}\,\frac{1}{\sqrt{2-\sqrt{2}}}, \quad Q_{32}^T = 0,\\
Q_{33}^T &= \frac{1}{\sqrt{2}}\,\frac{1}{\sqrt{2-\sqrt{2}}}
\end{aligned} \tag{1.188}$$

Hence, we obtain

$$\begin{aligned}
&Q_{11} = Q_{11}^T, \quad Q_{12} = 0, \quad Q_{13} = Q_{31}^T\\
&Q_{21} = 0, \quad Q_{22} = Q_{22}^T, \quad Q_{23} = 0,\\
&Q_{31} = Q_{13}^T, \quad Q_{32} = Q_{23}^T, \quad Q_{33} = Q_{33}^T
\end{aligned} \tag{1.189}$$

Since, by Eq. (1.184),

$$T_{ij}^* = Q_{ik}^T\, T_{ka}\, Q_{aj} \tag{1.190}$$

therefore, substituting T_{ka}, Q_{ik}^T, and Q_{aj} from Eqs. (1.183) in Problem 1.10, (1.188), and (1.189), respectively, into the RHS of (1.190), we obtain

$$\begin{aligned}
&T_{11}^* = \frac{2(3-2\sqrt{2})}{2-\sqrt{2}}, \quad T_{12}^* = T_{13}^* = 0\\
&T_{21}^* = 0, \quad T_{22}^* = 2, \quad T_{23}^* = 0\\
&T_{31}^* = 0, \quad T_{32}^* = 0, \quad T_{33}^* = \frac{2}{2-\sqrt{2}}
\end{aligned} \tag{1.191}$$

Equation (1.191) imply that $\mathbf{T}^*$ is represented by a diagonal matrix. In addition, it follows from (1.191) that (1.194) holds true, and this completes a solution to Problem 1.11.

Problem 1.12. Prove the following identities in which ϕ is a scalar field, $\mathbf{u}$ is a vector field, and $\mathbf{S}$ is a tensor field on a region $\mathrm{R} \subset E^3$. You may use the $\varepsilon - \delta$ relation.

$$\text{curl } \nabla\phi = \mathbf{0} \tag{1.192}$$

$$\text{div curl } \mathbf{u} = 0 \tag{1.193}$$

$$\text{curl curl } \mathbf{u} = \nabla \text{div } \mathbf{u} - \nabla^2 \mathbf{u} \tag{1.194}$$

$$\text{curl } \nabla\mathbf{u} = \mathbf{0} \tag{1.195}$$

$$\text{curl } (\nabla\mathbf{u}^{\mathrm{T}}) = \nabla \text{curl } \mathbf{u} \tag{1.196}$$

$$\text{If } \ \nabla\mathbf{u} = -\nabla\mathbf{u}^T \ \text{ then } \ \nabla\nabla\mathbf{u} = \mathbf{0} \tag{1.197}$$

$$\text{div curl } \mathbf{S} = \text{curl div } \mathbf{S}^{\mathrm{T}} \tag{1.198}$$

$$\text{div } (\text{curl } \mathbf{S})^{\mathrm{T}} = 0 \tag{1.199}$$

$$(\text{curl curl } \mathbf{S})^{\mathrm{T}} = \text{curl } (\text{curl } \mathbf{S}^{\mathrm{T}}) \tag{1.200}$$

$$\text{curl } (\phi\, \mathbf{1}) = -[\text{curl } (\phi\mathbf{1})]^{\mathrm{T}} \tag{1.201}$$

$$\text{div } (\mathbf{S}^{\mathrm{T}}\mathbf{u}) = \mathbf{u} \cdot \text{div } \mathbf{S} + \mathbf{S} \cdot \nabla\mathbf{u} \tag{1.202}$$

$$\text{tr } (\text{curl } \mathbf{S}) = 0 \ \text{ for every symmetric tensor } \mathbf{S} \tag{1.203}$$

If $\mathbf{S}$ is symmetric then

$$\text{curl curl } \mathbf{S} = -\nabla^2\mathbf{S} + 2\hat{\nabla}(\text{div } \mathbf{S}) - \nabla\nabla(\text{tr } \mathbf{S}) + \mathbf{1}[\nabla^2(\text{tr } \mathbf{S}) - \text{div div } \mathbf{S}] \tag{1.204}$$

If $\mathbf{S}$ is symmetric and $\mathbf{S} = \mathbf{G} - \mathbf{1}(\text{tr } \mathbf{G})$ then

$$\text{curl curl } \mathbf{S} = -\nabla^2\mathbf{G} + 2\hat{\nabla}(\text{div } \mathbf{G}) - \mathbf{1}\, \text{div div } \mathbf{G} \tag{1.205}$$

If $\mathbf{S}$ is skew and ω is its axial vector then

$$\text{curl } \mathbf{S} = \mathbf{1}\, (\text{div } \omega) - \nabla\omega \tag{1.206}$$

Solution. To show (1.192) we recall that for any vector field $\varphi = \varphi(\mathbf{x})$ the curl operator is defined by [see Eq. (1.68)]

$$(\text{curl } \varphi)_i = \varepsilon_{ijk}\, \varphi_{k,j} \tag{1.207}$$

By letting $\varphi = \nabla\phi$ in (1.207), we obtain

$$(\text{curl } \nabla\phi)_i = \varepsilon_{ijk}\, \phi_{,kj} \tag{1.208}$$

Since $\phi,_{kj}$ is a second order tensor that is symmetric with respect to the indexes k and j, while ε_{ijk} is asymmetric with respect to k, j, then by Eq. (1.42), the RHS of (1.208) vanishes, and this proves (1.192).

To show (1.193) we let $\varphi = \mathbf{u}$ in (1.207) and obtain

$$(\text{curl }\mathbf{u})_i = \varepsilon_{ijk}\, u_{k,j} \tag{1.209}$$

By taking the operator div on (1.209) we get

$$[(\text{curl }\mathbf{u})_i]_{,i} = \varepsilon_{ijk}\, u_{k,ij} \tag{1.210}$$

Since $u_{k,ij}$ is a third order tensor that is symmetric with respect to the indices i and j, while ε_{ijk} is asymmetric with respect to those indices, by Eq. (1.43), the RHS of (1.210) vanishes, and this proves (1.193).

To show (1.194) we write (1.194) in components

$$\varepsilon_{ida}\, \varepsilon_{abc}\, u_{c,bd} = u_{a,ai} - u_{i,aa} \tag{1.211}$$

Since, by the $\varepsilon - \delta$ relation [see Eq. (1.7)]

$$\varepsilon_{ida}\, \varepsilon_{abc} = \varepsilon_{ida}\, \varepsilon_{bca} = \delta_{ib}\, \delta_{\delta c} - \delta_{ic}\, \delta_{b\delta} \tag{1.212}$$

therefore

$$\varepsilon_{ida}\, \varepsilon_{abc}\, u_{c,bd} = (\delta_{ib}\, \delta_{dc} - \delta_{ic}\, \delta_{bd}) u_{c,bd} \tag{1.213}$$

and using the filtrating property of the Kronecker's delta

$$\delta_{ab}\, a_b = a_a \tag{1.214}$$

we obtain

$$\varepsilon_{ida}\, \varepsilon_{abc}\, u_{c,bd} = u_{c,ci} - u_{i,bb} \tag{1.215}$$

This proves (1.194).

To show (1.195) we note that

$$(\nabla \mathbf{u})_{ij} = u_{i,j} \tag{1.216}$$

and by the definition of curl of a second-order tensor field $\mathbf{T} = \mathbf{T}(\mathbf{x})$ [see Eq. (1.71)]

$$(\text{curl }\mathbf{T})_{ij} = \varepsilon_{ipq}\, T_{jq,p} \tag{1.217}$$

Substituting $\mathbf{T} = \nabla \mathbf{u}$ into (1.217) we get

$$(\text{curl }\nabla \mathbf{T})_{ij} = \varepsilon_{ipq}\, u_{j,qp} \tag{1.218}$$

Equation (1.218) together with Eq. (1.43) implies (1.195), and this completes a proof of (1.195).

To show (1.196) we replace $\mathbf{T}$ by $\mathbf{T}^{\mathrm{T}}$ in (1.217) and obtain

$$(\text{curl } \mathbf{T}^{\mathrm{T}})_{ij} = \varepsilon_{ipq}\, T^{T}_{jq,p} = \varepsilon_{ipq}\, T_{qj,p} \tag{1.219}$$

Next, by letting $\mathbf{T} = \nabla \mathbf{u}$ in Eq. (1.219) we obtain

$$(\text{curl } \nabla \mathbf{u}^{\mathrm{T}})_{ij} = \varepsilon_{ipq}\, u_{q,jp} = [\varepsilon_{ipq}\, u_{q,p}]_{,j} = (\nabla \text{ curl } \mathbf{u})_{ij} \tag{1.220}$$

This proves that (1.196) holds true.

To show (1.197) we need to prove that

$$u_{i,j} + u_{j,i} = 0 \Rightarrow u_{i,jk} = 0 \tag{1.221}$$

To this end we note that equation $u_{i,j} + u_{j,i} = 0$ implies

$$u_{i,jk} + u_{j,ki} = 0 \tag{1.222}$$

By replacing j by k and k by j in (1.222) we get

$$u_{i,jk} + u_{k,ji} = 0 \tag{1.223}$$

Now, if Eqs. (1.222) and (1.223) are added side by side we obtain

$$2u_{i,jk} + (u_{j,k} + u_{k,j})_{,i} = 0 \tag{1.224}$$

Since the second term on the LHS of (1.224) vanishes by the hypothesis, Eq. (1.224) implies (1.221).

To show (1.198), we write (1.198) in components and obtain [see Eq. (1.217)].

$$\varepsilon_{ipq}\, S_{jq,pj} = \varepsilon_{ipq}\, S^{T}_{qj,jp} \tag{1.225}$$

Since

$$S_{jq} = S^{T}_{qj} \tag{1.226}$$

Equation (1.225) is an identity, and this proves (1.198).

The relation (1.199) in components takes the form

$$\varepsilon_{jpq}\, S_{iq,pj} = 0 \tag{1.227}$$

Equation (1.227) represents an identity as ε_{jpq} is asymmetric with respect to indices p and j, and $S_{iq,pj}$ is symmetric with respect to p and j, and Eq. (1.43) holds true. This proves (1.199).

To show (1.200) introduce the notations

$$\text{curl}\,\mathbf{S} = \mathbf{A}, \quad \text{curl}\,\mathbf{S}^{\mathrm{T}} = \mathbf{B} \tag{1.228}$$

Then Eq. (1.200) is equivalent to

$$(\text{curl}\,\mathbf{A})^{\mathrm{T}} = \text{curl}\,\mathbf{B} \tag{1.229}$$

Eq. (1.229) in components takes the form

$$\varepsilon_{jpq}\, A_{iq,p} = \varepsilon_{ipq}\, B_{jq,p} \tag{1.230}$$

where

$$A_{iq} = \varepsilon_{iab}\, S_{qb,a} \tag{1.231}$$

$$B_{jq} = \varepsilon_{jab}\, S^{T}_{qb,a} = \varepsilon_{jab}\, S_{bq,a} \tag{1.232}$$

Substituting (1.231) and (1.232) into (1.230) we obtain

$$\varepsilon_{jpg}\, \varepsilon_{iab}\, S_{qb,ap} = \varepsilon_{ipq}\, \varepsilon_{jab}\, S_{bq.ap} \tag{1.233}$$

By letting $a = p,\ b = q$ in the RHS of (1.233) we arrive at an identity, and this proves (1.200).

To show (1.201), note that Eq. (1.201) in components takes the form

$$\varepsilon_{ipq}(\phi\, \delta_{jq}),_p = -\varepsilon_{jpq}(\phi\, \delta_{iq}),_p \tag{1.234}$$

or equivalently

$$\varepsilon_{ipj}\, \phi,_p = -\varepsilon_{jpi}\, \phi,_p \tag{1.235}$$

Since

$$-\,\varepsilon_{jpi} = +\varepsilon_{jip} = -\varepsilon_{ijp} = \varepsilon_{ipj} \tag{1.236}$$

therefore Eq. (1.235) is an identity, and this proves (1.201).

To show (1.202) we note that Eq. (1.202) in components reads

$$\left(S^{T}_{ij}\, u_j\right),_i = u_k\, S_{kj,j} + S_{ij}\, u_{i,j} \tag{1.237}$$

Since

$$\left(S^{T}_{ij}\, u_j\right),_i = (S_{ji}\, u_j),_i = S_{ji,i}\, u_j + S_{ji}\, u_{j,i} \tag{1.238}$$

therefore (1.237) is an identity, and this proves (1.202).

To prove (1.203) we note that Eq. (1.203) in components takes the form

$$\varepsilon_{ipq}\, S_{iq,p} = 0 \tag{1.239}$$

Equation (1.239) is an identity since $S_{iq} = S_{qi}$ and $\varepsilon_{ipq} = -\varepsilon_{qpi}$, and this completes proof of (1.203).

To show (1.204) note that the LHS of (1.204), written in components, takes the form

$$L_{ij} = \varepsilon_{ipq}\, \varepsilon_{jab}\, S_{qb,ap} = \varepsilon_{iqp}\, \varepsilon_{jba}\, S_{qb,pa} \tag{1.240}$$

Using the identity [see Eq. (1.9)]

$$\begin{aligned}\varepsilon_{iqp}\, \varepsilon_{jba} = \begin{vmatrix} \delta_{ij} & \delta_{ib} & \delta_{ia} \\ \delta_{qj} & \delta_{qb} & \delta_{qa} \\ \delta_{pj} & \delta_{pb} & \delta_{pa} \end{vmatrix} &= \delta_{ij}(\delta_{qb}\, \delta_{pa} - \delta_{qa}\, \delta_{pb}) - \delta_{ib}(\delta_{qj}\, \delta_{pa} - \delta_{pj}\, \delta_{qa}) \\ &\quad + \delta_{ia}(\delta_{qj}\, \delta_{pb} - \delta_{pj}\, \delta_{qb})\end{aligned} \tag{1.241}$$

as well as the filtrating property of the Kronecker symbol

$$\delta_{ab}\, a_b = a_a \tag{1.242}$$

where a_a is an arbitrary vector, we reduce Eq. (1.240) to the form

$$L_{ij} = \delta_{ij}(S_{qq,aa} - S_{ab,ab}) - \delta_{ib}(S_{jb,aa} - S_{ab,ja}) + \delta_{ia}(S_{jb,ab} - S_{bb,ja}) \tag{1.243}$$

or

$$L_{ij} = -S_{ij,aa} + S_{ia,aj} + S_{jb,bi} - S_{bb,ij} + \delta_{ij}(S_{qq,aa} - S_{ab,ab}) \tag{1.244}$$

Therefore

$$L_{ij} = R_{ij} \tag{1.245}$$

where R_{ij} is the RHS of (1.204) written in components, and this completes a proof of Eq. (1.204). Note that the symmetry of $\mathbf{S}$ was used to obtain (1.244) from (1.243).

To show (1.205) we substitute

$$S_{ij} = G_{ij} - \delta_{ij}\, G_{kk} \tag{1.246}$$

into the RHS of (1.244) and obtain

$$\begin{aligned} R_{ij} &= -(G_{ij} - \delta_{ij}\, G_{kk})_{,aa} + G_{ia,aj} + G_{jb,bi} - 2G_{kk,ij} + 2G_{aa,ij} - \delta_{ij}(G_{aa,bb} + G_{ab,ab}) \\ &= -G_{ij,aa} + 2G_{(ia,aj)} - \delta_{ij}\, G_{ab,ab} \end{aligned} \tag{1.247}$$

and this proves that (1.205) holds true.

Finally, to show (1.206) we recall the definition of an axial vector ω corresponding to a tensor $\mathbf{P}$ [see Eq. (1.24)]

$$\omega_i = -\frac{1}{2}\varepsilon_{ijk} P_{jk} \tag{1.248}$$

An equivalent form of (1.248) reads

$$P_{ij} = -\varepsilon_{ijk} \omega_k \tag{1.249}$$

By letting $P_{ij} = S_{ij}$ in (1.249) we obtain

$$S_{ij} = -\varepsilon_{ijk} \omega_k \tag{1.250}$$

Taking the curl of (1.250) we get

$$\varepsilon_{ipq} S_{jq,p} = -\varepsilon_{ipq} \varepsilon_{jqa} \omega_{a,p} \tag{1.251}$$

Since

$$-\varepsilon_{ipq} \varepsilon_{jqa} = \varepsilon_{ipq} \varepsilon_{jaq} = \delta_{ij} \delta_{pa} - \delta_{ia} \delta_{jp} \tag{1.252}$$

Eq. (1.251) implies

$$\varepsilon_{ipq} S_{jq,p} = \delta_{ij} \omega_{p,p} - \omega_{i,j} \tag{1.253}$$

and this proves (1.206).

As a result the solution to Problem 1.12 is complete.

Problem 1.13. Let f be a scalar field, **u** a vector field, and **T** a tensor field on a region $\mathrm{R} \subset E^3$. Let **n** be a unit outer normal vector to $\partial\mathrm{R}$, where $\partial\mathrm{R}$ stands for the boundary of R. Show that

$$\int_{\mathrm{R}} (\nabla \mathrm{f})\, dv = \int_{\partial\mathrm{R}} \mathrm{f}\, \mathbf{n}\, da \tag{1.254}$$

$$\int_{\mathrm{R}} (\mathrm{curl}\, \mathbf{u})\, dv = \int_{\partial\mathrm{R}} (\mathbf{n} \times \mathbf{u})\, da \tag{1.255}$$

$$\int_{\mathrm{R}} (\nabla \mathbf{u})\, dv = \int_{\partial\mathrm{R}} \mathbf{u} \otimes \mathbf{n}\, da \tag{1.256}$$

$$\int_{\mathrm{R}} [\mathbf{u} \otimes \mathrm{div}\, \mathbf{T} + (\nabla \mathbf{u})\mathbf{T}^{\mathrm{T}}]\, dv = \int_{\partial\mathrm{R}} \mathbf{u} \times \mathbf{T}\mathbf{n}\, da \tag{1.257}$$

Solution. To show (1.254) we use the formula [see Eq. (1.76) in which $T_{ij} = \hat{u}_j$ for a fixed index i]

$$\int_R \hat{u}_{k,k}\, dv = \int_{\partial R} \hat{u}_k\, n_k\, da \tag{1.258}$$

where $\hat{u}_k = \hat{u}_k(\mathbf{x})$ is an arbitrary vector field. By letting $\hat{u}_k = \delta_{ki} f$ into (1.258), where i is a fixed number from the set $\{1, 2, 3\}$, we obtain (1.254); and this completes proof of (1.254).

To show (1.255) we note that Eq. (1.255) in components reads

$$\int_R \varepsilon_{ijk}\, u_{k,j}\, dv = \int_{\partial R} \varepsilon_{ijk}\, n_j\, u_k\, da \tag{1.259}$$

By letting $\hat{u}_k = \varepsilon_{ika}\, u_a$ in (1.258), where i is a fixed number from the set $\{1, 2, 3\}$, we get

$$\int_R \varepsilon_{ika}\, u_{a,k}\, dv = \int_{\partial R} \varepsilon_{ika}\, n_k\, u_a\, da \tag{1.260}$$

Equation (1.260) is equivalent to Eq. (1.256), and this completes proof of (1.256).

To show (1.257), note that (1.257) written in components, takes the form

$$\int_R u_{i,j}\, dv = \int_{\partial R} u_i\, n_j\, da \tag{1.261}$$

By letting $\hat{u}_k = \delta_{jk}\, u_i$ in (1.258), where i and j are fixed numbers from the set $\{1, 2, 3\}$ we get

$$\int_R \delta_{jk}\, u_{i,k}\, dv = \int_{\delta R} \delta_{jk}\, u_i\, n_k\, da \tag{1.262}$$

or

$$\int_R u_{i,j}\, dv = \int_{\partial R} u_i\, n_j\, da \tag{1.263}$$

Equation (1.263) is equivalent to (1.257), and this completes proof of (1.257).

To show (1.258) we note that Eq. (1.258) in components takes the form

$$\int_R (u_i\, T_{jk,k} + u_{i,a}\, T_{aj}^T) dv = \int_{\partial R} u_i\, T_{jk}\, n_k\, da \tag{1.264}$$

Since

$$u_i\, T_{jk,k} + u_{i,a}\, T_{aj}^T = u_i\, T_{jk,k} + u_{i,k}\, T_{jk} = (u_i\, T_{jk})_{,k} \tag{1.265}$$

therefore, by letting $\hat{u}_k = u_i \, T_{jk}$ in (1.258), where i and j are fixed indices from the set $\{1, 2, 3\}$, we obtain (1.264); and this completes proof of (1.258).

Problem 1.14. Let $\mathbf{u}$ be a vector field on $\mathrm{R} \subset E^3$ subject to one of the conditions

$$\mathbf{u} = \mathbf{0} \quad \text{on} \quad \partial \mathrm{R} \tag{1.266}$$

or

$$\mathbf{n} \times \text{curl}\, \mathbf{u} = \mathbf{0} \quad \text{on} \quad \partial \mathrm{R} \tag{1.267}$$

Show that

$$\int_{\mathrm{R}} \mathbf{u} \cdot (\text{curl curl}\, \mathbf{u})\, dv = \int_{\mathrm{R}} (\text{curl}\, \mathbf{u})^2\, dv \tag{1.268}$$

Solution. Define a scalar field f, and a vector field $\mathbf{g}$ by

$$\mathrm{f} = \mathbf{u} \cdot (\text{curl}\, \mathbf{g}), \quad \mathbf{g} = \text{curl}\, \mathbf{u} \tag{1.269}$$

or in components

$$\mathrm{f} = u_k \, \varepsilon_{kab} \, g_{b,a}, \quad g_b = \varepsilon_{bcd} \, u_{d,c} \tag{1.270}$$

Also, note that

$$\begin{aligned} \mathrm{f} &= (u_k \, \varepsilon_{kab} \, g_b)_{,a} - u_{k,a} \, \varepsilon_{kab} \, g_b \\ &= (u_k \, \varepsilon_{kab} \, g_b)_{,a} + \varepsilon_{bak} \, u_{k,a} \, \varepsilon_{bcd} \, u_{d,c} \\ &= (u_k \, \varepsilon_{kab} \, g_b)_{,a} + (\text{curl}\, \mathbf{u})^2 \end{aligned} \tag{1.271}$$

By letting $\hat{u}_k = u_c \, \varepsilon_{ckb} \, g_b$ in Eq. (1.269) of Problem 1.13, we obtain

$$\int_{R} (u_c \, \varepsilon_{ckb} \, g_b)_{,k} \, dv = \int_{\partial R} u_c \, \varepsilon_{ckb} \, g_b \, n_k \, da \tag{1.272}$$

Therefore, if either (1.266) or (1.267) holds true, the RHS of (1.272) vanishes. Hence, integrating Eq. (1.271) over R and using (1.272), we find that Eq. (1.194) holds true provided either Eq. (1.192) or Eq. (1.193) is satisfied. This completes solution to Problem 1.14.

Problem 1.15. Let $\mathbf{u} = \mathbf{u}(\mathbf{x}, \mathrm{t})$ and $\mathbf{S} = \mathbf{S}(\mathbf{x}, \mathrm{t})$ denote a time-dependent vector field on $E^3 \times [0, \infty)$ and a time-dependent tensor field on $E^3 \times [0, \infty)$, respectively. Let $\rho = \rho(\mathbf{x})$ be a positive scalar field on E^3, and let the pair $[\mathbf{u}, \mathbf{S}]$ satisfy the differential equation

$$\text{div}\, \mathbf{S} - \rho\, \ddot{\mathbf{u}} = \mathbf{0} \quad \text{on} \quad E^3 \times [0, \infty) \tag{1.273}$$

subject to the conditions

$$\mathbf{u}(\mathbf{x},0)=\mathbf{u}_0(\mathbf{x}),\quad \dot{\mathbf{u}}(\mathbf{x},0)=\dot{\mathbf{u}}_0(\mathbf{x})\quad \text{for}\quad \mathbf{x}\in E^3 \tag{1.274}$$

where $\mathbf{u}_0$ and $\dot{\mathbf{u}}_0$ are prescribed vector fields on E^3. Show that $[\mathbf{u},\ \mathbf{S}]$ satisfies Eqs. (1.273) and (1.274) if and only if

$$\mathbf{u}=\rho^{-1}t*(\text{div}\ \mathbf{S})+\mathbf{u}_0+t\,\dot{\mathbf{u}}_0\quad \text{on}\quad E^3\times[0,\infty) \tag{1.275}$$

Here $*$ stands for the convolution product on the time-axis.

Solution. To show that (1.273) and (1.274) imply (1.275) we integrate Eq. (1.273) twice with respect to time over the interval $[0, t]$, and take into account the initial conditions (1.274).

To shows that (1.275) implies (1.273) and (1.274), we take the two steps:

(A) We let $t = 0$ in (1.194) to obtain $(b)_1$. Next, we differentiate Eq. (1.275) with respect to time and take the result at $t = 0$ to obtain $(b)_2$.

(B) We differentiate Eq. (1.275) twice with respect to time, take into account the formula

$$\frac{\partial^2}{\partial t^2}(t*f)=f \tag{1.276}$$

valid for an arbitrary function $f = f(x,t)$, and arrive at Eq. (1.273). This completes solution to Problem 1.15.

Chapter 2
Fundamentals of Linear Elasticity

In this chapter a number of concepts are introduced to describe a linear elastic body. In particular, the displacement vector, strain tensor, and stress tensor fields are introduced to define a linear elastic body which satisfies the strain-displacement relations, the equations of motion, and the constitutive relations. Also, the compatibility relations, the general solutions of elastostatics, and an alternative definition of the displacement field of elastodynamics are discussed. The stored energy of an elastic body, the positive definiteness and strong ellipticity of the elasticity fourth-order tensor, and the stress-strain-temperature relations for a thermoelastic body are also discussed.

2.1 Deformation of an Elastic Body

A material body B is defined as a set of elements **x**, called particles, for which there is a one-to-one correspondence with the points of a region $\kappa(\mathrm{B})$ of a physical space; while a *deformation* of B is a map κ of B onto a region $\kappa(\mathrm{B})$ in E^3 with det $(\nabla\kappa) > 0$. The point $\kappa(\mathbf{x})$ is the place occupied by the particle **x** in the deformation κ, and

$$\mathbf{u}(\mathbf{x}) = \kappa(\mathbf{x}) - \mathbf{x} \tag{2.1}$$

is the *displacement* of **x**.

If the mapping κ depends also on time $t \in [0, \infty)$, such a mapping defines a *motion* of B, and the displacement of **x** at time t is

$$\mathbf{u}(\mathbf{x}, \mathrm{t}) = \kappa(\mathbf{x}, t) - \mathbf{x} \tag{2.2}$$

M. Reza Eslami et al., *Theory of Elasticity and Thermal Stresses*, Solid Mechanics and Its Applications 197, DOI: 10.1007/978-94-007-6356-2_2,

By the *deformation gradient* and the *displacement gradient* we mean the tensor fields $\mathbf{F} = \nabla\kappa$ and $\nabla\mathbf{u}$, respectively. A *finite strain tensor* $\mathbf{D}$ is defined by

$$\mathbf{D} = \frac{1}{2}(\mathbf{F}^{\mathrm{T}}\mathbf{F} - \mathbf{1}) \tag{2.3}$$

or, equivalently, by

$$\mathbf{D} = \mathbf{E} + \frac{1}{2}(\nabla\mathbf{u})(\nabla\mathbf{u}^{\mathrm{T}}) \tag{2.4}$$

where

$$\mathbf{E} = \frac{1}{2}(\nabla\mathbf{u} + \nabla\mathbf{u}^{\mathrm{T}}) = \widehat{\nabla}\mathbf{u} \tag{2.5}$$

The tensor field $\mathbf{E}$ is called an *infinitesimal strain tensor.*

An *infinitesimal rigid displacement* of B is defined by

$$\mathbf{u}(\mathbf{x}) = \mathbf{u}_0 + \mathbf{W}(\mathbf{x} - \mathbf{x}_0) \tag{2.6}$$

where $\mathbf{u}_0$, $\mathbf{x}_0$ are constant vectors and $\mathbf{W}$ is a skew constant tensor.

An *infinitesimal volume change* of B is defined by

$$\delta\mathrm{v}(\mathrm{B}) = \int_{\mathrm{B}} \mathrm{div}\,\mathbf{u}\,\mathrm{dv} \tag{2.7}$$

while

$$\mathrm{div}\,\mathbf{u} = \mathrm{tr}\,\mathbf{E} \tag{2.8}$$

represents a *dilatation field.*

If a deformation is not accompanied by a change of volume, that is, if $\delta\mathrm{v}(\mathrm{P}) = 0$ for every $\mathrm{P} \subset \mathrm{B}$, the displacement $\mathbf{u}$ is called *isochoric.*

Kirchhoff Theorem. If two displacement fields $\mathbf{u}_1$ and $\mathbf{u}_2$ correspond to the same strain field $\mathbf{E}$ then

$$\mathbf{u}_1 - \mathbf{u}_2 = \mathbf{w} \tag{2.9}$$

where $\mathbf{w}$ is a rigid displacement field.

A *homogeneous displacement field* is defined by

$$\mathbf{u}(\mathbf{x}) = \mathbf{u}_0 + \mathbf{A}(\mathbf{x} - \mathbf{x}_0) \tag{2.10}$$

where $\mathbf{A}$ is an arbitrary constant tensor and $\mathbf{u}_0$, $\mathbf{x}_0$ are constant vectors. Clearly, if $\mathbf{A}$ is skew, (2.10) represents a rigid displacement, while for an arbitrary $\mathbf{A}$

$$\mathbf{u}(\mathbf{x}) = \mathbf{u}_1(\mathbf{x}) + \mathbf{u}_2(\mathbf{x}) \tag{2.11}$$

where $\mathbf{u}_1(\mathbf{x})$ is a rigid displacement field and $\mathbf{u}_2(\mathbf{x})$ is a displacement field corresponding to the strain $\mathbf{E} = \text{sym}\,\mathbf{A}$. The displacement $\mathbf{u}_2(\mathbf{x})$ of the form

$$\mathbf{u}_2(\mathbf{x}) = \mathbf{E}\,(\mathbf{x} - \mathbf{x}_0) \tag{2.12}$$

corresponds to a *pure strain from* $\mathbf{x}_0$.

Let $\mathrm{e} > 0$ and let $\mathbf{n}$ be a unit vector. Then by substituting $\mathbf{E} = \mathrm{e}\,\mathbf{n} \otimes \mathbf{n}$ into (2.12) we obtain a *simple extension of amount* e in the direction $\mathbf{n}$; and by substituting $\mathbf{E} = \mathrm{e}\,\mathbf{1}$ into (2.12) we obtain *uniform dilatation of amount* e. Finally, let $g > 0$ and let $\mathbf{m}$ be a unit vector perpendicular to $\mathbf{n}$. Then substituting $\mathbf{E} = \mathrm{g}\,[\mathbf{m} \otimes \mathbf{n} + \mathbf{n} \otimes \mathbf{m}\,]$ into (2.12) we obtain a *simple shear of amount* g with respect to the pair $(\mathbf{m},\mathbf{n})$.

Decomposition of a strain tensor E into spherical and deviatoric tensors

$$\mathbf{E} = \mathbf{E}^{(s)} + \mathbf{E}^{(d)} \tag{2.13}$$

where

$$\mathbf{E}^{(s)} = \frac{1}{3}(\mathrm{tr}\,\mathbf{E})\,\mathbf{1} \tag{2.14}$$

is called a *spherical part* of $\mathbf{E}$, and $\mathbf{E}^{(d)} = \mathbf{E} - \mathbf{E}^{(s)}$ is called a *deviatoric part* of $\mathbf{E}$. Clearly,

$$\mathrm{tr}\,(\mathbf{E}^{(d)}) = 0 \tag{2.15}$$

2.2 Compatibility

Theorem Let $\mathrm{B} \subset E^3$ be simply connected. If $\mathbf{u}$ is a displacement field corresponding to a strain field $\mathbf{E}$ on B, that is, if

$$\mathbf{E} = \frac{1}{2}(\nabla\mathbf{u} + \nabla\mathbf{u}^{\mathrm{T}}) \quad \text{on} \quad \mathrm{B} \tag{2.16}$$

then $\mathbf{E}$ satisfies the equations of compatibility

$$\mathrm{curl}\;\mathrm{curl}\;\mathbf{E} = \mathbf{0} \quad \text{on} \quad \mathrm{B} \tag{2.17}$$

Conversely, let $\mathbf{E}$ be a symmetric tensor field that satisfies the equations of compatibility (2.17), then there exists a displacement field $\mathbf{u}$ on B such that $\mathbf{u}$ and $\mathbf{E}$ satisfy (2.16).

In components the equations of compatibility (2.17) take the form

$$E_{ij,kl} + E_{kl,ij} - E_{ik,jl} - E_{jl,ik} = 0 \tag{2.18}$$

An alternative form of (2.17) reads

$$\nabla^2\mathbf{E} + \nabla\nabla(\text{tr}\,\mathbf{E}) - 2\widehat{\nabla}(\text{div}\,\mathbf{E}) = \mathbf{0} \tag{2.19}$$

2.3 Motion and Equilibrium

Let S be a surface in B with unit normal $\mathbf{n}$. Let B be subject to a deformation, and let $\mathbf{s}_\text{n} = \mathbf{s}_\text{n}(\mathbf{x}, t)$ denote a force per unit area at $\mathbf{x}$ and for $t \geq 0$ exerted by a portion of B on the side S toward which $\mathbf{n}$ points on a portion of B on the other side of S. The force $\mathbf{s}_\text{n}$ is called the *stress vector* at $(\mathbf{x}, t)$, while a second-order tensor field $\mathbf{S} = \mathbf{S}(\mathbf{x}, \text{t})$ such that

$$\mathbf{S}\,\mathbf{n} = \mathbf{s}_\text{n} \quad \text{on} \quad S \times [0, \infty) \tag{2.20}$$

is called a *time-dependent stress tensor field* on $S \times [0, \infty)$.

The equilibrium equations of elastostatics

$$\text{div}\,\mathbf{S} + \mathbf{b} = \mathbf{0} \tag{2.21}$$

$$\mathbf{S} = \mathbf{S}^\text{T} \tag{2.22}$$

Equation (2.21) expresses the *balance of forces*, and Eq. (2.22) expresses the *balance of moments;* and $\mathbf{b}$ in (2.21) is the *body force vector.*

The Beltrami representation of S

$$\mathbf{S} = \text{curl curl}\,\mathbf{A} \tag{2.23}$$

where $\mathbf{A}$ is a symmetric tensor field, or

$$\mathbf{S} = -\nabla^2\mathbf{G} + 2\widehat{\nabla}(\text{div}\,\mathbf{G}) - (\text{div div}\,\mathbf{G})\,\mathbf{1} \tag{2.24}$$

where $\mathbf{G}$ is a symmetric tensor field.

Self-equilibrated stress field

If $\mathbf{S} = \mathbf{S}^\text{T}$ on B, and

$$\int_\text{S} \mathbf{Sn}\,\text{da} = \mathbf{0} \tag{2.25}$$

$$\int_\text{S} \mathbf{x} \times (\mathbf{Sn})\,\text{da} = \mathbf{0} \tag{2.26}$$

for every closed surface S in B, then $\mathbf{S}$ is called a *self-equilibrated stress field.*

One can show that $\mathbf{S}$ given by (2.23) is a self-equilibrated stress field, and $\mathbf{S}$ *given by* (2.23) *is complete* in the sense that for any self-equilibrated $\mathbf{S}$ there is a symmetric tensor $\mathbf{A}$ such that (2.23) is satisfied.

The Beltrami-Schaefer representation of S

$$\mathbf{S} = \text{curl curl } \mathbf{A} + 2\widehat{\nabla}\mathbf{h} - (\text{div } \mathbf{h})\, \mathbf{1} \tag{2.27}$$

where **A** is a symmetric tensor field and **h** is a harmonic vector field on B.

2.4 Equations of Motion

$$\text{div } \mathbf{S} + \mathbf{b} = \rho\, \ddot{\mathbf{u}} \quad \text{on} \quad \text{B} \times [0, \infty) \tag{2.28}$$

where ρ is density and **b** is the body force vector field.

Kinetic energy of B for $t \geq 0$

$$\text{K(t)} = \frac{1}{2} \int_{\text{B}} \rho\, \dot{\mathbf{u}}^2 \text{dv} \tag{2.29}$$

Stress power of B for $t \geq 0$

$$\text{P(t)} = \int_{\text{B}} \mathbf{S} \cdot \dot{\mathbf{E}} \text{dv} \tag{2.30}$$

A dynamic process is identified with a triplet [**u**, **S**, **b**] that satisfies the equations of motion (2.28).

Theorem An array of functions [**u**, **S**, **b**] is a dynamic process consistent with the initial conditions

$$\mathbf{u}(\mathbf{x}, 0) = \mathbf{u}_0(\mathbf{x}), \quad \dot{\mathbf{u}}(\mathbf{x}, 0) = \dot{\mathbf{u}}_0(\mathbf{x}) \quad \text{for} \quad \mathbf{x} \in \overline{\text{B}} \tag{2.31}$$

if and only if

$$\text{i} * \text{div } \mathbf{S} + \mathbf{f} = \rho\, \mathbf{u} \quad \text{on} \quad \overline{\text{B}} \times [0, \infty) \tag{2.32}$$

where

$$\mathbf{f}(\mathbf{x}, t) = \text{i} * \mathbf{b}(\mathbf{x}, t) + \rho(\mathbf{x})\, [\mathbf{u}_0(\mathbf{x}) + t\, \dot{\mathbf{u}}_0(\mathbf{x})] \tag{2.33}$$

and

$$\text{i} = \text{i}(t) = t \tag{2.34}$$

The function $\mathbf{f} = \mathbf{f}(\mathbf{x}, t)$ given by (2.33) is called *pseudo-body force field.*

Clearly, since $\rho > 0$, Eq. (2.32) provides an alternative definition of the displacement vector $\mathbf{u} = \mathbf{u}(\mathbf{x}, t)$ related to the stress tensor $\mathbf{S} = \mathbf{S}(\mathbf{x}, t)$.

2.5 Constitutive Relations

A body B is said to be *linearly elastic* if for every point $\mathbf{x} \in$ B there is a linear transformation $\mathbf{C}$ from the space of all symmetric tensors $\mathbf{E}$ into the space of all symmetric tensors $\mathbf{S}$, or

$$\mathbf{S} = \mathbf{C}[\mathbf{E}] \tag{2.35}$$

In components

$$S_{ij} = C_{ijkl}E_{kl} \tag{2.36}$$

The tensor $\mathbf{C} = \mathbf{C}(\mathbf{x})$ is called the *elasticity tensor* field on B. It follows from Eq. (1.54) that

$$C_{ijkl} = (\mathbf{e}_i \otimes \mathbf{e}_j) \cdot \mathbf{C}[(\mathbf{e}_k \otimes \mathbf{e}_l)] \tag{2.37}$$

and, since $\mathbf{S}$ and $\mathbf{E}$ are symmetric, we postulate that

$$C_{ijkl} = C_{jikl} = C_{ijlk} \tag{2.38}$$

The elasticity tensor $\mathbf{C}$ is also assumed to be *invertible*, that means that a *restriction* of $\mathbf{C}$ to the space of all symmetric tensors is invertible. The elasticity tensor on the space of all tensors cannot be invertible since its value on every skew tensor is zero.

The *invertibility* of $\mathbf{C}$ means that there is a fourth-order tensor $\mathbf{K} = \mathbf{K}(\mathbf{x})$ such that

$$\mathbf{K} = \mathbf{C}^{-1} \tag{2.39}$$

Then equivalent form of (2.35) is

$$\mathbf{E} = \mathbf{K}[\mathbf{S}] \tag{2.40}$$

The tensor $\mathbf{K} = \mathbf{K}(\mathbf{x})$ is called the *compliance tensor*.

The fourth-order tensor $\mathbf{C}$ is *symmetric* if and only if

$$\mathbf{A} \cdot \mathbf{C}[\mathbf{B}] = \mathbf{B} \cdot \mathbf{C}[\mathbf{A}] \tag{2.41}$$

for any symmetric tensors $\mathbf{A}$ and $\mathbf{B}$.

In components the symmetry of $\mathbf{C}$ means that

$$C_{ijkl} = C_{klij} \tag{2.42}$$

The tensor $\mathbf{C}$ is *positive semi-definite* if

$$\mathbf{A} \cdot \mathbf{C}[\mathbf{A}] \geq 0 \tag{2.43}$$

for every symmetric tensor $\mathbf{A}$.

The tensor $\mathbf{C}$ is *positive definite* if

$$\mathbf{A} \cdot \mathbf{C}[\mathbf{A}] > 0 \tag{2.44}$$

for every symmetric nonzero tensor $\mathbf{A}$.

The compliance tensor $\mathbf{K}$ enjoys the properties similar to those of the elasticity tensor $\mathbf{C}$ [see, Eqs. (2.38) and (2.42)–(2.44)].

By an *anisotropic elastic body* we mean the body for which the tensor $\mathbf{C}$ possesses in general 21 different components.

2.6 Isotropic Elastic Body

For an *isotropic elastic body* the Eqs. (2.35) and (2.40), respectively, take the form

$$\mathbf{S} = 2\mu\,\mathbf{E} + \lambda\,(\mathrm{tr}\,\mathbf{E}) \tag{2.45}$$

and

$$\mathbf{E} = \frac{1}{2\mu}\left[\mathbf{S} - \frac{\lambda}{3\lambda + 2\mu}(\mathrm{tr}\,\mathbf{S})\,\mathbf{1}\right] \tag{2.46}$$

where λ and μ are Lamé moduli subject to the constitutive restrictions

$$\mu > 0, \quad 3\lambda + 2\mu > 0 \tag{2.47}$$

An alternative form of Eqs. (2.45) and (2.46), written in terms of Young's modulus E and Poisson's ratio ν, reads

$$\mathbf{S} = \frac{E}{1+\nu}\left[\mathbf{E} + \frac{\nu}{1-2\nu}(\mathrm{tr}\,\mathbf{E})\,\mathbf{1}\right] \tag{2.48}$$

$$\mathbf{E} = \frac{1}{E}\left[(1+\nu)\,\mathbf{S} - \nu\,(\mathrm{tr}\,\mathbf{S})\,\mathbf{1}\right] \tag{2.49}$$

where

$$E > 0 \quad \text{and} \quad -1 < \nu < 1/2 \tag{2.50}$$

Strain energy density of B

$$\mathrm{W}(\mathbf{E}) = \frac{1}{2}\mathbf{E} \cdot \mathbf{C}\,[\mathbf{E}] \tag{2.51}$$

Stress energy density of B

$$\widehat{\mathrm{W}}(\mathbf{S}) = \frac{1}{2}\mathbf{S} \cdot \mathbf{K}\,[\mathbf{S}] \tag{2.52}$$

The tensor C is said to be *strongly elliptic* if

$$\mathbf{A} \cdot \mathsf{C}[\mathbf{A}] > 0 \tag{2.53}$$

for every $\mathbf{A}$ of the form

$$\mathbf{A} = \mathbf{a} \otimes \mathbf{b} \tag{2.54}$$

where $\mathbf{a}$ and $\mathbf{b}$ are arbitrary nonzero vectors.

2.7 The Cauchy Relations

An anisotropic elastic body obeying, in addition to the symmetry relations (2.38) and (2.42), the restrictions

$$C_{ijkl} = C_{ikjl} \tag{2.55}$$

is said to be of the Cauchy type.

2.8 Constitutive Relations for a Thermoelastic Body

For an anisotropic body subject to an uneven heating the constitutive relations take the form

$$\mathbf{S} = \mathsf{C}\,[\mathbf{E}] + T\mathbf{M} \tag{2.56}$$

and

$$\mathbf{E} = \mathsf{K}\,[\mathbf{S}] + T\mathbf{A} \tag{2.57}$$

where

$$T = \theta - \theta_0, \quad \theta_0 > 0 \tag{2.58}$$

is a temperature change, $\mathbf{M} = \mathbf{M}^T$ is called the *stress-temperature tensor,* $\mathbf{A} = \mathbf{A}^T$ is called the *thermal expansion tensor,* θ is the *absolute temperature,* and θ_0 is a *reference temperature.*

Since relations (2.56) and (2.57) are equivalent

$$\mathsf{K} = \mathsf{C}^{-1} \quad \text{and} \quad \mathbf{A} = -\,\mathsf{K}[\mathbf{M}] \tag{2.59}$$

for an isotropic body

$$\mathbf{S} = 2\mu\,\mathbf{E} + \lambda\,(\text{tr}\,\mathbf{E}) - (3\lambda + 2\mu)\alpha\,T\,\mathbf{1} \tag{2.60}$$

and

$$\mathbf{E} = \frac{1}{2\mu}\left[\mathbf{S} - \frac{\lambda}{3\lambda + 2\mu}(\text{tr}\,\mathbf{S})\,\mathbf{1}\right] + \alpha\,T\,\mathbf{1} \tag{2.61}$$

where α is the *coefficient of thermal expansion*,
or

$$\mathbf{S} = \frac{E}{1+\nu}\left[\mathbf{E} + \frac{\nu}{1-2\nu}(\text{tr}\,\mathbf{E})\,\mathbf{1}\right] - \frac{E}{1-2\nu}\alpha\,T\,\mathbf{1} \tag{2.62}$$

and

$$\mathbf{E} = \frac{1}{E}\left[(1+\nu)\,\mathbf{S} - \nu\,(\text{tr}\,\mathbf{S})\,\mathbf{1}\right] + \alpha\,T\,\mathbf{1} \tag{2.63}$$

2.9 Problems and Solutions Related to the Fundamentals of Linear Elasticity

Problem 2.1. Show that if $\mathbf{u}$ is a pure strain from $\mathbf{x}_0$, then $\mathbf{u}$ admits the decomposition

$$\mathbf{u} = \mathbf{u}_1 + \mathbf{u}_2 + \mathbf{u}_3 \tag{2.64}$$

where $\mathbf{u}_1$, $\mathbf{u}_2$, and $\mathbf{u}_3$ are simple extensions in mutually perpendicular directions from $\mathbf{x}_0$.

Solution. Since $\mathbf{u}$ represents a pure strain from $\mathbf{x}_0$, $\mathbf{u}$ takes the form [see definition of $\mathbf{u}_2$ in (2.12)]

$$\mathbf{u} = \mathbf{E}(\mathbf{x} - \mathbf{x}_0) \tag{2.65}$$

where $\mathbf{E}$ is the strain tensor corresponding to $\mathbf{u}$. Now, by the decomposition spectral theorem [see Eq. (1.45) in which $\mathbf{T} = \mathbf{E}$ and $\lambda_i = e_i$]

$$\mathbf{E} = \sum_{i=1}^{3} e_i\,\mathbf{n}_i \otimes \mathbf{n}_i \tag{2.66}$$

where $\mathbf{n}_i$ is a principal direction corresponding to a principal value e_i of $\mathbf{E}$.

Substituting (2.66) into (2.65) we obtain

$$\mathbf{u} = \sum_{i=1}^{3} e_i(\mathbf{n}_i \otimes \mathbf{n}_i)(\mathbf{x} - \mathbf{x}_0) \tag{2.67}$$

Since for two arbitrary vectors $\mathbf{a}$ and $\mathbf{b}$

$$(\mathbf{a} \otimes \mathbf{a})\mathbf{b} = (\mathbf{a} \cdot \mathbf{b})\mathbf{a} \tag{2.68}$$

therefore, Eq. (2.67) is equivalent to

$$\mathbf{u} = \mathbf{u}_1 + \mathbf{u}_2 + \mathbf{u}_3 \tag{2.69}$$

where

$$\mathbf{u}_i = e_i[\mathbf{n}_i \cdot (\mathbf{x} - \mathbf{x}_0)]\mathbf{n}_i \text{ (no sum)} \tag{2.70}$$

Since $\mathbf{u}_i$ represents a simple extension of magnitude e_i in the direction of $\mathbf{n}_i$ [see Eq. (2.12)], and $\mathbf{n}_1$, $\mathbf{n}_2$, and $\mathbf{n}_3$ are orthogonal, Eq. (2.69) is equivalent to (2.64). This completes proof of (2.64).

Problem 2.2. Show that $\mathbf{u}$ in Problem 2.1 admits an alternative representation

$$\mathbf{u} = \mathbf{u}_{\mathrm{d}} + \mathbf{u}_{\mathrm{c}} \tag{2.71}$$

where $\mathbf{u}_{\mathrm{d}}$ is a uniform dilatation from $\mathbf{x}_0$, while $\mathbf{u}_{\mathrm{c}}$ is an isochoric pure strain from $\mathbf{x}_0$.

Solution. We rewrite $\mathbf{E}$ of Problem 2.1 as

$$\mathbf{E} = \mathbf{E}^{(s)} + \mathbf{E}^{(d)} \tag{2.72}$$

where

$$\mathbf{E}^{(s)} = \frac{1}{3}\mathbf{1}(\mathrm{tr}\ \mathbf{E}), \quad \mathbf{E}^{(d)} = \mathbf{E} - \frac{1}{3}\mathbf{1}(\mathrm{tr}\ \mathbf{E}) \tag{2.73}$$

Then Eq. (2.65) of Problem 2.1 takes the form

$$\mathbf{u} = \mathbf{u}_d + \mathbf{u}_c \tag{2.74}$$

where

$$\mathbf{u}_d = \mathbf{E}^{(s)}(\mathbf{x} - \mathbf{x}_0) \tag{2.75}$$

$$\mathbf{u}_c = \mathbf{E}^{(d)}(\mathbf{x} - \mathbf{x}_0) \tag{2.76}$$

It follows from Eqs. (2.73) and (2.75) that $\mathbf{u}_d$ represents a uniform dilatation of magnitude $e = \frac{1}{3}(\mathrm{tr}\ \mathbf{E})$, while the condition $\mathrm{tr}\ \mathbf{E}^{(d)} = 0$ implies that $\mathbf{u}_c$ represents an isochoric pure strain. This completes solution to Problem 2.2.

Problem 2.3. Show that if $\mathbf{u}$ is a simple shear of amount γ with respect to the pair ($\mathbf{m}$, $\mathbf{n}$), where $\mathbf{m}$ and $\mathbf{n}$ are perpendicular unit vectors, then $\mathbf{u}$ admits the decomposition

$$\mathbf{u} = \mathbf{u}_1 + \mathbf{u}_2 \tag{2.77}$$

where $\mathbf{u}_1$ is a simple extension of amount γ in the direction $\frac{1}{\sqrt{2}}(\mathbf{m} + \mathbf{n})$, and $\mathbf{u}_2$ is a simple extension of amount $-\gamma$ in the direction $\frac{1}{\sqrt{2}}(\mathbf{m} - \mathbf{n})$.

Solution. Since $\mathbf{u}$ represents a simple shear of amount γ with respect to $(\mathbf{m}, \mathbf{n})$, then the strain tensor corresponding to $\mathbf{u}$ takes the form [see the definition of a simple shear below Eq. (2.12)]

$$\mathbf{E} = \gamma\ (\mathbf{m} \otimes \mathbf{n} + \mathbf{n} \otimes \mathbf{m}) \tag{2.78}$$

Let λ and $\mathbf{a}$ denote a principal value and a principal vector of $\mathbf{E}$, respectively. Then

$$\gamma\ (\mathbf{n} \cdot \mathbf{a})\mathbf{m} + \gamma(\mathbf{m} \cdot \mathbf{a})\mathbf{n} - \lambda\mathbf{a} = \mathbf{0} \tag{2.79}$$

It is easy to check that Eq. (2.79) has the three eigensolutions

$$\mathbf{a}_1 = \mathbf{m} \times \mathbf{n}, \qquad \lambda_1 = 0 \tag{2.80}$$

$$\mathbf{a}_2 = \frac{1}{\sqrt{2}}(\mathbf{m} + \mathbf{n}), \quad \lambda_2 = \gamma \tag{2.81}$$

$$\mathbf{a}_3 = \frac{1}{\sqrt{2}}(\mathbf{m} - \mathbf{n}), \quad \lambda_3 = -\gamma \tag{2.82}$$

Therefore, using the solution (2.67) of Problem 2.1 we find that Eq. (2.77) holds true. This completes solution of Problem 2.3.

Problem 2.4. Let $\mathbf{u}$ and $\mathbf{E}$ denote a displacement vector field and the corresponding strain tensor field defined on $\overline{\text{B}}$. Show that the mean strain $\widehat{\mathbf{E}}(\text{B})$ is represented by the surface integral

$$\widehat{\mathbf{E}}(\text{B}) = \frac{1}{v(\text{B})} \int_{\partial\text{B}} \text{sym}\ (\mathbf{u} \otimes \mathbf{n})\, da \tag{2.83}$$

where $v(\text{B})$ is the volume of B.

Solution. The mean strain $\widehat{\mathbf{E}}(B)$ is defined by

$$\widehat{\mathbf{E}}(B) = \frac{1}{v(B)} \int_B \mathbf{E}\, dv \tag{2.84}$$

In components we obtain

$$\widehat{E}_{ij}(B) = \frac{1}{v(B)} \int_B E_{ij}\, dv \tag{2.85}$$

Since

$$E_{ij} = \frac{1}{2}(u_{i,j} + u_{j,i}) \tag{2.86}$$

therefore, by the divergence theorem,

$$\int_B E_{ij}\, dv = \frac{1}{2}\int_B (u_{i,j} + u_{j,i}) dv = \frac{1}{2}\int_{\partial B} (u_i\, n_j + u_j\, n_i) da \tag{2.87}$$

Equations (2.87) and (2.85) imply that Eq. (2.83) holds true, and this completes solution to Problem 2.4.

Problem 2.5. Show that if $\mathbf{u} = \mathbf{0}$ on ∂B then

$$\int_{\mathrm{B}} (\nabla \mathbf{u})^2 dv \leq 2 \int_{\mathrm{B}} |\mathbf{E}|^2\, dv \tag{2.88}$$

where $\mathbf{E}$ is the strain tensor field corresponding to a displacement field $\mathbf{u}$ on B.

Solution. We recall the relation

$$\nabla \mathbf{u} = \mathbf{E} + \mathbf{W} \tag{2.89}$$

where

$$\mathbf{E} = \frac{1}{2}(\nabla \mathbf{u} + \nabla \mathbf{u}^{\mathrm{T}}) \tag{2.90}$$

and

$$\mathbf{W} = \frac{1}{2}(\nabla \mathbf{u} - \nabla \mathbf{u}^{\mathrm{T}}) \tag{2.91}$$

Since $\mathbf{E} \cdot \mathbf{W} = 0$, Eq. (2.89) implies that

$$|\nabla \mathbf{u}|^2 = |\mathbf{E}|^2 + |\mathbf{W}|^2 \tag{2.92}$$

and it follows from Eqs. (2.90) and (2.91), respectively, that

$$|\mathbf{E}|^2 = \frac{1}{2}[(\nabla \mathbf{u})^2 + (\nabla \mathbf{u}) \cdot (\nabla \mathbf{u}^{\mathrm{T}})] \tag{2.93}$$

and

$$|\mathbf{W}|^2 = \frac{1}{2}[(\nabla \mathbf{u})^2 - (\nabla \mathbf{u}) \cdot (\nabla \mathbf{u}^{\mathrm{T}})] \tag{2.94}$$

Hence,

$$|\mathbf{E}|^2 - |\mathbf{W}|^2 = (\nabla \mathbf{u}) \cdot (\nabla \mathbf{u}^{\mathrm{T}}) \tag{2.95}$$

Now

$$\begin{aligned}
(\nabla \mathbf{u}) \cdot (\nabla \mathbf{u}^{\mathrm{T}}) &= u_{i,j}\, u_{i,j}^{T} = u_{i,j}\, u_{j,i} \\
&= (u_{i,j}\, u_j)_{,i} - u_{i,ji}\, u_j \\
&= (u_{i,j}\, u_j)_{,i} - (u_{i,i}\, u_j)_{,j} + (u_{i,i})^2
\end{aligned}$$

$$= (u_{j,i}\, u_i - u_{i,i}\, u_j)_{,j} + (u_{i,i})^2$$
$$= \text{div}[(\nabla \mathbf{u})\mathbf{u} - (\text{div}\, \mathbf{u})\mathbf{u}] + (\text{div}\, \mathbf{u})^2 \tag{2.96}$$

Therefore, integrating Eq. (2.96) over B, using the divergence theorem, and the homogeneous boundary condition: $\mathbf{u} = \mathbf{0}$ on ∂B, we obtain

$$\int_B (\nabla \mathbf{u}) \cdot (\nabla \mathbf{u}^{\mathrm{T}}) dv = \int_B (\text{div}\, \mathbf{u})^2 dv \tag{2.97}$$

Equations (2.95) and (2.97) imply that

$$\int_B (|\mathbf{E}|^2 - |\mathbf{W}|^2) dv = \int_B (\text{div}\, \mathbf{u})^2 dv \tag{2.98}$$

and it follows from Eq. (2.92) that

$$\int_B (|\mathbf{E}|^2 + |\mathbf{W}|^2) dv = \int_B |\nabla \mathbf{u}|^2 dv \tag{2.99}$$

Therefore, by adding Eqs. (2.98) and (2.99), we obtain

$$2 \int_B |\mathbf{E}|^2 dv = \int_B |\nabla \mathbf{u}|^2 dv + \int_B (\text{div}\, \mathbf{u})^2 dv \tag{2.100}$$

and Eq. (2.100) leads to the inequality

$$2 \int_B |\mathbf{E}|^2 dv \geq \int_B |\nabla \mathbf{u}|^2 dv \tag{2.101}$$

This completes solution of Problem 2.5.

Problem 2.6. (i) Let $\mathbf{E}$ be a strain tensor field on E^3 defined by the matrix

$$\mathbf{E} = \frac{N}{E} \begin{bmatrix} 1 & 0 & 0 \\ 0 & -\nu & 0 \\ 0 & 0 & -\nu \end{bmatrix} \tag{2.102}$$

where E, N, and ν are positive constants. Show that a solution $\mathbf{u}$ to the equation $\mathbf{E} = \widehat{\nabla} \mathbf{u}$ on E^3 subject to the condition $\mathbf{u}(\mathbf{0}) = \mathbf{0}$ takes the form

$$\mathbf{u} = \left[\frac{N}{E} x_1, \ -\nu \frac{N}{E} x_2, \ -\nu \frac{N}{E} x_3 \right]^{\mathrm{T}} \tag{2.103}$$

(ii) Let $\mathbf{E}$ be a strain tensor field on E^3 defined by the matrix

$$\mathbf{E} = \frac{M}{EI} x_1 \begin{bmatrix} \nu & 0 & 0 \\ 0 & \nu & 0 \\ 0 & 0 & -1 \end{bmatrix} \tag{2.104}$$

where M, E, I, and ν are positive constants. Show that a solution $\mathbf{u}$ to the equation $\mathbf{E} = \widehat{\nabla}\mathbf{u}$ on E^3 subject to the condition $\mathbf{u}(\mathbf{0}) = \mathbf{0}$ takes the form

$$\mathbf{u} = \frac{M}{EI} \left[\frac{1}{2}(x_3^2 + \nu x_1^2 - \nu x_2^2), \quad \nu x_1 x_2, \quad -x_1 x_3 \right]^{\mathrm{T}} \tag{2.105}$$

Solution. (i) Using (2.103) we find that $\mathbf{u}(\mathbf{0}) = \mathbf{0}$ and

$$\nabla\mathbf{u} = \frac{N}{E} \begin{bmatrix} 1 & 0 & 0 \\ 0 & -\nu & 0 \\ 0 & 0 & -\nu \end{bmatrix} \tag{2.106}$$

Since $\nabla\mathbf{u} = \nabla\mathbf{u}^{\mathrm{T}}$, the equation

$$\widehat{\nabla}\mathbf{u} = \mathbf{E} \tag{2.107}$$

in which $\mathbf{E}$ is given by (2.102) is identically satisfied. This completes a proof of (i).

(ii) Using (2.105) we obtain $\mathbf{u}(\mathbf{0}) = \mathbf{0}$ and

$$\nabla\mathbf{u} = \frac{M}{EI} \begin{bmatrix} \nu x_1 & -\nu x_2 & x_3 \\ \nu x_2 & \nu x_1 & 0 \\ -x_3 & 0 & -x_1 \end{bmatrix} \tag{2.108}$$

Hence

$$\nabla\mathbf{u}^{\mathrm{T}} = \frac{M}{EI} \begin{bmatrix} \nu x_1 & \nu x_2 & -x_3 \\ -\nu x_2 & \nu x_1 & 0 \\ x_3 & 0 & -x_1 \end{bmatrix} \tag{2.109}$$

and

$$\widehat{\nabla}\mathbf{u} = \frac{M}{EI} \begin{bmatrix} \nu x_1 & 0 & 0 \\ 0 & \nu x_1 & 0 \\ 0 & 0 & -x_1 \end{bmatrix} \tag{2.110}$$

Equation (2.110) implies that $\mathbf{u}$ given by (2.105) satisfies the equation

$$\widehat{\nabla}\mathbf{u} = \mathbf{E} \tag{2.111}$$

where $\mathbf{E}$ is given by (2.104). This completes proof of (ii).

Problem 2.7. Given a stress tensor $\mathbf{S}$ at a point A, find: (i) the stress vector $\mathbf{s}$ on a plane through A parallel to the plane $\mathbf{n} \cdot \mathbf{x} - \mathrm{vt} = 0 \quad (|\mathbf{n}| = 1,\ \mathrm{v} > 0,\ \mathrm{t} \geq 0)$, (ii) the magnitude of $\mathbf{s}$, (iii) the angle between $\mathbf{s}$ and the normal to the plane, and (iv) the normal and tangential components of the stress vector $\mathbf{s}$.

Answers. (i) $\mathbf{s} = \mathbf{Sn}$; (ii) $|\mathbf{s}| = |\mathbf{Sn}|$; (iii) $\cos\ \theta = \mathbf{s} \cdot \mathbf{n}/\ |\mathbf{s}|$; (iv) $\mathbf{s} = \mathbf{s}_{\mathrm{n}} + \mathbf{s}_{\tau}$, where $\mathbf{s}_{\mathrm{n}} = (\mathbf{n} \cdot \mathbf{s})\, \mathbf{n}$ and $\mathbf{s}_{\tau} = \mathbf{n} \times (\mathbf{s} \times \mathbf{n})$.

Solution. Solution to Problem 2.7 is presented by the answers (i)–(iv).

Problem 2.8. Let $\{\mathbf{e}_i\}$ be an orthonormal basis for a stress tensor $\mathbf{S}$, and let $\{\mathbf{e}_i^*\}$ be an orthonormal basis formed by the eigenvectors of $\mathbf{S}$. Then a tensor $\mathbf{S}^*$ obtained from $\mathbf{S}$ by the transformation formula from $\{\mathbf{e}_i\}$ to $\{\mathbf{e}_i^*\}$ takes the form

$$\mathbf{S}^* = \lambda_1\, \mathbf{e}_1^* \otimes \mathbf{e}_1^* + \lambda_2\, \mathbf{e}_2^* \otimes \mathbf{e}_2^* + \lambda_3\, \mathbf{e}_3^* \otimes \mathbf{e}_3^* \tag{2.112}$$

where λ_i is an eigenvalue of $\mathbf{S}$ corresponding to the eigenvector $\mathbf{e}_i^*$. Show that the function

$$g(\mathbf{n}^*) = \left|\mathbf{s}_\tau^*\right| = \left|\mathbf{n}^* \times (\mathbf{S}^*\mathbf{n}^* \times \mathbf{n}^*)\right| \tag{2.113}$$

representing the tangent stress vector magnitude with regard to a plane with a normal $\mathbf{n}^*$ in the $\{\mathbf{e}_i^*\}$ basis, assumes the extreme values

$$\left|\mathbf{s}_\tau^*\right|_1 = \frac{1}{2}\,|\lambda_2 - \lambda_3| \tag{2.114}$$

$$\left|\mathbf{s}_\tau^*\right|_2 = \frac{1}{2}\,|\lambda_3 - \lambda_1| \tag{2.115}$$

and

$$\left|\mathbf{s}_\tau^*\right|_3 = \frac{1}{2}\,|\lambda_1 - \lambda_2| \tag{2.116}$$

at

$$\mathbf{n}_1^* = [0,\ \ \pm 1/\sqrt{2},\ \ \pm 1/\sqrt{2}]^{\mathrm{T}} \tag{2.117}$$

$$\mathbf{n}_2^* = [\,\pm 1/\sqrt{2},\ \ 0,\ \ \pm 1/\sqrt{2}]^{\mathrm{T}} \tag{2.118}$$

and

$$\mathbf{n}_3^* = [\,\pm 1/\sqrt{2},\ \ \pm 1/\sqrt{2},\ 0]^{\mathrm{T}} \tag{2.119}$$

respectively. Hence, if $\lambda_1 > \lambda_2 > \lambda_3$ then the largest tangential stress vector magnitude is

$$\left|\mathbf{s}_\tau^*\right|_2 = \frac{1}{2}\,|\lambda_3 - \lambda_1| \tag{2.120}$$

and this extreme vector acts on the plane that bisects the angle between $\mathbf{e}_1^*$ and $\mathbf{e}_3^*$.

Solution. It follows from (iv) of Problem 2.7 that

$$\mathbf{s}^* = \mathbf{s}_n^* + \mathbf{s}_\tau^* \tag{2.121}$$

where

$$\mathbf{s}^* = \mathbf{S}^*\mathbf{n}^* \tag{2.122}$$

and

$$\mathbf{s}_n^* = (\mathbf{s}^* \cdot \mathbf{n}^*)\mathbf{n}, \quad \mathbf{s}_\tau^* = \mathbf{n}^* \times (\mathbf{s}^* \times \mathbf{n}) \tag{2.123}$$

Using (2.112), (2.122) and (2.123), we obtain

$$\mathbf{s}^* = \lambda_1 n_1^* \mathbf{e}_1^* + \lambda_2 n_2^* \mathbf{e}_2^* + \lambda_3 n_3^* \mathbf{e}_3^* \tag{2.124}$$

and

$$\mathbf{s}^* \cdot \mathbf{n}^* = \lambda_1 (n_1^*)^2 + \lambda_2 (n_2^*)^2 + \lambda_3 (n_3^*)^2 \tag{2.125}$$

Since $\mathbf{s}_n^* \cdot \mathbf{s}_\tau^* = 0$, by squaring (2.121), we get

$$|\mathbf{s}^*|^2 = \left|\mathbf{s}_n^*\right|^2 + \left|\mathbf{s}_\tau^*\right|^2 \tag{2.126}$$

Now, introduce the function

$$\begin{aligned} f(\mathbf{n}^*) = \left|\mathbf{s}_\tau^*\right|^2 = |\mathbf{s}^*|^2 - \left|\mathbf{s}_n^*\right|^2 &= \lambda_1^2 (n_1^*)^2 + \lambda_2^2 (n_2^*)^2 + \lambda_3^2 (n_3^*)^2 \\ &- \left[\lambda_1 (n_1^*)^2 + \lambda_2 (n_2^*)^2 + \lambda_3^2 (n_3^*)^2\right]^2 \end{aligned} \tag{2.127}$$

If there is an extremum of $f = f(\mathbf{n}^*)$, treated as a function of $n_1^*,\ n_2^*$, and n_3^*, it is also an extremum of $g = g(\mathbf{n}^*) = \sqrt{f(\mathbf{n}^*)}$.

To find the extreme values of $f = f(\mathbf{n}^*)$ subject to the condition $|\mathbf{n}^*| = 1$ we solve the algebraic equation for $\mathbf{n}^*$

$$\frac{\partial}{\partial n_i^*}\left[f(\mathbf{n}^*) - t(|\mathbf{n}^*|^2 - 1)\right] = 0 \tag{2.128}$$

where t is a Lagrangian multiplier. In expanded form Eq. (2.128) takes the form

$$\left[\lambda_1^2 - 2\lambda_1 (\mathbf{s}^* \cdot \mathbf{n}^*) - t\right] n_1^* = 0 \tag{2.129}$$

$$\left[\lambda_2^2 - 2\lambda_2 (\mathbf{s}^* \cdot \mathbf{n}^*) - t\right] n_2^* = 0 \tag{2.130}$$

$$\left[\lambda_3^2 - 2\lambda_3 (\mathbf{s}^* \cdot \mathbf{n}^*) - t\right] n_3^* = 0 \tag{2.131}$$

Fig. 2.1 The *wedge* region

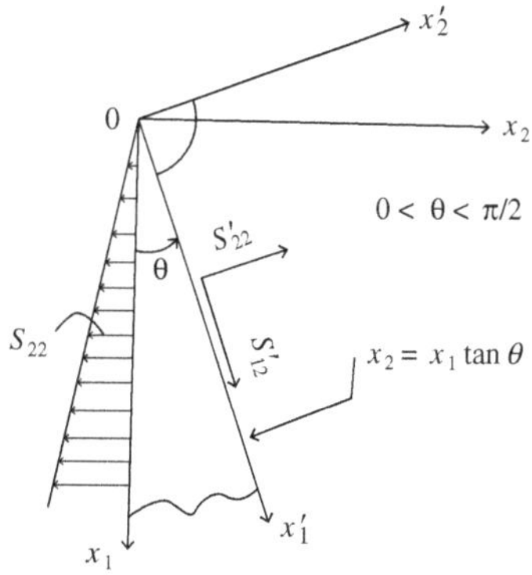

where $(\mathbf{s}^* \cdot \mathbf{n}^*)$ is given by (2.125). It can be verified that the unit vectors $\mathbf{n}_1^*$, $\mathbf{n}_2^*$, and $\mathbf{n}_3^*$, given by Eqs. (2.117), (2.118), and (2.119), respectively, satisfy Eqs. (2.129)–(2.131) with $t = \lambda_2\lambda_3$. In addition, by substituting $\mathbf{n}_1^*$, $\mathbf{n}_2^*$, and $\mathbf{n}_3^*$ into (2.127), we obtain Eqs. (2.114), (2.115), and (2.116), respectively. Also, the vector $\mathbf{n}_2^*$ that is normal to the surface element on which the largest tangential stress vector $(\mathbf{s}_\tau^*)_2$ acts bisects the angle between $\mathbf{e}_1^*$ and $\mathbf{e}_3^*$. This completes solution of Problem 2.8.

Problem 2.9. Let $D = \{\mathbf{x} : \ x_1 \geq 0, \quad x_1 \tan\theta \geq x_2 \geq 0\}$ be a two-dimensional wedge region shown in the Fig. 2.1, and let $S_{\alpha\beta} = S_{\alpha\beta}(\mathbf{x}), \ [\mathbf{x} = (x_1, x_2);\ \alpha, \beta = 1, 2]$ be a symmetric tensor field on D defined by

$$S_{11} = d\,x_2 + e\,x_1 - \rho\,g x_1, \quad S_{22} = -\gamma x_1, \quad S_{12} = S_{21} = -e\,x_2 \tag{2.132}$$

where d, e, g, ρ, and γ are constants [$g > 0,\ \rho > 0,\ \gamma > 0$]. (i) Show that

$$\text{div}\,\mathbf{S} + \mathbf{b} = \mathbf{0} \quad \text{on} \quad D \tag{2.133}$$

where

$$\mathbf{b} = [\rho\,g,\ 0]^{\mathrm{T}} \quad \text{on} \quad D \tag{2.134}$$

(ii) Using the transformation formula from the system x_α to the system x_α' [see Eq. (1.157) in Problem 1.8] find the components $S_{\alpha\beta}'$ in terms of $S_{\alpha\beta}$, and show that

$$S_{12}' = 0 \quad \text{and} \quad S_{22}' = 0 \quad \text{for} \quad x_2 = x_1 \tan\theta \tag{2.135}$$

provided

$$e = \frac{\gamma}{\tan^2\theta}, \quad \text{and} \quad d = \frac{\rho\,g}{\tan\theta} - \frac{2\gamma}{\tan^3\theta} \tag{2.136}$$

(iii) Give diagrams of S_{11} and S_{12} over a horizontal section $x_1 = x_1^0 =$ constant.
(iv) Give a diagram of S_{22} over the vertical section $x_2 = 0$.

Solution. To show (i) we note that $S_{\alpha\beta} = S_{\alpha\beta}(x_1, x_2)$ given by Eq. (2.132) satisfies the equilibrium equation

$$S_{\alpha\beta,\beta} + b_\alpha = 0 \quad \text{on } D \tag{2.137}$$

since

$$S_{1\beta,\beta} = -\rho\, g, \quad S_{2\beta,\beta} = 0 \quad \text{on } D \tag{2.138}$$

for arbitrary constants d, e, g, ρ, and γ. To show (ii) we use the transformation formulas [see Eq. (1.157) in Problem 1.8]

$$S'_{11} = S_{11} \cos^2\theta + S_{12} \sin 2\theta + S_{22} \sin^2\theta \tag{2.139}$$

$$S'_{12} = \frac{1}{2}(S_{22} - S_{11}) \sin 2\theta + S_{12} \cos 2\theta \tag{2.140}$$

$$S'_{22} = S_{11} \sin^2\theta - S_{12} \sin 2\theta + S_{22} \cos^2\theta \tag{2.141}$$

[see Fig. 2.1].

The components S_{11}, S_{12}, and S_{22} taken on the line $x_2 = x_1 \tan\theta$ assume the forms

$$S_{11}(x_1, x_1 \tan\theta) = (d \tan\theta + e - \rho g)x_1 \tag{2.142}$$

$$S_{12}(x_1, x_1 \tan\theta) = -(e \tan\theta)x_1 \tag{2.143}$$

$$S_{22}(x_1, x_1 \tan\theta) = -\gamma\, x_1 \tag{2.144}$$

Therefore, substituting (2.142)–(2.144) into the RHS$'$ of (2.140) and (2.141), and equating the results to zero, we obtain the algebraic equations for the unknown constants e and d, provided γ and ρg are prescribed

$$\begin{aligned} &e(\sin^2\theta + \tan\theta \sin 2\theta) + d \tan\theta \sin^2\theta \\ &\quad = \gamma \cos^2\theta + \rho g \sin^2\theta \end{aligned} \tag{2.145}$$

$$\begin{aligned} &e(\sin 2\theta + 2\tan\theta \cos 2\theta) + d \tan\theta \sin 2\theta \\ &\quad = -\gamma \sin 2\theta + \rho g \sin 2\theta \end{aligned} \tag{2.146}$$

Dividing Eq. (2.145) by $\sin^2\theta$ and Eq. (2.146) by $\sin 2\theta$ and introducing the notation

$$\tan\theta = u \tag{2.147}$$

we obtain

$$\begin{aligned} 3e + du &= \frac{\gamma}{u^2} + \rho g \\ (2 - u^2)e + du &= -\gamma + \rho g \end{aligned} \tag{2.148}$$

It is easy to check that a unique solution (e, d) of Eqs. (2.148) takes the form (2.136) that is

$$e = \gamma/u^2, \quad d = \rho g/u - 2\gamma/u^3 \tag{2.149}$$

This completes proof of (ii).

Finally, when $x_1 = x_1^0 = \text{const}$, S_{11} and S_{12} are represented by straight lines on the planes (x_2, S_{11}) and (x_2, S_{12}), respectively, and S_{22} at $x_2 = 0$ is represented by a straight line passing through the origin 0 as shown in Fig. 2.1. This completes solution to Problem 2.9.

Problem 2.10. Let B denote a cylinder of length l and of arbitrary cross section, suspended from the upper end and subject to its own weight ρg. Then the stress tensor $\mathbf{S} = \mathbf{S}(\mathbf{x})$ on B takes the form

$$\mathbf{S} = \begin{bmatrix} 0 & 0 & 0 \\ 0 & 0 & 0 \\ 0 & 0 & \rho g x_3 \end{bmatrix} \tag{2.150}$$

since, in this case, the body force vector field is given by $\mathbf{b} = [0, \ 0, \ -\rho\, g]^T$, and $\text{div}\, \mathbf{S} + \mathbf{b} = \mathbf{0}$ on B. The stress vector $\mathbf{s}$ associated with $\mathbf{S}$ on ∂B has the following properties: $\mathbf{s} = [0, 0, \rho\, g\, l\,]^T$ on the end plane $x_3 = l$; and $\mathbf{s} = \mathbf{0}$ on the plane $x_3 = 0$ and on the lateral surface of the cylinder since $\mathbf{n} = [n_1, n_2, 0]^T$ on the surface. Assuming that the cylinder is made of a homogeneous isotropic elastic material, the associated strain tensor field $\mathbf{E}$ takes the form [see Eqs. (2.49)]

$$\mathbf{E} = \frac{\rho\, g\, x_3}{E} \begin{bmatrix} -\nu & 0 & 0 \\ 0 & -\nu & 0 \\ 0 & 0 & 1 \end{bmatrix} \tag{2.151}$$

where E and ν are Young's modulus and Poisson's ratio, respectively.

(i) Show that a solution $\mathbf{u}$ of the equation

$$\mathbf{E} = \widehat{\nabla} \mathbf{u} \quad \text{on} \quad \text{B} \tag{2.152}$$

subject to the condition

$$\mathbf{u}(0, 0, l) = \mathbf{0} \tag{2.153}$$

takes the form

$$\mathbf{u} = \frac{\rho\, g}{E} \left[-\nu x_1 x_3, \quad -\nu x_2 x_3, \quad \frac{\nu}{2}(x_1^2 + x_2^2) + \frac{1}{2}(x_3^2 - l^2) \right]^T \tag{2.154}$$

(ii) Plot $u_3 = u_3(0, 0, x_3)$ over the range $0 \leq x_3 \leq l$.

Solution. To solve the problem we use (2.154) and obtain

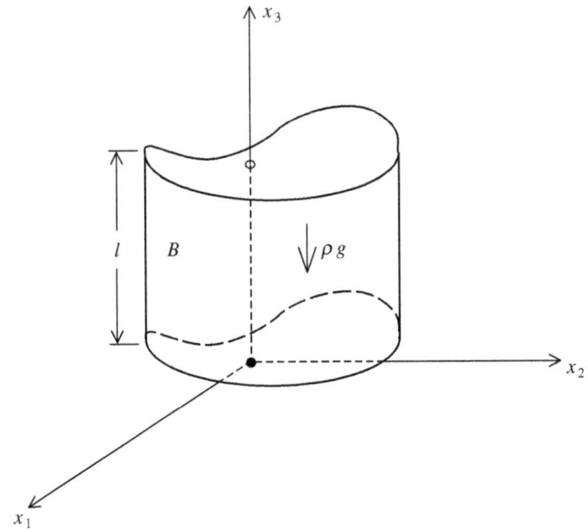

Fig. 2.2 The *cylinder* of arbitrary cross section

$$\nabla \mathbf{u} = \frac{\rho g}{E} \begin{bmatrix} -\nu x_3 & 0 & -\nu x_1 \\ 0 & -\nu x_3 & -\nu x_2 \\ \nu x_1 & \nu x_2 & x_3 \end{bmatrix} \tag{2.155}$$

and

$$\nabla \mathbf{u}^{\mathrm{T}} = \frac{\rho g}{E} \begin{bmatrix} -\nu x_3 & 0 & \nu x_1 \\ 0 & -\nu x_3 & \nu x_2 \\ -\nu x_1 & -\nu x_2 & x_3 \end{bmatrix} \tag{2.156}$$

Hence

$$\widehat{\nabla} \mathbf{u} = \frac{\rho g}{E} \begin{bmatrix} -\nu x_3 & 0 & 0 \\ 0 & -\nu x_3 & 0 \\ 0 & 0 & x_3 \end{bmatrix} \tag{2.157}$$

Therefore, **u** given by (2.154) satisfies (2.152). Also, it is easy to prove that **u** satisfies (2.153). Finally, $u_3 = u_3(0, 0, x_3)$ is represented by a parabolic curve restricted to the interval $0 \le x_3 \le \ell$. This completes solution to Problem 2.10.

Problem 2.11. For a transversely isotropic elastic body each material point posses an axis of rotational symmetry, which means that the elastic properties are the same in any direction on any plane perpendicular to the axis, but they are different than those in the direction of the axis. If the x_3 axis coincides with the axis of symmetry, then the stress-strain relation for such a body takes the form

$$\begin{bmatrix} S_{11} \\ S_{22} \\ S_{33} \\ S_{32} \\ S_{31} \\ S_{12} \end{bmatrix} = \begin{bmatrix} c_{11} & c_{12} & c_{13} & 0 & 0 & 0 \\ c_{12} & c_{11} & c_{13} & 0 & 0 & 0 \\ c_{13} & c_{13} & c_{33} & 0 & 0 & 0 \\ 0 & 0 & 0 & c_{44} & 0 & 0 \\ 0 & 0 & 0 & 0 & c_{44} & 0 \\ 0 & 0 & 0 & 0 & 0 & (c_{11}-c_{12})/2 \end{bmatrix} \begin{bmatrix} E_{11} \\ E_{22} \\ E_{33} \\ 2E_{32} \\ 2E_{31} \\ 2E_{12} \end{bmatrix} \tag{2.158}$$

where $\mathbf{S}$ and $\mathbf{E}$ are the stress and strain tensors, respectively, and five numerically independent moduli $c_{11}, c_{33}, c_{12}, c_{13},$ and c_{44} are related to the components C_{ijkl} of the fourth-order elasticity tensor C by [see Eq. (2.35)]

$$c_{11} = C_{1111}, \quad c_{12} = C_{1122}, \quad c_{13} = C_{1133}, \quad c_{33} = C_{3333}, \quad c_{44} = C_{1313} \tag{2.159}$$

Show that if the axis of symmetry of a transversely isotropic body coincides with the direction of an arbitrary unit vector $\mathbf{e}$, then the stress-strain relation takes the form

$$\begin{aligned} \mathbf{S} = \mathsf{C}[\mathbf{E}] &= (c_{11} - c_{12})\mathbf{E} + \{c_{12}(\operatorname{tr}\mathbf{E}) - (c_{12} - c_{13})\mathbf{e}\cdot(\mathbf{E}\mathbf{e})\}\,\mathbf{1} \\ &\quad - (c_{11} - c_{12} - 2c_{44})\{\mathbf{e}\otimes(\mathbf{E}\mathbf{e}) + (\mathbf{E}\mathbf{e})\otimes\mathbf{e}\} \\ &\quad - \{(c_{12} - c_{13})(\operatorname{tr}\mathbf{E}) - (c_{11} + c_{33} - 2c_{13} - 4c_{44})\mathbf{e}\cdot(\mathbf{E}\mathbf{e})\}\mathbf{e}\otimes\mathbf{e} \end{aligned} \tag{2.160}$$

Solution. For a transversely isotropic body in which the axis of symmetry coincides with an arbitrary unit vector $\mathbf{e}$, the stress–strain relation takes the form[1]

$$\mathbf{S} = \mathsf{C}[\mathbf{E}] \tag{2.161}$$

where $\mathbf{S}$ and $\mathbf{E}$ are the stress and strain tensors, respectively, and the elasticity tensor C is given by

$$\begin{aligned} \mathsf{C} = c_{33}\mathsf{C}^{(1)} + (c_{11} + c_{12})\mathsf{C}^{(2)} + \sqrt{2}\, c_{13}(\mathsf{C}^{(3)} + \mathsf{C}^{(4)}) \\ + (c_{11} - c_{12})\mathsf{C}^{(5)} + 2\, c_{44}\mathsf{C}^{(6)} \end{aligned} \tag{2.162}$$

In Eq. (2.162) the tensors $\mathsf{C}^{(a)}$, $a = 1, 2, 3, 4, 5, 6$, are defined by

$$\begin{aligned} C^{(1)}_{ijkl} &= A_{ij}\, A_{kl}, \quad C^{(2)}_{ijkl} = \frac{1}{2}\, B_{ij}\, B_{kl} \\ C^{(3)}_{ijkl} &= \frac{1}{\sqrt{2}}\, A_{ij}\, B_{kl}, \quad C^{(4)}_{ijkl} = \frac{1}{\sqrt{2}}\, B_{ij}\, A_{kl} \\ C^{(5)}_{ijkl} &= \frac{1}{2}(B_{ik}\, B_{jl} + B_{il}\, B_{jk} - B_{ij}\, B_{kl}) \\ C^{(6)}_{ijkl} &= \frac{1}{2}(A_{ik}\, B_{jl} + A_{il}\, B_{jk} + A_{jk}\, B_{il} + A_{jl}\, B_{ik}) \end{aligned} \tag{2.163}$$

[1] See P. Chadwick, Proc. R. Soc. London, A **422**, p. 26 (1989).

where

$$A_{ij} = e_i\, e_j, \quad B_{ij} = \delta_{ij} - e_i\, e_j \tag{2.164}$$

and e_i are the components of $\mathbf{e}$ in the coordinates $\{x_i\}$.

Using (2.163) and (2.164), we obtain

$$C^{(1)}_{ijkl}\, E_{kl} = e_i\, e_j\, e_k\, e_l\, E_{kl} \tag{2.165}$$

or in direct notation

$$\mathbf{C}^{(1)}[\mathbf{E}] = [\mathbf{e} \cdot (\mathbf{E}\mathbf{e})]\mathbf{e} \otimes \mathbf{e} \tag{2.166}$$

Similarly, by (2.163) and (2.164), we get

$$C^{(2)}_{ijkl}\, E_{kl} = \frac{1}{2}(\delta_{ij} - e_i\, e_j)(\delta_{kl} - e_k\, e_l)E_{kl} \tag{2.167}$$

or

$$\mathbf{C}^{(2)}[\mathbf{E}] = \frac{1}{2}(\mathbf{1} - \mathbf{e} \otimes \mathbf{e})[\mathrm{tr}\, \mathbf{E} - \mathbf{e} \cdot (\mathbf{E}\mathbf{e})] \tag{2.168}$$

Also, using (2.163) and (2.164), we obtain

$$\sqrt{2}\left(C^{(3)}_{ijkl} + C^{(4)}_{ijkl}\right)E_{kl} = e_i\, e_j(\delta_{kl} - e_k\, e_l)E_{kl} + (\delta_{ij} - e_i\, e_j)e_k\, e_l\, E_{kl} \tag{2.169}$$

or

$$\sqrt{2}(\mathbf{C}^{(3)}[\mathbf{E}] + \mathbf{C}^{(4)}[\mathbf{E}]) = [\mathbf{e} \cdot (\mathbf{E}\mathbf{e})]\mathbf{1} + [\mathrm{tr}\, \mathbf{E} - 2\mathbf{e} \cdot (\mathbf{E}\mathbf{e})]\mathbf{e} \otimes \mathbf{e} \tag{2.170}$$

and

$$\begin{aligned} C^{(5)}_{ijkl}\, E_{kl} &= E_{ij} - e_i\, e_k\, E_{jk} - e_j\, e_k\, E_{ik} + e_i\, e_j\, e_k\, e_l\, E_{kl} \\ &\quad - \frac{1}{2}(\delta_{ij} - e_i\, e_j)(E_{kk} - e_k\, e_l\, E_{kl}) \end{aligned} \tag{2.171}$$

or

$$\begin{aligned} \mathbf{C}^{(5)}[\mathbf{E}] &= \mathbf{E} - \mathbf{e} \otimes (\mathbf{E}\mathbf{e}) - (\mathbf{E}\mathbf{e}) \otimes \mathbf{e} + \mathbf{e} \otimes \mathbf{e}(\mathbf{e} \cdot \mathbf{E}\mathbf{e}) \\ &\quad - \frac{1}{2}(\mathbf{1} - \mathbf{e} \otimes \mathbf{e})(\mathrm{tr}\, \mathbf{E} - \mathbf{e} \cdot \mathbf{E}\mathbf{e}) \end{aligned} \tag{2.172}$$

and

$$C^{(6)}_{ijkl}\, E_{kl} = e_i\, E_{jk}\, e_k + E_{ik}\, e_k\, e_j - 2e_i\, e_j\, e_k\, e_l\, E_{kl} \tag{2.173}$$

or

$$\mathbf{C}^{(6)}[\mathbf{E}] = \mathbf{e} \otimes \mathbf{E}\mathbf{e} + (\mathbf{E}\mathbf{e}) \otimes \mathbf{e} - 2\mathbf{e} \otimes \mathbf{e}[\mathbf{e} \cdot (\mathbf{E}\mathbf{e})] \tag{2.174}$$

Therefore, substituting $\mathbf{C}$ from (2.162) into (2.161) and using (2.166), (2.168), (2.170), (2.172), and (2.174), we obtain (2.160). Note that the representation (2.160) coincides with (2.158) if $\mathbf{e} = (0, 0, 1)$. This can be proved by substituting $\mathbf{e} = (0, 0, 1)$ into (2.160).

Problem 2.12. Show that the stress-strain relation (2.160) in Problem 2.11 is invertible provided

$$c \equiv (c_{11} + c_{12})c_{33} - 2c_{13}^2 > 0, \quad c_{11} > |c_{12}|, \quad c_{44} > 0 \tag{2.175}$$

and that the strain-stress relation reads

$$\begin{aligned} \mathbf{E} = \mathbf{K}[\mathbf{S}] = (c_{11} - c_{12})^{-1}\mathbf{S} + \frac{1}{2}\,&\big[\{c^{-1}c_{33} - (c_{11} - c_{12})^{-1}\}\,(\mathrm{tr}\,\mathbf{S}) \\ &- \{c^{-1}(c_{33} + 2c_{13}) - (c_{11} - c_{12})^{-1}\}\,\mathbf{e}\cdot(\mathbf{S}\mathbf{e})\big]\,\mathbf{1} \\ &- \left\{(c_{11} - c_{12})^{-1} - \frac{1}{2}c_{44}^{-1}\right\}\{\mathbf{e}\otimes(\mathbf{S}\mathbf{e}) + (\mathbf{S}\mathbf{e})\otimes\mathbf{e}\} \\ &- \frac{1}{2}\,\big[\{c^{-1}(c_{33} + 2c_{13}) - (c_{11} - c_{12})^{-1}\}\,(\mathrm{tr}\,\mathbf{S}) \\ &- \{c^{-1}(2c_{11} + c_{33} + 2c_{12} + 4c_{13}) + (c_{11} - c_{12})^{-1} - 2c_{44}^{-1}\}\,\mathbf{e}\cdot(\mathbf{S}\mathbf{e})\big]\,\mathbf{e}\otimes\mathbf{e} \end{aligned} \tag{2.176}$$

Solution. To show that (2.160) in Problem 2.11 is invertible if the inequalities (2.175) are satisfied, and the inverted formula takes the form (2.176), consider the fourth-order tensor

$$\mathbf{A} = a_1\mathbf{C}^{(1)} + a_2\mathbf{C}^{(2)} + a_3(\mathbf{C}^{(3)} + \mathbf{C}^{(4)}) + a_5\mathbf{C}^{(5)} + a_6\mathbf{C}^{(6)} \tag{2.177}$$

where $a_1,\ a_2,\ a_3,\ a_5$, and a_6 are scalars that satisfy the inequalities

$$a_1 > 0, \quad a_2 > 0, \quad a_1a_2 - a_3^2 > 0, \quad a_5 > 0, \quad a_6 > 0 \tag{2.178}$$

and $\mathbf{C}^{(a)}$ $(a = 1, 2, 3, 4, 5, 6)$ are the fourth-order tensors defined by (2.163) and (2.164) in Problem 2.11.

Then, there is $\mathbf{A}^{-1}$ in the form

$$\mathbf{A}^{-1} = \left(a_1a_2 - a_3^2\right)^{-1}\{a_2\mathbf{C}^{(1)} + a_1\mathbf{C}^{(2)} - a_3(\mathbf{C}^{(3)} + \mathbf{C}^{(4)})\} + a_5^{-1}\mathbf{C}^{(5)} + a_6^{-1}\mathbf{C}^{(6)} \tag{2.179}$$

such that

$$\mathbf{A}\mathbf{A}^{-1} = \mathbf{A}^{-1}\mathbf{A} = \mathbf{1} \tag{2.180}$$

where $\mathbf{1}$ is the fourth-order identity tensor with components

$$I_{ijkl} = \frac{1}{2}(\delta_{ik}\,\delta_{jl} + \delta_{jk}\,\delta_{il}) \tag{2.181}$$

To prove (2.180) we use (2.163) and (2.164) of Problem 2.11 to obtain the 6×6 tensor matrix

$$[\mathbf{C}^{(a)}\,\mathbf{C}^{(b)}] = \begin{bmatrix} \mathbf{C}^{(1)} & \mathbf{0} & \mathbf{C}^{(3)} & \mathbf{0} & \mathbf{0} & \mathbf{0} \\ \mathbf{0} & \mathbf{C}^{(2)} & \mathbf{0} & \mathbf{C}^{(4)} & \mathbf{0} & \mathbf{0} \\ \mathbf{0} & \mathbf{C}^{(3)} & \mathbf{0} & \mathbf{C}^{(1)} & \mathbf{0} & \mathbf{0} \\ \mathbf{C}^{(4)} & \mathbf{0} & \mathbf{C}^{(2)} & \mathbf{0} & \mathbf{0} & \mathbf{0} \\ \mathbf{0} & \mathbf{0} & \mathbf{0} & \mathbf{0} & \mathbf{C}^{(5)} & \mathbf{0} \\ \mathbf{0} & \mathbf{0} & \mathbf{0} & \mathbf{0} & \mathbf{0} & \mathbf{C}^{(6)} \end{bmatrix} \tag{2.182}$$

as well as the identity

$$\mathbf{C}^{(1)} + \mathbf{C}^{(2)} + \mathbf{C}^{(5)} + \mathbf{C}^{(6)} = \mathbf{1} \tag{2.183}$$

Calculating the tensor $\mathbf{A}\,\mathbf{A}^{-1}$, by using Eqs. (2.177), (2.179), (2.182), and (2.183) we obtain

$$\mathbf{A}\,\mathbf{A}^{-1} = \mathbf{1} \tag{2.184}$$

Similarly, using Eqs. (2.177), (2.179), (2.182), and (2.183) we get

$$\mathbf{A}^{-1}\,\mathbf{A} = \mathbf{1} \tag{2.185}$$

Now, by letting

$$\begin{aligned} &a_1 = c_{33}, \quad a_2 = c_{11} + c_{12}, \quad a_3 = \sqrt{2}\,c_{13} \\ &a_5 = c_{11} - c_{12}, \quad a_6 = 2c_{44} \end{aligned} \tag{2.186}$$

in Eq. (2.177) we obtain $\mathbf{A} = \mathbf{C}$, and the inequalities (2.178) reduce to those of (2.175). Also, Eq. (2.179) reduces to

$$\begin{aligned} \mathbf{C}^{-1} = \mathbf{K} = c^{-1}\{(c_{11} + c_{12})\mathbf{C}^{(1)} + c_{33}\mathbf{C}^{(2)} - \sqrt{2}\,c_{13}(\mathbf{C}^{(3)} + \mathbf{C}^{(4)})\} \\ + (c_{11} - c_{12})^{-1}\mathbf{C}^{(5)} + 2^{-1}c_{44}^{-1}\mathbf{C}^{(6)} \end{aligned} \tag{2.187}$$

where

$$c = (c_{11} + c_{12})c_{33} - 2c_{13}^2 > 0 \tag{2.188}$$

Therefore, the strain–stress relation reads

$$\begin{aligned} \mathbf{E} = \mathbf{K}[\mathbf{S}] = c^{-1}\{(c_{11} + c_{12})\mathbf{C}^{(1)}[\mathbf{S}] + c_{33}\mathbf{C}^{(2)}[\mathbf{S}] - \sqrt{2}\,c_{13}(\mathbf{C}^{(3)}[\mathbf{S}] + \mathbf{C}^{(4)}[\mathbf{S}])\} \\ + (c_{11} - c_{12})^{-1}\mathbf{C}^{(5)}[\mathbf{S}] + 2^{-1}c_{44}^{-1}\mathbf{C}^{(6)}[\mathbf{S}] \end{aligned} \tag{2.189}$$

Finally, if Eqs. (2.166), (2.168), (2.170), (2.172), and (2.174) of Problem 2.11 in which $\mathbf{E}$ is replaced by $\mathbf{S}$ are taken into account, Eq. (2.189) reduces to (2.176).

Problem 2.13. Prove that the inequalities (2.175) in Problem 2.12 are necessary and sufficient conditions for the elasticity tensor C (compliance tensor K) to be positive definite. This means that the strain energy density (stress energy density) of a transversely isotropic body is positive definite if and only if the inequalities (2.175) in Problem 2.12 hold true.

Solution. Define the fourth-order tensor H by

$$\begin{aligned}\mathsf{H} = \sqrt{a_1}a_1'\left(a_1'^2 + a_3^2\right)^{-1/2}\mathsf{C}^{(1)} + \sqrt{a_2}a_2'\left(a_2'^2 + a_3^2\right)^{-1/2}\mathsf{C}^{(2)} \\ + a_3(a_1' + a_2')^{-1/2}(\mathsf{C}^{(3)} + \mathsf{C}^{(4)}) + \sqrt{a_5}\,\mathsf{C}^{(5)} + \sqrt{a_6}\,\mathsf{C}^{(6)}\end{aligned} \tag{2.190}$$

where

$$a_1' = a_1 + (a_1\,a_2 - a_3)^{1/2}, \quad a_2' = a_2 + (a_1\,a_2 - a_3)^{1/2} \tag{2.191}$$

and $a_1,\ a_2,\ a_3,\ a_5$, and a_6 satisfy the inequalities (2.178) of Problem 2.12.

Then using the matrix equation (2.182) of Problem 2.12 we obtain

$$\mathsf{H}\,\mathsf{H} = \mathsf{A} \tag{2.192}$$

where A is the fourth-order tensor given by (2.177) of Problem 2.12. Hence, we get

$$\mathsf{A}[\mathbf{E}] = \mathsf{H}(\mathsf{H}\,[\mathbf{E}]) \quad \forall\,\mathbf{E} = \mathbf{E}^{\mathrm{T}} \neq \mathbf{0} \tag{2.193}$$

and

$$\mathbf{E}\cdot\mathsf{A}[\mathbf{E}] = \mathbf{E}\cdot\mathsf{H}(\mathsf{H}\,[\mathbf{E}]) \tag{2.194}$$

or

$$\mathbf{E}\cdot\mathsf{A}[\mathbf{E}] = (\mathsf{H}\,[\mathbf{E}])\cdot(\mathsf{H}\,[\mathbf{E}]) \tag{2.195}$$

Since

$$(\mathsf{H}\,[\mathbf{E}])\cdot(\mathsf{H}\,[\mathbf{E}]) > 0 \quad \forall\,\mathbf{E} = \mathbf{E}^{\mathrm{T}} \neq \mathbf{0} \tag{2.196}$$

therefore, expressing a_1, a_2, a_3, a_5, and a_6 in terms of $c_{11}, c_{12}, c_{13}, c_{33}$, and c_{44} [see Eqs. (2.186) of Problem 2.12], we conclude that if (2.175) of Problem 2.12 is satisfied then

$$\mathbf{E}\cdot\mathsf{C}\,[\mathbf{E}] > 0 \quad \forall\,\mathbf{E} = \mathbf{E}^{\mathrm{T}} \neq \mathbf{0} \tag{2.197}$$

that is, the elasticity tensor C is positive definite.

To prove that (2.197) implies (2.175) of Problem 2.12, that is, that (2.175) of Problem 2.12 are also necessary conditions, we take advantage of the fact that $\mathbf{E}$

in (2.197) is an arbitrary second-order symmetric tensor, and select the following choices

$$\mathbf{E}_1 = \alpha \mathbf{e} \otimes \mathbf{e} + \frac{\beta}{\sqrt{2}}(\mathbf{1} - \mathbf{e} \otimes \mathbf{e}) \tag{2.198}$$

$$\mathbf{E}_2 = \mathbf{e} \otimes \mathbf{c} + \mathbf{c} \otimes \mathbf{e} \tag{2.199}$$

$$\mathbf{E}_3 = \mathbf{c} \otimes (\mathbf{e} \times \mathbf{c}) + (\mathbf{e} \times \mathbf{c}) \otimes \mathbf{c} \tag{2.200}$$

where α and β are the real numbers, and $\mathbf{c}$ is an arbitrary unit vector orthogonal to $\mathbf{e}$.

Then, substituting (2.198), (2.199), and (2.200) into (2.197), respectively, we obtain

$$\mathbf{E}_1 \cdot \mathbf{C}[\mathbf{E}_1] = c_{33}\alpha^2 + 2\sqrt{2}\, c_{13}\, \alpha\beta + (c_{11} + c_{12})\beta^2 \tag{2.201}$$

$$\mathbf{E}_2 \cdot \mathbf{C}[\mathbf{E}_2] = 4c_{44} \tag{2.202}$$

$$\mathbf{E}_3 \cdot \mathbf{C}[\mathbf{E}_3] = 2\,(c_{11} - c_{12}) \tag{2.203}$$

Since the RHS of (2.201) is positive for non-vanishing numbers α and β, we obtain

$$\Delta = 8c_{13}^2\,\alpha^2 - 4(c_{11} + c_{12})c_{33}\alpha^2 < 0 \tag{2.204}$$

or

$$c = (c_{11} + c_{12})c_{33} - 2c_{13}^2 > 0 \tag{2.205}$$

and

$$c_{11} + c_{12} > 0, \quad c_{33} > 0 \tag{2.206}$$

Also, Eqs. (2.202) and (2.203) together with the positiveness of $\mathbf{C}$ imply that

$$c_{44} > 0, \quad c_{11} - c_{12} > 0 \tag{2.207}$$

Since the inequalities (2.205)–(2.207) are equivalent to the inequalities (2.175) of Problem 2.12, the solution to Problem 2.13 is complete.

Problem 2.14. Consider a plane $\mathbf{n} \cdot \mathbf{x} - \mathrm{v}t = 0 \quad (|\mathbf{n}| = 1,\ \mathrm{v} > 0,\ \mathrm{t} \geq 0)$. Let $\mathbf{S}$ be the stress tensor obtained form Eq. (2.160) of Problem 2.11 in which $0 < \mathbf{e} \cdot \mathbf{n} < 1$, and the strain tensor $\mathbf{E}$ is defined by

$$\mathbf{E} = \mathrm{sym}\,(\mathbf{n} \otimes \mathbf{a}) \tag{2.208}$$

where $\mathbf{a}$ is an arbitrary vector orthogonal to $\mathbf{n}$. Let $\mathbf{S}^{\perp}$ and $\mathbf{S}^{||}$ represent the normal and tangential parts of $\mathbf{S}$ with respect to the plane [see Problem 1.4 in which $\mathbf{T}$ is replaced by $\mathbf{S}$]. Show that

$$\mathbf{S} = (c_{11} - c_{12})\,\mathrm{sym}\,(\mathbf{n} \otimes \mathbf{a}) - (c_{11} - c_{12} - 2c_{44})\,\cos\theta\,\mathrm{sym}\,(\mathbf{e} \otimes \mathbf{a}) \tag{2.209}$$

$$\mathbf{S}^{\perp} = [(c_{11} - c_{12})\,\sin^2\theta + 2c_{44}\cos^2\theta\,]\,\mathrm{sym}\,(\mathbf{n} \otimes \mathbf{a}) \tag{2.210}$$

$$\mathbf{S}^{||} = (c_{11} - c_{12} - 2c_{44})\,\cos\theta\,\mathrm{sym}\,[(\mathbf{n}\,\cos\theta - \mathbf{e}) \otimes \mathbf{a}] \tag{2.211}$$

where

$$\cos\theta = \mathbf{e} \cdot \mathbf{n}, \qquad 0 < \theta < \pi/2 \tag{2.212}$$

Solution. If we let $\mathbf{E} = \mathrm{sym}(\mathbf{n} \otimes \mathbf{a})$ into Eq. (2.160) of Problem 2.11, we obtain (2.209). To obtain (2.210) and (2.211) we use the formulas:

$$\mathbf{S}^{\perp} = 2\,\mathrm{sym}(\mathbf{n} \otimes \mathbf{S}\mathbf{n}) - (\mathbf{n} \cdot \mathbf{S}\mathbf{n})\mathbf{n} \otimes \mathbf{n} \tag{2.213}$$

and

$$\mathbf{S}^{||} = (\mathbf{1} - \mathbf{n} \otimes \mathbf{n})[\mathbf{S}(\mathbf{1} - \mathbf{n} \otimes \mathbf{n})] \tag{2.214}$$

By substituting $\mathbf{S}$ from Eq. (2.209) into Eqs. (2.213) and (2.214), we obtain (2.210) and (2.211), respectively. This completes solution to Problem 2.14.

Problem 2.15. Show that for a transversely isotropic elastic body the stress energy density

$$\widehat{\mathrm{W}}(\mathbf{S}) = \frac{1}{2}\mathbf{S} \cdot \mathbf{K}[\mathbf{S}] \tag{2.215}$$

corresponding to the stress tensor given by Eq. (2.209) of Problem 2.14 takes the form

$$\widehat{\mathrm{W}}(\mathbf{S}) = \frac{1}{2}\mathbf{a}^2 \left[\frac{1}{2}(c_{11} - c_{12})\,\sin^2\theta + c_{44}\cos^2\theta\right] \tag{2.216}$$

Also, show that

$$\widehat{\mathrm{W}}(\mathbf{S}) = \widehat{\mathrm{W}}(\mathbf{S}^{\perp}) - \widehat{\mathrm{W}}(\mathbf{S}^{||}) \tag{2.217}$$

where $\widehat{\mathrm{W}}(\mathbf{S}^{\perp})$ and $\widehat{\mathrm{W}}(\mathbf{S}^{||})$ represent the "normal" and "tangential" stress energies, respectively, given by

$$\widehat{\mathrm{W}}(\mathbf{S}^{\perp}) = \frac{1}{2}\mathbf{S}^{\perp} \cdot \mathbf{K}[\mathbf{S}^{\perp}] = \widehat{\mathrm{W}}(\mathbf{S})\left[1 + \frac{1}{8}c_{44}^{-1}(c_{11} - c_{12})^{-1}(c_{11} - c_{12} - 2c_{44})^2\sin^2 2\theta\right] \tag{2.218}$$

and

$$\widehat{\mathrm{W}}(\mathbf{S}^{||}) = \frac{1}{2}\mathbf{S}^{||} \cdot \mathbf{K}[\mathbf{S}^{||}] = \widehat{\mathrm{W}}(\mathbf{S})\left[\frac{1}{8}c_{44}^{-1}(c_{11} - c_{12})^{-1}(c_{11} - c_{12} - 2c_{44})^2\sin^2 2\theta\right] \tag{2.219}$$

Here, $\mathbf{S}^{\perp}$ and $\mathbf{S}^{||}$ are given by Eqs. (2.210) and (2.211), respectively, of Problem 2.14.

Solution. If we substitute **S** from Eq. (2.209) of Problem 2.14 into Eq. (2.215) we arrive at (2.216). Next, using Eqs. (2.208) and (2.210) of Problem 2.14 we obtain (2.218); and using Eqs. (2.208) and (2.211) of Problem 2.14, we get Eq. (2.219). Finally, subtracting (2.219) from (2.218) we obtain Eq. (2.217). This completes solution to Problem 2.15.

Problem 2.16. Let $\varphi(\theta) = \widehat{\mathrm{W}}(\mathbf{S}^{||})/\widehat{\mathrm{W}}(\mathbf{S}^{\perp})$, where $\widehat{\mathrm{W}}(\mathbf{S}^{\perp})$ and $\widehat{\mathrm{W}}(\mathbf{S}^{||})$ denote the normal and tangential stress energy densities, respectively, of Problem 2.15. Show that

$$\max_{\theta\in[0,\pi/2]}[\varphi(\theta)] = \varphi(\pi/4) = \frac{\mathrm{A}^2}{1+\mathrm{A}^2} \tag{2.220}$$

where

$$\mathrm{A} = \frac{1}{2\sqrt{2}}\frac{|c_{11}-c_{12}-2c_{44}|}{(c_{11}-c_{12})^{1/2}c_{44}^{1/2}} \tag{2.221}$$

Note. When the body is isotropic we have

$$c_{11} = c_{33} = \lambda + 2\mu, \quad c_{12} = c_{13} = \lambda, \quad c_{44} = \mu \tag{2.222}$$

where λ and μ are the Lamé material constants. In this case Eq. (2.221) reduces to $\mathrm{A} = 0$, which means that for an isotropic body the tangential stress energy corresponding to the stress (2.211) of Problem 2.14 vanishes.

Solution. Note that, by using Eqs. (2.218) and (2.219) of Problem 2.15, we obtain

$$\varphi(\theta) = \widehat{W}(\mathbf{S}^{||})/\widehat{W}(\mathbf{S}^{\perp}) = \frac{A^2\sin^2 2\theta}{(1+A^2\sin^2 2\theta)} \tag{2.223}$$

where A is given by Eq. (2.221). Equation (2.223) implies (2.220), and this completes solution of Problem 2.16.

Problem 2.17. Let **u**, **E**, and **S** denote the displacement vector, strain tensor, and stress tensor fields, respectively, corresponding to a body force **b** and a temperature change T. Suppose that the fields **u**, **E**, and **S** satisfy the equations

$$\mathbf{E} = \widehat{\nabla}\mathbf{u} \quad \text{on} \quad \mathrm{B} \tag{2.224}$$

$$\operatorname{div}\mathbf{S} + \mathbf{b} = \mathbf{0} \quad \text{on} \quad \mathrm{B} \tag{2.225}$$

$$\mathbf{S} = \mathsf{C}[\mathbf{E}] + T\mathbf{M} \quad \text{on} \quad \mathrm{B} \tag{2.226}$$

where B is a bounded domain in E^3; while C and **M** denote the elasticity and stress-temperature tensors, respectively, independent of $\mathbf{x} \in \overline{\mathrm{B}}$. Also, suppose that an alternative equation to Eq. (2.226) reads

$$\mathbf{E} = \mathsf{K}[\mathbf{S}] + T\mathbf{A} \quad \text{on} \quad \mathrm{B} \tag{2.227}$$

where **K** and **A** represent the compliance and thermal expansion tensors, respectively. Let $\widehat{\mathrm{f}} = \widehat{\mathrm{f}}(\mathrm{B})$ denote the mean value of a function $\mathrm{f} = \mathrm{f}(\mathbf{x})$ on B

$$\widehat{\mathrm{f}}(\mathrm{B}) = \frac{1}{v(\mathrm{B})} \int_{\mathrm{B}} \mathrm{f}(\mathbf{x})\, dv \tag{2.228}$$

where $v(\mathrm{B})$ stands for the volume of B. Show that

$$\widehat{\mathbf{E}}(\mathrm{B}) = \frac{1}{v(\mathrm{B})} \int_{\partial \mathrm{B}} \mathrm{sym}\,(\mathbf{u} \otimes \mathbf{n}) da \tag{2.229}$$

and

$$\widehat{\mathbf{S}}(\mathrm{B}) = \frac{1}{v(\mathrm{B})} \left[\int_{\partial \mathrm{B}} \mathrm{sym}\,(\mathbf{x} \otimes \mathbf{S}\mathbf{n})\, da + \int_{\mathrm{B}} \mathbf{sym}\,(\mathbf{u} \otimes \mathbf{b})\, dv \right] \tag{2.230}$$

where $\mathbf{n}$ is the outward unit normal on ∂B. Also, show that

$$\widehat{\mathbf{E}}(\mathrm{B}) = \mathbf{K}[\widehat{\mathbf{S}}(\mathrm{B})] + \widehat{T}(\mathrm{B})\mathbf{A} \tag{2.231}$$

and

$$\widehat{\mathbf{S}}(\mathrm{B}) = \mathbf{C}[\widehat{\mathbf{E}}(\mathrm{B})] + \widehat{T}(\mathrm{B})\mathbf{M} \tag{2.232}$$

Solution. Equation (2.229) is identical with Eq. (2.83) of Problem 2.4. Therefore, a proof of (2.229) is the same as that of (2.83) of Problem 2.4. To show (2.230), we note that Eq. (2.225) implies the tensorial equation

$$\mathbf{x} \otimes \mathrm{div}\, \mathbf{S} + \mathbf{x} \otimes \mathbf{b} = \mathbf{0} \tag{2.233}$$

or in components

$$x_i\, S_{jk,k} + x_i\, b_j = 0 \tag{2.234}$$

An equivalent form of Eq. (2.234) reads

$$(x_i\, S_{jk})_{,k} - S_{ji} + x_i\, b_j = 0 \tag{2.235}$$

Integrating Eq. (2.235) over B and using the divergence theorem we obtain

$$\int_{\partial B} x_i\, S_{jk}\, n_k\, da - \int_B S_{ji}\, dv + \int_B x_i\, b_j\, dv = 0 \tag{2.236}$$

Finally, taking into account the symmetry of S_{ij}, and applying the operator sym to Eq. (2.236); and dividing (2.236) by v(B), we obtain (2.230). Also applying the mean value operator to Eqs. (2.227) and (2.226), we obtain (2.231) and (2.232), respectively; since the fourth-order tensors $\mathbf{C}$ and $\mathbf{K}$, and the second-order tensors $\mathbf{M}$ and $\mathbf{A}$ are independent of $\mathbf{x}$. This completes solution of Problem 2.17.

Problem 2.18. The volume change $\delta v(\mathrm{B})$ associated with the fields $\mathbf{u}$, $\mathbf{E}$, and $\mathbf{S}$ in Problem 2.17 is defined by [see Eqs. (2.7)–(2.8)]

$$\delta v(\mathrm{B}) = v(\mathrm{B})\,\mathrm{tr}\,\widehat{\mathbf{E}}(\mathrm{B}) \tag{2.237}$$

Show that

(i)

$$\delta v(\mathrm{B}) = 0, \quad \widehat{\mathbf{S}}(\mathrm{B}) = \widehat{T}(\mathrm{B})\,\mathbf{M} \quad \text{if} \quad \mathbf{u} = \mathbf{0} \quad \text{on} \ \partial\mathrm{B} \tag{2.238}$$

and

(ii)

$$\widehat{\mathbf{S}}(\mathrm{B}) = \mathbf{0}, \quad \widehat{\mathbf{E}}(\mathrm{B}) = \widehat{T}(\mathrm{B})\mathbf{A}, \quad \delta v(\mathrm{B}) = v(\mathrm{B})\widehat{T}(\mathrm{B})\,\mathrm{tr}\,\mathbf{A}$$
$$\text{if} \ \mathbf{Sn} = \mathbf{0} \ \text{on} \ \partial\mathrm{B} \ \text{and} \ \mathbf{b} = \mathbf{0} \ \text{on} \ \mathrm{B} \tag{2.239}$$

Note. Equations (2.239) imply that the volume change $\delta v(\mathrm{B})$ of a homogeneous isotropic thermoelastic body with zero stress vector on $\partial\mathrm{B}$ and zero body force vector on B subject to a temperature change T on B is given by

$$\delta v(\mathrm{B}) = 3\,\alpha\,\widehat{T}(\mathrm{B})\,v(\mathrm{B}) \tag{2.240}$$

where α is the coefficient of linear thermal expansion of the body.

Solution. If $\mathbf{u} = \mathbf{0}$ on ∂B then it follows from Eq. (2.229) of Problem 2.17 that $\widehat{\mathbf{E}}(B) = \mathbf{0}$. This together with Eq. (2.232) of Problem 2.17 and Eq. (2.237) implies (i).

To show (ii) we note that if $\mathbf{Sn} = \mathbf{0}$ on ∂B and $\mathbf{b} = \mathbf{0}$ on $\overline{B}$ then, by virtue of (2.230) of Problem 2.17 we obtain

$$\widehat{\mathbf{S}}(B) = \mathbf{0} \tag{2.241}$$

Hence, using (2.231) of Problem 2.17 we get

$$\widehat{\mathbf{E}}(B) = \widehat{T}(B)\mathbf{A} \tag{2.242}$$

Finally, taking the trace of (2.234) and using (2.237) we obtain

$$\delta v(B) = v(B)\,\widehat{T}(B)\mathrm{tr}\,\mathbf{A} \tag{2.243}$$

This completes proof of (ii). The result (2.240) follows from the fact that in a homogeneous isotropic body

$$\mathrm{tr}\,\mathbf{A} = 3\alpha \tag{2.244}$$

This completes solution of Problem 2.18.

Chapter 3
Formulation of Problems of Elasticity

In this chapter both the basic boundary value problems of elastostatics and initial-boundary value problems of elastodynamics are recalled; in particular, the mixed boundary value problems of isothermal and nonisothermal elastostatics, as well as the pure displacement and the pure stress problems of classical elastodynamics are discussed. The Betti reciprocal theorem of elastostatics and Graffi's reciprocal theorem of elastodynamics together with the uniqueness theorems are also presented. An emphasis is made on a pure stress initial-boundary value problem of incompatible elastodynamics in which a body possesses initially distributed defects. [See also Chap. 16.]

3.1 Boundary Value Problems of Elastostatics

Field Equations of Isothermal Elastostatics

The strain-displacement relation

$$\mathbf{E} = \widehat{\nabla}\mathbf{u} = \frac{1}{2}(\nabla\mathbf{u} + \nabla\mathbf{u}^{\mathrm{T}}) \tag{3.1}$$

The equations of equilibrium

$$\operatorname{div}\mathbf{S} + \mathbf{b} = \mathbf{0}, \quad \mathbf{S} = \mathbf{S}^{\mathrm{T}} \tag{3.2}$$

The stress-strain relation

$$\mathbf{S} = \mathsf{C}\,[\mathbf{E}] \tag{3.3}$$

By eliminating $\mathbf{E}$ and $\mathbf{S}$ from Eqs. (3.1)–(3.3) we obtain the *displacement equation of equilibrium*

$$\operatorname{div}\mathsf{C}\,[\nabla\mathbf{u}] + \mathbf{b} = \mathbf{0} \tag{3.4}$$

M. Reza Eslami et al., *Theory of Elasticity and Thermal Stresses*, Solid Mechanics and Its Applications 197, DOI: 10.1007/978-94-007-6356-2_3,

For a homogeneous isotropic body the displacement equation of equilibrium (3.4) reduces to

$$\mu \nabla^2 \mathbf{u} + (\lambda + \mu) \nabla (\operatorname{div} \mathbf{u}) + \mathbf{b} = \mathbf{0} \tag{3.5}$$

or

$$\nabla^2 \mathbf{u} + \frac{1}{1 - 2\nu} \nabla (\operatorname{div} \mathbf{u}) + \frac{\mathbf{b}}{\mu} = \mathbf{0} \tag{3.6}$$

or

$$(\lambda + 2\mu) \nabla (\operatorname{div} \mathbf{u}) - \mu \operatorname{curl} \operatorname{curl} \mathbf{u} + \mathbf{b} = \mathbf{0} \tag{3.7}$$

An equivalent form of the stress-strain relation (3.3) reads

$$\mathbf{E} = \mathsf{K}[\mathbf{S}] \tag{3.8}$$

Therefore, by eliminating $\mathbf{u}$ and $\mathbf{E}$ from Eqs. (3.1), (3.2), and (3.8) the *stress equations of equilibrium* are obtained

$$\operatorname{div} \mathbf{S} + \mathbf{b} = \mathbf{0}, \quad \mathbf{S} = \mathbf{S}^T \tag{3.9}$$

$$\operatorname{curl} \operatorname{curl} \mathsf{K}[\mathbf{S}] = \mathbf{0} \tag{3.10}$$

For a homogeneous isotropic body, the stress equations of equilibrium (3.9)–(3.10) reduce to

$$\operatorname{div} \mathbf{S} + \mathbf{b} = \mathbf{0}, \quad \mathbf{S} = \mathbf{S}^T \tag{3.11}$$

$$\nabla^2 \mathbf{S} + \frac{1}{1 + \nu} \nabla \nabla (\operatorname{tr} \mathbf{S}) + \frac{\nu}{1 - \nu} (\operatorname{div} \mathbf{b}) \mathbf{1} + 2 \widehat{\nabla} \mathbf{b} = \mathbf{0} \tag{3.12}$$

Field Equations of nonisothermal Elastostatics

The strain-displacement relation

$$\mathbf{E} = \widehat{\nabla} \mathbf{u} = \frac{1}{2} (\nabla \mathbf{u} + \nabla \mathbf{u}^T) \tag{3.13}$$

The equations of equilibrium

$$\operatorname{div} \mathbf{S} + \mathbf{b} = \mathbf{0}, \quad \mathbf{S} = \mathbf{S}^T \tag{3.14}$$

The stress-strain-temperature relation

$$\mathbf{S} = \mathsf{C}[\mathbf{E}] + T \mathbf{M} \tag{3.15}$$

or, the strain-stress-temperature relation

$$\mathbf{E} = \mathsf{K}\,[\mathbf{S}] + T\,\mathbf{A} \tag{3.16}$$

In Eqs. (3.15) T stands for a temperature change; while $\mathbf{M} = \mathbf{M}^{\mathrm{T}}$ and $\mathbf{A} = \mathbf{A}^{\mathrm{T}}$ are the stress-temperature and thermal expansion tensors, respectively.

For an isotropic body Eqs. (3.15) and (3.16), respectively, take the form

$$\mathbf{S} = 2\mu\,\mathbf{E} + \lambda\,(\mathrm{tr}\,\mathbf{E})\,\mathbf{1} - (3\lambda + 2\mu)\,\alpha\,T\mathbf{1} \tag{3.17}$$

and

$$\mathbf{E} = \frac{1}{2\mu}\left[\mathbf{S} - \frac{\lambda}{3\lambda + 2\mu}(\mathrm{tr}\,\mathbf{S})\,\mathbf{1}\right] + \alpha\,T\,\mathbf{1} \tag{3.18}$$

By eliminating **E** and **S** from Eqs. (3.13)–(3.15) the *displacement-temperature equation of nonisothermal elastostatics* is obtained

$$\mathrm{div}\{\mathsf{C}\,[\nabla\mathbf{u}] + T\mathbf{M}\} + \mathbf{b} = \mathbf{0} \tag{3.19}$$

Also, by eliminating **u** and **E** from Eqs. (3.13), (3.14), and (3.16), the *stress-temperature equations of nonisothermal elastostatics* are obtained

$$\mathrm{div}\,\mathbf{S} + \mathbf{b} = \mathbf{0}, \quad \mathbf{S} = \mathbf{S}^{\mathrm{T}} \tag{3.20}$$

$$\mathrm{curl}\,\mathrm{curl}\,\{\mathsf{K}[\mathbf{S}] + T\mathbf{A}\} = \mathbf{0} \tag{3.21}$$

For an isotropic homogeneous body, Eqs. (3.19) and (3.20)–(3.21), respectively, take the form

$$\mu\,\nabla^2\mathbf{u} + (\lambda + \mu)\,\nabla(\mathrm{div}\,\mathbf{u}) - (3\lambda + 2\mu)\,\alpha\,\nabla\,T + \mathbf{b} = \mathbf{0} \tag{3.22}$$

and

$$\mathrm{div}\,\mathbf{S} + \mathbf{b} = \mathbf{0}, \quad \mathbf{S} = \mathbf{S}^{\mathrm{T}} \tag{3.23}$$

$$\nabla^2\mathbf{S} + \frac{1}{1+\nu}\nabla\nabla(\mathrm{tr}\,\mathbf{S}) + \frac{E\alpha}{1+\nu}\left(\nabla\nabla T + \frac{1+\nu}{1-\nu}\nabla^2 T\,\mathbf{1}\right) + \frac{\nu}{1-\nu}(\mathrm{div}\,\mathbf{b})\mathbf{1} + 2\widehat{\nabla}\mathbf{b} = \mathbf{0} \tag{3.24}$$

3.2 Concept of an Elastic State

An ordered array of functions $s = [\mathbf{u}, \mathbf{E}, \mathbf{S}]$ is called an *elastic state* corresponding to the body force **b** if the functions **u**, **E**, and **S** satisfy the system of fundamental field equations (3.1)–(3.3) on B.

An *external force system* for s is defined as a pair $[\mathbf{b}, \mathbf{s}]$ where $\mathbf{s} = \mathbf{S}\mathbf{n}$ with $\mathbf{n}$ being an outward unit vector normal to ∂B.

Theorem of work and energy. If $s = [\mathbf{u}, \mathbf{E}, \mathbf{S}]$ is an elastic state corresponding to the external force system $[\mathbf{b}, \mathbf{s}]$ then

$$\int_{\partial B} \mathbf{s} \cdot \mathbf{u}\, da + \int_{B} \mathbf{b} \cdot \mathbf{u}\, dv = 2U_C(\mathbf{E}) \tag{3.25}$$

where $U_C(\mathbf{E})$ is the total strain energy of body B

$$U_C(\mathbf{E}) = \frac{1}{2} \int_{B} \mathbf{E} \cdot \mathbf{C}[\mathbf{E}]\, dv \tag{3.26}$$

The Betti reciprocal theorem. Let the elasticity tensor $\mathbf{C}$ be symmetric, and let

$$s = [\mathbf{u}, \mathbf{E}, \mathbf{S}] \quad \text{and} \quad \widetilde{s} = [\widetilde{\mathbf{u}}, \widetilde{\mathbf{E}}, \widetilde{\mathbf{S}}] \tag{3.27}$$

be elastic states corresponding to the external force systems $[\mathbf{b}, \mathbf{s}]$ and $[\widetilde{\mathbf{b}}, \widetilde{\mathbf{S}}]$, respectively. Then the following reciprocity relation holds

$$\int_{\partial B} \mathbf{s} \cdot \widetilde{\mathbf{u}}\, da + \int_{B} \mathbf{b} \cdot \widetilde{\mathbf{u}}\, dv = \int_{\partial B} \widetilde{\mathbf{s}} \cdot \mathbf{u}\, da + \int_{B} \widetilde{\mathbf{b}} \cdot \mathbf{u}\, dv = \int_{B} \mathbf{S} \cdot \widetilde{\mathbf{E}}\, dv = \int_{B} \widetilde{\mathbf{S}} \cdot \mathbf{E}\, dv \tag{3.28}$$

3.3 Concept of a Thermoelastic State

An ordered array of functions $s = [\mathbf{u}, \mathbf{E}, \mathbf{S}]$ is called a *thermoelastic state* corresponding to an external force system $[\mathbf{b}, \mathbf{s}, T]$ if the functions $\mathbf{u}$, $\mathbf{E}$, and $\mathbf{S}$ satisfy the field equations of thermoelastostatics (3.13)–(3.15) on B.

Thermoelastic reciprocal theorem. Let $s = [\mathbf{u}, \mathbf{E}, \mathbf{S}]$ and $\widetilde{s} = [\widetilde{\mathbf{u}}, \widetilde{\mathbf{E}}, \widetilde{\mathbf{S}}]$ be thermoelastic states corresponding to the external force-temperature systems $[\mathbf{b}, \mathbf{s}, T]$ and $[\widetilde{\mathbf{b}}, \widetilde{\mathbf{s}}, \widetilde{T}]$, respectively. Then

$$\int_{\partial B} \mathbf{s} \cdot \widetilde{\mathbf{u}}\, da + \int_{B} \mathbf{b} \cdot \widetilde{\mathbf{u}}\, dv - \int_{B} T\mathbf{M} \cdot \widetilde{\mathbf{E}}\, dv = \int_{\partial B} \widetilde{\mathbf{s}} \cdot \mathbf{u}\, da + \int_{B} \widetilde{\mathbf{b}} \cdot \mathbf{u}\, dv - \int_{B} \widetilde{T}\mathbf{M} \cdot \mathbf{E}\, dv \tag{3.29}$$

3.4 Formulation of Boundary Value Problems

Mixed problems of elastostatics. By a mixed boundary value problem of elastostatics we mean the problem of finding an elastic state $s = [\mathbf{u}, \mathbf{E}, \mathbf{S}]$ corresponding to a body force $\mathbf{b}$ and satisfying the boundary conditions: the displacement condition

$$\mathbf{u} = \widehat{\mathbf{u}} \quad \text{on} \quad \partial \mathrm{B}_1 \tag{3.30}$$

and the traction condition

$$\mathbf{Sn} = \widehat{\mathbf{s}} \quad \text{on} \quad \partial \mathrm{B}_2 \tag{3.31}$$

where

$$\partial \mathrm{B}_1 \cup \partial \mathrm{B}_2 = \partial \mathrm{B}; \quad \partial \mathrm{B}_1 \cap \partial \mathrm{B}_2 = \varnothing \tag{3.32}$$

while $\widehat{\mathbf{u}}$ and $\widehat{\mathbf{s}}$ are prescribed functions.

An elastic state s that satisfies the boundary conditions (3.30)–(3.31) is called a *solution to the mixed problem.*

If $\partial \mathrm{B}_2 = \varnothing$, the mixed problem becomes a *displacement boundary value problem.* If $\partial \mathrm{B}_1 = \varnothing$, the mixed problem becomes a *traction boundary value problem.*

A *displacement field corresponding to a solution to a mixed problem* is a vector field $\mathbf{u}$ with the property that there are symmetric tensor fields $\mathbf{E}$ and $\mathbf{S}$ such that $s = [\mathbf{u}, \mathbf{E}, \mathbf{S}]$ is a solution to the mixed problem.

A *stress field corresponding to a solution to a mixed problem* is a tensor field $\mathbf{S}$ with the property that there are $\mathbf{u}$ and $\mathbf{E}$ such that $s = [\mathbf{u}, \mathbf{E}, \mathbf{S}]$ is a solution to the mixed problem.

Mixed Problem in Terms of Displacements

A vector field $\mathbf{u}$ corresponds to a solution to the mixed problem if and only if

$$\operatorname{div} \mathbf{C}[\nabla \mathbf{u}] + \mathbf{b} = \mathbf{0} \quad \text{on} \quad \mathrm{B} \tag{3.33}$$

$$\mathbf{u} = \widehat{\mathbf{u}} \quad \text{on} \quad \partial \mathrm{B}_1 \tag{3.34}$$

$$(\mathbf{C}[\nabla \mathbf{u}])\,\mathbf{n} = \widehat{\mathbf{s}} \quad \text{on} \quad \partial \mathrm{B}_2 \tag{3.35}$$

Displacement Problem in Terms of Displacements

A vector field $\mathbf{u}$ corresponds to a solution to the displacement problem if and only if

$$\operatorname{div} \mathbf{C}[\nabla \mathbf{u}] + \mathbf{b} = \mathbf{0} \quad \text{on} \quad \mathrm{B} \tag{3.36}$$

$$\mathbf{u} = \widehat{\mathbf{u}} \quad \text{on} \quad \partial \mathrm{B} \tag{3.37}$$

Traction Problem in Terms of Stresses

A tensor field $\mathbf{S}$ corresponds to a solution to the traction problem if and only if

$$\operatorname{div}\mathbf{S} + \mathbf{b} = \mathbf{0} \quad \text{on} \quad B \tag{3.38}$$

$$\operatorname{curl}\operatorname{curl}\mathbf{K}[\mathbf{S}] = \mathbf{0} \quad \text{on} \quad B \tag{3.39}$$

$$\mathbf{S}\mathbf{n} = \widehat{\mathbf{s}} \quad \text{on} \quad \partial B \tag{3.40}$$

3.5 Uniqueness

Uniqueness Theorem for the Mixed Problem. If the elasticity tensor $\mathbf{C}$ is positive definite, then any two solutions of the mixed problem of elastostatics are equal to within a rigid displacement. If $\partial B_1 \neq \varnothing$ then the rigid displacement vanishes.

3.6 Formulation of Problems of Nonisothermal Elastostatics

By a *mixed problem of nonisothermal elastostatics* we mean the problem of finding a thermoelastic state $s = [\mathbf{u}, \mathbf{E}, \mathbf{S}]$ that satisfies the field equations (3.13)–(3.15) on B subject to the boundary conditions (3.30)–(3.31).

Mixed Thermoelastic Problem in Terms of Displacements

A vector field $\mathbf{u}$ corresponds to a solution to the mixed thermoelastic problem if and only if

$$\operatorname{div}\{\mathbf{C}[\nabla\mathbf{u}] + T\mathbf{M}\} + \mathbf{b} = \mathbf{0} \quad \text{on} \quad B \tag{3.41}$$

$$\mathbf{u} = \widehat{\mathbf{u}} \quad \text{on} \quad \partial B_1 \tag{3.42}$$

$$(\mathbf{C}[\nabla\mathbf{u}] + T\mathbf{M})\,\mathbf{n} = \widehat{\mathbf{s}} \quad \text{on} \quad \partial B_2 \tag{3.43}$$

Traction Thermoelastic Problem in Terms of Stresses

A tensor field $\mathbf{S}$ corresponds to a solution to the traction thermoelastic problem if and only if

$$\operatorname{div}\mathbf{S} + \mathbf{b} = \mathbf{0} \quad \text{on} \quad B \tag{3.44}$$

$$\operatorname{curl}\operatorname{curl}\{\mathbf{K}[\mathbf{S}] + T\mathbf{A}\} = \mathbf{0} \quad \text{on} \quad B \tag{3.45}$$

$$\mathbf{S}\mathbf{n} = \widehat{\mathbf{s}} \quad \text{on} \quad \partial B \tag{3.46}$$

3.7 Initial-Boundary Value Problems of Elastodynamics

Field Equations of Isothermal Elastodynamics

The strain-displacement relation

$$\mathbf{E} = \widehat{\nabla}\mathbf{u} = \frac{1}{2}(\nabla\mathbf{u} + \nabla\mathbf{u}^{\mathrm{T}}) \tag{3.47}$$

The equations of motion

$$\operatorname{div}\mathbf{S} + \mathbf{b} = \rho\,\ddot{\mathbf{u}}, \quad \mathbf{S} = \mathbf{S}^{\mathrm{T}} \tag{3.48}$$

The stress-strain relation

$$\mathbf{S} = \mathbf{C}\,[\mathbf{E}] \tag{3.49}$$

or equivalently

$$\mathbf{E} = \mathbf{K}\,[\mathbf{S}] \tag{3.50}$$

By eliminating **E** and **S** from Eqs. (3.47)–(3.49) we obtain the *displacement equation of motion*

$$\operatorname{div}\,\mathbf{C}\,[\nabla\mathbf{u}] + \mathbf{b} = \rho\,\ddot{\mathbf{u}} \tag{3.51}$$

By eliminating **u** and **E** from Eqs. (3.47), (3.48), and (3.50) the *stress equation of motion* is obtained

$$\widehat{\nabla}[\rho^{-1}(\operatorname{div}\mathbf{S})] - \mathbf{K}[\ddot{\mathbf{S}}] = -\mathbf{B} \tag{3.52}$$

where

$$\mathbf{B} = \widehat{\nabla}(\rho^{-1}\mathbf{b}) \tag{3.53}$$

For a homogeneous isotropic elastic body Eqs. (3.51) and (3.52), respectively, reduce to

$$\mu\,\nabla^2\mathbf{u} + (\lambda + \mu)\,\nabla(\operatorname{div}\mathbf{u}) + \mathbf{b} = \rho\,\ddot{\mathbf{u}} \tag{3.54}$$

and

$$\widehat{\nabla}(\operatorname{div}\mathbf{S}) - \frac{\rho}{2\mu}\left[\ddot{\mathbf{S}} - \frac{\lambda}{3\lambda + 2\mu}(\operatorname{tr}\ddot{\mathbf{S}})\,\mathbf{1}\right] = -\widehat{\nabla}\,\mathbf{b} \tag{3.55}$$

Field Equations of nonisothermal Elastodynamics

The strain-displacement relation

$$\mathbf{E} = \widehat{\nabla}\mathbf{u} = \frac{1}{2}(\nabla\mathbf{u} + \nabla\mathbf{u}^{\mathrm{T}}) \tag{3.56}$$

The equations of motion

$$\operatorname{div}\mathbf{S} + \mathbf{b} = \rho\,\ddot{\mathbf{u}}, \quad \mathbf{S} = \mathbf{S}^{\mathrm{T}} \tag{3.57}$$

The stress-strain-temperature relation

$$\mathbf{S} = \mathsf{C}\,[\mathbf{E}] + T\,\mathbf{M} \tag{3.58}$$

or, the strain-stress-temperature relation

$$\mathbf{E} = \mathsf{K}\,[\mathbf{S}] + T\,\mathbf{A} \tag{3.59}$$

By eliminating $\mathbf{E}$ and $\mathbf{S}$ from Eqs. (3.56)–(3.58) we obtain the *displacement-temperature equation of motion*

$$\operatorname{div}\left(\mathsf{C}\,[\nabla\mathbf{u}] + T\mathbf{M}\right) + \mathbf{b} = \rho\,\ddot{\mathbf{u}} \tag{3.60}$$

By eliminating $\mathbf{u}$ and $\mathbf{E}$ from Eqs. (3.56), (3.57), and (3.59) the *stress-temperature equation of motion* is obtained

$$\widehat{\nabla}[\rho^{-1}(\operatorname{div}\mathbf{S})] - \mathsf{K}[\ddot{\mathbf{S}}] = -\widetilde{\mathbf{B}} \tag{3.61}$$

where

$$\widetilde{\mathbf{B}} = \widehat{\nabla}(\rho^{-1}\mathbf{b}) + \ddot{T}\mathbf{A} \tag{3.62}$$

Here, $\mathbf{M}$ and $\mathbf{A}$ are the stress-temperature and the thermal expansion tensors, respectively.

3.8 Concept of an Elastic Process

An *elastic process* corresponding to a body force $\mathbf{b}$ is defined as an ordered set of functions $p = [\mathbf{u}, \mathbf{E}, \mathbf{S}]$ that complies with the fundamental system of field equations of isothermal elastodynamics (3.47)–(3.49).

An *external force system for p* is defined as a pair $[\mathbf{b}, \mathbf{s}]$ in which $\mathbf{s} = \mathbf{S}\mathbf{n}$.

Graffi's Reciprocal Theorem. Let $p = [\mathbf{u}, \mathbf{E}, \mathbf{S}]$ be an elastic process corresponding to the external force system $[\mathbf{b}, \mathbf{s}]$ and to the initial data $[\mathbf{u}_0, \dot{\mathbf{u}}_0]$, Let $\widetilde{p} = [\widetilde{\mathbf{u}}, \widetilde{\mathbf{E}}, \widetilde{\mathbf{S}}]$ be another elastic process corresponding to $[\widetilde{\mathbf{b}}, \widetilde{\mathbf{s}}]$ and $[\widetilde{\mathbf{u}}_0, \dot{\widetilde{\mathbf{u}}}_0]$. Then the following integral relations hold true

$$\mathrm{i} * \int_{\partial B} \mathbf{s} * \widetilde{\mathbf{u}}\, da + \int_{B} \mathbf{f} * \widetilde{\mathbf{u}}\, dv = \mathrm{i} * \int_{\partial B} \widetilde{\mathbf{s}} * \mathbf{u}\, da + \int_{B} \widetilde{\mathbf{f}} * \mathbf{u}\, dv \tag{3.63}$$

$$\int_{\partial B} \mathbf{s} * \widetilde{\mathbf{u}}\, da + \int_{B} \mathbf{b} * \widetilde{\mathbf{u}}\, dv + \int_{B} \rho\, (\mathbf{u}_0 \cdot \dot{\widetilde{\mathbf{u}}} + \dot{\mathbf{u}}_0 \cdot \widetilde{\mathbf{u}})\, dv$$
$$= \int_{\partial B} \widetilde{\mathbf{s}} * \mathbf{u}\, da + \int_{B} \widetilde{\mathbf{b}} * \mathbf{u}\, dv + \int_{B} \rho\, (\widetilde{\mathbf{u}}_0 \cdot \dot{\mathbf{u}} + \dot{\widetilde{\mathbf{u}}}_0 \cdot \mathbf{u})\, dv \tag{3.64}$$

Here

$$\mathrm{i} = \mathrm{i}(t) = t, \quad t \geq 0 \tag{3.65}$$

and $\mathbf{f}$ and $\widetilde{\mathbf{f}}$ are pseudo-body forces corresponding to $p = [\mathbf{u}, \mathbf{E}, \mathbf{S}]$ and $\widetilde{p} = [\widetilde{\mathbf{u}}, \widetilde{\mathbf{E}}, \widetilde{\mathbf{S}}]$, respectively, defined by

$$\begin{aligned} \mathbf{f} &= \mathrm{i} * \mathbf{b}(\mathbf{x}, t) + \rho[\mathbf{u}_0(\mathbf{x}) + t\, \dot{\mathbf{u}}_0(\mathbf{x})] \\ \widetilde{\mathbf{f}} &= \mathrm{i} * \widetilde{\mathbf{b}}(\mathbf{x}, t) + \rho[\widetilde{\mathbf{u}}_0(\mathbf{x}) + t\, \dot{\widetilde{\mathbf{u}}}_0(\mathbf{x})] \end{aligned} \tag{3.66}$$

3.9 Formulation of Problems of Isothermal Elastodynamics

Mixed Problem in Terms of Displacements

A vector field $\mathbf{u}$ corresponds to a solution to a mixed problem of isothermal elastodynamics if and only if

$$\operatorname{div} \mathbf{C}\,[\nabla \mathbf{u}] + \mathbf{b} = \rho\, \ddot{\mathbf{u}} \quad \text{on} \quad B \times [0, \infty) \tag{3.67}$$

$$\mathbf{u}(\mathbf{x}, 0) = \mathbf{u}_0(\mathbf{x}), \quad \dot{\mathbf{u}}(\mathbf{x}, 0) = \dot{\mathbf{u}}_0(\mathbf{x}) \quad \text{on} \quad B \tag{3.68}$$

$$\mathbf{u} = \widehat{\mathbf{u}} \quad \text{on} \quad \partial B_1 \times [0, \infty) \tag{3.69}$$

$$(\mathbf{C}\,[\nabla \mathbf{u}]\,)\,\mathbf{n} = \widehat{\mathbf{s}} \quad \text{on} \quad \partial B_2 \times [0, \infty) \tag{3.70}$$

Traction Problem in Terms of Stresses

A tensor field $\mathbf{S}$ corresponds to a solution to a traction problem of isothermal elastodynamics if and only if

$$\widehat{\nabla}[\rho^{-1}(\operatorname{div} \mathbf{S})] - \mathbf{K}[\ddot{\mathbf{S}}] = -\mathbf{B} \quad \text{on} \quad B \times [0, \infty) \tag{3.71}$$

$$\mathbf{S}(\mathbf{x}, 0) = \mathbf{C}\,[\nabla \mathbf{u}_0], \quad \dot{\mathbf{S}}(\mathbf{x}, 0) = \mathbf{C}\,[\nabla \dot{\mathbf{u}}_0] \quad \text{on} \quad B \tag{3.72}$$

$$\mathbf{S}\mathbf{n} = \widehat{\mathbf{s}} \quad \text{on} \quad \partial B \times [0, \infty) \tag{3.73}$$

where

$$\mathbf{B} = \widehat{\nabla}(\rho^{-1}\mathbf{b}) \quad \text{on} \quad B \times [0, \infty) \tag{3.74}$$

Mixed Problem in Terms of Stresses

A tensor field **S** corresponds to a solution to a mixed problem of isothermal elastodynamics if and only if

$$\widehat{\nabla}[\rho^{-1}(i * \operatorname{div} \mathbf{S} + \mathbf{f})] - \mathbf{K}[\mathbf{S}] = \mathbf{0} \quad \text{on} \quad B \times [0, \infty) \tag{3.75}$$

$$\rho^{-1}(i * \operatorname{div} \mathbf{S} + \mathbf{f}) = \widehat{\mathbf{u}} \quad \text{on} \quad \partial B_1 \times [0, \infty) \tag{3.76}$$

$$\mathbf{Sn} = \widehat{\mathbf{s}} \quad \text{on} \quad \partial B_2 \times [0, \infty) \tag{3.77}$$

where **f** is the pseudo-body force given by $(3.66)_1$.

Nonconventional Traction Problem in Terms of Stresses (Uniqueness). Let **S** be a solution to the following initial-boundary value problem. Find a symmetric second-order tensor field **S** on $\overline{B} \times [0, \infty)$ that satisfies the field equation

$$\widehat{\nabla}[\rho^{-1}(\operatorname{div} \mathbf{S})] - \mathbf{K}[\ddot{\mathbf{S}}] = -\mathbf{F} \quad \text{on} \quad B \times [0, \infty) \tag{3.78}$$

the initial conditions

$$\mathbf{S}(\mathbf{x}, 0) = \mathbf{S}_0(\mathbf{x}), \quad \dot{\mathbf{S}}(\mathbf{x}, 0) = \dot{\mathbf{S}}_0(\mathbf{x}) \quad \text{for} \quad \mathbf{x} \in B \tag{3.79}$$

and the boundary condition

$$\mathbf{Sn} = \widehat{\mathbf{s}} \quad \text{on} \quad \partial B \times [0, \infty) \tag{3.80}$$

Here, **F** is an arbitrary symmetric second-order tensor field prescribed on $\overline{B} \times [0, \infty)$, $\mathbf{S}_0$ and $\dot{\mathbf{S}}_0$ are prescribed symmetric tensor fields on B, and $\widehat{\mathbf{s}}$ is a prescribed vector field on $\partial B \times [0, \infty)$. Then the problem described by Eqs. (3.78)–(3.80) has at most one solution (Uniqueness).

If $\mathbf{F} \neq \mathbf{B}$, where **B** is given by (3.74), $\mathbf{S}_0$ and $\dot{\mathbf{S}}_0$ are not given by Eqs. (3.72), then the problem (3.78)–(3.80) describes stress waves in an *elastic body with time-dependent continuously distributed defects.*

3.10 Problems and Solutions Related to the Formulation of Problems of Elasticity

Problem 3.1. For a homogeneous isotropic elastic body occupying a region $B \subset E^3$ subject to zero body forces, the displacement equation of equilibrium takes the form [see Eq. (3.6) with $\mathbf{b} = \mathbf{0}$]

$$\nabla^2 \mathbf{u} + \frac{1}{1-2\nu} \nabla (\operatorname{div} \mathbf{u}) = \mathbf{0} \quad \text{on} \quad \mathrm{B} \tag{3.81}$$

where ν is Poisson's ratio. Show that if $\mathbf{u} = \mathbf{u}(\mathbf{x})$ is a solution to Eq. (3.81) then $\mathbf{u}$ also satisfies the equation

$$\nabla^2 \left[\mathbf{u} + \frac{\mathbf{x}}{2(1-2\nu)} (\operatorname{div} \mathbf{u}) \right] = \mathbf{0} \quad \text{on} \quad \mathrm{B} \tag{3.82}$$

Solution. Equations (3.81) and (3.82) in components take the forms

$$u_{i,kk} + \frac{1}{1-2\nu} u_{k,ki} = 0 \quad \text{on } B \tag{3.83}$$

and

$$\left[u_i + \frac{x_i}{2(1-2\nu)} u_{k,k} \right]_{,jj} = 0 \quad \text{on } B \tag{3.84}$$

respectively.

It follows from (3.83) that

$$u_{i,ikk} + \frac{1}{1-2\nu} u_{k,kii} = 0 \quad \text{on } B \tag{3.85}$$

or

$$\frac{2(1-\nu)}{1-2\nu} \, u_{i,ikk} = 0 \quad \text{on } B \tag{3.86}$$

Since $-1 < \nu < 1/2 < 1$ [see Eq. (2.50)]

$$\frac{2-2\nu}{1-2\nu} > 0 \tag{3.87}$$

and (3.86) implies that

$$u_{i,ikk} = 0 \quad \text{on } B \tag{3.88}$$

In addition

$$(x_i \, u_{k,k})_{,jj} = (\delta_{ij} \, u_{k,k} + x_i \, u_{k,kj})_{,j} = 2\delta_{ij} \, u_{k,kj} + x_i \, u_{k,kjj} \tag{3.89}$$

Hence, it follows from Eqs. (3.88) and (3.89) that

$$(x_i \, u_{k,k})_{,jj} = 2u_{k,ki} \tag{3.90}$$

Substituting (3.90) into (3.84) we obtain (3.83), and this completes solution of Problem 3.1.

Problem 3.2. An alternative form of Eq. (3.81) in Problem 3.1 reads [see Eq. (3.7) in which $\lambda = 2\mu\nu/(1-2\nu)$ and $\mathbf{b} = \mathbf{0}$]

$$\nabla(\operatorname{div}\mathbf{u}) - \frac{1-2\nu}{2-2\nu}\operatorname{curl}\operatorname{curl}\mathbf{u} = \mathbf{0} \quad \text{on} \quad \text{B} \tag{3.91}$$

Show that if $\mathbf{u} = \mathbf{u}(\mathbf{x})$ is a solution to Eq. (3.91) then

$$\int_{\text{B}} \left[\operatorname{div}\mathbf{u})^2 + \frac{1-2\nu}{2-2\nu}(\operatorname{curl}\mathbf{u})^2\right] \text{dv}$$

$$= \int_{\partial\text{B}} \mathbf{u} \cdot \left[(\operatorname{div}\mathbf{u})\,\mathbf{n} + \frac{1-2\nu}{2-2\nu}(\operatorname{curl}\mathbf{u}) \times \mathbf{n}\right] \text{da} \tag{3.92}$$

where $\mathbf{n}$ is the unit outward normal vector field on ∂B.

Hint. Multiply Eq. (3.91) by $\mathbf{u}$ in the dot product sense, integrate the result over B, and use the divergence theorem.

Note. Since $-1 < \nu < 1/2$ [see Eq. (2.50)] then Eq. (3.92) implies that a displacement boundary value problem of homogeneous isotropic elastostatics may have at most one solution.

Solution. In components Eq. (3.91) takes the form

$$u_{k,ki} - \frac{1-2\nu}{2-2\nu}\varepsilon_{iab}\,\varepsilon_{bcd}\,u_{d,ca} = 0 \tag{3.93}$$

Since

$$u_i\,u_{k,ki} = (u_i\,u_{k,k})_{,i} - (u_{i,i})^2 \tag{3.94}$$

$$u_i\,\varepsilon_{iab}\,\varepsilon_{bcd}\,u_{d,ca} = \varepsilon_{iab}\,\varepsilon_{bcd}[(u_{d,c}\,u_i)_{,a} - u_{d,c}\,u_{i,a}] \tag{3.95}$$

and

$$\varepsilon_{iab}\,\varepsilon_{bcd}\,u_{d,c}\,u_{i,a} = -(\varepsilon_{bai}\,u_{i,a})(\varepsilon_{bcd}\,u_{d,c}) \tag{3.96}$$

therefore, multiplying (3.93) by u_i we obtain

$$(u_{i,i})^2 + \frac{1-2\nu}{2-2\nu}(\varepsilon_{bai}\,u_{i,a})(\varepsilon_{bcd}\,u_{d,c}) = \left(u_{k,k}\,u_a + \frac{1-2\nu}{2-2\nu}\varepsilon_{iab}\,\varepsilon_{bcd}\,u_{d,c}\,u_i\right)_{,a} \tag{3.97}$$

Finally, integrating (3.97) over B and using the divergence theorem we obtain

$$\int_B \left[(u_{i,i})^2 + \frac{1-2\nu}{2-2\nu} (\varepsilon_{bai}\, u_{i,a})(\varepsilon_{bcd}\, u_{d,c}) \right] dv$$

$$= \int_{\partial B} u_a \left[u_{k,k}\, n_a + \frac{1-2\nu}{2-2\nu} \varepsilon_{abi} (\varepsilon_{bcd}\, u_{d,c}) n_i \right] da \tag{3.98}$$

Equation (3.98) is equivalent to (3.92), and this completes solution of Problem 3.2.

Problem 3.3. Show that for a homogeneous isotropic infinite elastic body subject to a temperature change $T = T(\mathbf{x})$ its volume change is represented by the formula

$$\operatorname{tr} \mathbf{E}(\mathbf{x}) = \frac{1+\nu}{1-\nu} \alpha\, T(\mathbf{x}) \quad \text{for} \quad \mathbf{x} \in E^3 \tag{3.99}$$

where ν and α denote Poisson's ratio and coefficient of thermal expansion, respectively.

Hint. Apply the reciprocal relation (3.28) to the external force-temperature systems $[\mathbf{b}, \mathbf{s}, T] = [\mathbf{0}, \mathbf{0}, T]$ and $[\widetilde{\mathbf{b}}, \widetilde{\mathbf{s}}, \widetilde{T}] = [\mathbf{0}, \mathbf{0}, \delta(\mathbf{x} - \xi)]$ on E^3. Also note that for an isotropic body $\mathbf{M} = -(3\lambda + 2\mu)\alpha\, \mathbf{1}$ and $\operatorname{tr} \widetilde{\mathbf{E}} = [(3\lambda + 2\mu)\,/(\lambda + 2\mu)]\alpha\, \widetilde{T}$.

Solution. Let $s = [\mathbf{u}, \mathbf{E}, \mathbf{S}]$ and $\widetilde{s} = [\widetilde{\mathbf{u}}, \widetilde{\mathbf{E}}, \widetilde{\mathbf{S}}]$ be the thermoelastic states produced on E^3 by the external thermomechanical loads $[\mathbf{b}, \mathbf{s}, T] = [\mathbf{0}, \mathbf{0}, T(\mathbf{x})]$ and $[\widetilde{\mathbf{b}}, \widetilde{\mathbf{s}}, \widetilde{T}] = [\mathbf{0}, \mathbf{0}, \delta(\mathbf{x} - \xi)]$, respectively. Applying Eq. (3.29) to the states s and $\widetilde{S}$ we obtain

$$\int_{E^3} T\mathbf{M} \cdot \widetilde{\mathbf{E}}\, dv = \int_{E^3} \widetilde{T}\mathbf{M} \cdot \mathbf{E}\, dv \tag{3.100}$$

For a homogeneous isotropic thermoelastic solid

$$\mathbf{M} = -(3\lambda + 2\mu)\alpha\mathbf{1} \tag{3.101}$$

Therefore,

$$\mathbf{M} \cdot \widetilde{\mathbf{E}} = -(3\lambda + 2\mu)\alpha(\operatorname{tr} \widetilde{\mathbf{E}}) \tag{3.102}$$

and

$$\mathbf{M} \cdot \mathbf{E} = -(3\lambda + 2\mu)\alpha(\operatorname{tr} \mathbf{E}) \tag{3.103}$$

where λ and μ are Lamé constants. Substituting (3.102) and (3.103) into (3.100) we get

$$\int_{E^3} T(\operatorname{tr} \widetilde{\mathbf{E}}) dv = \int_{E^3} \widetilde{T}(\operatorname{tr} \mathbf{E}) dv \tag{3.104}$$

Since $\widetilde{s}$ is the thermoelastic state produced by the temperature $\widetilde{T}$ on E^3, $\widetilde{\mathbf{u}}$ takes the form

$$\widetilde{\mathbf{u}} = \nabla \widetilde{\phi} \tag{3.105}$$

where the thermoelastic potential $\widetilde{\phi}$ satisfies Poisson's equation

$$\nabla^2 \widetilde{\phi} = \frac{1+\nu}{1-\nu} \alpha \widetilde{T} \tag{3.106}$$

Hence

$$\text{tr}\, \widetilde{\mathbf{E}} = \nabla^2 \widetilde{\phi} = \frac{1+\nu}{1-\nu} \alpha \widetilde{T} \tag{3.107}$$

Therefore, substituting (3.107) into (3.104) we obtain

$$\frac{1+\nu}{1-\nu} \alpha \int\limits_{E^3} T(\xi) \delta(\mathbf{x}-\xi) dv(\xi) = \int\limits_{E^3} \delta(\mathbf{x}-\xi) [\text{tr}\, \mathbf{E}(\xi)] dv(\xi) \tag{3.108}$$

or using the filtrating property of the delta function we arrive at Eq. (3.99). This completes solution of Problem 3.3.

Problem 3.4. Assume T_0 to be a constant temperature, and let a_i (i $= 1, 2, 3$) be positive constants of the length dimension. Show that for a homogeneous isotropic infinite elastic body subject to the temperature change

$$\begin{aligned} T(\mathbf{x}) = T_0[H(x_1+a_1) - H(x_1-a_1)] \times [H(x_2+a_2) - H(x_2-a_2)] \\ \times [H(x_3+a_3) - H(x_3-a_3)] \end{aligned} \tag{3.109}$$

where $H = H(x)$ denotes the Heaviside function defined by: $H(x) = 1$ for $x > 0$ and $H(x) = 0$ for $x < 0$; the stress components S_{ij} are represented by the formulas

$$\begin{aligned} S_{ij}(\mathbf{x}) = \mathrm{A}_0 \int\limits_{-a_1}^{a_1} d\xi_1 \int\limits_{-a_2}^{a_2} d\xi_2 \int\limits_{-a_3}^{a_3} d\xi_3 \frac{\partial^2}{\partial \xi_i \partial \xi_j} \left[(x_1-\xi_1)^2 + (x_2-\xi_2)^2 + (x_3-\xi_3)^2 \right]^{-1/2} \\ + 4\pi \delta_{ij} \mathrm{A}_0 \int\limits_{-a_1}^{a_1} d\xi_1 \int\limits_{-a_2}^{a_2} d\xi_2 \int\limits_{-a_3}^{a_3} d\xi_3\, \delta(x_1-\xi_1) \delta(x_2-\xi_2) \delta(x_3-\xi_3) \end{aligned} \tag{3.110}$$

where

$$\mathrm{A}_0 = -\frac{\mu}{2\pi} \frac{1+\nu}{1-\nu} \alpha\, T_0 \tag{3.111}$$

Also, show that the integrals on the RHS of Eq. (3.110) can be calculated in terms of elementary functions, and for the exterior of the parallelepiped

$$|x_1| \le a_1, \quad |x_2| \le a_2 \quad |x_3| \le a_3 \tag{3.112}$$

we obtain

$$S_{12} = A_0 \ln \left[\frac{(x_3 + a_3 + r_{+1.+2.-3})}{(x_3 - a_3 + r_{+1.+2.+3})} \frac{(x_3 - a_3 + r_{+1.-2.+3})}{(x_3 + a_3 + r_{+1.-2.-3})} \right. \\ \left. \times \frac{(x_3 - a_3 + r_{-1.+2.+3})}{(x_3 + a_3 + r_{-1.+2.-3})} \frac{(x_3 + a_3 + r_{-1.-2.-3})}{(x_3 - a_3 + r_{-1.-2.+3})} \right] \tag{3.113}$$

$$S_{23} = A_0 \ln \left[\frac{(x_1 + a_1 + r_{-1.+2.+3})}{(x_1 - a_1 + r_{+1.+2.+3})} \frac{(x_1 - a_1 + r_{+1.+2.-3})}{(x_1 + a_1 + r_{-1.+2.-3})} \right. \\ \left. \times \frac{(x_1 - a_1 + r_{+1.-2.+3})}{(x_1 + a_1 + r_{-1.-2.+3})} \frac{(x_1 + a_1 + r_{-1.-2.-3})}{(x_1 - a_1 + r_{+1.-2.-3})} \right] \tag{3.114}$$

$$S_{31} = A_0 \ln \left[\frac{(x_2 + a_2 + r_{+1.-2.+3})}{(x_2 - a_2 + r_{+1.+2.+3})} \frac{(x_2 - a_2 + r_{-1.+2.+3})}{(x_2 + a_2 + r_{-1.-2.+3})} \right. \\ \left. \times \frac{(x_2 - a_2 + r_{+1.+2.-3})}{(x_2 + a_2 + r_{+1.-2.-3})} \frac{(x_2 + a_2 + r_{-1.-2.-3})}{(x_2 - a_2 + r_{-1.+2.-3})} \right] \tag{3.115}$$

and

$$S_{11} = A_0 \left[\tan^{-1}\left(\frac{x_2 + a_2}{x_1 - a_1} \frac{x_3 + a_3}{r_{+1.-2.-3}} \right) - \tan^{-1}\left(\frac{x_2 + a_2}{x_1 - a_1} \frac{x_3 - a_3}{r_{+1.-2.+3}} \right) \right. \\ - \tan^{-1}\left(\frac{x_2 - a_2}{x_1 - a_1} \frac{x_3 + a_3}{r_{+1.+2.-3}} \right) + \tan^{-1}\left(\frac{x_2 - a_2}{x_1 - a_1} \frac{x_3 - a_3}{r_{+1.+2.+3}} \right) \\ - \tan^{-1}\left(\frac{x_2 + a_2}{x_1 + a_1} \frac{x_3 + a_3}{r_{-1.-2.-3}} \right) + \tan^{-1}\left(\frac{x_2 + a_2}{x_1 + a_1} \frac{x_3 - a_3}{r_{-1.-2.+3}} \right) \\ \left. + \tan^{-1}\left(\frac{x_2 - a_2}{x_1 + a_1} \frac{x_3 + a_3}{r_{-1.+2.-3}} \right) - \tan^{-1}\left(\frac{x_2 - a_2}{x_1 + a_1} \frac{x_3 - a_3}{r_{-1.+2.+3}} \right) \right] \tag{3.116}$$

$$S_{22} = A_0 \left[\tan^{-1}\left(\frac{x_3 + a_3}{x_2 - a_2} \frac{x_1 + a_1}{r_{-1.+2.-3}} \right) - \tan^{-1}\left(\frac{x_3 + a_3}{x_2 - a_2} \frac{x_1 - a_1}{r_{+1.+2.-3}} \right) \right. \\ - \tan^{-1}\left(\frac{x_3 - a_3}{x_2 - a_2} \frac{x_1 + a_1}{r_{-1.+2.+3}} \right) + \tan^{-1}\left(\frac{x_3 - a_3}{x_2 - a_2} \frac{x_1 - a_1}{r_{+1.+2.+3}} \right) \\ - \tan^{-1}\left(\frac{x_3 + a_3}{x_2 + a_2} \frac{x_1 + a_1}{r_{-1.-2.-3}} \right) + \tan^{-1}\left(\frac{x_3 + a_3}{x_2 + a_2} \frac{x_1 - a_1}{r_{+1.-2.-3}} \right) \\ \left. + \tan^{-1}\left(\frac{x_3 - a_3}{x_2 + a_2} \frac{x_1 + a_1}{r_{-1.-2.+3}} \right) - \tan^{-1}\left(\frac{x_3 - a_3}{x_2 + a_2} \frac{x_1 - a_1}{r_{+1.-2.+3}} \right) \right] \tag{3.117}$$

$$
\begin{aligned}
S_{33} = A_0 \Big[& \tan^{-1}\left(\frac{x_1+a_1}{x_3-a_3}\frac{x_2+a_2}{r_{-1.-2.+3}}\right) - \tan^{-1}\left(\frac{x_1+a_1}{x_3-a_3}\frac{x_2-a_2}{r_{-1.+2.+3}}\right) \\
& - \tan^{-1}\left(\frac{x_1-a_1}{x_3-a_3}\frac{x_2+a_2}{r_{+1.-2.-3}}\right) + \tan^{-1}\left(\frac{x_1-a_1}{x_3-a_3}\frac{x_2-a_2}{r_{+1.+2.+3}}\right) \\
& - \tan^{-1}\left(\frac{x_1+a_1}{x_3+a_3}\frac{x_2+a_2}{r_{-1.-2.-3}}\right) + \tan^{-1}\left(\frac{x_1+a_1}{x_3+a_3}\frac{x_2-a_2}{r_{-1.+2.-3}}\right) \\
& + \tan^{-1}\left(\frac{x_1-a_1}{x_3+a_3}\frac{x_2+a_2}{r_{+1.-2.-3}}\right) - \tan^{-1}\left(\frac{x_1-a_1}{x_3+a_3}\frac{x_2-a_2}{r_{+1.+2.-3}}\right)\Big]
\end{aligned} \tag{3.118}
$$

where

$$
r_{\pm1.\pm2.\pm3} = [(x_1 \mp a_1)^2 + (x_2 \mp a_2)^2 + (x_3 \mp a_3)^2]^{1/2} \tag{3.119}
$$

Note that Eq. (3.114) follows from Eq. (3.113) by the transformation of indices

$$
1 \to 2, \quad 2 \to 3, \quad 3 \to 1
$$

and Eq. (3.115) follows from Eq. (3.114) by the transformation of indices

$$
2 \to 3, \quad 3 \to 1, \quad 1 \to 2
$$

Also, Eq. (3.117) follows from Eq. (3.116) by the transformation of indices

$$
1 \to 2, \quad 2 \to 3, \quad 3 \to 1
$$

and Eq. (3.118) follows from Eq. (3.117) by the transformation of indices

$$
2 \to 3, \quad 3 \to 1, \quad 1 \to 2.
$$

Hint. To find S_{12} use the formula

$$
\int \frac{du}{\sqrt{u^2+a^2}} = \ln\left(u + \sqrt{u^2+a^2}\right) \tag{3.120}
$$

and to calculate S_{11} take advantage of the formulas

$$
\int \frac{du}{(\sqrt{u^2+a^2})^3} = \frac{1}{a^2}\frac{u}{\sqrt{u^2+a^2}} \tag{3.121}
$$

and

$$
\int \frac{du}{(u^2+b^2)\sqrt{u^2+a^2}} = \frac{1}{b\sqrt{a^2-b^2}} \tan^{-1}\left(\frac{u\sqrt{a^2-b^2}}{b\sqrt{u^2+a^2}}\right) \tag{3.122}
$$

where a and b are constants subject to the conditions

$$a \neq 0, \quad b \neq 0, \quad |a| > |b| \tag{3.123}$$

Solution. To show (3.110) we note that

$$S_{ij}(\mathbf{x}) = \int_{-a_1}^{a_1} d\xi_1 \int_{-a_2}^{a_2} d\xi_2 \int_{-a_3}^{a_3} d\xi_3 \, S^*_{ij}(\mathbf{x}\xi) \tag{3.124}$$

where

$$S^*_{ij}(\mathbf{x}, \xi) = 2\mu \left(\phi^*_{,ij} - \delta_{ij} \phi^*_{,kk} \right) \tag{3.125}$$

and

$$\phi^*_{,kk} = \frac{1+\nu}{1-\nu} \alpha \, \delta(x_1 - \xi_1)\delta(x_2 - \xi_2)\delta(x_3 - \xi_3) \tag{3.126}$$

Since

$$\phi^*(\mathbf{x}, \xi) = -\frac{1}{4\pi} \frac{1+\nu}{1-\nu} \alpha \frac{1}{|\mathbf{x} - \xi|}$$

therefore

$$S^*_{ij}(\mathbf{x}\,\xi) = A_0 \left\{ \frac{\partial^2}{\partial \xi_i \partial \xi_j} \frac{1}{|\mathbf{x} - \xi|} + 4\pi \delta_{ij} \delta(x_1 - \xi_1)\delta(x_2 - \xi_2)\delta(x_3 - \xi_3) \right\}$$

where A_0 is given by (3.111). Hence, substituting S^*_{ij} into Eq. (3.124) we obtain (3.110).

To show (3.113)–(3.118) we note that for the exterior of the parallelepiped

$$|x_1| \leq a_1, \quad |x_2| \leq a_2, \quad |x_3| \leq a_3 \tag{3.127}$$

Equation (3.110) reduces to

$$\begin{aligned} S_{ij}(\mathbf{x}) = A_0 & \int_{-a_1}^{a_1} d\xi_1 \int_{-a_2}^{a_2} d\xi_2 \int_{-a_3}^{a_3} d\xi_3 \\ & \times \frac{\partial^2}{\partial \xi_i \, \partial \xi_j} [(x_1 - \xi_1)^2 + (x_2 - \xi_2)^2 + (x_3 - \xi_3)^2]^{-1/2} \\ & i, j = 1, \ 2, \ 3. \end{aligned} \tag{3.128}$$

Letting $i = 1, \ j = 2$ in (3.128) we obtain

$$
S_{12}(\mathbf{x}) = A_0 \int_{-a_3}^{a_3} d\xi_3 \int_{-a_2}^{a_2} d\xi_2 \, \frac{\partial}{\partial \xi_2}
\times \Big\{ [(x_1 - a_1)^2 + (x_2 - \xi_2)^2 + (x_3 - \xi_3)^2]^{-1/2}
- [(x_1 + a_1)^2 + (x_2 - \xi_2)^2 + (x_3 - \xi_3)^2]^{-1/2} \Big\} \quad (3.129)
$$

or

$$
\begin{aligned}
S_{12}(\mathbf{x}) = A_0 \int_{-a_3}^{a_3} d\xi_3 \times \Big\{ & [(x_1 - a_1)^2 + (x_2 - a_2)^2 + (x_3 - \xi_3)^2]^{-1/2} \\
& - [(x_1 - a_1)^2 + (x_2 + a_2)^2 + (x_3 - \xi_3)^2]^{-1/2} \\
& - [(x_1 + a_1)^2 + (x_2 - a_2)^2 + (x_3 - \xi_3)^2]^{-1/2} \\
& + [(x_1 + a_1)^2 + (x_2 + a_2)^2 + (x_3 - \xi_3)^2]^{-1/2} \Big\}
\end{aligned} \quad (3.130)
$$

Since for every $b > 0$

$$
\int_{-a_3}^{a_3} [b^2 + (x_3 - \xi_3)^2]^{-1/2} \, d\xi_3 = \int_{x_3 - a_3}^{x_3 + a_3} (b^2 + u^2)^{-1/2} \, du \quad (3.131)
$$

and by virtue of (3.121)

$$
\int (b^2 + u^2)^{-1/2} \, du = \ln \left(u + \sqrt{u^2 + b^2} \right) \quad (3.132)
$$

if follows from (3.130) that

$$
\begin{aligned}
S_{12} = A_0 \ln \Bigg\{ & \frac{(x_3 + a_3 + r_{+1.+2.-3})}{(x_3 - a_3 + r_{+1.+2.+3})} \, \frac{(x_3 - a_3 + r_{+1.-2.+3})}{(x_3 + a_3 + r_{+1.-2.-3})} \\
& \times \frac{(x_3 - a_3 + r_{-1.+2.+3})}{(x_3 + a_3 + r_{-1.+2.-3})} \, \frac{(x_3 + a_3 + r_{-1.-2.-3})}{(x_3 - a_3 + r_{-1.-2.+3})} \Bigg\}
\end{aligned} \quad (3.133)
$$

where $r_{\pm 1.\pm 2.\pm 3}$ is defined by (3.119). This completes proof of (3.113). The components S_{23} and S_{31}, respectively, are obtained from (3.113) by the transformation of the indices

$$
1 \to 2, \quad 2 \to 3, \quad 3 \to 1 \quad (3.134)
$$

and

$$
2 \to 3, \quad 3 \to 1, \quad 1 \to 2 \quad (3.135)
$$

By letting $i = 1,\ j = 1$ in (3.128) we obtain

$$S_{11}(\mathbf{x}) = A_0 \int_{-a_2}^{a_2} d\xi_2 \int_{-a_3}^{a_3} d\xi_3 \int_{-a_1}^{a_1} d\xi_1 \frac{\partial^2}{\partial \xi_1^2} \times [(x_2-\xi_2)^2 + (x_3-\xi_3)^2 + (x_1-\xi_1)^2]^{-1/2} \tag{3.136}$$

Since

$$\frac{\partial}{\partial \xi_1} [(x_1-\xi_1)^2 + \alpha^2]^{-1/2} = (x_1-\xi_1)[(x_1-\xi_1)^2 + \alpha^2]^{-3/2} \tag{3.137}$$

where

$$\alpha^2 = (x_2-\xi_2)^2 + (x_3-\xi_3)^2 \tag{3.138}$$

therefore Eq. (3.136) takes the form

$$S_{11}(\mathbf{x}) = A_0 \int_{-a_2}^{a_2} d\xi_2 \int_{-a_3}^{a_3} d\xi_3 \times \left\{ \frac{(x_1-a_1)}{[(x_1-a_1)^2+\alpha^2]^{3/2}} - \frac{(x_1+a_1)}{[(x_1+a_1)^2+\alpha^2]^{3/2}} \right\} \tag{3.139}$$

Now, because of (3.121),

$$\begin{aligned} \int_{-a_3}^{a_3} d\xi_3 \frac{1}{[(x_1-a_1)^2+\alpha^2]^{3/2}} &= \int_{x_3-a_3}^{x_3+a_3} \frac{du}{[(x_1-a_1)^2+(x_2-\xi_2)^2+u^2]^{3/2}} \\ &= \frac{1}{(x_1-a_1)^2+(x_2-\xi_2)^2} \left\{ \frac{(x_3+a_3)}{[(x_1-a_1)^2+(x_2-\xi_2)^2+(x_3+a_3)^2]^{1/2}} \right. \\ &\quad \left. - \frac{(x_3-a_3)}{[(x_1-a_1)^2+(x_2-\xi_2)^2+(x_3-a_3)^2]^{1/2}} \right\} \end{aligned} \tag{3.140}$$

Also, using (3.122), we obtain

$$\begin{aligned} &\int_{-a_2}^{a_2} d\xi_2 \frac{1}{(x_2-\xi_2)^2+(x_1-a_1)^2} \frac{1}{[(x_1-a_1)^2+(x_3+a_3)^2+(x_2-\xi_2)^2]^{1/2}} \\ &= \int_{-a_2}^{a_2} d\xi_2 \frac{1}{(x_2-\xi_2)^2+b^2} \frac{1}{[(x_2-\xi_2)^2+a^2]^{1/2}} = \int_{x_2-a_2}^{x_2+a_2} \frac{1}{u^2+b^2} \frac{1}{\sqrt{u^2+a^2}} du \\ &= \frac{1}{b\sqrt{a^2-b^2}} \left\{ \tan^{-1} \frac{u\sqrt{a^2-b^2}}{b\sqrt{u^2+a^2}} \right\}_{u=x_2-a_2}^{u=x_2+a_2} \end{aligned} \tag{3.141}$$

where

$$b^2 = (x_1 - a_1)^2, \quad a^2 = (x_1 - a_1)^2 + (x_3 + a_3)^2 \tag{3.142}$$

By letting $x_1 > a_1$ and $x_3 > -a_3$ we receive

$$b = x_1 - a_1, \quad \sqrt{a^2 - b^2} = x_3 + a_3 \tag{3.143}$$

and reduce Eq. (3.141) to

$$\int_{-a_2}^{a_2} d\xi_2 \frac{1}{(x_2 - \xi_2)^2 + b^2} \frac{1}{[(x_2 - \xi_2)^2 + a^2]^{1/2}}$$
$$= \frac{1}{(x_1 - a_1)(x_3 + a_3)} \left\{ \tan^{-1} \frac{x_2 + a_2}{x_1 - a_1} \frac{x_3 + a_3}{r_{+1,-2,-3}} - \tan^{-1} \frac{x_2 - a_2}{x_1 - a_1} \frac{x_3 + a_3}{r_{+1,+2,-3}} \right\} \tag{3.144}$$

It follows from Eq. (3.139) that

$$\begin{aligned} S_{11}(\mathbf{x}) = A_0 \int_{-a_2}^{a_2} & d\xi_2 \, (x_1 - a_1) \frac{1}{(x_1 - a_1)^2 + (x_2 - \xi_2)^2} \\ & \times \left\{ \frac{(x_3 + a_3)}{[(x_2 - \xi_2)^2 + (x_1 - a_1)^2 + (x_3 + a_3)^2]^{1/2}} \right. \\ & \left. - \frac{(x_3 - a_3)}{[(x_2 - \xi_2)^2 + (x_1 - a_1)^2 + (x_3 - a_3)^2]^{1/2}} \right\} \\ & - A_0 \int_{-a_2}^{a_2} d\xi_2 \, (x_1 + a_1) \frac{1}{(x_1 + a_1)^2 + \left(x_2 - \xi_2^2\right)^2} \\ & \times \left\{ \frac{(x_3 + a_3)}{[(x_2 - \xi_2)^2 + (x_1 + a_1)^2 + (x_3 + a_3)^2]^{1/2}} \right. \\ & \left. - \frac{(x_3 - a_3)}{[(x_2 - \xi_2)^2 + (x_1 + a_1)^2 + (x_3 - a_3)^2]^{1/2}} \right\} \end{aligned} \tag{3.145}$$

Therefore, using Eq. (3.141) as well as equations obtained from Eq. (3.141) by suitable choice of a and b, we obtain (3.116).

The components S_{22} and S_{33} are obtained from Eq. (3.116) by suitable transformation of indices.

Problem 3.5. Let $\mathbf{u} = \mathbf{u}(\mathbf{x}, t)$ be a solution of the vector equation

$$\nabla^2 \mathbf{u} - \frac{1}{c^2} \frac{\partial^2 \mathbf{u}}{\partial t^2} = -\frac{\mathbf{f}}{c^2} \quad \text{on} \quad B \times (0, \infty) \tag{3.146}$$

subject to the initial conditions

$$\mathbf{u}(\mathbf{x},0)=\mathbf{u}_0(\mathbf{x}),\quad \dot{\mathbf{u}}(\mathbf{x},0)=\dot{\mathbf{u}}_0(\mathbf{x})\quad \text{for}\quad \mathbf{x}\in\overline{\mathrm{B}} \tag{3.147}$$

where $\mathbf{f}=\mathbf{f}(\mathbf{x},t)$ is a prescribed vector field on $\overline{\mathrm{B}}\times[0,\infty)$; and $\mathbf{u}_0(\mathbf{x})$ and $\dot{\mathbf{u}}_0(\mathbf{x})$ are prescribed vector fields on $\overline{\mathrm{B}}$; and $c>0$.

Also, let $\widetilde{\mathbf{u}}=\widetilde{\mathbf{u}}(\mathbf{x},t)$ be a solution of the vector equation

$$\nabla^2\widetilde{\mathbf{u}}-\frac{1}{c^2}\frac{\partial^2\widetilde{\mathbf{u}}}{\partial t^2}=-\frac{\widetilde{\mathbf{f}}}{c^2}\quad \text{on}\quad \mathrm{B}\times(0,\infty) \tag{3.148}$$

subject to the initial conditions

$$\widetilde{\mathbf{u}}(\mathbf{x},0)=\widetilde{\mathbf{u}}_0(\mathbf{x}),\quad \dot{\widetilde{\mathbf{u}}}(\mathbf{x},0)=\dot{\widetilde{\mathbf{u}}}_0(\mathbf{x})\quad \text{for}\quad \mathbf{x}\in\overline{\mathrm{B}} \tag{3.149}$$

where $\widetilde{\mathbf{f}}=\widetilde{\mathbf{f}}(\mathbf{x},t)\neq\mathbf{f}(\mathbf{x},\mathrm{t})$, $\widetilde{\mathbf{u}}_0(\mathbf{x})\neq\mathbf{u}_0(\mathbf{x})$, and $\dot{\widetilde{\mathbf{u}}}_0(\mathbf{x})\neq\dot{\mathbf{u}}_0(\mathbf{x})$ are prescribed functions on $\overline{\mathrm{B}}\times[0,\infty)$, $\overline{\mathrm{B}}$, and $\overline{\mathrm{B}}$, respectively. Show that the following reciprocal relation holds true

$$\begin{aligned}\frac{1}{c^2}\int_{\mathrm{B}}(\mathbf{u}*\widetilde{\mathbf{f}}+\mathbf{u}\cdot\dot{\widetilde{\mathbf{u}}}_0+\dot{\mathbf{u}}\cdot\widetilde{\mathbf{u}}_0)\,dv+\int_{\partial\mathrm{B}}\mathbf{u}*\frac{\partial\widetilde{\mathbf{u}}}{\partial\mathrm{n}}da\\ =\frac{1}{c^2}\int_{\mathrm{B}}(\widetilde{\mathbf{u}}*\mathbf{f}+\widetilde{\mathbf{u}}\cdot\dot{\mathbf{u}}_0+\dot{\widetilde{\mathbf{u}}}\cdot\mathbf{u}_0)\,dv+\int_{\partial\mathrm{B}}\widetilde{\mathbf{u}}*\frac{\partial\mathbf{u}}{\partial\mathrm{n}}da\end{aligned} \tag{3.150}$$

where $*$ represents the *inner convolutional product*, that is, for any two vector fields $\mathbf{a}=\mathbf{a}(\mathbf{x},t)$ and $\mathbf{b}=\mathbf{b}(\mathbf{x},t)$ on $\overline{\mathrm{B}}\times[0,\infty)$

$$\mathbf{a}*\mathbf{b}=\int_0^t\mathbf{a}(\mathbf{x},t-\tau)\cdot\mathbf{b}(\mathbf{x},\tau)\,d\tau \tag{3.151}$$

Solution. Let $\overline{f}(\mathbf{x},p)$ denote the Laplace transform of a function $f=f(\mathbf{x},t)$ defined by

$$Lf\equiv\overline{f}(\mathbf{x},p)=\int_0^\infty e^{-pt}f(\mathbf{x},t)dt \tag{3.152}$$

Then

$$\overline{\dot{f}(\mathbf{x},t)}=p\,\overline{f}(\mathbf{x},p)-f(\mathbf{x},0) \tag{3.153}$$

$$\overline{\ddot{f}(\mathbf{x},t)}=p^2\,\overline{f}(\mathbf{x},p)-\dot{f}(\mathbf{x},0)-pf(\mathbf{x},0) \tag{3.154}$$

Now, Eqs. (3.146) and (3.147) in components take the forms

$$u_{i,kk} - \frac{1}{c^2}\ddot{u}_i = -\frac{f_i}{c^2} \tag{3.155}$$

and

$$u_i(\mathbf{x}, 0) = u_{0i}(\mathbf{x}),\ \ \dot{u}_i(\mathbf{x}, 0) = \dot{u}_{0i}(\mathbf{x}) \tag{3.156}$$

Taking the Laplace transform of Eq. (3.155) and using (3.156) we obtain

$$\overline{u}_{i,kk} - \frac{1}{c^2}(p^2\overline{u}_i - \dot{u}_{0i} - p\, u_{0i}) = -\frac{\overline{f_i}}{c^2} \tag{3.157}$$

Similarly, Eqs. (3.148) and (3.149) imply that

$$\overline{\tilde{u}}_{i,kk} - \frac{1}{c^2}(p^2\overline{\tilde{u}}_i - \dot{\tilde{u}}_{0i} - p\, \tilde{u}_{0i}) = \frac{\overline{\tilde{f}}_i}{c^2} \tag{3.158}$$

Multiplying (3.157) by $\overline{\tilde{u}_i}$ and (3.158) by $\overline{u_i}$, respectively, we obtain

$$\overline{\tilde{u}}_i\,\overline{u}_{i,kk} - \frac{1}{c^2}(p^2\overline{\tilde{u}}_i\overline{u}_i - \overline{\tilde{u}}_i\dot{u}_{0i} - p\,\overline{\tilde{u}}_i u_{0i}) = -\frac{\overline{f_i}\,\overline{\tilde{u}}_i}{c^2} \tag{3.159}$$

$$\overline{u}_i\,\overline{\tilde{u}}_{i,kk} - \frac{1}{c^2}(p^2\overline{u}_i\overline{\tilde{u}}_i - \overline{u}_i\dot{\tilde{u}}_{0i} - p\,\overline{u}_i\tilde{u}_{0i}) = -\frac{\overline{\tilde{f}}_i\,\overline{u}_i}{c^2} \tag{3.160}$$

Since

$$\overline{\tilde{u}}_i\overline{u}_{i,kk} = (\overline{\tilde{u}}_i\overline{u}_{i,k})_{,k} - \overline{\tilde{u}}_{i,k}\overline{u}_{i,k} \tag{3.161}$$

and

$$\overline{u}_i\overline{\tilde{u}}_{i,kk} = (\overline{u}_i\overline{\tilde{u}}_{i,k})_{,k} - \overline{u}_{i,k}\overline{\tilde{u}}_{i,k} \tag{3.162}$$

therefore, subtracting (3.160) from (3.159), and using the divergence theorem we obtain

$$\begin{aligned}\int\limits_{\partial B}(\overline{\tilde{u}}_i\overline{u}_{i,k}\,n_k - \overline{u}_i\overline{\tilde{u}}_{i,k}\,n_k)da + \frac{1}{c^2}\int\limits_B(\dot{u}_{0i}\overline{\tilde{u}}_i + p\,u_{0i}\overline{\tilde{u}}_i - \dot{\tilde{u}}_{0i}\overline{u}_i - p\tilde{u}_{0i}\overline{u}_i)dv \\ = -\frac{1}{c^2}\int\limits_B(\overline{f_i}\overline{\tilde{u}}_i - \overline{\tilde{f}}_i\overline{u}_i)dv\end{aligned} \tag{3.163}$$

or

$$\int\limits_{\partial B}\overline{u}_i\overline{\tilde{u}}_{i,k}n_k\,da + \frac{1}{c^2}\int\limits_B[\overline{u}_i\overline{\tilde{f}}_i + \overline{u}_i\dot{\tilde{u}}_{0i} + (p\overline{u}_i - u_{0i})\tilde{u}_{0i} + u_{0i}\tilde{u}_{0i}]dv$$

$$= \int_{\partial B} \overline{\widetilde{u}}_i \overline{u}_{i,k}\, n_k\, da + \frac{1}{c^2} \int_B [\overline{\widetilde{u}}_i \overline{f}_i + \overline{\widetilde{u}}_i \dot{u}_{0i} + (p\overline{\widetilde{u}}_i - \widetilde{u}_{0i}) u_{0i} + \widetilde{u}_{0i} u_{0i}] dv \tag{3.164}$$

Using the formula

$$L^{-1}(\overline{f}\,\overline{g}) = f * g \tag{3.165}$$

and applying the operator L^{-1} to Eq. (3.164) we arrive at Eq. (3.150). This completes solution of Problem 3.5.

Problem 3.6. Let $\widetilde{\mathbf{U}} = \widetilde{\mathbf{U}}(\mathbf{x}, t)$ be a symmetric second-order tensor field that satisfies the wave equation

$$\left(\nabla^2 - \frac{1}{c^2}\frac{\partial^2}{\partial t^2}\right)\widetilde{\mathbf{U}} = -\frac{\widetilde{\mathbf{F}}}{c^2} \quad \text{on} \quad \mathrm{B} \times (0, \infty) \tag{3.166}$$

subject to the initial conditions

$$\widetilde{\mathbf{U}}(\mathbf{x}, 0) = \widetilde{\mathbf{U}}_0(\mathbf{x}), \quad \dot{\widetilde{\mathbf{U}}}(\mathbf{x}, 0) = \dot{\widetilde{\mathbf{U}}}_0(\mathbf{x}) \quad \text{for} \quad \mathbf{x} \in \overline{\mathrm{B}} \tag{3.167}$$

where $\widetilde{\mathbf{F}} = \widetilde{\mathbf{F}}(\mathbf{x}, t)$, $\widetilde{\mathbf{U}}_0(\mathbf{x})$, and $\dot{\widetilde{\mathbf{U}}}_0(\mathbf{x})$ are prescribed functions on $\overline{\mathrm{B}} \times [0, \infty)$, $\overline{\mathrm{B}}$, and $\overline{\mathrm{B}}$, respectively. Let $\mathbf{u} = \mathbf{u}(\mathbf{x}, t)$ be a solution to Eq. (3.166) and (3.167) of Problem 3.5. Show that

$$\frac{1}{c^2}\int_{\mathrm{B}} (\widetilde{\mathbf{F}} * \mathbf{u} + \dot{\widetilde{\mathbf{U}}}_0 \mathbf{u} + \widetilde{\mathbf{U}}_0 \dot{\mathbf{u}})\, dv + \int_{\partial \mathrm{B}} \frac{\partial \widetilde{\mathbf{U}}}{\partial \mathrm{n}} * \mathbf{u}\, da$$
$$= \frac{1}{c^2}\int_{\mathrm{B}} (\widetilde{\mathbf{U}} * \mathbf{f} + \dot{\widetilde{\mathbf{U}}} \mathbf{u}_0 + \widetilde{\mathbf{U}} \dot{\mathbf{u}}_0)\, dv + \int_{\partial \mathrm{B}} \widetilde{\mathbf{U}} * \frac{\partial \mathbf{u}}{\partial \mathrm{n}} da \tag{3.168}$$

where for any tensor field $\mathbf{T} = \mathbf{T}(\mathbf{x}, t)$ on $\overline{\mathrm{B}} \times [0, \infty)$ and for any vector field $\mathbf{v} = \mathbf{v}(\mathbf{x}, t)$ on $\overline{\mathrm{B}} \times [0, \infty)$

$$\mathbf{T} * \mathbf{v} = \int_0^t \mathbf{T}(\mathbf{x}, t - \tau)\mathbf{v}(\mathbf{x}, \tau)\, d\tau \tag{3.169}$$

Solution. Equation (3.150) of Problem 3.5 in components takes the form

$$\frac{1}{c^2}\int_B (\widetilde{f}_i * u_i + \dot{\widetilde{u}}_{0i} u_i + \widetilde{u}_{0i} \dot{u}_i) dv + \int_{\partial B} \frac{\partial \widetilde{u}_i}{\partial n} * u_i da$$
$$= \frac{1}{c^2}\int_B (f_i * \widetilde{u}_i + \dot{u}_{0i} \widetilde{u}_i + u_{0i} \dot{\widetilde{u}}_i) dv + \int_{\partial B} \frac{\partial u_i}{\partial n} * \widetilde{u}_i da \tag{3.170}$$

It follows from the formulation of Problems 3.5 and 3.6 that Eq. (3.170) holds also true if for a fixed index j we let

$$\widetilde{u}_i = \widetilde{U}_{ij}, \quad \widetilde{u}_{0i} = \widetilde{U}_{0ij}, \quad \dot{\widetilde{u}}_{0i} = \dot{\widetilde{U}}_{0ij}, \quad \widetilde{f}_i = \widetilde{F}_{ij} \tag{3.171}$$

Therefore, substituting (3.171) into (3.170) we obtain

$$\begin{aligned} &\frac{1}{c^2}\int_B (\widetilde{F}_{ij} * u_i + \dot{\widetilde{U}}_{0ij} u_i + \widetilde{U}_{0ij}\dot{u}_i) dv + \int_B \frac{\partial \widetilde{U}_{ij}}{\partial n} * u_i da \\ &= \frac{1}{c^2}\int_B (f_i * \widetilde{U}_{ij} + \dot{u}_{0i}\, \widetilde{U}_{ij} + u_{0i}\, \dot{\widetilde{U}}_{ij}) dv + \int_B \frac{\partial u_i}{\partial n} * \widetilde{U}_{ij}\, da \end{aligned} \tag{3.172}$$

Finally, the symmetry of tensors $\widetilde{U}_{ij}$, $\widetilde{F}_{ij}$, $\widetilde{U}_{0ij}$, and $\dot{\widetilde{U}}_{0ij}$, as well as the relation

$$a * b = b * a \tag{3.173}$$

valid for arbitrary functions $a = a(\mathbf{x}, t)$ and $b = b(\mathbf{x}, t)$, imply that Eq. (3.172) is equivalent to Eq. (3.168). This completes solution to Problem 3.6.

Problem 3.7. Let $\mathbf{G} = \mathbf{G}(\mathbf{x}, \xi; t)$ be a symmetric second-order tensor field that satisfies the wave equation

$$\Box_0^2 \mathbf{G} = -\mathbf{1}\delta(\mathbf{x} - \xi)\delta(t) \quad \text{for} \quad \mathbf{x} \in E^3, \xi \in E^3, \ t > 0 \tag{3.174}$$

subject to the homogeneous initial conditions

$$\mathbf{G}(\mathbf{x}, \xi; 0) = \mathbf{0}, \quad \dot{\mathbf{G}}(\mathbf{x}, \xi; 0) = \mathbf{0} \quad \text{for} \quad \mathbf{x} \in E^3, \xi \in E^3 \tag{3.175}$$

where

$$\Box_0^2 = \frac{\partial^2}{\partial x_k \partial x_k} - \frac{1}{c^2}\frac{\partial^2}{\partial t^2} \quad (k = 1, 2, 3) \tag{3.176}$$

Show that a solution $\mathbf{u}$ to Eqs. (3.146) and (3.147) of Problem 3.5 admits the integral representation

$$\begin{aligned} \mathbf{u}(\mathbf{x}, t) = &\frac{1}{c^2}\int_B (\mathbf{G} * \mathbf{f} + \dot{\mathbf{G}}\mathbf{u}_0 + \mathbf{G}\dot{\mathbf{u}}_0)\, dv(\xi) \\ &+ \int_{\partial B} \left(\mathbf{G} * \frac{\partial \mathbf{u}}{\partial \mathrm{n}} - \frac{\partial \mathbf{G}}{\partial \mathrm{n}} * \mathbf{u}\right) da(\xi) \end{aligned} \tag{3.177}$$

Hint. Apply the reciprocal relation (3.168) of Problem 3.6 in which $\widetilde{\mathbf{F}}/c^2 = \mathbf{1}\delta(\mathbf{x} - \xi)\delta(t)$ and $\widetilde{\mathbf{U}} = \mathbf{G}(\mathbf{x}, \xi; t)$.

Solution. To solve this problem we let in Eq. (3.168) of Problem 3.6 the following

$$\widetilde{\mathbf{U}} = \mathbf{G}(\mathbf{x},\ \xi;\ t),\quad \widetilde{\mathbf{U}}_0 = \mathbf{G}(\mathbf{x},\ \xi;\ 0) = \mathbf{0}$$
$$\dot{\widetilde{\mathbf{U}}}_0 = \dot{\mathbf{G}}(\mathbf{x},\ \xi;\ 0) = \mathbf{0},\quad \widetilde{\mathbf{F}}/c^2 = \mathbf{1}\delta(\mathbf{x}-\xi)\,\delta(t) \tag{3.178}$$

and

$$dv = dv(\xi),\quad da = da(\xi) \tag{3.179}$$

and obtain

$$\int_B [\mathbf{1}\,\delta(\mathbf{x}-\xi)\delta(t)] * \mathbf{u}(\xi;\ t)dv(\xi) + \int_{\partial B} \left(\frac{\partial \mathbf{G}}{\partial n}\right) * \mathbf{u}\, da(\xi)$$
$$= \frac{1}{c^2}\int_B (\mathbf{G}*\mathbf{f} + \dot{\mathbf{G}}\,\mathbf{u}_0 + \mathbf{G}\,\dot{\mathbf{u}}_0)dv(\xi) + \int_{\partial B} \mathbf{G} * \frac{\partial \mathbf{u}}{\partial n} da(\xi) \tag{3.180}$$

Finally, using the filtrating property of the delta function we find that (3.180) is equivalent to (3.177). This completes solution to Problem 3.7.

Problem 3.8. Show that a unique solution to Eqs. (3.174) and (3.175) of Problem 3.7 takes the form

$$\mathbf{G}(\mathbf{x}, \xi; t) = \frac{1}{4\pi\,|\mathbf{x}-\xi|}\delta\left(t - \frac{|\mathbf{x}-\xi|}{c}\right)\mathbf{1} \tag{3.181}$$

and, hence, reduce Eq. (3.177) from Problem 3.7 to the Poisson-Kirchhoff integral representation

$$\mathbf{u}(\mathbf{x}, t) = \frac{1}{4\pi\, c^2}\int_B \frac{\mathbf{f}(\xi, t - |\mathbf{x}-\xi|\,/c)}{|\mathbf{x}-\xi|} dv(\xi) + \frac{\partial}{\partial t}[t\, M_{\mathbf{x},ct}(\mathbf{u}_0)] + t\, M_{\mathbf{x},ct}(\dot{\mathbf{u}}_0)$$
$$+ \frac{1}{4\pi}\int_{\partial B} \left\{ \frac{1}{|\mathbf{x}-\xi|}\frac{\partial \mathbf{u}}{\partial n}(\xi, t - |\mathbf{x}-\xi|\,/c) - \mathbf{u}(\xi, t - |\mathbf{x}-\xi|\,/c)\frac{\partial}{\partial n}\frac{1}{|\mathbf{x}-\xi|} \right.$$
$$\left. + \frac{1}{c\,|\mathbf{x}-\xi|}\left[\frac{\partial}{\partial n}|\mathbf{x}-\xi|\right]\left[\frac{\partial \mathbf{u}}{\partial t}(\xi, t - |\mathbf{x}-\xi|\,/c)\right]\right\} da(\xi) \tag{3.182}$$

where for any vector field $\mathbf{v} = \mathbf{v}(\mathbf{x})$ on $B \subset E^3$ the symbol $M_{\mathbf{x},ct}(\mathbf{v})$ represents the mean value of $\mathbf{v}$ over the spherical surface with a center at $\mathbf{x}$ and of radius ct, that is,

$$M_{\mathbf{x},ct}(\mathbf{v}) = \frac{1}{4\pi} \int_0^{2\pi} d\varphi \int_0^{\pi} d\theta \, \sin\theta$$
$$\times \, \mathbf{v}(x_1 + ct \, \sin\theta \, \cos\varphi, x_2 + ct \, \sin\theta \, \sin\varphi, x_3 + ct \, \cos\theta) \quad (3.183)$$

and we adopt the convention that all relevant quantities vanish for negative time arguments.

Note. If B $= E^3$ and $\mathbf{f} = \mathbf{0}$ on $E^3 \times [0, \infty)$ then Eq. (3.182) reduces to the form

$$\mathbf{u}(\mathbf{x}, t) = \frac{\partial}{\partial t}[t \, M_{\mathbf{x},ct}(\mathbf{u}_0)] + t \, M_{\mathbf{x},ct}(\dot{\mathbf{u}}_0) \tag{3.184}$$

Solution. Equation (3.174) of Problem 3.7 takes the form

$$\left(\nabla^2 - \frac{1}{c^2}\frac{\partial^2}{\partial t^2}\right) \mathbf{G} = -\mathbf{1} \, \delta(\mathbf{x} - \xi) \, \delta(t) \tag{3.185}$$

Applying the Laplace transform to this equation and using the homogeneous initial conditions (3.175) of Problem 3.7 we obtain

$$\left[\nabla^2 - \left(\frac{p}{c}\right)^2\right] \overline{\mathbf{G}} = -\mathbf{1} \, \delta(\mathbf{x} - \xi) \tag{3.186}$$

where

$$\overline{\mathbf{G}} = \overline{\mathbf{G}}(\mathbf{x}, \xi; \; p) = \int_0^{\infty} e^{-pt} \mathbf{G}(\mathbf{x}, \xi, t) dt \tag{3.187}$$

The only solution to Eq. (3.186) in E^3 that vanishes as $|\mathbf{x}| \to \infty$, $|\xi| < \infty$, takes the form $(p > 0)$

$$\overline{\mathbf{G}} = \frac{1}{4\pi} \frac{1}{|\mathbf{x} - \xi|} \, e^{-\frac{p}{c}|\mathbf{x} - \xi|} \mathbf{1} \tag{3.188}$$

Hence, applying the operator L^{-1} to (3.188) we obtain

$$\mathbf{G}(\mathbf{x}, \; \xi; \; t) = \frac{1}{4\pi \, |\mathbf{x} - \xi|} \, \delta\left(t - \frac{1}{c}|\mathbf{x} - \xi|\right) \mathbf{1} \tag{3.189}$$

This completes proof of (3.181). To show that (3.182) holds true, we split (3.177) of Problem 3.7 into the sum

$$\mathbf{u}(\mathbf{x}, t) = \mathbf{u}^{(1)}(\mathbf{x}, t) + \mathbf{u}^{(2)}(\mathbf{x}, t) + \mathbf{u}^{(3)}(\mathbf{x}, t) \tag{3.190}$$

where

$$\mathbf{u}^{(1)}(\mathbf{x}, t) = \frac{1}{c^2} \int_B \mathbf{G} * \mathbf{f} \, dv(\xi) \tag{3.191}$$

$$\mathbf{u}^{(2)}(\mathbf{x}, t) = \int_{\partial B} \left(\mathbf{G} * \frac{\partial \mathbf{u}}{\partial n} - \frac{\partial \mathbf{G}}{\partial n} * \mathbf{u} \right) da(\xi) \tag{3.192}$$

$$\mathbf{u}^{(3)}(\mathbf{x}, t) = \frac{1}{c^2} \int_B (\dot{\mathbf{G}} \, \mathbf{u}_0 + \mathbf{G} \, \dot{\mathbf{u}}_0) \, dv(\xi) \tag{3.193}$$

If $\mathbf{G}$ from Eq. (3.189) is substituted into Eq. (3.191) we obtain

$$\mathbf{u}^{(1)}(\mathbf{x}, t) = \frac{1}{4\pi c^2} \int_B dv(\xi) \int_0^t d\tau \frac{1}{|\mathbf{x} - \xi|} \mathbf{f}(\xi, t - \tau) \, \delta \left(\tau - \frac{1}{c} |\mathbf{x} - \xi| \right) \tag{3.194}$$

Using the filtrating property of the delta function

$$\int_0^t \delta(\tau - t_0) \, g(t - \tau) d\tau = g(t - t_0) H(t - t_0) \tag{3.195}$$

where $g = g(t)$ is an arbitrary function and $H = H(t)$ is the Heaviside function

$$H(t) = \begin{Bmatrix} 1 & t > 0 \\ 0 & t < 0 \end{Bmatrix} \tag{3.196}$$

we reduce Eq. (3.194) to the form

$$\mathbf{u}^{(1)}(\mathbf{x}, t) = \frac{1}{4\pi c^2} \int_B \frac{\mathbf{f}\left(\xi, t - \frac{R}{c}\right)}{R} H\left(t - \frac{R}{c}\right) dv(\xi) \tag{3.197}$$

where

$$R = |\mathbf{x} - \xi| \tag{3.198}$$

The function $\mathbf{u}^{(1)}(\mathbf{x}, t)$ given by (3.197) is identical to the first integral on the RHS of Eq. (3.182) when the convention that $\mathbf{f}(\mathbf{x}, t)$ vanishes for $t \le 0$ is adopted.

An alternative form of (3.197) reads

$$\mathbf{u}^{(1)}(\mathbf{x}, t) = \frac{1}{4\pi c^2} \int_{B \cap S(\mathbf{x}, ct)} \frac{1}{|\mathbf{x} - \xi|} \mathbf{f}\left(\xi, t - \frac{1}{c}|\mathbf{x} - \xi|\right) dv(\xi) \tag{3.199}$$

where

$$S(\mathbf{x}, ct) = \{\xi : |\xi - \mathbf{x}| < ct\}.$$

To show that $\mathbf{u}^{(2)}(\mathbf{x}, t)$ is identical to the last integral on the RHS of (3.182), we apply the Laplace transform to Eq. (3.192), use Eq. (3.188), and obtain

$$\overline{\mathbf{u}}^{(2)}(\mathbf{x}, p) = \frac{1}{4\pi} \int_{\partial B} \left\{ \frac{1}{R} e^{-\frac{p}{c}R} \frac{\partial \overline{\mathbf{u}}}{\partial n} - \left[\frac{\partial}{\partial n} \left(\frac{1}{R} \right) e^{-\frac{p}{c}R} - \frac{1}{cR} \frac{\partial R}{\partial n} e^{-\frac{p}{c}R} p \right] \overline{\mathbf{u}} \right\} da(\xi) \tag{3.200}$$

Applying the inverse Laplace transform to Eq. (3.200) we obtain

$$\begin{aligned} \mathbf{u}^{(2)}(\mathbf{x}, t) = \frac{1}{4\pi} \int_{\partial B} \left\{ \frac{1}{R} \frac{\partial \mathbf{u}}{\partial n} \left(\xi, t - \frac{R}{c} \right) - \frac{\partial}{\partial n} \left(\frac{1}{R} \right) \mathbf{u} \left(\xi, t - \frac{R}{c} \right) \right. \\ \left. + \frac{1}{cR} \frac{\partial R}{\partial n} \dot{\mathbf{u}} \left(\xi, t - \frac{R}{c} \right) \right\} H \left(t - \frac{R}{c} \right) da(\xi) \end{aligned} \tag{3.201}$$

The function $\mathbf{u}^{(2)}$ given by (3.201) is equivalent to the last integral on the RHS of (3.182), if the convention that $\mathbf{u}(\mathbf{x}, t) = \mathbf{0}$ for $t \leq 0$ is adopted. An equivalent form of (3.201) reads

$$\begin{aligned} \mathbf{u}^{(2)}(\mathbf{x}, t) = \frac{1}{4\pi} \int_{\partial B \cap S(\mathbf{x}, ct)} \left\{ \frac{1}{R} \frac{\partial \mathbf{u}}{\partial n} \left(\xi, t - \frac{R}{c} \right) - \frac{\partial}{\partial n} \left(\frac{1}{R} \right) \mathbf{u} \left(\xi, t - \frac{R}{c} \right) \right. \\ \left. + \frac{1}{cR} \frac{\partial R}{\partial n} \dot{\mathbf{u}} \left(\xi, t - \frac{R}{c} \right) \right\} da(\xi) \end{aligned} \tag{3.202}$$

To show that $\mathbf{u}^{(3)}(\mathbf{x}, t)$ given by (3.193) is equal to a sum of the second and third terms on the RHS of (3.182), consider the integral

$$\mathbf{h}(\mathbf{x}, t) = \frac{1}{c^2} \int_B \mathbf{G} \, \dot{\mathbf{u}}_0 \, dv(\xi) \tag{3.203}$$

Since

$$\frac{1}{R} \delta \left(t - \frac{R}{c} \right) = \frac{tc^2}{R^2} \delta(R - ct) \tag{3.204}$$

therefore, an alternative form of $\mathbf{G}$ given by (3.189) reads

$$\mathbf{G}(\mathbf{x}, \xi, t) = \frac{tc^2}{4\pi R^2} \delta(R - ct) \mathbf{1} \tag{3.205}$$

and the function $\mathbf{h} = \mathbf{h}(\mathbf{x}, t)$ takes the form

$$\mathbf{h}(\mathbf{x}, t) = \frac{t}{4\pi} \int_B \frac{\dot{\mathbf{u}}_0(\xi)}{|\mathbf{x} - \xi|^2} \delta(|\mathbf{x} - \xi| - ct) dv(\xi) \tag{3.206}$$

Next, we let

$$\mathbf{R} = \xi - \mathbf{x} \tag{3.207}$$

and introduce the spherical coordinates (R, φ, θ) with a center at $\mathbf{x}$

$$\begin{aligned} R_1 &= R \cos \varphi \sin \theta \\ R_2 &= R \sin \varphi \sin \theta \\ R_3 &= R \cos \theta \end{aligned} \tag{3.208}$$

$$0 \leq R < \infty, \quad 0 \leq \varphi \leq 2\pi, \quad 0 \leq \theta < \pi \tag{3.209}$$

Then the integral (3.206) takes the form

$$\mathbf{h}(\mathbf{x}, t) = \frac{t}{4\pi} \int_{B^*} \frac{\dot{\mathbf{u}}_0(\mathbf{R} + \mathbf{x})}{R^2} \delta(R - ct) dv(\mathbf{R}) \tag{3.210}$$

where

$$dv(\mathbf{R}) = R^2 \sin \theta \, d\varphi \, d\theta \, dR \tag{3.211}$$

and

$$B^* = \{(R, \varphi, \theta) : R_a < R < R_b; \ 0 \leq \varphi \leq 2\pi, 0 \leq \theta \leq \pi\} \tag{3.212}$$

The domain B^* is a mapping of B under the transformation defined by Eqs. (3.207) and (3.208), and R_a and R_b are uniquely defined nonnegative numbers. Hence, Eq. (3.210) can also be written as

$$\begin{aligned} \mathbf{h}(\mathbf{x}, t) = \frac{t}{4\pi} \int_{R_a}^{R_b} dR \, \delta(R - ct) \int_0^{2\pi} d\varphi \int_0^{\pi} d\theta \sin \theta \\ \times \dot{\mathbf{u}}_0(x_1 + R \cos \varphi \sin \theta, \ x_2 + R \sin \varphi \sin \theta, \ x_3 + R \cos \theta) \end{aligned} \tag{3.213}$$

or

$$\begin{aligned} \mathbf{h}(\mathbf{x}, t) = \frac{t}{4\pi} \int_0^{2\pi} d\varphi \int_0^{\pi} d\theta \sin \theta \\ \times \dot{\mathbf{u}}_0(x_1 + ct \cos \varphi \sin \theta, \ x_2 + ct \sin \varphi \sin \theta, \ x_3 + ct \cos \theta) \end{aligned} \tag{3.214}$$

or

$$\mathbf{h}(\mathbf{x}, t) = t \, \mathbf{M}_{\mathbf{x},ct}(\dot{\mathbf{u}}_0) \tag{3.215}$$

where $\mathbf{M}_{\mathbf{x},ct}(\dot{\mathbf{u}}_0)$ is defined by (3.183). Finally, if we note that

$$\mathbf{g}(\mathbf{x},t) \overset{\text{df}}{=} \frac{1}{c^2} \int_B \dot{\mathbf{G}}\mathbf{u}_0 \, dv(\xi) \tag{3.216}$$

can be written as

$$\mathbf{g}(\mathbf{x},t) = \frac{\partial}{\partial t} \left\{ \frac{1}{c^2} \int_B \mathbf{G}\mathbf{u}_0 \, dv(\xi) \right\} \tag{3.217}$$

then computing the integral on the RHS of (3.217) in a way similar to that of the integral $\mathbf{h} = \mathbf{h}(\mathbf{x},t)$, and taking into account Eq. (3.193) we obtain

$$\mathbf{u}^{(3)}(\mathbf{x},t) = \frac{\partial}{\partial t} \left[t\mathbf{M}_{\mathbf{x},ct}(\mathbf{u}_0) \right] + t\mathbf{M}_{\mathbf{x},ct}(\dot{\mathbf{u}}_0) \tag{3.218}$$

This completes proof of (3.182).

Problem 3.9. Let $\mathbf{G}^* = \mathbf{G}^*(\mathbf{x},\xi;t)$ be a solution to the initial-boundary value problem:

$$\Box_0^2 \mathbf{G}^* = -\mathbf{1}\delta(\mathbf{x}-\xi)\delta(t) \quad \text{for} \quad \mathbf{x}, \ \xi \in \mathrm{B}, \ \ t > 0 \tag{3.219}$$

$$\mathbf{G}^*(\mathbf{x},\xi;0) = \mathbf{0}, \quad \dot{\mathbf{G}}^*(\mathbf{x},\xi;0) = \mathbf{0} \quad \text{for} \quad \mathbf{x}, \xi \in \mathrm{B} \tag{3.220}$$

and

$$\mathbf{G}^*(\mathbf{x},\xi;t) = \mathbf{0} \quad \text{for} \quad \mathbf{x} \in \partial\mathrm{B}, \quad t > 0, \quad \xi \in \overline{\mathrm{B}} \tag{3.221}$$

and let $\mathbf{u} = \mathbf{u}(\mathbf{x},t)$ be a solution to the initial-boundary value problem

$$\Box_0^2 \mathbf{u} = -\frac{\mathbf{f}}{c^2} \quad \text{on} \quad \mathrm{B} \times (0,\infty) \tag{3.222}$$

$$\mathbf{u}(\mathbf{x},0) = \mathbf{u}_0(\mathbf{x}), \quad \dot{\mathbf{u}}(\mathbf{x},0) = \dot{\mathbf{u}}_0(\mathbf{x}) \quad \text{for} \quad \mathbf{x} \in \overline{\mathrm{B}} \tag{3.223}$$

$$\mathbf{u}(\mathbf{x},t) = \mathbf{g}(\mathbf{x},t) \quad \text{on} \quad \partial\mathrm{B} \times [0,\infty) \tag{3.224}$$

where the functions $\mathbf{f}$, $\mathbf{u}_0$, $\dot{\mathbf{u}}_0$, and $\mathbf{g}$ are prescribed. Use the representation formula (3.177) of Problem 3.7 to show that

$$\begin{aligned} \mathbf{u}(\mathbf{x},t) = {} & \frac{1}{c^2} \int_{\mathrm{B}} (\mathbf{G}^* * \mathbf{f} + \dot{\mathbf{G}}^* \mathbf{u}_0 + \mathbf{G}^* \dot{\mathbf{u}}_0) \, dv(\xi) \\ & - \int_{\partial\mathrm{B}} \left(\frac{\partial \mathbf{G}^*}{\partial \mathrm{n}} * \mathbf{g} \right) da(\xi) \end{aligned} \tag{3.225}$$

Solution. The representation formula (3.177) of Problem 3.7 reads

$$\mathbf{u}(\mathbf{x}, t) = \frac{1}{c^2} \int_{B} (\mathbf{G} * \mathbf{f} + \dot{\mathbf{G}}\mathbf{u}_0 + \mathbf{G}\dot{\mathbf{u}}_0)\, dv(\xi) + \int_{\partial B} \left(\mathbf{G} * \frac{\partial \mathbf{u}}{\partial n} - \frac{\partial \mathbf{G}}{\partial n} * \mathbf{u} \right) da(\xi) \tag{3.226}$$

where $\mathbf{G}$ satisfies Eqs. (3.174) and (3.175) of Problem 3.7, and $\mathbf{u}$ satisfies Eqs. (3.146) and (3.147) of Problem 3.5.

By letting $\mathbf{G} = \mathbf{G}^*$ in Eq. (3.226) and using the boundary conditions (3.221) and (3.224) we obtain (3.225).

This completes a solution to Problem 3.9.

Problem 3.10. A tensor field $\mathbf{S}$ corresponds to the solution of a traction problem of classical elastodynamics if and only if

$$\widehat{\nabla}[\rho^{-1}(\operatorname{div} \mathbf{S})] - \mathbf{K}[\ddot{\mathbf{S}}] = -\mathbf{B} \quad \text{on} \quad B \times [0, \infty) \tag{3.227}$$

$$\mathbf{S}(\mathbf{x}, 0) = \mathbf{S}^{(0)}(\mathbf{x}), \quad \dot{\mathbf{S}}(\mathbf{x}, 0) = \dot{\mathbf{S}}^{(0)}(\mathbf{x}) \quad \text{for} \quad \mathbf{x} \in B \tag{3.228}$$

$$\mathbf{S}\mathbf{n} = \widehat{\mathbf{s}} \quad \text{on} \quad \partial B \times [0, \infty) \tag{3.229}$$

[see Eqs. (3.71)–(3.73)] in which $\mathbf{B}$ is expressed in terms of a body force $\mathbf{b}$, and $\mathbf{S}^{(0)}$ and $\dot{\mathbf{S}}^{(0)}$ are defined in terms of two vector fields. A tensor field $\mathbf{S}$ corresponding to an external load $[\mathbf{B}, \mathbf{S}^{(0)}, \dot{\mathbf{S}}^{(0)}, \widehat{\mathbf{s}}]$ is said to be of a σ-type if $\mathbf{S}$ satisfies Eqs. (3.227) through (3.229) with an arbitrary symmetric second-order tensor field $\mathbf{B}$ and arbitrary symmetric initial tensor fields $\mathbf{S}^{(0)}$ and $\dot{\mathbf{S}}^{(0)}$, not necessarily related to the data of classic elastodynamics. Show that if $\mathbf{S}$ and $\widetilde{\mathbf{S}}$ are two different tensorial fields of σ-type corresponding to the external loads $[\mathbf{B}, \mathbf{S}^{(0)}, \dot{\mathbf{S}}^{(0)}, \widehat{\mathbf{s}}]$ and $[\widetilde{\mathbf{B}}, \widetilde{\mathbf{S}}^{(0)}, \dot{\widetilde{\mathbf{S}}}^{(0)}, \widehat{\widetilde{\mathbf{s}}}]$, respectively, then the following reciprocal relation holds true

$$\begin{aligned} &\int_{B} \left\{ \widetilde{\mathbf{B}} * \mathbf{S} + \widetilde{\mathbf{S}}^{(0)} \cdot \mathbf{K}[\dot{\mathbf{S}}] + \dot{\widetilde{\mathbf{S}}}^{(0)} \cdot \mathbf{K}[\mathbf{S}] \right\} dv + \int_{\partial B} \rho^{-1}(\operatorname{div} \widetilde{\mathbf{S}}) * (\mathbf{S}\mathbf{n})\, da \\ &= \int_{B} \left\{ \mathbf{B} * \widetilde{\mathbf{S}} + \mathbf{S}^{(0)} \cdot \mathbf{K}[\dot{\widetilde{\mathbf{S}}}] + \dot{\mathbf{S}}^{(0)} \cdot \mathbf{K}[\widetilde{\mathbf{S}}] \right\} dv + \int_{\partial B} \rho^{-1}(\operatorname{div} \mathbf{S}) * (\widetilde{\mathbf{S}}\mathbf{n})\, da \end{aligned} \tag{3.230}$$

Solution. The tensor fields S_{ij} and $\widetilde{S}_{ij}$, respectively, satisfy the equations

$$(\rho^{-1} S_{(ik,k),j}) - K_{ijkl}\, \ddot{S}_{kl} = -B_{ij} \quad \text{on } B \times (0, \infty) \tag{3.231}$$

$$S_{ij}(\mathbf{x}, 0) = S_{ij}^{(0)}(\mathbf{x}), \quad \dot{S}_{ij}(\mathbf{x}, 0) = \dot{S}_{ij}^{(0)}(\mathbf{x}) \quad \text{on } B \tag{3.232}$$

$$S_{ij} n_j = \widehat{s}_i \quad \text{on } \partial B \times (0, \infty) \tag{3.233}$$

and

$$(\rho^{-1}\tilde{S}_{(ik,k),j}) - K_{ijkl}\,\ddot{\tilde{S}}_{kl} = -\tilde{B}_{ij} \quad \text{on } B \times (0,\infty) \tag{3.234}$$

$$\tilde{S}_{ij}(\mathbf{x},0) = \tilde{S}^{(0)}_{ij}(\mathbf{x}), \quad \dot{\tilde{S}}_{ij}(\mathbf{x},0) = \dot{\tilde{S}}^{(0)}_{ij}(\mathbf{x}) \quad \text{on } B \tag{3.235}$$

$$\tilde{S}_{ij}n_j = \hat{\tilde{S}}_i \quad \text{on } \partial B \times (0,\infty) \tag{3.236}$$

Applying the Laplace transform to Eqs. (3.231) and (3.234), and using the initial conditions (3.232) and (3.235), respectively, we obtain

$$(\rho^{-1}\overline{S}_{(ik,k),j}) - K_{ijkl}\left(p^2\overline{S}_{kl} - \dot{S}^{(0)}_{kl} - pS^{(0)}_{kl}\right) = -\overline{B_{ij}} \tag{3.237}$$

and

$$(\rho^{-1}\overline{\tilde{S}}_{(ik,k),j}) - K_{ijkl}\left(p^2\overline{\tilde{S}}_{kl} - \dot{\tilde{S}}^{(0)}_{kl} - p\tilde{S}^{(0)}_{kl}\right) = -\overline{\tilde{B}_{ij}} \tag{3.238}$$

Next, multiplying (3.237) by $\overline{\tilde{S}}_{ij}$ and (3.238) by $-\overline{S}_{ij}$, and adding up the results we obtain

$$\begin{aligned}\overline{\tilde{S}}_{ij}(\rho^{-1}\overline{S}_{ik,k}),_j - \overline{\tilde{S}}_{ij}\,K_{ijkl}\left(p^2\overline{S}_{kl} - \dot{S}^{(0)}_{kl} - pS^{(0)}_{kl}\right) - \overline{S}_{ij}(\rho^{-1}\overline{\tilde{S}}_{ik,k}),_j \\ + \overline{S}_{ij}\,K_{ijkl}\left(p^2\overline{\tilde{S}}_{kl} - \dot{\tilde{S}}^{(0)}_{kl} - p\tilde{S}^{(0)}_{kl}\right) + \overline{B}_{ij}\,\overline{\tilde{S}}_{ij} - \overline{\tilde{B}}_{ij}\,\overline{S}_{ij} = 0\end{aligned} \tag{3.239}$$

Since

$$\overline{\tilde{S}}_{ij}(\rho^{-1}\overline{S}_{ik,k}),_j - \overline{S}_{ij}(\rho^{-1}\overline{\tilde{S}}_{ik,k}),_j = (\overline{\tilde{S}}_{ij}\rho^{-1}\overline{S}_{ik,k} - \overline{S}_{ij}\rho^{-1}\overline{\tilde{S}}_{ik,k}),_j \tag{3.240}$$

and

$$K_{ijkl}\,\overline{\tilde{S}}_{ij}\,\overline{S}_{kl} = K_{ijkl}\,\overline{S}_{ij}\,\overline{\tilde{S}}_{kl} \tag{3.241}$$

therefore, by integrating (3.239) over B and using the divergence theorem, we obtain

$$\begin{aligned}\int_B \left(\overline{B}_{ij}\,\overline{\tilde{S}}_{ij} - \overline{\tilde{B}}_{ij}\,\overline{S}_{ij}\right)dv(\xi) + \int_B K_{ijkl}\Big[\overline{\tilde{S}}_{ij}\,\dot{S}^{(0)}_{kl} + \left(p\overline{\tilde{S}}_{ij} - \tilde{S}^{(0)}_{ij}\right)S^{(0)}_{kl} + \tilde{S}^{0}_{ij}\,S^{(0)}_{kl} \\ - \overline{S}_{ij}\,\dot{\tilde{S}}^{(0)}_{kl} - \left(p\overline{S}_{ij} - S^{(0)}_{ij}\right)\tilde{S}^{(0)}_{kl} - S^{(0)}_{ij}\tilde{S}^{(0)}_{kl}\Big]dv \\ + \int_{\partial B} \rho^{-1}\left(\overline{\tilde{S}}_{ij}\,\overline{S}_{ik,k} - \overline{S}_{ij}\,\overline{\tilde{S}}_{ik,k}\right)n_j da(\xi) = 0\end{aligned} \tag{3.242}$$

Finally, applying the inverse Laplace transform to (3.242), using the convolution theorem

$$\overline{f * g} = \overline{f}\overline{g} \tag{3.243}$$

as well as the relation

$$K_{ijkl}\,\widetilde{S}_{ij}^{(0)}\,S_{kl}^{(0)} = K_{ijkl}\,S_{ij}^{(0)}\widetilde{S}_{kl}^{(0)} \tag{3.244}$$

we obtain

$$\int_B (\mathbf{B} * \widetilde{\mathbf{S}} - \widetilde{\mathbf{B}} * \mathbf{S})dv(\xi) + \int_B \left\{\widetilde{\mathbf{S}} \cdot \mathbf{K}[\dot{\mathbf{S}}^{(0)}] + \dot{\widetilde{\mathbf{S}}} \cdot \mathbf{K}[\mathbf{S}^{(0)}] - \mathbf{S} \cdot \mathbf{K}[\dot{\widetilde{\mathbf{S}}}^{(0)}] - \dot{\mathbf{S}} \cdot \mathbf{K}[\widetilde{\mathbf{S}}^{(0)}]\right\} dv(\xi)$$

$$+ \int_B \rho^{-1}\left[(\operatorname{div}\mathbf{S}) * (\widetilde{\mathbf{S}}\mathbf{n}) - (\operatorname{div}\widetilde{\mathbf{S}}) * (\mathbf{S}\mathbf{n})\right] da(\xi) = 0 \tag{3.245}$$

Since $\mathbf{K}$ is symmetric

$$\mathbf{A} \cdot \mathbf{K}[\mathbf{B}] = \mathbf{B} \cdot \mathbf{K}[\mathbf{A}] \quad \forall\, \mathbf{A} \text{ and } \mathbf{B} \tag{3.246}$$

therefore, Eq. (3.245) is equivalent to Eq. (3.230), and this completes a solution to Problem 3.10.

Problem 3.11. Let $S_{ij}^{(kl)} = S_{ij}^{(kl)}(\mathbf{x}, \xi; t)$ be a solution of the following equation

$$(\rho^{-1} S_{(ik,k),j}^{(kl)}) - K_{ijpq}\ddot{S}_{pq}^{(kl)} = 0$$
$$\text{for} \quad \mathbf{x} \in E^3, \quad \xi \in E^3; \quad \text{i, j, k, l} = 1, 2, 3 \tag{3.247}$$

subject to the initial conditions

$$S_{ij}^{(kl)}(\mathbf{x}, \xi; 0) = 0, \quad \dot{S}_{ij}^{(kl)}(\mathbf{x}, \xi; 0) = C_{ijkl}\delta(\mathbf{x} - \xi)$$
$$\text{for} \quad \mathbf{x} \in E^3, \quad \xi \in E^3; \quad \text{i, j, k, l} = 1, 2, 3 \tag{3.248}$$

where K_{ijkl} denotes the components of the compliance tensor $\mathbf{K}$, and C_{ijkl} stands for the components of elasticity tensor $\mathbf{C}$, that is,

$$C_{ijkl}K_{klmn} = \delta_{(im}\delta_{nj)} \tag{3.249}$$

Let $S_{ij} = S_{ij}(\mathbf{x}, t)$ be a solution of the equation

$$(\rho^{-1} S_{(ik,k),j}) - K_{ijkl}\ddot{S}_{kl} = 0 \quad \text{for} \quad \mathbf{x} \in \text{B}, \quad t > 0 \tag{3.250}$$

subject to the homogeneous initial conditions

$$S_{ij}(\mathbf{x}, 0) = 0, \quad \dot{S}_{ij}(\mathbf{x}, 0) = 0 \quad \text{for} \quad \mathbf{x} \in \text{B} \tag{3.251}$$

and the boundary condition

$$S_{ij}n_j = \widehat{s}_i \quad \text{on} \quad \partial\text{B} \times [0, \infty) \tag{3.252}$$

Use the reciprocal relation (3.230) of Problem 3.10 to show that

$$S_{kl}(\mathbf{x}, t) = \int_{\partial \mathrm{B}} \rho^{-1} \left(S_{im,m} * S_{ij}^{(kl)} n_j - \widehat{s}_i * S_{im,m}^{(kl)} \right) da(\xi) \tag{3.253}$$

Note. Equation (3.253) provides a solution to the traction initial-boundary value problem of classical elastodynamics if the field $S_{im,m}$ on $\partial \mathrm{B} \times [0, \infty)$ is found from an associated integral equation on $\partial \mathrm{B} \times [0, \infty)$. The idea of solving a traction problem of elastodynamics in terms of displacements through an associated boundary integral equation is due to V.D. Kupradze.

Solution. Note that for a fixed pair (k, l) $\widetilde{S}_{ij} = S_{ij}^{(kl)}(\mathbf{x}, \ \xi; \ t)$ is a tensor field of σ-type corresponding to the data: $\widetilde{B}_{ij} = 0$, $\widetilde{\widehat{s}}_i \neq 0$, $\widetilde{S}_{ij}^{(0)} = 0$, and $\dot{\widetilde{S}}_{ij}^{(0)} = C_{ijkl}\delta(\mathbf{x} - \xi)$; and $S_{ij} = S_{ij}(\mathbf{x}, t)$ is a tensor field of σ-type corresponding to the data: $B_{ij} = 0$, $\widehat{s}_i \neq 0$, $S_{ij}^{(0)} = 0$, and $\dot{S}_{ij}^{(0)} = 0$. Therefore, using the reciprocal relation (3.230) of Problem 3.10. in which $\widetilde{S}_{ij} = S_{ij}^{(kl)}(\mathbf{x}, \xi; t)$ and $S_{ij} = S_{ij}(\mathbf{x}, t)$, we obtain

$$\int_B \dot{\widetilde{S}}_{ij}^{(0)} K_{ijpq} \, S_{pq} dv(\xi) = \int_{\partial B} \rho^{-1} (S_{im,m} * \widetilde{S}_{ij} n_j - \widehat{s}_i * \widetilde{S}_{im,m}) \, da(\xi) \tag{3.254}$$

where

$$\dot{\widetilde{S}}_{ij}^{(0)} = \dot{S}_{ij}^{(kl)}(\mathbf{x}, \ \xi; 0) = C_{ijkl} \, \delta(\mathbf{x} - \xi) \tag{3.255}$$

$$\widetilde{S}_{ij} n_j = S_{ij}^{(kl)}(\mathbf{x}, \ \xi; t) n_j(\xi) \tag{3.256}$$

$$\widetilde{S}_{im,m} = S_{im,m}^{(kl)}(\mathbf{x}, \ \xi; \ t) \tag{3.257}$$

Since

$$\begin{aligned} \dot{\widetilde{S}}_{ij}^{(0)} K_{ijpq} \, S_{pq} &= C_{ijkl} \, K_{ijpq} \, S_{pq} \, \delta(\mathbf{x} - \xi) = C_{klij} \, K_{ijpq} \, S_{pq} \, \delta(\mathbf{x} - \xi) \\ &= \delta_{(kp} \, \delta_{ql)} \, S_{pq} \, \delta(\mathbf{x} - \xi) = S_{kl}(\xi, t) \delta(\mathbf{x} - \xi) \end{aligned} \tag{3.258}$$

therefore, Eq. (3.254) takes the form

$$\int_B S_{kl}(\xi, t) \delta(\mathbf{x} - \xi) dv(\xi) = \int_{\partial B} \rho^{-1} \left(S_{im,m} * S_{ij}^{(kl)} n_j - \widehat{s}_i * S_{im,m}^{(kl)} \right) da(\xi) \tag{3.259}$$

Finally, using the filtrating property of the delta function we obtain (3.253). This completes a solution to Problem 3.11.

Problem 3.12. Consider the pure stress initial-boundary value problem of linear elastodynamics for a homogeneous isotropic *incompressible* elastic body B [see Eq. (3.55) with $\mu > 0$ and $\lambda \to \infty$]. Find a tensor field $\mathbf{S} = \mathbf{S}(\mathbf{x}, t)$ on $B \times [0, \infty)$ that satisfies the equation

$$\widehat{\nabla}(\operatorname{div} \mathbf{S}) - \frac{\rho}{2\mu}\left[\ddot{\mathbf{S}} - \frac{1}{3}(\operatorname{tr} \ddot{\mathbf{S}})\,\mathbf{1}\right] = -\widehat{\nabla}\mathbf{b} \quad \text{on} \quad B \times [0, \infty) \tag{3.260}$$

subject to the initial conditions

$$\mathbf{S}(\mathbf{x}, 0) = \mathbf{S}_0(\mathbf{x}), \quad \dot{\mathbf{S}}(\mathbf{x}, 0) = \dot{\mathbf{S}}_0(\mathbf{x}) \quad \text{for} \quad \mathbf{x} \in B \tag{3.261}$$

and the traction boundary condition

$$\mathbf{Sn} = \widehat{\mathbf{s}} \quad \text{on} \quad \partial B \times [0, \infty) \tag{3.262}$$

Here, $\mathbf{b}, \widehat{\mathbf{s}}, \mathbf{S}_0,$ and $\dot{\mathbf{S}}_0$, are prescribed functions ($\mu > 0,\ \rho > 0$). Show that the problem (3.260) through (3.262) may have at most one solution.

Solution. We are to show that the field equation

$$S_{(ik,kj)} - \frac{\rho}{2\mu}\left(\ddot{S}_{ij} - \frac{1}{3}\ddot{S}_{kk}\,\delta_{ij}\right) = 0 \quad \text{on } B \times [0, \infty) \tag{3.263}$$

subject to the homogeneous initial conditions

$$S_{ij}(\mathbf{x},\ 0) = 0, \quad \dot{S}_{ij}(\mathbf{x},\ 0) = 0 \quad \text{on } B \tag{3.264}$$

and the homogeneous traction boundary condition

$$S_{ij}\, n_j = 0 \quad \text{on } \partial B \times [0, \infty) \tag{3.265}$$

imply that

$$S_{ij} = 0 \quad \text{on } \overline{B} \times [0, \infty) \tag{3.266}$$

To this end we multiply (3.263) by $\dot{S}_{ij}$ and obtain

$$S_{(ik,kj)}\,\dot{S}_{ij} - \frac{\rho}{2\mu}\left(\ddot{S}_{ij}\,\dot{S}_{ij} - \frac{1}{3}\ddot{S}_{kk}\,\dot{S}_{ii}\right) = 0 \tag{3.267}$$

Since

$$S_{(ik,kj)}\,\dot{S}_{ij} = S_{ik,kj}\,\dot{S}_{ij} = (S_{ik,k}\,\dot{S}_{ij}),_j - S_{ik,k}\,\dot{S}_{ij,j} \tag{3.268}$$

therefore (3.267) can be written in the form

$$(S_{ik,k}\,\dot{S}_{ij})_{,j} - S_{ik,k}\,\dot{S}_{ij,j} - \frac{\rho}{2\mu}\left[\frac{1}{2}\frac{\partial}{\partial t}(\dot{S}_{ij}\,\dot{S}_{ij}) - \frac{1}{3}\frac{1}{2}\frac{\partial}{\partial t}(\dot{S}_{aa})^2\right] = 0 \quad (3.269)$$

or

$$(S_{ik,k}\,\dot{S}_{ij})_{,j} - \frac{1}{2}\frac{\partial}{\partial t}(S_{ik,k}\,S_{ij,j}) - \frac{\rho}{2\mu}\left[\frac{1}{2}\frac{\partial}{\partial t}(\dot{S}_{ij}\,\dot{S}_{ij}) - \frac{1}{3}\frac{1}{2}\frac{\partial}{\partial t}(\dot{S}_{aa})^2\right] = 0 \quad (3.270)$$

Integrating Eq. (3.270) over the cartesian product $B \times [0, t]$, using the divergence theorem, the homogeneous initial conditions (3.264) as well as the boundary condition

$$\dot{S}_{ij}\,n_j = 0 \quad \text{on } \partial B \times [0, \infty) \quad (3.271)$$

obtained by differentiation of (3.265) with respect to time, we obtain

$$\int_B \left\{S_{ik,k}\,S_{ij,j} + \frac{\rho}{2\mu}\left[\dot{S}_{ij}\,\dot{S}_{ij} - \frac{1}{3}(\dot{S}_{aa})^2\right]\right\} dv = 0 \quad (3.272)$$

Since

$$\dot{S}_{ij} = \dot{S}_{ij}^{(d)} + \frac{1}{3}\dot{S}_{aa}\,\delta_{ij} \quad (3.273)$$

where

$$\dot{S}_{ij}^{(d)} = \dot{S}_{ij} - \frac{1}{3}\dot{S}_{aa}\,\delta_{ij} \quad (3.274)$$

and

$$\dot{S}_{ij}\,\dot{S}_{ij} = \dot{S}_{ij}^{(d)}\,\dot{S}_{ij}^{(d)} + \frac{1}{3}(\dot{S}_{aa})^2 \quad (3.275)$$

therefore, Eq. (3.272) can be written as

$$\int_B \left(S_{ik,k}\,S_{ij,j} + \frac{\rho}{2\mu}\dot{S}_{ij}^{(d)}\,\dot{S}_{ij}^{(d)}\right) dv = 0 \quad (3.276)$$

Equation (3.276) implies that

$$S_{ik,k} = 0, \quad \dot{S}_{ij}^{(d)} = 0 \quad \text{on } \overline{B} \times [0, \infty) \quad (3.277)$$

Equation $(3.277)_2$ together with the homogeneous initial conditions $(3.264)_1$ imply that

$$S_{ij} = \frac{1}{3}S_{aa}\,\delta_{ij} \quad \text{on } \overline{B} \times [0, \infty) \quad (3.278)$$

Equations (3.278) and $(3.277)_1$ imply that

$$S_{aa,i} = 0 \quad \text{on } \overline{B} \times [0, \infty) \tag{3.279}$$

which is equivalent to

$$S_{aa}(\mathbf{x}, t) = c(t) \quad \text{on } \overline{B} \times [0, \infty) \tag{3.280}$$

where $c = c(t)$ is an arbitrary function of time.

Finally, Eqs. (3.265), (3.278), and (3.280) lead to

$$c(t) n_i(\mathbf{x}) = 0 \quad \text{on } \partial B \times [0, \infty) \tag{3.281}$$

Since $|n_i\, n_i| = 1$ on ∂B, therefore Eq. (3.281) implies that

$$|c(t)| = 0 \tag{3.282}$$

Equation (3.282) together with Eq. (3.278) implies Eq. (3.266), and this completes a solution to Problem 3.12.

Chapter 4
Variational Formulation of Elastostatics

In this chapter the variational characterizations of a solution to a boundary value problem of elastostatics are recalled. They include the principle of minimum potential energy, the principle of minimum complementary energy, the Hu-Washizu principle, and the compatibility related principle for a traction problem. The variational principles are then used to solve typical problems of elastostatics.

4.1 Minimum Principles

To formulate the Principle of Minimum Potential Energy we recall the concept of the *strain energy,* of the *stress energy,* and of a *kinematically admissible state.*

By the *strain energy of a body B* we mean the integral

$$\mathrm{U_C}\{\mathbf{E}\} = \frac{1}{2}\int_{\mathrm{B}} \mathbf{E}\cdot\mathbf{C}[\mathbf{E}]\,dv \tag{4.1}$$

and by the *stress energy of a body* B we mean

$$\mathrm{U_K}\{\mathbf{S}\} = \frac{1}{2}\int_{\mathrm{B}} \mathbf{S}\cdot\mathbf{K}[\mathbf{S}]\,dv \tag{4.2}$$

Since $\mathbf{S} = \mathbf{C}[\mathbf{E}]$, therefore,

$$\mathrm{U_K}\{\mathbf{S}\} = \mathrm{U_C}\{\mathbf{E}\} \tag{4.3}$$

By a *kinematically admissible state* we mean a state $s = [\mathbf{u}, \mathbf{E}, \mathbf{S}]$ that satisfies

(1) the strain-displacement relation

$$\mathbf{E} = \widehat{\nabla}\mathbf{u} = \frac{1}{2}(\nabla\mathbf{u} + \nabla\mathbf{u}^{\mathrm{T}}) \quad \text{on} \quad \mathrm{B} \tag{4.4}$$

M. Reza Eslami et al., *Theory of Elasticity and Thermal Stresses*, Solid Mechanics and Its Applications 197, DOI: 10.1007/978-94-007-6356-2_4,

(2) the stress-strain relation

$$\mathbf{S} = \mathbf{C}\,[\mathbf{E}] \quad \text{on} \quad \mathrm{B} \tag{4.5}$$

(3) the displacement boundary condition

$$\mathbf{u} = \widehat{\mathbf{u}} \quad \text{on} \quad \partial \mathrm{B}_1 \tag{4.6}$$

where $\widehat{\mathbf{u}}$ is prescribed on $\partial \mathrm{B}_1$.

The Principle of Minimum Potential Energy is related to a mixed boundary value problem of elastostatics [see Chap. 3 on Formulation of Problems of Elasticity].

The Principle of Minimum Potential Energy

Let R be the set of all kinematically admissible states. Define a functional $\mathrm{F} = \mathrm{F}\{.\}$ on R by

$$\mathrm{F}\{s\} = \mathrm{U_C}\{\mathbf{E}\} - \int\limits_{\mathrm{B}} \mathbf{b} \cdot \mathbf{u}\, dv - \int\limits_{\partial \mathrm{B}_2} \widehat{\mathbf{s}} \cdot \mathbf{u}\, da \tag{4.7}$$

for every $s = [\mathbf{u},\ \mathbf{E},\ \mathbf{S}] \in \mathrm{R}$. Let s be a solution to the mixed problem of elastostatics. Then

$$\mathrm{F}\{s\} \le \mathrm{F}\{\tilde{s}\} \quad \text{for every} \quad \tilde{s} \in \mathrm{R} \tag{4.8}$$

and the equality holds true if s and $\tilde{s}$ differ by a rigid displacement.

By letting $\mathbf{E} = \widehat{\nabla}\mathbf{u}$ in (4.7) an *alternative form of the Principle of Minimum Potential Energy* is obtained.

Let R_1 denote a set of displacement fields that satisfy the boundary conditions (4.6), and define a functional $\mathrm{F}_1\{.\}$ on R_1 by

$$\mathrm{F}_1\{\mathbf{u}\} = \frac{1}{2}\int\limits_{\mathrm{B}} (\nabla\mathbf{u}) \cdot \mathbf{C}\,[\nabla\mathbf{u}]\, dv - \int\limits_{\mathrm{B}} \mathbf{b} \cdot \mathbf{u}\, dv - \int\limits_{\partial \mathrm{B}_2} \widehat{\mathbf{s}} \cdot \mathbf{u}\, da \quad \forall \mathbf{u} \in \mathrm{R}_1 \tag{4.9}$$

If $\mathbf{u}$ corresponds to a solution to the mixed problem, then

$$\mathrm{F}_1\{\mathbf{u}\} \le \mathrm{F}_1\{\tilde{\mathbf{u}}\} \quad \forall \tilde{\mathbf{u}} \in \mathrm{R}_1 \tag{4.10}$$

To formulate the Principle of Minimum Complementary Energy, we introduce a concept of a *statically admissible stress field*. By such a field we mean a symmetric second-order tensor field $\mathbf{S}$ that satisfies

(1) the equation of equilibrium

$$\operatorname{div}\mathbf{S} + \mathbf{b} = \mathbf{0} \quad \text{on} \quad \mathrm{B} \tag{4.11}$$

(2) the traction boundary condition

$$\mathbf{S}\mathbf{n} = \widehat{\mathbf{s}} \quad \text{on} \quad \partial \mathrm{B}_2 \tag{4.12}$$

The Principle of Minimum Complementary Energy

Let P denote a set of all statically admissible stress fields, and let $\mathrm{G} = \mathrm{G}\{.\}$ be a functional on P defined by

$$\mathrm{G}\{\mathbf{S}\} = \mathrm{U_K}\{\mathbf{S}\} - \int_{\partial \mathrm{B}_1} \mathbf{s} \cdot \widehat{\mathbf{u}}\, da \quad \forall \mathbf{S} \in P \tag{4.13}$$

If $\mathbf{S}$ is a stress field corresponding to a solution to the mixed problem, then

$$\mathrm{G}\{\mathbf{S}\} \le \mathrm{G}\{\tilde{\mathbf{S}}\} \quad \forall \tilde{\mathbf{S}} \in P \tag{4.14}$$

and the equality holds if $\mathbf{S} = \tilde{\mathbf{S}}$.

The Principle of Minimum Complementary Energy for Nonisothermal Elastostatics

The fundamental field equations of nonisothermal elastostatics may be written as

$$\mathbf{E} = \widehat{\nabla} \mathbf{u} = \frac{1}{2}(\nabla \mathbf{u} + \nabla \mathbf{u}^{\mathrm{T}}) \quad \text{on} \quad \mathrm{B} \tag{4.15}$$

$$\operatorname{div} \mathbf{S}' + \mathbf{b}' = \mathbf{0} \quad \text{on} \quad \mathrm{B} \tag{4.16}$$

$$\mathbf{S}' = \mathbf{C}\,[\mathbf{E}] \quad \text{on} \quad \mathrm{B} \tag{4.17}$$

where

$$\mathbf{b}' = \mathbf{b} + \operatorname{div}(T\mathbf{M}) \tag{4.18}$$

$$\mathbf{S}' = \mathbf{S} - T\mathbf{M} \tag{4.19}$$

$$\mathbf{s}' \equiv \mathbf{S}'\mathbf{n} \tag{4.20}$$

The Principle of Minimum Complementary Energy of nonisothermal Elastostatics reads: Let P denote a set of all statically admissible stress fields, and let $\mathrm{G_T} = \mathrm{G_T}\{.\}$ be a functional on P defined by

$$\mathrm{G_T}\{\mathbf{S}\} = \mathrm{U_K}\{\mathbf{S}'\} - \int_{\partial \mathrm{B}_1} \mathbf{s}' \cdot \widehat{\mathbf{u}}\, da \quad \forall \mathbf{S} \in P \tag{4.21}$$

If $\mathbf{S}$ is a stress field corresponding to a solution to the mixed problem of nonisothermal elastostatics, then

$$\mathrm{G_T}\{\mathbf{S}\} \le \mathrm{G_T}\{\tilde{\mathbf{S}}\} \quad \forall \tilde{\mathbf{S}} \in P \tag{4.22}$$

and the equality holds true if $\mathbf{S} = \tilde{\mathbf{S}}$.

Note. The functional $\mathrm{G_T} = \mathrm{G_T}\{.\}$ in Eq. (4.21) can be replaced by

$$\mathrm{G_T^*}\{\mathbf{S}\} = \mathrm{U_K}\{\mathbf{S}\} + \int_{\mathrm{B}} T\mathbf{S} \cdot \mathbf{A}\, dv - \int_{\partial \mathrm{B}_1} \mathbf{s} \cdot \widehat{\mathbf{u}}\, da \tag{4.23}$$

where $\mathbf{A}$ is the thermal expansion tensor.

4.2 The Rayleigh-Ritz Method

The functional $\mathrm{F_1} = \mathrm{F_1}\{\mathbf{u}\}$ [see Eq. (4.9)] can be minimized by looking for $\mathbf{u}$ in an approximate form

$$\mathbf{u} \cong \mathbf{u}^{(\mathrm{N})} = \widehat{\mathbf{u}}^{(\mathrm{N})} + \sum_{\mathrm{k}=1}^{\mathrm{N}} \mathrm{a_k} \mathbf{f}_\mathrm{k} \quad \text{on} \quad \overline{\mathrm{B}} \tag{4.24}$$

where $\widehat{\mathbf{u}}^{(\mathrm{N})}$ is a function on $\overline{\mathrm{B}}$ such that

$$\widehat{\mathbf{u}}^{(\mathrm{N})} = \widehat{\mathbf{u}} \quad \text{on} \quad \partial \mathrm{B}_1 \tag{4.25}$$

and $\{\mathbf{f}_\mathrm{k}\}$ stands for a set of functions on $\overline{\mathrm{B}}$ such that

$$\mathbf{f}_\mathrm{k} = \mathbf{0} \quad \text{on} \quad \partial \mathrm{B}_1 \tag{4.26}$$

and $\mathrm{a_k}$ are unknown constants to be determined from the condition that $\mathrm{F_1} = \mathrm{F_1}\{\mathbf{u}^{(\mathrm{N})}\} \equiv \varphi(\mathrm{a_1}, \mathrm{a_2}, \mathrm{a_3}, \dots, \mathrm{a_N})$ attains a minimum, that is, from the conditions

$$\frac{\partial \varphi}{\partial\, \mathrm{a_i}}(\mathrm{a_1}, \mathrm{a_2}, \mathrm{a_3}, \dots, \mathrm{a_N}) = 0 \quad i = 1, 2, 3, \dots, \mathrm{N} \tag{4.27}$$

One can show that Eqs. (4.27) represent a linear nonhomogeneous system of algebraic equations for which there is a unique solution $(\mathrm{a_1}, \mathrm{a_2}, \mathrm{a_3}, \dots, \mathrm{a_N})$.

Similarly, if $\partial \mathrm{B}_1 = \varnothing$, the functional $\mathrm{G} = \mathrm{G}\{.\}$ [see Eq. (4.13)] can be minimized by letting $\mathbf{S}$ in the form

$$\mathbf{S} \cong \mathbf{S}^{(\mathrm{N})} = \widehat{\mathbf{S}}^{(\mathrm{N})} + \sum_{\mathrm{k}=1}^{\mathrm{N}} \mathrm{a_k} \mathbf{S}_\mathrm{k} \quad \text{on} \quad \overline{\mathrm{B}} \tag{4.28}$$

where $\widehat{\mathbf{S}}^{(\mathrm{N})}$ is selected in such a way that

$$\operatorname{div} \widehat{\mathbf{S}}^{(\mathrm{N})} + \mathbf{b} = \mathbf{0} \quad \text{on} \quad \mathrm{B} \tag{4.29}$$

and

$$\widehat{\mathbf{S}}^{(N)}\mathbf{n} = \widehat{\mathbf{s}} \quad \text{on} \quad \partial B \tag{4.30}$$

while $\mathbf{S}_k$ are to satisfy the equations

$$\operatorname{div} \mathbf{S}_k = \mathbf{0} \quad \text{on} \quad B \tag{4.31}$$

and

$$\mathbf{S}_k\mathbf{n} = \mathbf{0} \quad \text{on} \quad \partial B \tag{4.32}$$

The unknown coefficients a_k are obtained by solving the linear algebraic equations

$$\frac{\partial \psi}{\partial\, a_i}(a_1, a_2, a_3, \ldots, a_N) = 0 \quad i = 1, 2, 3, \ldots, N \tag{4.33}$$

where

$$\psi(a_1, a_2, a_3, \ldots, a_N) \equiv G\{\mathbf{S}^{(N)}\} \tag{4.34}$$

The method of minimizing $F_1 = F_1\{\mathbf{u}\}$ and $G = G\{\mathbf{S}\}$ by postulating $\mathbf{u}$ and $\mathbf{S}$ by formulas (4.24) and (4.28), respectively, is called the *Rayleigh-Ritz Method.*

4.3 Variational Principles

Let $H\{s\}$ be a functional on A, where A is a set of admissible states $s = [\mathbf{u}, \mathbf{E}, \mathbf{S}]$. By the *first variation of* $H\{s\}$ we mean the number

$$\delta_{\tilde{s}} H\{s\} = \frac{d}{d\omega} H\{s + \omega\,\tilde{s}\}\Big|_{\omega=0} \tag{4.35}$$

where s and $\tilde{s} \in \mathsf{A}$, and $s + \omega\,\tilde{s} \in \mathsf{A}$ for every scalar ω, and we say that

$$\delta_{\tilde{s}} H\{s\} \equiv \delta\, H\{s\} = 0 \tag{4.36}$$

if $\delta_{\tilde{s}} H\{s\}$ exists and equals zero for any $\tilde{s}$ consistent with the relation $s + \omega\,\tilde{s} \in \mathsf{A}$.

Hu-Washizu Principle
Let A denote the set of all admissible states of elastostatics, and let $H\{s\}$ be the functional on A defined by

$$H\{s\} = U_C\{\mathbf{E}\} - \int_B \mathbf{S} \cdot \mathbf{E}\, dv - \int_B (\operatorname{div} \mathbf{S} + \mathbf{b}) \cdot \mathbf{u}\, dv + \int_{\partial B_1} \mathbf{s} \cdot \widehat{\mathbf{u}}\, da + \int_{\partial B_2} (\mathbf{s} - \widehat{\mathbf{s}}) \cdot \mathbf{u}\, da$$

$$\forall s = [\mathbf{u}, \mathbf{E}, \mathbf{S}] \in \mathsf{A} \tag{4.37}$$

Then

$$\delta\,\mathrm{H}\{\mathrm{s}\} = 0 \tag{4.38}$$

if and only if s is a solution to the mixed problem.

Note 1. If the set A in Hu-Washizu Principle is restricted to the set of all kinematically admissible states R [see the Principle of Minimum Potential Energy] then Hu-Washizu Principle reduces to that of Minimum Potential Energy.

Hellinger-Reissner Principle
Let A_1 denote the set of all admissible states that satisfy the strain-displacement relation, and let $\mathrm{H}_1 = \mathrm{H}_1\{\mathrm{s}\}$ be the functional on A_1 defined by

$$\mathrm{H}_1\{\mathrm{s}\} = \mathrm{U_K}\{\mathbf{S}\} - \int_{\mathrm{B}} \mathbf{S}\cdot\mathbf{E}\,dv + \int_{\mathrm{B}} \mathbf{b}\cdot\mathbf{u}\,dv + \int_{\partial\mathrm{B}_1} \mathbf{s}\cdot(\mathbf{u}-\widehat{\mathbf{u}})\,da + \int_{\partial\mathrm{B}_2} \widehat{\mathbf{s}}\cdot\mathbf{u}\,da$$
$$\forall\,\mathrm{s} = [\mathbf{u},\mathbf{E},\mathbf{S}] \in \mathsf{A}_1 \tag{4.39}$$

Then

$$\delta\,\mathrm{H}_1\{\mathrm{s}\} = 0 \tag{4.40}$$

if and only if s is a solution to the mixed problem.

Note 2. By restricting A_1 to the set $\mathsf{A}_2 = \mathsf{A}_1 \cap \mathrm{P}$, where P is the set of all statically admissible states, we reduce Hellinger-Reissner Principle to that of the Principle of Minimum Complementary Energy.

4.4 Compatibility-Related Principle

Consider a traction problem for a body B subject to an external load $[\mathbf{b},\widehat{\mathbf{s}}]$. Let Q denote the set of all admissible states that satisfy the equation of equilibrium, the stress-strain relations, and the traction boundary condition; and let I{.} be the functional on Q defined by

$$\mathrm{I}\{\mathrm{s}\} = \mathrm{U_K}\{\mathbf{S}\} = \frac{1}{2}\int_{\mathrm{B}} \mathbf{S}\cdot\mathbf{K}[\mathbf{S}]\,dv \quad \forall\,\mathrm{s} = [\mathbf{u},\mathbf{E},\mathbf{S}] \in \mathrm{Q} \tag{4.41}$$

Then

$$\delta\,\mathrm{I}\{\mathrm{s}\} = 0 \tag{4.42}$$

if and only if s is a solution to the mixed problem.

A proof of the above variational principles is based on the *Fundamental Lemma of Calculus of Variations* which states that for every smooth function $\tilde{\mathrm{g}} = \tilde{\mathrm{g}}(\mathbf{x})$ on

$\overline{B}$ that vanishes near ∂B, and for a fixed continuous function $f = f(\mathbf{x})$ on $\overline{B}$, the condition $\int_B f(\mathbf{x})\tilde{g}(\mathbf{x})\, dv(\mathbf{x}) = 0$ is equivalent to $f(\mathbf{x}) = 0$ on $\overline{B}$.

4.5 Problems and Solutions Related to Variational Formulation of Elastostatics

Problem 4.1. Consider a generalized plane stress traction problem of homogeneous isotropic elastostatics for a region C_0 of (x_1, x_2) plane (see Sect. 7). For such a problem the stress energy is represented by the integral

$$\overline{U}_K\{\overline{\mathbf{S}}\} = \frac{1}{2}\int_{C_0} \overline{\mathbf{S}} \cdot \mathbf{K}[\overline{\mathbf{S}}]\, da \tag{4.43}$$

where $\overline{\mathbf{S}}$ is the stress tensor corresponding to a solution $\overline{s} = [\overline{\mathbf{u}}, \overline{\mathbf{E}}, \overline{\mathbf{S}}]$ of the traction problem, and

$$\overline{\mathbf{E}} = \mathbf{K}[\overline{\mathbf{S}}] = \frac{1}{2\mu}\left[\overline{\mathbf{S}} - \frac{\nu}{1+\nu}(\mathrm{tr}\,\overline{\mathbf{S}})\,\mathbf{1}\right] \quad \text{on} \quad C_0 \tag{4.44}$$

$$\mathrm{div}\,\overline{\mathbf{S}} + \overline{\mathbf{b}} = \mathbf{0} \quad \text{on} \quad C_0 \tag{4.45}$$

$$\overline{\mathbf{E}} = \widehat{\nabla}\overline{\mathbf{u}} \quad \text{on} \quad C_0 \tag{4.46}$$

and

$$\overline{\mathbf{S}}\mathbf{n} = \widehat{\mathbf{s}} \quad \text{on} \quad \partial C_0 \tag{4.47}$$

Let $\overline{Q}$ denote the set of all admissible states that satisfy Eq. (4.44) through (4.47) except for Eq. (4.46). Define the functional $\overline{I}\{.\}$ on $\overline{Q}$ by

$$\overline{I}\{\overline{s}\} = U_K\{\overline{\mathbf{S}}\} \quad \text{for every } \overline{s} \in \overline{Q} \tag{4.48}$$

Show that

$$\delta\,\overline{I}\{\overline{s}\} = 0 \tag{4.49}$$

if and only if $\overline{\mathbf{s}}$ is a solution to the traction problem.

Hint: The proof is similar to that of the compatibility-related principle of Sect. 4.4. First, we note that if $\overline{s} \in \overline{Q}$ and $\tilde{s} \in \overline{Q}$ then $\overline{s} + \omega\tilde{s} \in \overline{Q}$ for every scalar ω, and

$$\delta\,\overline{I}\{\overline{s}\} = \int_{C_0} \tilde{\mathbf{S}} \cdot \overline{\mathbf{E}}\, da \tag{4.50}$$

Next, by letting

$$\tilde{S}_{\alpha\beta} = \varepsilon_{\alpha\gamma 3}\, \varepsilon_{\beta\delta 3}\, \tilde{F}_{,\gamma\delta} \tag{4.51}$$

where $\tilde{F}$ is an Airy stress function such that $\tilde{F}$, $\tilde{F}_{,\alpha}$, and $\tilde{F}_{,\alpha\beta}$ $(\alpha, \beta = 1, 2)$ vanish near $\partial\, \mathrm{C}_0$, w find that

$$\delta\, \overline{I}\{\overline{\mathrm{s}}\} = \int_{\mathrm{C}_0} \tilde{F}\, \varepsilon_{\alpha\gamma 3}\, \varepsilon_{\beta\delta 3}\, \overline{E}_{\alpha\beta,\gamma\delta}\, da \tag{4.52}$$

The proof then follows from (4.52).

Solution. We are to show that

(A) If $\overline{s}$ is a solution to the traction problem then

$$\delta \overline{I}(\overline{s}) = 0 \tag{4.53}$$

and

(B) If

$$\delta \overline{I}(\overline{s}) = 0 \quad \text{for } \overline{s} \in \overline{Q} \tag{4.54}$$

then $\overline{s}$ is a solution to the traction problem.

Proof of (A). Using (4.52) we obtain

$$\delta \overline{I}(\overline{s}) = \int_{C_0} \tilde{F} \varepsilon_{\alpha\gamma 3}\, \varepsilon_{\beta\delta 3}\, \overline{E}_{\alpha\beta,\gamma\delta}\, da \tag{4.55}$$

Since $\overline{s} = [\overline{\mathbf{u}}, \overline{\mathbf{E}}, \overline{\mathbf{S}}]$ is a solution to the fraction problem, Eqs. (4.44)–(4.47) are satisfied, and in particular

$$\overline{E}_{\alpha\beta} = \overline{u}_{(\alpha,\beta)} \tag{4.56}$$

Substituting (4.56) into the RHS of (4.55) we obtain (4.53), and this completes proof of (A).

Proof of (B). We assume that

$$\delta \overline{I}(\overline{s}) = 0 \quad \text{for } \overline{s} \in \overline{Q} \tag{4.57}$$

or

$$\int_{C_0} \tilde{F} \varepsilon_{\alpha\gamma 3}\, \varepsilon_{\beta\delta 3}\, \overline{E}_{\alpha\beta,\gamma\delta}\, da = 0 \tag{4.58}$$

where $\tilde{F}$ is an arbitrary function on C_0 that vanishes near ∂C_0, and $\overline{E}_{\alpha\beta}$ is a symmetric second order tensor field on C_0 that complies with Eqs. (4.44), (4.45), and (4.47). It follows from (4.58) and the Fundamental Lemma of calculus of variations that

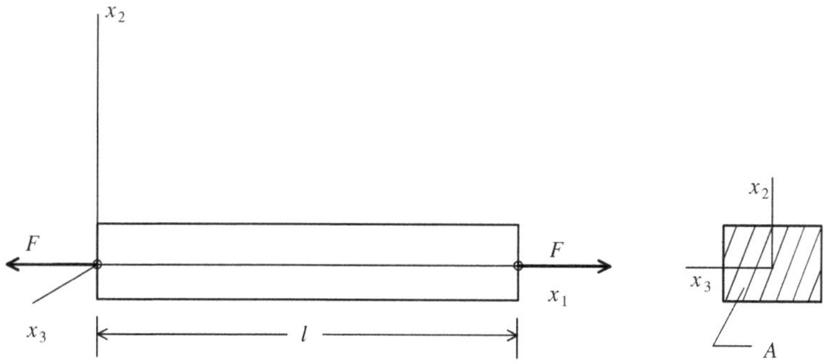

Fig. 4.1 The prismatic bar in simple tension

$$\varepsilon_{\alpha\gamma 3}\, \varepsilon_{\beta\delta 3}\, \overline{E}_{\alpha\beta,\gamma\delta} = 0 \quad \text{on} \quad C_0 \tag{4.59}$$

This implies that there is $\overline{u}_\alpha$ such that

$$\overline{E}_{\alpha\beta} = \overline{u}_{(\alpha,\beta)} \tag{4.60}$$

As a result $\overline{s} = [\overline{\mathbf{u}}, \overline{\mathbf{E}}, \overline{\mathbf{S}}]$ satisfies Eqs. (4.44)–(4.47), that is, $\overline{s}$ is a solution to the traction problem. This completes proof of (B).

Problem 4.2. Consider an elastic prismatic bar in simple tension shown in Fig. 4.1.
The stress energy of the bar takes the form

$$\mathrm{U_K}\{\mathbf{S}\} = \int_0^l \left(\int_A \frac{1}{2E} S_{11}^2 \, da \right) dx_1 = \frac{1}{2E} \int_0^l \left(\frac{F}{A} \right)^2 A \, dx = \frac{F^2 l}{2EA} \tag{4.61}$$

where A is the cross section of the bar, and E denotes Young's modulus.
The strain energy of the bar is obtained from

$$\mathrm{U_C}\{\mathbf{E}\} = \mathrm{U_K}\{\mathbf{S}\} = \frac{EAe^2}{2l} \tag{4.62}$$

where e is an elongation of the bar produced by the force $F = AEE_{11} = AEe/l$. The elastic state of the bar is then represented by

$$s = [u_1, E_{11}, S_{11}] = [e, e/l, F/A] \tag{4.63}$$

(i) Define a potential energy of the bar as $\widehat{\mathrm{F}}\{\mathrm{s}\} \equiv \varphi(e)$ and show that the relation

$$\delta\varphi(e) = 0 \tag{4.64}$$

is equivalent to the condition

$$\frac{\partial U_C}{\partial e} = F \tag{4.65}$$

(ii) Define a complementary energy of the bar as $\widehat{G}\{s\} \equiv \psi(F)$ and show that the condition

$$\delta\psi(F) = 0 \tag{4.66}$$

is equivalent to the equation

$$\frac{\partial U_K}{\partial F} = e \tag{4.67}$$

Hint: The functions $\varphi = \varphi(e)$ and $\psi = \psi(F)$ are given by

$$\varphi(e) = \frac{EA}{2l}e^2 - Fe$$

and

$$\psi(F) = \frac{l}{2EA}F^2 - Fe$$

respectively.

Note: Equations (4.65) and (4.67) constitute the *Castigliano theorem.*

Solution. The potential energy of the bar is given by

$$\varphi(e) = U_c(e) - Fe \tag{4.68}$$

where

$$U_c(e) = \frac{EAe^2}{2l} \tag{4.69}$$

Hence, the relation

$$\delta\varphi(e) = \varphi'(e) = 0 \tag{4.70}$$

takes the form

$$\frac{\partial U_c}{\partial e} = F \tag{4.71}$$

Equations (4.69) and (4.71) imply that

$$F = \frac{EAe}{l} \tag{4.72}$$

which is consistent with the definition of F. This shows that (i) holds true. To prove (ii) we define the complementary energy of the bar as

$$\psi(F) = U_k(F) - Fe \tag{4.73}$$

where

$$U_k(F) = \frac{F^2 l}{2EA} \tag{4.74}$$

and from the relation

$$\delta\psi(F) = \psi'(F) = 0 \tag{4.75}$$

we obtain

$$\frac{\partial U_k}{\partial F} = e \tag{4.76}$$

Equations (4.74) and (4.76) imply that

$$e = \frac{Fl}{EA} \tag{4.77}$$

which is consistent with the definition of e. This shows that (ii) holds true. Hence, a solution to Problem 4.2 is complete.

Problem 4.3. The complementary energy of a cantilever beam loaded at the end by force P takes the form (see Fig. 4.2)

$$\begin{aligned}\psi(P) &= \frac{1}{2E}\int_B S_{11}^2\, dv - Pu_2(l)\\ &= \frac{1}{2E}\int_0^l \left\{\int_A \frac{M^2(x_1)}{I^2} x_2^2\, dA\right\} dx_1 - Pu_2(l)\end{aligned} \tag{4.78}$$

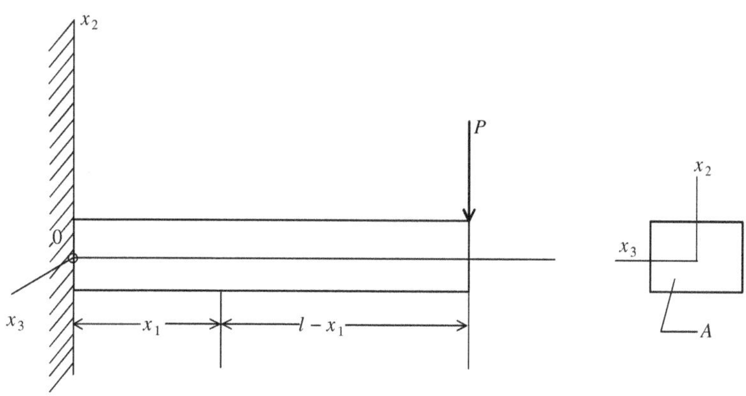

Fig. 4.2 The cantilever beam loaded at the end

where $M = M(x_1)$ and I stand for the bending moment and the moment of inertia of the area A with respect to the x_3 axis, respectively, given by

$$M(x_1) = P(l - x_1), \quad I = \int_A x_2^2 \, da \tag{4.79}$$

Use the minimum complementary energy principle for the cantilever beam in the form

$$\delta\psi(P) = 0 \tag{4.80}$$

to show that the magnitude of deflection at the end of the beam is

$$u_2(l) = \frac{Pl^3}{3EI} \tag{4.81}$$

Solution. Substituting $M = M(x_1)$ and I from (4.79) into (4.78) and performing the integration we obtain.

$$\psi(P) = \frac{P^2}{2EI}\frac{l^3}{3} - P\, u_2(l) \tag{4.82}$$

Finally, using the minimum complementary energy principle

$$\delta\psi(P) = \psi'(P) = 0 \tag{4.83}$$

we arrive at (4.81), and this completes a solution to Problem 4.3.

Problem 4.4. An elastic beam which is clamped at one end and simply supported at the other end is loaded at an internal point $x_1 = \xi$ by force P (see Fig. 4.3)

The potential energy of the beam, treated as a functional depending on a deflection of the beam $u_2 = u_2(x_1)$, takes the form

$$\varphi\{u_2\} = \frac{EI}{2}\int_0^l \left(\frac{d^2u_2}{dx_1^2}\right)^2 dx_1 - Pu_2(\xi) \tag{4.84}$$

and $u_2 \in \tilde{P} = \{u_2 = u_2(x_1) : u_2(0) = u_2'(0) = 0; \quad u_2(l) = u_2''(l) = 0\}$. Let $u_2 = u_2(x_1)$ be a solution of the equation

$$EI\,\frac{d^4u_2}{dx_1^4} = P\delta(x_1 - \xi) \quad \text{for} \quad 0 < x_1 < l \tag{4.85}$$

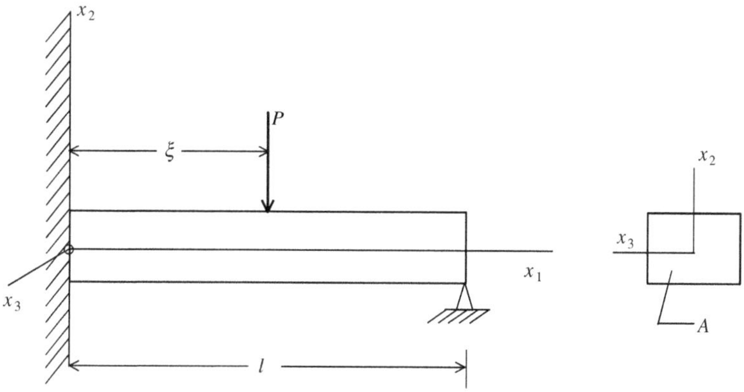

Fig. 4.3 The beam clamped at one end and simply supported at the other end

subject to the conditions

$$u_2(0) = u_2'(0) = 0; \quad u_2(l) = u_2''(l) = 0 \tag{4.86}$$

Show that

$$\delta\varphi\{u_2\} = 0 \tag{4.87}$$

if and only if u_2 is a solution to the boundary value problem (4.85)–(4.86).

Solution. Since

$$\delta\varphi\{u_2\} = \frac{d}{d\omega}\varphi\{u_2 + \omega\tilde{u}_2\}\bigg|_{\omega=0} \tag{4.88}$$

where

$$\tilde{u}_2(0) = \tilde{u}_2'(0) = 0 \tag{4.89}$$

and

$$\tilde{u}_2(l) = \tilde{u}_2''(l) = 0 \tag{4.90}$$

therefore, Eq. (4.88) takes the form

$$\delta\varphi\{u_2\} = EI\int_0^l u_2''(x)\tilde{u}_2''(x)dx - P\,\tilde{u}_2(\xi) \tag{4.91}$$

Integrating by parts we obtain

$$\int_0^l u_2''(x)\tilde{u}_2''(x)dx = u_2''(x)\tilde{u}_2'(x)\Big|_{x=0}^{x=l} - u_2'''(x)\tilde{u}_2(x)\Big|_{x=0}^{x=l} + \int_0^l u_2^{(4)}(x)\tilde{u}_2(x)dx \tag{4.92}$$

Since $u_2 \in \tilde{P}$ and $\tilde{u}_2 \in \tilde{P}$, Eq. (4.92) reduces to

$$\int_0^l u_2''(x)\tilde{u}_2''(x)dx = \int_0^l u_2^{(4)}(x)\tilde{u}_2(x)dx \tag{4.93}$$

and Eq. (4.91) takes the form

$$\delta\varphi\{u_2\} = \int_0^l \Big[EI\, u_2^{(4)}(x) - P\delta(x-\xi)\Big]\tilde{u}_2(x)dx \tag{4.94}$$

Equation (4.94) together with the Fundamental Lemma of calculus of variations imply that Eq. (4.87) is satisfied if and only if u_2 is a solution to problem (4.85)–(4.86). And this completes a solution to Problem 4.4.

Problem 4.5. Use the Rayleigh-Ritz method to show that an approximate deflection of the beam of Problem 4.4 takes the form ($x_1 = x$)

$$u_2(x) = -\mathrm{c}l^3 \left(\frac{x}{l}\right)^2 \left(1 - \frac{x}{l}\right)\left(1 - \frac{2}{3}\frac{x}{l}\right) \tag{4.95}$$

where

$$\mathrm{c} = -\frac{5}{4}\frac{P}{EI}\left(\frac{\xi}{l}\right)^2 \left(1 - \frac{\xi}{l}\right)\left(1 - \frac{2}{3}\frac{\xi}{l}\right) \tag{4.96}$$

Also, show that for $\xi = l/2$ we obtain

$$u_2(l/2) = 0.0086\frac{l^3 P}{EI} \tag{4.97}$$

Solution. Note that $u_2 = u_2(x)$ given by Eq. (4.95) can be written in the form

$$u_2(x) = -cl^3 f\left(\frac{x}{l}\right) \tag{4.98}$$

where

$$f\left(\frac{x}{l}\right) = \left(\frac{x}{l}\right)^2 \left(1 - \frac{x}{l}\right)\left(1 - \frac{2}{3}\frac{x}{l}\right) \tag{4.99}$$

and

$$f(0) = f'(0) = f(1) = f''(1) = 0 \tag{4.100}$$

Hence

$$u_2(x) \in \tilde{P} \tag{4.101}$$

where $\tilde{P}$ is the domain of the functional $\varphi\{u_2\}$ from Problem 4.4, and substituting (4.98) into Eq. (4.95) of Problem 4.4 we obtain

$$\varphi\{u_2\} = l^3 \left\{ \frac{EI}{2} c^2 \int_0^1 [f''(u)]^2 du + P\, c\, f\left(\frac{\xi}{l}\right) \right\} \equiv \psi(c) \tag{4.102}$$

The condition

$$\delta\varphi\{u_2\} = \psi'(c) = 0 \tag{4.103}$$

is satisfied if and only if

$$c \int_0^1 (f'')^2 du = -\frac{P}{EI}\, f\left(\frac{\xi}{l}\right) \tag{4.104}$$

Since

$$f''(u) = 2(4u^2 - 5u + 1) \tag{4.105}$$

and

$$\int_0^1 (f'')^2 du = \frac{4}{5} \tag{4.106}$$

it follows from Eq. (4.104) that c is given by Eq. (4.96). Finally, by letting $x = l/2$ and $\xi = l/2$ in Eqs. (4.95) and (4.96), respectively, we obtain (4.97). This completes a solution to Problem 4.5.

Problem 4.6. The potential energy of a rectangular thin elastic membrane fixed at its boundary and subject to a vertical load $f = f(x_1, x_2)$ is

$$I\{u\} = \int_{-a_1}^{a_1} \int_{-a_2}^{a_2} \left(\frac{T_0}{2} u_{,\alpha} u_{,\alpha} - f\, u \right) dx_1 dx_2 \tag{4.107}$$

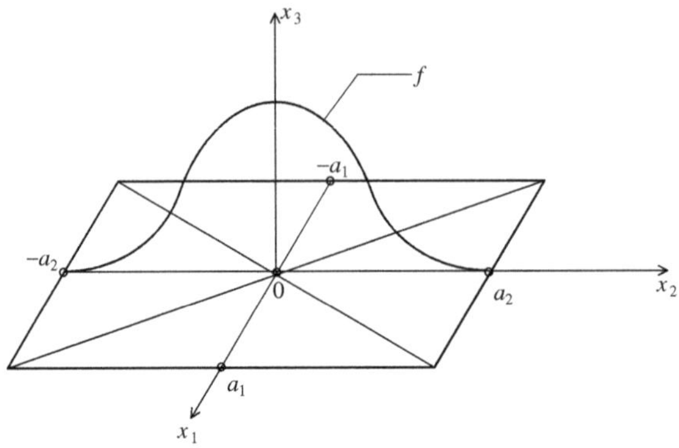

Fig. 4.4 The thin membrane fixed at its boundary

where $u \in \widehat{P}$, and

$$\widehat{P} = \{u = u(x_1, x_2) \; : \; u(\pm a_1, x_2) = 0 \quad \text{for} \quad |x_2| < a_2; \quad u(x_1, \pm a_2) = 0 \quad \text{for} \quad |x_1| < a_1\} \tag{4.108}$$

Here, $u = u(x_1, x_2)$ is a deflection of the membrane in the x_3 direction, and T_0 is a uniform tension of the membrane (see Fig. 4.4). Let the load function $f = f(x_1, x_2)$ be represented by the series

$$f(x_1, x_2) = \sum_{m,n=1}^{\infty} f_{mn} \sin \frac{m\pi(x_1 - a_1)}{2a_1} \sin \frac{n\pi(x_2 - a_2)}{2a_2} \tag{4.109}$$

Use the Rayleigh-Ritz method to show that the functional $I\{u\}$ attains a minimum over $\widehat{P}$ at

$$u(x_1, x_2) = \sum_{m,n=1}^{\infty} u_{mn} \sin \frac{m\pi(x_1 - a_1)}{2a_1} \sin \frac{n\pi(x_2 - a_2)}{2a_2} \tag{4.110}$$

where

$$u_{mn} = \frac{1}{T_0} \frac{f_{mn}}{[(m\pi/2a_1)^2 + (n\pi/2a_2)^2]} \qquad m, n = 1, 2, 3, \ldots \tag{4.111}$$

Solution. Let C_0 stand for the interior of rectangular region

$$C_0 = \{(x_1, \; x_2) : |x_1| < a_1, \; |x_2| < a_2\} \tag{4.112}$$

and let ∂C_0 denote its boundary.

Then

$$\widehat{P} = \{u : u = 0 \quad \text{on} \quad \partial C_0\} \tag{4.113}$$

let $u \in \widehat{P}$ and $u + \omega\tilde{u} \in \widehat{P}$, where ω is a scalar. Then

$$\tilde{u} \in \widehat{P}, \text{ that is, } \tilde{u} = 0 \text{ on } \partial C_0 \tag{4.114}$$

Computing the first variation of $I\{u\}$ we obtain

$$\delta I\{u\} = \frac{d}{d\omega} I\{u + \omega\tilde{u}\}|_{\omega=0} = \int_{C_0} (T_0 u_{,\alpha}\ \tilde{u}_{,\alpha} - f\tilde{u}) da \tag{4.115}$$

Since

$$u_{,\alpha}\ \tilde{u}_{,\alpha} = (u_{,\alpha}\ \tilde{u})_{,\alpha} - u_{,\alpha\alpha}\ \tilde{u} \tag{4.116}$$

therefore, using the divergence theorem, from Eqs. (4.115) and (4.116) we obtain

$$\delta I\{u\} = -\int_{C_0} (T_0 u_{,\alpha\alpha} + f)\tilde{u}\ da \tag{4.117}$$

and

$$\delta I\{u\} = 0 \quad \text{for every } u \in \widehat{P} \tag{4.118}$$

if and only if $u = u(x_1, x_2)$ is a solution to the boundary value problem

$$u_{,\alpha\alpha} = -\frac{1}{T_0} f \quad \text{on } C_0 \tag{4.119}$$

$$u = 0 \quad \text{on } \partial C_0 \tag{4.120}$$

Therefore, the Rayleigh Ritz method applied to the functional $I = I\{u\}$ leads to a solution of problem (4.119)–(4.120). It is easy to show, by substituting (4.110) into Eq. (4.119), that $u = u(x_1, x_2)$ given by (4.110) is a solution to problem (4.119)–(4.120).

To obtain the formula (4.110) by the Rayleigh Ritz method we look for $u = u(x_1, x_2)$ that minimizes $I\{u\}$ in the form

$$u(x_1, x_2) = \sum_{mn} c_{mn} \varphi_m(x_1)\ \psi_n(x_2) \tag{4.121}$$

where

$$\varphi_m(x_1) = \sin \frac{m\pi(x_1 - a_1)}{2a_1} \tag{4.122}$$

and

$$\psi_n(x_2) = \sin \frac{n\pi(x_2 - a_2)}{2a_2} \tag{4.123}$$

Substituting u from (4.121) into (4.107) and using f given by (4.109) we obtain

$$I\{u\} \equiv F(c_{mn}) = \int_{-a_1}^{a_1} dx_1 \int_{-a_2}^{a_2} dx_2 \left\{ \frac{T_0}{2} \left[\sum_{mn} c_{mn} \varphi'_m(x_1)\, \psi_n(x_2) \right]^2 \right.$$
$$+ \frac{T_0}{2} \left[\sum_{mn} c_{mn} \varphi_m(x_1)\, \psi'_n(x_2) \right]^2 - \left[\sum_{mn} c_{mn} \varphi_m(x_1)\, \psi_n(x_2) \right]$$
$$\left. \times \left[\sum_{pq} f_{pq} \varphi_p(x_1)\, \psi_q(x_2) \right] \right\} \tag{4.124}$$

The conditions

$$\frac{\partial F}{\partial c_{mn}} = 0 \quad m, n = 1, 2, \dots \tag{4.125}$$

together with the orthogonality relations

$$\frac{1}{a_1} \int_{-a_1}^{a_1} \varphi_m(x_1)\, \varphi_k(x_1) dx_1 = \delta_{mk} \tag{4.126}$$

$$\frac{1}{a_2} \int_{-a_2}^{a_2} \psi_m(x_2)\, \psi_k(x_2) dx_2 = \delta_{mk} \tag{4.127}$$

lead to the simple algebraic equation for c_{mn}

$$T_0 c_{mn} [(m\pi/2a_1)^2 + (n\pi/2a_2)^2] - f_{mn} = 0 \tag{4.128}$$

Therefore, $c_{mn} = u_{mn}$, where u_{mn} is given by (4.111). This completes a solution to Problem 4.6.

Problem 4.7. Use the solution obtained in Problem 4.6 to find the deflection of a square membrane of side a that is held fixed at its boundary and is vertically loaded by a load f of the form

$$f(x_1, x_2) = f_0 [H(x_1 + \varepsilon) - H(x_1 - \varepsilon)][H(x_2 + \varepsilon) - H(x_2 - \varepsilon)] \tag{4.129}$$

where $H = H(x)$ is the Heaviside function, and f_0 and ε are positive constants $(0 < \varepsilon < a)$. Also, compute a deflection of the square membrane at its center when $\varepsilon = a/8$.

Solution. Let f be a function represented by the double series [see (4.109) of Problem 4.6]

$$f(x_1, x_2) = \sum_{mn} f_{mn} \varphi_m(x_1)\, \psi_n(x_2) \tag{4.130}$$

where φ_m and ψ_n are given by Eqs. (4.122) and (4.123), respectively, of Problem 4.6. Using the orthogonality conditions (4.126) and (4.127) of Problem 4.6, we find that

$$f_{mn} = \frac{1}{a_1 a_2} \int_{-a_1}^{a_1} dx_1 \int_{-a_2}^{a_2} dx_2\, f(x_1, x_2)\, \varphi_m(x_1)\, \psi_n(x_2) \tag{4.131}$$

For a square membrane of side a

$$a_1 = a_2 = a \tag{4.132}$$

and

$$\varphi_m(x_1) = \sin \frac{m\pi(x_1 - a)}{2a} \tag{4.133}$$

$$\psi_n(x_2) = \sin \frac{n\pi(x_2 - a)}{2a} \tag{4.134}$$

Substituting f from (4.129) into (4.131) we obtain

$$f_{mn} = \frac{f_0}{a^2} \int_{-\varepsilon}^{\varepsilon} dx_1 \int_{-\varepsilon}^{\varepsilon} dx_2\, \varphi_m(x_1)\psi_n(x_2) \tag{4.135}$$

$$= \frac{16}{\pi^2} f_0 \frac{1}{mn} \sin\left(\frac{m\pi}{2}\right) \sin\left(\frac{m\pi}{2}\frac{\varepsilon}{a}\right) \times \sin\left(\frac{n\pi}{2}\right) \sin\left(\frac{n\pi}{2}\frac{\varepsilon}{a}\right) \tag{4.136}$$

Therefore, for a load f of the form (4.129) the deflection of the membrane is given by

$$u(x_1, x_2) = \sum_{m,n=1}^{\infty} u_{mn}\varphi_m(x_1)\, \varphi_n(x_2) \tag{4.137}$$

where

$$u_{mn} = \frac{1}{T_0} \frac{4a^2}{\pi^2} \frac{f_{mn}}{m^2 + n^2} \tag{4.138}$$

and f_{mn} is given by (4.136).

Letting $x_1 = 0$ and $x_2 = 0$ in (4.137) we obtain

$$u(0,0) = \frac{64a^2}{\pi^4} \frac{f_0}{T_0} \times \sum_{m,n=1}^{\infty} \frac{1}{mn(m^2+n^2)} \sin^2 \frac{m\pi}{2} \sin^2 \frac{m\pi}{2} \left(\frac{\varepsilon}{a}\right) \times \sin^2 \frac{n\pi}{2} \sin \frac{n\pi}{2} \left(\frac{\varepsilon}{a}\right) \tag{4.139}$$

Since

$$\sin^2 \frac{m\pi}{2} = \frac{1}{2}(1 - \cos m\pi) = \frac{1-(-)^m}{2} \tag{4.140}$$

and

$$\sin^2 \frac{n\pi}{2} = \frac{1}{2}(1 - \cos n\pi) = \frac{1-(-)^n}{2} \tag{4.141}$$

therefore, (4.139) can be written as

$$u(0,0) = \frac{64a^2}{\pi^4}\frac{f_0}{T_0} \times \sum_{m,n=1,3,5,\ldots}^{\infty} \frac{1}{mn(m^2+n^2)} \sin\frac{m\pi}{2}\left(\frac{\varepsilon}{a}\right) \sin\frac{n\pi}{2}\left(\frac{\varepsilon}{a}\right) \tag{4.142}$$

Using the orthogonality relations

$$\int_0^1 \sin m\pi\zeta \,\sin n\pi\zeta \, d\zeta = \frac{1}{2}\,\delta_{mn} \tag{4.143}$$

it is easy to show that

$$\frac{\pi}{4n^2}\left[1 - \frac{\cos\text{h}\left[\frac{n\pi}{2}(1-2\zeta)\right]}{\cos\text{h}\,\frac{n\pi}{2}}\right] = \sum_{m=1,3,5,\ldots}^{\infty} \frac{\sin m\pi\zeta}{m(m^2+n^2)} \quad \text{for } 0<\zeta<1 \tag{4.144}$$

Since

$$\varepsilon < 2a$$

therefore, letting $\zeta = \varepsilon/2a < 1$ into (4.144) we reduce the double series (4.142) to the single one

$$u(0,0) = \frac{16a^2}{\pi^3}\frac{f_0}{T_0} \times \sum_{n=1,3,5,\ldots}^{\infty} \frac{1}{n^3}\sin\left(\frac{n\pi}{2}\frac{\varepsilon}{a}\right)\left[1 - \frac{\cos\text{h}\frac{n\pi}{2}\left(1-\frac{\varepsilon}{a}\right)}{\cosh\frac{n\pi}{2}}\right] \tag{4.145}$$

Finally, letting $\varepsilon/a = 1/8$ in (4.145) we get

$$u(0,0) = \frac{16a^2}{\pi^3}\frac{f_0}{T_0} \sum_{n=1,3,\ldots}^{\infty} \frac{1}{n^3}\sin\frac{n\pi}{16} \times \left[1 - \frac{\cos\text{h}\left(\frac{7}{16}n\pi\right)}{\cos\text{h}\left(\frac{1}{2}n\pi\right)}\right] \tag{4.146}$$

This completes a solution to Problem 4.7.

Problem 4.8. The potential energy of a rectangular thin elastic plate that is simply supported along all the edges and is vertically loaded by a force P at a point (ξ_1, ξ_2)

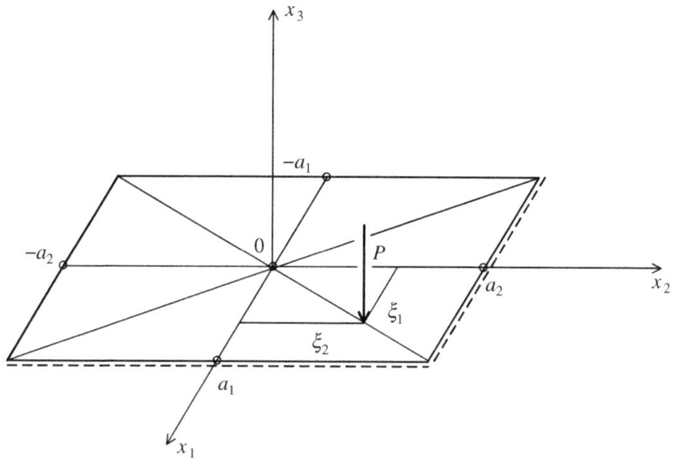

Fig. 4.5 The *rectangular* thin plate simply supported along all edges

takes the form

$$\widehat{I}\{w\} = \frac{1}{2}D \int_{-a_1}^{a_1} \int_{-a_2}^{a_2} (\nabla^2 w)^2 dx_1 dx_2 - P\, w(\xi_1, \xi_2) \tag{4.147}$$

where $w \in \tilde{P}$, and

$$\begin{aligned} \tilde{P} = \{w = w(x_1, x_2) : \; & w(\pm a_1, x_2) = 0, \quad \nabla^2 w(\pm a_1, x_2) = 0 \quad \text{for } |x_2| < a_2; \\ & w(x_1, \pm a_2) = 0, \quad \nabla^2 w(x_1, \pm a_2) = 0 \quad \text{for } |x_1| < a_1\} \end{aligned} \tag{4.148}$$

Here $w = w(x_1, x_2)$ is a deflection of the plate, and D is the bending rigidity of the plate (see Fig. 4.5).

Show that a minimum of the functional $\widehat{I}\{.\}$ over $\tilde{P}$ is attained at a function $w = w(x_1, x_2)$ represented by the series

$$w(x_1, x_2) = \sum_{m,n=1}^{\infty} w_{mn} \sin \frac{m\pi(x_1 - a_1)}{2a_1} \sin \frac{n\pi(x_2 - a_2)}{2a_2} \tag{4.149}$$

where

$$w_{mn} = \frac{P}{Da_1 a_2} \frac{\sin \dfrac{m\pi}{2a_1}(\xi_1 - a_1) \sin \dfrac{n\pi}{2a_2}(\xi_2 - a_2)}{[(m\pi/2a_1)^2 + (n\pi/2a_2)^2]^2} \quad m, n = 1, 2, 3, \ldots \tag{4.150}$$

Hint: Use the series representation of the concentrated load P

$$
\begin{aligned}
&P\delta(x_1-\xi_1)\delta(x_2-\xi_2)\\
&\quad=\frac{P}{a_1a_2}\sum_{m,n=1}^{\infty}\sin\frac{m\pi}{2a_1}(\xi_1-a_1)\sin\frac{n\pi}{2a_2}(\xi_2-a_2)\sin\frac{m\pi}{2a_1}(x_1-a_1)\\
&\qquad\times\sin\frac{n\pi}{2a_2}(x_2-a_2)\\
&\qquad\text{for every } |x_1|<a_1,\quad |x_2|<a_2,\quad |\xi_1|<a_1,\quad |\xi_2|<a_2.
\end{aligned}
\tag{4.151}
$$

Solution. Let $w\in\tilde{P}$ and $\tilde{w}\in\tilde{P}$. Then $w+\omega\tilde{w}\in\tilde{P}$, and the first variation of $\widehat{I}\{w\}s$ takes the form

$$
\begin{aligned}
\delta\widehat{I}\{w\}=&\frac{d}{d\omega}\widehat{I}\,\{w+\omega\tilde{w}\}_{|\omega=0}\\
=&D\int_{-a_1}^{a_1}dx_1\int_{-a_2}^{a_2}dx_2(\nabla^2w)(\nabla^2\tilde{w})-P\tilde{w}(\xi)
\end{aligned}
\tag{4.152}
$$

Let C_0 be an interior of the rectangular region, and let ∂C_0 denote its boundary. Then Eq. (4.152) can be written as

$$
\delta\widehat{I}\{w\}=D\int_{C_0}w,_{\alpha\alpha}\ \tilde{w},_{\beta\beta}\ da-P\tilde{w}(\xi)
\tag{4.153}
$$

Since

$$
\begin{aligned}
w,_{\alpha\alpha}\ \tilde{w},_{\beta\beta}&=(w,_{\alpha\alpha}\ \tilde{w},_{\beta}),_{\beta}-w,_{\alpha\alpha\beta}\ \tilde{w},_{\beta}\\
&=(w,_{\alpha\alpha}\ \tilde{w},_{\beta}-w,_{\alpha\alpha\beta}\ \tilde{w}),_{\beta}+w,_{\alpha\alpha\beta\beta}\ \tilde{w}
\end{aligned}
\tag{4.154}
$$

therefore, integrating (4.154) over C_0, using the divergence theorem, and the relations

$$
w,_{\alpha\alpha}=0,\quad \tilde{w}=0\quad \text{on } \partial C_0
\tag{4.155}
$$

we reduce (4.153) to the form

$$
\delta\widehat{I}\{w\}=\int_{C_0}[D\nabla^4w-P\delta(\mathbf{x}-\xi)]\tilde{w}(\xi)da
\tag{4.156}
$$

A minimum of the functional $\widehat{I}\{w\}$ over $\tilde{P}$ is attained at w that satisfies the field equation

$$
\nabla^4w=\frac{P}{D}\delta(\mathbf{x}-\xi)\quad \text{on } C_0
\tag{4.157}
$$

subject to the homogeneous b conditions

$$w = 0, \quad \nabla^2 w = 0 \quad \text{on } \partial C_0 \tag{4.158}$$

To obtain a solution to problem (4.157)–(4.158) we use the representation of $\delta(\mathbf{x}-\xi)$

$$\delta(\mathbf{x} - \xi) = \frac{1}{a_1 a_2} \sum_{m,n=1}^{\infty} \varphi_m(x_1)\, \varphi_m(\xi_1)\, \psi_n(x_2)\, \psi_n(\xi_2) \tag{4.159}$$

where $\varphi_m(x_1)$ and $\psi_n(x_2)$, respectively, are given by Eqs. (4.122) and (4.123) of Problem 4.6 Since

$$\nabla^2 \varphi_m(x_1)\, \psi_n(x_2) = -\left[\left(\frac{m\pi}{2a_1}\right)^2 + \left(\frac{n\pi}{2a_2}\right)^2\right] \varphi_m(x_1)\, \psi_n(x_2) \tag{4.160}$$

therefore, by looking for a solution of Eq. (4.157) in the form

$$w(x_1, x_2) = \sum_{m,n=1}^{\infty} w_{mn}\, \varphi_m(x_1)\, \psi_n(x_2) \tag{4.161}$$

and substituting (4.159) and (4.161) into (4.157) we find that

$$w_{mn} \left[\left(\frac{m\pi}{2a_1}\right)^2 + \left(\frac{n\pi}{2a_2}\right)^2\right]^2 = \frac{P}{D a_1 a_2}\, \varphi_m(\xi_1)\, \psi_n(\xi_2) \tag{4.162}$$

This completes a solution to Problem 4.8.

Problem 4.9. Show that the central deflection of a square plate of side a that is simply supported along all the edges, and is loaded by a force P at its center, takes the form

$$w(0, 0) \approx 0.0459 \frac{P a^2}{D} \tag{4.163}$$

Hint: Use the result obtained in Problem 4.8 when $\xi_1 = \xi_2 = 0$, $x_1 = x_2 = 0$, $a_1 = a_2 = a$

$$w(0, 0) = \frac{16 P a^2}{D \pi^4} \sum_{m,n=1}^{\infty} \frac{1}{[(2m-1)^2 + (2n-1)^2]^2} \tag{4.164}$$

Also, by taking advantage of the formula

$$\sum_{m=1}^{\infty} \frac{1}{[(2m-1)^2+x^2]^2} = \frac{\pi}{8x^3}\left(\tan\text{h}\frac{\pi x}{2} - \frac{\pi x}{2}\sec\text{h}^2\frac{\pi x}{2}\right) \quad \text{for every } x > 0 \tag{4.165}$$

which is obtained by differentiating with respect to x the formula

$$\sum_{k=1}^{\infty} \frac{1}{(2k-1)^2+x^2} = \frac{\pi}{4x}\tan\text{h}\frac{\pi x}{2} \tag{4.166}$$

we reduce Eq. (4.164) to the simple form

$$w(0,0) = \frac{2Pa^2}{D\pi^3}\sum_{n=1}^{\infty}\frac{1}{(2n-1)^3}\left[\tan\text{h}\frac{\pi}{2}(2n-1) - \frac{\pi}{2}(2n-1)\sec\text{h}^2\frac{\pi}{2}(2n-1)\right] \tag{4.167}$$

The result (4.163) then follows by truncating the series (4.167).

Solution. By letting $a_1 = a_2 = a,\ x_1 = x_2 = 0,\ \xi_1 = \xi_2 = 0$ in Eq. (4.165) of Problem 4.8 we obtain

$$w(0,0) = \sum_{m,n=1}^{\infty} w_{mn}\ \sin\left(\frac{m\pi}{2}\right)\ \sin\left(\frac{n\pi}{2}\right) \tag{4.168}$$

where

$$w_{mn} = \frac{P}{Da^2}\,\frac{\sin\left(\frac{m\pi}{2}\right)\sin\left(\frac{n\pi}{2}\right)}{(m^2\pi^2/4a^2+n^2\pi^2/4a^2)^2} \tag{4.169}$$

Hence

$$w(0,0) = \frac{16Pa^2}{D\pi^4}\sum_{m,n=1}^{\infty}\frac{\sin^2\left(\frac{m\pi}{2}\right)\sin^2\left(\frac{n\pi}{2}\right)}{(m^2+n^2)^2} \tag{4.170}$$

or

$$w(0,0) = \frac{16Pa^2}{D\pi^4}\sum_{m,n=1}^{\infty}\frac{1}{[(2m-1)^2+(2n-1)^2]^2} \tag{4.171}$$

which is equivalent to Eq. (4.164).

Finally, using (4.165) with $x = 2n-1$, we reduce (4.171) to the single series formula (4.167). This completes solution to Problem 4.9.

Chapter 5
Variational Principles of Elastodynamics

In this chapter both the classical Hamilton-Kirchhoff Principle and a convolutional variational principle of Gurtin's type that describes completely a solution to an initial-boundary value problem of elastodynamics are used to solve a number of typical problems of elastodynamics.

5.1 The Hamilton-Kirchhoff Principle

To formulate H-K principle we introduce a notion of *kinematically admissible process,* and by this we mean an admissible process that satisfies the strain-displacement relation, the stress-strain relation, and the displacement boundary condition.

(H-K) **The Hamilton-Kirchhoff Principle.** Let P denote the set of all kinematically admissible processes $p = [\mathbf{u}, \mathbf{E}, \mathbf{S}]$ on $\overline{\mathrm{B}} \times [0, \infty)$ satisfying the conditions

$$\mathbf{u}(\mathbf{x}, \mathrm{t}_1) = \mathbf{u}_1(\mathbf{x}), \quad \mathbf{u}(\mathbf{x}, \mathrm{t}_2) = \mathbf{u}_2(\mathbf{x}) \quad \text{on} \quad \overline{\mathrm{B}} \tag{5.1}$$

where t_1 and t_2 are two arbitrary points on the t-axis such that $0 \leq t_1 < t_2 < \infty$, and $\mathbf{u}_1(\mathbf{x})$ and $\mathbf{u}_2(\mathbf{x})$ are prescribed fields on $\overline{\mathrm{B}}$. Let $\mathsf{K} = \mathsf{K}\{p\}$ be the functional on P defined by

$$\mathsf{K}\{p\} = \int_{t_1}^{t_2} [\mathrm{F}(t) - \mathrm{K}(t)]\, dt \tag{5.2}$$

where

$$F(t) = \mathrm{U_C}\{\mathbf{E}\} - \int_{\mathrm{B}} \mathbf{b} \cdot \mathbf{u}\, dv - \int_{\partial \mathrm{B}_2} \hat{\mathbf{s}} \cdot \mathbf{u}\, da \tag{5.3}$$

M. Reza Eslami et al., *Theory of Elasticity and Thermal Stresses*, Solid Mechanics and Its Applications 197, DOI: 10.1007/978-94-007-6356-2_5,

and

$$K(t) = \frac{1}{2} \int\limits_{B} \rho \, \dot{\mathbf{u}}^2 dv \tag{5.4}$$

for every $p = [\mathbf{u}, \mathbf{E}, \mathbf{S}] \in P$. Then

$$\delta \, \mathsf{K}\{p\} = 0 \tag{5.5}$$

if and only if p satisfies the equation of motion and the traction boundary condition.

Clearly, in the (H-K) principle a displacement vector $\mathbf{u} = \mathbf{u}(\mathbf{x}, \mathrm{t})$ needs to be prescribed at two points t_1 and t_2 of the time axis. If $t_1 = 0$, then $\mathbf{u}(\mathbf{x}, 0)$ may be identified with the initial value of the displacement vector in the formulation of an initial-boundary value problem, however, the value $\mathbf{u}(\mathbf{x}, \mathrm{t}_2)$ is not available in this formulation. This is the reason why the (H-K) principle can not be used to describe the initial-boundary value problem. A full variational characterization of an initial-boundary value problem of elastodynamics is due to Gurtin, and it has the form of a convolutional variational principle. The idea of a convolutional variational principle of elastodynamics is now explained using a traction initial-boundary value problem of incompatible elastodynamics. In such a problem we are to find a symmetric second-order tensor field $\mathbf{S} = \mathbf{S}(\mathbf{x}, t)$ on $\overline{\mathrm{B}} \times [0, \infty)$ that satisfies the field equation

$$\hat{\nabla}[\rho^{-1}(\mathrm{div}\, \mathbf{S})] - \mathbf{K}[\ddot{\mathbf{S}}] = -\mathbf{B} \quad \text{on} \quad \mathrm{B} \times [0, \infty) \tag{5.6}$$

subject to the initial conditions

$$\mathbf{S}(\mathbf{x}, 0) = \mathbf{S}_0(\mathbf{x}), \quad \dot{\mathbf{S}}(\mathbf{x}, 0) = \dot{\mathbf{S}}_0(\mathbf{x}) \quad \text{for} \quad \mathbf{x} \in \mathrm{B} \tag{5.7}$$

and the boundary condition

$$\mathbf{s} = \mathbf{S}\mathbf{n} = \hat{\mathbf{s}} \quad \text{on} \quad \partial\mathrm{B} \times [0, \infty) \tag{5.8}$$

Here $\mathbf{S}_0$ and $\dot{\mathbf{S}}_0$ are arbitrary symmetric tensor fields on B, and $\mathbf{B}$ is a prescribed symmetric second-order tensor field on $\overline{\mathrm{B}} \times [0, \infty)$. Moreover, ρ, $\mathbf{K}$, and $\hat{\mathbf{s}}$ have the same meaning as in classical elastodynamics.

First, we note that the problem is equivalent to the following one. Find a symmetric second-order tensor field on $\overline{\mathrm{B}} \times [0, \infty)$ that satisfies the integro-differential equation

$$\hat{\nabla}[\rho^{-1} t * (\mathrm{div}\, \mathbf{S})] - \mathbf{K}[\mathbf{S}] = -\hat{\mathbf{B}} \quad \text{on} \quad \mathrm{B} \times [0, \infty) \tag{5.9}$$

subject to the boundary condition

$$\mathbf{s} = \mathbf{S}\mathbf{n} = \hat{\mathbf{s}} \quad \text{on} \quad \partial\mathrm{B} \times [0, \infty) \tag{5.10}$$

where

$$\hat{\mathbf{B}} = t * \mathbf{B} + \mathbf{K}[\mathbf{S}_0 + t\,\dot{\mathbf{S}}_0] \tag{5.11}$$

and $*$ stands for the convolution product, that is, for any two scalar functions $a = a(\mathbf{x}, t)$ and $b = b(\mathbf{x}, t)$

$$(a * b)(\mathbf{x}, t) = \int_0^t a(\mathbf{x}, \tau) b(\mathbf{x}, t - \tau)\, d\tau \tag{5.12}$$

Next, the convolutional variational principle is formulated for the problem described by Eqs. (5.9)–(5.10).

Principle of Incompatible Elastodynamics. Let N denote the set of all symmetric second-order tensor fields $\mathbf{S}$ on $\overline{\mathrm{B}} \times [0, \infty)$ that satisfy the traction boundary condition (5.8) $\equiv$ (5.10). Let $\mathrm{I}_t\{\mathbf{S}\}$ be the functional on N defined by

$$\mathrm{I}_t\{\mathbf{S}\} = \frac{1}{2} \int_{\mathrm{B}} \{\rho^{-1} t * (\operatorname{div}\mathbf{S}) * (\operatorname{div}\mathbf{S}) + \mathbf{S} * \mathbf{K}[\mathbf{S}] - 2\,\mathbf{S} * \hat{\mathbf{B}}\}\, dv \tag{5.13}$$

Then

$$\delta\, \mathrm{I}_t\{\mathbf{S}\} = 0 \tag{5.14}$$

at a particular $\mathbf{S} \in \mathrm{N}$ if and only if $\mathbf{S}$ is a solution to the traction problem described by Eqs. (5.6)–(5.8).

Note. When the fields $\mathbf{B}$, $\mathbf{S}_0$, and $\dot{\mathbf{S}}_0$ are arbitrarily prescribed, the principle of incompatible elastodynamics may be useful in a study of elastic waves in bodies with various types of defects.

5.2 Problems and Solutions Related to Variational Principles of Elastodynamics

Problem 5.1. A symmetrical elastic beam of flexural rigidity EI, density ρ, and length L, is acted upon by: (i) the transverse force $F = F(x_1, t)$, (ii) the end shear forces V_0 and V_L, and (iii) the end bending moments M_0 and M_L shown in Fig. 5.1. The strain energy of the beam is

$$F(t) = \frac{1}{2} \int_0^L E\, I (u_2'')^2\, dx_1 \tag{5.15}$$

the kinetic energy of the beam is

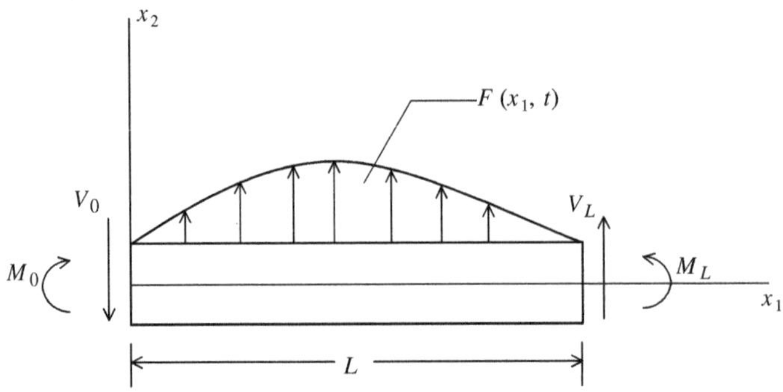

Fig. 5.1 The symmetrical beam

$$K(t) = \frac{1}{2} \int_0^L \rho \, (\dot{u}_2)^2 \, dx_1 \tag{5.16}$$

and the energy of external forces is

$$\begin{aligned} V(t) = &- \int_0^L F(x_1, t) \, u_2(x_1, t) \, dx_1 + V_0 u_2(0, t) \\ &+ M_0 u_2'(0, t) - V_L u_2(L, t) - M_L u_2'(L, t) \end{aligned} \tag{5.17}$$

where the prime denotes differentiation with respect to x_1. Let U be the set of functions $u_2 = u_2(x_1, t)$ that satisfies the conditions

$$u_2(x_1, t_1) = u(x_1), \quad u_2(x_1, t_2) = v(x_1) \tag{5.18}$$

where t_1 and t_2 are two arbitrary points on the t-axis such that $0 \le t_1 < t_2 < \infty$, and $u(x_1)$ and $v(x_1)$ are prescribed fields on $[0, L]$. Define a functional $\hat{K}\{u_2\}$ on U by

$$\hat{K}\{u_2\} = \int_{t_1}^{t_2} [F(t) + V(t) - K(t)] \, dt \tag{5.19}$$

Show that

$$\delta \hat{K}\{u_2\} = 0 \tag{5.20}$$

if and only if u_2 satisfies the equation of motion

$$(EI u_2'')'' + \rho \, \ddot{u}_2 = F \quad \text{on} \quad [0, L] \times [0, \infty) \tag{5.21}$$

and the boundary conditions

$$[(EI\,u_2'')'](0,t) = -V_0 \quad \text{on} \quad [0,\infty) \tag{5.22}$$
$$[(EI\,u_2'')](0,t) = M_0 \quad \text{on} \quad [0,\infty) \tag{5.23}$$
$$[(EI\,u_2'')'](L,t) = -V_L \quad \text{on} \quad [0,\infty) \tag{5.24}$$
$$[(EI\,u_2'')](L,t) = M_L \quad \text{on} \quad [0,\infty) \tag{5.25}$$

The field equaton (5.21) and the boundary conditions (5.22) through (5.25) describe flexural waves in the beam.

Solution. Introduce the notation

$$u_2(x_1,t) \equiv u(x,t) \tag{5.26}$$

Then the functional $\hat{K}\{u_2\}$ takes the form

$$\hat{K}\{u\} = \frac{1}{2}\int_{t_1}^{t_2} dt \int_0^L dx\,[EI(u'')^2 - \rho(\dot{u})^2]$$
$$+ \int_{t_1}^{t_2} \left\{ -\int_0^L Fu\,dx + V_0 u(0,t) + M_0 u'(0,t) - V_L u(L,t) - M_L u'(L,t) \right\} dt \tag{5.27}$$

Let $u \in U$ and $u + \omega\tilde{u} \in U$. Then

$$\tilde{u}(x,t_1) = \tilde{u}(x,t_2) = 0 \qquad x \in [0,L] \tag{5.28}$$

Computing $\delta\hat{K}\{u\}$ we obtain

$$\delta\hat{K}\{u\} = \frac{d}{d\omega}\,\hat{K}\{u+\omega\tilde{u}\}\bigg|_{\omega=0} = \int_{t_1}^{t_2} dt \int_0^L dx\,[EIu''\,\tilde{u}'' - \rho\,\dot{u}\,\dot{\tilde{u}}]$$
$$+ \int_{t_1}^{t_2} dt \left\{ -\int_0^L F\tilde{u}\,dx + V_0\tilde{u}(0,t) + M_0\tilde{u}'(0,t) \right.$$
$$\left. -V_L\tilde{u}(L,t) - M_L\tilde{u}'(L,t) \right\} \tag{5.29}$$

Next, note that integrating by parts we obtain

$$\int_0^L dx(EIu''\tilde{u}'') = (EIu'')\tilde{u}'\Big|_{x=0}^{x=L} - \int_0^L dx(EIu'')'\tilde{u}'$$

$$= (EIu'')\tilde{u}'\Big|_{x=0}^{x=L} - (EIu'')'\tilde{u}\Big|_{x=0}^{x=L} + \int_0^L dx(EIu'')''\tilde{u} \quad (5.30)$$

and

$$-\int_{t_1}^{t_2} \rho\dot{u}\dot{\tilde{u}}\,dt = -\rho\dot{u}\tilde{u}\Big|_{t=t_1}^{t=t_2} + \int_{t_1}^{t_2} \rho\ddot{u}\tilde{u}\,dt \quad (5.31)$$

Hence, using the homogeneous conditions (5.28) we reduce (5.29) to the form

$$\delta\hat{K}\{u\} = \int_{t_1}^{t_2} dt \int_0^L dx\,[(EIu'')'' + \rho\ddot{u} - F]\tilde{u}(x,t)$$
$$+ \int_{t_1}^{t_2} dt\{[V_0 + (EIu'')'(0,t)]\tilde{u}(0,t) - [V_L + (EIu'')'(L,t)]\tilde{u}(L,t)$$
$$+ [M_0 - (EIu'')(0,t)]\tilde{u}'(0,t) - [M_L - (EIu'')(L,t)]\tilde{u}'(L,t)\} \quad (5.32)$$

Now, if $u = u(x,t)$ satisfies (5.21)–(5.25) then $\delta\hat{K}\{u\} = 0$. Conversely, if $\delta\hat{K}\{u\} = 0$ then selecting $\tilde{u} = \tilde{u}(x,t)$ in such a way that $\tilde{u} = \tilde{u}(x,t)$ is an arbitrary smooth function on $[0, L] \times [t_1, t_2]$ and such that $\tilde{u}(0,t) = \tilde{u}(L,t) = 0$ on $[t_1, t_2]$ and $\tilde{u}'(0,t) = \tilde{u}'(L,t) = 0$ on $[t_1, t_2]$, from Eq. (5.32) we obtain

$$\int_{t_1}^{t_2}\int_0^L [(EIu'')'' + \rho\ddot{u} - F]\tilde{u}\,dt dx = 0 \quad (5.33)$$

and by the Fundamental Lemma of the calculus of variations we obtain

$$(EIu'')'' + \rho\ddot{u} = F \quad (5.34)$$

Next, by selecting $\tilde{u} = \tilde{u}(x,t)$ in such a way that $\tilde{u}$ is an arbitrary smooth function on $[0, L] \times [t_1, t_2]$ that complies with the conditions $\tilde{u}(0,t) \neq 0$ on $[t_1, t_2]$, $\tilde{u}(L,t) = 0$, $\tilde{u}'(0,t) = \tilde{u}'(L,t) = 0$ on $[t_1, t_2]$, and by using (5.32) and (5.34), we obtain

$$\int_{t_1}^{t_2} [V_0 + (EIu'')'(0,t)]\tilde{u}(0,t) = 0 \quad (5.35)$$

This together with the Fundamental Lemma of calculus of variations yields

$$(EIu'')'(0,t) = -V_0 \tag{5.36}$$

Next, by selecting $\tilde{u}$ to be an arbitrary smooth function on $[0, L]\times[t_1, t_2]$ that satisfies the conditions $\tilde{u}(L,t) \neq 0$ on $[t_1, t_2]$, $\tilde{u}\,'(0,t) = 0$, and $\tilde{u}\,'(L,t) = 0$ on $[t_1, t_2]$, we find from Eqs. (5.34), (5.36), and (5.32) that

$$\int_{t_1}^{t_2} [V_L + (EIu'')'(L,t)]\tilde{u}(L,t)dt = 0 \tag{5.37}$$

Equation (5.37) together with the Fundamental Lemma of calculus of variations imply that

$$(EIu'')'(L,t) = -V_L \tag{5.38}$$

Next, by selecting $\tilde{u}$ to be an arbitrary smooth function on $[0, L] \times [t_1, t_2]$ that meets the conditions $\tilde{u}\,'(0,t) \neq 0$ on $[t_1, t_2]$, and $\tilde{u}\,'(L,t) = 0$ on $[t_1, t_2]$, by virtue of Eqs. (5.34), (5.36), (5.38), and (5.32), we obtain

$$\int_{t_1}^{t_2} [M_0 - (EIu'')(0,t)]\tilde{u}\,'(0,t)dt = 0 \tag{5.39}$$

This together with the Fundamental Lemma of calculus of variations yields

$$(EIu'')(0,t) = M_0 \tag{5.40}$$

Finally, by letting $\tilde{u}$ to be an arbitrary smooth function on $[0, L] \times [t_1, t_2]$ and such that $\tilde{u}\,'(L,t) \neq 0$, from Eqs. (5.34), (5.36), (5.38), (5.40), and (5.32) we obtain

$$\int_{t_1}^{t_2} [M_L - (EIu'')(L,t)]\tilde{u}\,'(L,t) = 0 \tag{5.41}$$

Equation (5.41) together with the Fundamental Lemma of calculus of variations yields

$$(EIu'')(L,t) = M_L \tag{5.42}$$

This completes a solution to Problem 5.1.

Problem 5.2. A thin elastic membrane of uniform area density $\hat{\rho}$ is stretched to a uniform tension $\hat{T}$ over a region C_0 of the x_1, x_2 plane. The membrane is subject to a vertical load $f = f(\mathbf{x}, t)$ on $C_0 \times [0, \infty)$ and the initial conditions

$$u(\mathbf{x}, 0) = u_0(\mathbf{x}), \quad \dot{u}(\mathbf{x}, 0) = \dot{u}_0(\mathbf{x}) \quad \text{for} \quad \mathbf{x} \in C_0$$

where $u = u(\mathbf{x}, t)$ is a vertical deflection of the membrane on $\overline{C}_0 \times [0, \infty)$, and $u_0(\mathbf{x})$ and $\dot{u}_0(\mathbf{x})$ are prescribed functions on C_0. Also, $u = u(\mathbf{x}, t)$ on $\partial C_0 \times [0, \infty)$ is represented by a given function $g = g(\mathbf{x}, t)$. The strain energy of the membrane is

$$F(t) = \frac{\hat{T}}{2} \int_{C_0} u_{,\alpha} u_{,\alpha} \, da \tag{5.43}$$

The kinetic energy of the membrane is

$$K(t) = \frac{\hat{\rho}}{2} \int_{C_0} (\dot{u})^2 \, da \tag{5.44}$$

The external load energy is

$$V(t) = -\int_{C_0} f \, u \, da \tag{5.45}$$

Let U be the set of functions $u = u(\mathbf{x}, t)$ on $C_0 \times [0, \infty)$ that satisfy the conditions

$$u(\mathbf{x}, t_1) = a(\mathbf{x}), \quad u(\mathbf{x}, t_2) = b(\mathbf{x}) \quad \text{for} \quad \mathbf{x} \in C_0 \tag{5.46}$$

and

$$u(\mathbf{x}, t) = g(\mathbf{x}, t) \quad \text{on} \quad \partial C_0 \times [0, \infty) \tag{5.47}$$

where t_1 and t_2 have the same meaning as in Problem 5.1, and $a(\mathbf{x})$ and $b(\mathbf{x})$ are prescribed functions on C_0. Define a functional $\hat{K}\{.\}$ on U by

$$\hat{K}\{u\} = \int_{t_1}^{t_2} [F(t) + V(t) - K(t)] \, dt \tag{5.48}$$

Show that the condition

$$\delta \hat{K}\{u\} = 0 \quad \text{on} \quad U \tag{5.49}$$

implies the wave equation

$$\left(\nabla^2 - \frac{1}{c^2}\frac{\partial^2}{\partial t^2}\right) u = -\frac{f}{\hat{T}} \quad \text{on} \quad C_0 \times [0, \infty) \tag{5.50}$$

where

$$c = \sqrt{\frac{\hat{T}}{\hat{\rho}}} \tag{5.51}$$

Note that $[\hat{T}] = [\text{Force} \times L^{-1}]$, $[\hat{\rho}] = [\text{Density} \times L]$, $[c] = [LT^{-1}]$, where L and T are the length and time units, respectively.

Solution. The functional $\hat{K} = \hat{K}\{u\}$ takes the form

$$\hat{K}\{u\} = \int_{t_1}^{t_2} dt \int_{C_0} \left(\frac{\widehat{T}}{2} u_{,\alpha}\, u_{,\alpha} - \frac{\hat{\rho}}{2} \dot{u}^2 - f u \right) da \quad \text{for every} \quad u \in U \tag{5.52}$$

Let $u \in U$ and $u + \omega\tilde{u} \in U$. Then

$$\tilde{u}(\mathbf{x}, t_1) = \tilde{u}(\mathbf{x}, t_2) = 0 \quad \text{for } \mathbf{x} \in C_0 \tag{5.53}$$

and

$$\tilde{u}(\mathbf{x}, t) = 0 \quad \text{on } \partial C_0 \times [0, \infty) \tag{5.54}$$

Computing $\delta \widehat{K}\{u\}$ we obtain

$$\begin{aligned} \delta \widehat{K}\{u\} &= \frac{d}{d\omega} \widehat{K}\{u + \omega\tilde{u}\}|_{\omega=0} \\ &= \int_{t_1}^{t_2} dt \int_{C_0} (\widehat{T} u_{,\alpha}\, \tilde{u}_{,\alpha} - \hat{\rho} \dot{u} \dot{\tilde{u}} - f \tilde{u}) da \end{aligned} \tag{5.55}$$

Since

$$u_{,\alpha}\, \tilde{u}_{,\alpha} = (u_{,\alpha}\, \tilde{u})_{,\alpha} - u_{,\alpha\alpha}\, \tilde{u} \tag{5.56}$$

and

$$\dot{u}\dot{\tilde{u}} = (\dot{u}\tilde{u})^{\cdot} - \ddot{u}\tilde{u} \tag{5.57}$$

therefore, using the divergence theorem and the homogeneous conditions (5.53) and (5.54), we reduce (5.55) into the form

$$\delta \widehat{K}\{u\} = \int_{t_1}^{t_2} dt \int_{C_0} (-\widehat{T} u_{,\alpha\alpha} + \hat{\rho} \ddot{u} - f) \tilde{u}\, da \tag{5.58}$$

Hence, the condition

$$\delta \widehat{K}\{u\} = 0 \quad \text{on } U \tag{5.59}$$

together with the Fundamental Lemma of calculus of variations imply Eq. (5.50). This completes a solution to Problem 5.2.

Problem 5.3. Transverse waves propagating in a thin elastic membrane are described by the field equation (see Problem 5.2.)

$$\left(\nabla^2 - \frac{1}{c^2}\frac{\partial^2}{\partial t^2}\right) u = -\frac{f}{\hat{T}} \quad \text{on} \quad C_0 \times [0, \infty) \tag{5.60}$$

the initial conditions

$$u(\mathbf{x}, 0) = u_0(\mathbf{x}), \quad \dot{u}(\mathbf{x}, 0) = \dot{u}_0(\mathbf{x}) \quad \text{for} \quad \mathbf{x} \in C_0 \tag{5.61}$$

and the boundary condition

$$u(\mathbf{x}, t) = g(\mathbf{x}, t) \quad \text{on} \quad \partial C_0 \times [0, \infty) \tag{5.62}$$

Let $\hat{U}$ be a set of functions $u = u(\mathbf{x}, t)$ on $C_0 \times [0, \infty)$ that satisfy the boundary condition (5.62). Define a functional $\mathcal{F}_t\{.\}$ on $\hat{U}$ in such a way that

$$\delta F_t\{u\} = 0 \tag{5.63}$$

if and only if $u = u(\mathbf{x}, t)$ is a solution to the initial-boundary value problem (5.60) through (5.62).

Solution. By transforming the initial-boundary value problem (5.60)–(5.62) to an equivalent integro-differential boundary-value problem in a way similar to that of the Principle of Incompatible Elastodynamics [see Eqs. (5.6)–(5.12)] we find that the functional $\mathcal{F}_t\{u\}$ on $\hat{U}$ takes the form

$$\mathcal{F}_t\{u\} = \frac{1}{2}\int_{C_0} (i * u_{,\alpha} * u_{,\alpha} + \frac{1}{c^2} u * u - 2g * u) da \tag{5.64}$$

where

$$i = i(t) = t \tag{5.65}$$

and

$$g = i * \frac{f}{\hat{T}} + \frac{1}{c^2}(u_0 + t\dot{u}_0) \tag{5.66}$$

The associated variational principle reads:

$$\delta\mathcal{F}_t\{u\} = 0 \quad \text{on} \quad \hat{U} \tag{5.67}$$

if and only if u is a solution to the initial-boundary value problem (5.60)–(5.62). This completes a solution to Problem 5.3.

Problem 5.4. A homogeneous isotropic thin elastic plate defined over a region C_0 of the x_1, x_2 plane, and clamped on its boundary ∂C_0, is subject to a transverse load $p = p(\mathbf{x}, t)$ on $C_0 \times [0, \infty)$. The strain energy of the plate is

$$F(t) = \frac{D}{2} \int_{C_0} (\nabla^2 w)^2 \, da \tag{5.68}$$

The kinetic energy of the plate is

$$K(t) = \frac{\hat{\rho}}{2} \int_{C_0} (\dot{w})^2 \, da \tag{5.69}$$

The external energy is

$$V(t) = -\int_{C_0} p \, w \, da \tag{5.70}$$

Here, $w = w(\mathbf{x}, t)$ is a transverse deflection of the plate on $C_0 \times [0, \infty)$, D is the bending rigidity of the plate ($[D] = [\text{Force} \times \text{Length}]$), and $\hat{\rho}$ is the area density of the plate ($[\hat{\rho}] = [\text{Density} \times \text{Length}]$).

Let W be the set of functions $w = w(\mathbf{x}, t)$ on $C_0 \times [0, \infty)$ that satisfy the conditions

$$w(\mathbf{x}, t_1) = a(\mathbf{x}), \quad w(\mathbf{x}, t_2) = b(\mathbf{x}) \quad \text{for} \quad \mathbf{x} \in C_0 \tag{5.71}$$

and

$$w = 0, \quad \frac{\partial w}{\partial n} = 0 \quad \text{on} \quad \partial C_0 \times [0, \infty) \tag{5.72}$$

where t_1, t_2, $a(\mathbf{x})$ and $b(\mathbf{x})$ have the same meaning as in Problem 5.2, and $\partial/\partial n$ is the normal derivative on ∂C_0. Define a functional $\hat{K}\{.\}$ on W by

$$\hat{K}\{w\} = \int_{t_1}^{t_2} [F(t) + V(t) - K(t)] \, dt \tag{5.73}$$

Show that

$$\delta \hat{K}\{w\} = 0 \quad \text{on} \quad W \tag{5.74}$$

if and only if $w = w(\mathbf{x}, t)$ satisfies the differential equation

$$\nabla^2 \nabla^2 w + \frac{\hat{\rho}}{D} \frac{\partial^2 w}{\partial t^2} = \frac{p}{D} \quad \text{on} \quad C_0 \times [0, \infty) \tag{5.75}$$

and the boundary conditions

$$w = 0, \quad \frac{\partial w}{\partial n} = 0 \quad \text{on} \quad \partial C_0 \times [0, \infty) \tag{5.76}$$

Solution. The functional $\hat{K} = \hat{K}\{w\}$ on W takes the form

$$\widehat{K}\{w\} = \frac{1}{2}\int_{t_1}^{t_2} dt \int_{C_0} [D(\nabla^2 w)^2 - \hat{\rho}\dot{w}^2 - 2pw]da \tag{5.77}$$

Let $w \in W$, $w + \omega\tilde{w} \in W$. Then

$$\tilde{w}(\mathbf{x}, t_1) = \tilde{w}(\mathbf{x}, t_2) = 0 \quad \text{for } \mathbf{x} \in C_0 \tag{5.78}$$

and

$$\tilde{w} = 0, \quad \frac{\partial \tilde{w}}{\partial n} = 0 \quad \text{on } \partial C_0 \times [0, \infty) \tag{5.79}$$

Hence, we obtain

$$\begin{aligned} \delta\widehat{K}\{w\} &= \frac{d}{d\omega}\widehat{K}\{w + \omega\tilde{w}\}|_{\omega=0} \\ &= \int_{t_1}^{t_2} dt \int_{C_0} [D(\nabla^2 w)(\nabla^2 \tilde{w}) - \hat{\rho}\dot{w}\dot{\tilde{w}} - p\tilde{w}]da \end{aligned} \tag{5.80}$$

Since

$$\begin{aligned} &(\nabla^2 w)(\nabla^2 \tilde{w}) = w_{,\alpha\alpha}\, \tilde{w}_{,\beta\beta} = (w_{,\alpha\alpha}\, \tilde{w}_{,\beta})_{,\beta} \\ &-w_{,\alpha\alpha\beta}\, \tilde{w}_{,\beta} = (w_{,\alpha\alpha}\, \tilde{w}_{,\beta} - w_{,\alpha\alpha\beta}\, \tilde{w})_{,\beta} + w_{,\alpha\alpha\beta\beta}\, \tilde{w} \end{aligned} \tag{5.81}$$

and

$$\dot{w}\dot{\tilde{w}} = (\dot{w}\tilde{w})^{\cdot} - \ddot{w}\tilde{w} \tag{5.82}$$

therefore, using the divergence theorem as well as the homogeneous conditions (5.78) and (5.79), we reduce (5.80) to the form

$$\delta\widehat{K}\{w\} = \int_{t_1}^{t_2} dt \int_{C_0} (D\nabla^4 w + \hat{\rho}\ddot{w} - p)\tilde{w}da \tag{5.83}$$

Hence, by virtue of the Fundamental Lemma of calculus of variations

$$\delta \widehat{K}\{w\} = 0 \quad \text{on } W \tag{5.84}$$

if and only if w satisfies the differential equation

$$\nabla^2 \nabla^2 w + \frac{\hat{\rho}}{D} \frac{\partial^2 w}{\partial t^2} = \frac{p}{D} \quad \text{on } C_0 \times [0, \infty) \tag{5.85}$$

and the boundary conditions

$$w = \frac{\partial w}{\partial n} = 0 \quad \text{on } \partial C_0 \times [0, \infty) \tag{5.86}$$

This completes a solution to Problem 5.4.

Problem 5.5. Transverse waves propagating in a clamped thin elastic plate are described by the equations (see Problem 5.4)

$$\nabla^2 \nabla^2 w + \frac{\hat{\rho}}{D} \frac{\partial^2 w}{\partial t^2} = \frac{p}{D} \quad \text{on} \quad C_0 \times [0, \infty) \tag{5.87}$$

$$w(\mathbf{x}, 0) = w_0(\mathbf{x}), \quad \dot{w}(\mathbf{x}, 0) = \dot{w}_0(\mathbf{x}) \quad \text{for} \quad \mathbf{x} \in C_0 \tag{5.88}$$

and

$$w = 0, \quad \frac{\partial w}{\partial n} = 0 \quad \text{on} \quad \partial C_0 \times [0, \infty) \tag{5.89}$$

where $w_0(\mathbf{x})$ and $\dot{w}_0(\mathbf{x})$ are prescribed functions on C_0. Let W^* denote the set of functions $w = w(\mathbf{x}, t)$ that satisfy the homogeneous boundary conditions (5.89). Find a functional $\hat{\mathcal{F}}_t\{.\}$ on W^* with the property that

$$\delta \, \hat{\mathcal{F}}_t\{w\} = 0 \quad \text{on} \quad W^* \tag{5.90}$$

if and only if w is a solution to the initial-boundary value problem (5.87) through (5.89).

Solution. First, we note that the initial-boundary value problem (5.87)–(5.89) is equivalent to the following boundary-value problem. Find $w = w(\mathbf{x}, t)$ on $C_0 \times [0, \infty)$ that satisfies the integro-differential equation.

$$i * \nabla^4 w + \frac{1}{c^2} w = h \quad \text{on} \quad C_0 \times [0, \infty) \tag{5.91}$$

subject to the boundary conditions

$$w = \frac{\partial w}{\partial n} = 0 \quad \text{on} \quad \partial C_0 \times [0, \infty) \tag{5.92}$$

Here,

$$i = i(t) = t, \quad h(\mathbf{x}, t) = i * \frac{p}{D} + \frac{1}{c^2}(w_0 + t\dot{w}_0),$$
$$\text{and} \quad \frac{1}{c^2} = \frac{\hat{\rho}}{D} \tag{5.93}$$

Next, we define a functional $\hat{\mathcal{F}}_t\{w\}$ on W^* by

$$\hat{\mathcal{F}}_t\{w\} = \frac{1}{2}\int_{C_0}\left(i * \nabla^2 w * \nabla^2 w + \frac{1}{c^2} w * w - 2h * w\right) da \tag{5.94}$$

By computing $\delta\hat{\mathcal{F}}_t\{w\}$, we obtain

$$\delta\hat{\mathcal{F}}_t\{w\} = \int_{C_0}\left(i * \nabla^4 w + \frac{1}{c^2} w - h\right) * \tilde{w} da \tag{5.95}$$

where $\tilde{w}$ is an arbitrary smooth function on C_0 such that

$$\tilde{w} = \frac{\partial \tilde{w}}{\partial n} = 0 \quad \text{on } \partial C_0 \times [0, \infty) \tag{5.96}$$

Therefore, using the Fundamental Lemma of calculus of variations, it follows from Eq. (5.95) that the condition

$$\delta\hat{\mathcal{F}}_t\{w\} = 0 \quad \text{on } W^* \tag{5.97}$$

holds true if and only if w is a solution to the initial-boundary value problem (5.87)–(5.89). This completes a solution to Problem 5.5.

Problem 5.6. Free longitudinal vibrations of an elastic bar are defined as solutions of the form

$$u(x, t) = \phi(x) \sin(\omega t + \gamma) \tag{5.98}$$

to the homogeneous wave equation

$$\frac{\partial}{\partial x}\left(E\frac{\partial u}{\partial x}\right) - \rho\frac{\partial^2 u}{\partial t^2} = 0 \quad \text{on} \quad [0, L] \times [0, \infty) \tag{5.99}$$

subject to the homogeneous boundary conditions

$$u(0, t) = u(L, t) = 0 \quad \text{on} \quad [0, \infty) \tag{5.100}$$

or

$$\frac{\partial u}{\partial x}(0, t) = \frac{\partial u}{\partial x}(L, t) = 0 \quad \text{on} \quad [0, \infty) \tag{5.101}$$

Here, ω is a circular frequency of vibrations, γ is a dimensionless constant, and $\phi = \phi(x)$ is an unknown function that complies with Eqs. (5.99) and (5.100), or Eqs. (5.99) and (5.101). Substituting $u = u(\mathbf{x}, t)$ from Eq. (5.98) into (5.99) through (5.101) we obtain

$$\frac{d}{dx}\left(E\frac{d\phi}{dx}\right) + \lambda\phi = 0 \quad \text{on} \quad [0, L] \tag{5.102}$$

$$\phi(0) = \phi(L) = 0 \tag{5.103}$$

or

$$\phi'(0) = \phi'(L) = 0 \tag{5.104}$$

where the prime stands for derivative with respect to x, and

$$\lambda = \rho\omega^2 \tag{5.105}$$

Therefore, introduction of (5.98) into (5.99) through (5.101) results in an eigenproblem in which an eigenfunction $\phi = \phi(x)$ corresponding to an eigenvalue λ is to be found. An eigenproblem that covers both boundary conditions (5.100) and (5.101) can be written as

$$\frac{d}{dx}\left(E\frac{d\phi}{dx}\right) + \lambda\phi = 0 \quad \text{on} \quad [0, L] \tag{5.106}$$

$$\phi'(0) - \alpha\,\phi(0) = 0, \quad \phi'(L) + \beta\,\phi(L) = 0 \tag{5.107}$$

where $|\alpha| + |\beta| > 0$. Let U be the set of functions $\phi = \phi(x)$ on $[0, L]$ that satisfy the boundary conditions (5.107). Define a functional $\pi\{.\}$ on U by

$$\pi\{\phi\} = \frac{1}{2}\int_0^L \left[E\left(\frac{d\phi}{dx}\right)^2 - \lambda\phi^2\right] dx + \frac{1}{2}\alpha\, E(0)\, [\phi(0)]^2 + \frac{1}{2}\beta\, E(L)\, [\phi(L)]^2 \tag{5.108}$$

Show that

$$\delta\,\pi\{\phi\} = 0 \quad \text{over} \quad U \tag{5.109}$$

if and only if $\phi = \phi(x)$ is an eigenfunction corresponding to an eigenvalue λ in the eigenproblem (5.106) and (5.107).

Solution. Let $\phi \in U$ and $\phi + \omega\,\tilde{\phi} \in U$. Then

$$\tilde{\phi}'(0) - \alpha\tilde{\phi}(0) = 0, \quad \tilde{\phi}'(L) + \beta\tilde{\phi}(L) = 0 \tag{5.110}$$

and

$$\pi\{\phi+\omega\tilde{\phi}\}=\frac{1}{2}\int_0^L[E(\phi'+\omega\tilde{\phi}')^2-\lambda(\phi+\omega\tilde{\phi})^2]dx$$
$$+\frac{1}{2}\alpha E(0)[\phi(0)+\omega\tilde{\phi}(0)]^2+\frac{1}{2}\beta E(L)[\phi(L)+\omega\tilde{\phi}(L)]^2 \tag{5.111}$$

Hence, we obtain

$$\delta\pi\{\phi\}=\frac{d}{d\omega}\pi\{\phi+\omega\tilde{\phi}\}\bigg|_{\omega=0}$$
$$=\int_0^L[E\phi'\tilde{\phi}'-\lambda\phi\tilde{\phi}]\,dx+\alpha E(0)\,\phi(0)\,\tilde{\phi}(0)+\beta E(L)\,\phi(L)\,\tilde{\phi}(L) \tag{5.112}$$

Since

$$\int_0^L E\phi'\tilde{\phi}'dx=E\phi'\tilde{\phi}\bigg|_{x=0}^{x=L}-\int_0^L(E\phi')'\tilde{\phi}dx \tag{5.113}$$

therefore, Eq. (5.112) takes the form

$$\delta\pi\{\phi\}=-\int_0^L[(E\phi')'+\lambda\phi]\tilde{\phi}dx-E(0)[\phi'(0)-\alpha\phi(0)]\,\tilde{\phi}(0)$$
$$+E(L)[\phi'(L)+\beta\phi(L)]\tilde{\phi}(L) \tag{5.114}$$

Now, if $\phi=\phi(x)$ is an eigenfunction corresponding to an eigenvalue λ in the problem (5.106)–(5.107), then by virtue of (5.114) $\delta\pi\{\phi\}=0$ over U. Conversely, if $\delta\pi\{\phi\}=0$ then selecting $\tilde{\phi}=\tilde{\phi}(x)$ to be a smooth function on $[0, L]$ such that $\tilde{\phi}(0)=\tilde{\phi}(L)=0$, and using the Fundamental Lemma of calculus of variations, we obtain

$$(E\phi')'+\lambda\phi=0 \quad \text{on} \quad [0, L] \tag{5.115}$$

Next, if $\delta\pi\{\phi\}=0$ then selecting $\tilde{\phi}=\tilde{\phi}(x)$ to be a smooth function on $[0, L]$ and such that $\tilde{\phi}(L)=0$, and $\tilde{\phi}(0)\neq 0$, by virtue of (5.115), we obtain

$$E(0)[\phi'(0)-\alpha\phi(0)]\tilde{\phi}(0)=0 \tag{5.116}$$

Since

$$E(0)>0 \tag{5.117}$$

Equation (5.116) implies that $\phi = \phi(x)$ satisfies the boundary condition

$$\phi'(0) - \alpha\phi(0) = 0 \tag{5.118}$$

Finally, if $\delta\pi\{\phi\} = 0$ then selecting $\tilde{\phi}$ to be a smooth function on $[0, L]$ and such that $\tilde{\phi}(L) \neq 0$, by virtue of (5.115) and (5.118), we obtain

$$E(L)[\phi'(L) + \beta\phi(L)]\tilde{\phi}(L) = 0 \tag{5.119}$$

Since $E(L) > 0$, Eq. (5.119) implies that

$$\phi'(L) + \beta\phi(L) = 0 \tag{5.120}$$

This shows that if Eq. (5.110) holds true then (ϕ, λ) is an eigenpair for the problem (5.106)–(5.107). This completes a solution to Problem 5.6.

Problem 5.7. Free lateral vibrations of an elastic bar clamped at the end $x = 0$ and supported by a spring of stiffness k at the end $x = L$ are defined as solutions of the form

$$u(x, t) = \phi(x)\,\sin(\omega t + \gamma) \tag{5.121}$$

to the equation [see Problem 5.1, Eq. (5.127) in which $u_2 = u$, and $F = 0$]

$$\frac{\partial^2}{\partial x^2}\left(EI\frac{\partial^2 u}{\partial x^2}\right) + \rho\frac{\partial^2 u}{\partial t^2} = 0 \quad \text{on} \quad [0, L] \times [0, \infty) \tag{5.122}$$

subject to the boundary conditions

$$u(0, t) = u'(0, t) = 0 \quad \text{on} \quad [0, \infty) \tag{5.123}$$

$$u''(L, t) = 0, \quad (EI\,u'')'(L, t) - k\,u(L, t) = 0 \quad \text{on} \quad [0, \infty) \tag{5.124}$$

Let $\rho = \text{const}$, and $\lambda = \rho\,\omega^2$. Then the associated eigenproblem reads

$$(EI\,\phi'')'' - \lambda\phi = 0 \quad \text{on} \quad [0, L] \tag{5.125}$$

$$\phi(0) = \phi'(0) = 0 \tag{5.126}$$

$$\phi''(L) = 0, \quad (EI\phi'')'(L) - k\phi(L) = 0 \tag{5.127}$$

Let V denote the set of functions $\phi = \phi(x)$ on $[0, L]$ that satisfy the boundary conditions (5.126) and (5.127). Define a functional $\pi\{.\}$ on V by

$$\pi\{\phi\} = \frac{1}{2}\int_0^L EI\,(\phi'')^2 dx + \frac{1}{2}k[\phi(L)]^2 - \frac{\lambda}{2}\int_0^L \phi^2 dx \tag{5.128}$$

Show that

$$\delta\,\pi\{\phi\} = 0 \quad \text{over} \quad V \tag{5.129}$$

if and only if (λ, ϕ) is a solution to the eigenproblem (5.125) through (5.127).

Solution. Let $\phi \in V$ and $\phi + \omega\tilde{\phi} \in V$. Then

$$\tilde{\phi}(0) = 0, \quad \tilde{\phi}'(0) = 0 \tag{5.130}$$

Computing the first variation of the functional $\pi\{\phi\}$ given by (5.128), we obtain

$$\begin{aligned}\delta\pi\{\phi\} &= \frac{d}{d\omega}\pi\ \{\phi + \omega\tilde{\phi}\}\Big|_{\omega=0} \\ &= \int_0^L (EI\,\phi''\tilde{\phi}'' - \lambda\phi\tilde{\phi})dx + k\phi(L)\,\tilde{\phi}(L)\end{aligned} \tag{5.131}$$

Since

$$\int_0^L EI\,\phi''\tilde{\phi}''dx = (EI\,\phi'')\tilde{\phi}'\Big|_{x=0}^{x=L} - (EI\,\phi'')'\tilde{\phi}\Big|_{x=0}^{x=L} + \int_0^L (EI\,\phi'')''\,\tilde{\phi}dx \tag{5.132}$$

therefore, using (5.130) we reduce (5.131) into the form

$$\delta\pi\{\phi\} = \int_0^L [(EI\,\phi'')''-\lambda\phi]\tilde{\phi}dx + (EI\,\phi'')(L)\tilde{\phi}'(L) - [(EI\,\phi'')'(L) - k\phi(L)]\tilde{\phi}(L) \tag{5.133}$$

Now, if (λ, ϕ) is a solution to the eigenproblem (5.125)–(5.127), then $\delta\pi\{\phi\} = 0$. Conversely, if $\delta\pi\{\phi\} = 0$ over V, then selecting $\tilde{\phi}$ to be an arbitrary smooth function on $[0, L]$ such that $\tilde{\phi}(x) \not\equiv 0$ for $x \in (0, L)$, $\tilde{\phi}'(L) = 0$, $\tilde{\phi}(L) = 0$, we obtain

$$\int_0^L [(EI\,\phi'')'' - \lambda\phi]\,\tilde{\phi}\,dx = 0 \tag{5.134}$$

Equation (5.134) together with the Fundamental Lemma of calculus of variations implies

$$(EI\,\phi'')'' - \lambda\phi = 0 \quad \text{on} \quad [0, L] \tag{5.135}$$

Next, by selecting $\tilde{\phi}$ on $[0, L]$ in such a way that

$$\tilde{\phi}'(L) \neq 0, \quad \tilde{\phi}(L) = 0 \tag{5.136}$$

we find that the condition $\delta\pi\{\phi\} = 0$ and Eq. (5.135) imply that

$$(EI\,\phi'')(L) = 0 \tag{5.137}$$

Since

$$E(L) > 0, \quad I(L) > 0 \tag{5.138}$$

we obtain

$$\phi''(L) = 0 \tag{5.139}$$

Finally, by selecting $\tilde{\phi}$ on $[0, L]$ in such a way that

$$\tilde{\phi}(L) \neq 0 \tag{5.140}$$

we conclude that the condition $\delta\pi\{\phi\} = 0$ together with Eqs. (5.135), and (5.139) lead to the boundary condition

$$(EI\,\phi'')'(L) - k\,\phi(L) = 0 \tag{5.141}$$

This completes a solution to Problem 5.7.

Problem 5.8. Show that the eigenvalues $\lambda_{\rm i}$ and the eigenfunctions $\phi_i = \phi_i(x)$ for the longitudinal vibrations of a uniform elastic bar having one end clamped and the other end free are given by the relations

$$\omega_{\rm i} = \sqrt{\frac{\lambda_{\rm i}}{\rho}} = \frac{(2i-1)}{2L}\sqrt{\frac{E}{\rho}}$$

$$\phi_{\rm i}(x) = \sin\frac{(2i-1)\pi x}{2L}, \quad i = 1, 2, 3, \ldots, 0 \leq x \leq L$$

(see Problem 5.6).

Solution. For an elastic bar that is clamped at $x = 0$ and free at $x = L$ the eigenproblem reads

$$E\phi''(x) + \lambda\phi(x) = 0 \qquad x \in [0, L] \tag{5.142}$$

$$\phi(0) = 0, \quad \phi'(L) = 0 \tag{5.143}$$

where

$$\lambda = \omega^2\rho \tag{5.144}$$

There is an infinite sequence of eigensolutions (λ_i, ϕ_i) to the problem (5.142)–(5.143) of the form

$$\lambda_i = \frac{(2i-1)^2\pi^2}{4L^2}E \tag{5.145}$$

$$\phi_i(x) = \sin\frac{(2i-1)\pi x}{2L}, \quad i = 1, 2, 3, \ldots \tag{5.146}$$

This can be shown by substituting (5.145) and (5.146) into (5.142), and by showing that $\phi_i(x)$ satisfies (5.143). By combining (5.144) and (5.145) we obtain

$$\omega_i \equiv \sqrt{\frac{\lambda_i}{\rho}} = \frac{(2i-1)\pi}{2L}\sqrt{\frac{E}{\rho}} \tag{5.147}$$

This completes a solution to Problem 5.8.

Problem 5.9. Show that the eigenvalues λ_i and the eigenfunctions $\phi_i = \phi_i(x)$ for the lateral vibrations of a uniform, simply supported elastic beam are given by the relations

$$\omega_i = \sqrt{\frac{\lambda_i}{\rho}} = \frac{\pi^2 i^2}{L^2}\sqrt{\frac{EI}{\rho}}$$

$$\phi_i(x) = \sin\frac{i\pi x}{L}, \quad i = 1, 2, 3\ldots, 0 \le x \le L$$

(see Problem 5.1).

Solution. For a uniform, simply supported beam with the lateral vibrations, the eigenproblem takes the form

$$EI\phi^{(4)} - \lambda\phi = 0 \quad \text{on} \quad [0, L] \tag{5.148}$$

$$\phi(0) = \phi''(0) = 0, \quad \phi(L) = \phi''(L) = 0 \tag{5.149}$$

where

$$\lambda = \omega^2\rho \tag{5.150}$$

There is an infinite sequence of eigensolutions (λ_i, ϕ_i) to the problem (5.148)–(5.149) of the form

$$\lambda_i = EI\left(\frac{i\pi}{L}\right)^4 \tag{5.151}$$

$$\phi_i(x) = \sin\frac{i\pi x}{L} \quad i = 1, 2, 3, \ldots \tag{5.152}$$

To prove that (λ_i, ϕ_i) given by (5.151)–(5.152) satisfies Eqs. (5.148)–(5.149), we note that

$$\phi_i''(x) = -\left(\frac{i\pi}{L}\right)^2 \phi_i(x) \tag{5.153}$$

and

$$\phi_i^{(4)}(x) = \left(\frac{i\pi}{L}\right)^4 \phi_i(x) \tag{5.154}$$

Substituting (5.151) and (5.152) into (5.148) and using (5.154) we find that $\phi_i = \phi_i(x)$ satisfies Eq. (5.148) on $[0, L]$. Also, it follows from Eqs. (5.152) and (5.153) that the boundary conditions (5.149) are satisfied; and Eqs. (5.150) and (5.151) imply that

$$\omega_i = \sqrt{\frac{\lambda i}{\rho}} = \frac{\pi^2 i^2}{L^2}\sqrt{\frac{EI}{\rho}} \tag{5.155}$$

These steps complete a solution to Problem 5.9.

Problem 5.10. Show that the eigenvalues λ_{mn} and the eigenfunctions $\phi_{mn} = \phi_{mn}(x)$ for the transversal vibrations of a rectangular elastic membrane: $0 \le x_1 \le a_1,\ 0 \le x_2 \le a_2$, that is clamped on its boundary, are given by

$$\omega_{mn} = \sqrt{\frac{\lambda_{mn}}{\hat{\rho}}} = \pi\sqrt{\frac{\hat{T}}{\hat{\rho}}\left(\frac{m^2}{a_1^2} + \frac{n^2}{a_2^2}\right)}$$

$$\phi_{mn}(x_1, x_2) = \sin\frac{m\pi x_1}{a_1}\sin\frac{n\pi x_2}{a_2},$$
$$m, n = 1, 2, 3, \ldots, 0 \le x_1 \le a_1,\ 0 \le x_2 \le a_2$$

(See Problem 5.2).

Solution. Let C_0 denote the rectangular region

$$0 < x_1 < a_1, \quad 0 < x_2 < a_2 \tag{5.156}$$

and let ∂C_0 be its boundary. Then the associated eigenproblem reads. Find an eigenpair (λ, ϕ) such that

$$\hat{T}\,\nabla^2\phi + \lambda\phi = 0 \quad \text{on } C_0 \tag{5.157}$$

and

$$\phi = 0 \quad \text{on } \partial C_0 \tag{5.158}$$

where

$$\lambda = \omega^2 \hat{\rho} \tag{5.159}$$

There is an infinite number of eigenpairs $(\lambda_{mn}, \phi_{mn}),\ m, n = 1, 2, 3, \ldots$ that satisfy Eqs. (5.157) and (5.158), and they are given by Equation

$$\lambda_{mn} = \pi^2 \hat{T} \left(\frac{m^2}{a_1^2} + \frac{n^2}{a_2^2} \right) \tag{5.160}$$

$$\phi_{mn}(x_1, x_2) = \sin\left(\frac{m\pi x_1}{a_1}\right) \sin\left(\frac{n\pi x_2}{a_2}\right) \tag{5.161}$$

This can be proved by substituting (5.152) and (5.153) into (5.149) and (5.150).

Also, the eigenvalues λ_{mn} generate the eigenfrequencies ω_{mn} by the formulas

$$\omega_{mn} = \sqrt{\frac{\lambda_{mn}}{\hat{\rho}}} = \pi \sqrt{\frac{\hat{T}}{\hat{\rho}}} \sqrt{\left(\frac{m^2}{a_1^2}\right) + \left(\frac{n^2}{a_2^2}\right)} \tag{5.162}$$

This completes a solution to Problem 5.10.

Problem 5.11. Show that the eigenvalues λ_{mn} and the eigenfunctions $\phi_{mn} = \phi_{mn}(x_1, x_2)$ for the transversal vibrations of a thin elastic rectangular plate: $0 \le x_1 \le a_1,\ 0 \le x_2 \le a_2$, that is simply supported on its boundary are given by the relations

$$\omega_{mn} = \sqrt{\frac{\lambda_{mn}}{\hat{\rho}}} = \pi^2 \left(\frac{m^2}{a_1^2} + \frac{n^2}{a_2^2} \right) \sqrt{\frac{D}{\hat{\rho}}}$$

$$\phi_{mn}(x_1, x_2) = \sin \frac{m\,\pi\,x_1}{a_1} \sin \frac{n\,\pi\,x_2}{a_2},$$

$$m, n = 1, 2, 3, \ldots, 0 \le x_1 \le a_1,\ 0 \le x_2 \le a_2$$

(See Problem 5.4).

Solution. The eigenproblem associated with the transversal vibrations of a thin elastic rectangular plate that is simply supported on its boundary, reads [see Eq. (5.85) of Problem 5.4]

$$D\,\nabla^2 \nabla^2 \phi - \lambda \phi = 0 \quad \text{on } C_0 \tag{5.163}$$

$$\phi = \nabla^2 \phi = 0 \quad \text{on } \partial C_0 \tag{5.164}$$

where

$$\lambda = \omega^2 \hat{\rho} \tag{5.165}$$

and C_0 and ∂C_0 are the same as in Problem 5.10.

There are an infinite number of eigenpairs $(\lambda_{mn}, \phi_{mn})$ that satisfy Eqs. (5.163) and (5.164), and the eigenpairs are given by

$$\lambda_{mn} = \pi^4 D \left(\frac{m^2}{a_1^2} + \frac{n^2}{a_2^2} \right)^2 \tag{5.166}$$

$$\phi_{mn}(x_1, x_2) = \sin\left(\frac{m\pi x_1}{a_1}\right) \sin\left(\frac{n\pi x_2}{a_2}\right) \quad m, n = 1, 2, 3, \ldots \tag{5.167}$$

This is proved by substituting (5.166) and (5.167) into (5.163) and (5.164).

Also, by using (5.165) the eigenfrequencies ω_{mn} are obtained

$$\omega_{mn} = \sqrt{\frac{\lambda_{mn}}{\hat{\rho}}} = \pi^2 \left(\frac{m^2}{a_1^2} + \frac{n^2}{a_2^2} \right) \sqrt{\frac{D}{\hat{\rho}}} \tag{5.168}$$

This completes a solution to Problem 5.11.

Chapter 6
Complete Solutions of Elasticity

In this chapter general solutions of the homogeneous isotropic elastostatics and elastodynamics are discussed. The general solutions are related to both the displacement and stress governing equations, and emphasis is made on completeness of the solutions [See also Chap. 16].

6.1 Complete Solutions of Elastostatics

A vector field $\mathbf{u} = \mathbf{u}(\mathbf{x})$ on B that satisfies the displacement equation of equilibrium

$$\nabla^2 \mathbf{u} + \frac{1}{1-2\nu} \nabla(\text{div}\, \mathbf{u}) + \frac{\mathbf{b}}{\mu} = \mathbf{0} \tag{6.1}$$

is called *an elastic displacement field corresponding to* **b**.

Boussinesq-Papkovitch-Neuber (B-P-N) Solution. Let

$$\mathbf{u} = \psi - \frac{1}{4(1-\nu)} \nabla(\mathbf{x} \cdot \psi + \varphi) \tag{6.2}$$

where φ and ψ are fields on B that satisfy Poisson's equations

$$\nabla^2 \psi = -\frac{1}{\mu} \mathbf{b} \tag{6.3}$$

and

$$\nabla^2 \varphi = \frac{1}{\mu} \mathbf{b} \cdot \mathbf{x} \tag{6.4}$$

Then **u** is an elastic displacement field corresponding to **b**.

M. Reza Eslami et al., *Theory of Elasticity and Thermal Stresses*, Solid Mechanics and Its Applications 197, DOI: 10.1007/978-94-007-6356-2_6,

Boussinesq-Somigliana-Galerkin (B-S-G) Solution. Let $\mathbf{u}$ be a vector field given by

$$\mathbf{u} = \nabla^2\mathbf{g} - \frac{1}{2(1-\nu)}\nabla(\operatorname{div}\mathbf{g}) \tag{6.5}$$

where

$$\nabla^2\nabla^2\mathbf{g} = -\frac{1}{\mu}\mathbf{b} \tag{6.6}$$

Then $\mathbf{u}$ is an elastic displacement field corresponding to $\mathbf{b}$.

We say that a *representation for the displacement* $\mathbf{u}$ *expressed in terms of auxiliary functions is complete if these auxiliary functions exist for any* $\mathbf{u}$ *that satisfies the displacement equation of equilibrium* (6.1).

For B-P-N solution such auxiliary functions are the fields φ and ψ; while for B-S-G solution an auxiliary function is the field $\mathbf{g}$.

Completeness of B-P-N and B-S-G Solutions. Let $\mathbf{u}$ be a solution to the displacement equation of equilibrium with the body force $\mathbf{b}$. Then there exists a field $\mathbf{g}$ on B that satisfies Eqs. (6.5)–(6.6). Also, there exist fields φ and ψ that satisfy Eqs. (6.2)–(6.4).

B-P-N solution for axial symmetry. For an axially symmetric problem with $\mathbf{b} = \mathbf{0}$ in which $x_3 = z$ is the axis symmetry of a body, the displacement vector field $\mathbf{u} = \mathbf{u}(r, z)$ referred to the cylindrical coordinates (r, θ, z) takes the form

$$\mathbf{u} = \psi\mathbf{k} - \frac{1}{4(1-\nu)}\nabla(z\psi + \varphi) \tag{6.7}$$

where

$$z = \mathbf{x}\cdot\mathbf{k} \tag{6.8}$$

with $\mathbf{k}$ being a unit vector along the x_3 axis, and with scalar-valued harmonic functions $\varphi = \varphi(r, z)$ and $\psi = \psi(r, z)$. In components we obtain

$$\mathbf{u} = [u_r(r, z),\ 0,\ u_z(r, z)] \tag{6.9}$$

where

$$u_r = -\frac{1}{4(1-\nu)}\frac{\partial}{\partial r}(z\psi + \varphi) \tag{6.10}$$

$$u_z = \psi - \frac{1}{4(1-\nu)}\frac{\partial}{\partial z}(z\psi + \varphi) \tag{6.11}$$

and

$$\nabla^2\varphi = 0, \quad \nabla^2\psi = 0 \tag{6.12}$$

with

$$\nabla^2 = \frac{\partial^2}{\partial r^2} + \frac{1}{r}\frac{\partial}{\partial r} + \frac{\partial^2}{\partial z^2} \tag{6.13}$$

B-S-G solution for axial symmetry with $\mathbf{b} = \mathbf{0}$ and $\mathbf{g} = \chi\mathbf{k}$ is also called Love's solution.

$$\mathbf{u} = (\nabla^2\chi)\mathbf{k} - \frac{1}{2(1-\nu)}\nabla(\mathbf{k}\cdot\nabla\chi) \tag{6.14}$$

where

$$\nabla^2\nabla^2\chi = 0 \tag{6.15}$$

In cylindrical coordinates (r, θ, z)

$$u_r = -\frac{1}{2(1-\nu)}\frac{\partial^2}{\partial r\partial z}\chi \tag{6.16}$$

$$u_\theta = 0 \tag{6.17}$$

$$u_z = \frac{1}{2(1-\nu)}\left[2(1-\nu)\nabla^2 - \frac{\partial^2}{\partial z^2}\right]\chi \tag{6.18}$$

6.2 Complete Solutions of Elastodynamics

The displacement equation of motion for a homogeneous isotropic elastic body takes the form

$$\Box_2^2\mathbf{u} + \left[\left(\frac{c_1}{c_2}\right)^2 - 1\right]\nabla(\operatorname{div}\mathbf{u}) + \frac{\mathbf{b}}{\mu} = \mathbf{0} \tag{6.19}$$

where

$$\Box_2^2 = \nabla^2 - \frac{1}{c_2^2}\frac{\partial^2}{\partial t^2}, \quad \frac{1}{c_1^2} = \frac{\rho}{\lambda+2\mu}, \quad \frac{1}{c_2^2} = \frac{\rho}{\mu} \tag{6.20}$$

The body force $\mathbf{b}$ is represented by Helmholtz's decomposition formula

$$\mathbf{b} = -\nabla h - \operatorname{curl}\mathbf{k}, \quad \operatorname{div}\mathbf{k} = 0 \tag{6.21}$$

A solution $\mathbf{u}$ on $\overline{B}\times[0,\infty)$ to Eq. (6.19) will be called an *elastic motion corresponding to* $\mathbf{b}$.

Green-Lame (G-L) Solution. Let

$$\mathbf{u} = \nabla\varphi + \operatorname{curl}\psi \tag{6.22}$$

where φ and ψ satisfy, respectively, the equations

$$\Box_1^2 \varphi = \frac{h}{\lambda + 2\mu} \tag{6.23}$$

and

$$\Box_2^2 \psi = \frac{\mathbf{k}}{\mu} \tag{6.24}$$

Then **u** is an elastic motion corresponding to **b** given by Eq. (6.21). In Eq. (6.23)

$$\Box_1^2 = \nabla^2 - \frac{1}{c_1^2} \frac{\partial^2}{\partial t^2} \tag{6.25}$$

Cauchy-Kovalevski-Somigliana (C-K-S) Solution. Let

$$\mathbf{u} = \Box_1^2 \mathbf{g} + \left(\frac{c_2^2}{c_1^2} - 1 \right) \nabla (\text{div } \mathbf{g}) \tag{6.26}$$

where **g** satisfies the inhomogeneous biwave equation

$$\Box_1^2 \Box_2^2 \mathbf{g} = -\frac{\mathbf{b}}{\mu} \tag{6.27}$$

Then **u** is an elastic motion corresponding to **b**.

Note. Both G-L and C-K-S solutions are complete.

6.3 Complete Stress Solution of Elastodynamics

The stress equation of motion for a homogeneous isotropic elastic body takes the form

$$\widehat{\nabla}(\text{div } \mathbf{S}) - \frac{\rho}{2\mu} \left[\ddot{\mathbf{S}} - \frac{\nu}{1+\nu} (\text{tr } \ddot{\mathbf{S}}) \mathbf{1} \right] = -\widehat{\nabla} \mathbf{b} \tag{6.28}$$

A solution **S** on $\bar{\mathrm{B}} \times [0, \infty)$ to Eq. (6.28) will be called a *stress motion corresponding to* **b**.

Stress Solution of Galerkin Type. Let

$$\mathbf{S} = \left[\left(\nabla\nabla - \nu \mathbf{1} \Box_2^2 \right) \text{tr } \mathbf{G} - 2(1-\nu) \Box_1^2 \mathbf{G} \right] \tag{6.29}$$

where **G** is a symmetric second-order tensor field on $\bar{\mathrm{B}} \times [0, \infty)$ that satisfies the equations

$$\Box_1^2 \Box_2^2 \mathbf{G} = \frac{1}{1-\nu} \widehat{\nabla} \mathbf{b} \tag{6.30}$$

and

$$\nabla^2\mathbf{G} + \nabla\nabla(\text{tr}\,\mathbf{G}) - 2\widehat{\nabla}(\text{div}\,\mathbf{G}) = \mathbf{0} \tag{6.31}$$

Then **S** is a stress motion corresponding to **b**, that is, **S** satisfies Eq. (6.28).

Completeness of the Stress Solution of Galerkin Type.

The stress solution of Galerkin type corresponding to homogeneous initial conditions is complete, that is, for any stress motion **S** corresponding to **b** there exists a second-order symmetric tensor field **G** such that Eqs. (6.29)–(6.31) are satisfied.

6.4 Problems and Solutions Related to Complete Solutions of Elasticity

Problem 6.1. The displacement $\mathbf{u} = \mathbf{u}(\mathbf{x}, \xi)$ at a point $\mathbf{x}$ due to a concentrated force $\boldsymbol{l}$ applied at a point ξ of a homogeneous isotropic infinite elastic body is given by $(\mathbf{x} \neq \xi)$

$$\mathbf{u}(\mathbf{x}, \xi) = \mathbf{U}(\mathbf{x}, \xi)\boldsymbol{l}$$

where

$$\mathbf{U}(\mathbf{x}, \xi) = \frac{1}{16\pi\mu(1-\nu)}\frac{1}{\mathrm{R}}\left[(3-4\nu)\mathbf{1} + \frac{(\mathbf{x}-\xi)\otimes(\mathbf{x}-\xi)}{\mathrm{R}^2}\right]$$

with

$$\mathrm{R} = |\mathbf{x} - \xi|$$

Use the stress-displacement relation to show that the associated stress $\mathbf{S} = \mathbf{S}(\mathbf{x}, \xi)$ takes the form

$$\mathbf{S}(\mathbf{x}, \xi) = -\frac{1}{8\pi(1-\nu)}\frac{1}{\mathrm{R}^3}\left\{\frac{3}{\mathrm{R}^2}[(\mathbf{x}-\xi)\cdot\boldsymbol{l}](\mathbf{x}-\xi)\otimes(\mathbf{x}-\xi) \right.$$
$$\left. +(1-2\nu)\{(\mathbf{x}-\xi)\otimes\boldsymbol{l} + \boldsymbol{l}\otimes(\mathbf{x}-\xi) - [(\mathbf{x}-\xi)\cdot\boldsymbol{l}]\,\mathbf{1}\}\right\}$$

Solution. The displacement **u** in components takes the form

$$u_i = U_{ik}\,\ell_k \tag{6.32}$$

where

$$U_{ik} = \frac{A}{2\mu}R^{-1}\left[(3-4\nu)\delta_{ik} + (x_i - \xi_i)(x_k - \xi_k)R^{-2}\right] \tag{6.33}$$

and

$$A = \frac{1}{8\pi(1-\nu)} \tag{6.34}$$

The stress tensor S_{ij} is computed from the stress-strain relation

$$S_{ij} = 2\mu\left(E_{ij} + \frac{\nu}{1-2\nu} E_{kk}\,\delta_{ij}\right) \tag{6.35}$$

where

$$E_{ij} = u_{(i,j)} = U_{(ik,j)}\,\ell_k \tag{6.36}$$

Calculating $U_{ik,j}$, by using Eq. (6.33), we obtain

$$U_{ik,j} = -\frac{A}{2\mu}R^{-3}\Big[(3-4\nu)(x_j-\xi_j)\delta_{ki} + 3(x_i-\xi_i)(x_j-\xi_j)(x_k-\xi_k)R^{-2} \\ -(x_k-\xi_k)\delta_{ij} - (x_i-\xi_i)\delta_{kj}\Big] \tag{6.37}$$

Hence, taking the trace of (6.37) with respect to the indices i and j, we get

$$\begin{aligned} U_{ik,i} &= -\frac{A}{2\mu}R^{-3}[(3-4\nu)(x_k-\xi_k) + 3(x_k-\xi_k) - 3(x_k-\xi_k) - (x_k-\xi_k)] \\ &= -\frac{A}{2\mu}R^{-3} \times 2(1-2\nu)(x_k-\xi_k) \end{aligned} \tag{6.38}$$

Since

$$E_{ii} = U_{ik,i}\,\ell_k \tag{6.39}$$

therefore,

$$E_{ii} = E_{kk} = -\frac{A}{2\mu}R^{-3} \times 2(1-2\nu)(x_k-\xi_k)\ell_k \tag{6.40}$$

Also, by taking the symmetric part of (6.37) with respect to the indices i and j we obtain

$$U_{(ik,j)} = -\frac{A}{2\mu}R^{-3}\Big\{(1-2\nu)[(x_j-\xi_j)\delta_{ki} + (x_i-\xi_i)\delta_{kj}] \\ +3(x_i-\xi_i)(x_j-\xi_j)(x_k-\xi_k)R^{-2} - (x_k-\xi_k)\delta_{ij}\Big\} \tag{6.41}$$

Hence, because of (6.36), we get

$$E_{ij} = -\frac{A}{2\mu}R^{-3}\Big\{(1-2\nu)[(x_j-\xi_j)\ell_i + (x_i-\xi_i)\ell_j] + 3(x_i-\xi_i)(x_j-\xi_j)(x_k-\xi_k)\ell_k R^{-2} \\ -(x_k-\xi_k)\ell_k\,\delta_{ij}\Big\} \tag{6.42}$$

Finally, substituting E_{ij} from (6.42) and E_{kk} from (6.40), respectively, into (6.35), we obtain

$$S_{ij} = -AR^{-3}\left\{3R^{-2}(x_i - \xi_i)(x_j - \xi_j)(x_k - \xi_k)\ell_k \right.$$
$$\left. +(1-2\nu)[(x_i - \xi_i)\ell_j + (x_j - \xi_j)\ell_i - (x_k - \xi_k)\ell_k\delta_{ij}\right\} \tag{6.43}$$

Equation. (6.43) is equivalent to the stress formula of Problem 6.1. This completes a solution to Problem 6.1.

Problem 6.2. The displacement equation of thermoelastostatics for a homogeneous isotropic body subject to a temperature change $T = T(\mathbf{x})$ takes the form

$$\nabla^2\mathbf{u} + \frac{1}{1-2\nu}\nabla(\text{div}\,\mathbf{u}) - \frac{2+2\nu}{1-2\nu}\alpha\,\nabla T = \mathbf{0} \tag{6.44}$$

Let

$$\mathbf{u} = \psi - \frac{1}{4(1-\nu)}\nabla(\mathbf{x}\cdot\psi + \widehat{\varphi}) \tag{6.45}$$

where

$$\nabla^2\psi = \mathbf{0} \tag{6.46}$$

and

$$\nabla^2\widehat{\varphi} = -4(1+\nu)\alpha\,T \tag{6.47}$$

Show that $\mathbf{u}$ given by Eqs. (6.45) through (6.46) satisfies Eq. (6.44).

Solution. Eqs. (6.44)–(6.47) in components take the form

$$u_{i,kk} + \frac{1}{1-2\nu}u_{k,ki} - \frac{2+2\nu}{1-2\nu}\alpha\,T_{,i} = 0 \tag{6.48}$$

$$u_i = \psi_i - \frac{1}{4(1-\nu)}(x_a\psi_a + \widehat{\varphi})_{,i} \tag{6.49}$$

where

$$\psi_{i,aa} = 0 \tag{6.50}$$

and

$$\widehat{\varphi}_{,kk} = -4(1+\nu)\alpha T \tag{6.51}$$

Taking the gradient of (6.49) we obtain

$$u_{i,k} = \psi_{i,k} - \frac{1}{4(1-\nu)}(x_a\psi_a + \widehat{\varphi})_{,ik} \tag{6.52}$$

Hence, from (6.50),

$$u_{i,kk} = -\frac{1}{4(1-\nu)}(x_a\psi_a + \widehat{\varphi})_{,ikk} \tag{6.53}$$

and

$$u_{k,k} = \psi_{k,k} - \frac{1}{4(1-\nu)}(x_a\psi_a + \widehat{\varphi}),_{kk} \tag{6.54}$$

$$u_{k,ki} = \left[\psi_{k,k} - \frac{1}{4(1-\nu)}(x_a\psi_a + \widehat{\varphi}),_{kk}\right],_i \tag{6.55}$$

Using the relation

$$(x_a\psi_a),_{kk} = 2\psi_{k,k} + x_a\psi_{a,kk} = 2\psi_{k,k} \tag{6.56}$$

we reduce (6.53) and (6.55) into

$$u_{i,kk} = -\frac{1}{4(1-\nu)}(2\psi_{k,k} + \widehat{\varphi},_{kk}),_i \tag{6.57}$$

and

$$u_{k,ki} = \left[\psi_{k,k} - \frac{1}{4(1-\nu)}(2\psi_{k,k} + \widehat{\varphi},_{kk})\right],_i \tag{6.58}$$

Therefore, substituting (6.57) and (6.58) into the LHS of (6.48) we obtain

$$\left\{-\frac{1}{4(1-\nu)}(2\psi_{k,k} + \widehat{\varphi},_{kk}) + \frac{1}{1-2\nu}\left[\psi_{k,k} - \frac{1}{4(1-\nu)}(2\psi_{k,k} + \widehat{\varphi},_{kk})\right] - \frac{2+2\nu}{1-2\nu}\alpha T\right\},_i$$
$$= -\frac{1}{2(1-2\nu)}\{\widehat{\varphi},_{kk} + 4(1+\nu)\alpha T\},_i \tag{6.59}$$

Equation (6.59) together with Eq. (6.51) imply that u_i given by (6.49) meets (6.48). This completes a solution to Problem 6.2.

Problem 6.3. The temperature change T of a homogeneous isotropic infinite elastic body is represented by

$$T(\mathbf{x}) = \widehat{T}\,\delta(\mathbf{x}) \tag{6.60}$$

where

$$\delta(\mathbf{x}) = \delta(x_1)\delta(x_2)\delta(x_3) \tag{6.61}$$

$\delta = \delta(x_i)$, i $= 1, 2, 3$, is a one dimensional Dirac delta function, and $\widehat{T}$ is a constant with the dimension $[\widehat{T}]$ = [Temperature × Volume]. Show that an elastic displacement $\mathbf{u} = \mathbf{u}(\mathbf{x})$ and stress $\mathbf{S} = \mathbf{S}(\mathbf{x})$ corresponding to $T = T(\mathbf{x})$ are given by

$$\mathbf{u}(\mathbf{x}) = -\frac{1}{4\pi}\frac{1+\nu}{1-\nu}\alpha\,\widehat{T}\,\nabla\,\frac{1}{|\mathbf{x}|} \tag{6.62}$$

and

$$\mathbf{S}(\mathbf{x}) = -\frac{\mu}{2\pi}\frac{1+\nu}{1-\nu}\alpha\,\widehat{T}\,(\nabla\nabla - \mathbf{1}\,\nabla^2)\,\frac{1}{|\mathbf{x}|} \tag{6.63}$$

Hint. Use the representation (6.61) through (6.63) of Problem 6.2 in which $\psi = \mathbf{0}$ and $T = \widehat{T}\,\delta(\mathbf{x})$. Also, note that

$$\mathbf{S} = -\frac{\mu}{2(1-\nu)}(\nabla\nabla - \mathbf{1}\,\nabla^2)\,\widehat{\varphi} \tag{6.64}$$

Solution. By letting $\psi = \mathbf{0}$ and

$$\widehat{\varphi} = -4(1-\nu)\phi \tag{6.65}$$

in Eqs. (6.61)–(6.63) of Problem 6.2 we obtain

$$u_i = \phi_{,i} \tag{6.66}$$

where

$$\nabla^2\phi = m\,T \tag{6.67}$$

and

$$m = \frac{1+\nu}{1-\nu}\alpha \tag{6.68}$$

Also,

$$S_{ij} = 2\mu(\phi_{,ij} - \nabla^2\phi\delta_{ij}) \tag{6.69}$$

If $T(\mathbf{x}) = \widehat{T}\delta(\mathbf{x})$, a solution to (6.67) in E^3 takes the form

$$\phi = -\frac{m\widehat{T}}{4\pi}\frac{1}{|\mathbf{x}|} \tag{6.70}$$

Substituting ϕ from (6.70) into (6.66) and (6.69), respectively, we obtain (6.62) and (6.63). This completes a solution to Problem 6.3.

Problem 6.4. A solution $\varphi = \varphi(\mathbf{x}, t)$ to the nonhomogeneous wave equation

$$\square_0^2\varphi(\mathbf{x}, t) = -F(\mathbf{x}, t) \quad \text{on} \quad E^3 \times [0, \infty) \tag{6.71}$$

subject to the homogeneous initial conditions

$$\varphi(\mathbf{x}, 0) = 0, \quad \dot{\varphi}(\mathbf{x}, 0) = 0 \qquad \text{on} \qquad E^3 \tag{6.72}$$

takes the form

$$\varphi(\mathbf{x}, t) = \frac{1}{4\pi}\int\limits_{|\mathbf{x}-\xi|\le ct} \frac{F(\xi, t - |\mathbf{x}-\xi|\,/\,c)}{|\mathbf{x}-\xi|}\,dv(\xi) \quad \text{on} \quad E^3 \times [0, \infty) \tag{6.73}$$

Here

$$\Box_0^2 = \nabla^2 - \frac{1}{c^2}\frac{\partial^2}{\partial t^2} \tag{6.74}$$

Show that an equivalent form of Eq. (6.73) reads

$$\varphi(\mathbf{x}, t) = -\frac{c^2 t^2}{4\pi} \int\limits_{|\xi|\leq 1} \frac{F[\mathbf{x} - c\,t\xi, (1 - |\xi|)t]}{|\xi|}\, dv(\xi) \quad \text{on} \quad E^3 \times [0, \infty) \tag{6.75}$$

Solution. Introduce the transformation of variables

$$\mathbf{x} - \xi = c\,t\zeta, \quad c\,t > 0 \tag{6.76}$$

Then

$$dv(\xi) = d\xi_1\, d\xi_2\, d\xi_3 = -c^3 t^3 d\zeta_1 d\zeta_2 d\zeta_3 \tag{6.77}$$

and

$$dv(\xi) = -c^3 t^3 dv(\zeta) \tag{6.78}$$

Since

$$|\mathbf{x} - \xi| = ct|\zeta| \leq ct \tag{6.79}$$

therefore,

$$|\zeta| \leq 1 \tag{6.80}$$

and the integral (6.73) reduces to (6.75). This completes a solution to Problem 6.4.

Problem 6.5. Let $\mathbf{S} = \mathbf{S}(\mathbf{x}, t)$ be a solution to the stress equation of motion of a homogeneous anisotropic elastodynamics [see Eq. (3.51) in which ρ and K are constants]

$$\widehat{\nabla}(\operatorname{div} \mathbf{S}) - \rho\, \mathbf{K}[\ddot{\mathbf{S}}] = -\mathbf{B} \quad \text{on} \quad B \times [0, \infty) \tag{6.81}$$

subject to the initial conditions

$$\mathbf{S}(\mathbf{x}, 0) = \mathbf{S}_0(\mathrm{x}), \quad \dot{\mathbf{S}}(\mathbf{x}, 0) = \dot{\mathbf{S}}_0(\mathbf{x}) \quad \text{for} \quad \mathbf{x} \in B \tag{6.82}$$

Here, $\mathbf{B} = \mathbf{B}(\mathbf{x}, t)$, $\mathbf{S}_0 = \mathbf{S}_0(\mathbf{x})$, and $\dot{\mathbf{S}}_0 = \dot{\mathbf{S}}_0(\mathbf{x})$ are prescribed functions. Show that the compatibility condition

$$\operatorname{curl}\operatorname{curl} \mathbf{K}[\mathbf{S}] = 0 \quad \text{on} \quad B \times [0, \infty) \tag{6.83}$$

is satisfied if and only if there exists a vector field $\mathbf{u} = \mathbf{u}(\mathbf{x}, t)$ on $B \times [0, \infty)$ such that

$$\widehat{\nabla}\ddot{\mathbf{u}} = -\rho^{-1}\mathbf{B} \quad \text{on} \quad B \times [0, \infty) \tag{6.84}$$

and

$$\mathbf{S}_0(\mathbf{x}) = \mathbf{K}^{-1}\,[\widehat{\nabla}\mathbf{u}(\mathbf{x},0)], \quad \dot{\mathbf{S}}_0(\mathbf{x}) = \mathbf{K}^{-1}\,[\widehat{\nabla}\dot{\mathbf{u}}(\mathbf{x},0)] \quad \text{on} \quad B \tag{6.85}$$

Note that $\mathbf{B}$ in Eq. (6.81) represents an arbitrary second-order symmetric tensor field on $B \times [0, \infty)$, while $\mathbf{S}_0$ and $\dot{\mathbf{S}}_0$ in Eq. (6.82) stand for arbitrary second-order symmetric tensor fields on B.

Solution. A solution to Problem 6.5 is based on the following

Lemma. A symmetric tensor field $\mathbf{E}$ on $B \times [0, \infty)$ satisfies the condition

$$\text{curl curl } \mathbf{E} = \mathbf{0} \quad \text{on } B \times [0, \infty) \tag{6.86}$$

if and only if there is a vector field $\mathbf{u}$ on $B \times [0, \infty)$ such that

$$\mathbf{E} = \widehat{\nabla}\mathbf{u} \quad \text{on } B \times (0, \infty) \tag{6.87}$$

Proof of Lemma. The proof is split into two parts
(i) (6.87) ⇒ (6.86), (ii) (6.86) ⇒ (6.87).

To show (i) we substitute (6.87) into the LHS of Eq. (6.86) and find that Eq. (6.86) holds true. To show (ii), we note that Eq. (6.86) implies that there is a vector field $\mathbf{a}$ such that

$$\text{curl } \mathbf{E} = \nabla\mathbf{a} \tag{6.88}$$

Since, by Helmholtz's theorem, there are a scalar field φ and a vector field $\mathbf{b}$ such that

$$\mathbf{a} = \nabla\varphi + \text{curl } \mathbf{b}, \quad \text{div } \mathbf{b} = 0 \tag{6.89}$$

Equation (6.88) can be written as

$$\text{curl } \mathbf{E} = \nabla\nabla\varphi + \nabla \text{ curl } \mathbf{b} \tag{6.90}$$

or

$$\varepsilon_{iab}\, E_{jb,a} = \varphi,_{ij} + \varepsilon_{iab}\, b_{b,aj} \tag{6.91}$$

By taking the trace of (6.91), it is, by letting $i = j$ in (6.91), we obtain

$$\varphi,_{ii} = 0 \quad \text{or} \quad \text{div } \nabla\varphi = 0 \tag{6.92}$$

This implies that there is a vector field $\mathbf{c}$ on $B \times [0, \infty)$ such that

$$\nabla\varphi = \text{curl } \mathbf{c} \tag{6.93}$$

Substituting (6.93) into (6.90) we obtain

$$\text{curl } \mathbf{E} = \nabla \text{ curl } (\mathbf{c} + \mathbf{b}) \tag{6.94}$$

Since for any vector $\mathbf{v}$

$$\nabla \text{ curl } \mathbf{v} = \text{curl } (\nabla \mathbf{v}^T) \tag{6.95}$$

or

$$\varepsilon_{iab}\, v_{b,aj} = \varepsilon_{iab}(v_{j,b})^T_{,a} \tag{6.96}$$

therefore, Eqs. (6.94) and (6.95) imply that

$$\text{curl } (\mathbf{E} - \nabla \mathbf{v}^T) = \mathbf{0} \tag{6.97}$$

where

$$\mathbf{v} = \mathbf{b} + \mathbf{c} \tag{6.98}$$

Next, it follows from Eq. (6.97) that there is a vector field $\mathbf{e}$ on $B \times [0, \infty)$ such that

$$\mathbf{E} - \nabla \mathbf{v}^T = \nabla\, \mathbf{e} \tag{6.99}$$

By taking the transpose of (6.99) and using the symmetry of $\mathbf{E}$ ($\mathbf{E} = \mathbf{E}^T$) we obtain

$$\mathbf{E} - \nabla \mathbf{v} = \nabla \mathbf{e}^T \tag{6.100}$$

By adding Eqs. (6.99) and (6.100) we get

$$\mathbf{E} = \widehat{\nabla}(\mathbf{v} + \mathbf{e}) \tag{6.101}$$

Hence, if we let

$$\mathbf{u} = \mathbf{v} + \mathbf{e} \tag{6.102}$$

in Eq. (6.101) we obtain (6.87). This shows (ii), and proof of Lemma is complete.

To show that the compatibility condition (6.83) is satisfied if and only if there exists a vector field $\mathbf{u}$ on $B \times [0, \infty)$ such that (6.84) and (6.85) hold true, we note that Eqs. (6.81) and (6.82) are satisfied if and only if

$$\mathsf{K}[\mathbf{S}] = \mathsf{K}[\mathbf{S}_0 + t\dot{\mathbf{S}}_0] + t * \rho^{-1}(\widehat{\nabla}\text{div}\, \mathbf{S} + \mathbf{B}) \tag{6.103}$$

Applying curl curl to this equation we obtain

$$\text{curl curl } \mathsf{K}[\mathbf{S}] = \text{curl curl } \left\{t * \rho^{-1}\mathbf{B} + \mathsf{K}[\mathbf{S}_0 + t\dot{\mathbf{S}}_0]\right\} \tag{6.104}$$

Hence, the compatibility condition (6.83) is equivalent to

$$\text{curl curl } \{t * \rho^{-1}\mathbf{B} + \mathsf{K}[\mathbf{S}_0 + t\dot{\mathbf{S}}_0]\} = \mathbf{0} \tag{6.105}$$

Using the Lemma we find that Eq. (6.105) holds true if and only if there is a vector field **u** such that

$$t * \rho^{-1}\mathbf{B} + \mathbf{K}[\mathbf{S}_0 + t\dot{\mathbf{S}}_0] = \widehat{\nabla}\mathbf{u} \tag{6.106}$$

If **B**, $\mathbf{S}_0$, and $\dot{\mathbf{S}}_0$ are given by (6.84)–(6.85), then Eq. (6.106) is identically satisfied. Conversely, by differentiating twice (6.106) with respect to time we obtain (6.84). Also, by differentiating (6.106) with respect to time and letting $t = 0$ we obtain

$$\dot{\mathbf{S}}_0 = \mathbf{K}^{-1}[\widehat{\nabla}\dot{\mathbf{u}}(\mathbf{x}, 0)] \tag{6.107}$$

Finally, by letting $t = 0$ in (6.106) we get

$$\mathbf{S}_0 = \mathbf{K}^{-1}[\widehat{\nabla}\mathbf{u}(\mathbf{x}, 0)] \tag{6.108}$$

This completes a solution to Problem 6.5.

Problem 6.6. Consider a homogeneous isotropic elastic body occupying a region B. Let $\mathbf{S} = \mathbf{S}(\mathbf{x}, t)$ be a tensor field defined by

$$\mathbf{S}(\mathbf{x}, t) = \left[\left(\nabla\nabla - \nu\mathbf{1}\Box_2^2\right) \operatorname{tr}\chi - 2(1-\nu)\Box_1^2\chi\right] \quad \text{on} \quad \overline{B} \times [0, \infty) \tag{6.109}$$

where $\chi = \chi(\mathbf{x}, t)$ is a symmetric second-order tensor field that satisfies the equations

$$\Box_1^2\Box_2^2\chi = \frac{1}{1-\nu}\widehat{\nabla}\mathbf{b} \quad \text{on} \quad \mathrm{B} \times [0, \infty) \tag{6.110}$$

and

$$\nabla^2\chi + \nabla\nabla(\operatorname{tr}\chi) - 2\widehat{\nabla}(\operatorname{div}\chi) = \mathbf{0} \quad \text{on} \quad B \times [0, \infty) \tag{6.111}$$

Show that $\mathbf{S} = \mathbf{S}(\mathbf{x}, t)$ satisfies the stress equation of motion [see Eq. (6.28)]

$$\widehat{\nabla}(\operatorname{div}\mathbf{S}) - \frac{\rho}{2\mu}\left[\ddot{\mathbf{S}} - \frac{\nu}{1+\nu}(\operatorname{tr}\ddot{\mathbf{S}})\,\mathbf{1}\right] = -\widehat{\nabla}\mathbf{b} \quad \text{on} \quad B \times [0, \infty) \tag{6.112}$$

Note. The stress field **S** in the form of Eqs. (6.109) through (6.111) is a tensor solution of the homogeneous isotropic elastodynamics of the Galerkin type. To show this we let $\chi = -[\mu/(1-\nu)]\widehat{\nabla}\mathbf{g}$, where **g** is the Galerkin vector satisfying Eq. (6.27). Then Eqs. (6.110 and 6.111) are satisfied identically, and Eq. (6.109) reduces to

$$\mathbf{S} = \mu\left[2\Box_1^2\widehat{\nabla}\mathbf{g} - \frac{1}{1-\nu}\nabla\nabla(\operatorname{div}\mathbf{g}) + \frac{\nu}{1-\nu}\mathbf{1}\,\Box_2^2(\operatorname{div}\mathbf{g})\right] \tag{6.113}$$

The stress field **S** given by (6.111) corresponds to a solution of C-K-S [or Galerkin] type defined by Eqs. (6.26)–(6.27).

Solution. Eqs. (6.109)–(6.111), respectively, in components take the form

$$S_{ij} = \chi_{aa,ij} - \nu\Box_2^2\chi_{aa}\,\delta_{ij} - 2(1-\nu)\Box_1^2\chi_{ij} \tag{6.114}$$

$$\Box_1^2\Box_2^2\chi_{ij} = \tfrac{1}{1-\nu}b_{(i,j)} \tag{6.115}$$

and

$$\chi_{ij,aa} + \chi_{aa,ij} - \chi_{ik,kj} - \chi_{jk,ki} = 0 \tag{6.116}$$

The stress equation of motion (6.112) is rewritten as

$$S_{(ik,kj)} - \frac{\rho}{2\mu}\left(\ddot{S}_{ij} - \frac{\nu}{1+\nu}\ddot{S}_{kk}\,\delta_{ij}\right) = -b_{(i,j)} \tag{6.117}$$

In Eqs. (6.114)–(6.115)

$$\Box_1^2 = \nabla^2 - \frac{1}{c_1^2}\frac{\partial^2}{\partial t^2}, \quad \Box_2^2 = \nabla^2 - \frac{1}{c_2^2}\frac{\partial^2}{\partial t^2} \tag{6.118}$$

and

$$\frac{1}{c_2^2} = \frac{\rho}{\mu}, \quad \frac{1}{c_1^2} = \frac{1}{c_2^2}\frac{1-2\nu}{2-2\nu} \tag{6.119}$$

By taking the trace of (6.114) we obtain

$$S_{aa} = -(1+\nu)\,\Box_2^2\,\chi_{aa} \tag{6.120}$$

Hence, an alternative form of (6.114) reads

$$S_{ij} - \frac{\nu}{1+\nu}S_{kk}\,\delta_{ij} = \chi_{aa,ij} - 2(1-\nu)\Box_1^2\,\chi_{ij} \tag{6.121}$$

Next, using (6.114) we obtain

$$S_{(ik,kj)} = \chi_{aa,kkij} - \nu\Box_2^2\,\chi_{aa,ij} - 2(1-\nu)\Box_1^2\,\chi_{(ik,kj)} \tag{6.122}$$

Since, from (6.116)

$$2\chi_{(ik,kj)} = \chi_{ij,aa} + \chi_{aa,ij} \tag{6.123}$$

therefore, (6.122) can be written as

$$S_{(ik,kj)} = \nabla^2\chi_{aa,ij} - \nu\Box_2^2\,\chi_{aa,ij} - (1-\nu)\Box_1^2\,\chi_{aa,ij} - (1-\nu)\Box_1^2\,\chi_{ij,aa} \tag{6.124}$$

or

$$S_{(ik,kj)} = \frac{1}{2c_2^2}\ddot{\chi}_{aa,ij} - (1-\nu)\Box_1^2\,\chi_{ij,aa} \tag{6.125}$$

Substituting (6.121) and (6.125) into the LHS of (6.117) we obtain

$$S_{(ik,kj)} - \frac{1}{2c_2^2}\left(\ddot{S}_{ij} - \frac{\nu}{1+\nu}\ddot{S}_{kk}\delta_{ij}\right) = \frac{1}{2c_2^2}\ddot{\chi}_{aa,ij} - (1-\nu)\Box_1^2\,\chi_{ij,aa}$$
$$-\frac{1}{2c_2^2}\left[\ddot{\chi}_{aa,ij} - 2(1-\nu)\Box_1^2\,\ddot{\chi}_{ij}\right] = -(1-\nu)\Box_1^2\,\Box_2^2\,\chi_{ij} \tag{6.126}$$

Since χ_{ij} satisfies Eq. (6.115), therefore, by virtue of (6.126), S_{ij} given by (6.114) meets (6.117). This completes a solution to Problem 6.6.

Problem 6.7. Let $\mathbf{S}$ be the tensor solution of homogeneous isotropic elastodynamics of Problem 6.6 corresponding to homogeneous initial conditions. Show that the solution is complete, that is, there exists a second-order symmetric tensor field χ such that Eqs. (6.109) through (6.111) of Problem 6.6 are satisfied.

Solution. To solve Problem 6.7 we prove the two Lemmas.

Lemma 1. Let $\mathbf{S}$ satisfy the field equation

$$\widehat{\nabla}(\operatorname{div}\mathbf{S}) - \frac{1}{2c_2^2}\left(\ddot{\mathbf{S}} - \frac{\nu}{1+\nu}(\operatorname{tr}\ddot{\mathbf{S}})\mathbf{1}\right) = -\widehat{\nabla}\mathbf{b} \quad \text{on } B\times[0,\infty) \tag{6.127}$$

subject to the homogeneous initial conditions

$$\mathbf{S}(\mathbf{x},0) = \mathbf{0}, \quad \dot{\mathbf{S}}(\mathbf{x},0) = 0 \quad \text{on } B \tag{6.128}$$

Then

$$\operatorname{curl}\operatorname{curl}\mathbf{E} = \mathbf{0} \quad \text{on } B\times[0,\infty) \tag{6.129}$$

where

$$\mathbf{E} = \frac{1}{2\mu}\left(\mathbf{S} - \frac{\nu}{1+\nu}(\operatorname{tr}\mathbf{S})\mathbf{1}\right) \tag{6.130}$$

Lemma 2. Let $\mathbf{S}$ satisfy the hypotheses of Lemma 1, and let $\tilde{\mathbf{S}}$ be a continuation of $\mathbf{S}$ on $E^3\times[0,\infty)$ such that

$$\operatorname{curl}\operatorname{curl}\tilde{\mathbf{E}} = \mathbf{0} \quad \text{on } E^3\times[0,\infty) \tag{6.131}$$

where

$$\tilde{\mathbf{E}} = \frac{1}{2\mu}\left(\tilde{\mathbf{S}} - \frac{\nu}{1+\nu}(\operatorname{tr}\tilde{\mathbf{S}})\mathbf{1}\right) \quad \text{on } E^3\times[0,\infty) \tag{6.132}$$

Define a second-order tensor field χ on $\bar{B}\times[0,\infty)$ such that

$$-2(1-\nu)\chi = 2\mu\,\frac{c_1^2t^2}{4\pi}\int\limits_{|\xi|\le 1}\frac{dv(\xi)}{|\xi|}\tilde{\mathbf{E}}(\mathbf{x}-c_1t\xi,(1-|\xi|)t) + \frac{2\mu}{1-2\nu}\frac{c_1^2c_2^2t^4}{16\pi^2}\nabla\nabla$$

$$\times \int_{|\xi|\le 1} \frac{dv(\xi)}{|\xi|} \int_{|\eta|\le 1} \frac{dv(\eta)}{|\eta|} (\text{tr}\, \tilde{\mathbf{E}})$$
$$\times (\mathbf{x} - c_1 t\xi - c_2 t\eta,\ (1 - |\xi|)(1 - |\eta|)t) \tag{6.133}$$

Then

$$\Box_2^2(\text{tr}\, \chi) = -\frac{2\mu}{1-2\nu}(\text{tr}\, \mathbf{E}) \tag{6.134}$$

Notes

(1) Equations (6.129) and (6.131), respectively, are equivalent to

$$2\widehat{\nabla}(\text{div}\, \mathbf{S}) - \nabla^2 \mathbf{S} + \frac{1}{1+\nu}(\nu \mathbf{1}\nabla^2 - \nabla\nabla)(\text{tr}\, \mathbf{S}) = \mathbf{0} \tag{6.135}$$

and

$$2\widehat{\nabla}(\text{div}\, \tilde{\mathbf{S}}) - \nabla^2 \tilde{\mathbf{S}} + \frac{1}{1+\nu}(\nu \mathbf{1}\nabla^2 - \nabla\nabla)(\text{tr}\, \tilde{\mathbf{S}}) = \mathbf{0} \tag{6.136}$$

To prove that (6.135) $\Leftrightarrow$ (6.129) we use the identity [see Problem 1.12, Eq. (1.204) in which **S** is replaced by **E**]

$$\text{curl curl}\, \mathbf{E} = 2\widehat{\nabla}(\text{div}\, \mathbf{E}) - \nabla^2 \mathbf{E} - \nabla\nabla\, (\text{tr}\, \mathbf{E}) + \mathbf{1}[\nabla^2(\text{tr}\, \mathbf{E}) - \text{div div}\, \mathbf{E}] \tag{6.137}$$

Equation (6.129) implies that the LHS of (6.137) vanishes which written in components means

$$E_{ik,kj} + E_{jk,ki} - E_{ij,kk} - E_{aa,ij} + \delta_{ij}(E_{aa,bb} - E_{ab,ab}) = 0 \tag{6.138}$$

By letting $i = j$ in (6.138) we obtain

$$2E_{ik,ki} - 2E_{aa,bb} + 3(E_{aa,bb} - E_{ab,ab}) = 0 \tag{6.139}$$

or

$$E_{aa,bb} - E_{ab,ab} = 0 \tag{6.140}$$

Hence (6.129) is equivalent to

$$2\widehat{\nabla}(\text{div}\, \mathbf{E}) - \nabla^2 \mathbf{E} - \nabla\nabla(\text{tr}\, \mathbf{E}) = \mathbf{0} \tag{6.141}$$

Substituting **E** from (6.130) into (6.141), and using the relations

$$\text{div}\, \mathbf{E} = \frac{1}{2\mu}\left[\text{div}\, \mathbf{S} - \frac{\nu}{1+\nu}\nabla(\text{tr}\, \mathbf{S})\right] \tag{6.142}$$

$$(\text{tr}\,\mathbf{E}) = \frac{1}{2\mu}\,\frac{1-2\nu}{1+\nu}(\text{tr}\,\mathbf{S}) \tag{6.143}$$

$$2\widehat{\nabla}\nabla(\text{tr}\,\mathbf{S}) = 2\nabla\nabla(\text{tr}\,\mathbf{S}) \tag{6.144}$$

we obtain

$$\begin{aligned}
&2\widehat{\nabla}(\text{div}\,\mathbf{S}) - \nabla^2\mathbf{S} - \frac{2\nu}{1+\nu}\,\nabla\nabla(\text{tr}\,\mathbf{S}) + \frac{\nu}{1+\nu}\mathbf{1}\nabla^2(\text{tr}\,\mathbf{S}) - \frac{1-2\nu}{1+\nu}\nabla\nabla(\text{tr}\,\mathbf{S}) \\
&= 2\widehat{\nabla}(\text{div}\,\mathbf{S}) - \nabla^2\mathbf{S} + \frac{1}{1+\nu}[\nu\mathbf{1}\nabla^2(\text{tr}\,\mathbf{S}) - \nabla\nabla(\text{tr}\,\mathbf{S})] = \mathbf{0}
\end{aligned} \tag{6.145}$$

Therefore, (6.129) $\Leftrightarrow$ (6.135) $\Leftrightarrow$ (6.145).

(2) An alternative form of (6.133) reads

$$\begin{aligned}
-2(1-\nu)\chi &= \frac{c_1^2t^2}{4\pi}\int\limits_{|\xi|\le 1}\frac{dv(\xi)}{|\xi|}\left[\tilde{\mathbf{S}} - \frac{\nu}{1+\nu}(\text{tr}\,\tilde{\mathbf{S}})\mathbf{1}\right](\mathbf{x}-c_1t\xi,\ t(1-|\xi|)) \\
&\quad + \frac{1}{1+\nu}\nabla\nabla\left(\frac{c_1^2c_2^2t^4}{16\pi^2}\right)\int\limits_{|\xi|\le 1}\frac{dv(\xi)}{|\xi|}\int\limits_{|\eta|\le 1}\frac{dv(\eta)}{|\eta|}(\text{tr}\,\tilde{\mathbf{S}}) \\
&\quad \times (\mathbf{x}-c_1t\xi - c_2t\eta,\ t(1-|\xi|)(1-|\eta|))
\end{aligned} \tag{6.146}$$

To prove that (6.133) $\Leftrightarrow$ (6.146) we substitute $\tilde{\mathbf{E}}$ from (6.132) into (6.133) and obtain (6.146).

(3) A solution $\widehat{\varphi} = \widehat{\varphi}(\mathbf{x}, t)$ of the biwave equation

$$\Box_1^2\,\Box_2^2\widehat{\varphi} = f \quad \text{on } \bar{B}\times[0,\infty) \tag{6.147}$$

subject to the homogeneous initial conditions

$$\widehat{\varphi}^{(k)}(\mathbf{x}, 0) = 0 \quad \text{on } B, \quad k = 0, 1, 2, 3 \tag{6.148}$$

takes the form of iterated retarded potential

$$\widehat{\varphi}(\mathbf{x}, t) = \frac{c_1^2c_2^2t^4}{16\pi^2}\int\limits_{|\xi|\le 1}\frac{dv(\xi)}{|\xi|}\int\limits_{|\eta|\le 1}\frac{dv(\eta)}{|\eta|}f(\mathbf{x}-c_1t\xi - c_2t\eta,\ t(1-|\xi|)(1-|\eta|)) \tag{6.149}$$

To show that $\widehat{\varphi}$ given by (6.149) satisfies Eq. (6.147), note that because of the solution to Problem 6.4, a solution to the equation

$$\Box_1^2 u = f \quad \text{on } B\times[0,\infty) \tag{6.150}$$

subject to the homogeneous initial conditions

$$u(\mathbf{x}, 0) = 0, \ \dot{u}(\mathbf{x}, 0) = 0 \quad \text{on } B \tag{6.151}$$

takes the form

$$u(\mathbf{x}, t) = \frac{c_1^2 t^2}{4\pi} \int\limits_{|\xi| \leq 1} \frac{dv(\xi)}{|\xi|} f(\mathbf{x} - c_1 t \xi, \ t(1 - |\xi|)) \tag{6.152}$$

Similarly, a solution to the equation

$$\Box_2^2 \widehat{\varphi} = u \quad \text{on } B \times [0, \infty) \tag{6.153}$$

in which u is prescribed, and subject to the conditions

$$\widehat{\varphi}(\mathbf{x}, 0) = 0, \quad \dot{\widehat{\varphi}}(\mathbf{x}, 0) = 0 \quad \text{on } B \tag{6.154}$$

takes the form

$$\widehat{\varphi}(\mathbf{x}, t) = \frac{c_2^2 t^2}{4\pi} \int\limits_{|\eta| \leq 1} \frac{dv(\eta)}{|\eta|} u(\mathbf{x} - c_2 t \eta, \ t(1 - |\eta|)) \tag{6.155}$$

Note that substituting u from (6.153) into (6.150) we obtain

$$\Box_1^2 \, \Box_2^2 \, \widehat{\varphi} = f \tag{6.156}$$

and it follows from (6.152) that

$$\begin{aligned} & u(\mathbf{x} - c_2 t \eta, \ t(1 - |\eta|)) \\ & \quad = \frac{c_1^2 t^2}{4\pi} \int\limits_{|\xi| \leq 1} \frac{dv(\xi)}{|\xi|} \times f(\mathbf{x} - c_1 t \xi - c_2 t \eta, \ t(1 - |\xi|)(1 - |\eta|)) \end{aligned} \tag{6.157}$$

Therefore, substituting (6.157) into (6.155) we find that a solution of (6.156) takes the form (6.149). Also, by differentiating (6.149) with respect to time we obtain

$$\widehat{\varphi}(\mathbf{x}, 0) = \dot{\widehat{\varphi}}(\mathbf{x}, 0) = \ddot{\widehat{\varphi}}(\mathbf{x}, 0) = \dddot{\widehat{\varphi}}(\mathbf{x}, 0) = 0 \tag{6.158}$$

This completes the proof that $\widehat{\varphi}$ given by (6.149) satisfies (6.147) and (6.148).

Proof of Lemma 1. Applying the operator curl curl to Eq. (6.127) and using the relation

$$\text{curl curl } \widehat{\nabla}(\text{div } \mathbf{S} + \mathbf{b}) = \mathbf{0} \tag{6.159}$$

we obtain

$$\text{curl curl } \ddot{\mathbf{E}} = \mathbf{0} \tag{6.160}$$

where $\mathbf{E}$ is given by Eq. (6.130).

From (6.128) and (6.130)

$$\mathbf{E}(\mathbf{x}, 0) = \mathbf{0}, \quad \dot{\mathbf{E}}(\mathbf{x}, 0) = \mathbf{0} \tag{6.161}$$

Therefore, integrating (6.160) twice with respect to time, and using (6.161), we arrive at Eq. (6.129). This completes the proof of Lemma 1.

Proof of Lemma 2. Introduce $\tilde{\varphi} = \tilde{\varphi}(\mathbf{x}, t)$ on $E^3 \times [0, \infty)$ by the formula

$$\tilde{\varphi}(\mathbf{x}, t) = -\frac{c_2^2 t^2}{4\pi} \frac{1}{1+\nu} \int_{|\xi| \leq 1} \frac{dv(\xi)}{|\xi|} (\text{tr } \tilde{\mathbf{S}})(\mathbf{x} - c_2 t \xi, t(1 - |\xi|)) \tag{6.162}$$

and let $\tilde{\chi} = \tilde{\chi}(\mathbf{x}, t)$ be an extension of χ on $E^3 \times [0, \ \infty)$.

Then

$$\begin{array}{ll} \tilde{\varphi}(\mathbf{x}, 0) = 0, & \dot{\tilde{\varphi}}(\mathbf{x}, 0) = 0 \\ \tilde{\chi}(\mathbf{x}, 0) = 0, & \dot{\tilde{\chi}}(\mathbf{x}, 0) = 0 \end{array} \quad \text{on } E^3 \tag{6.163}$$

and

$$\Box_2^2 \, \tilde{\varphi} = -\frac{1}{1+\nu} (\text{tr } \tilde{\mathbf{S}}) \quad \text{on } E^3 \times [0, \infty) \tag{6.164}$$

Also, using Note 3, and applying the wave operator $\Box_1^2$ to Eq. (6.146), extended to $E^3 \times [0, \infty)$, we obtain

$$\begin{aligned} -2(1-\nu)\Box_1^2 \, \tilde{\chi} = {} & \left[\tilde{\mathbf{S}} - \frac{\nu}{1+\nu} (\text{tr } \tilde{\mathbf{S}}) \mathbf{1} \right] + \nabla\nabla \frac{c_2^2 t^2}{4\pi} \frac{1}{1+\nu} \\ & \times \int_{|\eta| \leq 1} \frac{dv(\eta)}{|\eta|} (\text{tr } \tilde{\mathbf{S}})(\mathbf{x} - c_2 t \eta, \ t(1 - |\eta|)) \end{aligned} \tag{6.165}$$

Taking the trace of (6.165) and using definition of $\tilde{\varphi}$ [see (6.162)] we get

$$-2(1-\nu)\Box_1^2 (\text{tr } \tilde{\chi}) = \frac{1-2\nu}{1+\nu} (\text{tr } \tilde{\mathbf{S}}) - \nabla^2 \tilde{\varphi} \tag{6.166}$$

Next, multiplying (6.164) by $(1 - 2\nu)$, and using the identity

$$(1 - 2\nu)\Box_2^2 = 2(1-\nu)\Box_1^2 - \nabla^2 \tag{6.167}$$

we obtain

$$2(1-\nu)\Box_1^2\tilde{\varphi} = -\frac{1-2\nu}{1+\nu}(\text{tr}\,\tilde{\mathbf{S}}) + \nabla^2\tilde{\varphi} \tag{6.168}$$

By addition of Eqs. (6.166) and (6.168) we get

$$\Box_1^2(\tilde{\varphi} - \text{tr}\,\tilde{\chi}) = 0 \tag{6.169}$$

Since, in view of (6.163),

$$(\tilde{\varphi} - \text{tr}\,\tilde{\chi})(\mathbf{x}, 0) = 0 \quad \text{on } E^3 \tag{6.170}$$

and

$$(\dot{\tilde{\varphi}} - \text{tr}\,\dot{\tilde{\chi}})(\mathbf{x}, 0) = 0 \quad \text{on } E^3 \tag{6.171}$$

therefore, it follows from (6.169) that

$$\tilde{\varphi} = \text{tr}\,\tilde{\chi} \quad \text{on } E^3 \times [0, \infty) \tag{6.172}$$

Substituting $\tilde{\varphi}$ from (6.172) into (6.162) and applying the operator $\nabla\nabla$ we get

$$\nabla\nabla(\text{tr}\,\tilde{\chi}) = -\frac{c_2^2 t^2}{4\pi}\frac{1}{1+\nu}\nabla\nabla\int\limits_{|\xi|\leq 1}\frac{dv(\xi)}{|\xi|}(\text{tr}\,\tilde{\mathbf{S}})(\mathbf{x} - c_2 t\xi, t(1-|\xi|)) \tag{6.173}$$

Also, substituting $\tilde{\varphi}$ from (6.172) into (6.164) we obtain

$$\Box_2^2(\text{tr}\,\tilde{\chi}) = -\frac{1}{1+\nu}(\text{tr}\,\tilde{\mathbf{S}}) \tag{6.174}$$

Since, because of (6.132),

$$(\text{tr}\,\tilde{\mathbf{E}}) = \frac{1}{2\mu}\frac{1-2\nu}{1+\nu}(\text{tr}\,\tilde{\mathbf{S}}) \tag{6.175}$$

Equation (6.174) is equivalent to (6.134). A restriction of (6.134) to $\bar{B} \times [0, \infty)$ leads to

$$\Box_2^2(\text{tr}\,\chi) = -\frac{1}{1+\nu}(\text{tr}\,\mathbf{S}) \tag{6.176}$$

This completes proof of Lemma 2.

Solution to Problem 6.7. We are to show that χ introduced by Lemma 2 [see Eq. (6.133)] satisfies Eqs. (6.109)–(6.111) of Problem 6.6.

By Lemma 2 [see also Eq. (6.176)]

$$\Box_2^2(\text{tr}\,\chi) = -\frac{1}{1+\nu}(\text{tr}\,\mathbf{S}) \tag{6.177}$$

An equivalent form to Eq. (6.177) reads

$$\operatorname{tr}\chi = -\frac{c_2^2 t^2}{4\pi}\frac{1}{1+\nu}\int\limits_{|\xi|\leq 1}\frac{dv(\xi)}{|\xi|}(\operatorname{tr}\mathbf{S})(\mathbf{x}-c_2 t\xi, t(1-|\xi|)) \tag{6.178}$$

By applying the operator $\nabla\nabla$ to this equation we obtain

$$\nabla\nabla(\operatorname{tr}\chi) = -\frac{c_2^2 t^2}{4\pi}\nabla\nabla\frac{1}{1+\nu}\int\limits_{|\xi|\leq 1}\frac{dv(\xi)}{|\xi|}(\operatorname{tr}\mathbf{S})(\mathbf{x}-c_2 t\xi,\; t(1-|\xi|)) \tag{6.179}$$

It follows from Eq. (6.165), restricted to $\bar{B}\times[0,\;\infty)$, and from Eq. (6.179) that

$$-2(1-\nu)\Box_1^2\chi = \mathbf{S} - \frac{\nu}{1+\nu}(\operatorname{tr}\mathbf{S})\mathbf{1} - \nabla\nabla(\operatorname{tr}\chi) \tag{6.180}$$

Also, from Eqs. (6.177) and (6.180), we obtain

$$\mathbf{S} = \nabla\nabla(\operatorname{tr}\chi) - \mathbf{1}\nu\Box_2^2(\operatorname{tr}\chi) - 2(1-\nu)\Box_1^2\chi \tag{6.181}$$

Therefore, χ introduced by Lemma 2 satisfies Eq. (6.109) of Problem 6.6.

Next, applying the operator $\Box_1^2\Box_2^2$ to Eq. (6.146) we obtain

$$\begin{aligned}-2(1-\nu)\Box_1^2\Box_2^2\chi &= \Box_2^2\left(\mathbf{S} - \frac{\nu}{1+\nu}(\operatorname{tr}\mathbf{S})\mathbf{1}\right) + \frac{1}{1+\nu}\nabla\nabla(\operatorname{tr}\mathbf{S})\\ &= \frac{1}{1+\nu}\nabla\nabla(\operatorname{tr}\mathbf{S}) + \nabla^2\mathbf{S} - \frac{\nu}{1+\nu}\nabla^2(\operatorname{tr}\mathbf{S})\mathbf{1}\\ &\quad - \frac{1}{c_2^2}\left(\ddot{\mathbf{S}} - \frac{\nu}{1+\nu}(\operatorname{tr}\ddot{\mathbf{S}})\mathbf{1}\right)\end{aligned} \tag{6.182}$$

It follows from (6.145) that

$$\nabla^2\mathbf{S} + \frac{1}{1+\nu}\left[\nabla\nabla(\operatorname{tr}\mathbf{S}) - \nu\mathbf{1}\nabla^2(\operatorname{tr}\mathbf{S})\right] = 2\widehat{\widehat{\nabla}}(\operatorname{div}\mathbf{S}) \tag{6.183}$$

Therefore, Eq. (6.182) is reduced to

$$-2(1-\nu)\Box_1^2\Box_2^2\chi = 2\widehat{\nabla}(\operatorname{div}\mathbf{S}) - \frac{1}{c_2^2}\left(\ddot{\mathbf{S}} - \frac{\nu}{1+\nu}(\operatorname{tr}\ddot{\mathbf{S}})\mathbf{1}\right) \tag{6.184}$$

and, since $\mathbf{S}$ is a solution to Eq. (6.127), we obtain

$$\Box_1^2\Box_2^2\chi = \frac{1}{1-\nu}\widehat{\nabla}\mathbf{b} \tag{6.185}$$

This shows that χ introduced by Lemma 2 satisfies Eq. (6.110) of Problem 6.6. Finally, introduce the notation

$$\psi = -2(1-\nu)\chi \tag{6.186}$$

then using Eqs. (6.146) and (6.178) we obtain

$$\begin{aligned}
\left(\nabla^2 - 2\widehat{\nabla}\mathrm{div} + \nabla\nabla\mathrm{tr}\right)\psi = {} & \frac{c_1^2 t^2}{4\pi}\int\limits_{|\xi|\le 1}\frac{dv(\xi)}{|\xi|}\Big\{\nabla^2\tilde{\mathbf{S}} - 2\widehat{\nabla}(\mathrm{div}\,\tilde{\mathbf{S}}) + \frac{1}{1+\nu}\Big[\nabla\nabla(\mathrm{tr}\,\tilde{\mathbf{S}}) \\
& -\nu\mathbf{1}\nabla^2(\mathrm{tr}\,\tilde{\mathbf{S}})\Big]\Big\}\times(\mathbf{x}-c_1 t\xi, t(1-|\xi|)) \\
& -\frac{c_1^2 t^2}{4\pi}\frac{(1-2\nu)}{1+\nu}\int\limits_{|\xi|\le 1}\frac{dv(\xi)}{|\xi|}\nabla\nabla(\mathrm{tr}\,\tilde{\mathbf{S}})(\mathbf{x}-c_1 t\xi, t(1-|\xi|)) \\
& -\nabla\nabla\nabla^2\left(\frac{c_1^2 c_2^2 t^4}{16\pi^2}\right)\frac{1}{1+\nu}\int\limits_{|\xi|\le 1}\frac{dv(\xi)}{|\xi|}\int\limits_{|\eta|\le 1}\frac{dv(\eta)}{|\eta|}(\mathrm{tr}\,\tilde{\mathbf{S}}) \\
& \times(\mathbf{x}-c_1 t\xi - c_2 t\eta,\ t(1-|\xi|)(1-|\eta|)) \\
& +\nabla\nabla\frac{2(1-\nu)}{1+\nu}\frac{c_2^2 t^2}{4\pi}\int\limits_{|\eta|\le 1}\frac{dv(\eta)}{|\eta|}(\mathrm{tr}\,\tilde{\mathbf{S}}) \\
& \times(\mathbf{x}-c_2 t\eta,\ t(1-|\eta|))
\end{aligned} \tag{6.187}$$

Since [see (6.167)]

$$\nabla^2 = 2(1-\nu)\Box_1^2 - (1-2\nu)\Box_2^2 \tag{6.188}$$

therefore

$$\begin{aligned}
& \nabla^2\left(\frac{c_1^2 c_2^2 t^4}{16\pi^4}\right)\int_{|\xi|\le 1}\frac{dv(\xi)}{|\xi|}\int\limits_{|\eta|\le 1}\frac{dv(\eta)}{|\eta|}(\mathrm{tr}\,\tilde{\mathbf{S}})(\mathbf{x}-c_1 t\xi - c_2 t\eta,\ t(1-|\xi|)(1-|\eta|)) \\
& = 2(1-\nu)\frac{c_2^2 t^2}{4\pi}\int\limits_{|\eta|\le 1}\frac{dv(\eta)}{|\eta|}(\mathrm{tr}\,\tilde{\mathbf{S}})(\mathbf{x}-c_2 t\eta,\ t(1-|\eta|)) \\
& \quad -(1-2\nu)\frac{c_1^2 t^2}{4\pi}\int\limits_{|\xi|<1}\frac{dv(\xi)}{|\xi|}(\mathrm{tr}\,\tilde{\mathbf{S}})(\mathbf{x}-c_1 t\xi,\ t(1-|\xi|))
\end{aligned} \tag{6.189}$$

Substituting (6.189) into the RHS of (6.187) we obtain

$$\nabla^2\psi - 2\widehat{\nabla}(\operatorname{div}\psi) + \nabla\nabla(\operatorname{tr}\psi) = \frac{c_1^2 t^2}{4\pi} \int\limits_{|\xi|\leq 1} \frac{dv(\xi)}{|\xi|} \left\{\nabla^2 \tilde{\mathbf{S}} - 2\widehat{\nabla}(\operatorname{div}\tilde{\mathbf{S}}) + \frac{1}{1+\nu} \times [\nabla\nabla(\operatorname{tr}\tilde{\mathbf{S}}) - \nu\mathbf{1}\nabla^2(\operatorname{tr}\tilde{\mathbf{S}})]\right\} \times (\mathbf{x} - c_1 t\xi, t(1-|\xi|)) \tag{6.190}$$

Because of (6.136) the integrand on the RHS of (6.190) vanishes. Therefore, if follows from Eqs. (6.186) and (6.190) that

$$\nabla^2\chi - 2\widehat{\nabla}(\operatorname{div}\chi) + \nabla\nabla(\operatorname{tr}\chi) = 0 \tag{6.191}$$

This means that χ introduced by Lemma 2 satisfies Eq. (6.111) of Problem 6.6. Therefore, χ meets Eqs. (6.109)–(6.111), and a solution to Problem 6.7 is complete.

Problem 6.8. Consider the stress equation of motion

$$\widehat{\nabla}(\operatorname{div}\mathbf{S}) - \frac{\rho}{2\mu}\left[\ddot{\mathbf{S}} - \frac{\nu}{1+\nu}(\operatorname{tr}\ddot{\mathbf{S}})\,\mathbf{1}\right] = -\widehat{\nabla}\mathbf{b} \quad \text{on} \quad B \times [0,\infty) \tag{6.192}$$

subject to the initial conditions

$$\mathbf{S}(\mathbf{x},0) = \mathbf{S}_0(\mathbf{x}), \quad \ddot{\mathbf{S}}(\mathbf{x},0) = \ddot{\mathbf{S}}_0(\mathbf{x}) \quad \text{for} \quad \mathbf{x} \in B \tag{6.193}$$

where $\mathbf{b}$, $\mathbf{S}_0$, and $\dot{\mathbf{S}}_0$ are prescribed functions. Define a scalar field $\alpha = \alpha(\mathbf{x}, t)$ and a vector field $\beta = \beta(\mathbf{x}, t)$ by

$$\alpha(\mathbf{x},t) = \frac{1}{4\pi c_1^2} \operatorname{div} \int\limits_B \frac{\gamma(\mathbf{y},t)}{|\mathbf{x}-\mathbf{y}|} dv(\mathbf{y}) \tag{6.194}$$

and

$$\beta(\mathbf{x},t) = -\frac{1}{4\pi c_2^2} \operatorname{curl} \int\limits_B \frac{\gamma(\mathbf{y},t)}{|\mathbf{x}-\mathbf{y}|} dv(\mathbf{y}) \tag{6.195}$$

where

$$\gamma(\mathbf{x},t) = \mathbf{b}(\mathbf{x},t) + \operatorname{div}[S_0(\mathbf{x}) + t\,\dot{\mathbf{S}}_0(\mathbf{x})] \tag{6.196}$$

Let ϕ and ω satisfy the equations

$$\Box_1^2\phi = \alpha \quad \text{on} \quad B \times [0,\infty) \tag{6.197}$$

and

$$\Box_2^2\omega = \beta, \quad \operatorname{div}\omega = 0 \quad \text{on} \quad B \times [0,\infty) \tag{6.198}$$

subject to the homogeneous initial conditions

$$\begin{aligned} \phi(\mathbf{x},0) = \dot{\phi}(\mathbf{x},0) = 0 \\ \omega(\mathbf{x},0) = \dot{\omega}(\mathbf{x},0) = 0 \end{aligned} \quad \text{on } B \tag{6.199}$$

Let

$$\mathbf{S}(\mathbf{x},t) = \mathbf{S}_0(\mathbf{x}) + t\,\dot{\mathbf{S}}_0(\mathbf{x}) + 2c_2^2\,[\nabla\nabla\phi + \widehat{\nabla}(\text{curl}\,\boldsymbol{\omega})] + \left(c_1^2 - 2c_2^2\right)\nabla^2\phi\,\mathbf{1} \tag{6.200}$$

Show that $\mathbf{S}$ satisfies Eqs. (6.192) and (6.193).

Note. The solution (i), in which ϕ and $\boldsymbol{\omega}$ satisfy Equations (6.197) through (6.199), represents a tensor solution of homogeneous isotropic elastodynamics of the Lame-type [see Eqs. (6.22)–(6.24)].

Solution. Let $\phi = \phi(\mathbf{x},t)$ and $\omega = \omega(\mathbf{x},t)$ satisfy the equations

$$\Box_1^2\phi = \alpha \quad \text{on } B \times [0,\infty) \tag{6.201}$$

$$\Box_2^2\omega = \beta, \quad \text{div}\,\omega = 0 \quad \text{on } B \times [0,\infty) \tag{6.202}$$

subject to the homogeneous initial conditions

$$\phi(\mathbf{x},0) = 0, \quad \dot{\phi}(\mathbf{x},0) = 0 \quad \text{on } B \tag{6.203}$$

$$\omega(\mathbf{x},0) = \mathbf{0}, \quad \dot{\omega}(\mathbf{x},0) = \mathbf{0} \quad \text{on } B \tag{6.204}$$

where α and β are defined by

$$\alpha(\mathbf{x},t) = +\frac{1}{4\pi c_1^2}\,\text{div} \int_B \frac{\gamma(\mathbf{y},t)}{|\mathbf{x}-\mathbf{y}|}\,dv(\mathbf{y}) \tag{6.205}$$

$$\beta(\mathbf{x},t) = -\frac{1}{4\pi c_2^2}\,\text{curl} \int_B \frac{\gamma(\mathbf{y},t)}{|\mathbf{x}-\mathbf{y}|}\,dv(\mathbf{y}) \tag{6.206}$$

and

$$\gamma(\mathbf{x},t) = \mathbf{b}(\mathbf{x},t) + \text{div}[\mathbf{S}_0(\mathbf{x}) + t\,\dot{\mathbf{S}}_0(\mathbf{x})] \tag{6.207}$$

Define $\mathbf{S} = \mathbf{S}(\mathbf{x},t)$ on $\bar{B} \times [0,\infty)$ by

$$\mathbf{S}(\mathbf{x},t) = \mathbf{S}_0(\mathbf{x}) + t\,\dot{\mathbf{S}}_0(\mathbf{x}) + 2c_2^2\,\widehat{\nabla}\,(\nabla\phi + \text{curl}\,\omega) + \left(c_1^2 - 2c_2^2\right)\nabla^2\phi\mathbf{1} \tag{6.208}$$

we are to show that

$$\widehat{\nabla}(\text{div}\,\mathbf{S}) - \frac{1}{2c_2^2}\left(\ddot{\mathbf{S}} - \frac{\nu}{1+\nu}(\text{tr}\,\ddot{\mathbf{S}})\mathbf{1}\right) = -\widehat{\nabla}\,\mathbf{b} \quad \text{on } B \times [0,\infty) \tag{6.209}$$

and

$$\mathbf{S}(\mathbf{x},0)=\mathbf{S}_0(\mathbf{x}),\quad \dot{\mathbf{S}}(\mathbf{x},0)=\dot{\mathbf{S}}_0(\mathbf{x})\quad \text{on } B. \tag{6.210}$$

To this end we note first that due to the homogeneous initial conditions (6.203) and (6.204), **S** given by (6.208) meets the nonhomogeneous initial conditions (6.210). To show that **S** satisfies the field Eq. (6.209) we make the following steps.

Applying the identity

$$\text{curl curl } \mathbf{u}=\nabla \text{ div } \mathbf{u}-\nabla^2\mathbf{u} \tag{6.211}$$

to the field

$$\mathbf{u}(\mathbf{x},t)=-\frac{1}{4\pi}\int_B \frac{\gamma(\mathbf{y},t)}{|\mathbf{x}-\mathbf{y}|}dv(\mathbf{y}) \tag{6.212}$$

that satisfies Poisson's equation

$$\nabla^2\mathbf{u}=\gamma \tag{6.213}$$

and using the definitions of α and β [see Eqs. (6.205) and (6.206)] we obtain

$$-c_1^2\,\nabla\alpha-c_2^2\,\text{curl }\beta=\gamma \tag{6.214}$$

Next, by using the relations

$$\frac{1}{c_2^2}=\frac{\rho}{\mu},\ \frac{1}{c_1^2}=\frac{1}{c_2^2}\frac{1-2\nu}{2-2\nu} \tag{6.215}$$

and differentiating (6.208) twice with respect to time we obtain

$$\ddot{\mathbf{S}}=2\,c_2^2\left[\widehat{\nabla}(\nabla\,\ddot{\phi}+\text{curl }\ddot{\omega})+\frac{\nu}{1-2\nu}\,\nabla^2\ddot{\phi}\,\mathbf{1}\right] \tag{6.216}$$

Since

$$\text{tr }\widehat{\nabla}\,(\nabla\,\ddot{\phi})=\nabla^2\ddot{\phi},\quad \text{tr }\widehat{\nabla}(\text{curl }\ddot{\omega})=0 \tag{6.217}$$

therefore, taking the trace of (6.216) we get

$$\text{tr }\ddot{\mathbf{S}}=c_2^2\,\frac{2(1+\nu)}{1-2\nu}\,\nabla^2\ddot{\phi} \tag{6.218}$$

and it follows from (6.216) and (6.218) that

$$\frac{1}{2\mu}\left[\ddot{\mathbf{S}}-\frac{\nu}{1+\nu}(\text{tr }\ddot{\mathbf{S}})\mathbf{1}\right]=\frac{1}{\rho}\widehat{\nabla}(\nabla\ddot{\phi}+\text{curl }\ddot{\omega}) \tag{6.219}$$

In addition, if follows from (6.208) that

$$\text{div}\,\mathbf{S} + \mathbf{b} = \text{div}\left\{\mathbf{S}_0 + t\,\dot{\mathbf{S}}_0 + 2c_2^2\,\widehat{\nabla}(\nabla\phi + \text{curl}\,\omega) + (c_1^2 - 2c_2^2)\nabla^2\phi\mathbf{1}\right\} + \mathbf{b} \tag{6.220}$$

$$= \text{div}(\mathbf{S}_0 + t\,\dot{\mathbf{S}}_0) + \mathbf{b}(\mathbf{x}, t) + \left[c_2^2\,\nabla^2 + (c_1^2 - c_2^2)\nabla\,\text{div}\right](\nabla\phi + \text{curl}\,\omega) \tag{6.221}$$

Hence, in view of (6.207) and (6.214) we get

$$\text{div}\,\mathbf{S} + \mathbf{b} = -c_1^2\,\nabla\alpha - c_2^2\,\text{curl}\,\beta + \nabla^2\,(c_1^2\nabla\,\phi + c_2^2\,\text{curl}\,\omega) \tag{6.222}$$

Finally, applying the operator $\widehat{\nabla}$ to (6.222) we obtain

$$\begin{aligned}\widehat{\nabla}(\text{div}\,\mathbf{S} + \mathbf{b}) &= \widehat{\nabla}[c_1^2\nabla(\nabla^2\phi - \alpha) + c_2^2\text{curl}(\nabla^2\omega - \beta)]\\ &= \widehat{\nabla}\left[c_1^2\nabla\left(\Box_1^2\phi - \alpha + c_1^{-2}\ddot{\phi}\right) + c_2^2\text{curl}\left(\Box_2^2\omega - \beta + c_2^{-2}\ddot{\boldsymbol{\omega}}\right)\right]\end{aligned} \tag{6.223}$$

Since ϕ and ω satisfy Eqs. (6.201) and (6.202), respectively, therefore dividing (6.223) by ρ and taking into account (6.219) we find that $\mathbf{S} = \mathbf{S}(\mathbf{x}, t)$ satisfies the stress equation of motion in the form

$$\rho^{-1}\widehat{\nabla}\,(\text{div}\,\mathbf{S} + \mathbf{b}) - \frac{1}{2\mu}\left[\ddot{\mathbf{S}} - \frac{\nu}{1+\nu}(\text{tr}\,\ddot{\mathbf{S}})\mathbf{1}\right] = 0 \tag{6.224}$$

This completes solution to Problem 6.8.

Problem 6.9. Let $\mathbf{S}$ be a symmetric second-order tensor field on $B \times [0, \infty)$ that satisfies the stress equation of motion

$$\widehat{\nabla}(\text{div}\,\mathbf{S}) - \frac{\rho}{2\mu}\left[\ddot{\mathbf{S}} - \frac{\nu}{1+\nu}(\text{tr}\,\ddot{\mathbf{S}})\,\mathbf{1}\right] = -\widehat{\nabla}\mathbf{b} \quad \text{on} \quad B \times [0, \infty) \tag{6.225}$$

subject to the initial conditions

$$\mathbf{S}(\mathbf{x}, 0) = \mathbf{S}_0(\mathbf{x}), \quad \dot{\mathbf{S}}(\mathbf{x}, 0) = \dot{\mathbf{S}}_0(\mathbf{x}) \quad \text{for} \quad \mathbf{x} \in B \tag{6.226}$$

Show that there are a scalar field $\phi = \phi(\mathbf{x}, t)$ and a vector field $\omega = \omega(\mathbf{x}, t)$ such that

$$\mathbf{S} = \mathbf{S}_0 + t\,\dot{\mathbf{S}}_0 + 2c_2^2\,\left[\nabla\nabla\phi + \widehat{\nabla}(\text{curl}\,\omega)\right] + (c_1^2 - 2c_2^2)\nabla^2\phi\,\mathbf{1} \tag{6.227}$$

$$\Box_1^2\phi = \alpha, \quad \phi(\mathbf{x}, 0) = \dot{\phi}(\mathbf{x}, 0) = 0 \tag{6.228}$$

$$\Box_2^2\omega = \beta, \quad \text{div}\,\omega = 0, \quad \omega(\mathbf{x}, 0) = \dot{\omega}(\mathbf{x}, 0) = \mathbf{0} \tag{6.229}$$

where the fields α and β are given by Eqs. (6.227) and (6.228), respectively, of Problem 6.8.

Note. Solution to Problem 6.9 implies that the tensor solution of Lame-type [Eqs. (6.227) through (6.229)] is complete.

Solution. In this problem the fields $\mathbf{S}_0$, $\dot{\mathbf{S}}_0$, and $\mathbf{b}$ are prescribed. Therefore, the fields α and β, given by Eqs. (6.234) and (6.235), respectively, of Problem 6.8 are given.

Let $\phi^{(0)}$ and $\omega^{(0)}$ be solutions to the equations [see Eqs. (6.230) and (6.231) of Problem 6.8]

$$\Box_1^2\,\phi^{(0)} = \alpha \quad \text{on } B \times [0, \infty) \tag{6.230}$$

$$\Box_2^2\,\omega^{(0)} = \beta, \quad \text{div}\,\omega = 0 \quad \text{on } B \times [0, \infty) \tag{6.231}$$

subject to the homogeneous initial conditions

$$\phi^{(0)}(\mathbf{x}, 0) = 0, \qquad \dot{\phi}^{(0)}(\mathbf{x}, 0) = 0 \tag{6.232}$$

$$\omega^{(0)}(\mathbf{x}, 0) = \mathbf{0}, \qquad \dot{\omega}^{(0)}(\mathbf{x}, 0) = 0 \tag{6.233}$$

For example, $\phi^{(0)}$ and $\omega^{(0)}$ can be taken in the form of retarded potentials. Then, it follows from the solution to Problem 6.8 that the tensor field $\mathbf{S}^{(0)}$ defined by

$$\mathbf{S}^{(0)}(\mathbf{x}, t) = \mathbf{S}_0(\mathbf{x}) + t\dot{\mathbf{S}}_0(\mathbf{x}) + 2c_2^2\widehat{\nabla}\left(\nabla\phi^{(0)} + \text{curl}\,\omega^{(0)}\right) + \left(c_1^2 - 2c_2^2\right)\nabla^2\phi^{(0)}\mathbf{1} \tag{6.234}$$

satisfies the equation

$$\widehat{\nabla}\left(\text{div}\,\mathbf{S}^{(0)} + \mathbf{b}\right) - \frac{\rho}{2\mu}\left[\ddot{\mathbf{S}}^{(0)} - \frac{\nu}{1+\nu}(\text{tr}\,\ddot{\mathbf{S}}^{(0)})\mathbf{1}\right] = \mathbf{0} \tag{6.235}$$

In addition, because of (6.232) and (6.233)

$$\mathbf{S}^{(0)}(\mathbf{x}, 0) = \mathbf{S}_0(\mathbf{x}), \quad \dot{\mathbf{S}}^{(0)}(\mathbf{x}, 0) = \dot{\mathbf{S}}_0(x) \tag{6.236}$$

Introduce the notation

$$\mathbf{R} = \mathbf{S} - \mathbf{S}^{(0)} \tag{6.237}$$

where $\mathbf{S}$ satisfies Eqs. (6.225) and (6.226) of the Problem 6.9. Then $\mathbf{R}$ satisfies the equations

$$\widehat{\nabla}(\text{div}\,\mathbf{R}) - \frac{\rho}{2\mu}\left[\ddot{\mathbf{R}} - \frac{\nu}{1+\nu}(\text{tr}\,\ddot{\mathbf{R}})\mathbf{1}\right] = \mathbf{0} \tag{6.238}$$

and

$$\mathbf{R}(\mathbf{x}, 0) = \mathbf{0}, \quad \dot{\mathbf{R}}(\mathbf{x}, 0) = \mathbf{0} \tag{6.239}$$

An equivalent form of (6.238) reads

$$\ddot{\mathbf{R}} = \rho^{-1} \left\{ 2\mu \widehat{\nabla}(\text{div } \mathbf{R}) + \frac{2\mu\nu}{1-2\nu} \times [\text{div}(\text{div } \mathbf{R})]\mathbf{1} \right\} \tag{6.240}$$

In components (6.240) reads

$$\ddot{R}_{ij} = \rho^{-1} \left\{ \mu(R_{ik,kj} + R_{jk,ki}) + \frac{2\mu\nu}{1-2\nu} R_{ab,ab}\, \delta_{ij} \right\} \tag{6.241}$$

Hence

$$\ddot{R}_{ia,a} = \rho^{-1} \left\{ \mu(R_{ik,kaa} + R_{ak,kia}) + \frac{2\mu\nu}{1-2\nu} R_{mn,mni} \right\} \tag{6.242}$$

Taking into account the relations

$$\nabla^2 = \nabla \text{ div} - \text{curl curl} \tag{6.243}$$

$$c_2^2 = \frac{\mu}{\rho}, \quad c_1^2 = \frac{2(1-\nu)}{1-2\nu} c_2^2 \tag{6.244}$$

and using direct notation, we reduce (6.242) to the form

$$\text{div } \ddot{\mathbf{R}} = \left(c_1^2\, \nabla \text{ div} - c_2^2 \text{ curl curl} \right) \text{ div } \mathbf{R} \tag{6.245}$$

Let ξ and $\mathbf{r}$ be defined by

$$\xi = c_1^2 \text{ div div } \mathbf{R} * t \tag{6.246}$$

$$\mathbf{r} = -c_2^2 \text{ curl div } \mathbf{R} * t \tag{6.247}$$

where $*$ represents the convolution product. Then

$$\ddot{\xi} = c_1^2 \text{ div div } \mathbf{R} \tag{6.248}$$

$$\xi(\mathbf{x}, 0) = 0, \quad \dot{\xi}(\mathbf{x}, 0) = 0 \tag{6.249}$$

and

$$\ddot{\mathbf{r}} = -c_2^2 \text{ curl div } \mathbf{R} \tag{6.250}$$

$$\mathbf{r}(\mathbf{x}, 0) = 0, \quad \dot{\mathbf{r}}(\mathbf{x}, 0) = 0 \tag{6.251}$$

$$\text{div } \mathbf{r}(\mathbf{x}, t) = 0 \tag{6.252}$$

It follows from (6.239) and (6.245)–(6.251) that

$$\text{div}\,\mathbf{R} = \nabla\xi + \text{curl}\,\mathbf{r} \tag{6.253}$$

By taking the div operator on (6.253) we obtain

$$\text{div div}\,\mathbf{R} = \nabla^2\xi \tag{6.254}$$

and applying the curl to (6.253) we get

$$\text{curl div}\,\mathbf{R} = \text{curl curl}\,\mathbf{r} \tag{6.255}$$

Hence, in view of (6.243) and (6.252)

$$\text{curl div}\,\mathbf{R} = -\nabla^2\mathbf{r} \tag{6.256}$$

Also, because of Eqs. (6.248) and (6.254) we obtain

$$\Box_1^2\xi = 0 \tag{6.257}$$

and using Eqs. (6.250) and (6.256) we obtain

$$\Box_2^2\mathbf{r} = \mathbf{0} \tag{6.258}$$

Substituting div **R** from (6.253) into (6.240) we obtain

$$\ddot{\mathbf{R}} = \frac{\mu}{\rho}\left\{2\widehat{\nabla} + \mathbf{1}\frac{2\nu}{1-2\nu}\text{div}\right\}(\nabla\xi + \text{curl}\,\mathbf{r}) \tag{6.259}$$

Let

$$\phi^{(1)} = \xi * t, \quad \omega^{(1)} = \mathbf{r} * t \tag{6.260}$$

Then

$$\ddot{\phi}^{(1)} = \xi, \quad \phi^{(1)}(\mathbf{x},0) = \dot{\phi}^{(1)}(\mathbf{x},0) = 0 \tag{6.261}$$

and

$$\ddot{\omega}^{(1)} = \mathbf{r}, \quad \omega^{(1)}(\mathbf{x},0) = \dot{\omega}^{(1)}(\mathbf{x},0) = \mathbf{0}, \quad \text{div}\,\omega^{(1)} = 0 \tag{6.262}$$

Integrating (6.259) twice with respect to time and taking into account the homogeneous initial conditions (6.239), (6.261) and (6.262) we obtain

$$\mathbf{R} = \rho^{-1}\mu\left(2\widehat{\nabla} + \mathbf{1}\frac{2\nu}{1-2\nu}\,\text{div}\right)\left(\nabla\phi^{(1)} + \text{curl}\,\omega^{(1)}\right) \tag{6.263}$$

where, because of (6.257) and (6.258), $(6.261)_1$ and $(6.262)_1$,

$$\Box_1^2 \ddot{\phi}^{(1)} = 0 \tag{6.264}$$

and

$$\Box_2^2 \ddot{\omega}^{(1)} = \mathbf{0} \tag{6.265}$$

Since, in view of (6.249), (6.251), $(6.261)_2$ and $(6.262)_2$

$$\phi^{(1)}(\mathbf{x}, 0) = \dot{\phi}^{(1)}(\mathbf{x}, 0) = \ddot{\phi}^{(1)}(\mathbf{x}, 0) = \dddot{\phi}^{(1)}(\mathbf{x}, 0) = 0 \tag{6.266}$$

$$\omega^{(1)}(\mathbf{x}, 0) = \dot{\omega}^{(1)}(\mathbf{x}, 0) = \ddot{\omega}^{(1)}(\mathbf{x}, 0) = \dddot{\omega}^{(1)}(\mathbf{x}, 0) = 0 \tag{6.267}$$

therefore, integrating twice Eqs. (6.264) and (6.265), with respect to time, we obtain

$$\Box_1^2 \, \phi^{(1)} = 0 \tag{6.268}$$

and

$$\Box_1^2 \, \omega^{(1)} = \mathbf{0} \tag{6.269}$$

Also, note that an alternative form of (6.263) reads

$$\mathbf{R} = 2c_2^2 \widehat{\nabla} \left(\nabla \phi^{(1)} + \operatorname{curl} \omega^{(1)} \right) + \left(c_1^2 - 2c_2^2 \right) \left(\nabla^2 \phi^{(1)} \right) \mathbf{1} \tag{6.270}$$

Therefore, because of (6.234), (6.237), and (6.270)

$$\begin{aligned} \mathbf{S} = \mathbf{S}_0 + \mathbf{R} = \mathbf{S}_0(\mathbf{x}) + t\dot{\mathbf{S}}_0(\mathbf{x}) + 2c_2^2 \widehat{\nabla} \left[\nabla \left(\phi^{(0)} + \phi^{(1)} \right) + \operatorname{curl} \left(\omega^{(0)} + \omega^{(1)} \right) \right] \\ + \left(c_1^2 - 2c_2^2 \right) \left[\nabla^2 \left(\phi^{(0)} + \phi^{(1)} \right) \right] \mathbf{1} \end{aligned} \tag{6.271}$$

The fields ϕ and ω are defined by

$$\phi = \phi^{(0)} + \phi^{(1)}, \quad \omega = \omega^{(0)} + \omega^{(1)} \tag{6.272}$$

where $(\phi^{(0)}, \ \omega^{(0)})$ and $(\phi^{(1)}, \ \omega^{(1)})$ have been defined before.

If the definition (6.272) of ϕ and ω is taken into account, Eq. (6.271) reduces to Eq. (6.225) of Problem 6.9.

Finally, if we note that the pair $(\phi^{(0)}, \ \omega^{(0)})$ satisfies Eqs. (6.230)–(6.233) and the pair $(\phi^{(1)}, \ \omega^{(1)})$ satisfies Eqs. (6.261)–(6.262), and (6.268)–(6.269), we find that the pair $(\phi, \ \omega)$ meets Eqs. (6.228)–(6.229). This completes a solution to Problem 6.9.

Problem 6.10. Consider the stress equation of motion in the form

$$\widehat{\nabla}(\operatorname{div} \mathbf{S}) - \rho \, \mathbf{K}[\ddot{\mathbf{S}}] = -\mathbf{B} \quad \text{on} \quad B \times [0, \infty) \tag{6.273}$$

subject to the homogeneous initial conditions

$$\mathbf{S}(\mathbf{x}, 0) = \mathbf{0}, \quad \dot{\mathbf{S}}(\mathbf{x}, \mathbf{0}) = \mathbf{0} \quad \text{for} \quad \mathbf{x} \in B \tag{6.274}$$

where

$$\mathbf{K}[\mathbf{S}] = \frac{1}{2\mu}\left[\mathbf{S} - \frac{\nu}{1+\nu}(\text{tr}\,\mathbf{S})\,\mathbf{1}\right] \tag{6.275}$$

and $\mathbf{B} = \mathbf{B}(\mathbf{x}, t)$ is an arbitrary symmetric second-order tensor field on $B \times [0, \infty)$. Define a vector field $\mathbf{v} = \mathbf{v}(\mathbf{x}, t)$ by

$$\mathbf{v}(\mathbf{x}, t) = -\frac{c_2^2 t^2}{4\pi} \int\limits_{|\xi| \le 1} \frac{\mathbf{f}[\mathbf{x} - c_2 t\xi,\ (1 - |\xi|)t]}{|\xi|}\, dv(\xi) \tag{6.276}$$

where

$$\mathbf{f}(\mathbf{x}, t) = \left\{\left(\frac{c_1^2}{c_2^2} - 1\right)\nabla g + \frac{1}{\rho\, c_2^2}\text{div}\,\mathbf{K}^{-1}[\mathbf{B}]\right\}(\mathbf{x}, t) \tag{6.277}$$

$$\text{g}(\mathbf{x}, t) = -\frac{c_1^2 t^2}{4\pi} \int\limits_{|\xi| \le 1} \frac{\text{h}[\mathbf{x} - c_1 t\xi,\ (1 - |\xi|)t]}{|\xi|}\, dv(\xi) \tag{6.278}$$

and

$$\text{h}(\mathbf{x}, t) = \frac{1}{\rho\, c_1^2}\text{div}\,\text{div}\,\mathbf{K}^{-1}[\mathbf{B}](\mathbf{x}, t) \tag{6.279}$$

Let

$$\mathbf{S}(\mathbf{x}, t) = \frac{1}{\rho}\mathbf{K}^{-1}[\widehat{\nabla}\mathbf{v} + \mathbf{B}] * t \tag{6.280}$$

Show that $\mathbf{S}$ satisfies Eqs. (6.273) and (6.274).
Hint. Use the result of Problem 6.4 that the function

$$\varphi(\mathbf{x}, t) = -\frac{c^2 t^2}{4\pi} \int\limits_{|\xi| \le 1} \frac{F[\mathbf{x} - ct\xi,\ (1 - |\xi|)t]}{|\xi|}\, dv(\xi) \quad \text{on} \quad E^3 \times [0, \infty) \tag{6.281}$$

satisfies the inhomogeneous wave equation

$$\left(\nabla^2 - \frac{1}{c^2}\frac{\partial^2}{\partial t^2}\right)\varphi = -F \quad \text{on} \quad E^3 \times [0, \infty) \tag{6.282}$$

subject to the homogeneous initial conditions

$$\varphi(\mathbf{x}, 0) = \dot{\varphi}(\mathbf{x}, 0) = 0 \tag{6.283}$$

Solution. To show that $\mathbf{S}$ given by

$$\mathbf{S}(\mathbf{x}, t) = \rho^{-1}\mathbf{K}^{-1}[\widehat{\nabla}\mathbf{v} + \mathbf{B}] * t \tag{6.284}$$

satisfies Eqs. (6.273)–(6.274), we note that

$$\mathbf{S}(\mathbf{x}, 0) = \mathbf{0}, \quad \dot{\mathbf{S}}(\mathbf{x}, 0) = \mathbf{0} \tag{6.285}$$

Hence, **S** given by (6.284), satisfies Eqs. (6.274). To show that **S** given by (6.284) satisfies (6.273) we substitute **S** given by (6.284) to (6.273) and obtain

$$\widehat{\nabla}(\text{div } \mathbf{S}) - \rho\mathbf{K}[\ddot{\mathbf{S}}] = \widehat{\nabla}(\text{div } \mathbf{S}) - \widehat{\nabla}\mathbf{v} - \mathbf{B} = -\mathbf{B} \tag{6.286}$$

In the following we prove that

$$\mathbf{v} = \text{div } \mathbf{S} \tag{6.287}$$

This implies that **S** given by (6.284) meets (6.273). To this end we note that from Eqs. (6.276)–(6.277) we obtain

$$\Box_2^2\mathbf{v} = -\mathbf{f} = -\left\{\left(\frac{c_1^2}{c_2^2} - 1\right)\nabla g + \frac{1}{\rho c_2^2}\text{div } \mathbf{K}^{-1}[\mathbf{B}]\right\} \tag{6.288}$$

$$\mathbf{v}(\mathbf{x}, 0) = \mathbf{0}, \quad \dot{\mathbf{v}}(\mathbf{x}, 0) = \mathbf{0} \tag{6.289}$$

Also, Eqs. (6.278)–(6.280) imply that

$$\Box_1^2 g = -h = -\frac{1}{\rho c_1^2}\text{div div } \mathbf{K}^{-1}[\mathbf{B}] \tag{6.290}$$

$$g(\mathbf{x}, 0) = \dot{g}(\mathbf{x}, 0) = 0 \tag{6.291}$$

By taking the div operator of (6.288) we get

$$\Box_2^2\text{div } \mathbf{v} = -\left\{\left(\frac{c_1^2}{c_2^2} - 1\right)\nabla^2 g + \frac{1}{\rho c_2^2}\text{ div div } \mathbf{K}^{-1}[\mathbf{B}]\right\} \tag{6.292}$$

By eliminating div div $\mathbf{K}^{-1}[\mathbf{B}]$ from Eqs. (6.290) and (6.292), we obtain

$$\begin{aligned}\Box_2^2(\text{div } \mathbf{v}) &= -\left\{\left(\frac{c_1^2}{c_2^2} - 1\right)\nabla^2 g - \frac{c_1^2}{c_2^2}\Box_1^2 g\right\} \\ &= -\left\{-\nabla^2 g + \frac{1}{c_2^2}\ddot{g}\right\} = +\Box_2^2 g\end{aligned} \tag{6.293}$$

Hence

$$\Box_2^2(\text{div } \mathbf{v} - g) = 0 \tag{6.294}$$

Since, $\mathbf{v}$ and g satisfy the homogeneous initial conditions (6.289) and (6.291), respectively, Eq. (6.294) implies that

$$\text{div } \mathbf{v} = g \tag{6.295}$$

Substituting g from (6.295) into the RHS of (6.288), we obtain

$$\ddot{\mathbf{v}} = c_2^2 \nabla^2 \mathbf{v} + \left(c_1^2 - c_2^2\right) \nabla \text{div } \mathbf{v} + \rho^{-1} \text{div } \mathsf{K}^{-1}[\mathbf{B}] \tag{6.296}$$

Since

$$\nabla^2 \mathbf{v} = 2 \,\text{div } (\widehat{\nabla} \mathbf{v}) - \nabla \text{div } \mathbf{v} \tag{6.297}$$

and

$$\nabla \text{div } \mathbf{v} = \text{div } [\mathbf{1} \text{ tr } (\widehat{\nabla} \mathbf{v})] \tag{6.298}$$

therefore, Eq. (6.296) can be written as

$$\ddot{\mathbf{v}} = \text{div } \left\{2c_2^2 (\widehat{\nabla} \mathbf{v}) + \left(c_1^2 - 2c_2^2\right) \mathbf{1} \text{ tr } (\widehat{\nabla} \mathbf{v}) + \rho^{-1} \mathsf{K}^{-1}[\mathbf{B}]\right\} \tag{6.299}$$

or, in view of (6.284),

$$\ddot{\mathbf{v}} = \text{div } \ddot{\mathbf{S}} \tag{6.300}$$

Integrating (6.300) with respect to time twice, and using the homogeneous initial conditions for $\mathbf{v}$ and $\mathbf{S}$, given by Eqs. (6.285) and (6.289), respectively, we arrive at Eq. (6.287). This completes a solution to Problem 6.10.

Note that the solution to Problem 6.10 provides an effective solution of the incompatible elastodynamics when $\mathbf{B}$ represents a space-time distribution of defects on $B \times [0, \infty)$.

Chapter 7
Formulation of Two-Dimensional Problems

In this chapter a class of problems is discussed in which an elastic state depends on two space variables only, or an elastic process depends on two space variables and time only. In particular, problems related to a plane strain state and a generalized plane stress state of homogeneous isotropic elastostatics, a plane strain process, and a generalized plane stress process of homogeneous isotropic elastodynamics are discussed. The problems related to a two-dimensional homogeneous isotropic elastodynamics described in terms of stresses only are also considered. [See also Chaps. 16 and 17].

7.1 Two-Dimensional Problems of Isothermal Elastostatics

A state of plane strain. An elastic body is said to be in a state of plane strain corresponding to a body force $\mathbf{b} = (b_1, b_2, 0)$ if the elastic state $s = [\mathbf{u}, \mathbf{E}, \mathbf{S}]$ complies with the two-dimensional field equations

$$u_\alpha = u_\alpha(x_1, x_2) \quad \text{for} \quad (x_1, x_2) \in C_0 \tag{7.1}$$

$$E_{\alpha\beta} = \frac{1}{2}(u_{\alpha,\beta} + u_{\beta,\alpha}) \quad \text{on} \quad C_0 \tag{7.2}$$

$$S_{\alpha\beta,\beta} + b_\alpha = 0 \quad \text{on} \quad C_0 \tag{7.3}$$

$$S_{\alpha\beta} = 2\mu\, E_{\alpha\beta} + \lambda\, E_{\gamma\gamma}\delta_{\alpha\beta} \quad \text{on} \quad C_0 \tag{7.4}$$

In Eqs. (7.1)–(7.4), and in all plane problems, the Greek subscripts α, β, and γ take values 1 and 2; and C_0 is a domain in the x_1, x_2 plane. In Eq. (7.3) $b_\alpha = b_\alpha(x_1, x_2)$, and the remaining components of $s = [\mathbf{u}, \mathbf{E}, \mathbf{S}]$ are given by

M. Reza Eslami et al., *Theory of Elasticity and Thermal Stresses*, Solid Mechanics and Its Applications 197, DOI: 10.1007/978-94-007-6356-2_7,

$$u_3 = 0, \quad E_{13} = E_{23} = E_{33} = 0 \quad \text{on} \quad C_0 \tag{7.5}$$
$$S_{13} = S_{23} = 0, \quad S_{33} = \nu\, S_{\alpha\alpha} \quad \text{on} \quad C_0 \tag{7.6}$$

An alternative form of the constitutive relation (7.4) is

$$E_{\alpha\beta} = \frac{1}{2\mu}(S_{\alpha\beta} - \nu\, S_{\gamma\gamma}\delta_{\alpha\beta}) \quad \text{on} \quad C_0 \tag{7.7}$$

By eliminating the fields $E_{\alpha\beta}$ and $S_{\alpha\beta}$ from Eqs. (7.2)–(7.4) we obtain *the displacement field equations for a plane strain problem*

$$\mu\, u_{\alpha,\gamma\gamma} + (\lambda + \mu)u_{\gamma,\gamma\alpha} + b_\alpha = 0 \quad \text{on} \quad C_0 \tag{7.8}$$

A generalized plane stress state. A generalized plane stress state $\overline{s} = [\overline{\mathbf{u}}, \overline{\mathbf{E}}, \overline{\mathbf{S}}]$ corresponding to a body force $\overline{\mathbf{b}} = (\overline{b}_1, \overline{b}_2, 0)$ is defined as an elastic state which complies with the two-dimensional field equations

$$\overline{u}_\alpha = \overline{u}_\alpha(x_1, x_2) \quad \text{for} \quad (x_1, x_2) \in C_0 \tag{7.9}$$
$$\overline{E}_{\alpha\beta} = \frac{1}{2}(\overline{u}_{\alpha,\beta} + \overline{u}_{\beta,\alpha}) \quad \text{on} \quad C_0 \tag{7.10}$$
$$\overline{S}_{\alpha\beta,\beta} + \overline{b}_\alpha = 0 \quad \text{on} \quad C_0 \tag{7.11}$$
$$\overline{S}_{\alpha\beta} = 2\mu\, \overline{E}_{\alpha\beta} + \overline{\lambda}\, \overline{E}_{\gamma\gamma}\delta_{\alpha\beta} \quad \text{on} \quad C_0 \tag{7.12}$$

The remaining components of $\overline{s} = [\overline{\mathbf{u}}, \overline{\mathbf{E}}, \overline{\mathbf{S}}]$ are given by

$$\overline{u}_3 = 0, \quad \overline{E}_{13} = \overline{E}_{23} = 0, \quad \overline{E}_{33} = -\frac{\lambda}{\lambda + 2\mu}\overline{E}_{\gamma\gamma} \quad \text{on} \quad C_0 \tag{7.13}$$
$$\overline{S}_{13} = \overline{S}_{23} = \overline{S}_{33} = 0 \quad \text{on} \quad C_0 \tag{7.14}$$

In Eq. (7.12)

$$\overline{\lambda} = \frac{2\mu\,\lambda}{\lambda + 2\mu} \tag{7.15}$$

and an alternative form of (7.12) reads

$$\overline{E}_{\alpha\beta} = \frac{1}{2\mu}(\overline{S}_{\alpha\beta} - \overline{\nu}\, \overline{S}_{\gamma\gamma}\delta_{\alpha\beta}) \quad \text{on} \quad C_0 \tag{7.16}$$

By eliminating the fields $\overline{E}_{\alpha\beta}$ and $\overline{S}_{\alpha\beta}$ from Eqs. (7.10)–(7.12) we obtain *the displacement equations for a body subject to generalized plane stress conditions*

$$\mu\, \overline{u}_{\alpha,\gamma\gamma} + (\overline{\lambda} + \mu)\, \overline{u}_{\gamma,\gamma\alpha} + \overline{b}_\alpha = 0 \quad \text{on} \quad C_0 \tag{7.17}$$

7.2 Two-Dimensional Problems of Nonisothermal Elastostatics

A nonisothermal plane strain state in the x_1, x_2 plane corresponding to zero body forces and a temperature change $T = T(x_1, x_2)$ is defined as a *thermoelastic state* $s = [\mathbf{u}, \mathbf{E}, \mathbf{S}]$ that complies with the two-dimensional field equations

$$u_\alpha = u_\alpha(x_1, x_2) \quad \text{for} \quad (x_1, x_2) \in C_0 \tag{7.18}$$

$$E_{\alpha\beta} = \frac{1}{2}(u_{\alpha,\beta} + u_{\beta,\alpha}) \quad \text{on} \quad C_0 \tag{7.19}$$

$$S_{\alpha\beta,\beta} = 0 \quad \text{on} \quad C_0 \tag{7.20}$$

$$S_{\alpha\beta} = 2\mu\, E_{\alpha\beta} + \lambda\, E_{\gamma\gamma}\delta_{\alpha\beta} - (3\lambda + 2\mu)\alpha T\delta_{\alpha\beta} \quad \text{on} \quad C_0 \tag{7.21}$$

The remaining components of $s = [\mathbf{u}, \mathbf{E}, \mathbf{S}]$ are given by

$$u_3 = 0, \quad E_{13} = E_{23} = E_{33} = 0 \quad \text{on} \quad C_0 \tag{7.22}$$

$$S_{13} = S_{23} = 0, \quad S_{33} = \nu\, S_{\alpha\alpha} - (3\lambda + 2\mu)\alpha\, T \quad \text{on} \quad C_0 \tag{7.23}$$

and an alternative form of (7.21) is

$$E_{\alpha\beta} = \frac{1}{2\mu}(S_{\alpha\beta} - \nu\, S_{\gamma\gamma}\delta_{\alpha\beta}) + (1+\nu)\alpha\, T\delta_{\alpha\beta} \quad \text{on} \quad C_0 \tag{7.24}$$

By eliminating the fields $E_{\alpha\beta}$ and $S_{\alpha\beta}$ from Eqs. (7.19)–(7.21) we obtain *the displacement-temperature field equations for a body under plane strain conditions*

$$\mu\, u_{\alpha,\gamma\gamma} + (\lambda + \mu)u_{\gamma,\gamma\alpha} - \gamma T_{,\alpha} = 0 \quad \text{on} \quad C_0 \tag{7.25}$$

where

$$\gamma = (3\lambda + 2\mu)\alpha \tag{7.26}$$

A nonisothermal generalized plane stress state in the x_1, x_2 plane corresponding to zero body forces and a temperature change $\overline{T} = \overline{T}(x_1, x_2)$ is defined as a thermoelastic state $\overline{s} = [\overline{\mathbf{u}}, \overline{\mathbf{E}}, \overline{\mathbf{S}}]$ that complies with the two-dimensional field equations

$$\overline{u}_\alpha = \overline{u}_\alpha(x_1, x_2) \quad \text{for} \quad (x_1, x_2) \in C_0 \tag{7.27}$$

$$\overline{E}_{\alpha\beta} = \frac{1}{2}(\overline{u}_{\alpha,\beta} + \overline{u}_{\beta,\alpha}) \quad \text{on} \quad C_0 \tag{7.28}$$

$$\overline{S}_{\alpha\beta,\beta} = 0 \quad \text{on} \quad C_0 \tag{7.29}$$

$$\overline{S}_{\alpha\beta} = 2\mu\, \overline{E}_{\alpha\beta} + \overline{\lambda}\, \overline{E}_{\gamma\gamma}\delta_{\alpha\beta} - 2\mu\frac{3\lambda + 2\mu}{\lambda + 2\mu}\alpha\, \overline{T}\delta_{\alpha\beta} \quad \text{on} \quad C_0 \tag{7.30}$$

The remaining components of $\overline{s} = [\overline{\mathbf{u}}, \overline{\mathbf{E}}, \overline{\mathbf{S}}]$ are given by

$$\overline{u}_3 = 0, \quad \overline{E}_{13} = \overline{E}_{23} = 0, \quad \overline{E}_{33} = -\frac{\lambda}{\lambda + 2\mu}\overline{E}_{\gamma\gamma} + \frac{3\lambda + 2\mu}{\lambda + 2\mu}\alpha\overline{T} \quad \text{on} \quad C_0 \tag{7.31}$$

$$\overline{S}_{13} = \overline{S}_{23} = \overline{S}_{33} = 0 \quad \text{on} \quad C_0 \tag{7.32}$$

and an alternative form of Eq. (7.30) reads

$$\overline{E}_{\alpha\beta} = \frac{1}{2\mu}(\overline{S}_{\alpha\beta} - \overline{\nu}\,\overline{S}_{\gamma\gamma}\delta_{\alpha\beta}) + \alpha\,\overline{T}\delta_{\alpha\beta} \quad \text{on} \quad C_0 \tag{7.33}$$

where

$$\overline{\nu} = \frac{\nu}{1+\nu} \tag{7.34}$$

By eliminating the fields $\overline{E}_{\alpha\beta}$ and $\overline{S}_{\alpha\beta}$ from Eqs. (7.28)–(7.30) we obtain *the displacement-temperature equations for a body subject to generalized plane stress conditions*

$$\mu\,\overline{u}_{\alpha,\gamma\gamma} + (\overline{\lambda} + \mu)\,\overline{u}_{\gamma,\gamma\alpha} - \overline{\gamma}\,\overline{T}_{,\alpha} = 0 \quad \text{on} \quad C_0 \tag{7.35}$$

where

$$\overline{\gamma} = 2\mu\frac{3\lambda + 2\mu}{\lambda + 2\mu}\alpha \tag{7.36}$$

7.3 Two-Dimensional Problems of Elastodynamics

Two-Dimensional Problems of Isothermal Elastodynamics

A plane strain process $p = [\mathbf{u}, \mathbf{E}, \mathbf{S}]$ complies with the two-dimensional field equations

$$u_\alpha = u_\alpha(\mathbf{x}; t) \quad \text{for} \quad (\mathbf{x}, t) \in C_0 \times [0, \infty) \tag{7.37}$$

$$E_{\alpha\beta} = \frac{1}{2}(u_{\alpha,\beta} + u_{\beta,\alpha}) \quad \text{on} \quad C_0 \times [0, \infty) \tag{7.38}$$

$$S_{\alpha\beta,\beta} + b_\alpha = \rho\,\ddot{u}_\alpha \quad \text{on} \quad C_0 \times [0, \infty) \tag{7.39}$$

$$S_{\alpha\beta} = 2\mu\,E_{\alpha\beta} + \lambda\,E_{\gamma\gamma}\delta_{\alpha\beta} \quad \text{on} \quad C_0 \times [0, \infty) \tag{7.40}$$

where ρ is the density of the body, and $\mathbf{x} = (x_1, x_2) \in C_0$.

By eliminating the fields $E_{\alpha\beta}$ and $S_{\alpha\beta}$ from Eqs. (7.38)–(7.40) we obtain *the displacement equations of isothermal elastodynamics for a body subject to plane strain conditions*

$$\mu\,u_{\alpha,\gamma\gamma} + (\lambda + \mu)u_{\gamma,\gamma\alpha} + b_\alpha = \rho\,\ddot{u}_\alpha \quad \text{on} \quad C_0 \times [0, \infty) \tag{7.41}$$

A generalized plane stress process $\overline{p} = [\overline{\mathbf{u}}, \overline{\mathbf{E}}, \overline{\mathbf{S}}]$ complies with the two-dimensional field equations

$$\overline{u}_\alpha = \overline{u}_\alpha(\mathbf{x}, t) \quad \text{for} \quad (\mathbf{x}, t) \in C_0 \times [0, \infty) \tag{7.42}$$

$$\overline{E}_{\alpha\beta} = \frac{1}{2}(\overline{u}_{\alpha,\beta} + \overline{u}_{\beta,\alpha}) \quad \text{on} \quad C_0 \times [0, \infty) \tag{7.43}$$

$$\overline{S}_{\alpha\beta,\beta} + \overline{b}_\alpha = \rho\,\ddot{\overline{u}}_\alpha \quad \text{on} \quad C_0 \times [0, \infty) \tag{7.44}$$

$$\overline{S}_{\alpha\beta} = 2\mu\,\overline{E}_{\alpha\beta} + \overline{\lambda}\,\overline{E}_{\gamma\gamma}\delta_{\alpha\beta} \quad \text{on} \quad C_0 \times [0, \infty) \tag{7.45}$$

An alternative form of Eqs. (7.40) and (7.45), respectively, is

$$E_{\alpha\beta} = \frac{1}{2\mu}(S_{\alpha\beta} - \nu\, S_{\gamma\gamma}\delta_{\alpha\beta}) \quad \text{on} \quad C_0 \times [0, \infty) \tag{7.46}$$

and

$$\overline{E}_{\alpha\beta} = \frac{1}{2\mu}(\overline{S}_{\alpha\beta} - \overline{\nu}\,\overline{S}_{\gamma\gamma}\delta_{\alpha\beta}) \quad \text{on} \quad C_0 \times [0, \infty) \tag{7.47}$$

where

$$\overline{\nu} = \frac{\nu}{1+\nu} \tag{7.48}$$

By eliminating the fields $\overline{E}_{\alpha\beta}$ and $\overline{S}_{\alpha\beta}$ from Eqs. (7.43)–(7.45) we obtain *the displacement equations of isothermal elastodynamics for a body subject to generalized plane stress conditions*

$$\mu\,\overline{u}_{\alpha,\gamma\gamma} + (\overline{\lambda} + \mu)\,\overline{u}_{\gamma,\gamma\alpha} + \overline{b}_\alpha = \rho\,\ddot{\overline{u}}_\alpha \quad \text{on} \quad C_0 \times [0, \infty) \tag{7.49}$$

Also, by eliminating the fields u_α and $E_{\alpha\beta}$ from Eqs. (7.38), (7.39), and (7.46) *the stress equation of motion for a body under plane strain conditions* is obtained

$$S_{(\alpha\gamma,\gamma\beta)} - \frac{\rho}{2\mu}(\ddot{S}_{\alpha\beta} - \nu\,\ddot{S}_{\gamma\gamma}\delta_{\alpha\beta}) = -b_{(\alpha,\beta)} \quad \text{on} \quad C_0 \times [0, \infty) \tag{7.50}$$

And, by eliminating the fields $\overline{u}_\alpha$ and $\overline{E}_{\alpha\beta}$ from Eqs. (7.43), (7.44), and (7.47) *the stress equation of motion for a body under generalized plane stress conditions* is obtained

$$\overline{S}_{(\alpha\gamma,\gamma\beta)} - \frac{\rho}{2\mu}(\ddot{\overline{S}}_{\alpha\beta} - \overline{\nu}\,\ddot{\overline{S}}_{\gamma\gamma}\delta_{\alpha\beta}) = -\overline{b}_{(\alpha,\beta)} \quad \text{on} \quad C_0 \times [0, \infty) \tag{7.51}$$

7.4 Two-Dimensional Problems of Nonisothermal Elastodynamics

A nonisothermal plane strain process corresponding to zero body forces and a temperature change $T = T(\mathbf{x}, t)$ is defined as a thermoelastic process $p = [\mathbf{u}, \mathbf{E}, \mathbf{S}]$ that complies with the two-dimensional field equations

$$u_\alpha = u_\alpha(\mathbf{x}; t) \quad \text{for} \quad (\mathbf{x}, t) \in C_0 \times [0, \infty) \tag{7.52}$$

$$E_{\alpha\beta} = \frac{1}{2}(u_{\alpha,\beta} + u_{\beta,\alpha}) \quad \text{on} \quad C_0 \times [0, \infty) \tag{7.53}$$

$$S_{\alpha\beta,\beta} = \rho\, \ddot{u}_\alpha \quad \text{on} \quad C_0 \times [0, \infty) \tag{7.54}$$

$$S_{\alpha\beta} = 2\mu\, E_{\alpha\beta} + \lambda\, E_{\gamma\gamma}\delta_{\alpha\beta} - (3\lambda + 2\mu)\alpha\, T\delta_{\alpha\beta} \quad \text{on} \quad C_0 \times [0, \infty) \tag{7.55}$$

or

$$E_{\alpha\beta} = \frac{1}{2\mu}(S_{\alpha\beta} - \nu\, S_{\gamma\gamma}\delta_{\alpha\beta}) + (1 + \nu)\alpha\, T\delta_{\alpha\beta} \quad \text{on} \quad C_0 \times [0, \infty) \tag{7.56}$$

Similarly, *a nonisothermal generalized plane stress process* corresponding to zero body forces and a temperature change $\overline{T} = \overline{T}(\mathbf{x}, t)$ is defined as a process $\overline{p} = [\overline{\mathbf{u}}, \overline{\mathbf{E}}, \overline{\mathbf{S}}]$ that complies with the two-dimensional field equations

$$\overline{u}_\alpha = \overline{u}_\alpha(\mathbf{x}, t) \quad \text{for} \quad (\mathbf{x}, t) \in C_0 \times [0, \infty) \tag{7.57}$$

$$\overline{E}_{\alpha\beta} = \frac{1}{2}(\overline{u}_{\alpha,\beta} + \overline{u}_{\beta,\alpha}) \quad \text{on} \quad C_0 \times [0, \infty) \tag{7.58}$$

$$\overline{S}_{\alpha\beta,\beta} = \rho\, \ddot{\overline{u}}_\alpha \quad \text{on} \quad C_0 \times [0, \infty) \tag{7.59}$$

$$\overline{S}_{\alpha\beta} = 2\mu\, \overline{E}_{\alpha\beta} + \overline{\lambda}\, \overline{E}_{\gamma\gamma}\delta_{\alpha\beta} - 2\mu\frac{3\lambda + 2\mu}{\lambda + 2\mu}\alpha\, \overline{T}\delta_{\alpha\beta} \quad \text{on} \quad C_0 \times [0, \infty) \tag{7.60}$$

or

$$\overline{E}_{\alpha\beta} = \frac{1}{2\mu}(\overline{S}_{\alpha\beta} - \overline{\nu}\, \overline{S}_{\gamma\gamma}\delta_{\alpha\beta}) + \alpha\, \overline{T}\delta_{\alpha\beta} \quad \text{on} \quad C_0 \times [0, \infty) \tag{7.61}$$

By eliminating the fields $E_{\alpha\beta}$ and $S_{\alpha\beta}$ from Eqs. (7.53)–(7.55), *the displacement-temperature equations for a body subject to plane strain conditions* are obtained

$$\mu\, u_{\alpha,\gamma\gamma} + (\lambda + \mu)u_{\gamma,\gamma\alpha} - \gamma T_{,\alpha} = \rho\, \ddot{u}_\alpha \quad \text{on} \quad C_0 \times [0, \infty) \tag{7.62}$$

Similarly, by eliminating the fields $\overline{E}_{\alpha\beta}$ and $\overline{S}_{\alpha\beta}$ from Eqs. (7.58)–(7.60), *the displacement-temperature equations of motion for a body subject to generalized plane stress conditions* are obtained

$$\mu\, \overline{u}_{\alpha,\gamma\gamma} + (\overline{\lambda} + \mu)\, \overline{u}_{\gamma,\gamma\alpha} - \overline{\gamma}\, \overline{T}_{,\alpha} = \rho\, \ddot{\overline{u}}_\alpha \quad \text{on} \quad C_0 \times [0, \infty) \tag{7.63}$$

Also, by eliminating the fields u_α and $E_{\alpha\beta}$ from Eqs. (7.53), (7.54), and (7.56) *the stress-temperature equation of motion for a body under plane strain conditions* is obtained

$$S_{(\alpha\gamma,\gamma\beta)} - \frac{\rho}{2\mu}(\ddot{S}_{\alpha\beta} - \nu\, \ddot{S}_{\gamma\gamma}\delta_{\alpha\beta}) - \rho(1 + \nu)\alpha\, \ddot{T}\delta_{\alpha\beta} = 0 \quad \text{on} \quad C_0 \times [0, \infty) \tag{7.64}$$

Similarly, by eliminating the fields $\overline{u}_\alpha$ and $\overline{E}_{\alpha\beta}$ from Eqs. (7.58), (7.59), and (7.61) *the stress-temperature equation of motion for a body under generalized plane stress conditions* is obtained

$$\overline{S}_{(\alpha\gamma,\gamma\beta)} - \frac{\rho}{2\mu}(\ddot{\overline{S}}_{\alpha\beta} - \overline{\nu}\,\ddot{\overline{S}}_{\gamma\gamma}\delta_{\alpha\beta}) - \rho\,\alpha\,\ddot{\overline{T}}\delta_{\alpha\beta} = 0 \quad \text{on} \quad C_0 \times [0,\infty) \tag{7.65}$$

It is easy to show that *a particular solution* $S_{\alpha\beta}$ to Eq. (7.64) takes the form

$$S_{\alpha\beta} = 2\mu\,(\phi_{,\alpha\beta} - \phi_{,\gamma\gamma}\delta_{\alpha\beta}) + \rho\,\ddot{\phi}\delta_{\alpha\beta} \tag{7.66}$$

where $\phi = \phi(\mathbf{x}, t)$ satisfies the wave equation

$$\Box_1^2\phi = m\,T \tag{7.67}$$

in which

$$\Box_1^2 = \nabla^2 - \frac{1}{c_1^2}\frac{\partial^2}{\partial t^2}, \quad \text{and} \quad m = \frac{1+\nu}{1-\nu}\alpha \tag{7.68}$$

7.5 Problems and Solutions Related to Formulation of Two-Dimensional Problems

Problem 7.1. The displacement equations of isothermal elastostatics for a body subject to plane strain conditions take the form [see Eq. (7.8)]

$$\mu\,u_{\alpha,\gamma\gamma} + (\lambda+\mu)u_{\gamma,\gamma\alpha} + b_\alpha = 0 \tag{7.69}$$

where u_α is the displacement corresponding to the body force b_α $(\alpha, \gamma = 1, 2)$.

Let

$$u_\alpha = \psi_\alpha - \frac{1}{4(1-\nu)}(x_\gamma\psi_\gamma + \varphi)_{,\alpha} \tag{7.70}$$

where

$$\psi_{\alpha,\gamma\gamma} = -\frac{b_\alpha}{\mu}, \quad \varphi_{,\gamma\gamma} = \frac{x_\gamma b_\gamma}{\mu} \tag{7.71}$$

Show that u_α defined by Eqs. (7.70) and (7.71) satisfies Eq. (7.69).

Solution. Since

$$\lambda = \frac{2\mu\nu}{1-2\nu} \tag{7.72}$$

an equivalent form of (7.69) reads

$$u_{\alpha,\gamma\gamma} + \frac{1}{1-2\nu}u_{\gamma,\gamma\alpha} + \frac{b_\alpha}{\mu} = 0 \tag{7.73}$$

Also, an equivalent form of (7.70) takes the form

$$u_\alpha = \psi_\alpha - \frac{1}{4(1-\nu)}(x_\beta \psi_\beta + \varphi),_\alpha \tag{7.74}$$

Computing $u_{\alpha,\gamma}$ and $u_{\alpha,\gamma\gamma}$, respectively, we obtain

$$u_{\alpha,\gamma} = \psi_{\alpha,\gamma} - \frac{1}{4(1-\nu)}(x_\beta \psi_\beta + \varphi),_{\alpha\gamma} \tag{7.75}$$

and

$$u_{\alpha,\gamma\gamma} = \psi_{\alpha,\gamma\gamma} - \frac{1}{4(1-\nu)}(x_\beta \psi_\beta + \varphi),_{\alpha\gamma\gamma} \tag{7.76}$$

Hence, it follows from (7.75) that

$$u_{\gamma,\gamma} = \psi_{\gamma,\gamma} - \frac{1}{4(1-\nu)}(x_\beta \psi_\beta + \varphi),_{\gamma\gamma} \tag{7.77}$$

Since

$$(x_\beta \psi_\beta),_{\gamma\gamma} = (\psi_\gamma + x_\beta \psi_{\beta,\gamma}),_\gamma = 2\psi_{\gamma,\gamma} + x_\beta \psi_{\beta,\gamma\gamma} \tag{7.78}$$

and

$$(x_\beta \psi_\beta),_{\gamma\gamma\alpha} = 2\psi_{\gamma,\gamma\alpha} + \psi_{\alpha,\gamma\gamma} + x_\beta \psi_{\beta,\gamma\gamma\alpha} \tag{7.79}$$

therefore, (7.76) and (7.77), respectively, reduce to

$$u_{\alpha,\gamma\gamma} = \frac{1}{4(1-\nu)}[(3-4\nu)\psi_{\alpha,\gamma\gamma} - \varphi,_{\gamma\gamma\alpha} - 2\psi_{\gamma,\gamma\alpha} - x_\beta \psi_{\beta,\gamma\gamma\alpha}] \tag{7.80}$$

and

$$u_{\gamma,\gamma} = \frac{1}{4(1-\nu)}[2(1-2\nu)\psi_{\gamma,\gamma} - x_\beta \psi_{\beta,\gamma\gamma} - \varphi,_{\gamma\gamma}] \tag{7.81}$$

Next, if we take into account Eq. (7.71)

$$\psi_{\alpha,\gamma\gamma} = -\frac{b_\alpha}{\mu}, \quad \varphi,_{\gamma\gamma} = \frac{x_\beta b_\beta}{\mu} \tag{7.82}$$

we obtain

$$\psi_{\beta,\gamma\gamma\alpha} = \frac{-b_{\beta,\alpha}}{\mu} \tag{7.83}$$

and

$$\varphi,_{\gamma\gamma\alpha} = \frac{1}{\mu}(x_\beta b_\beta),_\alpha = \frac{b_\alpha}{\mu} + \frac{x_\beta b_{\beta,\alpha}}{\mu} \tag{7.84}$$

and (7.80) and (7.81), respectively, reduce to

$$u_{\alpha,\gamma\gamma} = \frac{1}{4(1-\nu)}\left[-(3-4\nu)\frac{b_\alpha}{\mu} - \frac{b_\alpha}{\mu} - x_\beta\frac{b_{\beta,\alpha}}{\mu} + x_\beta\frac{b_{\beta,\alpha}}{\mu} - 2\psi_{\gamma,\gamma\alpha}\right] \quad (7.85)$$

and

$$u_{\gamma,\gamma} = \frac{1}{4(1-\nu)}\left[2(1-2\nu)\psi_{\gamma,\gamma} + x_\beta\frac{b_\beta}{\mu} - x_\beta\frac{b_\beta}{\mu}\right] \quad (7.86)$$

Hence

$$u_{\alpha,\gamma\gamma} = \frac{1}{4(1-\nu)}\left[-4(1-\nu)\frac{b_\alpha}{\mu} - 2\psi_{\gamma,\gamma\alpha}\right] \quad (7.87)$$

and

$$u_{\gamma,\gamma\alpha} = \frac{1-2\nu}{2(1-\nu)}\psi_{\gamma,\gamma\alpha} \quad (7.88)$$

Finally, substituting (7.87) and (7.88) into LHS of (7.73) we find that u_α given by (7.70) and (7.71) satisfies (7.73). This completes a solution to Problem 7.1.

Problem 7.2. The displacement equations of equilibrium for a body subject to generalized plane stress conditions take the form [see Eq. (7.17)]

$$\mu\,\overline{u}_{\alpha,\gamma\gamma} + (\overline{\lambda} + \mu)\,\overline{u}_{\gamma,\gamma\alpha} + \overline{b}_\alpha = 0 \quad (7.89)$$

where $\overline{u}_\alpha$ is the displacement corresponding to the body force $\overline{b}_\alpha$ $(\alpha, \gamma = 1, 2)$, and

$$\overline{\lambda} = \frac{2\mu}{\lambda + 2\mu}\,\lambda \quad (7.90)$$

Let

$$\overline{u}_\alpha = \overline{\psi}_{,\alpha} - \frac{1}{4(1-\nu)}(x_\gamma\overline{\psi}_\gamma + \overline{\varphi})_{,\alpha} \quad (7.91)$$

where

$$\overline{\psi}_{\alpha,\gamma\gamma} = -\frac{\overline{b}_\alpha}{\mu}, \quad \overline{\varphi}_{,\gamma\gamma} = \frac{x_\gamma\overline{b}_\gamma}{\mu} \quad (7.92)$$

and

$$\overline{\nu} = \frac{\nu}{1+\nu} \quad (7.93)$$

Show that $\overline{u}_\alpha$ defined by Eqs. (7.91) through (7.93) satisfies the equilibrium equation (7.89).

Solution. To solve Problem 7.2 we let

$$\overline{\lambda} = 2\mu \frac{\overline{\nu}}{1 - 2\overline{\nu}} \tag{7.94}$$

Then $\overline{u}_\alpha$ given by (7.91), where $\overline{\psi}_\alpha$ and $\overline{\varphi}$ satisfy (7.92), in view of the solution to Problem 7.1, satisfies (7.89). To obtain (7.93) of Problem 7.2, note that because of (7.94) and (7.90)

$$\frac{\overline{\nu}}{1 - 2\overline{\nu}} = \frac{\lambda}{\lambda + 2\mu} = \frac{\nu}{1 - \nu}, \tag{7.95}$$

Hence

$$(1 - \nu)\overline{\nu} = \nu(1 - 2\overline{\nu}) \tag{7.96}$$

and

$$\overline{\nu} = \frac{\nu}{1 + \nu} \tag{7.97}$$

This completes a solution to Problem 7.2.

Problem 7.3. The displacement equations of thermoelastostatics for a body under plane strain conditions subject to a temperature change $T = T(\mathbf{x})$ take the form [see Eq. (7.25)]

$$u_{\alpha,\gamma\gamma} + \frac{1}{1 - 2\nu} u_{\gamma,\gamma\alpha} - \frac{\gamma}{\mu} T_{,\alpha} = 0 \tag{7.98}$$

where

$$\gamma = 2\mu \frac{1 + \nu}{1 - 2\nu} \alpha \tag{7.99}$$

Let

$$u_\alpha = \nabla^2 g_\alpha - \frac{1}{2(1 - \nu)} g_{\gamma,\gamma\alpha} \tag{7.100}$$

where

$$\nabla^2 \nabla^2 g_\alpha = \frac{\gamma}{\mu} T_{,\alpha} \tag{7.101}$$

Show that u_α given by Eqs. (7.100) and (7.101) satisfies Eq. (7.98).

Solution. From Eq. (7.100) we obtain

$$u_{\alpha,\gamma\gamma} = \nabla^2 \nabla^2 g_\alpha - \frac{1}{2(1 - \nu)} \nabla^2 g_{\beta,\beta\alpha} \tag{7.102}$$

and

$$u_{\gamma,\gamma\alpha} = \nabla^2 g_{\gamma,\gamma\alpha} - \frac{1}{2(1 - \nu)} \nabla^2 g_{\gamma,\gamma\alpha} = \frac{1 - 2\nu}{2(1 - \nu)} \nabla^2 g_{\gamma,\gamma\alpha} \tag{7.103}$$

Therefore,

$$u_{\alpha,\gamma\gamma} + \frac{1}{1-2\nu} u_{\gamma,\gamma\alpha} = \nabla^2\nabla^2 g_\alpha \tag{7.104}$$

Since, because of (7.101),

$$\nabla^2\nabla^2 g_\alpha = \frac{\gamma}{\mu} T_{,\alpha} \tag{7.105}$$

it follows from Eqs. (7.104) and (7.105) that u_α given by (7.100) and (7.101) satisfies (7.98). This completes solution to Problem 7.3.

Problem 7.4. The displacement equations of thermoelastostatics for a body under generalized plane stress conditions subject to a temperature change $\overline{T} = \overline{T}(\mathbf{x})$ take the form [see Eq. (7.35)]

$$\overline{u}_{\alpha,\gamma\gamma} + \frac{1}{1-2\overline{\nu}} \overline{u}_{\gamma,\gamma\alpha} - \frac{\overline{\gamma}}{\mu} \overline{T}_{,\alpha} = 0 \tag{7.106}$$

where

$$\overline{\nu} = \frac{\nu}{1+\nu}, \quad \overline{\gamma} = 2\mu \frac{1+\nu}{1-\nu} \alpha \tag{7.107}$$

Let

$$\overline{u}_\alpha = \nabla^2 \overline{g}_\alpha - \frac{1}{2(1-\overline{\nu})} \overline{g}_{\gamma,\gamma\alpha} \tag{7.108}$$

where

$$\nabla^2\nabla^2 \overline{g}_\alpha = \frac{\overline{\gamma}}{\mu} T_{,\alpha} \tag{7.109}$$

Show that $\overline{u}_\alpha$ given by Eqs. (7.108) and (7.109) satisfies Eq. (7.106).

Solution. Solution to this problem follows directly from the solution to Problem 7.3 in which we replace λ by $\overline{\lambda}$ given by

$$\overline{\lambda} = 2\mu \frac{\overline{\nu}}{1-2\overline{\nu}} = 2\mu \frac{\nu}{1-\nu} \tag{7.110}$$

and u_α, g_α, γ, and T by $\overline{u}_\alpha$, $\overline{g}_\alpha$, $\overline{\gamma}$, and $\overline{T}$, respectively.

This completes solution to Problem 7.4.

Problem 7.5. The displacement equations of isothermal elastodynamics for a body subject to plane strain conditions take the form [see Eq. (7.41)]

$$\mu\, u_{\alpha,\gamma\gamma} + (\lambda+\mu) u_{\gamma,\gamma\alpha} + b_\alpha = \rho\, \ddot{u}_\alpha \tag{7.111}$$

Let

$$b_\alpha = -h_{,\alpha} - \varepsilon_{\alpha\beta 3} k_{,\beta} \tag{7.112}$$

where $h = h(\mathbf{x})$ and $k = k(\mathbf{x})$ are prescribed scalar fields.

Let

$$u_\alpha = \varphi_{,\alpha} + \varepsilon_{\alpha\beta 3}\psi_{,\beta} \tag{7.113}$$

where φ *and* ψ satisfy, respectively, the equations

$$\Box_1^2\varphi = \frac{h}{\lambda + 2\mu} \tag{7.114}$$

and

$$\Box_2^2\psi = \frac{k}{\mu} \tag{7.115}$$

Show that u_α given by Eqs. (7.113) through (7.115) satisfies Eq. (7.111).

Solution. Using (7.113) we obtain

$$u_{\alpha,\gamma\gamma} = (\nabla^2\varphi)_{,\alpha} + \varepsilon_{\alpha\beta 3}(\nabla^2\psi)_{,\beta} \tag{7.116}$$

and

$$u_{\gamma,\gamma\alpha} = (\nabla^2\varphi)_{,\alpha} \tag{7.117}$$

Therefore,

$$\begin{aligned} u_{\alpha,\gamma\gamma} &+ \frac{\lambda+\mu}{\mu}u_{\gamma,\gamma\alpha} - \frac{1}{c_2^2}\ddot{u}_\alpha \\ &= \frac{\lambda+2\mu}{\mu}(\nabla^2\varphi)_{,\alpha} - \frac{1}{c_2^2}\ddot{\varphi}_{,\alpha} + \varepsilon_{\alpha\beta 3}\left[(\nabla^2\psi)_{,\beta} - \frac{1}{c_2^2}\ddot{\psi}_{,\beta}\right] \\ &= \frac{\lambda+2\mu}{\mu}\left(\nabla^2\varphi - \frac{1}{c_1^2}\ddot{\varphi}\right)_{,\alpha} + \varepsilon_{\alpha\beta 3}\left(\nabla^2\psi - \frac{1}{c_2^2}\ddot{\psi}\right)_{,\beta} \end{aligned} \tag{7.118}$$

Substituting (7.118) into (7.111) divided by μ, and using (7.112), (7.114), and (7.115), we obtain

$$\begin{aligned} &\frac{\lambda+2\mu}{\mu}\left(\nabla^2\varphi - \frac{1}{c_1^2}\ddot{\varphi}\right)_{,\alpha} + \varepsilon_{\alpha\beta 3}\left(\nabla^2\psi - \frac{1}{c_2^2}\ddot{\psi}\right)_{,\beta} - \frac{1}{\mu}h_{,\alpha} - \frac{1}{\mu}\varepsilon_{\alpha\beta 3}k_{,\beta} \\ &= \frac{\lambda+2\mu}{\mu}\left(\nabla^2\varphi - \frac{1}{c_1^2}\ddot{\varphi} - \frac{1}{\lambda+2\mu}h\right)_{,\alpha} + \varepsilon_{\alpha\beta 3}\left(\nabla^2\psi - \frac{1}{c_2^2}\ddot{\psi} - \frac{1}{\mu}k\right)_{,\beta} = 0 \end{aligned} \tag{7.119}$$

This completes solution to Problem 7.5.

Problem 7.6. The displacement equations of isothermal elastodynamics for a body subject to generalized plane stress conditions take the form [see Eq. (7.49)]

$$\mu \overline{u}_{\alpha,\gamma\gamma} + (\overline{\lambda} + \mu)\overline{u}_{\gamma,\gamma\alpha} + \overline{b}_\alpha = \rho \ddot{\overline{u}}_\alpha \tag{7.120}$$

Let

$$\overline{b}_\alpha = -\overline{h}_{,\alpha} - \varepsilon_{\alpha\beta 3}\overline{k}_{,\beta} \tag{7.121}$$

where $\overline{h} = \overline{h}(\mathbf{x})$ *and* $\overline{k} = \overline{k}(\mathbf{x})$ are prescribed scalar fields.

Let

$$\overline{u}_\alpha = \overline{\varphi}_{,\alpha} + \varepsilon_{\alpha\beta 3}\overline{\psi}_{,\beta} \tag{7.122}$$

where $\overline{\varphi}$ and $\overline{\psi}$ satisfy, respectively, the equations

$$\Box_1^2 \overline{\varphi} = \frac{\overline{h}}{\overline{\lambda} + 2\mu} \tag{7.123}$$

and

$$\Box_2^2 \overline{\psi} = \frac{\overline{k}}{\mu} \tag{7.124}$$

Here,

$$\Box_1^2 = \nabla^2 - \frac{1}{\overline{c}_1^2}\frac{\partial^2}{\partial t^2}, \quad \frac{1}{\overline{c}_1^2} = \frac{\rho}{\overline{\lambda} + 2\mu} \tag{7.125}$$

Show that u_α given by Eqs. (7.122) through (7.125) satisfies Eq. (7.120).

Solution. Solution to Problem 7.6 is obtained from the solution to Problem (7.5) in which we replace λ by $\overline{\lambda}$ (c_1 by $\overline{c}_1$), and u_α, b_α, (h and k), φ, and ψ, by $\overline{u}_\alpha$, $\overline{b}_\alpha$ ($\overline{h}$ and $\overline{k}$) $\overline{\varphi}$ and $\overline{\psi}$ respectively.

This completes solution to Problem 7.6.

Problem 7.7. The displacement equations of isothermal elastodynamics for a body subject to plane strain conditions take the form [see Eq. (7.41)]

$$\mu\, u_{\alpha,\gamma\gamma} + (\lambda + \mu)u_{\gamma,\gamma\alpha} + b_\alpha = \rho\, \ddot{u}_\alpha \tag{7.126}$$

Let u_α be a vector field defined by

$$u_\alpha = \Box_1^2 g_\alpha + \left(\frac{c_2^2}{c_1^2} - 1\right) g_{\gamma,\gamma\alpha} \tag{7.127}$$

where

$$\Box_1^2 \Box_2^2 g_\alpha = -\frac{b_\alpha}{\mu} \tag{7.128}$$

Show that u_α given by Eqs. (7.127) and (7.128) satisfies Eq. (7.126).

Solution. To show that u_α given by (7.127) and (7.128) meets (7.126), we note that (7.127) implies

$$u_{\gamma,\gamma\alpha} = \Box_1^2\, g_{\gamma,\gamma\alpha} + \left(\frac{c_2^2}{c_1^2} - 1\right)\nabla^2\, g_{\gamma,\gamma\alpha} \tag{7.129}$$

Also, note that (7.126) can be written as

$$\Box_2^2 u_\alpha + \left(\frac{c_1^2}{c_2^2} - 1\right) u_{\gamma,\gamma\alpha} + \frac{b_\alpha}{\mu} = 0 \tag{7.130}$$

Therefore, substituting u_α given by (7.127) and (7.128) into LHS of (7.130) we obtain

$$\begin{aligned} \text{LHS of (3)} &= \Box_1^2\Box_2^2\, g_\alpha + \frac{b_\alpha}{\mu} + \left(\frac{c_2^2}{c_1^2} - 1\right)\Box_2^2\, g_{\gamma,\gamma\alpha} \\ &\quad + \left(\frac{c_1^2}{c_2^2} - 1\right)\left[\Box_1^2\, g_{\gamma,\gamma\alpha} + \left(\frac{c_2^2}{c_1^2} - 1\right)\nabla^2\, g_{\gamma,\gamma\alpha}\right] \end{aligned} \tag{7.131}$$

Since

$$\frac{c_2^2}{c_1^2} - 1 + \frac{c_1^2}{c_2^2} - 1 + \left(\frac{c_1^2}{c_2^2} - 1\right)\left(\frac{c_2^2}{c_1^2} - 1\right) = \left(\frac{c_2^2}{c_1^2} - 1\right)\left(1 + \frac{c_1^2}{c_2^2} - 1\right) + \frac{c_1^2}{c_2^2} - 1 = 0 \tag{7.132}$$

and

$$\frac{1}{c_2^2}\left(1 - \frac{c_2^2}{c_1^2}\right) + \frac{1}{c_1^2}\left(1 - \frac{c_1^2}{c_2^2}\right) = 0 \tag{7.133}$$

$$\text{LHS of (3)} = \Box_1^2\, \Box_2^2\, g_\alpha + \frac{b_\alpha}{\mu} \tag{7.134}$$

Finally, because of (7.128) and (7.134), we obtain

$$\text{LHS of (3)} = 0 \tag{7.135}$$

This completes solution to Problem 7.7.

Problem 7.8. The stress equations of isothermal elastodynamics for a body subject to plane strain conditions take the form [see Eq. (7.50)]

$$S_{(\alpha\gamma,\gamma\beta)} - \frac{\rho}{2\mu}(\ddot{S}_{\alpha\beta} - \nu\, \ddot{S}_{\gamma\gamma}\delta_{\alpha\beta}) = -b_{(\alpha,\beta)} \tag{7.136}$$

Let $S_{\alpha\beta}$ be a tensor field defined by

$$S_{\alpha\beta} = 2\mu \left[\Box_1^2 \chi_{\alpha\beta} - \frac{1}{2(1-\nu)} (\chi_{\gamma\gamma,\alpha\beta} - \nu \, \delta_{\alpha\beta} \Box_2^2 \chi_{\gamma\gamma}) \right] \tag{7.137}$$

where $\chi_{\alpha\beta}$ is a symmetric second-order tensor field that satisfies the equations

$$\Box_1^2 \Box_2^2 \chi_{\alpha\beta} = -\frac{b_{(\alpha,\beta)}}{\mu} \tag{7.138}$$

and

$$\chi_{\alpha\beta,\gamma\gamma} + \chi_{\gamma\gamma,\alpha\beta} - 2\chi_{(\alpha\gamma,\gamma\beta)} = 0 \tag{7.139}$$

Show that $S_{\alpha\beta}$ given by Eqs. (7.137) through (7.139) satisfies the tensorial equation (7.136).

Note. If $\chi_{\alpha\beta} = g_{(\alpha,\beta)}$, where g_α is the vector field of Galerkin type from Problem 7.7, then $\chi_{\alpha\beta}$ satisfies Eq. (7.139) identically, and $S_{\alpha\beta}$ is the stress tensor corresponding to the displacement vector u_α of Problem 7.7.

Solution. To show that $S_{\alpha\beta}$ given by Eqs. (7.137)–(7.139) meets (7.136), we rewrite Eqs. (7.137)–(7.139) and obtain

$$S_{\alpha\beta} = 2\mu \left[\Box_1^2 \chi_{\alpha\beta} - \frac{1}{2(1-\nu)} \left(\chi_{\gamma\gamma,\alpha\beta} - \nu \delta_{\alpha\beta} \Box_2^2 \chi_{\gamma\gamma} \right) \right] \tag{7.140}$$

$$\Box_1^2 \Box_2^2 \chi_{\alpha\beta} = -\frac{1}{\mu} \, b_{(\alpha,\beta)} \tag{7.141}$$

$$\chi_{\alpha\beta,\gamma\gamma} + \chi_{\gamma\gamma,\alpha\beta} - 2\chi_{(\alpha\gamma,\gamma\beta)} = 0 \tag{7.142}$$

The stress equation of motion (7.136) is rewritten as

$$S_{(\alpha\gamma,\gamma\beta)} - \frac{\rho}{2\mu} (\ddot{S}_{\alpha\beta} - \nu \, \ddot{S}_{\gamma\gamma} \, \delta_{\alpha\beta}) = -b_{(\alpha,\beta)} \tag{7.143}$$

By taking the trace of (7.140), and using the identity

$$\Box_1^2 - \frac{1}{2(1-\nu)} \left(\nabla^2 - 2\nu \Box_2^2 \right) = \frac{1}{2(1-\nu)} \, \Box_2^2 \tag{7.144}$$

we obtain

$$S_{\gamma\gamma} = \frac{\mu}{1-\nu} \Box_2^2 \, \chi_{\gamma\gamma} \tag{7.145}$$

Hence, an alternative form of (7.140) reads

$$S_{\alpha\beta} - \nu\, S_{\gamma\gamma}\, \delta_{\alpha\beta} = 2\mu \left[\Box_1^2 \chi_{\alpha\beta} - \frac{1}{2(1-\nu)} \chi_{\gamma\gamma,\alpha\beta} \right] \tag{7.146}$$

Next, using (7.140) we obtain

$$S_{\alpha\gamma} = 2\mu \left[\Box_1^2 \chi_{\alpha\gamma} - \frac{1}{2(1-\nu)} \left(\chi_{\delta\delta,\alpha\gamma} - \nu \delta_{\alpha\gamma}\, \Box_2^2 \chi_{\delta\delta} \right) \right] \tag{7.147}$$

$$S_{\alpha\gamma,\gamma} = 2\mu \left[\Box_1^2 \chi_{\alpha\gamma,\gamma} - \frac{1}{2(1-\nu)} \left(\nabla^2 \chi_{\delta\delta,\alpha} - \nu\, \Box_2^2 \chi_{\delta\delta,\alpha} \right) \right] \tag{7.148}$$

$$S_{(\alpha\gamma,\gamma\beta)} = 2\mu \left[\Box_1^2 \chi_{(\alpha\gamma,\gamma\beta)} - \frac{1}{2(1-\nu)} \left(\nabla^2 - \nu\, \Box_2^2 \right) \chi_{\delta\delta,\alpha\beta} \right] \tag{7.149}$$

Since, because of (7.142),

$$2\chi_{(\alpha\gamma,\gamma\beta)} = \nabla^2 \chi_{\alpha\beta} + \chi_{\delta\delta,\alpha\beta} \tag{7.150}$$

therefore, (7.149) can be written as

$$S_{(\alpha\gamma,\gamma\beta)} = \mu \left\{ \Box_1^2 (\nabla^2 \chi_{\alpha\beta} + \chi_{\delta\delta,\alpha\beta}) - \frac{1}{1-\nu} \left(\nabla^2 - \nu \Box_2^2 \right) \chi_{\delta\delta,\alpha\beta} \right\} \tag{7.151}$$

Also, note that

$$\begin{aligned} \Box_1^2 - \frac{1}{1-\nu} \left(\nabla^2 - \nu \Box_2^2 \right) &= -\frac{1}{c_1^2} - \frac{\nu}{1-\nu} \frac{1}{c_2^2} \\ &= -\frac{1}{c_2^2} \frac{1-2\nu}{2-2\nu} - \frac{2\nu}{2-2\nu} \frac{1}{c_2^2} = -\frac{1}{2(1-\nu)} \frac{1}{c_2^2} \end{aligned} \tag{7.152}$$

Hence (7.151) reduces to

$$S_{(\alpha\gamma,\gamma\beta)} = \mu \left[\Box_1^2 \nabla^2 \chi_{\alpha\beta} - \frac{1}{2(1-\nu)} \frac{1}{c_2^2} \ddot{\chi}_{\delta\delta,\alpha\beta} \right] \tag{7.153}$$

Substituting (7.146) and (7.153) into LHS of (7.143) we obtain

$$\begin{aligned} LHS\ of\ (4) = {} & \mu \left[\Box_1^2\, \nabla^2 \chi_{\alpha\beta} - \frac{1}{2(1-\nu)} \frac{1}{c_2^2}\, \ddot{\chi}_{\delta\delta,\alpha\beta} \right] \\ & - \rho \left[\Box_1^2\, \ddot{\chi}_{\alpha\beta} - \frac{1}{2(1-\nu)}\, \ddot{\chi}_{\gamma\gamma,\alpha\beta} \right] \\ = {} & \mu \Box_1^2\, \Box_2^2\, \chi_{\alpha\beta} \end{aligned} \tag{7.154}$$

Finally, it follows from (7.141) and (7.154) that $S_{\alpha\beta}$ satisfies (7.136) $\Leftrightarrow$ (7.143). This completes a solution to Problem 7.8.

Problem 7.9. Let $S_{\alpha\beta}$ be a solution to the stress equations of elastodynamics [see Eq. (7.50)]

$$S_{(\alpha\gamma,\gamma\beta)} - \frac{\rho}{2\mu}(\ddot{S}_{\alpha\beta} - \nu\,\ddot{S}_{\gamma\gamma}\delta_{\alpha\beta}) = -b_{(\alpha,\beta)} \quad \text{on} \quad C_0 \times [0,\infty) \tag{7.155}$$

subject to the homogeneous initial conditions

$$S_{\alpha\beta}(\mathbf{x},0) = 0, \quad \dot{S}_{\alpha\beta}(\mathbf{x},0) = 0 \quad \text{for} \quad \mathbf{x} \in C_0 \tag{7.156}$$

Show that there is a second-order symmetric tensor field $\chi_{\alpha\beta}$ that satisfies Eqs. (7.137) through (7.139) of Problem 7.8, that is, the stress representation of Problem 7.8 is complete.

Solution. To solve the problem we use the following three Lemmas.

Lemma 1. Let $f = f(\boldsymbol{x}, t)$ be a prescribed function on $E^2 \times [0,\infty)$. Then a solution $g = g(\boldsymbol{x}, t)$ of the wave equation

$$\left(\nabla^2 - \frac{1}{c^2}\frac{\partial^2}{\partial t^2}\right) g = f \quad \text{on} \quad E^2 \times [0,\infty) \tag{7.157}$$

subject to the homogeneous initial conditions

$$g(\mathbf{x},0) = \dot{g}(\mathbf{x},0) = 0 \quad \text{on} \quad E^2 \tag{7.158}$$

takes the form

$$g(\mathbf{x},t) = -\frac{c}{2\pi}\int_0^t \int_{|\mathbf{x}-\xi|\le c(t-\tau)} \frac{f(\xi,\tau)da(\xi)d\tau}{\sqrt{c^2(t-\tau)^2 - |\mathbf{x}-\xi|^2}} \tag{7.159}$$

Proof of Lemma 1. Is omitted.

Before formulating Lemma 2, we introduce a symmetric second-order tensor field χ^* that is related to χ by the formula

$$2\mu\chi = -2(1-\nu)\chi^* \tag{7.160}$$

Then Eqs. (7.137)–(7.139) of Problem 7.8 take the form

$$\mathbf{S} = \left(\nabla\nabla - \nu\mathbf{1}\,\Box_2^2\right) \text{tr}\,\chi^* - 2(1-\nu)\Box_1^2\,\chi^* \tag{7.161}$$

$$\Box_1^2\,\Box_2^2\chi^* = \frac{1}{1-\nu}\hat{\nabla}\mathbf{b} \tag{7.162}$$

$$\nabla^2\chi^* + \nabla\nabla\,\text{tr}\,\chi^* - 2\hat{\nabla}\,\text{div}\,\chi^* = \mathbf{0} \tag{7.163}$$

Also, note that (7.163) is equivalent to

$$\nabla^2(\text{tr}\,\chi^*) - \text{div}\,\text{div}\,\chi^* = 0 \tag{7.164}$$

To prove that (7.163) ⇔ (7.164) we write (7.163) and (7.164) in components, and find that two of Eqs. (7.163) are identical with (7.164), while the third one is identically satisfied.

In the following, for simplicity $*$ in Eqs. (7.161)–(7.164) is omitted. It follows from Eqs. (7.137)–(7.139) of Problem 7.8 and Eqs. (7.161)–(7.164) that the representation **S** of Problem 7.8 is equivalent to Eqs. (7.161), (7.162) and (7.164).

Lemma 2. If **S** satisfies the stress equation of motion

$$\hat{\nabla}(\text{div}\,\mathbf{S}) - \frac{1}{2c_2^2}[\ddot{\mathbf{S}} - \nu\mathbf{1}(\text{tr}\,\ddot{\mathbf{S}})] = -\hat{\nabla}\mathbf{b} \tag{7.165}$$

subject to the homogeneous initial conditions

$$\mathbf{S}(\mathbf{x}, 0) = \mathbf{0}, \quad \dot{\mathbf{S}}(\mathbf{x}, 0) = \mathbf{0} \tag{7.166}$$

then **S** satisfies the equation

$$(1-\nu)\nabla^2(\text{tr}\,\mathbf{S}) - \text{div}\,\text{div}\,\mathbf{S} = 0 \tag{7.167}$$

Proof of Lemma 2. Since for any vector field **v** on $E^2 \times [0, \infty)$

$$(\nabla^2\text{tr} - \text{div}\,\text{div})\hat{\nabla}\mathbf{v} = 0 \tag{7.168}$$

therefore, applying the operator $(\nabla^2\text{tr} - \text{div}\,\text{div})$ to Eq. (7.165) and using the relation

$$(\nabla^2\text{tr} - \text{div}\,\text{div})\mathbf{1}a = \nabla^2 a \tag{7.169}$$

valid for an arbitrary scalar field a on $E^2 \times [0, \infty)$, we obtain

$$(1-\nu)\nabla^2(\text{tr}\,\ddot{\mathbf{S}}) - \text{div}\,\text{div}\,\ddot{\mathbf{S}} = 0 \tag{7.170}$$

Finally, integrating (7.170) twice with respect to time and using (7.166) we obtain (7.167). This completes proof of Lemma 2.

Lemma 3. Let $\mathbf{S}$ be a solution to Eq. (7.165) on $B \times [0, \infty)$ subject to the initial conditions (7.166) on B ($B \subset E^2$). Let $\tilde{\mathbf{S}}$ be an extension of $\mathbf{S}$ to $E^2 \times [0, \infty)$ preserving Eq. (7.167). Then there is a symmetric tensor field $\tilde{\Lambda}$ on $E^2 \times [0, \infty)$ such that

$$-2(1-\nu)\Box_1^2 \Box_2^2 \tilde{\Lambda} = \tilde{\mathbf{S}} \tag{7.171}$$

and

$$[(1-\nu)\nabla^2 \text{tr} - \text{div div}]\tilde{\Lambda} = 0 \tag{7.172}$$

Proof of Lemma 3. First, we note that Eq. (7.167) extended to $E^2 \times [0, \infty)$ is a necessary condition for solvability of Eqs. (7.171), (7.172). Next, define the components of $\tilde{\Lambda}$ in terms of $\tilde{\mathbf{S}}$ by the formulas:

$$\tilde{\Lambda}_{\alpha\alpha}(\boldsymbol{x}, t) = -\frac{c_2}{2\pi} \int\limits_0^t \int\limits_{|\boldsymbol{x}-\xi| \le c_2(t-\tau)} \frac{\varphi_\alpha(\xi, \tau) da(\xi) d\tau}{\left[c_2^2(t-\tau) - |\boldsymbol{x} - \xi|^2\right]^{1/2}} \tag{7.173}$$

where

$$\varphi_\alpha(\boldsymbol{x}, t) = \frac{1}{2(1-\nu)} \frac{c_1}{2\pi} \int\limits_0^t \int\limits_{|\boldsymbol{x}-\xi| \le c_1(t-\tau)} \frac{\tilde{S}_{\alpha\alpha}(\xi, \tau) da(\xi) d\tau}{\left[c_1^2(t-\tau)^2 - |\boldsymbol{x} - \xi|^2\right]^{1/2}} \tag{7.174}$$

($\alpha = 1, 2$, do not sum on α)
and

$$\tilde{\Lambda}_{12}(\boldsymbol{x}, t) = \phi(\boldsymbol{x}, t) - \psi(\boldsymbol{x}, t) \tag{7.175}$$

where ϕ and ψ are given by

$$\begin{aligned}\phi(\boldsymbol{x}, t) &= \phi\left(x_1^0, x_2, t\right) + \phi\left(x_1, x_2^0, t\right) - \phi\left(x_1^0, x_2^0, t\right) \\ &\quad + \frac{1}{2} \int\limits_{x_1^0}^{x_1} \int\limits_{x_2^0}^{x_2} \left[(1-\nu)\nabla^2(\tilde{\Lambda}_{11} + \tilde{\Lambda}_{22}) - \partial_1^2 \tilde{\Lambda}_{11} - \partial_2^2 \tilde{\Lambda}_{22}^2\right](\xi, t) d\xi_1 d\xi_2\end{aligned} \tag{7.176}$$

and

$$\psi(\boldsymbol{x}, t) = -\frac{c_2}{2\pi} \int\limits_0^t \int\limits_{|\boldsymbol{x}-\xi| \le c_2(t-\tau)} \frac{\omega(\xi, \tau) da(\xi) d\tau}{\left[c_2^2(t-\tau)^2 - |\boldsymbol{x} - \xi|^2\right]^{1/2}} \tag{7.177}$$

in which

$$\omega(\mathbf{x},t) = -\frac{c_1}{2\pi}\int_0^t \int_{|\mathbf{x}-\xi|\le c_1(t-\tau)} \frac{\left[\square_1^2\square_2^2\phi + \tilde{S}_{12}/2(1-\nu)\right]}{\left[c_1^2(t-\tau)^2 - |\mathbf{x}-\xi|^2\right]^{1/2}} da(\xi)d\tau \tag{7.178}$$

In Eq. (7.176) $\mathbf{x}^0$ is a fixed point of B, and $\phi(x_1^0, x_2, t)$, $\phi(x_1, x_2^0, t)$, and $\phi(x_1^0, x_2^0, t)$ are arbitrary functions.

In the following we show that the tensor field $\tilde{\Lambda}$ defined by (7.173)–(7.178) satisfies Eqs. (7.171) and (7.172), that is, Lemma 3 holds true. To this end we first show that

(i) The tensor field $\tilde{\Lambda}$ satisfies Eq. (7.171).

To show (i) we note that because of Lemma 1, the functions $\tilde{\Lambda}_{\alpha\alpha}$ and φ_α given by (7.173) and (7.174), respectively, satisfy the equations

$$\square_2^2\tilde{\Lambda}_{\alpha\alpha} = \varphi_\alpha \tag{7.179}$$

and

$$\square_1^2\varphi_\alpha = -\frac{1}{2(1-\nu)}\tilde{S}_{\alpha\alpha} \tag{7.180}$$

($\alpha = 1, 2$; do not sum on α).

Hence, we obtain

$$-2(1-\nu)\square_1^2\square_2^2\tilde{\Lambda}_{\alpha\alpha} = \tilde{S}_{\alpha\alpha} \tag{7.181}$$

($\alpha = 1, 2$; do not sum on α).

This means that $\tilde{\Lambda}_{11}$ and $\tilde{\Lambda}_{22}$ satisfy Eqs. $(7.171)_1$ and $(7.171)_2$, respectively. Also, from Lemma 1, the functions ψ and ω, defined by (7.177) and (7.178), respectively, satisfy the equations

$$\square_2^2\psi = \omega \tag{7.182}$$

and

$$\square_1^2\omega = \square_1^2\square_2^2\phi + \frac{1}{2(1-\nu)}\tilde{S}_{12} \tag{7.183}$$

Hence, using (7.175), (7.182), and (7.183) we find that $\tilde{\Lambda}_{12}$ satisfies Eq. $(7.171)_3$, that is,

$$-2(1-\nu)\square_1^2\square_2^2\tilde{\Lambda}_{12} = \tilde{S}_{12} \tag{7.184}$$

This completes proof of (i).

Next, we are to show that

(ii) The tensor field $\tilde{\Lambda}$ satisfies Eq. (7.172).

By applying the operator $\partial^2/\partial x_1\partial x_2$ to Eq. (7.176) we obtain

$$2\partial_1\partial_2\phi = (1-\nu)\nabla^2(\tilde{\Lambda}_{11} + \tilde{\Lambda}_{22}) - \partial_1^2\tilde{\Lambda}_{11} - \partial_2^2\tilde{\Lambda}_{22} \tag{7.185}$$

or using (7.175), we get

$$[(1-\nu)\nabla^2 \text{tr} - \text{div div}]\tilde{\Lambda} = 2\partial_1\partial_2\psi \tag{7.186}$$

Therefore, to show (ii) it is sufficient to prove that

$$\partial_1\partial_2\psi = 0 \tag{7.187}$$

To this end we let $\xi - \boldsymbol{x} = z$ in the integrals (7.177) and (7.178), respectively, and obtain

$$\psi(\boldsymbol{x}, t) = -\frac{c_2}{2\pi}\int_0^t \int_{|z|\le c_2(t-\tau)} \frac{\omega(z+\boldsymbol{x}, \tau)da(z)d\tau}{\left[c_2^2(t-\tau)^2 - |z|^2\right]^{1/2}} \tag{7.188}$$

and

$$\omega(\boldsymbol{x}, t) = -\frac{c_1}{2\pi}\int_0^t \int_{|z|\le c_1(t-\tau)} \frac{\left[\square_1^2\square_2^2\phi + \tilde{S}_{12}/2(1-\nu)\right](z+\boldsymbol{x}, \tau)}{\left[c_1^2(t-\tau)^2 - |z|^2\right]^{1/2}} da(z)d\tau \tag{7.189}$$

By applying the operator $\square_1^2\square_2^2$ to (7.185) and using (7.171) we obtain

$$2\partial_1\partial_2\square_1^2\square_2^2\phi = -\frac{1}{2(1-\nu)}[(1-\nu)\nabla^2(\text{tr}\,\tilde{\mathbf{S}}) - \partial_1^2\tilde{S}_{11} - \partial_2^2\tilde{S}_{22}] \tag{7.190}$$

Next, applying the operator $\partial_1\partial_2$ to (7.189) and using (7.190) we obtain

$$\begin{aligned} 2\partial_1\partial_2\,\omega(x, t) = &-\frac{1}{2(1-\nu)}\frac{c_1}{2\pi}\int_0^t \int_{|z|\le c_1(t-\tau)} da(z)d\tau \\ &\times \frac{[\text{div div} - (1-\nu)\nabla^2\text{tr}]\tilde{\mathbf{S}}(x+z, \tau)}{[c_1^2(t-\tau)^2 - |z|^2]^{1/2}} \end{aligned} \tag{7.191}$$

Hence, and from Eq. (7.167) of Lemma 2 we obtain

$$2\,\partial_1\partial_2\omega = 0 \tag{7.192}$$

and applying the operator $\partial_1\partial_2$ to (7.188) we obtain

$$\partial_1\partial_2\psi = 0 \tag{7.193}$$

Therefore, (7.193) $\Leftrightarrow$ (7.187) holds true, and this completes proof of (ii).

The conclusions (i) and (i) imply that Lemma 3 holds true.

Proof of the completeness of the stress representation in Problem 7.9. We are to show that the tensor field χ, defined in terms of Λ from Lemma 3, by

$$\chi = 2\left[\hat{\nabla}(\mathrm{div}\Lambda) - \frac{1}{2c_2^2}(\ddot{\Lambda} - \nu \mathbf{1}\,\mathrm{tr}\,\ddot{\Lambda})\right] \tag{7.194}$$

satisfies Eqs. (7.161), (7.162), and (7.164) in which asterisk is omitted.

First, we show that χ satisfies (7.162), that is,

$$\Box_1^2\Box_2^2\chi = \frac{1}{1-\nu}\hat{\nabla}\mathbf{b} \tag{7.195}$$

Applying the operator $\Box_1^2\Box_2^2$ to (7.194) and using (7.171) we obtain

$$\Box_1^2\Box_2^2\chi = -\frac{1}{1-\nu}\left[\hat{\nabla}(\mathrm{div}\,\mathbf{S}) - \frac{1}{2c_2^2}(\ddot{\mathbf{S}} - \nu\mathbf{1}\,\mathrm{tr}\,\ddot{\mathbf{S}})\right] \tag{7.196}$$

Since $\mathbf{S}$ is a solution to the stress equation of motion (7.165), Eq. (7.196) implies that χ satisfies (7.195).

Next, we show that χ satisfies (7.164), that is,

$$\nabla^2(\mathrm{tr}\,\chi) - \mathrm{div}\,\mathrm{div}\,\chi = 0 \tag{7.197}$$

To this end we take the trace of (7.194) and obtain

$$\mathrm{tr}\,\chi = 2\left[\mathrm{div}\,\mathrm{div}\,\Lambda - \frac{1-2\nu}{2c_2^2}(\mathrm{tr}\,\ddot{\Lambda})\right] \tag{7.198}$$

Applying the operator ($\nabla^2\mathrm{tr}$ – div div) to (7.194) and using (7.198) we get

$$\begin{aligned}\nabla^2(\mathrm{tr}\,\chi) - \mathrm{div}\,\mathrm{div}\,\chi &= 2\left[\nabla^2(\mathrm{div}\,\mathrm{div}\Lambda) - \frac{1-2\nu}{2c_2^2}\nabla^2(\mathrm{tr}\,\ddot{\Lambda})\right]\\ &\quad - 2\left\{\nabla^2(\mathrm{div}\,\mathrm{div}\Lambda) - \frac{1}{2c_2^2}[\mathrm{div}\,\mathrm{div}\ddot{\Lambda} - \nu\nabla^2(\mathrm{tr}\,\ddot{\Lambda})]\right\}\\ &= -\frac{1}{c_2^2}[(1-\nu)\nabla^2(\mathrm{tr}\,\ddot{\Lambda}) - \mathrm{div}\,\mathrm{div}\ddot{\Lambda}]\end{aligned} \tag{7.199}$$

Because of Eq. (7.172) restricted to $B \times [0, \infty)$, the RHS of (7.199) vanishes: this shows that (7.197) $\Leftrightarrow$ (7.164) holds true.

Finally, we are to show that if χ from Eq. (7.194) is substituted into the RHS of Eq. (7.161) in which asterisk is omitted we obtain $\mathbf{S}$, that is, Eq. (7.161) holds true.

It follows from the relation

$$\frac{1}{c_1^2} = \frac{1}{c_2^2}\frac{1-2\nu}{2-2\nu} \tag{7.200}$$

and from Eqs. (7.172) restricted to $B \times [0, \infty)$ and (7.198) that

$$\text{tr}\,\chi = 2[(1-\nu)\nabla^2(\text{tr}\,\Lambda) - \frac{1-2\nu}{2c_2^2}(\text{tr}\,\ddot{\Lambda})] = 2(1-\nu)\Box_1^2(\text{tr}\,\Lambda) \tag{7.201}$$

Applying the operator $2(1-\nu)\Box_1^2$ to Eq. (7.194) we obtain

$$2(1-\nu)\Box_1^2\chi = 2(1-\nu)\Box_1^2\left[2\hat{\nabla}(\text{div}\Lambda) + \left(\Box_2^2 - \nabla^2\right)(\Lambda - \nu\mathbf{1}\,\text{tr}\,\Lambda)\right] \tag{7.202}$$

and applying the operator $(\nabla\nabla - \nu\mathbf{1}\Box_2^2)$ to (7.201) we get

$$(\nabla\nabla - \nu\mathbf{1}\Box_2^2)\text{tr}\,\chi = 2(1-\nu)\Box_1^2(\nabla\nabla - \nu\mathbf{1}\Box_2^2)\text{tr}\,\Lambda \tag{7.203}$$

Next, subtracting (7.203) from (7.202) we get

$$\begin{aligned}&\left[\left(\nabla\nabla - \nu\mathbf{1}\Box_2^2\right)\text{tr} - 2(1-\nu)\Box_1^2\right]\chi \\ &= -2(1-\nu)\Box_1^2\left[\Box_2^2 + 2\hat{\nabla}\text{div} - \nabla^2 - (\nabla\nabla - \nu\mathbf{1}\nabla^2)\text{tr}\right]\Lambda\end{aligned} \tag{7.204}$$

Since, from (7.172),

$$[(1-\nu)\nabla^2\text{tr} - \text{div div}]\Lambda = 0 \tag{7.205}$$

and (7.205) is equivalent to the tensorial equation

$$[2\hat{\nabla}\text{div} - \nabla^2 - (\nabla\nabla - \nu\mathbf{1}\nabla^2)\text{tr}]\Lambda = \mathbf{0} \tag{7.206}$$

therefore, (7.204) takes the form

$$\left[\left(\nabla\nabla - \nu\mathbf{1}\Box_2^2\right)\text{tr} - 2(1-\nu)\Box_1^2\right]\chi = -2(1-\nu)\Box_1^2\Box_2^2\Lambda \tag{7.207}$$

This together with (7.171) restricted to $B \times [0, \infty)$ implies that χ satisfies (7.161). This completes a solution to Problem 7.9.

Problem 7.10. A homogeneous isotropic infinite elastic body under plane strain conditions and initially at rest is subject to a temperature change $T = T(\mathbf{x}, t)$ for every $(\mathbf{x}, t) \in E^2 \times [0, \infty)$. Show that the dynamic thermal stresses $S_{\alpha\beta} = S_{\alpha\beta}(\mathbf{x}, t)$ corresponding to the temperature $T = T(\mathbf{x}, t)$ are represented by the formulas [see Eqs. (7.66), (7.67)]

$$S_{\alpha\beta} = 2\mu\,(\phi_{,\alpha\beta} - \phi_{,\gamma\gamma}\delta_{\alpha\beta}) + \rho\,\ddot{\phi}\delta_{\alpha\beta} \tag{7.208}$$

where

$$\phi(\mathbf{x}, t) = -\frac{m\, c_1}{2\pi} \int_0^t d\tau \int\limits_{|\mathbf{x}-\xi|<c_1\tau} \frac{T(\xi, t-\tau)\, da(\xi)}{\sqrt{c_1^2\tau^2 - |\mathbf{x}-\xi|^2}} \tag{7.209}$$

and

$$m = \frac{1+\nu}{1-\nu}\alpha, \quad \frac{1}{c_1} = \sqrt{\frac{\rho}{\lambda + 2\mu}} \tag{7.210}$$

Solution. In the plane-strain case when $T = T(\mathbf{x}, t)$ is prescribed on $E^2 \times [0, \infty)$, the dynamic thermal stresses $S_{\alpha\beta} = S_{\alpha\beta}(\mathbf{x}, t)$ are computed from the formula [see Eqs. (7.66) and (7.67)]

$$S_{\alpha\beta} = 2\mu(\phi,_{\alpha\beta} - \phi,_{\gamma\gamma}\, \delta_{\alpha\beta}) + \rho\ddot{\phi}\delta_{\alpha\beta} \tag{7.211}$$

where $\phi = \phi(\mathbf{x}, t)$ is a solution to Poisson's equation

$$\Box_1^2\phi = mT \quad \text{on } E^2 \times [0, \infty) \tag{7.212}$$

By Lemma 1 from the solution to Problem 7.9, a unique solution to Eq. (7.212) subject to the homogeneous initial conditions

$$\phi(\mathbf{x}, 0) = 0, \quad \dot{\phi}(\mathbf{x}, 0) = 0 \quad \text{on } E^2 \tag{7.213}$$

takes the form

$$\phi(\mathbf{x}, t) = -\frac{mc_1}{2\pi} \int_0^t \int\limits_{|x-\xi|\le c_1(t-\tau)} \frac{T(\xi, \tau)}{\left[c_1^2(t-\tau)^2 - |\mathbf{x}-\xi|^2\right]^{1/2}} da(\xi) d\tau \tag{7.214}$$

By introducing the new integration variable: ($v = t - \tau$) we find that ϕ is equivalent to (7.209). This completes a solution to Problem 7.10.

Chapter 8
Solutions to Particular Three-Dimensional Boundary Value Problems of Elastostatics

In this chapter the boundary value problems related to torsion of a prismatic bar bounded by a cylindrical lateral surface and by a pair of planes normal to the lateral surface, are discussed. It is assumed that a resultant torsion moment is applied at one of the bases while the other is subject to a warping and the lateral surface is stress free. In each of the problems an approximate three-dimensional formulation is reduced to a two-dimensional one for Laplace's or Poisson's equation on the cross section of the bar.

8.1 Torsion of Circular Bars

We consider a circular prismatic bar of length l and radius a referred to the Cartesian coordinates (x_1, x_2, x_3) in such a way that x_3 coincides with the axis of the bar, the bar is fixed at $x_3 = 0$ in the (x_1, x_2) plane, while at $x_3 = l$ a torsion moment M_3 is applied. This moment causes the bar to be twisted, and the generators of the circular cylinder deform into helical curves.

An elastic state $s = [\mathbf{u}, \mathbf{E}, \mathbf{S}]$ in the bar is approximated by $\widetilde{s} = [\widetilde{\mathbf{u}}, \widetilde{\mathbf{E}}, \widetilde{\mathbf{S}}]$, where

$$\widetilde{u}_1 = -\alpha\, x_2\, x_3, \quad \widetilde{u}_2 = \alpha\, x_1\, x_3, \quad \widetilde{u}_3 = 0 \tag{8.1}$$

and α is the angle of twist per unit length along the x_3 axis

$$\begin{aligned} &\widetilde{E}_{11} = \widetilde{E}_{22} = \widetilde{E}_{33} = \widetilde{E}_{12} = 0 \\ &\widetilde{E}_{23} = \frac{1}{2}\alpha\, x_1, \quad \widetilde{E}_{31} = -\frac{1}{2}\alpha\, x_2 \end{aligned} \tag{8.2}$$

and

$$\begin{aligned} &\widetilde{S}_{11} = \widetilde{S}_{22} = \widetilde{S}_{33} = \widetilde{S}_{12} = 0 \\ &\widetilde{S}_{23} = \mu\alpha\, x_1, \quad \widetilde{S}_{31} = -\mu\alpha\, x_2 \end{aligned} \tag{8.3}$$

M. Reza Eslami et al., *Theory of Elasticity and Thermal Stresses*, Solid Mechanics and Its Applications 197, DOI: 10.1007/978-94-007-6356-2_8,

The torsion moment M_3 is

$$M_3 = \int_A (x_1 \widetilde{S}_{23} - x_2 \widetilde{S}_{31}) dx_1 dx_2 = \mu\alpha \int_A (x_1^2 + x_2^2) dx_1 dx_2 \equiv \mu\,\alpha\,J \tag{8.4}$$

where A is the area of the cross section : $A = \{(x_1, x_2) : \sqrt{x_1^2 + x_2^2} \le a\}$, and J is the polar moment of inertia of the cross section about its center. The product μJ is called the *torsional rigidity of the bar*. Also, since

$$n_\alpha = x_\alpha / a, \quad n_3 = 0 \quad \text{on} \;\; \partial A \quad (\alpha = 1, 2) \tag{8.5}$$

therefore, because of Eq. (8.3)

$$\widetilde{S}_{ij} n_j = 0 \quad \text{on} \quad \partial A \;\; (i, j = 1, 2, 3) \tag{8.6}$$

that is, the elastic state $\widetilde{s} = [\widetilde{\mathbf{u}}, \widetilde{\mathbf{E}}, \widetilde{\mathbf{S}}]$ satisfies the homogeneous boundary conditions: (i) $\widetilde{\mathbf{u}} = \mathbf{0}$ on $x_3 = 0$, and (ii) $\widetilde{\mathbf{S}}\mathbf{n} = \mathbf{0}$ on the lateral surface $\partial A \times [0, l]$. A shear stress boundary condition on the plane $x_3 = l$ is replaced by application of the resultant moment M_3 on this plane.

Torsion of Noncircular Prismatic Bars

A noncircular prismatic bar of length l is fixed at $x_3 = 0$ in the sense that the displacement components in the (x_1, x_2) plane vanish while the axial displacement is subject to a warping, and the other end $x_3 = l$ is twisted by a moment M_3; the lateral surface of the bar is stress free and no body forces are present. Therefore, an elastic state $s = [\mathbf{u}, \mathbf{E}, \mathbf{S}]$ in the bar is approximated by $\widetilde{s} = [\widetilde{\mathbf{u}}, \widetilde{\mathbf{E}}, \widetilde{\mathbf{S}}]$ in which

$$\widetilde{u}_1 = -\alpha\, x_2\, x_3, \quad \widetilde{u}_2 = \alpha\, x_1\, x_3, \quad \widetilde{u}_3 = \alpha\, \psi(x_1,\, x_2) \tag{8.7}$$

where $\psi = \psi(x_1, x_2)$ is called a *warping function*. The strain-displacement relations, the equilibrium equations with zero body forces, and the constitutive stress-strain relations, respectively, associated with the displacements (8.7), take the forms

$$\begin{aligned} &\widetilde{E}_{11} = \widetilde{E}_{22} = \widetilde{E}_{33} = \widetilde{E}_{12} = 0 \\ &\widetilde{E}_{23} = \frac{1}{2}\alpha(\psi_{,2} + x_1), \quad \widetilde{E}_{31} = \frac{1}{2}\alpha\,(\psi_{,1} - x_2) \end{aligned} \tag{8.8}$$

$$\psi_{,11} + \psi_{,22} = 0 \tag{8.9}$$

and

$$\begin{aligned} &\widetilde{S}_{11} = \widetilde{S}_{22} = \widetilde{S}_{33} = \widetilde{S}_{12} = 0 \\ &\widetilde{S}_{23} = \mu\alpha(\psi_{,2} + x_1), \quad \widetilde{S}_{31} = \mu\alpha(\psi_{,1} - x_2) \end{aligned} \tag{8.10}$$

The torsion moment M_3 takes the form

$$M_3 = \int_A (x_1\widetilde{S}_{23} - x_2\widetilde{S}_{31})dx_1dx_2 = \alpha D \tag{8.11}$$

where

$$D = \mu \int_A (x_1^2 + x_2^2 + x_1\psi_{,2} - x_2\psi_{,1})dx_1dx_2 \tag{8.12}$$

is called the *torsional rigidity of the bar.*

The boundary conditions are satisfied in the following sense. The bases $x_3 = 0$ and $x_3 = l$ of the bar are the resultant force free, that is,

$$F_1 = \int_A \widetilde{S}_{31}dx_1dx_2 = 0, \quad F_2 = \int_A \widetilde{S}_{32}dx_1dx_2 = 0 \tag{8.13}$$

and the distribution of shear stresses on the base $x_3 = l$ is represented by the torsion moment M_3. To satisfy the stress free lateral surface boundary condition, we postulate that

$$\frac{\partial\psi}{\partial n} = x_2n_1 - x_1n_2 \quad \text{on} \quad \partial A \tag{8.14}$$

As a result, the torsion problem of a noncircular prismatic bar has been solved once a warping function $\psi = \psi(x_1, x_2)$ that satisfies the harmonic equation

$$\nabla^2\psi = 0 \quad \text{on} \quad A \tag{8.15}$$

subject to the boundary condition

$$\frac{\partial\psi}{\partial n} = x_2n_1 - x_1n_2 \quad \text{on} \quad \partial A \tag{8.16}$$

has been found.

For example, for an elliptic bar with semi-axes a and b and with the center at the origin, we obtain

$$\psi(x_1, x_2) = \frac{b^2 - a^2}{b^2 + a^2}x_1x_2 \tag{8.17}$$

and

$$D = \frac{\pi\,\mu\,a^3b^3}{a^2 + b^2}, \quad M_3 = \alpha D \tag{8.18}$$

$$\widetilde{S}_{13} = -\frac{2M_3}{\pi\,ab^3}x_2, \quad \widetilde{S}_{13} = \frac{2M_3}{\pi\,a^3b}x_1 \tag{8.19}$$

Prandtl's Stress Function
Prandtl's stress function $\phi = \phi(x_1, x_2)$ is defined in terms of the warping function $\psi = \psi(x_1, x_2)$ by the formulas

$$\begin{aligned} \phi_{,2} &= \mu\,\alpha(\psi_{,1} - x_2) = \widetilde{S}_{13} \\ \phi_{,1} &= -\mu\,\alpha(\psi_{,2} + x_1) = -\widetilde{S}_{23} \end{aligned} \tag{8.20}$$

One can show that the boundary value problem for the warping function $\psi = \psi(x_1, x_2)$, described by Eqs. (8.15) and (8.16), is equivalent to finding a Prandtl's stress function $\phi = \phi(x_1, x_2)$ that satisfies Poisson's equation

$$\nabla^2\phi = -2\mu\,\alpha \quad \text{on} \quad A \tag{8.21}$$

subject to the homogeneous boundary condition

$$\phi = 0 \quad \text{on} \quad \partial A \tag{8.22}$$

while the torsion moment M_3 is calculated from the formula

$$M_3 = 2\int_A \phi(x_1, x_2)dx_1 dx_2 \tag{8.23}$$

8.2 Problems and Solutions Related to Particular Three-Dimensional Boundary Value Problems of Elastostatics—Torsion Problems

Problem 8.1. Show that the warping function ψ = const solves the torsion problem of a circular bar.

Solution. By letting $\psi = 0$ in Eqs. (8.7)–(8.10) we obtain $\widetilde{s} = [\widetilde{\mathbf{u}}, \widetilde{\mathbf{E}}, \widetilde{\mathbf{S}}]$, where $\widetilde{\mathbf{u}}$, $\widetilde{\mathbf{E}}$ and $\widetilde{\mathbf{S}}$ are given by Eqs. (8.1)–(8.3), respectively, that describes a solution to the torsion problem of a circular bar.

Problem 8.2. Show that in the torsion problem of an elliptic bar, the resultant shear stress $\widetilde{S}_t$ at points on a given diameter of the ellipse is parallel to the tangent at the point of intersection of the diameter and the ellipse [see Fig. 8.1].

Solution. For an elliptic bar subject to a torsion moment M_3, the stresses $\widetilde{S}_{13}$ and $\widetilde{S}_{23}$, respectively, are given by [see Eqs. (8.19)]

$$\widetilde{S}_{13} = -\frac{2\,M_3}{\pi\,ab^3}x_2 \tag{8.24}$$

and

$$\widetilde{S}_{23} = \frac{2\,M_3}{\pi\,a^3 b} x_1 \tag{8.25}$$

The resultant shear stress magnitude is then computed from the formula

$$\widetilde{S}_t = \left(\widetilde{S}_{13}^2 + \widetilde{S}_{23}^2\right)^{1/2} = \frac{2\,M_3}{\pi\,ab}\left(\frac{x_1^2}{a^4} + \frac{x_2^2}{b^4}\right)^{1/2} \tag{8.26}$$

Equations (8.24)–(8.26) hold true for any point of the elliptical cross section of the bar. In particular, for any such point, because of (8.24) and (8.25)

$$\frac{\widetilde{S}_{13}}{\widetilde{S}_{23}} = -\frac{a^2}{b^2}\frac{x_2}{x_1} \tag{8.27}$$

Therefore, the ratio $\widetilde{S}_{13}/\widetilde{S}_{23}$ is constant along the diameter of the ellipse shown in Fig. of Problem 8.2 represented by the equation

$$-\frac{a^2}{b^2}\frac{x_2}{x_1} = c = \text{const } (c > 0) \tag{8.28}$$

As a result, the resultant shear stress vector $\widetilde{\boldsymbol{\tau}} = \widetilde{S}_{13}\,\mathbf{e}_1 + \widetilde{S}_{23}\,\mathbf{e}_2$, where $\mathbf{e}_1 = (1,0)^T$, $\mathbf{e}_2 = (0,1)^T$, coincides with the tangent vector at the point of intersection of the diameter and the ellipse. Substituting x_2 from (8.28) into (8.26) we obtain

$$\widetilde{S}_t = \frac{2\,M_3}{\pi\,ab}\sqrt{1+c^2}\;\frac{|x_1|}{a^2} \tag{8.29}$$

This formula shows that for $x_1 > 0$ $\widetilde{S}_t$ is a linear function of x_1 along the diameter.

This completes a solution to Problem 8.2.

Problem 8.3. Show that the torsion moment in terms of Prandtl's stress function $\phi = \phi(x_1, x_2)$ is expressed by

$$M_3 = 2\int_A \phi(x_1, x_2)dx_1 dx_2$$

Solution. A solution to this problem is obtained from Eqs. (8.11)–(8.12), (8.15)–(8.16), and (8.20)–(8.22).

Problem 8.4. Show that Prandtl's stress function $\phi = \phi(x_1, x_2)$ given by

$$\phi(x_1, x_2) = \frac{32\mu\,\alpha\,a_1^2}{\pi^3}\sum_{n=1,3,5,\ldots}^{\infty}\frac{\sin\left(\frac{n\pi}{2}\right)}{n^3}\left[1 - \frac{\cosh\left(\frac{n\pi\,x_2}{2a_1}\right)}{\cosh\left(\frac{n\pi\,a_2}{2a_1}\right)}\right]\cos\left(\frac{n\pi\,x_1}{2a_1}\right)$$

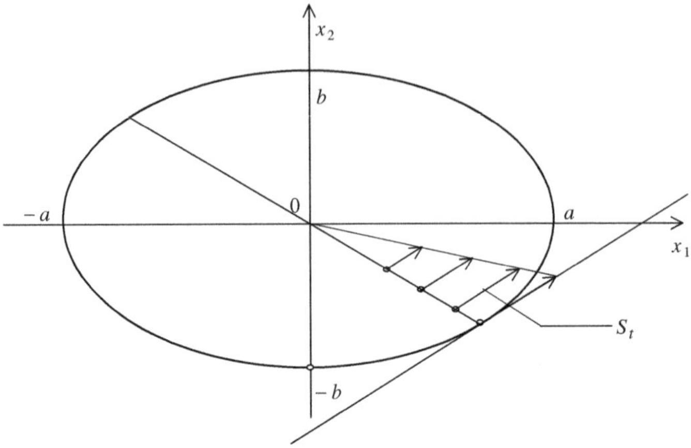

Fig. 8.1 The cross section of an elliptic bar in torsion

solves the torsion problem of a bar with the rectangular cross section: $|x_1| \leq a_1$, $|x_2| \leq a_2$. Also, show that in this case the torsion moment

$$M_3 = 2 \int_{-a_1}^{a_1} \int_{-a_2}^{a_2} \phi(x_1, x_2) dx_1 dx_2 = \mu\, \alpha (2a_1)^3 (2a_2)\, k^*$$

where

$$k^* = \frac{1}{3} \left[1 - \frac{192}{\pi^5} \left(\frac{a_1}{a_2} \right) \sum_{n=1,3,5,\ldots}^{\infty} \frac{1}{n^5} \tanh \left(\frac{n\pi\, a_2}{2a_1} \right) \right]$$

Solution. For the rectangular cross section $C_0 : \ |x_1| \leq a_1, \ |x_2| \leq a_2$, Prandtl's stress function $\phi = \phi(x_1, x_2)$ satisfies Poisson's equation

$$\nabla^2 \phi = -2\mu\alpha \quad \text{on } C_0 \tag{8.30}$$

subject to the homogeneous boundary condition

$$\phi = 0 \quad \text{on } \partial C_0 \tag{8.31}$$

Since

$$\cos \left(\frac{n\pi}{2} \right) = 0 \quad \text{for } n = 1, 3, 5, \ldots \tag{8.32}$$

therefore, $\phi = \phi(x_1, x_2)$ given by

$$\phi(x_1, x_2) = \frac{32\,\mu\alpha\, a_1^2}{\pi^3} \times \sum_{n=1,3,5,\ldots} \frac{\sin\left(\frac{n\pi}{2}\right)}{n^3}\left[1 - \frac{ch\left(\frac{n\pi x_2}{2a_1}\right)}{ch\left(\frac{n\pi a_2}{a_1}\right)}\right]\cos\left(\frac{n\pi x_1}{2a_1}\right) \tag{8.33}$$

satisfies the homogeneous boundary condition (8.31).

In addition, applying ∇^2 to (8.33) and using the identity

$$\nabla^2\left[\cos\left(\frac{n\pi x_1}{2a_1}\right) ch\left(\frac{n\pi x_2}{2a_1}\right)\right] = 0 \tag{8.34}$$

we obtain

$$\nabla^2\phi = -\frac{8\,\mu\alpha}{\pi} \sum_{n=1,3,5,\ldots} \frac{\sin\left(\frac{n\pi}{2}\right)\cos\left(\frac{n\pi x_1}{2a_1}\right)}{n} \tag{8.35}$$

Hence, ϕ given by (8.33) satisfies (8.30) if the function 1 on $|x_1| \le a_1$ can be represented by the Fourier's series

$$1 = \frac{4}{\pi} \sum_{n=1,3,5,\ldots} \frac{\sin\left(\frac{n\pi}{2}\right)\cos\left(\frac{n\pi x_1}{2a_1}\right)}{n} \qquad |x_1| \le a_1 \tag{8.36}$$

To show (8.36) we multiply (8.36) by $\cos\left(\frac{k\pi x_1}{2a_1}\right)$ and integrate over $|x_1| \le a_1$, and obtain

$$\int_{-a_1}^{a_1} \cos\left(\frac{k\pi x_1}{2a_1}\right) dx_1 = \frac{4}{\pi} \sum_{n=1,3,5,\ldots} \frac{\sin\left(\frac{n\pi}{2}\right)}{n} \times \int_{-a_1}^{a_1} \cos\left(\frac{k\pi x_1}{2a_1}\right)\cos\left(\frac{n\pi x_1}{2a_1}\right) dx_1 \tag{8.37}$$

Since

$$\int_{-a_1}^{a_1} \cos\left(\frac{k\pi x_1}{2a_1}\right)\cos\left(\frac{n\pi x_1}{2a_1}\right) dx_1 = a_1\delta_{kn} \quad \text{for } n, k = 1, 3, 5, \ldots \tag{8.38}$$

and

$$\int_{-a_1}^{a_1} \cos\left(\frac{k\pi x_1}{2a_1}\right) dx_1 = \frac{4a_1}{k\pi}\sin\left(\frac{k\pi}{2}\right) \tag{8.39}$$

therefore, Eq. (8.37) is an identity. This proves that the expansion (8.36) holds true, and as a result ϕ given by (8.33) solves the torsion problem of a bar with the rectangular cross section.

To calculate the torsion moment we use the formula

$$M_3 = 2 \int_{-a_1}^{a_1} \int_{-a_2}^{a_2} \phi(x_1, x_2) dx_1 dx_2 \tag{8.40}$$

Substituting ϕ from (8.33) into (8.40) we obtain

$$\begin{aligned} M_3 &= \frac{64\,\mu\alpha a_1^2}{\pi^3} \times \int_{-a_1}^{a} \int_{-a_2}^{a_2} \sum_{n=1,3,5,\ldots} \frac{\sin\left(\frac{n\pi}{2}\right)}{n^3} \left[1 - \frac{\text{ch}\left(\frac{n\pi x_2}{2a_1}\right)}{\text{ch}\left(\frac{n\pi a_2}{2a_1}\right)}\right] \cos\left(\frac{n\pi x_1}{2a_1}\right) dx_1 dx_2 \\ &= \frac{32\,\mu\alpha (2a_1)^3 (2a_2)}{\pi^4} \sum_{n=1,3,5,\ldots} \frac{1}{n^4} - \frac{64\,\mu\alpha (2a_1)^4}{\pi^5} \sum_{n=1,3,5,\ldots} \frac{1}{n^5} \tanh\left(\frac{n\pi a_2}{2a_1}\right) \end{aligned} \tag{8.41}$$

Since

$$\sum_{n=1,3,5,\ldots} \frac{1}{n^4} = \frac{\pi^4}{96} \tag{8.42}$$

therefore, substituting (8.42) into (8.41) we obtain

$$M_3 = \frac{1}{3}\,\mu\alpha (2a_1)^3 2a_2 \times \left[1 - \frac{192}{\pi^5} \frac{a_1}{a_2} \sum_{n=1,3,5,\ldots} \frac{1}{n^5} \tanh\left(\frac{n\pi a_2}{2a_1}\right)\right] \tag{8.43}$$

This completes a solution to Problem 8.4.

Problem 8.5. Show that Prandtl's stress function

$$\phi(r, \theta) = \frac{\mu\alpha}{2}(r^2 - b^2)\left(\frac{2a\cos\theta}{r} - 1\right)$$

defined over the region

$$0 < b \le r \le 2a - b, \quad -\cos^{-1}\left(\frac{b}{2a}\right) \le \theta \le \cos^{-1}\left(\frac{b}{2a}\right)$$

solves the torsion problem of the circular shaft with a circular groove shown in Fig. 8.2; in particular, find the stresses $\widetilde{S}_{13}$ and $\widetilde{S}_{23}$ on the boundary of the shaft.

Hint. Use the polar coordinates

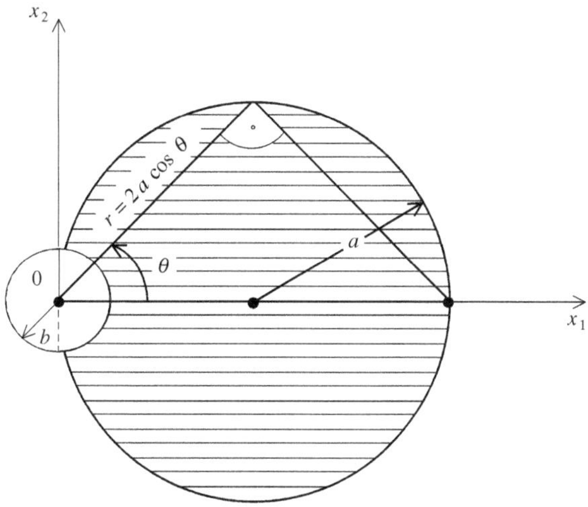

Fig. 8.2 The cross section of a circular bar with a circular groove

$$x_1 = r \cos\theta, \quad x_2 = r \sin\theta$$

Solution. First, we note that the function

$$\phi(r,\theta) = \frac{\mu\alpha}{2}(r^2 - b^2)\left(\frac{2a\cos\theta}{r} - 1\right) \tag{8.44}$$

vanishes on the boundary of the circular shaft with a circular groove shown in Fig. of Problem 8.5.

Next, using ∇^2 in the form

$$\nabla^2 = \frac{\partial^2}{\partial r^2} + \frac{1}{r}\frac{\partial}{\partial r} + \frac{1}{r^2}\frac{\partial^2}{\partial \theta^2} \tag{8.45}$$

and the equations

$$\nabla^2(r\cos\theta) = \nabla^2(r^{-1}\cos\theta) = 0 \tag{8.46}$$

and applying ∇^2 to (8.44) we obtain

$$\nabla^2\phi = -2\,\mu\alpha \tag{8.47}$$

Therefore, ϕ solves the torsion problem of the circular shaft with a circular groove. In particular, the stresses $\widetilde{S}_{13}$ and $\widetilde{S}_{23}$ are computed from the formulas [see Eqs. (8.20)]

$$\widetilde{S}_{13} = \phi_{,2}, \quad \widetilde{S}_{23} = -\phi_{,1} \tag{8.48}$$

Substituting ϕ from (8.44) into (8.48), and using the polar coordinates, we obtain

$$\widetilde{S}_{13} = \mu\alpha x_2 \left(2a\, x_1 \frac{b^2}{r^4} - 1\right) \tag{8.49}$$

and

$$\widetilde{S}_{23} = -\mu\alpha \left[a\left(1 - \frac{b^2}{r^2}\right) - x_1 + 2a\, x_1^2 \frac{b^2}{r^4}\right] \tag{8.50}$$

In Eqs. (8.49) and (8.50)

$$x_1 = r\cos\theta \quad x_2 = r\sin\theta \tag{8.51}$$

By letting $r = b$ in (8.49) and (8.50) we get

$$\widetilde{S}_{13}|_{r=b} = \mu\alpha(2a\cos\theta - b)\sin\theta \tag{8.52}$$

$$\widetilde{S}_{23}|_{r=b} = -\mu\alpha(2a\cos\theta - b)\cos\theta \tag{8.53}$$

Hence, the resultant shear stress magnitude for $r = b$ takes the form

$$\widetilde{S}_t = \left(\widetilde{S}_{13}^2 + \widetilde{S}_{23}^2\right)^{1/2} = \mu\alpha(2a\cos\theta - b) \tag{8.54}$$

Since

$$\frac{\partial \widetilde{S}_t}{\partial\theta} = 0, \quad \frac{\partial^2 \widetilde{S}_t}{\partial\theta^2} < 0 \quad \text{at } \theta = 0 \tag{8.55}$$

the function $\widetilde{S}_t = \widetilde{S}_t(\theta)$ attains a maximum at $\theta = 0$. Hence, the resultant shear stress attains a maximum at the point $(x_1, x_2) = (b, 0)$ and

$$\widetilde{S}_t(\theta = 0) = \mu\alpha(2a - b) \tag{8.56}$$

If $b \to 0$, the RHS of (8.56) $\to 2\mu\alpha a$. Hence, for a small groove radius the maximum resultant shear stress doubles that of a bar with a circular cross section [see Eq. (8.3)].

This completes a solution to Problem 8.5.

Chapter 9
Solutions to Particular Two-Dimensional Boundary Value Problems of Elastostatics

In this chapter a number of two-dimensional boundary value problems for a body under plane strain conditions or under generalized plane stress conditions are solved. The problems include: (i) a semispace subject to an internal concentrated body force, (ii) an elastic wedge subject to a concentrated load at its tip, and (iii) an infinite elastic strip subject to a discontinuous temperature field. To solve the problems a two-dimensional version of the Boussinesq-Papkovitch-Neuber solution as well as an Airy stress function method, are used.

9.1 The Two-Dimensional Version of Boussinesq-Papkovitch-Neuber Solution for a Body Under Plane Strain Conditions

An elastic state $s = [\mathbf{u}, \mathbf{E}, \mathbf{S}]$ corresponding to a body under plane strain conditions is described by the equations [see Eqs. 7.70 and 7.71 in Problem 7.1.]

$$u_\alpha = \psi_\alpha - \frac{1}{4(1-\nu)}(x_\gamma \psi_\gamma + \varphi)_{,\alpha} \tag{9.1}$$

where

$$\psi_{\alpha,\gamma\gamma} = -\frac{b_\alpha}{\mu} \tag{9.2}$$

and

$$\varphi_{,\gamma\gamma} = \frac{x_\gamma b_\gamma}{\mu} \tag{9.3}$$

The strains $E_{\alpha\beta}$ and stresses $S_{\alpha\beta}$, associated with u_α, are given, respectively, by

M. Reza Eslami et al., *Theory of Elasticity and Thermal Stresses*, Solid Mechanics and Its Applications 197, DOI: 10.1007/978-94-007-6356-2_9,

$$E_{\alpha\beta} = \frac{1}{4(1-\nu)}[2(1-2\nu)\,\psi_{(\alpha,\beta)} - x_\gamma\,\psi_{\gamma,\alpha\beta} - \varphi_{,\alpha\beta}] \tag{9.4}$$

and

$$S_{\alpha\beta} = \frac{\mu}{4(1-\nu)}[2(1-2\nu)\,\psi_{(\alpha,\beta)} - x_\gamma\,\psi_{\gamma,\alpha\beta} + 2\nu\,\psi_{\gamma,\gamma}\delta_{\alpha\beta} - \varphi_{,\alpha\beta}] \tag{9.5}$$

If a concentrated force P_0 normal to the boundary of a semispace $|x_1| < \infty, \quad x_2 \geq 0$ is applied at the point $(x_1, x_2) = (0, 0)$, and suitable asymptotic conditions are imposed on $s = [\mathbf{u}, \mathbf{E}, \mathbf{S}]$ at infinity, then a suitable choice of the pair (φ, ψ_α) leads to the stress tensor $S_{\alpha\beta}$ in the form

$$S_{11} = -\frac{2P_0}{\pi\, r^4}x_1^2 x_2, \quad S_{22} = -\frac{2P_0}{\pi\, r^4}x_2^3, \quad S_{12} = -\frac{2P_0}{\pi\, r^4}x_1 x_2^2 \tag{9.6}$$

where

$$r = |\mathbf{x}| = \sqrt{x_1^2 + x_2^2} \tag{9.7}$$

In polar coordinates (r, φ) related to the Cartesian coordinates (x_1, x_2) by

$$x_1 = r\,\cos\varphi, \quad x_2 = r\,\sin\varphi \tag{9.8}$$

we obtain

$$S_{rr} = -\frac{2P_0}{\pi\, r}\sin\varphi, \quad S_{\varphi\varphi} = S_{r\varphi} = 0 \tag{9.9}$$

Clearly, it follows from (9.6) and (9.9) that

$$|S| \to 0 \quad \text{as} \quad r \to \infty \tag{9.10}$$

Similarly, if a concentrated force T_0 tangent to the boundary of a semispace $|x_1| < \infty, x_2 \geq 0$ is applied at the point $(x_1, x_2) = (0, 0)$, and suitable asymptotic conditions are imposed on $s = [\mathbf{u}, \mathbf{E}, \mathbf{S}]$ at infinity, then a suitable choice of the pair (φ, ψ_α) leads to the stress tensor $S_{\alpha\beta}$ in the form

$$S_{11} = -\frac{2T_0}{\pi\, r^4}x_1^3, \quad S_{22} = -\frac{2T_0}{\pi\, r^4}x_1 x_2^2, \quad S_{12} = -\frac{2T_0}{\pi\, r^4}x_1^2 x_2 \tag{9.11}$$

In polar coordinates (r, φ) we obtain

$$S_{rr} = -\frac{2T_0}{\pi\, r}\cos\varphi, \quad S_{\varphi\varphi} = S_{r\varphi} = 0 \tag{9.12}$$

and it follows from Eqs. (9.11) and (9.12) that

$$|\mathbf{S}| \to 0 \quad \text{as} \quad r \to \infty \tag{9.13}$$

9.2 Problems and Solutions Related to Particular Two-Dimensional Boundary Value Problems of Elastostatics

Problem 9.1. Find an elastic state $s = [\mathbf{u}, \mathbf{E}, \mathbf{S}]$ corresponding to a concentrated body force in an interior of a homogeneous and isotropic semispace $|x_1| < \infty$, $x_2 \geq 0$, under plane strain conditions, when the boundary of semispace is stress free and the elastic state satisfies suitable asymptotic conditions at infinity.

Solution. We confine ourselves to the case when the semispace: $|x_1| < \infty$, $x_2 \geq 0$ with stress free boundary $x_2 = 0$ is subject to the body force of the form

$$b_\alpha = b_0\, \delta_{\alpha 2}\, \delta(x_1)\, \delta(x_2 - \xi_2) \tag{9.14}$$

where b_0 represents intensity of the force and $\xi_2 > 0$. This means that the semispace is subject to an internal force that is normal to its boundary and concentrated at the point $(0, \xi_2)$.

A solution $s = [\mathbf{u}, \mathbf{E}, \mathbf{S}]$ to the problem is to be found by using a restricted form of Boussinesq–Papkowitch–Neuber solution [see Eqs. (9.1)–(9.5) in which we let $\psi_1 = 0$, $\psi_2 = \psi$, $\varphi = \varphi$]

$$u_1 = -\frac{1}{4(1-\nu)}(x_2\psi_{,1} + \varphi_{,1}) \tag{9.15}$$

$$u_2 = \frac{1}{4(1-\nu)}[(3-4\nu)\,\psi - x_2\psi_{,2} - \varphi_{,2}] \tag{9.16}$$

where $\psi = \psi(x_1, x_2)$ and $\varphi = \varphi(x_1, x_2)$ satisfy Poisson's equations

$$\psi_{,rr} = -\frac{1}{\mu} b_2 \tag{9.17}$$

and

$$\varphi_{,rr} = \frac{1}{\mu} x_2 b_2 \tag{9.18}$$

The strain and stress fields are then given, respectively, by

$$E_{11} = \frac{1}{4(1-\nu)}[-x_2\psi_{,11} - \varphi_{,11}] \tag{9.19}$$

$$E_{22} = \frac{1}{4(1-\nu)}[2(1-2\nu)\psi_{,2} - x_2\psi_{,22} - \varphi_{,22}] \tag{9.20}$$

$$E_{12} = \frac{1}{4(1-\nu)}[(1-2\nu)\psi_{,1} - x_2\psi_{,12} - \varphi_{,12}] \tag{9.21}$$

and

$$S_{11} = \frac{\mu}{2(1-\nu)} [-x_2\psi,_{11} + 2\nu\psi,_2 - \varphi,_{11}] \tag{9.22}$$

$$S_{22} = \frac{\mu}{2(1-\nu)} [2(1-\nu)\psi,_2 - x_2\psi,_{22} - \varphi,_{22}] \tag{9.23}$$

$$S_{12} = \frac{\mu}{2(1-\nu)} [(1-2\nu)\psi,_1 - x_2\psi,_{12} - \varphi,_{12}] \tag{9.24}$$

The boundary conditions take the form

$$S_{12}(x_1, 0) = S_{22}(x_1, 0) = 0 \quad \text{for } |x_1| < \infty \tag{9.25}$$

In addition, we assume suitable vanishing conditions at infinity, and suitable restrictions on $\mathbf{u}$ to obtain a unique solution to the problem.

To this end we look for a solution $s = [\mathbf{u}, \mathbf{E}, \mathbf{S}]$ in the form

$$s = s^{(0)} + s^{(1)} \tag{9.26}$$

where $s^{(0)} = [\mathbf{u}^{(0)}, \mathbf{E}^{(0)}, \mathbf{S}^{(0)}]$ is a solution for an infinite plane $|x_1| < \infty$, $|x_2| < \infty$ subject to the body force (9.14), and $s^{(1)} = [\mathbf{u}^{(1)}, \mathbf{E}^{(1)}, \mathbf{S}^{(1)}]$ is a solution for a semispace $|x_1| < \infty$, $x_2 \geq 0$ subject to the boundary conditions

$$S_{12}^{(1)}(x_1, 0) = -S_{12}^{(0)}(x_1, 0) \tag{9.27}$$

and

$$S_{22}^{(1)}(x_1, 0) = -S_{22}^{(0)}(x_1, 0) \tag{9.28}$$

This amounts to looking for a pair (ψ, φ) in the form

$$\psi = \psi^{(0)} + \psi^{(1)} \tag{9.29}$$

and

$$\varphi = \varphi^{(0)} + \varphi^{(1)} \tag{9.30}$$

where

$$\nabla^2\psi^{(0)} = -\frac{1}{\mu} b_2 \tag{9.31}$$

and

$$\nabla^2\varphi^{(0)} = \frac{1}{\mu} x_2 b_2 \tag{9.32}$$

and

$$\nabla^2\psi^{(1)} = 0, \quad \nabla^2\varphi^{(1)} = 0 \tag{9.33}$$

Substituting b_2 from (9.14) into (9.31) and (9.32), respectively, we obtain

$$\nabla^2 \psi^{(0)} = -\frac{b_0}{\mu}\delta(x_1)\delta(x_2 - \xi_2) \tag{9.34}$$

and

$$\nabla^2 \varphi^{(0)} = \frac{b_0 \xi_2}{\mu}\delta(x_1)\delta(x_2 - \xi_2) \tag{9.35}$$

where we used the identity

$$x_2\, \delta(x_2 - \xi_2) = \xi_2\, \delta(x_2 - \xi_2) \tag{9.36}$$

Equations (9.34) and (9.35) are to be satisfied for every $(x_1, x_2) \in E^2$ and for a fixed positive ξ_2. The unique solutions to Eqs. (9.34) and (9.35) are then given, respectively, by

$$\psi^{(0)} = -\frac{b_0}{2\pi\mu}\ln\left(\frac{r_1}{L}\right) \tag{9.37}$$

and

$$\varphi^{(0)} = \frac{b_0 \xi_2}{2\pi\mu}\ln\left(\frac{r_1}{L}\right) \tag{9.38}$$

where

$$r_1 = \sqrt{x_1^2 + (x_2 - \xi_2)^2} \tag{9.39}$$

and L is a positive constant of the length dimension. The solution $s^{(0)}$ is obtained by letting $\psi = \psi^{(0)}$ and $\varphi = \varphi^{(0)}$ into Eqs. (9.15)–(9.16), (9.19)–(9.21), and (9.22)–(9.24). Therefore, we obtain

$$u_1^{(0)} = \frac{b_0}{8\pi\mu(1-\nu)}(x_2 - \xi_2)\frac{\partial}{\partial x_1}\ln\left(\frac{r_1}{L}\right) \tag{9.40}$$

$$u_2^{(0)} = -\frac{b_0}{8\pi\mu(1-\nu)}\left[(3-4\nu) - (x_2 - \xi_2)\frac{\partial}{\partial x_2}\right]\ln\left(\frac{r_1}{L}\right) \tag{9.41}$$

and

$$E_{11}^{(0)} = \frac{b_0}{8\pi\mu(1-\nu)}(x_2 - \xi_2)\frac{\partial^2}{\partial x_1^2}\ln\left(\frac{r_1}{L}\right) \tag{9.42}$$

$$E_{22}^{(0)} = -\frac{b_0}{8\pi\mu(1-\nu)}\left[2(1-2\nu)\frac{\partial}{\partial x_2} - (x_2 - \xi_2)\frac{\partial^2}{\partial x_2^2}\right]\ln\left(\frac{r_1}{L}\right) \tag{9.43}$$

$$E_{12}^{(0)} = -\frac{b_0}{8\pi\mu(1-\nu)}\left[(1-2\nu)\frac{\partial}{\partial x_1} - (x_2 - \xi_2)\frac{\partial^2}{\partial x_1 \partial x_2}\right]\ln\left(\frac{r_1}{L}\right) \tag{9.44}$$

The stress components $S_{11}^{(0)}$, $S_{22}^{(0)}$, and $S_{12}^{(0)}$ are given by

$$S_{11}^{(0)} = -\frac{b_0}{4\pi(1-\nu)}\left[2\nu\frac{\partial}{\partial x_2} - (x_2-\xi_2)\frac{\partial^2}{\partial x_1^2}\right]\ln\left(\frac{r_1}{L}\right) \tag{9.45}$$

$$S_{22}^{(0)} = -\frac{b_0}{4\pi(1-\nu)}\left[2(1-\nu)\frac{\partial}{\partial x_2} - (x_2-\xi_2)\frac{\partial^2}{\partial x_2^2}\right]\ln\left(\frac{r_1}{L}\right) \tag{9.46}$$

$$S_{12}^{(0)} = -\frac{b_0}{4\pi(1-\nu)}\left[(1-2\nu)\frac{\partial}{\partial x_1} - (x_2-\xi_2)\frac{\partial^2}{\partial x_1 \partial x_2}\right]\ln\left(\frac{r_1}{L}\right) \tag{9.47}$$

The solution $s^{(1)}$ is to be found by letting $\psi = \psi^{(1)}$ and $\varphi = \varphi^{(1)}$ into Eqs.(9.15)–(9.16), (9.19)–(9.21), and (9.22)–(9.24), where $\psi^{(1)}$ and $\varphi^{(1)}$ satisfy Eqs. $(9.33)_1$ and $(9.33)_2$, respectively, for $|x_1| < \infty, x_2 > 0$, subject to suitable boundary conditions at $x_2 = 0$. A hint as to how $\psi^{(1)}$ and $\varphi^{(1)}$ could be found comes from the boundary conditions (9.27) and (9.28) written in terms of the pairs $(\psi^{(0)}, \varphi^{(0)})$ and $(\psi^{(1)}, \varphi^{(1)})$:

$$\left[(1-2\nu)\psi^{(1)} - \varphi,_2^{(1)}\right],_1(x_1,0) = -f(x_1) \tag{9.48}$$

$$\left[(2-2\nu)\psi^{(1)} - \varphi,_2^{(1)}\right],_2(x_1,0) = -g(x_1) \tag{9.49}$$

where

$$f(x_1) = \left[(1-2\nu)\psi^{(0)} - \varphi,_2^{(0)}\right],_1(x_1,0) \tag{9.50}$$

and

$$g(x_1) = \left[(2-2\nu)\psi^{(0)} - \varphi,_2^{(0)}\right],_2(x_1,0) \tag{9.51}$$

Since

$$\frac{\partial}{\partial x_1}\ln\left(\frac{r_1}{L}\right) = \int_0^\infty e^{-\alpha|x_2-\xi_2|}\sin\alpha x_1 d\alpha \tag{9.52}$$

and

$$\frac{\partial}{\partial x_2}\ln\left(\frac{r_1}{L}\right) = \int_0^\infty e^{-\alpha|x_2-\xi_2|}\cos\alpha x_1 d\alpha \tag{9.53}$$

therefore, because of (9.37) and (9.38) we obtain

$$\psi,_1^{(0)} = -\frac{b_0}{2\pi\mu}\int_0^\infty e^{-\alpha|x_2-\xi_2|}\sin\alpha x_1 d\alpha \tag{9.54}$$

and

$$\varphi,_2^{(0)} = \frac{b_0 \xi_2}{2\pi \mu} \int_0^\infty e^{-\alpha|x_2 - \xi_2|} \cos \alpha x_1 d\alpha \tag{9.55}$$

It follows from (9.55) that

$$\varphi,_{12}^{(0)} = -\frac{b_0 \xi_2}{2\pi \mu} \int_0^\infty e^{-\alpha|x_2 - \xi_2|} \alpha \sin \alpha x_1 d\alpha \tag{9.56}$$

and an extension of the RHS of (9.50) to include arbitrary point (x_1, x_2) reads

$$\begin{aligned} &\left[(1 - 2\nu)\psi,_1^{(0)} - \varphi,_{12}^{(0)}\right](x_1, x_2) \\ &= -\frac{b_0}{2\pi \mu} \int_0^\infty e^{-\alpha|x_2 - \xi_2|} \left[(1 - 2\nu) - \alpha \xi_2\right] \sin \alpha x_1 d\alpha \end{aligned} \tag{9.57}$$

Hence

$$f(x_1) = -\frac{b_0}{2\pi \mu} \int_0^\infty e^{-\alpha \xi_2} \left[(1 - 2\nu) - \alpha \xi_2\right] \sin \alpha x_1 d\alpha \tag{9.58}$$

Similarly, we obtain

$$\psi,_2^{(0)} = -\frac{b_0}{2\pi \mu} \int_0^\infty e^{-\alpha|x_2 - \xi_2|} \cos \alpha x_1 d\alpha \tag{9.59}$$

and

$$\varphi,_{22}^{(0)} = \frac{b_0 \xi_2}{2\pi \mu} \frac{\partial}{\partial x_2} \int_0^\infty e^{-\alpha|x_2 - \xi_2|} \cos \alpha x_1 d\alpha \tag{9.60}$$

For $0 \le x_2 < \xi_2$ Eq. (9.60) takes the form

$$\varphi,_{22}^{(0)} = \frac{b_0 \xi_2}{2\pi \mu} \int_0^\infty e^{-\alpha(\xi_2 - x_2)} \alpha \cos \alpha x_1 d\alpha \tag{9.61}$$

Hence, using (9.59) and (9.61) we reduce (9.51) to the form

$$g(x_1) = -\frac{b_0}{2\pi\mu} \int_0^\infty e^{-\alpha\xi_2}[(2-2\nu) + \alpha\xi_2] \cos\alpha x_1 d\alpha \tag{9.62}$$

Since

$$\nabla^2[e^{-\alpha x_2} \cos\alpha x_1] = 0 \tag{9.63}$$

an inspection of Eqs. (9.33) and of the boundary conditions (9.48) and (9.49) in which f and g are given by the integrals (9.58) and (9.62), respectively, leads to the integral form of $\psi^{(1)}$ and $\varphi^{(1)}$ for $|x_1| < \infty$, $x_2 > 0$:

$$\psi^{(1)}(x_1, x_2) = \int_0^\infty A(\alpha) e^{-\alpha x_2} \cos\alpha x_1 d\alpha \tag{9.64}$$

and

$$\varphi^{(1)}(x_1, x_2) = \int_0^\infty B(\alpha) e^{-\alpha x_2} \cos\alpha x_1 d\alpha \tag{9.65}$$

where $A(\alpha)$ and $B(\alpha)$ are arbitrary functions on $[0, \infty)$ to be selected in such a way that the boundary conditions (9.48) and (9.49) are satisfied. For the partial derivatives of $\psi^{(1)}$ and $\varphi^{(1)}$ that come into the boundary conditions (9.48) and (9.49) we obtain

$$\psi,_1^{(1)} = -\int_0^\infty A(\alpha) e^{-\alpha x_2} \alpha \sin\alpha x_1 d\alpha \tag{9.66}$$

$$\varphi,_2^{(1)} = -\int_0^\infty B(\alpha) e^{-\alpha x_2} \alpha \cos\alpha x_1 d\alpha \tag{9.67}$$

$$\varphi,_{21}^{(1)} = \int_0^\infty B(\alpha) e^{-\alpha x_2} \alpha^2 \sin\alpha x_1 d\alpha \tag{9.68}$$

$$\psi,_2^{(1)} = -\int_0^\infty A(\alpha) e^{-\alpha x_2} \alpha \cos\alpha x_1 d\alpha \tag{9.69}$$

$$\varphi,_{22}^{(1)} = \int_0^\infty B(\alpha) e^{-\alpha x_2} \alpha^2 \cos\alpha x_1 d\alpha \tag{9.70}$$

Therefore, substituting (9.66) and (9.68) into (9.48), and (9.69) and (9.70) into (9.49), and using f and g in the forms (9.58) and (9.62), respectively, we find that the functions $A = A(\alpha)$ and $B = B(\alpha)$ must satisfy the linear algebraic equations

$$(1-2\nu)A+\alpha B=-\frac{b_0}{2\pi\mu}e^{-\alpha\xi_2}\frac{1}{\alpha}[1-2\nu-\alpha\xi_2]$$
$$(2-2\nu)A+\alpha B=-\frac{b_0}{2\pi\mu}e^{-\alpha\xi_2}\frac{1}{\alpha}[2-2\nu+\alpha\xi_2] \tag{9.71}$$

and the only solution (A, B) to (9.71) takes the form

$$A=-\frac{b_0}{2\pi\mu}e^{-\alpha\xi_2}\frac{(1+2\alpha\xi_2)}{\alpha} \tag{9.72}$$
$$B=\frac{b_0\xi_2}{2\pi\mu}e^{-\alpha\xi_2}\frac{(3-4\nu)}{\alpha} \tag{9.73}$$

It follows from Eqs. (9.64)–(9.65) and (9.72)–(9.73) that the integral representations of $\psi^{(1)}$ and $\varphi^{(1)}$ are divergent, however, all partial derivatives of $\psi^{(1)}$ and $\varphi^{(1)}$ are represented by the convergent integrals. This implies that the integral representations of $\mathbf{E}^{(1)}$ and $\mathbf{S}^{(1)}$ are convergent. In the following we are to obtain first the integral forms of $\mathbf{E}^{(1)}$ and $\mathbf{S}^{(1)}$, as well as of $u_1^{(1)}$ and $u_{2,1}^{(1)}$, and next the integral representation of $u_{2,1}^{(1)}$ is used to recover $u_2^{(1)}$ by integration. Note that an alternative form of Eqs. (9.19)–(9.21) and (9.22)–(9.24), respectively, taken at $\psi=\psi^{(1)}$ and $\varphi=\varphi^{(1)}$, reads

$$E_{11}^{(1)}+E_{22}^{(1)}=\frac{1-2\nu}{2-2\nu}\psi,_2^{(1)} \tag{9.74}$$

$$E_{11}^{(1)}-E_{22}^{(1)}=-\frac{1-2\nu}{2-2\nu}\psi,_2^{(1)}-\frac{1}{4(1-\nu)}\left[x_2\left(\psi,_{11}^{(1)}-\psi,_{22}^{(1)}\right)+\varphi,_{11}^{(1)}-\varphi,_{22}^{(1)}\right] \tag{9.75}$$
$$E_{12}^{(1)}=\frac{1-2\nu}{4(1-\nu)}\psi,_1^{(1)}-\frac{1}{4(1-\nu)}\left(x_2\psi,_{12}^{(1)}+\varphi,_{12}^{(1)}\right) \tag{9.76}$$

and

$$S_{11}^{(1)}+S_{22}^{(1)}=\frac{\mu}{1-\nu}\psi,_2^{(1)} \tag{9.77}$$

$$S_{11}^{(1)}-S_{22}^{(1)}=-\mu\frac{1-2\nu}{1-\nu}\psi,_2^{(1)}-\frac{\mu}{2(1-\nu)}\left[x_2\left(\psi,_{11}^{(1)}-\psi,_{22}^{(1)}\right)+\varphi,_{11}^{(1)}-\varphi,_{22}^{(1)}\right] \tag{9.78}$$
$$S_{12}^{(1)}=\frac{\mu(1-2\nu)}{2(1-\nu)}\psi,_1^{(1)}-\frac{\mu}{2(1-\nu)}\left(x_2\psi,_{12}^{(1)}+\varphi,_{12}^{(1)}\right) \tag{9.79}$$

Also, it follows from (9.15) and (9.16), respectively, taken at $\psi=\psi^{(1)}$ and $\varphi=\varphi^{(1)}$ that

$$u_1^{(1)}=-\frac{1}{4(1-\nu)}\left(x_2\psi,_1^{(1)}+\varphi,_1^{(1)}\right) \tag{9.80}$$

and

$$u^{(1)}_{2,1} = \frac{1}{4(1-\nu)}\left[(3-4\nu)\psi,^{(1)}_{1} - x_2\psi,^{(1)}_{12} - \varphi,^{(1)}_{12}\right] \tag{9.81}$$

Therefore, substituting $\psi^{(1)}$ and $\varphi^{(1)}$ from (9.64) and (9.65), respectively, where A and B are given by (9.72) and (9.73), respectively into Eqs. (9.74)–(9.81), we obtain

$$E^{(1)}_{11} + E^{(1)}_{22} = \frac{b_0}{2\pi\mu}\frac{1-2\nu}{2-2\nu}\int_0^\infty e^{-\alpha(x_2+\xi_2)}(1+2\alpha\xi_2)\cos\alpha x_1 d\alpha \tag{9.82}$$

$$\begin{aligned} E^{(1)}_{11} - E^{(1)}_{22} = &-\frac{b_0}{2\pi\mu}\frac{1-2\nu}{2-2\nu}\int_0^\infty e^{-\alpha(x_2+\xi_2)}(1+2\alpha\xi_2)\cos\alpha x_1 d\alpha \\ &- \frac{b_0}{4\pi\mu(1-\nu)}\left\{\int_0^\infty e^{-\alpha(x_2+\xi_2)}\alpha[x_2(1+2\alpha\xi_2)-\xi_2(3-4\nu)]\cos\alpha x_1 d\alpha\right\} \end{aligned} \tag{9.83}$$

$$\begin{aligned} E^{(1)}_{12} = &\frac{b_0}{2\pi\mu}\frac{1-2\nu}{4(1-\nu)}\int_0^\infty e^{-\alpha(x_2+\xi_2)}(1+2\alpha\xi_2)\sin\alpha x_1 d\alpha \\ &+ \frac{b_0}{2\pi\mu}\frac{1}{4(1-\nu)}\int_0^\infty e^{-\alpha(x_2+\xi_2)}[x_2-(3-4\nu)\xi_2+2\alpha\xi_2 x_2]\times\alpha\sin\alpha x_1 d\alpha \end{aligned} \tag{9.84}$$

and

$$S^{(1)}_{11} + S^{(1)}_{22} = \frac{b_0}{2\pi(1-\nu)}\int_0^\infty e^{-\alpha(x_2+\xi_2)}(1+2\alpha\xi_2)\cos\alpha x_1 d\alpha \tag{9.85}$$

$$\begin{aligned} S^{(1)}_{11} - S^{(1)}_{22} = &-\frac{b_0(1-2\nu)}{2\pi(1-\nu)}\int_0^\infty e^{-\alpha(x_2+\xi_2)}(1+2\alpha\xi_2)\cos\alpha x_1 d\alpha \\ &- \frac{b_0}{2\pi(1-\nu)}\int_0^\infty e^{-\alpha(x_2+\xi_2)}\alpha[x_2-(3-4\nu)\xi_2+2\alpha\xi_2 x_2]\cos\alpha x_1 d\alpha \end{aligned} \tag{9.86}$$

$$S_{12}^{(1)} = \frac{b_0(1-2\nu)}{4\pi(1-\nu)} \int_0^\infty e^{-\alpha(x_2+\xi_2)}(1+2\alpha\xi_2)\sin\alpha x_1 d\alpha$$
$$+ \frac{b_0}{4\pi(1-\nu)} \int_0^\infty e^{-\alpha(x_2+\xi_2)}[x_2-(3-4\nu)\xi_2+2\alpha\xi_2 x_2]\alpha \sin\alpha x_1 d\alpha \tag{9.87}$$

In addition, Eqs. (9.80) and (9.81), respectively, imply that

$$u_1^{(1)} = -\frac{b_0}{8\pi\mu(1-\nu)} \int_0^\infty e^{-\alpha(x_2+\xi_2)}[x_2-(3-4\nu)\xi_2+2\alpha\xi_2 x_2]\sin\alpha x_1 d\alpha \tag{9.88}$$

and

$$u_{2,1}^{(1)} = \frac{b_0}{8\pi\mu(1-\nu)} \int_0^\infty e^{-\alpha(x_2+\xi_2)}\{3-4\nu+[(3-4\nu)\xi_2+x_2]\alpha+2\xi_2 x_2\alpha^2\}$$
$$\times \sin\alpha x_1 d\alpha \tag{9.89}$$

It follows from Eqs. (9.82)–(9.89), respectively, that $\mathbf{E}^{(1)}$, $\mathbf{S}^{(1)}$, $u_1^{(1)}$, and $u_{2,1}^{(1)}$ are represented by the convergent integrals for any point of the semispace: $|x_1|<\infty$, $x_2\geq 0$. In addition, by using the formulas [see (9.52) and (9.53)]

$$\int_0^\infty e^{-\alpha u}\cos\alpha x_1 d\alpha = \frac{\partial}{\partial u}\ln\left(\frac{R}{L}\right) \tag{9.90}$$
$$\int_0^\infty e^{-\alpha u}\sin\alpha x_1 d\alpha = \frac{\partial}{\partial x_1}\ln\left(\frac{R}{L}\right) \tag{9.91}$$

and the formulas obtained from (9.90) and (9.91) by differentiation

$$\int_0^\infty e^{-\alpha u}\alpha\sin\alpha x_1 d\alpha = -\frac{\partial^2}{\partial x_1\partial u}\ln\left(\frac{R}{L}\right) \tag{9.92}$$
$$\int_0^\infty e^{-\alpha u}\alpha^2\sin\alpha x_1 d\alpha = \frac{\partial^3}{\partial x_1\partial u^2}\left[\ln\left(\frac{R}{L}\right)\right] \tag{9.93}$$
$$\int_0^\infty e^{-\alpha u}\alpha\cos\alpha x_1 d\alpha = -\frac{\partial^2}{\partial u^2}\left[\ln\left(\frac{R}{L}\right)\right] \tag{9.94}$$
$$\int_0^\infty e^{-\alpha u}\alpha^2\cos\alpha x_1 d\alpha = \frac{\partial^3}{\partial u^3}\left[\ln\left(\frac{R}{L}\right)\right] \tag{9.95}$$

where

$$R = \sqrt{x_1^2 + u^2}, \quad u > 0 \tag{9.96}$$

the fields $\mathbf{E}^{(1)}$, $\mathbf{S}^{(1)}$, $u_1^{(1)}$, and $u_{2,1}^{(1)}$ can be obtained in terms of elementary functions.

For example, by using (9.91) and (9.92), the closed form of $u_1^{(1)}$ is obtained

$$u_1^{(1)} = -\frac{b_0}{8\pi\mu(1-\nu)} \left\{ [x_2 - (3-4\nu)\xi_2] \frac{\partial}{\partial x_1} \left[\ln\left(\frac{r_2}{L}\right) \right] -2x_2\xi_2 \frac{\partial^2}{\partial x_1 \partial x_2} \left[\ln\left(\frac{r_2}{L}\right) \right] \right\} \tag{9.97}$$

where

$$r_2 = \sqrt{x_1^2 + (x_2 + \xi_2)^2} \tag{9.98}$$

To obtain a closed-form of $u_2^{(1)}$ we integrate (9.89) with respect to x_1 over the interval $[0, x_1]$ and obtain

$$u_2^{(1)}(x_1, x_2) - u_2^{(1)}(0, x_2) = \frac{b_0}{8\pi\mu(1-\nu)} \int_0^\infty e^{-\alpha(x_2+\xi_2)} \times \{3 - 4\nu + [(3-4\nu)\xi_2 + x_2]\alpha + 2\xi_2 x_2 \alpha^2\} \times \frac{1 - \cos\alpha x_1}{\alpha} d\alpha \tag{9.99}$$

By letting

$$u_2^{(1)}(0, x_2) = 0 \quad \text{for } x_2 > 0 \tag{9.100}$$

Equation (9.99) can be written as

$$u_2^{(1)} = \frac{b_0}{8\pi\mu(1-\nu)} \left\{ (3-4\nu) \int_0^\infty e^{-\alpha(x_2+\xi_2)} \frac{(1-\cos\alpha x_1)}{\alpha} d\alpha + [(3-4\nu)\xi_2 + x_2] \int_0^\infty e^{-\alpha(x_2+\xi_2)} (1 - \cos\alpha x_1) d\alpha + 2\xi_2 x_2 \int_0^\infty e^{-\alpha(x_2+\xi_2)} \alpha (1 - \cos\alpha x_1) d\alpha \right\} \tag{9.101}$$

By integrating (9.91) with respect to x_1 we obtain

$$\int_0^\infty e^{-\alpha u} \frac{1-\cos \alpha x_1}{\alpha} d\alpha = \ln \left(\frac{R}{u} \right) \tag{9.102}$$

Hence

$$\int_0^\infty e^{-\alpha(x_2+\xi_2)} \frac{1-\cos \alpha x_1}{\alpha} d\alpha = \ln \frac{r_2}{x_2+\xi_2} \tag{9.103}$$

and by differentiation of (9.103) with respect to x_2, we obtain

$$\int_0^\infty e^{-\alpha(x_2+\xi_2)} (1-\cos \alpha x_1) d\alpha = -\frac{\partial}{\partial x_2} \ln \left(\frac{r_2}{x_2+\xi_2} \right) \tag{9.104}$$

and

$$\int_0^\infty e^{-\alpha(x_2+\xi_2)} \alpha (1-\cos \alpha x_1) d\alpha - \frac{\partial^2}{\partial x_2^2} \ln \left(\frac{r_2}{x_2+\xi_2} \right) \tag{9.105}$$

Finally, substituting (9.103), (9.104), and (9.105) into (9.101) we obtain $u_2^{(1)}$ in the form

$$u_2^{(1)} = \frac{b_0}{8\pi\mu(1-\nu)} \left\{ (3-4\nu) \ln \left(\frac{r_2}{x_2+\xi_2} \right) - [(3-4\nu)\xi_2 + x_2] \frac{\partial}{\partial x_2} \ln \left(\frac{r_2}{x_2+\xi_2} \right) \right. \\ \left. + 2\xi_2 x_2 \frac{\partial^2}{\partial x_2^2} \ln \left(\frac{r_2}{x_2+\xi_2} \right) \right\} \tag{9.106}$$

This completes a solution to Problem 9.1 in which the semispace is subject to an internal force that is normal to its boundary and concentrated at the point $(0, \xi_2)$.

In a similar way a solution to Prob. 9.1 in which the semispace is subject to a force that is parallel to its boundary and concentrated at $(0, \xi_2)$, may be obtained.

Let $U_{\alpha 2}$ and $U_{\alpha 1}$, respectively, denote the displacement of the semispace corresponding to the unit normal and parallel forces at $(0, \xi_2)$, and let $\boldsymbol{l} = (l_1, l_2)$ be an arbitrary force concentrated at $(0, \xi_2)$. Then the displacement u_α corresponding to a solution to Problem 9.1 in which the semispace is subject to the concentrated force $\boldsymbol{l}$ at $(0, \xi_2)$ takes the form

$$u_\alpha = U_{\alpha\beta} l_\beta \tag{9.107}$$

This completes a solution to Problem 9.1 in which the semispace with stress free boundary is subject to a concentrated force $\boldsymbol{l}$ at $(0, \xi_2)$.

Problem 9.2. Find an elastic state $s = [\mathbf{u}, \mathbf{E}, \mathbf{S}]$ corresponding to a concentrated body force in an interior of a homogeneous and isotropic semispace $|x_1| < \infty$,

$x_2 \geq 0$, under plane strain conditions, when the boundary of semispace is clamped and the elastic state vanishes at infinity.

Solution. Let the semispace $|x_1| < \infty$, $x_2 \geq 0$ with a clamped boundary $x_2 = 0$ be subject to the body force

$$b_\alpha = b_0 \delta_{\alpha 2} \delta(x_1) \delta(x_2 - \xi_2) \tag{9.108}$$

An elastic state $s = [\mathbf{u}, \mathbf{E}, \mathbf{S}]$ corresponding to (9.108) and satisfying the boundary conditions

$$u_1(x_1, 0) = u_2(x_1, 0) = 0 \quad |x_1| < \infty \tag{9.109}$$

and suitable vanishing conditions at infinity may be found in a way similar to that of Problem 9.1. To this end we use Eqs. (9.109)–(9.119) of Problem 9.1 to obtain **u**, **E**, and **S**, respectively.

In particular, u_1 and u_2 are to be found from the equations

$$u_1 = -\frac{1}{4(1-\nu)}(x_2 \psi_{,1} + \varphi_{,1}) \tag{9.110}$$

$$u_2 = \frac{1}{4(1-\nu)}[(3-4\nu)\psi - x_2 \psi_{,2} - \varphi_{,2}] \tag{9.111}$$

where ψ and φ satisfy Poisson's equations

$$\nabla^2 \psi = -\frac{1}{\mu} b_2 \tag{9.112}$$

and

$$\nabla^2 \varphi = \frac{1}{\mu} x_2 b_2 \tag{9.113}$$

To find a pair (ψ, φ) that generates (u_1, u_2) by Eqs. (9.110)–(9.111) in such a way that Eqs. (9.109) are satisfied, we let

$$\psi = \psi^{(0)} + \psi^{(1)} \quad |x_1| < \infty, x_2 \geq 0 \tag{9.114}$$

and

$$\varphi = \varphi^{(0)} + \varphi^{(1)} \quad |x_1| < \infty, x_2 \geq 0 \tag{9.115}$$

where

$$\nabla^2 \psi^{(0)} = -\frac{1}{\mu} b_2 \quad |x_1| < \infty, x_2 \geq 0 \tag{9.116}$$

and

$$\nabla^2 \varphi^{(0)} = \frac{1}{\mu} x_2 b_2 \quad |x_1| < \infty, x_2 \geq 0 \tag{9.117}$$

and

$$\nabla^2\psi^{(1)} = 0, \quad \nabla^2\varphi^{(1)} = 0 \tag{9.118}$$

and the harmonic functions $\psi^{(1)}$ and $\varphi^{(1)}$ defined for $|x_1| < \infty, \quad x_2 \geq 0$ are selected in such a way that u_1 and u_2 vanish at $x_2 = 0$. To obtain a pair $(\psi^{(0)}, \varphi^{(0)})$ we extend Eqs. (9.116)–(9.117) to the whole plane E^2 in which a normal force of intensity b_0 is concentrated at $(0, \xi_2)$ and a normal force of intensity—b_0 is concentrated at $(0, -\xi_2)$. This amounts to solving the equations

$$\nabla^2\psi^{(0)} = -\frac{b_0}{\mu}\delta(x_1)[\delta(x_2 - \xi_2) - \delta(x_2 + \xi_2)] \tag{9.119}$$

and

$$\nabla^2\varphi^{(0)} = \frac{1}{\mu}b_0 x_2[\delta(x_2 - \xi_2) - \delta(x_2 + \xi_2)] \quad \text{for } |x_1| < \infty, \ |x_2| < \infty \tag{9.120}$$

Note that a restriction of Eqs. (9.119)–(9.120) to the semispace $|x_1| < \infty, x_2 \geq 0$ leads to Eqs. (9.116)–(9.117), and an extension of $(\psi^{(0)}, \varphi^{(0)})$ is denoted in the same way as its restriction.

Since

$$x_2\delta(x_2 - \xi_2) = \xi_2\delta(x_2 - \xi_2) \tag{9.121}$$

then

$$-x_2\delta(x_2 - \xi_2) = -\xi_2\delta(x_2 - \xi_2) \tag{9.122}$$

and replacing ξ_2 by $-\xi_2$ in (9.122) we get

$$-x_2\delta(x_2 + \xi_2) = \xi_2\delta(x_2 + \xi_2) \tag{9.123}$$

Hence, Eq. (9.120) can be written as

$$\nabla^2\varphi^{(0)} = \frac{b_0\xi_2}{\mu}[\delta(x_2 - \xi_2) + \delta(x_2 + \xi_2)] \quad \text{for } |x_1| < \infty, |x_2| < \infty \tag{9.124}$$

Proceeding in a way similar to that of solving Eqs. (9.129) and (9.130) of Problem 9.1, from Eqs. (9.119) and (9.124), respectively, we obtain

$$\psi^{(0)} = -\frac{b_0}{2\pi\mu}\left[\ln\left(\frac{r_1}{L}\right) - \ln\left(\frac{r_2}{L}\right)\right] \tag{9.125}$$

and

$$\varphi^{(0)} = \frac{b_0\xi_2}{2\pi\mu}\left[\ln\left(\frac{r_1}{L}\right) + \ln\left(\frac{r_2}{L}\right)\right] \tag{9.126}$$

where

$$r_{1.2} = \sqrt{x_1^2 + (x_2 \mp \xi_2)^2} \tag{9.127}$$

It follows from (9.125)–(9.127) that

$$\psi^{(0)}(x_1, 0) = 0, \quad |x_1| < \infty \tag{9.128}$$

and

$$\varphi^{(0)}(x_1, 0) = \frac{b_0 \xi_2}{\pi \mu} \ln\left(\frac{r_0}{L}\right), \quad |x_1| < \infty \tag{9.129}$$

where

$$r_0 = \sqrt{x_1^2 + \xi_2^2} \tag{9.130}$$

Also, using (9.110) and (9.111) in which $\psi = \psi^{(0)}$ and $\varphi = \varphi^{(0)}$, and letting $x_2 = 0$ we obtain

$$u_1^{(0)}(x_1, 0) = -\frac{b_0 \xi_2}{4\pi \mu(1-\nu)} \frac{\partial}{\partial x_1} \ln\left(\frac{r_0}{L}\right) \tag{9.131}$$

and

$$u_2^{(0)}(x_1, 0) = 0 \tag{9.132}$$

The displacements $u_1^{(1)}$ and $u_2^{(1)}$ are represented by

$$u_1^{(1)}(x_1, x_2) = -\frac{1}{4(1-\nu)} \left(x_2 \psi,_1^{(1)} + \varphi,_1^{(1)}\right) \tag{9.133}$$

$$u_2^{(1)}(x_1, x_2) = \frac{1}{4(1-\nu)} \left[(3-4\nu)\psi^{(1)} - x_2 \psi,_2^{(1)} - \varphi,_2^{(1)}\right] \tag{9.134}$$

where $\psi^{(1)}$ and $\varphi^{(1)}$ are harmonic on the semispace $|x_1| < \infty$, $x_2 > 0$.

It is easy to check that the function $\varphi^{(1)} = \varphi^{(1)}(x_1, x_2)$ given by

$$\varphi^{(1)}(x_1, x_2) = -\frac{b_0 \xi_2}{\pi \mu} \ln\left(\frac{r_2}{L}\right) \tag{9.135}$$

satisfies the Laplace's equation

$$\nabla^2 \varphi^{(1)} = 0 \quad \text{for } |x_1| < \infty, \ x_2 > 0 \tag{9.136}$$

and complies with the boundary condition

$$u_1(x_1, 0) = u_1^{(0)}(x_1, 0) + u_2^{(1)}(x_1, 0) = 0 \tag{9.137}$$

To find $\psi^{(1)}$, note that because of (9.132) and (9.134), $\psi^{(1)}$ must satisfy the Laplace equation for $|x_1| < \infty, x_2 > 0$ subject to the boundary condition

$$(3-4\nu)\psi^{(1)}(x_1,0) = \varphi_{,2}^{(1)}(x_1,0) = -\frac{b_0\xi_2}{\pi\mu}\frac{\partial}{\partial x_2}\ln\left(\frac{r_2}{L}\right)\Big|_{x_2=0}$$
$$= -\frac{b_0\xi_2}{\pi\mu}\frac{\partial}{\partial \xi_2}\ln\left(\frac{r_0}{L}\right) \tag{9.138}$$

Since

$$\frac{\partial}{\partial \xi_2}\ln\left(\frac{r_0}{L}\right) = \int_0^\infty e^{-\alpha\xi_2}\cos\alpha x_1 d\alpha \tag{9.139}$$

therefore, (9.138) takes the form

$$(3-4\nu)\psi^{(1)}(x_1,0) = -\frac{b_0\xi_2}{\pi\mu}\int_0^\infty e^{-\alpha\xi_2}\cos\alpha x_1 d\alpha \tag{9.140}$$

and we find that for any point of the semispace $|x_1| < \infty, x_2 \geq 0$

$$\psi^{(1)}(x_1,x_2) = -\frac{b_0\xi_2}{\pi\mu(3-4\nu)}\frac{\partial}{\partial x_2}\ln\left(\frac{r_2}{L}\right) \tag{9.141}$$

As a result, because of (9.114) and (9.115), (9.125) and (9.126), and (9.135) and (9.141), we obtain

$$\psi(x_1,x_2) = -\frac{b_0}{2\pi\mu(3-4\nu)}\left[(3-4\nu)\ln\left(\frac{r_1}{r_2}\right) + 2\xi_2\frac{\partial}{\partial x_2}\ln\left(\frac{r_2}{L}\right)\right] \tag{9.142}$$

and

$$\varphi(x_1,x_2) = \frac{b_0\xi_2}{2\pi\mu}\ln\left(\frac{r_1}{r_2}\right) \tag{9.143}$$

Next, substituting ψ and φ from Eqs. (9.142) and (9.143), respectively, to Eqs. (9.110)–(9.111), we obtain a closed-form displacement vector **u** corresponding to the solution $s = [\mathbf{u}, \mathbf{E}, \mathbf{S}]$. The associated fields **E** and **S**, respectively, are obtained by substituting ψ and φ from (9.142) and (9.143), into Eqs. (9.113)–(9.115) and (9.116)–(9.119) of Problem 9.1.

This completes a solution to Problem 9.2 when the concentrated body force is normal to the clamped boundary $x_2 = 0$.

If the concentrated body force is parallel to the clamped boundary $x_2 = 0$, Problem 9.2 may be solved in a similar way. When the body force is arbitrarily oriented with regard to the clamped boundary $x_2 = 0$, a solution to Problem 9.2 is a linear combination of the solutions corresponding to the normal and parallel directions of the concentrated body force. This completes a solution to Problem 9.2.

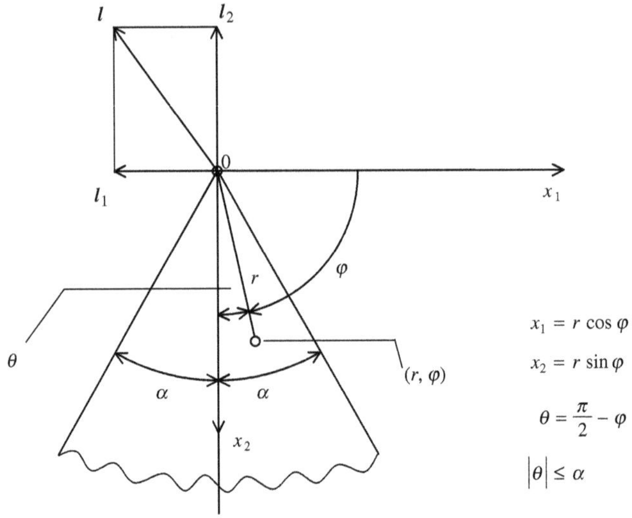

Fig. 9.1 The infinite wedge loaded by a concentrated force

Problem 9.3. Suppose that a homogeneous isotropic infinite elastic wedge, subjected to generalized plane stress conditions, is loaded in its plane by a concentrated force $\boldsymbol{l}$ applied at its tip (see Fig. 9.1)

$$x_1 = r\,\cos\varphi,\quad \theta = \frac{\pi}{2} - \varphi$$
$$x_2 = r\,\sin\varphi,\quad |\theta| \le \alpha$$

Show that the stress components $\overline{S}_{rr}$, $\overline{S}_{r\varphi}$, and $\overline{S}_{\varphi\varphi}$ corresponding to the force $\boldsymbol{l}$ and vanishing at infinity take the form

$$\overline{S}_{rr}(r,\varphi) = \frac{2l_1}{r}\frac{\cos\varphi}{(2\alpha - \sin 2\alpha)} + \frac{2l_2}{r}\frac{\sin\varphi}{(2\alpha + \sin 2\alpha)}$$
$$\overline{S}_{r\varphi}(r,\varphi) = \overline{S}_{\varphi\varphi}(r,\varphi) = 0$$

for every $0 < r < \infty$., $\pi/2 - \alpha \le \varphi \le \pi/2 + \alpha$. Note that $l_1 < 0$ and $l_2 < 0$, and $\left|\overline{S}_{rr}\right| \to \infty$ for $\alpha \to 0$ and $r > 0$.

Solution. A solution to this problem is to be given in the two cases

(i) $l_1 \ne 0,\quad l_2 = 0$
(ii) $l_1 = 0,\quad l_2 \ne 0$

The case (i). The stress components $S_{rr}^{||}$, $S_{r\varphi}^{||}$, and $S_{\varphi\varphi}^{||}$ in a semi-infinite disk $|x_1| < \infty, x_2 \geq 0$ subject to a tangent concentrated force T_0 at (0, 0) take the form [see Eq. (9.12)]

$$S_{rr}^{||}(r,\varphi) = -\frac{2T_0}{\pi r}\cos\varphi = -\frac{2T_0}{\pi r}\sin\theta \tag{9.144}$$

$$S_{r\varphi}^{||}(r,\varphi) = S_{\varphi\varphi}^{||}(r,\varphi) = 0 \tag{9.145}$$

where $0 \leq r < \infty, 0 \leq \varphi < 2\pi$.

A restriction of Eqs. (9.144) and (9.145) to the wedge region shown in the Figure provides a solution to Problem 9.3 in case (i) if a resultant tangent force at the tip of the wedge is equal to l_1, that is, if for every $r \geq 0$

$$\int_{-\alpha}^{\alpha} S_{rr}^{||}\, r\sin\theta d\theta = l_1 \tag{9.146}$$

Substituting $S_{rr}^{||}$ from (9.144) into (9.146) we obtain

$$-\frac{2T_0}{\pi}\times 2\int_0^{\alpha}\sin^2\theta\, d\theta = -\frac{2T_0}{\pi}\int_0^{\alpha}(1-\cos 2\theta)\, d\theta = -\frac{T_0}{\pi}(2\alpha - \sin 2\alpha) = l_1 \tag{9.147}$$

Hence

$$T_0 = -\frac{l_1\pi}{2\alpha - \sin 2\alpha} \tag{9.148}$$

and substituting T_0 from (9.148) into (9.144) we obtain

$$S_{rr}^{*}(r,\varphi) = \frac{2l_1\cos\varphi}{r(2\alpha - \sin 2\alpha)} \tag{9.149}$$

$$S_{r\varphi}^{*}(r,\varphi) = 0, \quad S_{\varphi\varphi}^{*}(r,\varphi) = 0 \ \text{ for } 0 < r < \infty, \ \left|\varphi - \frac{\pi}{2}\right| \leq \alpha \tag{9.150}$$

The stress components S_{rr}^{*}, $S_{r\varphi}^{*}$, and $S_{\varphi\varphi}^{*}$ represent a solution to Problem 9.3 in case (i).

The case (ii). The stress components $S_{rr}^{\perp}$, $S_{r\varphi}^{\perp}$, and $S_{\varphi\varphi}^{\perp}$ in a semi-infinite disk $|x_1| < \infty, x_2 \geq 0$ subject to a normal force P_0 concentrated at (0, 0) take the form [see Eqs. (9.9)]

$$S_{rr}^{\perp}(r,\varphi) = -\frac{2P_0}{\pi r}\sin\varphi = -\frac{2P_0}{\pi r}\cos\theta \tag{9.151}$$

$$S_{r\varphi}^{\perp}(r,\varphi) = S_{\varphi\varphi}^{\perp}(r,\varphi) = 0 \tag{9.152}$$

for every $0 < r < \infty,\ 0 < \varphi \le 2\pi$.

Similarly as in the case (i), a restriction of Eqs. (9.151)–(9.152) to the wedge region provides a solution to Prob. 9.3 in case (ii) if a resultant normal force at the tip of wedge is equal to l_2, that is, if for every $r \ge 0$

$$\int_{-\alpha}^{\alpha} S_{rr}^{\perp}\, r\cos\theta d\theta = l_2 \tag{9.153}$$

Substituting $S_{rr}^{\perp}$ from (9.151) into (9.153) and integrating, we obtain

$$-\frac{2P_0}{\pi}\int_{0}^{\alpha}(1+\cos 2\theta)d\theta = l_2 \tag{9.154}$$

or

$$P_0 = -\frac{l_2\pi}{2\alpha + \sin 2\alpha} \tag{9.155}$$

Finally, substituting P_0 from (9.155) into (9.151) we obtain

$$S_{rr}^{**}(r,\varphi) = \frac{2l_2\sin\varphi}{r(2\alpha+\sin 2\alpha)} \tag{9.156}$$

$$S_{r\varphi}^{**}(r,\varphi) = S_{\varphi\varphi}^{**}(r,\varphi) = 0 \tag{9.157}$$

for every $0 < r < \infty,\ \left|\frac{\pi}{2} - \varphi\right| \le \alpha$.

The stress components S_{rr}^{**}, $S_{r\varphi}^{**}$, and $S_{\varphi\varphi}^{**}$, represent a solution to Prob. 9.3 in case (ii).

A solution to Problem 9.3 takes the form

$$\overline{S}_{rr} = S_{rr}^{*} + S_{rr}^{**}, \quad \overline{S}_{r\varphi} = \overline{S}_{\varphi\varphi} = 0 \tag{9.158}$$

This completes a solution to Problem 9.3

Problem 9.4. Show that for a homogeneous isotropic infinite elastic wedge under generalized plane stress conditions loaded by a concentrated moment M at its tip (see Fig. 9.2) the stress components $\overline{S}_{rr}$, $\overline{S}_{r\varphi}$, and $\overline{S}_{\varphi\varphi}$ vanishing at infinity take the form

$$\overline{S}_{rr}(\mathrm{r},\varphi) = \frac{2M}{r^2}\frac{\sin(2\varphi-\alpha)}{\sin\alpha - \alpha\cos\alpha}$$

$$\overline{S}_{r\varphi}(\mathrm{r},\varphi) = -\frac{M}{r^2}\frac{\cos(2\varphi-\alpha)-\cos\alpha}{\sin\alpha - \alpha\cos\alpha}$$

$$\overline{S}_{\varphi\varphi}(\mathrm{r},\varphi) = 0 \quad \text{for every } r > 0,\ 0 < \varphi < \alpha$$

where

$$M = -r \int_0^{\alpha} (\overline{S}_{r\varphi} r) \, \mathrm{d}\varphi$$

Note that the stress components $\overline{S}_{rr}$ and $\overline{S}_{r\varphi}$ become unbounded for $\alpha = \alpha^*$, where α^* is the only root of the equation

$$\sin \alpha^* - \alpha^* \cos \alpha^* = 0$$

that is, for $\alpha^* = 257.4°$. Hence, the solution makes sense for an elastic wedge that obeys the condition $0 < \alpha < \alpha^*$.

Solution. The stress components $\overline{S}_{rr}$, $\overline{S}_{r\varphi}$, and $\overline{S}_{\varphi\varphi}$ produced in the wedge by a moment M at its tip (see Figure) are to be found using an Airy stress function $\overline{F} = \overline{F}(r, \varphi)$

$$\overline{S}_{rr} = \left(\nabla^2 - \frac{\partial^2}{\partial r^2} \right) \overline{F} \tag{9.159}$$

$$\overline{S}_{\varphi\varphi} = \frac{\partial^2}{\partial r^2} \overline{F} \tag{9.160}$$

$$\overline{S}_{r\varphi} = -\frac{\partial}{\partial r} \left(\frac{1}{r} \frac{\partial \overline{F}}{\partial \varphi} \right) \tag{9.161}$$

where

$$\nabla^2 \nabla^2 \overline{F} = 0 \tag{9.162}$$

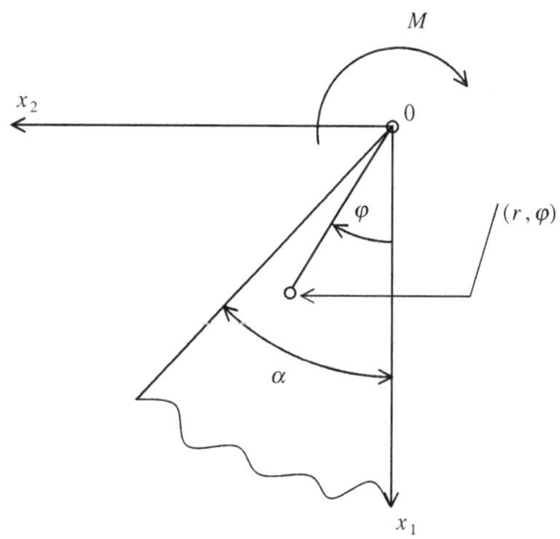

Fig. 9.2 The infinite wedge loaded by a concentrated moment

and

$$\nabla^2 = \frac{\partial^2}{\partial r^2} + \frac{1}{r}\frac{\partial}{\partial r} + \frac{1}{r^2}\frac{\partial^2}{\partial \varphi^2} \tag{9.163}$$

By inspection of (9.160) and (9.161), and of the definition of moment M, we conclude that a biharmonic function $\overline{F} = \overline{F}(\varphi)$ is to solve the problem. In the following we are to show that a solution to Problem 9.4 is obtained if $\bar{F}$ takes the form

$$\overline{F}(r, \varphi) = c_1(2\varphi - \alpha) + c_2 \sin(2\varphi - \alpha) \tag{9.164}$$

where c_1 and c_2 are constants to be determined from the stress free boundary conditions at $\varphi = 0$ and $\varphi = \alpha$ for $r > 0$, and from the definition of M:

$$M = -r \int_0^\alpha \overline{S}_{r\varphi}\, r d\varphi \tag{9.165}$$

Note that $\overline{F}$ given by (9.164) satisfies Eq. (9.162) since

$$\nabla^2\, \nabla^2 \overline{F} = c_2 \nabla^2 [r^{-2} \sin(2\varphi - \alpha)] = 0 \quad \text{for } 0 < r < \infty,\ 0 \le \varphi \le \alpha \tag{9.166}$$

Substituting $\overline{F}$ from (9.164) into (9.159), (9.160), and (9.161), respectively, we obtain

$$\overline{S}_{rr} = -4c_2 \frac{\sin(2\varphi - \alpha)}{r^2} \tag{9.167}$$

$$\overline{S}_{\varphi\varphi} = 0 \tag{9.168}$$

$$\overline{S}_{r\varphi} = \frac{2}{r^2}\, [c_1 + c_2 \cos(2\varphi - \alpha)] \tag{9.169}$$

Also note that the boundary conditions

$$\overline{S}_{r\varphi} = 0 \quad \text{at } \varphi = 0 \text{ and } \varphi = \alpha \tag{9.170}$$

are satisfied if

$$c_1 = -c_2 \cos \alpha \tag{9.171}$$

Therefore, substituting (9.171) into (9.164) and (9.169), respectively, we obtain

$$\overline{F} = c_2\, [\sin(2\varphi - \alpha) - (2\varphi - \alpha) \cos \alpha] \tag{9.172}$$

and

$$\overline{S}_{r\varphi} = \frac{2c_2}{r^2}\, [\cos(2\varphi - \alpha) - \cos \alpha] \tag{9.173}$$

Hence, substituting (9.173) into (9.165) we obtain

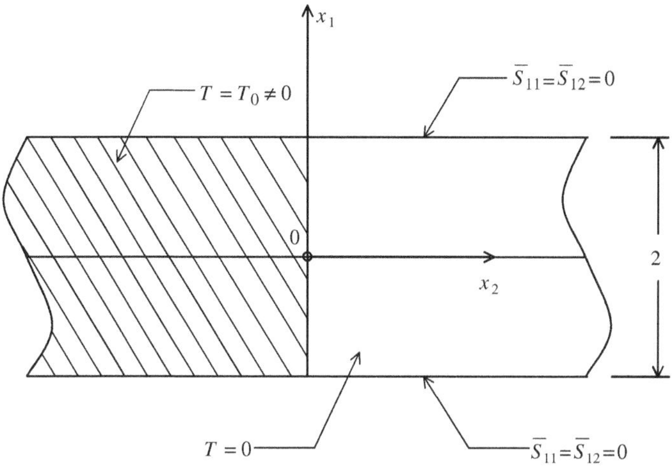

Fig. 9.3 The infinite strip

$$M = -2c_2 \int_0^{\alpha} [\cos(2\varphi - \alpha) - \cos\alpha] d\varphi \tag{9.174}$$

or

$$c_2 = -\frac{M}{2(\sin\alpha - \alpha\cos\alpha)} \tag{9.175}$$

Finally, by substituting c_2 from (9.175) into (9.167) and (9.173), respectively, we obtain the required result. This completes a solution to Problem 9.4.

Problem 9.5. Consider a homogeneous isotropic infinite elastic strip under generalized plane stress conditions: $|x_1| \leq 1,\ |x_2| < \infty$ subject to the temperature field of the form

$$\overline{T}(x_1, x_2) = T_0[1 - H(x_2)] \tag{9.176}$$

where T_0 is a constant temperature and $H = H(x)$ is the Heaviside function

$$H(x) = \begin{cases} 1 & \text{for} \quad x > 0 \\ 1/2 & \text{for} \quad x = 0 \\ 0 & \text{for} \quad x < 0 \end{cases} \tag{9.177}$$

Note that in this case, we complemented the definition of the Heaviside function by specifying its value at $x = 0$ (Fig. 9.3).

Show that the stress tensor field $\overline{S} = \overline{S}(x_1, x_2)$ corresponding to the discontinuous temperature (9.176) is represented by the sum

$$\overline{S} = \overline{S}^{(1)} + \overline{S}^{(2)} \tag{9.178}$$

where

$$\overline{S}_{11}^{(1)} = -E\alpha T_0[1 - H(x_2)], \quad \overline{S}_{22}^{(1)} = \overline{S}_{12}^{(1)} = 0 \tag{9.179}$$

and

$$\overline{S}_{11}^{(2)} = F_{,22}, \quad \overline{S}_{22}^{(2)} = F_{,11}, \quad \overline{S}_{12}^{(2)} = -F_{,12} \tag{9.180}$$

where the biharmonic function $F = F(x_1, x_2)$ is given by

$$F(x_1, x_2) = E\alpha T_0 \left[\frac{x_2^2}{4} + \frac{2}{\pi} \int_0^\infty (A \cosh \beta x_1 + B \beta x_1 \sinh \beta x_1) \frac{\sin \beta x_2}{\beta^3} d\beta \right] \tag{9.181}$$

$$A = \frac{\sinh \beta + \beta \cosh \beta}{\sinh 2\beta + 2\beta}, \quad B = -\frac{\sinh \beta}{\sinh 2\beta + 2\beta} \tag{9.182}$$

Hint. Note that

$$\begin{aligned} \overline{S}_{11}^{(2)}(\pm 1, x_2) &= -\overline{S}_{11}^{(1)}(\pm 1, x_2) \\ &= \frac{E\alpha T_0}{2} \left(1 - \frac{2}{\pi} \int_0^\infty \frac{\sin \beta x_2}{\beta} d\beta \right) \quad \text{for} \quad |x_2| < \infty \end{aligned}$$

Solution. To solve the problem we recall the integral representation of the Heaviside step function

$$H(x_2) = \left\{ \begin{array}{l} 0 \text{ for } x_2 < 0 \\ \frac{1}{2} \text{ for } x_2 = 0 \\ 1 \text{ for } x_2 > 0 \end{array} \right\} = \frac{1}{2} + \frac{1}{\pi} \int_0^\infty \frac{\sin \beta x_2}{\beta} d\beta \tag{9.183}$$

The temperature $\overline{T}(x_1, x_2)$ on the strip $|x_1| \leq 1, |x_2| < \infty$ is then represented by

$$\overline{T}(x_1, x_2) = T_0 \, [1 - H(x_2)] = \frac{T_0}{2} \left[1 - \frac{2}{\pi} \int_0^\infty \frac{\sin \beta x_2}{\beta} d\beta \right] \tag{9.184}$$

We are to find a stress tensor $\overline{\mathbf{S}}$ corresponding to a solution $s = [\overline{\mathbf{u}}, \overline{\mathbf{E}}, \overline{\mathbf{S}}]$ of a problem of thermo-elastostatics for the strip subject to the discontinuous temperature (9.184) when the strip boundaries $x_1 = 1$ and $x_1 = -1$ are stress free, that is, when

$$\overline{S}_{11}(\pm 1, x_2) = \overline{S}_{12}(\pm 1, x_2) = 0 \quad \text{for } |x_2| < \infty \tag{9.185}$$

To this end we look for s in the form

$$s = s_1 + s_2 \tag{9.186}$$

where $s_1 = [\mathbf{u}^{(1)}, \mathbf{E}^{(1)}, \mathbf{S}^{(1)}]$ is generated from a displacement potential $\overline{\phi} = \overline{\phi}(x_2)$ by the formulas

$$\mathbf{u}^{(1)} = \nabla\overline{\phi} \tag{9.187}$$

$$\mathbf{E}^{(1)} = \nabla\nabla\overline{\phi} \tag{9.188}$$

$$\mathbf{S}^{(1)} = 2\mu(\nabla\nabla\overline{\phi} - \nabla^2\overline{\phi}\mathbf{1}) \tag{9.189}$$

in which $\overline{\phi}$ satisfies Poisson's equation

$$\nabla^2\overline{\phi} = m_0\overline{T}, \quad m_0 = (1+\nu)\alpha \tag{9.190}$$

where $\overline{T}$ is given by Eq. (9.184) and $s_2 = [\mathbf{u}^{(2)}, \mathbf{E}^{(2)}, \mathbf{S}^{(2)}]$ is a solution of isothermal elastostatics for the strip that complies with the boundary conditions

$$\begin{aligned} S_{11}^{(2)}(\pm 1, x_2) &= -S_{11}^{(1)}(\pm 1, x_2) \\ S_{12}^{(2)}(\pm 1, x_2) &= -S_{12}^{(1)}(\pm 1, x_2) \end{aligned} \tag{9.191}$$

The stress components $S_{11}^{(2)}$, $S_{22}^{(2)}$, and $S_{12}^{(2)}$ are to be computed from an Airy stress function $\overline{F} = \overline{F}(x_1, x_2)$ by the formulas

$$S_{11}^{(2)} = \overline{F}_{,22}\,, \; S_{22}^{(2)} = \overline{F}_{,11}\,, \; S_{12}^{(2)} = -\overline{F}_{,12} \tag{9.192}$$

$$\nabla^2\nabla^2\overline{F} = 0 \tag{9.193}$$

Since $\overline{\phi} = \overline{\phi}(x_2)$, it follows from Eqs. (9.189) and (9.190) that the stress components $S_{11}^{(1)}$, $S_{22}^{(1)}$, and $S_{12}^{(1)}$ are given by

$$S_{11}^{(1)} = -2\mu m_0\overline{T} = -E\alpha T_0[1 - H(x_2)] \tag{9.194}$$

$$S_{22}^{(1)} = S_{12}^{(1)} = 0, \quad |x_1| \le 1, \; |x_2| < \infty \tag{9.195}$$

and the problem is reduced to that of finding a biharmonic function $\overline{F} = \overline{F}(x_1, x_2)$ on the strip region: $|x_1| \le 1$, $|x_2| < \infty$ that complies with the boundary conditions

$$\overline{F}_{,22}\,(\pm 1, x_2) = 2\mu m_0\overline{T} \tag{9.196}$$

and

$$\overline{F}_{,12}\,(\pm 1, x_2) = 0 \tag{9.197}$$

An alternative form of (9.196) reads

$$\overline{F}_{,22}(\pm 1, x_2) = \frac{E\alpha T_0}{2}\left[1 - \frac{2}{\pi}\int_0^\infty \frac{\sin\beta x_2}{\beta} d\beta\right] \tag{9.198}$$

Since

$$\nabla^2(\cos h\beta x_1 \ \sin\beta x_2) = 0 \tag{9.199}$$

$$\nabla^2\nabla^2(\beta x_1 \ \sinh\beta x_1 \ \sin\beta x_2) = 0 \tag{9.200}$$

and

$$\overline{F}(x_1, x_2) = \overline{F}(-x_1, x_2) \tag{9.201}$$

the function $\overline{F}$ is postulated in the form [see Eq. 9.181 of Problem 9.5]

$$\overline{F}(x_1, x_2) = E\alpha T_0\left[\frac{x_2^2}{4} + \frac{2}{\pi}\int_0^\infty (A\cosh\beta x_1 + B\beta x_1 \sinh\beta x_1)\frac{\sin\beta x_2}{\beta^3} d\beta\right] \tag{9.202}$$

where A and B are arbitrary functions on $[0, \infty)$ that make the integral (9.202) to converge for $|x_1| \leq 1$, $|x_2| < \infty$.

Substituting (9.202) into the boundary conditions (9.197) and (9.198), respectively, we obtain

$$A\sinh\beta + B(\sinh\beta + \beta\cosh\beta) = 0$$

and

$$A\cosh\beta + B\beta\sinh\beta = 1/2 \tag{9.203}$$

A unique solution to Eqs. (9.203) takes the form

$$A = \frac{\sinh\beta + \beta\cosh\beta}{\sinh 2\beta + 2\beta} \tag{9.204}$$

$$B = -\frac{\sinh\beta}{\sinh 2\beta + 2\beta} \tag{9.205}$$

Hence substituting $\overline{F} = \overline{F}(x_1, x_2)$ given by (9.202) in which A and B are given by (9.204) and (9.205), respectively, into (9.192) we obtain the integral representation of the stress tensor $\mathbf{S}^{(2)}$. A solution to Problem 9.5 is obtained in the form

$$\overline{\mathbf{S}} = \mathbf{S}^{(1)} + \mathbf{S}^{(2)} \tag{9.206}$$

where $\mathbf{S}^{(1)}$ and $\mathbf{S}^{(2)}$ are given by Eqs. (9.194)–(9.195) and (9.192), respectively.

Chapter 10
Solutions to Particular Three-Dimensional Initial-Boundary Value Problems of Elastodynamics

In this chapter a number of spherically symmetric initial-boundary value problems of the dynamic theory of thermal stresses for a homogeneous isotropic infinite elastic body are solved. The problems include: (i) the dynamic thermal stresses due to an instantaneous temperature distributed on a spherical surface in E^3, (ii) the dynamic thermal stresses due to a time-dependent spherically symmetric temperature field that satisfies a parabolic heat conduction equation in E^3, and (iii) the dynamic thermal stresses propagating in an infinite body with a stress free spherical cavity, corresponding to an instantaneous temperature distributed on a spherical surface lying inside the body. To solve the problems a method of the dynamic thermoelastic displacement potential in spherical coordinates is used.

10.1 A Spherically Symmetric Green's Function of the Dynamic Theory of Thermal Stresses for an Infinite Elastic Body

In dimensionless spherical coordinates $(R, \varphi, \theta) : [R \geq 0, \quad 0 \leq \varphi \leq 2\pi, \quad 0 \leq \theta \leq \pi]$, a spherically symmetric Green's function of the dynamic theory of thermal stresses for a homogeneous isotropic infinite elastic body, corresponding to a spherical time-dependent temperature field of the form

$$T^*(R, R_0; \tau) = \delta(R - R_0)\delta(\tau) \quad \text{for} \quad R > 0,\ R_0 > 0,\ \tau > 0 \tag{10.1}$$

can be identified with a solution to the following initial-boundary value problem. Find a function $\Phi^* = \Phi^*(R, R_0; \tau)$ for $R > 0,\ R_0 > 0,\ \tau > 0$, that satisfies the inhomogeneous wave equation

$$\left(\nabla^2 - \frac{\partial^2}{\partial \tau^2}\right) \Phi^* = T^* \quad \text{for} \quad R > 0,\ R_0 > 0,\ \tau > 0 \tag{10.2}$$

M. Reza Eslami et al., *Theory of Elasticity and Thermal Stresses*, Solid Mechanics and Its Applications 197, DOI: 10.1007/978-94-007-6356-2_10,

the initial conditions

$$\Phi^*(R, R_0; 0) = \dot{\Phi}^*(R, R_0; 0) = 0 \quad \text{for} \quad R > 0,\ R_0 > 0 \tag{10.3}$$

and the boundary conditions at infinity

$$\Phi^*(R, R_0; \tau) \to 0 \quad \text{as} \quad R \to \infty \quad \text{and} \quad R_0 > 0,\ \tau > 0 \tag{10.4}$$

The function $\Phi^* = \Phi^*(R, R_0; \tau)$ generates the radial displacement $\widehat{\mathrm{u}}_{\mathrm{R}}^* = \widehat{\mathrm{u}}_{\mathrm{R}}^*(R, R_0; \tau)$ and the stresses $\widehat{S}_{\mathrm{RR}}^* = \widehat{S}_{\mathrm{RR}}^*(R, R_0; \tau)$ and $\widehat{S}_{\theta\theta}^* = \widehat{S}_{\theta\theta}^*(R, R_0; \tau)$ by the formulas

$$u_{\mathrm{R}}^*(R, R_0; \tau) = \frac{\partial \Phi^*}{\partial R} \quad \text{for} \quad R > 0, R_0 > 0,\ \tau > 0 \tag{10.5}$$

and

$$\begin{aligned} \mathrm{S}_{\mathrm{RR}}^*(R, R_0; \tau) &= \frac{\partial^2 \Phi^*}{\partial \tau^2} - \frac{2(1-2\nu)}{1-\nu} \frac{1}{R} \frac{\partial \Phi^*}{\partial R} \\ \mathrm{S}_{\varphi\varphi}^*(R, R_0; \tau) &= S_{\theta\theta}^*(R, R_0; \tau) \qquad\qquad \text{for} \quad R > 0,\ R_0 > 0,\ \tau > 0 \\ &= \frac{\nu}{1-\nu} \frac{\partial^2 \Phi^*}{\partial \tau^2} + \frac{(1-2\nu)}{1-\nu} \left(\frac{1}{R} \frac{\partial \Phi^*}{\partial R} - T^* \right) \end{aligned} \tag{10.6}$$

A unique solution $\Phi^* = \Phi^*(R, R_0; \tau)$ to Eqs. (10.2)–(10.4) is obtained by using an integral representation of $\delta = \delta(R - R_0)$. To this end, we introduce the Laplace transform of a function $f = f(R, \tau)$ by

$$Lf = \bar{f}(R, p) = \int_0^\infty e^{-p\tau} f(R, \tau) d\tau \tag{10.7}$$

and apply the operator L to Eq. (10.2). Then, because of (10.3) we obtain

$$(\nabla^2 - p^2)\overline{\Phi}^* = \delta(R - R_0) \tag{10.8}$$

while the asymptotic condition (10.4), in the transform domain, takes the form

$$\overline{\Phi}^*(R, R_0; p) \to 0 \quad \text{as} \quad R \to \infty, \quad R_0 > 0 \tag{10.9}$$

Since

$$R\nabla^2 \overline{\Phi}^* = R \left(\frac{\partial^2}{\partial R}^2 + \frac{2}{R} \frac{\partial}{\partial R} \right) \overline{\Phi}^* = \frac{\partial^2}{\partial R^2} (R\overline{\Phi}^*) \tag{10.10}$$

therefore multiplying Eq. (10.8) by R we obtain

$$\left(\frac{\partial^2}{\partial R^2} - p^2\right)(R\overline{\Phi}^*) = R\delta(R - R_0) = R_0\delta(R - R_0) \tag{10.11}$$

Since

$$R_0\delta(R - R_0) = \frac{2}{\pi}R_0 \int_0^\infty \sin\alpha R_0 \sin\alpha R \, d\alpha \tag{10.12}$$

therefore, a solution to Eq. (10.11) may be represented by the integral

$$R\overline{\Phi}^*(R, R_0; p) = -\frac{2}{\pi}R_0 \int_0^\infty \frac{\sin\alpha R \sin\alpha R_0}{\alpha^2 + p^2} d\alpha \tag{10.13}$$

or, using the identity,

$$2\sin x \sin y = \cos(x - y) - \cos(x + y) \tag{10.14}$$

by the formula

$$R\overline{\Phi}^*(R, R_0; p) = -\frac{1}{\pi}R_0 \left[\int_0^\infty \frac{\cos\alpha(R - R_0)}{\alpha^2 + p^2} d\alpha - \int_0^\infty \frac{\cos\alpha(R + R_0)}{\alpha^2 + p^2} d\alpha\right] \tag{10.15}$$

Since

$$\int_0^\infty \frac{\cos\alpha x}{\alpha^2 + p^2} d\alpha = \frac{\pi}{2}\frac{e^{-p|x|}}{p} \quad \text{for} \quad p > 0 \tag{10.16}$$

hence, Eq. (10.15) reduces to the form

$$R\overline{\Phi}^*(R, R_0; p) = -\frac{R_0}{2p}\left[e^{-p|R-R_0|} - e^{-p(R+R_0)}\right] \tag{10.17}$$

By applying the operator L^{-1} to Eq. (10.17) we obtain

$$\Phi^*(R, R_0; \tau) = -\frac{R_0}{2R}\{H[\tau - |R - R_0|] - H[\tau - (R + R_0)]\} \tag{10.18}$$

where $H = H(x)$ is the Heaviside function

$$H(x) = \begin{cases} 1 & \text{for} \quad x > 0 \\ 0 & \text{for} \quad x < 0 \end{cases} \tag{10.19}$$

10.2 A Spherically Symmetric Green's Function for a Parabolic Heat Conduction Equation in an Infinite Space

A spherically symmetric Green's function $\widehat{T} = \widehat{T}(R, \tau)$ for a parabolic heat conduction equation for an infinite space can be identified with a solution to the dimensionless initial-boundary value problem. Find a function $\widehat{T} = \widehat{T}(R, \tau)$ that satisfies the inhomogeneous parabolic heat conduction equation

$$\left(\nabla^2 - \frac{\partial}{\partial \tau}\right) \widehat{T} = -\frac{\delta(R)}{4\pi R^2}\delta(\tau) \quad \text{for} \quad R > 0, \quad \tau > 0 \tag{10.20}$$

the initial condition

$$\widehat{T}(R, 0) = 0 \quad \text{for} \quad R > 0 \tag{10.21}$$

and the vanishing condition

$$\widehat{T}(R, \tau) \to 0 \quad \text{as} \quad R \to \infty \quad \text{and} \quad \tau > 0 \tag{10.22}$$

By applying the Laplace transform to Eq. (10.20) and using the initial condition (10.21) we obtain

$$\left(\nabla^2 - p\right) \overline{\widehat{T}} = -\frac{\delta(R)}{4\pi R^2} \quad \text{for} \quad R > 0 \quad \text{and} \quad p > 0 \tag{10.23}$$

while Eq. (10.22) is reduced to

$$\overline{\widehat{T}}(R, p) \to 0 \quad \text{as} \quad R \to \infty \quad \text{and} \quad p > 0 \tag{10.24}$$

Since

$$\left(\nabla^2 - p\right)^{-1} \frac{\delta(R)}{4\pi R^2} = -\frac{e^{-R\sqrt{p}}}{4\pi R} \quad \text{for} \quad R > 0 \quad \text{and} \quad p > 0 \tag{10.25}$$

therefore, a unique solution $\overline{\widehat{T}} = \overline{\widehat{T}}(R, p)$ to Eq. (10.23) subject to the asymptotic condition (10.24) takes the form

$$\overline{\widehat{T}}(R, p) = \frac{e^{-R\sqrt{p}}}{4\pi R} \tag{10.26}$$

By applying the operator L^{-1} to Eq. (10.26) we obtain

$$\widehat{T}(R, \tau) = \frac{1}{4\pi} \frac{1}{\sqrt{4\pi \tau^3}} e^{-\frac{R^2}{4\tau}} \quad \text{for} \quad R > 0, \quad \tau > 0 \tag{10.27}$$

The function $\widehat{T} = \widehat{T}(R, \tau)$ is the Green's function for a parabolic heat conduction equation corresponding to an instantaneous concentrated heat source in an infinite space.

10.3 Problems and Solutions Related to Particular Three-Dimensional Initial-Boundary Value Problems of Elastodynamics

Problem 10.1. A dimensionless temperature field $\widehat{T}^* = \widehat{T}^*(R, R_0; \tau)$ is assumed in the form

$$\widehat{T}^*(R, R_0; \tau) = \delta(R - R_0)\delta(\tau) \tag{10.28}$$

for any point of an infinite body referred to a spherical system of coordinates $(R, \varphi, \theta) : [R \geq 0,\ 0 \leq \varphi \leq 2\pi,\ 0 \leq \theta \leq \pi]$ and for $R_0 > 0$, and $\tau \geq 0$. Show that a solution $\widehat{\Phi}^* = \widehat{\Phi}^*(R, R_0; \tau)$ to the nonhomogeneous wave equation

$$\left(\nabla^2 - \frac{\partial^2}{\partial \tau^2}\right) \widehat{\Phi}^* = \widehat{T}^* \quad \text{for} \quad R > 0,\ R_0 > 0,\ \tau > 0 \tag{10.29}$$

subject to the initial conditions

$$\widehat{\Phi}^*(R, R_0; 0) = \dot{\widehat{\Phi}}^*(R, R_0; 0) = 0 \quad \text{for} \quad R > 0,\ R_0 > 0 \tag{10.30}$$

and vanishing conditions at infinity takes the form

$$\begin{aligned}\widehat{\Phi}^*(R, R_0; \tau) = &-\frac{R_0}{2R}\{H[\tau - (R_0 - R)] - H[\tau - (R_0 + R)]\}\, H(R_0 - R)\\ &-\frac{R_0}{2R}\{H[\tau - (R - R_0)] - H[\tau - (R + R_0)]\}\, H(R - R_0)\end{aligned} \tag{10.31}$$

for every $R > 0,\ R_0 > 0,\ \tau > 0$.

Solution. A solution $\widehat{\phi}^* = \widehat{\phi}^*(R, R_0, \tau)$ to the nonhomogeneous wave equation (10.29)

$$\left(\nabla^2 - \frac{\partial^2}{\partial \tau^2}\right) \widehat{\phi}^* = \delta(R - R_0)\delta(\tau) \tag{10.32}$$

subject to the homogeneous initial conditions (10.30)

$$\widehat{\phi}^*(R, R_0; 0) = \dot{\widehat{\phi}}^*(R, R_0; 0) = 0 \tag{10.33}$$

and suitable vanishing conditions at $R \to \infty$, takes the form

[For the derivation see Eqs. (10.1)–(10.4) and (10.7)–(10.18)]

$$\widehat{\phi}^*(R, R_0; \tau) = -\frac{R_0}{2R}\{H[\tau - |R - R_0|] - H[\tau - (R + R_0)]\} \tag{10.34}$$

Since Eq. (10.34) is equivalent to (10.31) of Problem 10.1, a solution to the problem is complete.

Problem 10.2. Compute the radial displacements $\widehat{u}_R^* = \widehat{u}_R^*(R, R_0; \tau)$, and the stress components $\widehat{S}_{RR}^* = \widehat{S}_{RR}^*(R, R_0; \tau)$ and $\widehat{S}_{\theta\theta}^* = \widehat{S}_{\theta\theta}^*(R, R_0; \tau)$ generated by the potential $\widehat{\Phi}^* = \widehat{\Phi}^*(R, R_0; \tau)$ of the Problem 10.1.

Hint. Use Eqs. (10.5)–(10.6).

Solution. The radial displacement $\widehat{u}_R^*$, and the stress components $\widehat{S}_{RR}^*$, and $\widehat{S}_{\theta\theta}^*$ generated by $\widehat{\phi}^*$ of Problem 10.1 are given by

$$\widehat{u}_R^* = \frac{\partial \widehat{\phi}^*}{\partial R} \tag{10.35}$$

$$\widehat{S}_{RR}^* = \frac{\partial^2 \widehat{\phi}^*}{\partial \tau^2} - \frac{2(1-2\nu)}{1-\nu}\frac{\widehat{u}_R^*}{R} \tag{10.36}$$

$$\widehat{S}_{\theta\theta}^* = \widehat{S}_{\varphi\varphi}^* = \frac{\nu}{1-\nu}\frac{\partial^2 \widehat{\phi}^*}{\partial \tau^2} + \frac{1-2\nu}{1-\nu}\left(\frac{\widehat{u}_R^*}{R} - \widehat{T}^*\right) \tag{10.37}$$

where

$$\widehat{T}^* = \delta(R - R_0)\delta(\tau) \tag{10.38}$$

Applying the Laplace transform to Eqs. (10.35)–(10.38), respectively, we obtain

$$\overline{\widehat{u}}_R^* = \frac{\partial \overline{\widehat{\phi}}^*}{\partial R} \tag{10.39}$$

$$\overline{\widehat{S}}_{RR}^* = p^2\overline{\widehat{\phi}} - \frac{2(1-2\nu)}{1-\nu}\frac{\overline{\widehat{u}}_R^*}{R} \tag{10.40}$$

$$\overline{\widehat{S}}_{\theta\theta}^* = \frac{\nu}{1-\nu}p^2\overline{\widehat{\phi}}^* - \frac{1-2\nu}{1-\nu}\left(\frac{\overline{\widehat{u}}_R^*}{R} - \overline{\widehat{T}}^*\right) \tag{10.41}$$

where

$$\overline{\widehat{T}}^* = \delta(R - R_0) \tag{10.42}$$

and p is the Laplace transform parameter. Also, applying the Laplace transform to Eq. (10.37) ⇔ (10.31) of Problem 10.1, we get

$$\overline{\widehat{\phi}}^* = -\frac{R_0}{2Rp}\left[e^{-p|R-R_0|} - e^{-p(R+R_0)}\right] \tag{10.43}$$

Hence

$$p^2\overline{\widehat{\phi}}^* = -\frac{R_0}{2R}\left[pe^{-p|R-R_0|} - pe^{-p(R+R_0)}\right] \tag{10.44}$$

and

$$\frac{\partial \overline{\widehat{\phi}}^*}{\partial R} = \frac{R_0}{2R^2}\left\{\frac{e^{-p|R-R_0|} - e^{-p(R+R_0)}}{p}\right\} + \frac{R_0}{2R}\left\{\text{sign}(R-R_0)e^{-p|R-R_0|} - e^{-p(R+R_0)}\right\} \tag{10.45}$$

where

$$\text{sign}\, x = \left\{\begin{array}{l} 1 \ \text{ if } x > 0 \\ -1 \text{ if } x < 0 \end{array}\right\} \tag{10.46}$$

Since

$$\int_0^\infty e^{-pt} H(t-\alpha)dt = \frac{e^{-p\alpha}}{p}, \ \alpha > 0 \tag{10.47}$$

therefore, by differentiation of (10.47) with respect to α we obtain

$$\int_0^\infty e^{-pt}\delta(t-\alpha)dt = e^{-p\alpha} \tag{10.48}$$

and

$$\int_0^\infty e^{-pt}\delta'(t-\alpha)dt = pe^{-\alpha p} \tag{10.49}$$

where δ is the delta function.

Next, applying the inverse Laplace transform to (10.44) and (10.45), respectively in view of Eqs. (10.47) and (10.48), we obtain

$$\frac{\partial^2 \widehat{\phi}^*}{\partial \tau^2} = -\frac{R_0}{2R}\{\delta'[\tau - |R-R_0|] - \delta'[\tau - (R+R_0)]\} \tag{10.50}$$

and

$$\frac{\partial \widehat{\phi}^*}{\partial R} = \frac{R_0}{2R^2}\{H[\tau - |R-R_0|] - H[\tau - (R+R_0)]\} + \frac{R_0}{2R}\{\text{sign}(R-R_0)\delta[\tau - |R-R_0|] - \delta[\tau - (R+R_0)]\} \tag{10.51}$$

Finally, substituting (10.50) and (10.51) into (10.35), (10.36), and (10.37), respectively, we obtain $\widehat{u}_R^*$, $\widehat{S}_{RR}^*$, and $\widehat{S}_{\theta\theta}^*$. This completes a solution to Problem 10.2.

Problem 10.3. Show that if $\widehat{T} = \widehat{T}(R, \tau)$ satisfies a parabolic heat conduction equation for $R \geq 0$, $\tau > 0$, then the radial displacement $\widehat{u}_R = \widehat{u}_R(R; \tau)$ and the stresses $\widehat{S}_{RR} = \widehat{S}_{RR}(R; \tau)$ and $\widehat{S}_{\theta\theta} = \widehat{S}_{\theta\theta}(R; \tau)$ produced by the temperature $\widehat{T} = \widehat{T}(R, \tau)$ in an infinite body $R \geq 0$ and for $\tau > 0$ are generated by the potential $\widehat{\Phi} = \widehat{\Phi}(R, \tau)$ given by the double integral

$$\widehat{\Phi}(R, \tau) = \int_0^\tau \int_0^\infty \widehat{\Phi}^*(R, R_0; \tau - u)\widehat{T}(R_0, u)\, dR_0 du \tag{10.52}$$

where $\widehat{\Phi}^* = \widehat{\Phi}^*(R, R_0; \tau)$ is the thermoelastic displacement potential of Problem 10.1.

Solution. The function $\widehat{\phi}^*(R, R_0; \tau - u)$ for $R \geq 0,\ R_0 > 0,\ \tau - u \geq 0$ with u fixed, satisfies the equation [see Eq. (10.29) of Problem 10.1]

$$\left(\nabla^2 - \frac{\partial^2}{\partial \tau^2}\right)\widehat{\phi}^*(R, R_0; \tau - u) = \delta(R - R_0)\delta(\tau - u) \tag{10.53}$$

Let $\widehat{T} = \widehat{T}(R, \tau)$ be a solution to a parabolic heat conduction equation for $R \geq 0$, $\tau > 0$.

Since

$$\int_0^\infty \int_0^\infty \delta(R - R_0)\delta(\tau - u)\widehat{T}(R_0, u) dR_0 du = \widehat{T}(R, \tau) \tag{10.54}$$

therefore, by multiplying (10.53) by $\widehat{T}(R_0, u)$ and integrating over the intervals: $0 < R_0 < \infty,\ 0 < u < \infty$, and taking into account the property

$$\widehat{\phi}^*(R, R_0; \tau - u) = 0 \quad \text{for} \quad \tau - u \leq 0 \tag{10.55}$$

we find that the integral

$$\begin{aligned} \widehat{\phi}(R, \tau) &= \int_0^\infty \int_0^\infty \widehat{\phi}^*(R, R_0; \tau - u)\widehat{T}(R_0, u) dR_0 du \\ &= \int_0^\tau \int_0^\infty \widehat{\phi}^*(R, R_0; \tau - u)\widehat{T}(R_0, u) dR_0 du \end{aligned} \tag{10.56}$$

satisfies the wave equation

$$\left(\nabla^2 - \frac{\partial^2}{\partial \tau^2}\right)\widehat{\phi} = \widehat{T}(R, \tau) \tag{10.57}$$

The radial displacement $\widehat{u}_R$ and the stresses $\widehat{S}_{RR}$ and $\widehat{S}_{\theta\theta}$ corresponding to the temperature $\widehat{T}$ are then computed from Eqs. (10.35), (10.36), and (10.37), respectively, of the solution to Problem 10.2 in which asterisk is omitted. This completes a solution to Problem 10.3.

Problem 10.4. The Laplace transform of the temperature $\widehat{T} = \widehat{T}(R,\tau)$ due to an instantaneous concentrated source of heat in an infinite body $R \geq 0$ takes the form [see Eq. (10.26)]

$$\overline{\widehat{T}}(R,p) = \frac{e^{-R\sqrt{p}}}{4\pi R} \tag{10.58}$$

where p is the Laplace transform parameter. By applying the Laplace transform to Eq. (10.52) of Problem 10.3, show that

$$\overline{\widehat{\Phi}}(R,p) = \frac{1}{4\pi\, p(p-1)} \frac{e^{-Rp} - e^{-R\sqrt{p}}}{R} \tag{10.59}$$

where $\overline{\widehat{\Phi}} = \overline{\widehat{\Phi}}(R,p)$ is the Laplace transform of $\widehat{\Phi} = \widehat{\Phi}(R,\tau)$ associated with $\widehat{T} = \widehat{T}(R,\tau)$.

Solution. Applying the Laplace transform to Eq. (10.56) of the solution to Problem 10.3 we obtain

$$\overline{\widehat{\phi}}(R,p) = \int_0^\infty \overline{\widehat{\phi}}^*(R,R_0,p)\overline{\widehat{T}}(R_0,p)dR_0 \tag{10.60}$$

where $\overline{\widehat{T}}(R,p)$ is given by Eq. (10.58) and $\overline{\widehat{\phi}}^*(R,R_0;p)$ is given by Eq. (10.43) of the solution to Problem 10.2.

Hence

$$\begin{aligned}
\widehat{\phi}(R,p) &= -\frac{1}{2Rp}\frac{1}{4\pi}\int_0^\infty e^{-R_0\sqrt{p}} \times \left[e^{-p|R-R_0|} - e^{-p(R+R_0)}\right] dR_0 \\
&= -\frac{1}{2Rp}\frac{1}{4\pi}\left\{\int_0^R e^{-pR} \times e^{-R_0(\sqrt{p}-p)}dR_0 + \int_R^\infty e^{pR} \times e^{-R_0(\sqrt{p}+p)}dR_0\right. \\
&\qquad \left. - \int_0^\infty e^{-pR} \times e^{-R_0(\sqrt{p}+p)}dR_0\right\}
\end{aligned} \tag{10.61}$$

Since

$$\int e^{-R_0\alpha}\, dR_0 = -\frac{1}{\alpha} e^{-R_0\alpha} \quad \alpha \neq 0 \tag{10.62}$$

therefore, computing the integrals on RHS of (10.61) we obtain

$$\begin{aligned}
\widehat{\phi}(R,p) &= -\frac{1}{2Rp}\frac{1}{4\pi}\left\{e^{-pR}\times\frac{1-e^{-R(\sqrt{p}-p)}}{\sqrt{p}-p} + e^{pR}\times\frac{e^{-R(\sqrt{p}+p)}}{\sqrt{p}+p} - e^{-pR}\frac{1}{\sqrt{p}+p}\right\}\\
&= -\frac{1}{2Rp}\frac{1}{4\pi}\left\{\frac{e^{-pR}-e^{-R\sqrt{p}}}{\sqrt{p}-p} - \frac{e^{-Rp}-e^{-R\sqrt{p}}}{\sqrt{p}+p}\right\}\\
&= \frac{1}{4\pi p(p-1)}\left\{\frac{e^{-Rp}-e^{-R\sqrt{p}}}{R}\right\}
\end{aligned} \tag{10.63}$$

This completes a solution to Problem 10.4.

Problem 10.5. Find a thermoelastic displacement potential $\widehat{\Phi}^{**} = \widehat{\Phi}^{**}(R, R_0; \tau)$ for an infinite body with a stress free spherical cavity of radius $R = a$ corresponding to the temperature field

$$\widehat{T}^{**}(R, R_0; \tau) = \delta(R - R_0)\delta(\tau) \quad \text{for} \quad R \geq a,\ R_0 > a,\ \tau > 0 \tag{10.64}$$

Solution. The function $\widehat{\phi}^{**} = \widehat{\phi}^{**}(R, R_0; \tau)$, defined for $R \geq a, R_0 > a,\ \tau \geq 0$, is to meet the nonhomogeneous wave equation

$$\left(\nabla^2 - \frac{\partial^2}{\partial\tau^2}\right)\widehat{\phi}^{**} = \delta(R - R_0)\delta(\tau) \quad \text{for} \quad R > a,\ \tau > 0 \tag{10.65}$$

subject to the homogeneous initial conditions

$$\widehat{\phi}^{**}(R, R_0; 0) = \dot{\widehat{\phi}}^{**}(R, R_0; 0) = 0 \quad \text{for} \quad R > a,\ R_0 > a \tag{10.66}$$

the homogenous boundary condition

$$\widehat{S}^{**}_{RR}(a, R_0; \tau) \equiv \frac{\partial^2\widehat{\phi}^{**}}{\partial\tau^2}(a, R_0; \tau) - \frac{2(1-2\nu)}{1-\nu}\frac{1}{a}\frac{\partial\widehat{\phi}^{**}}{\partial R}(a, R_0; \tau) = 0 \quad \text{for } R_0 > a,\ \tau > 0 \tag{10.67}$$

and suitable vanishing conditions as $R \to \infty$.

Applying the Laplace transform to (10.65) and (10.67), respectively, and using (10.66), we obtain

$$(\nabla^2 - p^2)\overline{\widehat{\phi}}^{**} = \delta(R - R_0) \tag{10.68}$$

and

$$p^2\bar{\hat{\phi}}^{**}(a,R_0;p)-\frac{2(1-2\nu)}{1-\nu}\,\frac{1}{a}\,\frac{\partial\bar{\hat{\phi}}^{**}}{\partial R}(a,R_0;p)=0 \tag{10.69}$$

A solution to the problem is obtained if we let

$$\bar{\hat{\phi}}^{**}(R,R_0;p)=\bar{\hat{\phi}}^{*}(R,R_0;p)+\bar{\hat{\varphi}}^{*}(R,R_0;p) \tag{10.70}$$

where $\bar{\hat{\phi}}^{*}$ is given by [see Problem 10.1]

$$\bar{\hat{\phi}}^{*}(R,R_0;p)=-\frac{R_0}{2Rp}[e^{-p|R-R_0|}-e^{-p(R+R_0)}] \tag{10.71}$$

and

$$\bar{\hat{\varphi}}^{*}(R,R_0;p)=\frac{A}{R}e^{-pR} \tag{10.72}$$

In Eq. (10.72) A is a constant to be determined from the boundary condition (10.69).

Note that $\bar{\hat{\phi}}^{*}$ satisfies the nonhomogeneous wave equation (10.68) for $R > a$ and $\bar{\hat{\varphi}}^{*}$ satisfies the homogeneous wave equation

$$(\nabla^2-p^2)\bar{\hat{\varphi}}^{*}=0 \quad R>a \tag{10.73}$$

Also, $\bar{\hat{\phi}}^{*}$ and $\bar{\hat{\varphi}}^{*}$ satisfy vanishing conditions as $R\to\infty$ if $p>0$. Since, because of Eq. (10.45) from the solution to Problem 10.2,

$$\begin{aligned}\frac{\partial\bar{\hat{\phi}}^{*}}{\partial R}&=\frac{R_0}{2R^2}\left\{\frac{e^{-p|R-R_0|}-e^{-p(R+R_0)}}{p}\right\}\\&\quad+\frac{R_0}{2R}\left\{\operatorname{sign}(R-R_0)e^{-p|R-R_0|}-e^{-p(R+R_0)}\right\}\end{aligned} \tag{10.74}$$

therefore, substituting $\bar{\hat{\phi}}^{**}$ from (10.70) into the boundary condition (10.69), and using Eqs. (10.71), (10.72), and (10.74), we obtain

$$\begin{aligned}A\left[(ap)^2+\frac{2(1-2\nu)}{1-\nu}(1+pa)\right]&=\frac{R_0}{2p}\left\{\left[e^{-p(R_0-2a)}-e^{-pR_0}\right](pa)^2\right\}\\&\quad+\frac{R_0}{2p}\,\frac{2(1-2\nu)}{1-\nu}\left\{\left[e^{-p(R_0-2a)}-e^{-pR_0}\right]-\left[e^{-p(R_0-2a)}+e^{-pR_0}\right](pa)\right\}\end{aligned} \tag{10.75}$$

Substituting A from (10.75) into (10.72) we obtain

$$\overline{\widehat{\varphi}}^{*}(R, R_0; p) = \frac{R_0}{2R}\frac{1}{p} \times \left\{ \frac{(ap)^2 - \frac{2(1-2\nu)}{1-\nu}(ap) + \frac{2(1-2\nu)}{1-\nu}}{(ap)^2 + \frac{2(1-2\nu)}{1-\nu}(ap) + \frac{2(1-2\nu)}{1-\nu}} e^{-p(R+R_0-2a)} - e^{-p(R+R_0)} \right\} \tag{10.76}$$

By letting

$$x = ap; \quad \alpha = \frac{1-2\nu}{1-\nu}; \quad \beta = \frac{\sqrt{1-2\nu}}{1-\nu} \tag{10.77}$$

into Eq. (10.76) we get

$$\overline{\widehat{\varphi}}^{*} = \frac{R_0}{2R}\left\{ \frac{1}{p}\left[e^{-p(R+R_0-2a)} - e^{-p(R+R_0)} \right] - \frac{4\alpha a}{(x+\alpha)^2 + \beta^2} e^{-p(R+R_0-2a)} \right\} \tag{10.78}$$

An alternative form of (10.78) reads

$$\overline{\widehat{\varphi}}^{*} = \frac{R_0}{2R}\left\{ \frac{e^{-p(R+R_0-2a)} - e^{-p(R+R_0)}}{p} - 4\frac{\alpha}{\beta}\frac{1}{2i}\left[\frac{1}{p - \frac{1}{a}(-\alpha + i\beta)} - \frac{1}{p - \frac{1}{a}(-\alpha - i\beta)} \right] \times e^{-p(R+R_0-2a)} \right\} \tag{10.79}$$

Since

$$L^{-1}\left\{ \frac{1}{p-\alpha^*} e^{-p\beta^*} \right\} = e^{\alpha^*(\tau-\beta^*)} H(\tau - \beta^*) \tag{10.80}$$

where α^* and β^* are independent of p, therefore, applying the operator L^{-1} to Eq. (10.79) we obtain

$$\begin{aligned} \widehat{\varphi}^{*}(R, R_0; \tau) = \frac{R_0}{2R}\Bigg\{ \Bigg[1 - 4\sqrt{1-2\nu}\exp\left[-\frac{1-2\nu}{1-\nu}\frac{\tau - (R+R_0-2a)}{a} \right] \\ \times \sin\left[\frac{\sqrt{1-2\nu}}{1-\nu}\frac{\tau - (R+R_0-2a)}{a} \right] \Bigg] H[\tau - (R+R_0-2a)] \\ - H[\tau - (R+R_0)] \Bigg\} \end{aligned} \tag{10.81}$$

Finally, applying L^{-1} to Eq. (10.70) and using $\widehat{\phi}^{*}$ in the form of Eq. (10.34) of the solution to Problem 10.1, and Eq. (10.81) we arrive at

$$\widehat{\phi}^{**}(R, R_0, \tau) = \frac{R_0}{2R}\left\{1 - 4\sqrt{1-2\nu}\exp\left[-\frac{1-2\nu}{1-\nu}\frac{\tau-(R+R_0-2a)}{a}\right]\right.$$
$$\left.\times \sin\left[\frac{\sqrt{1-2\nu}}{1-\nu}\frac{\tau-(R+R_0-2a)}{a}\right]\right\}H[\tau-(R+R_0-2a)]$$
$$-\frac{R_0}{2R}H[\tau-|R-R_0|] \tag{10.82}$$

This completes a solution to Problem 10.5.

Problem 10.6. The temperature field $\widehat{T} = \widehat{T}(R, \tau)$ produced by a sudden heating of a spherical cavity of radius $R = a$ in an infinite body $R \geq a$ to a constant temperature T_0 is given by the formula

$$\widehat{T}(R, \tau) = \frac{T_0 a}{R}\,\text{erfc}\left(\frac{R-a}{\sqrt{4\tau}}\right) \quad \text{for} \quad R \geq a \quad \text{and} \quad \tau \geq 0 \tag{10.83}$$

Show that the associated potential $\widehat{\Phi} = \widehat{\Phi}(R, \tau)$ that generates the displacement $\widehat{u}_R = \widehat{u}_R(R; \tau)$, and the stresses $\widehat{S}_{RR} = \widehat{S}_{RR}(R; \tau)$ and $\widehat{S}_{\theta\theta} = \widehat{S}_{\theta\theta}(R; \tau)$ in an infinite body $R \geq a$ subject to the stress free boundary condition

$$\widehat{S}_{RR}(a; \tau) = 0 \quad \text{for} \quad \tau > 0 \tag{10.84}$$

and zero stresses at infinity, admits the integral representation

$$\widehat{\Phi}(R, \tau) = T_0 a \int_a^{\infty}\int_0^{\tau} \text{erfc}\left(\frac{R_0-a}{\sqrt{4u}}\right)\frac{1}{R_0}\widehat{\Phi}^{**}(R, R_0; \tau-u)\,du\,dR_0 \tag{10.85}$$

where $\widehat{\Phi}^{**} = \widehat{\Phi}^{**}(R, R_0; \tau)$ is the potential obtained in Problem 10.5.

Solution. The function $\widehat{\phi}^{**}(R, R_0; \tau-u)$ for $R \geq a,\ R_0 > a,\ \tau - u \geq 0$ with u fixed, satisfies the equation [see Eq. (10.65) from the solution to Problem 10.5]

$$\left(\nabla^2 - \frac{\partial^2}{\partial\tau^2}\right)\widehat{\phi}^{**}(R, R_0; \tau-u) = \delta(R-R_0)\delta(\tau-u) \tag{10.86}$$

Let $\hat{\text{T}}$ be given by Eq. (10.83). Multiplying (10.86) by $\widehat{T}(R_0, u)$, integrating over the intervals $a < R_0 < \infty,\ 0 < u < \infty$, and using the property

$$\widehat{\phi}^{**}(R, R_0; \tau-u) = 0 \quad \text{for}\ \tau - u \leq 0 \tag{10.87}$$

we find that the integral

$$\widehat{\phi}(R,\tau) = \int_a^\infty \int_0^\infty \widehat{\phi}^{**}(R, R_0; \tau - u)\widehat{T}(R_0, u) dR_0 du$$

$$= \int_0^\tau \int_a^\infty \widehat{\phi}^{**}(R, R_0; \tau - u)\widehat{T}(R_0, u) du\, dR_0$$

$$= T_0 a \int_a^\infty \int_0^\tau \operatorname{erfc}\left(\frac{R_0 - a}{\sqrt{4u}}\right) \frac{1}{R_0} \widehat{\phi}^{**}(R, R_0, \tau - u) du\, dR_0 \tag{10.88}$$

solves Problem 10.6, since $\widehat{\phi} = \widehat{\phi}(R, t)$ satisfies equation

$$\left(\nabla^2 - \frac{\partial^2}{\partial \tau^2}\right)\widehat{\phi} = \widehat{T}(R, \tau) \quad \text{for} \quad R > a,\ \tau > 0 \tag{10.89}$$

and $\widehat{\phi}$ generates the radial stress $\widehat{S}_{RR} = \widehat{S}_{RR}(R, \tau)$ that meets the stress free boundary condition

$$\widehat{S}_{RR}(a, \tau) = 0 \quad \text{for} \quad \tau \geq 0 \tag{10.90}$$

In addition, $\widehat{\phi}$ generates a solution $s = [\mathbf{u}, \mathbf{E}, \mathbf{S}]$ to the problem that vanishes as $R \to \infty$ and $t > 0$. This completes a solution to Problem 10.6.

Chapter 11
Solutions to Particular Two-Dimensional Initial-Boundary Value Problems of Elastodynamics

The particular solutions discussed in this chapter include: (i) dynamic thermal stresses in an infinite elastic sheet subject to a discontinuous temperature field, and (ii) dynamic thermal stresses produced by a concentrated heat source in an infinite elastic body subject to plane strain conditions.

11.1 Dynamic Thermal Stresses in an Infinite Elastic Body Under Plane Strain Conditions Subject to a Time-Dependent Temperature Field

The dynamic thermal stresses in a homogeneous isotropic infinite elastic body under plane strain conditions and initially at rest, and subject to a time-dependent temperature field $T = T(\mathbf{x}, t)$ are computed from the formulas

$$S_{\alpha\beta} = 2\mu\,(\phi_{,\alpha\beta} - \phi_{,\gamma\gamma}\delta_{\alpha\beta}) + \rho\,\ddot{\phi}\delta_{\alpha\beta} \quad \text{on} \quad E^2 \times [0, \infty) \tag{11.1}$$

where $\phi = \phi(x, t)$ satisfies the field equation

$$\left(\nabla^2 - \frac{1}{c^2}\frac{\partial^2}{\partial t^2}\right)\phi = m\,T \quad \text{on} \quad E^2 \times [0, \infty) \tag{11.2}$$

subject to the homogeneous initial conditions

$$\phi(\mathbf{x}, 0) = \dot{\phi}(\mathbf{x}, 0) = 0 \quad \text{for} \quad \mathbf{x} \in E^2 \tag{11.3}$$

where

$$\frac{1}{c^2} = \frac{\rho}{\lambda + 2\mu} \quad \text{and} \quad m = \frac{1+\nu}{1-\nu}\alpha \tag{11.4}$$

M. Reza Eslami et al., *Theory of Elasticity and Thermal Stresses*, Solid Mechanics and Its Applications 197, DOI: 10.1007/978-94-007-6356-2_11,

11.2 Dynamic Thermal Stresses in an Infinite Elastic Body Under Generalized Plane Stress Conditions Subject to a Time-Dependent Temperature Field

The dynamic thermal stresses in a homogeneous isotropic infinite elastic body under generalized plane stress conditions are computed from the formulas similar to those of a plane strain state. The stresses $\overline{S}_{\alpha\beta} = \overline{S}_{\alpha\beta}(\mathbf{x}, t)$ produced by a temperature $\overline{T} = \overline{T}(\mathbf{x}, t)$ on $E^2 \times [0, \infty)$ and corresponding to a body initially at rest are given by

$$\overline{S}_{\alpha\beta} = 2\mu\,(\overline{\phi}_{,\alpha\beta} - \overline{\phi}_{,\gamma\gamma}\delta_{\alpha\beta}) + \rho\,\ddot{\overline{\phi}}\delta_{\alpha\beta} \quad \text{on} \quad E^2 \times [0, \infty) \tag{11.5}$$

where $\overline{\phi} = \overline{\phi}(\mathbf{x}, t)$ satisfies the field equation

$$\left(\nabla^2 - \frac{1}{\overline{c}^2}\frac{\partial^2}{\partial t^2}\right)\overline{\phi} = m\,\overline{T} \quad \text{on} \quad E^2 \times [0, \infty) \tag{11.6}$$

subject to the homogeneous initial conditions

$$\overline{\phi}(\mathbf{x}, 0) = \dot{\overline{\phi}}(\mathbf{x}, 0) = 0 \quad \text{for} \quad \mathbf{x} \in E^2 \tag{11.7}$$

where

$$\frac{1}{\overline{c}^2} = \frac{\rho}{\overline{\lambda} + 2\mu}, \quad \overline{\lambda} = \frac{2\mu\lambda}{\lambda + 2\mu} \tag{11.8}$$

and

$$\overline{m} = (1 + \nu)\alpha \tag{11.9}$$

11.3 Problems and Solutions Related to Particular Two-Dimensional Initial-Boundary Value Problems of Elastodynamics

Problem 11.1. Find the dynamic thermal stresses in an infinite elastic sheet with a quiescent past subject to the temperature $\overline{T}^* = \overline{T}^*(\mathbf{x}, t)$ of the form

$$\overline{T}^*(\mathbf{x}, t) = T_0\delta(x_1)\delta(x_2)\delta(t)$$

where T_0 is a constant temperature and $\delta = \delta(x)$ is the delta function.

Solution. The dynamic thermal stresses in a homogeneous isotropic infinite sheet initially at rest are computed from the formulas [see Eqs. (11.5)–(11.9) with $c \equiv \overline{c}$ and $m \equiv \overline{m} = (1 + \nu)\alpha$]

$$\overline{S}_{\alpha\beta} = 2\,\mu(\overline{\phi}_{,\alpha\beta} - \nabla^2\overline{\phi}\delta_{\alpha\beta}) + \rho\ddot{\overline{\phi}}\delta_{\alpha\beta} \quad \text{on } E^2 \times [0, \infty) \tag{11.10}$$

where $\overline{\phi} = \overline{\phi}(\mathbf{x}, t)$ satisfies the field equation

$$\left(\nabla^2 - \frac{1}{c^2}\frac{\partial^2}{\partial t^2}\right)\overline{\phi} = mT_0\,\delta(x_1)\,\delta(x_2)\,\delta(t) \quad \text{on } E^2 \times (0, \infty) \tag{11.11}$$

subject to the homogeneous initial conditions

$$\overline{\phi}(\mathbf{x}, 0) = \dot{\overline{\phi}}(\mathbf{x}, 0) = 0, \quad \mathbf{x} \in E^2 \tag{11.12}$$

Since the problem is radially symmetric

$$\delta(x_1)\delta(x_2) = \frac{1}{2\pi r}\delta(r) \tag{11.13}$$

where r is the radial coordinate of the polar coordinates (r, φ)

$$0 < r < \infty, \quad 0 < \varphi \le 2\pi \tag{11.14}$$

a unique solution to Eq. (11.11) that satisfies Eq. (11.12) takes the form

$$\overline{\phi}(r, t) = -\frac{mT_0}{2\pi}H\left(t - \frac{r}{c}\right)\left(t^2 - \frac{r^2}{c^2}\right)^{-1/2} \tag{11.15}$$

In the polar coordinates (r, φ)

$$\nabla^2 = \frac{\partial^2}{\partial r^2} + \frac{1}{r}\frac{\partial}{\partial r} + \frac{1}{r^2}\frac{\partial^2}{\partial \varphi^2} \tag{11.16}$$

Hence, for the function $\overline{\phi}$ that depends on r and t only, because of (11.10), we obtain

$$\overline{S}_{rr}(r, t) = -2\mu\,\frac{1}{r}\frac{\partial\overline{\phi}}{\partial r} + \rho\,\ddot{\overline{\phi}} \tag{11.17}$$

$$\overline{S}_{\varphi\varphi}(r, t) = -2\mu\,\frac{\partial^2\overline{\phi}}{\partial r^2} + \rho\,\ddot{\overline{\phi}} \tag{11.18}$$

$$S_{r\varphi}(r, t) = 0, \quad 0 < r < \infty, t \ge 0 \tag{11.19}$$

Substituting $\overline{\phi}$ from Eq. (11.15) into Eqs. (11.17) and (11.18), respectively, we get

$$\overline{S}_{rr} = -\frac{mT_0}{2\pi}H\left(t - \frac{r}{c}\right)\frac{1}{(t^2 - r^2/c^2)^{5/2}} \times \left[\left(2t^2 + \frac{r^2}{c^2}\right)\rho - \frac{2\mu}{c^2}\left(t^2 - \frac{r^2}{c^2}\right)\right] \tag{11.20}$$

and

$$\overline{S}_{\varphi\varphi} = -\frac{mT_0}{2\pi} H\left(t - \frac{r}{c}\right) \frac{1}{(t^2 - r^2/c^2)^{5/2}} \times \left[\left(2t^2 + \frac{r^2}{c^2}\right)\rho - \frac{2\mu}{c^2}\left(t^2 + 2\frac{r^2}{c^2}\right)\right] \tag{11.21}$$

It follows from Eqs. (11.20) and (11.21) that

$$\overline{S}_{rr} \to 0 \quad \text{and} \quad \overline{S}_{\varphi\varphi} \to 0 \quad \text{as } t \to \infty$$

which means that the dynamic thermal stresses produced by an instantaneous concentrated nucleus of thermoelastic strain [that is, by the strain of the form $E^0_{\alpha\beta} = \alpha T_0 \delta_{\alpha\beta}\delta(x_1)\delta(x_2)\delta(t)$] in an infinite sheet have a transient character. Also, it follows from Eqs. (11.20) and (11.21) that such stress waves are represented by the cylindrical waves propagating from the origin $r = 0$ to infinity with the velocity c, and they become unbounded on the wave front

$$t^2 - r^2/c^2 = 0 \tag{11.22}$$

This completes a solution to Problem 11.1.

Problem 11.2. Find the dynamic thermal stresses in an infinite elastic sheet with a quiescent past subject to the temperature $\widehat{T}^* = \widehat{T}^*(\mathbf{x}, t)$ of the form

$$\widehat{T}^*(\mathbf{x}, t) = T_0 \delta(x_1)\delta(x_2)H(t)$$

where T_0 is a constant temperature, while $\delta = \delta(x)$ and $H = H(t)$ represent the Dirac delta and Heaviside functions, respectively.

Solution. A solution to the problem is obtained by integrating the solution of Problem 11.1 with respect to time. In particular, the potential $\overline{\phi}$, denoted by the same symbol as in Problem 11.1, is obtained from the formula

$$\begin{aligned}\overline{\phi}(\mathbf{x}, t) &= -\frac{mT_0}{2\pi} \int_0^t H\left(\tau - \frac{r}{c}\right)\left(\tau^2 - \frac{r^2}{c^2}\right)^{-1/2} d\tau \\ &= -\frac{mT_0}{2\pi} H\left(t - \frac{r}{c}\right) \int_{r/c}^t \left(\tau^2 - \frac{r^2}{c^2}\right)^{-1/2} d\tau \\ &= -\frac{mT_0}{2\pi} H\left(t - \frac{r}{c}\right)\left[\ln\left(t + \sqrt{t^2 - r^2/c^2}\right) - \ln(r/c)\right]\end{aligned} \tag{11.23}$$

Substituting $\overline{\phi}$ from (11.23) into Eqs. (11.17) and (11.18) of the solution to Problem 11.1, respectively, we obtain

$$\overline{S}_{rr} = \frac{mT_0}{2\pi} H(t - r/c)$$
$$\times \left[\rho \frac{t}{\left(\sqrt{t^2 - r^2/c^2}\right)^3} - 2\mu \left(\frac{1}{c^2} \frac{1}{\sqrt{t^2 - c^2/r^2}} \frac{1}{t + \sqrt{t^2 - r^2/c^2}} + \frac{1}{r^2} \right) \right] \tag{11.24}$$

and

$$\overline{S}_{\varphi\varphi} = \frac{mT_0}{2\pi} H(t - r/c)$$
$$\times \left\{ \rho \frac{t}{\left(\sqrt{t^2 - r^2/c^2}\right)^3} - 2\mu \left[\frac{2}{c^2} \frac{1}{\sqrt{t^2 - r^2/c^2}} \frac{1}{t + \sqrt{t^2 - r^2/c^2}} \right.\right.$$
$$\left.\left. \times \left(1 + \frac{r^2}{c^2} \frac{t + 2\sqrt{t^2 - r^2/c^2}}{t + \sqrt{t^2 - r^2/c^2}} \frac{1}{t^2 - r^2/c^2} \right) - \frac{1}{r^2} \right] \right\} \tag{11.25}$$

If $t \to \infty$, Eqs. (11.24) and (11.25) imply a solution corresponding to a static nucleus:

$$\overline{S}_{rr} \to \overline{S}_{rr}^{0}, \quad \overline{S}_{\varphi\varphi} \to \overline{S}_{\varphi\varphi}^{0} \quad \text{as } t \to \infty \tag{11.26}$$

where

$$\overline{S}_{rr}^{(0)} = -\frac{mT_0\mu}{\pi} \frac{1}{r^2}, \quad \overline{S}_{\varphi\varphi}^{(0)} = \frac{mT_0\mu}{\pi} \frac{1}{r^2} \tag{11.27}$$

This completes a solution to Problem 11.2.

Problem 11.3. Find the dynamic shear stress $\overline{S}_{12} = \overline{S}_{12}(\mathbf{x}, t)$ in an infinite elastic sheet with a quiescent past subject to a temperature $\overline{T} = \overline{T}(\mathbf{x}, t)$ of the form

$$\overline{T}(\mathbf{x}, t) = T_0[H(x_1 + a_1) - H(x_1 - a_1)][H(x_2 + a_2) - H(x_2 - a_2)]\, H(t)$$

where T_0 is a constant temperature, $H = H(t)$ is the Heaviside function, while a_1 and a_2 are positive parameters of the length dimension.

Solution. The dynamic shear stress $\overline{S}_{12}$ produced by the rectangular discontinuous temperature field

$$\overline{T}(\mathbf{x}, t) = T_0[H(x_1 + a_1) - H(x_1 - a_1)] \times [H(x_2 + a_2) - H(x_2 - a_2)]H(t) \tag{11.28}$$

in an infinite elastic sheet:

$$|x_1| < \infty, \quad |x_2| < \infty, \quad a_1 > 0, \quad a_2 > 0$$

is obtained from the formula

$$\overline{S}_{12}(\mathbf{x}, t) = 2\mu\overline{\phi}_{,12}(\mathbf{x}, t) \tag{11.29}$$

where

$$\overline{\phi}(\mathbf{x}, t) = \int_{-a_1}^{a_1}\int_{-a_2}^{a_2} \overline{\psi}(\mathbf{x}, \xi; t)d\xi_1 d\xi_2 \tag{11.30}$$

and

$$\begin{aligned}\overline{\psi}(\mathbf{x}, \xi; t) = &-\frac{mT_0}{2\pi} H\left(t - \frac{1}{c}|\mathbf{x} - \xi|\right)\\ &\times \left[\ln\left(t + \sqrt{t^2 - |\mathbf{x} - \xi|^2/c^2}\right) - \ln(|\mathbf{x} - \xi|/c)\right]\end{aligned} \tag{11.31}$$

[Note that Eq. (11.29) is obtained from Eq. (11.10) of the solution to Problem 11.1, and (11.31) is implied by Eq. (11.28) of the solution to Problem 11.2].

Since

$$\frac{\partial^2}{\partial x_1 \partial x_2}\overline{\psi}(\mathbf{x}, \xi; t) = \frac{\partial^2}{\partial \xi_1 \partial \xi_2}\overline{\psi}(\mathbf{x}, \xi; t) \tag{11.32}$$

the shear stress $\overline{S}_{12}$ may be written as

$$\overline{S}_{12} = 2\mu \int_{-a_1}^{a_1} \frac{\partial}{\partial \xi_1}\left[\int_{-a_2}^{a_2} \frac{\partial}{\partial \xi_2}\overline{\psi}(\mathbf{x}, \xi; t)d\xi_2\right] d\xi_1 \tag{11.33}$$

or

$$\begin{aligned}\overline{S}_{12} = &-\frac{mT_0\mu}{\pi}\int_{-a_1}^{a_1} \frac{\partial}{\partial \xi_1}\left\{H\left[t - \sqrt{(x_1 - \xi_1)^2 + (x_2 - a_2)^2}/c\right]\right.\\ &\times \left[\ln\left(t + \sqrt{t^2 - (x_1 - \xi_1)^2/c^2 - (x_2 - a_2)^2/c^2}\right)\right.\\ &\left. - \ln\left(\sqrt{(x_1 - \xi_1)^2 + (x_2 - a_2)^2}/c\right)\right]\\ &- H\left[t - \sqrt{(x_1 - \xi_1)^2 + (x_2 + a_2)^2}/c\right]\\ &\times \left[\ln\left(t + \sqrt{t^2 - (x_1 - \xi_1)^2/c^2 - (x_2 + a_2)^2/c^2}\right)\right.\\ &\left.\left. - \ln\left(\sqrt{(x_1 - \xi_1)^2 + (x_2 + a_2)^2}/c\right)\right]\right\}\end{aligned}$$

$$= -\frac{mT_0\mu}{\pi}\Big\{H(t - r_{+1+2}/c) \times \Big[\ln\Big(t + \sqrt{t^2 - r_{+1+2}^2/c^2}\Big) - \ln(r_{+1+2}/c)\Big]$$
$$- H(t - r_{+1-2}/c)\Big[\ln\Big(t + \sqrt{t^2 - r_{+1-2}^2/c^2}\Big) - \ln(r_{+1-2}/c)\Big]$$
$$- H(t - r_{-1+2}/c) \times \Big[\ln\Big(t + \sqrt{t^2 - r_{-1+2}^2/c^2}\Big) - \ln(r_{-1+2}/c)\Big]$$
$$+ H(t - r_{-1-2}/c) \times \Big[\ln\Big(t + \sqrt{t^2 - r_{-1-2}^2/c^2}\Big) - \ln(r_{-1-2}/c)\Big]\Big\} \quad (11.34)$$

where

$$r_{\pm1\pm2} = \sqrt{(x_1 \mp a_1)^2 + (x_2 \mp a_2)^2} \quad (11.35)$$

Equation (11.34) provides a closed-form representation of the shear stress wave produced by the rectangular discontinuous temperature field in an infinite elastic sheet. For $t \to \infty$ $\overline{S}_{12}$ attains a steady-state, that is,

$$\overline{S}_{12}(\mathbf{x}, t) \to \frac{mT_0\mu}{\pi} \ln \frac{r_{+1+2}\, r_{-1-2}}{r_{+1-2}\, r_{-1+2}} \quad \text{as } t \to \infty \quad (11.36)$$

This completes a solution to Problem 11.3.

Problem 11.4. An infinite elastic body described by the inequalities

$$0 \le r < \infty, \quad 0 \le \varphi \le 2\pi, \quad |x_3| < \infty \quad (11.37)$$

is subject to a line heat source of the form

$$Q(r, t) = \frac{Q_0 H(t)\delta(r)}{2\pi r} \quad (11.38)$$

Use a parabolic heat conduction equation for the temperature $T = T(r, t)$ together with a wave equation for the thermoelastic displacement potential $\Phi = \Phi(r, t)$ to find the temperature $T = T(r, t)$ and associated thermal stresses $S_{rr} = S_{rr}(r, t)$ and $S_{\varphi\varphi} = S_{\varphi\varphi}(r, t)$ produced by the line heat source. Assume that the body is initially at rest, which means that $T(r, 0) = 0$, $u_r(r, 0) = 0$, and $\dot{u}_r(r, 0) = 0$, where $u_r = u_r(r, t)$ is the radial displacement corresponding to the heat source. Also, note that the stress components are generated by the potential $\Phi = \Phi(r, t)$ through Eqs. (11.1)–(11.4) restricted to the radial symmetry.

Solution. The line heat source

$$Q(r, t) = Q_0 H(t)\delta(r)/2\pi r \quad (11.39)$$

corresponds to a plane strain thermoelastodynamics in which a temperature field $T = T(r, t)$ satisfies the field equation

$$\left(\nabla_r^2 - \frac{1}{\varkappa}\frac{\partial}{\partial t}\right) T = -\frac{Q_0}{\varkappa}\frac{\delta(r)}{2\pi r} H(t) \quad \text{for } r > 0, t > 0 \tag{11.40}$$

the initial condition

$$T(r, 0) = 0 \quad \text{for } r > 0 \tag{11.41}$$

and the vanishing condition

$$T(r, t) \to 0 \quad \text{as } r \to \infty, t > 0 \tag{11.42}$$

In Eq. (11.40)

$$\nabla_r^2 = \frac{\partial^2}{\partial r^2} + \frac{1}{r}\frac{\partial}{\partial r} \tag{11.43}$$

a dynamic thermoelastic process $s = [\mathbf{u}, \mathbf{E}, \mathbf{S}]$ corresponding to $T = T(r, t)$ is described by the relations

$$\mathbf{u}(r, t) = [\partial\phi/\partial r, 0] \tag{11.44}$$

$$E_{rr} = \frac{\partial^2 \phi}{\partial r^2}, \quad E_{\varphi\varphi} = \frac{1}{r}\frac{\partial \phi}{\partial r} \tag{11.45}$$

$$S_{rr} = -2\mu \frac{1}{r}\frac{\partial \phi}{\partial r} + \rho\ddot{\phi} \tag{11.46}$$

$$S_{\varphi\varphi} = -2\mu \frac{\partial^2 \phi}{\partial r^2} + \rho\ddot{\phi} \tag{11.47}$$

where $\phi = \phi(r, t)$ satisfies the wave equation

$$\left(\nabla_r^2 - \frac{1}{c^2}\frac{\partial^2}{\partial t^2}\right)\phi = mT \quad \text{for } r > 0, t > 0 \tag{11.48}$$

subject to the conditions

$$\phi(r, 0) = 0, \quad \dot{\phi}(r, 0) = 0 \quad r > 0 \tag{11.49}$$

and suitable vanishing conditions as $r \to \infty$.

In Eq. (11.48)

$$\frac{1}{c^2} = \frac{\rho}{\lambda + 2\mu}, \quad m = \frac{1+\nu}{1-\nu}\alpha \tag{11.50}$$

To find $s = [\mathbf{u}, \mathbf{E}, \mathbf{S}]$ we proceed in the following way.

Let $\overline{f}(p)$ denote the Laplace transform of a function $f = f(t)$ on $[0, \infty)$, that is,

$$Lf \equiv \overline{f}(p) = \int_0^\infty e^{-pt} f(t) dt \tag{11.51}$$

where p is the Laplace transform parameter.

Applying the Laplace transform to solve the problem (11.40)–(11.42) we obtain

$$\overline{T}(r,p)=\frac{Q_0}{2\pi\varkappa}\frac{1}{p}K_0\left(r\sqrt{\frac{p}{\varkappa}}\right) \tag{11.52}$$

Hence

$$L^{-1}\overline{T}=T(r,t)=\frac{Q_0}{4\pi\varkappa}\int\limits_0^t\frac{e^{-\frac{r^2}{4\varkappa\tau}}}{\tau}d\tau \tag{11.53}$$

Similarly, solving the problem (11.48)–(11.49) by the Laplace transform technique we obtain

$$\overline{\phi}(r,p)=\frac{A}{p}\left(\frac{1}{p-c^2/\varkappa}-\frac{1}{p}\right)\left[K_0\left(\frac{r}{c}p\right)-K_0\left(\frac{r}{\sqrt{\varkappa}}\sqrt{p}\right)\right] \tag{11.54}$$

whcrc

$$A=\frac{mQ_0}{2\pi} \tag{11.55}$$

An alternative form of (11.54) reads

$$\overline{\phi}(r,p)=A\left\{\frac{\varkappa}{c^2}\left(\frac{1}{p-c^2/\varkappa}-\frac{1}{p}\right)-\frac{1}{p^2}\right\}\times\left[K_0\left(\frac{r}{c}p\right)-K_0\left(\frac{r}{\sqrt{\varkappa}}\sqrt{p}\right)\right] \tag{11.56}$$

Since

$$L^{-1}K_0(a\sqrt{p})=\frac{1}{2t}e^{-\frac{a^2}{4t}},\quad a>0 \tag{11.57}$$

and

$$L^{-1}\,K_0(bp)=H(t-b)(t^2-b^2)^{-1/2},\quad b>0 \tag{11.58}$$

therefore, applying L^{-1} to (11.56) and using the convolution theorem we get

$$\phi(r,t)=\phi_w(r,t)+\phi_d(r,t) \tag{11.59}$$

where ϕ_w and ϕ_d, respectively, represent a wave part of ϕ and a diffusive part of ϕ, given by

$$\phi_w(r,t)=AH\left(t-\frac{r}{c}\right)\int\limits_{r/c}^{t}\frac{1}{\sqrt{\tau^2-r^2/c^2}}\times\left\{\frac{\varkappa}{c^2}[e^{(c^2/\varkappa)(t-\tau)}-1]-(t-\tau)\right\}d\tau \tag{11.60}$$

and

$$\phi_d(r,t) = -\frac{A}{2}\int_0^t \frac{e^{-\frac{r^2}{4\varkappa\tau}}}{\tau} \times \left\{\frac{\varkappa}{c^2}[e^{(c^2/\varkappa)(t-\tau)} - 1] - (t-\tau)\right\} d\tau \tag{11.61}$$

By substituting ϕ from Eq. (11.59) into Eqs. (11.44), (11.46), and (11.47), a closed-form of the functions $u_r = u_r\,(r,t)$, $S_{rr} = S_{rr}(r,t)$, and $S_{\varphi\varphi} = S_{\varphi\varphi}(r,t)$, respectively, is obtained.

In particular, for $u_r = u_r(r,t)$ we obtain

$$u_r(r,t) = u_r^{(w)}(r,t) + u_r^{(d)}(r,t) \tag{11.62}$$

where

$$u_r^{(w)}(r,t) = -\frac{A}{r}H\left(t-\frac{r}{c}\right)\int_{r/c}^t [e^{(c^2/\varkappa)(t-\tau)} - 1] \times \frac{\tau}{\sqrt{\tau^2 - r^2/c^2}}d\tau \tag{11.63}$$

and

$$u_r^{(d)}(r,t) = \frac{A}{r}\int_0^t [e^{(c^2/\varkappa)(t-\tau)} - 1] \times e^{-r^2/4\varkappa\tau}\,d\tau \tag{11.64}$$

Also, for $\ddot{\phi} = \ddot{\phi}(r,t)$ we get

$$\ddot{\phi}(r,t) = A\frac{c^2}{\varkappa}\left\{H\left(t-\frac{r}{c}\right)\int_{r/c}^t e^{(c^2/\varkappa)(t-\tau)}\frac{d\tau}{\sqrt{\tau^2 - r^2/c^2}} - \frac{1}{2}\int_0^t e^{(c^2/\varkappa)(t-\tau)}\frac{e^{-r^2/4\varkappa\tau}}{\tau}d\tau\right\} \tag{11.65}$$

Substituting u_r from (11.62) and $\ddot{\phi}$ from (11.65) into the relation

$$S_{rr}(r,t) = -2\mu\frac{1}{r}u_r(r,t) + \rho\,\ddot{\phi}(r,t) \tag{11.66}$$

a closed-form of the radial stress is obtained.

In a similar way a closed-form of $S_{\varphi\varphi} = S_{\varphi\varphi}(r,t)$ may be obtained. This completes a solution to Problem 11.4.

Problem 11.5. An infinite elastic body described by the inequalities

$$0 \le r < \infty, \quad 0 \le \varphi \le 2\pi, \quad |x_3| < \infty \tag{11.67}$$

is subject to a time-periodic line heat source of the form

$$Q(r,t) = \frac{Q_0 e^{i\omega t}\delta(r)}{2\pi r}, \quad i = \sqrt{-1} \tag{11.68}$$

where $\omega > 0$ is the frequency. Show that the temperature $T = T(r,t)$ and the thermoelastic displacement potential $\Phi = \Phi(r,t)$ corresponding to the heat source take the form

$$T(r,t) = \frac{Q_0 e^{i\omega t}}{2\pi\kappa} K_0(r\sqrt{i\omega/\kappa}) \tag{11.69}$$

and

$$\Phi(r,t) = \frac{Q_0 m e^{i\omega t}}{2\pi\kappa(i\omega/\kappa + \omega^2/c_1^2)} \left[K_0(r\sqrt{i\omega/\kappa}) - K_0(ir\omega/c_1)\right] \tag{11.70}$$

where $K_0 = K_0(z)$ is the modified Bessel function of the second kind and zero order; κ stands for the thermal diffusivity, c_1 is the longitudinal velocity, and

$$m = \frac{1+\nu}{1-\nu}\alpha \tag{11.71}$$

Here, ν and α are Poisson's ratio and the coefficient of linear thermal expansion, respectively.

Solution. Note that the temperature field $T(r,t)$ given by Eq. (11.69) of the problem can be written in the form

$$T(r,t) = \overline{T}(r,i\omega)i\omega\, e^{i\omega t} \tag{11.72}$$

where $\overline{T} = \overline{T}(r,p)$ is given by Eq. (11.52) of the solution to Problem 11.4.

Also we check that $T(r,t)$ given by (11.72) satisfies the heat conduction equation

$$\left(\nabla_r^2 - \frac{1}{\varkappa}\frac{\partial}{\partial t}\right) T = -\frac{Q_0}{\varkappa}\frac{\delta(r)}{2\pi r} e^{i\omega t} \quad \text{for } r > 0,\ t > 0 \tag{11.73}$$

and represents an outgoing thermal wave from $r = 0$ to $r = \infty$ if $\sqrt{i\omega/\varkappa}$ in Eq. (11.69) is properly selected.

As a result, the thermoelastic displacement potential $\phi = \phi(r,t)$ corresponding to $T = T(r,t)$ is obtained from the formula

$$\phi(r,t) = \overline{\phi}(r,i\omega)i\omega\, e^{i\omega t} \tag{11.74}$$

where $\overline{\phi} = \overline{\phi}(r, p)$ is given by Eq. (11.54) of the solution to Problem 11.4, that is,

$$\phi(r,t) = -A\, e^{i\omega t} \left(\frac{1}{i\omega - c^2/\varkappa} - \frac{1}{i\omega} \right) \times \left[K_0\left(\frac{ir\omega}{c}\right) - K_0\left(r\sqrt{\frac{i\omega}{\varkappa}}\right) \right]$$

$$= \frac{Q_0 m\, e^{i\omega t}}{2\pi\,\varkappa(i\omega/\varkappa + \omega^2/c^2)} \times [K_0(r\sqrt{i\omega/\varkappa}) - K_0(ir\omega/c)] \tag{11.75}$$

By letting

$$\sqrt{i} = \frac{1}{\sqrt{2}}(1+i) \tag{11.76}$$

and using the asymptotic formula

$$K_0(z) \approx \sqrt{\frac{\pi}{2z}} e^{-z} \quad \text{as } |z| \to \infty \tag{11.77}$$

we find that $\phi(r,t)$ given by Eq. (11.75) generates an outgoing thermoelastic wave from $r = 0$ to $r = \infty$.

This completes a solution to Problem 11.5.

Chapter 12
One-Dimensional Solutions of Elastodynamics

In this chapter a number of typical one-dimensional initial-boundary value problems of homogeneous isotropic isothermal and nonisothermal elastodynamics are solved in a closed-form using the Laplace transform technique. The isothermal solutions include: (a) one-dimensional displacement waves in a semispace subject to a uniform dynamic boundary pressure, (b) one-dimensional displacement waves in a semispace subject to the initial disturbances, and (c) one-dimensional stress waves in an infinite space composed of two homogeneous isotropic elastic semispaces of different material properties. The nonisothermal solutions include: (i) one-dimensional dynamic thermal stresses produced by a plane source of heat that varies harmonically with time in an infinite elastic solid, (ii) one-dimensional dynamic thermal stresses produced by a plane nucleus of thermoelastic strain in an infinite elastic solid, and (iii) one-dimensional dynamic thermal stresses in a semispace due to the action of a plane internal nucleus of thermoelastic strain.

12.1 One-Dimensional Field Equations of Isothermal Elastodynamics

The one-dimensional field equations of isothermal elastodynamics describe an elastic process $p = [\mathbf{u}, \mathbf{E}, \mathbf{S}]$ that depends on a single space variable $x = x_1$ and on time t. Such a process corresponds to the data that depend on x and t only. In the following we let

$$\mathbf{u}(\mathbf{x}, t) = [u(x, t), 0, 0] \tag{12.1}$$

and

$$\mathbf{b}(\mathbf{x}, t) = [b(x, t), 0, 0] \tag{12.2}$$

where $\mathbf{b} = \mathbf{b}(\mathbf{x}, t)$ is the body force vector field.

M. Reza Eslami et al., *Theory of Elasticity and Thermal Stresses*, Solid Mechanics and Its Applications 197, DOI: 10.1007/978-94-007-6356-2_12,

The strain tensor $\mathbf{E}$ has only one component: $E_{11} = E_{11}(x,t)$, and the strain-displacement relation reads

$$E_{11} = \frac{\partial u}{\partial x} \tag{12.3}$$

The equation of motion takes the form

$$\frac{\partial S_{11}}{\partial x} + b = \rho \frac{\partial^2 u}{\partial t^2} \tag{12.4}$$

The constitutive relation reads

$$S_{11} = (\lambda + 2\mu)\frac{\partial u}{\partial x} \tag{12.5}$$

provided the body is homogeneous and isotropic.

By letting $S_{11} = S(x,t)$ and eliminating $E_{11} = E_{11}(x,t)$ and $S = S(x,t)$ from Eqs. (12.3)–(12.5), the one-dimensional displacement equation of motion is obtained

$$\left(\frac{\partial^2}{\partial x^2} - \frac{1}{c^2}\frac{\partial^2}{\partial t^2}\right) u = -\frac{b}{\lambda + 2\mu} \tag{12.6}$$

where

$$c = \sqrt{\frac{\lambda + 2\mu}{\rho}} \tag{12.7}$$

Also, by eliminating $u = u(x,t)$ and $E_{11} = E_{11}(x,t)$ from Eqs. (12.3)–(12.5), the one-dimensional stress equation of motion is obtained

$$\left(\frac{\partial^2}{\partial x^2} - \frac{1}{c^2}\frac{\partial^2}{\partial t^2}\right) S = -\frac{\partial b}{\partial x} \tag{12.8}$$

12.2 One-Dimensional Field Equations of Nonisothermal Elastodynamics

A one-dimensional thermoelastic process $p = [\mathbf{u}, \mathbf{E}, \mathbf{S}]$ can be associated with a pair (Φ, T) in which $\Phi = \Phi(x,t)$ represents a thermoelastic displacement potential and $T = T(x,t)$ is a temperature field.

In a dimensionless setting, the displacement $u = u(x,t)$ and the stress $S = S(x,t)$, respectively, are computed from the formulas

$$u(x,t) = \frac{\partial \Phi}{\partial x} \tag{12.9}$$

and

$$S(x,t) = \frac{\partial^2 \Phi}{\partial t^2} \tag{12.10}$$

where

$$\left(\frac{\partial^2}{\partial x^2} - \frac{\partial^2}{\partial t^2}\right) \Phi = T \tag{12.11}$$

and

$$\left(\frac{\partial^2}{\partial x^2} - \frac{\partial}{\partial t}\right) T = -Q \tag{12.12}$$

In Eq. (12.12) Q represents a dimensionless heat source field.

By eliminating $\Phi = \Phi(x,t)$ from Eqs. (12.10) and (12.11), the one-dimensional stress-temperature equation is obtained

$$\left(\frac{\partial^2}{\partial x^2} - \frac{\partial^2}{\partial t^2}\right) S = \frac{\partial^2 T}{\partial t^2} \tag{12.13}$$

Finally, by applying the operator $(\partial^2/\partial x^2 - \partial/\partial t)$ to Eq. (12.13) and using Eq. (12.12), we obtain the one-dimensional stress equation of nonisothermal elastodynamics

$$\left(\frac{\partial^2}{\partial x^2} - \frac{\partial}{\partial t}\right)\left(\frac{\partial^2}{\partial x^2} - \frac{\partial^2}{\partial t^2}\right) S = -\frac{\partial^2 Q}{\partial t^2} \tag{12.14}$$

Green's Functions of One-Dimensional Nonisothermal Elastodynamics

Consider the following initial-boundary value problem of one-dimensional nonisothermal elastodynamics. Find a temperature $T^* = T^*(x,t)$ that satisfies the parabolic heat conduction equation

$$\left(\frac{\partial^2}{\partial x^2} - \frac{\partial}{\partial t}\right) T^* = -\delta(x)\delta(t) \quad \text{for} \quad |x| < \infty, \quad t > 0 \tag{12.15}$$

the initial condition

$$T^*(x,0) = 0 \quad \text{for} \quad |x| < \infty \tag{12.16}$$

and the asymptotic condition

$$T^*(x,t) \to 0 \quad \text{as} \quad |x| \to \infty \quad \text{and} \quad t > 0 \tag{12.17}$$

Applying the Laplace transform to Eq. (12.15) and using the initial condition (12.16), we obtain

$$\left(\frac{\partial^2}{\partial x^2} - p\right) \overline{T}^* = -\delta(x) \quad \text{for} \quad |x| < \infty \tag{12.18}$$

where $\overline{T}^* = \overline{T}^*(x, p)$ is the Laplace transform of $T^* = T^*(x, t)$, and p is the transform parameter.

Since

$$\left(\frac{\partial^2}{\partial x^2} - k^2\right)^{-1} \delta(x) = -\frac{e^{-k|x|}}{2k} \quad \text{for} \quad |x| < \infty \quad \text{and} \quad k > 0 \tag{12.19}$$

and the asymptotic condition (12.17) in the transform domain takes the form

$$\overline{T}^*(x, p) \to 0 \quad \text{as} \quad |x| \to \infty \quad \text{and} \quad p > 0 \tag{12.20}$$

therefore, a unique solution to Eq. (12.18) subject to (12.20), is

$$\overline{T}^*(x, p) = \frac{e^{-|x|\sqrt{p}}}{2\sqrt{p}} \quad \text{for} \quad |x| < \infty \quad \text{and} \quad p > 0 \tag{12.21}$$

Finally, by applying the inverse Laplace transform to Eq. (12.21), we obtain

$$T^*(x, t) = \frac{1}{2\sqrt{\pi t}} e^{-x^2/4t} \quad \text{for} \quad |x| < \infty \quad \text{and} \quad t \geq 0 \tag{12.22}$$

The function $T^* = T^*(x, t)$, satisfying Eqs. (12.15)–(12.17), is a Green's function for the one-dimensional parabolic heat conduction equation of the nonisothermal elastodynamics.

A Green's function related to the one-dimensional stress equation of nonisothermal elastodynamics is a solution to the following initial-boundary value problem. Find a function $S^* = S^*(x, t)$ that satisfies the field equation

$$\left(\frac{\partial^2}{\partial x^2} - \frac{\partial}{\partial t}\right)\left(\frac{\partial^2}{\partial x^2} - \frac{\partial^2}{\partial t^2}\right) S^* = -\delta(x)\frac{\partial^2}{\partial t^2}\delta(t) \quad \text{for} \quad |x| < \infty, \quad t > 0 \tag{12.23}$$

the initial conditions

$$S^*(x, 0) = \frac{\partial}{\partial t} S^*(x, 0) = \frac{\partial^2}{\partial t^2} S^*(x, 0) = 0 \quad \text{for} \quad |x| < \infty \tag{12.24}$$

and the asymptotic condition

$$S^*(x, t) \to 0 \quad \text{as} \quad |x| \to \infty \quad \text{and} \quad t > 0 \tag{12.25}$$

By applying the Laplace transform to Eqs. (12.23)–(12.25) and using the homogeneous initial conditions (12.24), we obtain

$$\left(\frac{\partial^2}{\partial x^2} - p\right)\left(\frac{\partial^2}{\partial x^2} - p^2\right) \overline{S}^* = -p^2\delta(x) \quad \text{for} \quad |x| < \infty \tag{12.26}$$

and

$$S^*(x, p) \to 0 \quad \text{as} \quad |x| \to \infty \quad \text{and} \quad p > 0 \tag{12.27}$$

Since

$$\left[\left(\frac{\partial^2}{\partial x^2} - p\right)\left(\frac{\partial^2}{\partial x^2} - p^2\right)\right]^{-1} = \frac{1}{p(p-1)}\left[\left(\frac{\partial^2}{\partial x^2} - p^2\right)^{-1} - \left(\frac{\partial^2}{\partial x^2} - p\right)^{-1}\right] \tag{12.28}$$

therefore, using Eq. (12.19), we obtain a unique solution to Eq. (12.26) subject to the asymptotic condition (12.27) in the form

$$\overline{S}^*(x, p) = \frac{1}{2}\frac{p}{p-1}\left[\frac{e^{-|x|p}}{p} - \frac{e^{-|x|\sqrt{p}}}{\sqrt{p}}\right] \quad \text{for} \quad |x| < \infty, \quad p > 0 \tag{12.29}$$

An alternative form of this equation reads

$$\overline{S}^*(x, p) = \frac{1}{2}\left(\frac{e^{-|x|p}}{p-1} + \frac{\partial}{\partial |x|}\frac{e^{-|x|\sqrt{p}}}{p-1}\right) \quad \text{for} \quad |x| < \infty, \quad p > 0 \tag{12.30}$$

Since

$$L^{-1}\left\{\frac{e^{-|x|p}}{p-1}\right\} = e^{t-|x|}H(t - |x|) \tag{12.31}$$

and

$$L^{-1}\left\{\frac{e^{-|x|\sqrt{p}}}{p-1}\right\} = U(|x|, t) \tag{12.32}$$

where

$$U(|x|, t) = \frac{1}{2}e^t\left[e^{-|x|}\text{erfc}\left(\frac{|x|}{2\sqrt{t}} - \sqrt{t}\right) + e^{|x|}\text{erfc}\left(\frac{|x|}{2\sqrt{t}} + \sqrt{t}\right)\right] \tag{12.33}$$

therefore, applying the operator L^{-1} to Eq. (12.30) we obtain

$$\begin{aligned} S^*(x, t) &= \frac{1}{2}\left[e^{t-|x|}H(t - |x|) + \frac{\partial}{\partial |x|}U(|x|, t)\right] \\ &= \frac{1}{2}\left[e^{t-|x|}H(t - |x|) - \frac{1}{\sqrt{\pi t}}e^{-x^2/4t} - V(|x|, t)\right] \end{aligned} \tag{12.34}$$

where

$$V(|x|, t) = \frac{1}{2}e^t\left[e^{-|x|}\text{erfc}\left(\frac{|x|}{2\sqrt{t}} - \sqrt{t}\right) - e^{|x|}\text{erfc}\left(\frac{|x|}{2\sqrt{t}} + \sqrt{t}\right)\right] \tag{12.35}$$

The function $S^* = S^*(x,t)$ represents Green's function of one-dimensional non-isothermal elastodynamics for $|x| < \infty$, $t \geq 0$ corresponding to an instantaneous heat source distributed on the plane $x = 0$.

12.3 Problems and Solutions Related to One-Dimensional Initial-Boundary Value Problems of Elastodynamics

Problem 12.1. Using the displacement characterization of one-dimensional homogeneous isotropic elastodynamics, find the displacement $u = u(x,t)$ in a semispace $x \geq 0$ that is initially at rest and subject to a uniform dynamic pressure $s = s(t)$ on the boundary $x = 0$. Also, show that the particle velocity $\dot{u} = \dot{u}(x,t)$ is related to the stress $S = S(x,t)$ by the equation $S(x,t) = -\rho\, c\, \dot{u}(x,t)$, where ρ and c are the density and the longitudinal velocity of the body, respectively.

Solution. We are to find a solution $u = u(x,t)$ of the equation

$$\left(\frac{\partial^2}{\partial x^2} - \frac{1}{c^2}\frac{\partial^2}{\partial t^2}\right) u = 0, \quad \frac{1}{c^2} = \frac{\rho}{\lambda + 2\,\mu} \quad \text{for } x > 0,\ t > 0 \tag{12.36}$$

subject to the initial conditions

$$u(x,0) = 0, \quad \dot{u}(x,0) = 0, \quad x > 0 \tag{12.37}$$

the boundary condition

$$(\lambda + 2\,\mu)\frac{\partial u}{\partial x}(0,t) = -s(t), \quad t > 0 \tag{12.38}$$

and suitable vanishing conditions at $x = \infty$. The stress field $S = S(x,t)$ is related to the displacement field $u = u(x,t)$ by

$$S(x,t) = (\lambda + 2\,\mu)\frac{\partial u}{\partial x}(x,t) \tag{12.39}$$

Let $\overline{f} = \overline{f}(p)$ be the Laplace transform of a function $f = f(t)$ on $[0,\infty)$, that is,

$$Lf \equiv \overline{f}(p) = \int_0^\infty e^{-pt} f(t)dt \tag{12.40}$$

Applying the Laplace transform to Eqs. (12.36), (12.38), and (12.39), respectively, and using the initial conditions (12.37) we obtain

$$\left(\frac{\partial^2}{\partial x^2} - \frac{p^2}{c^2}\right)\overline{u} = 0, \quad x > 0 \tag{12.41}$$

$$(\lambda + 2\,\mu)\frac{\partial \overline{u}}{\partial x}(0, p) = -\overline{s}(p) \tag{12.42}$$

and

$$\overline{S}(x, p) = (\lambda + 2\,\mu)\frac{\partial \overline{u}}{\partial x}(x, p) \quad x \geq 0 \tag{12.43}$$

Hence, a solution $\overline{u} = \overline{u}(x, p)$ of Eq. (12.41) that satisfies the boundary condition (12.42) and vanishes at infinity takes the form

$$\overline{u}(x, p) = \frac{c}{\lambda + 2\,\mu}\frac{\overline{s}(p)}{p}e^{-\frac{p}{c}x} \quad x \geq 0 \tag{12.44}$$

Also, substituting $\overline{u}$ from (12.44) into (12.43) we obtain

$$\overline{S}(x, p) = -\overline{s}(p)\, e^{-\frac{p}{c}x} \quad x \geq 0 \tag{12.45}$$

and it follows from Eqs. (12.44) and (12.45) that

$$\overline{S}(x, p) = -\rho\, c\, p\, \overline{u}(x, p) \quad x \geq \dot{0} \tag{12.46}$$

Finally, applying the operator L^{-1} to Eqs. (12.44)–(12.46), respectively, and using the homogeneous initial condition $(12.37)_1$ and the convolution theorem, we obtain

$$u(x, t) = \frac{c}{\lambda + 2\,\mu}H\left(t - \frac{x}{c}\right)\int\limits_0^{t-x/c} s(\tau)d\tau \tag{12.47}$$

$$S(x, t) = -H\left(t - \frac{x}{c}\right)s\left(t - \frac{x}{c}\right) \tag{12.48}$$

and

$$S(x, t) = -\rho c\, \dot{u}(x, t) \tag{12.49}$$

This completes a solution to Problem 12.1.

Problem 12.2. Solve a one-dimensional initial-boundary value problem for a semi-space with fixed boundary subject to the initial disturbances $u(x, 0) = u_0(x)$, $\dot{u}(x, 0) = \dot{u}_0(x)$ for $x \geq 0$, where $u_0(x)$ and $\dot{u}_0(x)$ are prescribed functions on $[0, \infty)$.

Solution. The problem reads:

Find $u = u(x, t)$ that satisfies the field equation

$$\left(\frac{\partial^2}{\partial x^2} - \frac{1}{c^2}\frac{\partial^2}{\partial t^2}\right)u = 0 \quad x > 0,\ t > 0 \tag{12.50}$$

the initial conditions

$$u\ (x,0) = u_0(x), \quad \dot{u}(x,0) = \dot{u}_0(x) \tag{12.51}$$

the boundary condition

$$u(0,t) = 0, \quad t \geq 0 \tag{12.52}$$

and the vanishing condition

$$u(x,t) \to 0 \ \text{ as } x \to \infty, \ t > 0 \tag{12.53}$$

By using the Laplace transform technique to solve the initial-boundary value problem (12.50)–(12.53), we obtain

$$\begin{aligned} u(x,t) = {} & \frac{1}{2}\left[H(x-ct)u_0(x-ct) + u_0(x+ct) - H(ct-x)u_0(ct-x)\right] \\ & + \frac{1}{2c}\left[H(x-ct)\int_{x-ct}^{x} \dot{u}_0(\xi)d\xi + H(ct-x)\int_{0}^{x} \dot{u}_0(\xi)d\xi \right. \\ & \left. + \int_{x}^{x+ct} \dot{u}_0(\xi)d\xi - H(ct-x)\int_{0}^{ct-x} \dot{u}_0(\xi)d\xi \right] \quad \text{for } x \geq 0, \ t \geq 0 \end{aligned} \tag{12.54}$$

It is easy to check that u given by (12.54) satisfies (12.50)–(12.53), and this completes a solution to Problem 12.2.

Problem 12.3. Find the displacement $u = u(x,t)$ in a homogeneous isotropic elastic semispace $x \geq 0$ with free boundary subject to the initial disturbances as in Problem 12.2.

Solution. We are to find a solution $u = u(x,t)$ to the equation

$$\left(\frac{\partial^2}{\partial x^2} - \frac{1}{c^2}\frac{\partial^2}{\partial t^2}\right) u = 0 \quad x > 0, \ t > 0 \tag{12.55}$$

subject to the conditions

$$u(x,0) = u_0(x), \quad \dot{u}(x,0) = \dot{u}_0(x), \quad x > 0 \tag{12.56}$$

$$\frac{\partial u}{\partial x}(0,t) = 0, \quad t \geq 0 \tag{12.57}$$

and the vanishing condition

$$u(x,t) \to 0 \ \text{ as } x \to \infty, \ t > 0 \tag{12.58}$$

By applying the Laplace transform to Eqs. (12.55) and (12.57), respectively, and using (12.56) we obtain

$$\left(\frac{\partial^2}{\partial x^2} - \frac{p^2}{c^2}\right)\overline{u} = -\frac{1}{c^2}[pu_0(x) + \dot{u}_0(x)] \quad x > 0, \ p > 0 \tag{12.59}$$

and

$$\frac{\partial \overline{u}}{\partial x}(0, p) = 0, \quad p > 0 \tag{12.60}$$

where p is the transform parameter.

Let $G = G(x, \xi; t)$ be Green's function defined by the equations:

$$\left(\frac{\partial^2}{\partial x^2} - \frac{1}{c^2}\frac{\partial^2}{\partial t^2}\right) G = -[\delta(x-\xi) + \delta(x+\xi)]\delta(t) \quad x \geq 0, \ \xi > 0, \ t > 0 \tag{12.61}$$

$$G(x, \xi; 0) = 0, \quad \dot{G}(x, \xi; 0) = 0 \quad x > 0, \ \xi > 0 \tag{12.62}$$

$$\frac{\partial G}{\partial x}(0, \xi; t) = 0, \quad \xi > 0, \ t > 0 \tag{12.63}$$

and

$$G(x, \xi; t) \to 0 \quad \text{as } x \to \infty, \ t > 0 \tag{12.64}$$

By applying the Laplace transform to Eqs. (12.61), (12.63), and (12.64), respectively, and using (12.62) we get

$$\left(\frac{\partial^2}{\partial x^2} - \frac{p^2}{c^2}\right)\overline{G} = -[\delta(x-\xi) + \delta(x+\xi)] \tag{12.65}$$

$$\frac{\partial \overline{G}}{\partial x}(0, \xi; p) = 0 \tag{12.66}$$

and

$$\overline{G}(x, \xi; p) \to 0 \quad \text{as } x \to +\infty \tag{12.67}$$

It is easy to show that $\overline{G}$ takes the form

$$\overline{G}(x, \xi; p) = \frac{c}{2p}\left[e^{-\frac{p}{c}|x-\xi|} + e^{-\frac{p}{c}(x+\xi)}\right] \quad p > 0 \tag{12.68}$$

Therefore, applying L^{-1} to (12.68) we get

$$G(x, \xi; t) = \frac{c}{2}\left[H\left(t - \frac{|x-\xi|}{c}\right) + H\left(t - \frac{x+\xi}{c}\right)\right] \quad \text{for } x \geq 0, \ t \geq 0, \ \xi > 0. \tag{12.69}$$

Also, it follows from Eqs. (12.59)–(12.60) and (12.65)–(12.66) that

$$\overline{u}(x, p) = \frac{1}{c^2} \int_0^\infty \overline{G}(x, \xi; p)[pu_0(\xi) + \dot{u}_0(\xi)]d\xi \tag{12.70}$$

Hence, substituting $\overline{G}$ from (12.68) into (12.70) we obtain

$$\overline{u}(x, p) = \frac{1}{2c} \left\{ \int_0^\infty \left[e^{-\frac{p}{c}|x-\xi|} + e^{-\frac{p}{c}(x+\xi)} \right] u_0(\xi) d\xi \right.$$
$$\left. + \int_0^\infty \frac{1}{p} \left[e^{-\frac{p}{c}|x-\xi|} + e^{-\frac{p}{c}(x+\xi)} \right] \dot{u}_0(\xi) d\xi \right\} \tag{12.71}$$

and applying L^{-1} to (12.71) we get

$$u(x, t) = \frac{1}{2c} \left\{ \int_0^\infty u_0(\xi) \left[\delta\left(t - \frac{|x-\xi|}{c}\right) + \delta\left(t - \frac{x+\xi}{c}\right) \right] d\xi \right.$$
$$\left. + \int_0^\infty \dot{u}_0(\xi) \left[H\left(t - \frac{|x-\xi|}{c}\right) + H\left(t - \frac{x+\xi}{c}\right) \right] d\xi \right\} \tag{12.72}$$

or

$$u(x, t) = \frac{1}{2} \{H(x - ct)u_0(x - ct) + u_0(x + ct) + H(ct - x)u_0(ct - x)\}$$
$$+ \frac{1}{2c} \left\{ H(x - ct) \int_{x-ct}^{x} \dot{u}_0(\xi) d\xi + \int_x^{x+ct} \dot{u}_0(\xi) d\xi \right.$$
$$\left. + H(ct - x) \int_0^x \dot{u}_0(\xi) d\xi + H(ct - x) \int_0^{ct-x} \dot{u}_0(\xi) d\xi \right\} \tag{12.73}$$

It is easy to check that (12.73) satisfies Eq. (12.55) for $x > 0$, $t > 0$, the initial conditions (12.56), the boundary condition (12.57) as well as the vanishing condition (12.58).

This completes a solution to Problem 12.3.

Problem 12.4. Find the displacement $u = u(x, t)$ in a homogeneous isotropic elastic finite strip $0 \le x \le l$ fixed at $x = 0$ and subject to an impulsive unit traction at $x = l$ [that is, $S(l, t) = \delta(t)$, where $S = S(x, t)$ is the stress associated with

$u = u(x, t)$, and $\delta = \delta(t)$ is the delta function]. The initial conditions are assumed to be homogeneous, that is, $u(x, 0) = 0, \quad \dot{u}(x, 0) = 0$ for $0 \le x \le l$.

Solution. We are to find a solution to the equation

$$\left(\frac{\partial^2}{\partial x^2} - \frac{1}{c^2} \frac{\partial^2}{\partial t^2} \right) u = 0, \quad 0 < x < l, \ t > 0 \tag{12.74}$$

subject to the conditions

$$u(x, 0) = 0, \quad \dot{u}(x, 0) = 0, \quad 0 < x < l \tag{12.75}$$

$$u(0, t) = 0, \quad \frac{\partial u}{\partial x}(l, t) = \frac{\delta(t)}{\lambda + 2\mu}, \quad t > 0 \tag{12.76}$$

A solution $u = u(x, t)$ is sought in the form

$$u(x, t) = \phi(x, t) + \sum_{n=1}^{\infty} \widehat{\phi}_n(t) \psi_n(x) \tag{12.77}$$

where $\phi = \phi(x, t)$ is a solution to the quasi-static problem:

$$\frac{\partial^2 \phi}{\partial x^2}(x, t) = 0, \quad 0 < x < l, \ t \ge 0 \tag{12.78}$$

$$\phi(0, t) = 0, \quad \frac{\partial \phi}{\partial x}(l, t) = \frac{\delta(t)}{\lambda + 2\mu}, \quad t > 0 \tag{12.79}$$

$\psi_n = \psi_n(x)$ are the eigenfunctions satisfying the equation

$$\psi_n''(x) + \left(\frac{\omega_n}{c} \right)^2 \psi_n(x) = 0, \quad 0 < x < l \tag{12.80}$$

subject to the boundary conditions

$$\psi_n(0) = 0, \quad \psi_n'(l) = 0 \tag{12.81}$$

and $\widehat{\phi}_n = \widehat{\phi}_n(t)$ are selected in such a way that $u = u(x, t)$ satisfies (12.74)–(12.76). It is easy to check that a pair $(\omega_n, \ \psi_n)$ is represented by

$$\omega_n = \frac{(2n - 1)\pi c}{2l}, \quad n = 1, 2, 3, \ldots \tag{12.82}$$

$$\psi_n(x) = \left(\frac{2}{l} \right)^{1/2} \sin \frac{(2n - 1)\pi x}{2l} \tag{12.83}$$

and

$$\int_0^l \psi_n(x)\psi_m(x)dx = \delta_{nm} \quad n, m = 1, 2, 3, \ldots \tag{12.84}$$

Substituting $u(x, t)$ from (12.77) into (12.74)–(12.76) and taking into account (12.78)–(12.83) we find that $\widehat{\phi}_n = \widehat{\phi}_n(t)$ satisfies the relation

$$\sum_{n=1}^{\infty} \psi_n(x)\left[\ddot{\widehat{\phi}}_n(t) + \omega_n{}^2\widehat{\phi}_n(t)\right] = -\ddot{\phi}(x, t) \tag{12.85}$$

By multiplying (12.85) by $\psi_m(x)$ and integrating over $[0, l]$, and using the orthogonality conditions (12.84) we obtain

$$\ddot{\widehat{\phi}}_m(t) + \omega_m^2\widehat{\phi}_m(t) = \ddot{\phi}_m(t) \tag{12.86}$$

where

$$\phi_m(t) = -\int_0^l \phi(x, t)\psi_m(x)dx \tag{12.87}$$

Since $\phi = \phi(x, t)$ that satisfies (12.78)–(12.79) takes the form

$$\phi(x, t) = \frac{\delta(t)}{\lambda + 2\mu}x, \quad 0 \le x \le l \tag{12.88}$$

we obtain

$$\begin{aligned} \phi_m(t) &= -\frac{\delta(t)}{\lambda + 2\mu}\left(\frac{2}{l}\right)^{1/2}\int_0^l x \sin\frac{(2m-1)\pi x}{2l}dx \\ &= \frac{\delta(t)}{\lambda + 2\mu}\frac{4(2l)^{1/2}l}{\pi^2}\frac{(-)^m}{(2m-1)^2} = \frac{2(2l)^{1/2}\delta(t)(-)^m}{\pi\rho c\omega_m(2m-1)} \end{aligned} \tag{12.89}$$

Also note that

$$\widehat{\phi}_m(0) = \dot{\widehat{\phi}}_m(0) = 0 \tag{12.90}$$

and

$$\phi_m(0) = 0, \quad \dot{\phi}_m(0) = 0 \tag{12.91}$$

Therefore, applying the Laplace transform to (12.86) and using (12.90) and (12.91) we obtain

$$\left(p^2 + \omega_m{}^2\right)\overline{\widehat{\phi}}_m(p) = p^2\overline{\phi}_m(p) \tag{12.92}$$

Hence

$$\overline{\overline{\phi}}_m(p) = \left(1 - \frac{\omega_m^2}{p^2 + \omega_m^2}\right) \overline{\phi}_m(p) \tag{12.93}$$

and

$$\widehat{\phi}_m(t) = \phi_m(t) - \omega_m \int_0^t \phi_m(\tau) \sin \omega_m (t - \tau) d\tau \tag{12.94}$$

Substituting $\widehat{\phi}_m$ from (12.94) into (12.77) we obtain

$$u(x, t) = \phi(x, t) + \sum_{n=1}^{\infty} \phi_n(t)\psi_n(x) - \sum_{n=1}^{\infty} \omega_n \left[\int_0^t \phi_n(\tau) \sin \omega_n (t - \tau) d\tau \right] \psi_n(x) \tag{12.95}$$

Since

$$\phi(x, t) = -\sum_{n=1}^{\infty} \phi_n(t)\psi_n(x) \tag{12.96}$$

[see Eq. (12.87)], therefore substituting $\phi_n(t)$ from (12.89) into (12.95) we obtain

$$u(x, t) = \frac{4}{\pi \rho c} \sum_{n=1}^{\infty} \frac{(-)^{n-1}}{2n - 1} \sin \left[\frac{(2n - 1)\pi c t}{2l} \right] \times \sin \left[\frac{(2n - 1)\pi x}{2l} \right] \tag{12.97}$$

This completes a solution to Problem 12.4.

Problem 12.5. Let $x = 0$ be an interface between two homogeneous isotropic elastic semispaces of different material properties, and assume that $(\rho_-, \lambda_-, \mu_-)$ and $(\rho_+, \lambda_+, \mu_+)$ denote the material properties of the $(-\infty, 0)$ and $(0, \infty)$ semispaces, respectively. Also, assume that an incident stress wave $S^{(i)} = S^{(i)}(t - x/c_-)$, where $S^{(i)}(s) = 0$ for $s < 0$ and $c_- = \sqrt{(\lambda_- + 2\mu_-)/\rho_-}$, strikes the interface $x = 0$ in such a way that it is completely reflected. Let $S^{(r)} = S^{(r)}(x, t)$ be the reflected wave. Show that the total stress $S = S(x, t)$ on $|x| < \infty$ and $t > 0$ is represented by

$$S(x, t) = \begin{cases} S^{(i)}(t - x/c_-) + S^{(r)}(x, t) & \text{for} \quad x \le 0 \\ 0 & \text{for} \quad x \ge 0 \end{cases}$$

where

$$S^{(r)}(x, t) = -S^{(i)}(t + x/c_-)$$

This case occurs if the semispace $[0, \infty)$ is a vacuum.

Solution. A stress wave $S^{(r)} = S^{(r)}(x, t)$ reflected from the interface $x = 0$ propagates in the negative direction of the x-axis. Therefore it takes the form

$$S^{(r)}(x, t) = Af\left(t + \frac{x}{c_-}\right) \tag{12.98}$$

where A is an arbitrary constant, and $f = f(s)$ is an arbitrary function. Since the interface $x = 0$ is stress free

$$S^{(i)}(0,t) + S^{(r)}(0,t) = 0 \tag{12.99}$$

where

$$S^{(i)}(x,t) = S^{(i)}(t - x/c_-) \tag{12.100}$$

therefore, (12.99) takes the form

$$S^{(i)}(t) + A\ f(t) = 0 \quad t \geq 0 \tag{12.101}$$

Equation (12.101) is satisfied if and only if

$$f(t) = S^{(i)}(t) \text{ and } A = -1 \tag{12.102}$$

This implies that

$$S^{(r)}(x,t) = -S^{(i)}(t + x/c_{-1}) \tag{12.103}$$

A complete reflection of the striking wave from $x = 0$ also means that $S(x,t) = 0$ for $x \geq 0,\ t > 0$. This completes a solution to Problem 12.5.

Problem 12.6. Assume that an incident stress wave $S^{(i)} = S^{(i)}(t - x/c_-)$ strikes the interface $x = 0$ between the two semispaces introduced in Problem 12.5 in such a way that a part of $S^{(i)}$ is reflected and a part is transmitted across the interface. This case occurs when the total stress and the particle velocity are continuous at $x = 0$ for every $t > 0$. Show that the total stress wave propagating in the two semispace solid is represented by the formula

$$S(x,t) = \begin{cases} S^{(i)}(t - x/c_-) + S^{(r)}(t + x/c_-) & \text{for} \quad x \leq 0 \\ S^{(t)}(t - x/c_+) & \text{for} \quad x \geq 0 \end{cases}$$

where $S^{(r)} = S^{(r)}(t + x/c_-)$ and $S^{(t)} = S^{(t)}(t - x/c_+)$ are the reflected and transmitted stress waves, respectively, given by

$$S^{(r)}(t + x/c_-) = c^{(r)} S^{(i)}(t + x/c_-)$$
$$S^{(t)}(t - x/c_+) = c^{(t)} S^{(i)}(t - x/c_+)$$

and

$$c^{(r)} = -\frac{1 - (\rho_+ c_+)/(\rho_- c_-)}{1 + (\rho_+ c_+)/(\rho_- c_-)}$$
$$c^{(t)} = 2\frac{(\rho_+ c_+)/(\rho_- c_-)}{1 + (\rho_+ c_+)/(\rho_- c_-)}$$

$$c_+ = \sqrt{\frac{\lambda_+ + 2\mu_+}{\rho_+}}, \quad c_- = \sqrt{\frac{\lambda_- + 2\mu_-}{\rho_-}}$$

The dimensionless constants $c^{(r)}$ and $c^{(t)}$ are called the reflection and transmission coefficients for the one-dimensional stress wave motion in the nonhomogeneous two semispace medium.
Hint. Use the relation $S(x,t) = -\rho\, c\, \dot{u}(x,t)$ from Problem 12.1 for a stress wave propagating in the positive direction of x with a velocity c to satisfy the particle velocity continuity condition at the interface $x = 0$.

Solution. Let $S = S(x,t)$ be defined by

$$S(x,t) = \begin{cases} S^{(i)}(t - x/c_-) + S^r(t + x/c_-), & x \le 0 \\ S^{(t)}(t - x/c_+) & x \ge 0 \end{cases} \tag{12.104}$$

where $S^{(i)} = S^{(i)}(s)$ is a prescribed function on $[0, \infty)$ such that $S^{(i)}(s) = 0$ for $s < 0$, and

$$S^{(r)}(t + x/c_-) = c^{(r)}\, S^{(i)}(t + x/c_-) \tag{12.105}$$

$$S^{(t)}(t - x/c_+) = c^{(t)}\, S^{(i)}(t - x/c_+) \tag{12.106}$$

where the reflection and transmission coefficients $c^{(r)}$ and $c^{(t)}$, are given by

$$c^{(r)} = -\frac{1 - \varkappa}{1 + \varkappa}, \quad c^{(t)} = 2\frac{\varkappa}{1 + \varkappa} \tag{12.107}$$

with

$$\varkappa = \frac{\rho_+ c_+}{\rho_- c_-} \tag{12.108}$$

For a stress wave propagating in the positive direction of x with a velocity c the particle velocity formula reads

$$\dot{u}(x,t) = -\frac{1}{\rho c} S(x,t) \tag{12.109}$$

Hence

$$\dot{u}^{(i)}(x,t) = -\frac{1}{\rho_- c_-} S^{(i)}(x,t) \tag{12.110}$$

$$\dot{u}^{(r)}(x,t) = +\frac{1}{\rho_- c_-} S^{(r)}(x,t) \tag{12.111}$$

and

$$\dot{u}^{(t)}(x,t) = -\frac{1}{\rho_+ c_+} S^{(t)}(x,t) \tag{12.112}$$

where $\dot{u}^{(i)}$, $\dot{u}^{(r)}$, and $\dot{u}^{(t)}$, respectively, are the particle velocity associated with the stress waves $S^{(i)}$, $S^{(r)}$, and $S^{(t)}$. Therefore, to solve Problem 12.6 it is sufficient to show that S given by (12.104)–(12.108) satisfies the interface conditions at $x = 0$

$$S^{(i)}(t) + S^{(r)}(t) = S^{(t)}(t), \ t \geq 0 \tag{12.113}$$

$$\dot{u}^{(i)}(t) + \dot{u}^{(r)}(t) = \dot{u}^{(t)}(t), \ t \geq 0 \tag{12.114}$$

Substituting $S^{(r)}$ and $S^{(t)}$, respectively, from (12.105) and (12.106) into (12.113) and (12.114) in which (12.110)–(12.112) are taken into account and dividing by $S^{(i)}(t) \neq 0$, we obtain

$$\begin{gathered} 1 + c^{(r)} = c^{(t)} \\ -\frac{1}{\rho_- c_-} + \frac{1}{\rho_- c_-} c^{(r)} = -\frac{1}{\rho_+ c_+} c^{(t)} \end{gathered} \tag{12.115}$$

or

$$\begin{gathered} 1 + c^{(r)} = c^{(t)} \\ \varkappa\,(1 - c^{(r)}) = c^{(t)} \end{gathered} \tag{12.116}$$

where $\varkappa$ is defined by Eq. (12.108).

Finally, substituting $c^{(r)}$ and $c^{(t)}$ from (12.107) into (12.116) we find that Eq. (12.116) are identically satisfied. Also note that for $\varkappa = 0$, that is, when $x = 0$ is a free surface, $c^{(t)} = 0$ and $c^{(r)} = -1$, which corresponds to the situation discussed in Problem 12.5. If $\varkappa = 1$, $c^{(r)} = 0$, that is, the pulse is completely transmitted.

This completes a solution to Problem 12.6.

Problem 12.7. Find the dynamic thermal stresses produced by a plane source of heat that varies harmonically with time in an infinite elastic solid.

Hint. Assume the heat source function as $Q(x,t) = e^{i\omega t}\delta(x)$, where $\omega > 0$ is a prescribed frequency, and use Eqs. (12.12) and (12.14).

Solution. Equations (12.12) and (12.14), respectively, take the forms

$$\left(\frac{\partial^2}{\partial x^2} - \frac{\partial}{\partial t}\right) T = -Q \quad \text{for } |x| < \infty, \ t > 0 \tag{12.117}$$

and

$$\left(\frac{\partial^2}{\partial x^2} - \frac{\partial}{\partial t}\right)\left(\frac{\partial^2}{\partial x^2} - \frac{\partial^2}{\partial t^2}\right) S = -\frac{\partial^2 Q}{\partial t^2} \quad \text{for } |x| < \infty, \ t > 0 \tag{12.118}$$

If

$$Q(x,t) = e^{i\omega t}\delta(x) \tag{12.119}$$

a solution (T, S) to Eqs. (12.117) and (12.118) takes the form

$$(T, S) = e^{i\omega t}(\widetilde{T}(x,\omega),\ \widetilde{S}(x,\omega)) \tag{12.120}$$

where $\widetilde{T}$ and $\widetilde{S}$, respectively, satisfy the equations

$$\left(\frac{\partial^2}{\partial x^2} - i\omega\right)\widetilde{T} = -\delta(x) \tag{12.121}$$

and

$$\left(\frac{\partial^2}{\partial x^2} - i\omega\right)\left[\frac{\partial^2}{\partial x^2} - (i\omega)^2\right]\widetilde{S} = -(i\omega)^2\delta(x) \tag{12.122}$$

Proceeding in a way similar to that of section on Green's functions of one-dimensional nonisothermal elastodynamics [see Eqs. (12.15)–(12.35)], we find that

$$\widetilde{T}(x,w) = \frac{1}{2}\frac{e^{-|x|\sqrt{i\omega}}}{\sqrt{i\omega}} \tag{12.123}$$

and

$$\widetilde{S}(x,w) = \frac{1}{2}\frac{i\omega}{i\omega - 1}\left[\frac{e^{-i\omega|x|}}{i\omega} - \frac{e^{-|x|\sqrt{i\omega}}}{\sqrt{i\omega}}\right] \tag{12.124}$$

Note that Eqs. (12.123) and (12.124) are identical to Eqs. (12.21) and (12.29), respectively, when $p = i\omega$.

Therefore

$$T(x,t) = \frac{1}{2\sqrt{i\omega}}e^{i\omega t - |x|\sqrt{i\omega}} \tag{12.125}$$

and

$$S(x,t) = \frac{1}{2}\frac{i\omega}{i\omega - 1}e^{i\omega t}\left[\frac{e^{-i\omega|x|}}{i\omega} - \frac{e^{-|x|\sqrt{i\omega}}}{\sqrt{i\omega}}\right] \tag{12.126}$$

By letting

$$\sqrt{i} = \frac{1}{\sqrt{2}}(1+i) \tag{12.127}$$

and substituting (12.127) into (12.125) and (12.126), respectively, we obtain

$$T(x,t) = \frac{1}{2\sqrt{\omega}}e^{-|x|\sqrt{\frac{\omega}{2}}} \times \exp\left[i\left(\omega t - |x|\sqrt{\frac{\omega}{2}} - \frac{\pi}{4}\right)\right] \tag{12.128}$$

and

$$S(x,t) = -\frac{1}{2(\omega^2+1)}\left\{(1+i\omega)\exp[i\omega(t-|x|)] + \sqrt{\omega}(\omega - i) \times \exp\left[-|x|\sqrt{\frac{\omega}{2}} + i\left(\omega t - |x|\sqrt{\frac{\omega}{2}} - \frac{\pi}{4}\right)\right]\right\} \tag{12.129}$$

If the heat source function is assumed in the form

$$Q^*(x,t) = Re[Q(x,t)] \tag{12.130}$$

where Re z denote the real part of z, then by taking the real parts of (12.128) and (12.129), respectively, we obtain

$$T^*(x,t) = \frac{1}{2\sqrt{\omega}} e^{-|x|\sqrt{\frac{\omega}{2}}} \times \cos\left(\omega t - |x|\sqrt{\frac{\omega}{2}} - \frac{\pi}{4}\right) \tag{12.131}$$

and

$$S^*(x,t) = -\frac{1}{2(1+\omega^2)}\left\{\cos\omega(t-|x|) - \omega\sin\omega(t-|x|) + \sqrt{\omega}e^{-|x|\sqrt{\frac{\omega}{2}}} \times \left[\omega\cos\left(\omega t - |x|\sqrt{\frac{\omega}{2}} - \frac{\pi}{4}\right) + \sin\left(\omega t - |x|\sqrt{\frac{\omega}{2}} - \frac{\pi}{4}\right)\right]\right\} \tag{12.132}$$

Equation (12.132) describes the dynamic thermal stress produced by the plane heat source of the form

$$Q^*(x,t) = \cos\omega t\,\delta(x) \tag{12.133}$$

in an infinite elastic solid $|x| < \infty$. This completes a solution to Problem 12.7.

Problem 12.8. Find the dynamic thermal stresses produced by a plane heat source of the form $Q(x,t) = H(t)\delta(x)$ in an infinite elastic solid which is initially at rest. Here $H = H(t)$ and $\delta = \delta(x)$ are the Heaviside and delta functions, respectively.

Solution. The dynamic thermal stress produced by a plane heat source of the form $Q(x,t) = H(t)\,\delta(x)$ in an infinite elastic solid that is initially at rest, satisfies the equation [see Eq. (12.14)]

$$\left(\frac{\partial^2}{\partial x^2} - \frac{\partial}{\partial t}\right)\left(\frac{\partial^2}{\partial x^2} - \frac{\partial^2}{\partial t^2}\right) S = -\frac{\partial^2}{\partial t^2}[H(t)\,\delta(x)] \quad \text{on } (-\infty, +\infty) \times (0, \infty) \tag{12.134}$$

subject to the initial conditions

$$S\,(x,0) = \frac{\partial}{\partial t} S\,(x,0) = \frac{\partial^2}{\partial t^2} S\,(x,0) = 0, \quad |x| < \infty \tag{12.135}$$

and the vanishing condition

$$S(x,t) \to 0 \quad \text{as } |x| \to \infty,\ t > 0 \tag{12.136}$$

Let $\overline{f} = \overline{f}(x, p)$ denote the Laplace transform of a function $f = f(x,t)$

$$Lf \equiv \overline{f}(x,p) = \int_0^\infty e^{-pt} f(x,t)dt \tag{12.137}$$

where p is the transform parameter. Applying the operator L to (12.134), and taking into account the conditions (12.135) we obtain

$$\left(\frac{\partial^2}{\partial x^2} - p\right)\left(\frac{\partial^2}{\partial x^2} - p^2\right)\overline{S} = -p\,\delta(x) \quad \text{for } |x| < \infty,\ p > 0 \tag{12.138}$$

and

$$\overline{S}(x,p) \to 0 \quad \text{as } |x| \to \infty, p > 0 \tag{12.139}$$

Next, proceeding in a way similar to that of section on Green's functions of one-dimensional nonisothermal elastodynamics [see Eqs. (12.15)–(12.35)], we obtain, [see Eq. (12.29) divided by p]

$$\overline{S}(x,p) = \frac{1}{2p}\left[\frac{e^{-|x|p}}{p-1} + \frac{\partial}{\partial |x|}\frac{e^{-|x|\sqrt{p}}}{p-1}\right] \tag{12.140}$$

Since

$$\frac{1}{p(p-1)} = \frac{1}{p-1} - \frac{1}{p} \tag{12.141}$$

Equation (12.140) can also be written as

$$\overline{S}(x,p) = \frac{1}{2}\left(\frac{e^{-|x|p}}{p-1} - \frac{e^{-|x|p}}{p}\right) + \frac{1}{2}\frac{\partial}{\partial |x|}\left(\frac{e^{-|x|\sqrt{p}}}{p-1} - \frac{e^{-|x|\sqrt{p}}}{p}\right) \tag{12.142}$$

Now

$$L^{-1}\frac{e^{-|x|p}}{p-1} = e^{t-|x|}\,H(t-|x|) \tag{12.143}$$

$$L^{-1}\frac{e^{-|x|p}}{p} = H(t-|x|) \tag{12.144}$$

$$L^{-1}\frac{e^{-|x|\sqrt{p}}}{p-1} = U(|x|,\ t) \tag{12.145}$$

where

$$U(x,t) = \frac{1}{2}e^t \left\{ e^{-|x|}\text{erfc}\left[\frac{|x|}{2\sqrt{t}} - \sqrt{t}\right] + e^{|x|}\text{erfc}\left[\frac{|x|}{2\sqrt{t}} + \sqrt{t}\right]\right\} \tag{12.146}$$

and

$$L^{-1}\frac{e^{-|x|\sqrt{p}}}{p} = \text{erfc}\left(\frac{|x|}{2\sqrt{t}}\right) \tag{12.147}$$

In Eqs. (12.146) and (12.147)

$$\text{erfc}\,(z) = \frac{2}{\sqrt{\pi}} \int_z^\infty e^{-\xi^2} d\xi \tag{12.148}$$

Hence, by applying the operator L^{-1} to Eq. (12.142), we obtain

$$S(x,t) = \frac{1}{2}\{[e^{t-|x|} - 1]H(t-|x|) - V(|x|,t)\} \tag{12.149}$$

where

$$V(|x|,t) = \frac{1}{2}e^t \left\{ e^{-|x|}\text{erfc}\left[\frac{|x|}{2\sqrt{t}} - \sqrt{t}\right] - e^{|x|}\text{erfc}\left[\frac{|x|}{2\sqrt{t}} + \sqrt{t}\right]\right\} \tag{12.150}$$

It follows from (12.149) that

$$S(x,t) = S_w(x,t) + S_d(x,t) \tag{12.151}$$

where S_w and S_d represent the wave and diffusive parts of S, respectively, defined by

$$S_w(x,t) = \frac{1}{2}\left[e^{t-|x|} - 1\right] H(t-|x|) \tag{12.152}$$

and

$$S_d(x,t) = -\frac{1}{2}V(|x|,\ t) \tag{12.153}$$

The function S_w represents a wave with the two plane fronts propagating with a unit velocity in the opposite directions: $t + x = 0$ and $t - x = 0$ for every $|x| < \infty$ and $t > 0$; and the wave enters an undisturbed region in a continuous manner, that is.

$$S_w(x,\ |x|+0) - S_w(x,\ |x|-0) = 0 \tag{12.154}$$

On the other hand, the diffusive part S_d is felt instantaneously at any distance $|x|$ from the heat source plane $x = 0$.

Finally, using the limit formula

$$\overline{S}(x, p)p \to S(x, +\infty) \quad \text{as } p \to 0 \tag{12.155}$$

we find

$$S(x, +\infty) = -\frac{1}{2} \quad \text{for } |x| < \infty \tag{12.156}$$

which means that the dynamic stress attains a steady state as time goes to infinity.

This completes a solution to Problem 12.8.

Problem 12.9. Find the dynamic thermal stresses in a semispace $x \geq 0$ subject to the boundary heating $T(0, t) = T_0 t^2 \exp(-a\,t)$ $(a > 0,\ T_0 > 0)$ when the boundary is stress free, the body is initially at rest, and both the temperature $T = T(x, t)$ and the induced stress $S = S(x, t)$ vanish as $x \to \infty$ for $t > 0$.

Solution. A temperature $T = T(x, t)$ is to satisfy the field equation [see Eq. (12.12) with $Q = 0$]

$$\left(\frac{\partial^2}{\partial x^2} - \frac{\partial}{\partial t}\right) T = 0 \quad \text{for } x > 0,\ t > 0 \tag{12.157}$$

the initial condition

$$T(x, 0) = 0 \quad \text{for } x > 0 \tag{12.158}$$

the boundary condition

$$T(0, t) = T_0\, t^2 e^{-at}, \quad a > 0,\ t > 0 \tag{12.159}$$

and the vanishing condition at infinity

$$T(x, t) \to 0 \quad \text{as } x \to \infty,\ t > 0 \tag{12.160}$$

The associated stress $S = S(x, t)$ is to satisfy the field equation [see Eq. (12.13)]

$$\left(\frac{\partial^2}{\partial x^2} - \frac{\partial^2}{\partial t^2}\right) S = \frac{\partial^2 T}{\partial t^2} \quad \text{for } x > 0,\ t > 0 \tag{12.161}$$

subject to the conditions

$$S(x, 0) = 0, \quad \frac{\partial S}{\partial t}(x, 0) = 0 \quad \text{for } x > 0 \tag{12.162}$$

$$S(0, t) = 0 \quad \text{for } t \geq 0 \tag{12.163}$$

and

$$S(x, t) \to 0 \quad \text{as } x \to \infty,\ t > 0 \tag{12.164}$$

Let $\overline{f}(p) = \overline{f}$ denote the Laplace transform of a function $f = f(t)$

$$Lf \equiv \overline{f}(p) = \int_0^\infty e^{-pt} f(t)dt \tag{12.165}$$

where p is the transform parameter. Then, applying the Laplace transform to Eqs. (12.157), (12.159)–(12.161), (12.163), and (12.164), respectively, and using (12.158) and (12.162), we obtain

$$\left(\frac{\partial^2}{\partial x^2} - p\right)\overline{T} = 0 \quad \text{for } x > 0 \tag{12.166}$$

$$\overline{T}(0, p) = \frac{2T_0}{(p+a)^3}, \quad p > 0, \tag{12.167}$$

$$\overline{T}(x, p) \to 0 \quad \text{as } x \to \infty, \ p > 0 \tag{12.168}$$

$$\left(\frac{\partial^2}{\partial x^2} - p^2\right)\overline{S} = p^2\overline{T}, \quad x > 0, \ p > 0 \tag{12.169}$$

$$\overline{S}(0, p) = 0 \quad p > 0 \tag{12.170}$$

and

$$\overline{S}(x, p) \to 0 \quad \text{as } x \to \infty, \ p > 0 \tag{12.171}$$

Hence, we obtain

$$\overline{T}(x, p) = T_0 \frac{\partial^2}{\partial a^2}\left(\frac{e^{-x\sqrt{p}}}{p+a}\right) \tag{12.172}$$

and

$$\overline{S}(x, p) = T_0 \frac{\partial^2}{\partial a^2}\left[\frac{p(e^{-xp} - e^{-x\sqrt{p}})}{(p-1)(p+a)}\right] \tag{12.173}$$

or by letting $a = -\omega > 0$ in (12.172) and (12.173), respectively, we get

$$\overline{T}(x, p) = T_0 \frac{\partial^2}{\partial \omega^2}\left(\frac{e^{-x\sqrt{p}}}{p-\omega}\right) \tag{12.174}$$

and

$$\overline{S}(x, p) = T_0 \frac{\partial^2}{\partial \omega^2}\left[\frac{p(e^{-xp} - e^{-x\sqrt{p}})}{(p-1)(p-\omega)}\right] \tag{12.175}$$

Since

$$\frac{p}{(p-1)(p-\omega)} = \frac{1}{1-\omega}\left(\frac{1}{p-1} - \frac{\omega}{p-\omega}\right) \quad \omega \neq 1 \tag{12.176}$$

therefore, an alternative form of $\overline{S}$ reads

$$\overline{S}(x,p) = T_0 \frac{\partial^2}{\partial\omega^2}\left\{\frac{1}{1-\omega}\left(\frac{1}{p-1} - \frac{\omega}{p-\omega}\right)\right\} \times (e^{-xp} - e^{-x\sqrt{p}}) \tag{12.177}$$

Now

$$L^{-1}\left(\frac{e^{-xp}}{p-\omega}\right) = H(t-x)e^{\omega(t-x)} \tag{12.178}$$

and

$$L^{-1}\left(\frac{e^{-x\sqrt{p}}}{p-\omega}\right) = U(x,t;\omega) \tag{12.179}$$

where $U(x,t,w)$ is defined by

$$U(x,t;\omega) = \frac{1}{2}e^{\omega t}\left\{e^{-x\sqrt{\omega}}\text{erfc}\left[\frac{x}{2\sqrt{t}} - \sqrt{\omega t}\right] + e^{x\sqrt{\omega}}\text{erfc}\left[\frac{x}{2\sqrt{t}} + \sqrt{\omega t}\right]\right\} \tag{12.180}$$

Hence, applying the operator L^{-1} to Eqs. (12.174) and (12.177), respectively, we obtain

$$T\,(x,t) = T_0 \frac{\partial^2 U}{\partial\omega^2}(x,t;\omega) \tag{12.181}$$

and

$$\begin{aligned} S(x,t) = T_0 \frac{\partial^2}{\partial\omega^2}\Bigg\{&\frac{1}{1-\omega}[e^{t-x} - \omega\, e^{\omega(t-x)}]H(t-x) \\ &- \frac{1}{1-\omega}[U(x,t;1) - \omega U(x,t;\omega)]\Bigg\} \end{aligned} \tag{12.182}$$

Substituting T from (12.181) into Eqs. (12.157)–(12.160) we find that T is a solution to the heat conduction problem (12.157)–(12.160). Also, substituting S from (12.182) into Eqs. (12.161)–(12.164) we check that S is a solution to Problem (12.161)–(12.164).

This completes a solution to problem 12.9.

Problem 12.10. An instantaneous nucleus of thermoelastic strain distributed over the plane $x = 0$ in an infinite solid can be identified with the temperature

$$T^*(x,t) = \delta(x)\delta(t) \quad \text{for} \quad |x| < \infty, \quad t > 0$$

Find the dynamic thermal stresses $S^* = S^*(x,t)$ produced by the nucleus provided the infinite body is initially at rest, that is, $S^*(x,0) = 0, \;\; \dot{S}^*(x,0) = 0 \;\; \text{for} \;\; |x| < \infty$.

Solution. The dynamic thermal stress $S^* = S^*(x,t)$ produced by the nucleus of thermoelastic strain in an infinite body, that is initially at rest, satisfies the equation [see Eq. (12.13)] with $T = T^*$ and $S = S^*$]

$$\left(\frac{\partial^2}{\partial x^2} - \frac{\partial^2}{\partial t^2}\right) S^* = \frac{\partial^2}{\partial t^2} [\delta(x)\, \delta(t)] \quad \text{for } |x| < \infty,\ t > 0 \tag{12.183}$$

subject to the initial conditions

$$S^*(x, 0) = 0, \quad \frac{\partial S^*}{\partial t}(x, 0) = 0, \quad |x| < \infty \tag{12.184}$$

and the vanishing condition at infinity

$$S^*(x, t) \to 0 \quad \text{as } |x| \to \infty,\ t > 0 \tag{12.185}$$

Let $\overline{S}^*(x, p)$ be the Laplace transform of $S^*(x, t)$, that is,

$$LS^* \equiv \overline{S}^*(x, p) = \int_0^\infty S^*(x, t)\, e^{-pt}\, dt \tag{12.186}$$

where p is the transform parameter. Applying the operator L to Eq. (12.183) and using the conditions (12.184) we obtain

$$\left(\frac{\partial^2}{\partial x^2} - p^2\right) \overline{S}^* = p^2\, \delta(x) \quad \text{for } |x| < \infty \tag{12.187}$$

Now using the integral representation of $\delta(x)$

$$\delta(x) = \frac{1}{\pi} \int_0^\infty \cos \alpha\, x\, d\alpha \tag{12.188}$$

we find that the only solution $\overline{S}^*$ to Eq. (12.187) that vanishes at $|x| = \infty$ takes the form

$$\overline{S}^*(x, p) = -\frac{p^2}{\pi} \int_0^\infty \frac{\cos \alpha x}{\alpha^2 + p^2} d\alpha = -\frac{1}{2}\, p\, e^{-|x|p}, \quad p > 0 \tag{12.189}$$

Finally, applying the operator L^{-1} to Eq. (12.189) we obtain

$$S^*(x, t) = -\frac{1}{2}\, \delta'(t - |x|), \quad |x| < \infty, \quad t > 0 \tag{12.190}$$

This completes a solution to Problem 12.10.

Problem 12.11. Find the dynamic thermal stresses in a semispace $x \geq 0$ due to the action of an instantaneous nucleus of thermoelastic strain distributed over the plane $x = x_0 > 0$, when the boundary $x = 0$ is stress free, and the semispace is initially at rest.

Solution. We are to find a solution $S = S(x, x_0; t)$ to the field equation [see Eq. (12.13) with $T = \delta(x - x_0)\,\delta(t)$]

$$\left(\frac{\partial^2}{\partial x^2} - \frac{\partial^2}{\partial t^2}\right) S = \frac{\partial^2}{\partial t^2}\,[\delta(x - x_0)\,\delta(t)] \quad \text{for } x \geq 0,\ x_0 > 0,\ t > 0 \tag{12.191}$$

subject to the homogeneous initial conditions

$$S(x, x_0; 0) = 0, \quad \frac{\partial S}{\partial t}(x, x_0; 0) = 0 \quad \text{for } x \geq 0,\ x_0 > 0 \tag{12.192}$$

the homogeneous boundary condition

$$S(0, x_0; t) = 0 \quad \text{for } x_0 > 0,\ t > 0 \tag{12.193}$$

and the vanishing condition at infinity

$$S(x, x_0; t) \to 0 \quad \text{as } x \to \infty,\ x_0 > 0,\ t > 0 \tag{12.194}$$

Let $\overline{S} = \overline{S}(x, x_0;\, p)$ be the Laplace transform of $S = S(x, x_0; t)$ defined by

$$LS = \overline{S}(x, x_0;\, p) = \int_0^\infty e^{-pt}\, S(x, x_0; t)dt \tag{12.195}$$

where p is the transform parameter. Applying the L operator to Eqs. (12.191), (12.193) and (12.194), respectively, and using (12.192) we obtain

$$\left(\frac{\partial^2}{\partial x^2} - p^2\right) \overline{S} = p^2 \delta(x - x_0) \quad \text{for } x \geq 0,\ x_0 > 0,\ p > 0 \tag{12.196}$$

$$\overline{S}(0, x_0;\, p) = 0, \quad x_0 > 0,\ p > 0 \tag{12.197}$$

and

$$\overline{S}(x, x_0;\, p) \to 0 \quad \text{as } x \to \infty,\ x_0 > 0,\ p > 0 \tag{12.198}$$

It follows from the solution to Problem 12.10 that

$$\left(\frac{\partial^2}{\partial x^2} - p^2\right) \left\{-\frac{1}{2} p\, e^{-|x - x_0| p}\right\} = p^2 \delta(x - x_0) \tag{12.199}$$

therefore, a solution $\overline{S}$ to problem (12.196)–(12.198) is sought in the form

$$\overline{S}(x, x_0; p) = A\, e^{-xp} - \frac{1}{2} p\, e^{-|x-x_0|p} \tag{12.200}$$

where A is a constant to be selected in such a way that $\overline{S}$ meets (12.197).

Substituting $\overline{S}$ from (12.200) into (12.197) we obtain

$$A = \frac{1}{2} p\, e^{-x_0 p} \tag{12.201}$$

As a result, Eq. (12.200) takes the form

$$\overline{S}(x, x_0; p) = -\frac{1}{2} p\, [e^{-|x-x_0|p} - e^{-(x+x_0)p}] \tag{12.202}$$

Finally, applying the operator L^{-1} to (12.202) we obtain

$$S(x, x_0; t) = -\frac{1}{2} \left\{\delta'[t - |x - x_0|] - \delta'[t - (x + x_0)]\right\} \quad \text{for } x \geq 0,\ x_0 > 0,\ t \geq 0 \tag{12.203}$$

If we note that

$$\delta(t) \sim \frac{1}{\varepsilon\sqrt{\pi}} e^{-\left(\frac{t^2}{\varepsilon^2}\right)} \quad \text{as } \varepsilon \to 0 \tag{12.204}$$

and

$$\delta'(t) \sim -\frac{2t}{\varepsilon^3\sqrt{\pi}} e^{-\left(\frac{t^2}{\varepsilon^2}\right)} \quad \text{as } \varepsilon \to 0 \tag{12.205}$$

the asymptotic form of S is obtained

$$\begin{aligned} S(x, x_0; t) \sim \frac{1}{\sqrt{\pi}} \frac{1}{\varepsilon^3} \Big\{ &[t - |x - x_0|] \exp\left[-(t - |x - x_0|)^2 / \varepsilon^2\right] \\ &- [t - (x + x_0)] \exp\left[-(t - (x + x_0))^2 / \varepsilon^2\right]\Big\} \\ &\text{for } x \geq 0,\ x_0 > 0,\ t > 0 \end{aligned} \tag{12.206}$$

By letting ε to be a small dimensionless number in (12.206) a graph of S for a fixed $x_0 > 0$ and for $x \geq 0,\ t \geq 0$ may be obtained.

This completes a solution to Problem 12.11.

Problem 12.12. Let $S = S(x, x_0; t)$ be the stress field obtained in Problem 12.11. Show that the dynamic thermal stress $S = S(x, t)$ produced in a semispace with free boundary by an arbitrary temperature field $T = T(x, t)$ on $[0, \infty) \times [0, \infty)$ takes the form

$$S(x,t)=\int_0^t\int_0^\infty S(x,x_0;t-\tau)\,T(x_0,\tau)dx_0d\tau$$

Also, use the formula to get the closed-form solution to Problem 12.9.

Solution. Let

$$S(x,t)=\int_0^t\int_0^\infty S(x,x_0;t-\tau)T(x_0,\tau)dx_0d\tau \quad x\geq 0,\ t\geq 0 \tag{12.207}$$

Applying the L operator to (12.207) we obtain

$$\overline{S}(x,p)=\int_0^\infty \overline{S}(x,x_0;p)\overline{T}(x_0,p)dx_0 \tag{12.208}$$

We are to show that $\overline{S}$ given by (12.208) satisfies the equation

$$\left(\frac{\partial^2}{\partial x^2}-p^2\right)\overline{S}(x,p)=p^2\overline{T}(x,p) \quad \text{for } x\geq 0,\ p>0 \tag{12.209}$$

subject to the conditions

$$\overline{S}(0,p)=0 \tag{12.210}$$

and

$$\overline{S}(x,p)\to 0 \quad \text{as } x\to\infty,\ p>0 \tag{12.211}$$

Multiplying Eq. (12.212) of the solution to Problem 12.11 by $\overline{T}(x_0,p)$ we obtain

$$\left(\frac{\partial^2}{\partial x^2}-p^2\right)\overline{S}(x,x_0;p)\,\overline{T}(x_0,p)=p^2\delta(x-x_0)\,\overline{T}\,(x,p) \tag{12.212}$$

Next, integrating (12.212) with respect to x_0 from $x_0=0$ to $x_0=\infty$, we get

$$\left(\frac{\partial^2}{\partial x^2}-p^2\right)\int_0^\infty \overline{S}(x,x_0;p)\overline{T}(x_0,p)dx_0=p^2\overline{T}(x,p) \tag{12.213}$$

Therefore, $\overline{S}$ given by (12.208) meets (12.209). In addition, since $\overline{S}(x,x_0;p)$ satisfies the boundary condition

$$\overline{S}(0,x_0;p)=0, \quad x_0>0,\ p>0 \tag{12.214}$$

and the asymptotic condition

$$\overline{S}(x, x_0; p) \to 0 \quad \text{as } x \to \infty \quad x_0 > 0, \ p > 0 \tag{12.215}$$

$\overline{S}$ given by (12.208) also satisfies (12.210) and (12.211). This shows that $S = S(x,t)$ produced in a semispace with free boundary by an arbitrary temperature $T = T(x,t)$ and corresponding to the homogeneous initial conditions takes the form (12.207).

To obtain the closed-form solution to Problem 12.9 [see Eq. (12.182) of the solution to Problem 12.9], using (12.208), we substitute $\overline{S}(x, x_0; p)$ and $\overline{T}(x_0, p)$ from Eq. (12.218) of the solution to Problem 12.11 and (12.174) of the solution to Problem 12.9, respectively, into (12.208), and obtain

$$\begin{aligned}\overline{S}(x, p) &= -\frac{T_0}{2} p \int_0^\infty [e^{-|x-x_0|p} - e^{-(x+x_0)p}] \times \frac{\partial^2}{\partial \omega^2}\left(\frac{e^{-x_0\sqrt{p}}}{p-\omega}\right) dx_0 \\ &= -\frac{T_0}{2}\frac{\partial^2}{\partial \omega^2}\frac{p}{p-\omega}\int_0^\infty [e^{-|x-x_0|p} - e^{-(x+x_0)p}] e^{-x_0\sqrt{p}} dx_0.\end{aligned} \tag{12.216}$$

Calculating the integral on the RHS of (12.216) we obtain

$$\int_0^\infty [e^{-|x-x_0|p - x_0\sqrt{p}} - e^{-(x+x_0)p - x_0\sqrt{p}}] dx_0 = -\frac{2}{p-1}(e^{-xp} - e^{-x\sqrt{p}}) \tag{12.217}$$

Finally, substituting (12.217) into the RHS of (12.216) we get

$$\overline{S}(x, p) = T_0 \frac{\partial^2}{\partial \omega^2}\left\{\frac{p(e^{-xp} - e^{-x\sqrt{p}})}{(p-1)(p-\omega)}\right\} \tag{12.218}$$

Equation (12.218) is identical to Eq. (12.175) of the solution to Problem 12.9. Therefore, applying the operator L^{-1} to (12.218) we arrive at the closed-form solution of Problem 12.9 [see Eq. (12.182) of the solution to Problem 12.9].

This completes a solution to Problem 12.2.

Part II
Thermal Stresses

Chapter 13
Thermal Stresses in Bars

In this chapter the concept of thermal stresses in bars is introduced for the simple case of a perfectly clamped bar subjected to arbitrary temperature change. The problems and solutions related to thermal stresses in bars are: a perfectly clamped bar, a clamped bar with a small gap, a clamped circular frustum, a bar with variable cross-sectional area, two bars attached to each other, three bars fastened to each other, truss of three bars, and three bars hanging from a rigid plate.

13.1 Thermal Stresses in Bars

When the temperature of a circular bar of length l changes from an initial temperature T_0 to its final temperature T_1, the free thermal elongation λ_T of the bar is defined by

$$\lambda_T = \alpha(T_1 - T_0)l = \alpha\tau l \tag{13.1}$$

where α is the coefficient of linear thermal expansion which is measured in one per one degree of the temperature 1/K, and τ denotes the temperature change given by

$$\tau = T_1 - T_0 \tag{13.2}$$

The free thermal strain is given by

$$\epsilon_T = \frac{\lambda_T}{l} = \alpha\tau \tag{13.3}$$

When an internal force and the temperature change act simultaneously in the bar, the normal strain is given by

$$\epsilon = \epsilon_s + \epsilon_T \tag{13.4}$$

M. Reza Eslami et al., *Theory of Elasticity and Thermal Stresses*, Solid Mechanics and Its Applications 197, DOI: 10.1007/978-94-007-6356-2_13,

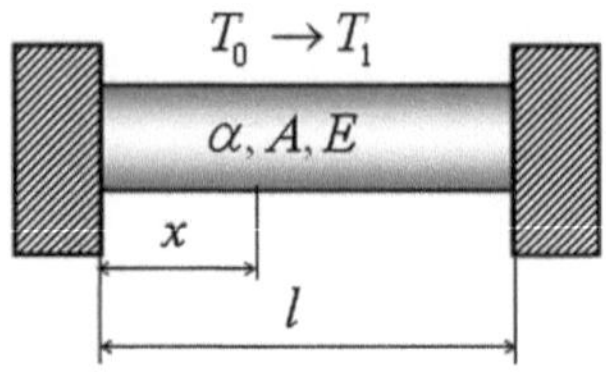

Fig. 13.1 A perfectly clampled bar

where ϵ_s denotes the strain produced by the internal force. The strain ϵ_s produced by the internal force is proportional to the normal stress σ

$$\epsilon_s = \frac{\sigma}{E} \tag{13.5}$$

where E denotes Young's modulus.

Hooke's law with the temperature change is

$$\epsilon = \frac{\sigma}{E} + \alpha\tau \tag{13.6}$$

When a perfectly clamped bar with length l and cross-sectional area A, shown in Fig. 13.1, is subjected to the uniform temperature change τ, the thermal stress is

$$\sigma = -\alpha E\tau \tag{13.7}$$

If the temperature change $\tau(x)$ is a function of the position x, the free thermal elongation λ_T of the bar of length l is

$$\lambda_T = \int d\lambda_T = \int_0^l \alpha\tau(x)\,dx = \alpha\int_0^l \tau(x)\,dx \tag{13.8}$$

The thermal strain ϵ_T is

$$\epsilon_T = \frac{\lambda_T}{l} = \frac{\alpha}{l}\int_0^l \tau(x)\,dx \tag{13.9}$$

The thermal stress in the perfectly clamped bar is

$$\sigma = -\frac{\alpha E}{l}\int_0^l \tau(x)\,dx \tag{13.10}$$

13.2 Problems and Solutions Related to Thermal Stresses in Bars

Problem 13.1. If the temperature in a mild steel rail with length 25 m is raised to 50 K, and the coefficient of linear thermal expansion for mild steel is 11.2×10^{-6} 1/K, what elongation is produced in the rail?

Solution. The elongation λ_T is from Eq. (13.1)

$$\lambda_T = \alpha \tau l = 11.2 \times 10^{-6} \times 50 \times 25 = 14 \times 10^{-3}\,\text{m} = 14\,\text{mm} \qquad \text{(Answer)}$$

Problem 13.2. The temperature of a bar of length 1 m of mild steel is kept at 300 K. If the temperature at one end of the bar is raised to 380 K and at the other end to 480 K, and the temperature distribution is linear along the bar, what elongation is produced in the bar? The coefficient of linear thermal expansion for mild steel is 11.2×10^{-6} 1/K.

Solution. The temperature rise $\tau(x) = T_1(x) - T_0$ is

$$\tau(x) = T_1(x) - T_0 = \left[380 + (480 - 380)\frac{x}{1}\right] - 300 = 80 + 100x \tag{13.11}$$

The free thermal elongation λ_T is

$$\begin{aligned}\lambda_T &= \int_0^1 \alpha\tau(x)dx = \alpha\int_0^1 (80 + 100x)dx \\ &= 11.2 \times 10^{-6} \times \left[80x + 50x^2\right]_0^1 = 1.456 \times 10^{-3}\,\text{m} = 1.46\,\text{mm} \qquad \text{(Answer)}\end{aligned} \tag{13.12}$$

Problem 13.3. A bar of mild steel at 300 K is clamped between two walls. Calculate the thermal stress produced in the bar when the bar is heated to 360 K. The coefficient of linear thermal expansion and Young's modulus are $\alpha = 11.2 \times 10^6$ 1/K and $E = 206$ GPa, respectively.

Solution. The thermal stress σ is from Eq. (13.7)

$$\sigma = -\alpha E \tau = -138.4 \times 10^6\,\text{Pa} = -138\,\text{MPa} \qquad \text{(Answer)}$$

Problem 13.4. In Problem 13.2, calculate the thermal stress produced in the bar if it is clamped between two walls. The coefficient of linear thermal expansion and Young's modulus are $\alpha = 11.2 \times 10^6$ 1/K and $E = 206$ GPa, respectively.

Solution. As the summation of the free thermal elongation λ_T and the elongation λ_s due to the stress is zero, we get

$$\sigma = \frac{E\lambda_s}{l} = -\frac{E\lambda_T}{l} = -\frac{206 \times 10^9 \times 1.456 \times 10^{-3}}{1} = -300\,\text{MPa} \quad \text{(Answer)}$$

Problem 13.5. A bar of mild steel at 300 K is clamped between two walls in such a way that the initial stress is zero. Calculate the temperature when the thermal stress in the bar reaches the compressive strength ($\sigma_{BC} = 400\,\text{MPa}$). The coefficient of linear thermal expansion and Young's modulus are $\alpha = 11.2 \times 10^6$ 1/K and $E = 206\,\text{GPa}$, respectively.

Solution. The compressive thermal stress σ is given by Eq. (13.7). Therefore, the temperature rise τ is

$$\tau = \frac{\sigma_{BC}}{\alpha E} = \frac{400 \times 10^6}{11.2 \times 10^{-6} \times 206 \times 10^9} = 173.37 \tag{13.13}$$

Then

$$T_1 = T_0 + \tau = 300 + 173.37 = 473.37\,\text{K} = 473\,\text{K} \quad \text{(Answer)}$$

Problem 13.6. The temperature of a bar with a small gap $e = 1$ mm, shown in Fig. 13.2 is kept at 300 K. If the temperature at one end of the bar is raised to 380 K and at the other end to 480 K, and the temperature distribution is linear along the bar, calculate the thermal stress. Where length of the bar is 1 m, and the coefficient of linear thermal expansion and Young's modulus are $\alpha = 11.2 \times 10^6$ 1/K and $E = 206\,\text{GPa}$, respectively.

Solution. The free thermal elongation is assumed to be longer than the gap. The summation of elongations due to the free thermal elongation and the elongation due to the stress is equal to the small gap e

$$\int_0^l \alpha\tau(x)dx + \frac{\sigma l}{E} = e \tag{13.14}$$

Then, we get

$$\sigma = -\frac{E}{l}\left[\alpha \int_0^l \tau(x)dx - e\right] \tag{13.15}$$

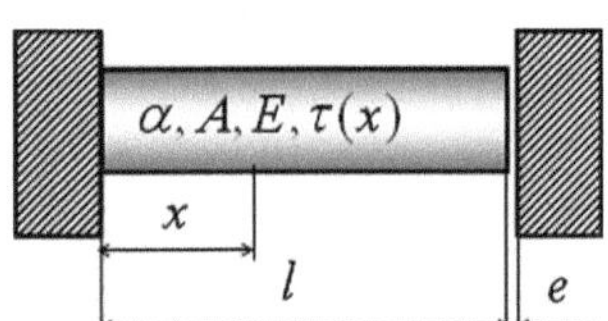

Fig. 13.2 A bar with a small gap

The free thermal expansion is given by Eq. (13.12). Therefore,

$$\sigma = -\frac{206 \times 10^9}{1}\left(1.46 \times 10^{-3} - 1 \times 10^{-3}\right) = -94.8 \times 10^6 = -94.8\,\text{MPa}$$

(Answer)

Problem 13.7. If a clamped circular frustum of mild steel with $d_0 = 1\,\text{cm}$, $d_1 = 2\,\text{cm}$, and $l = 2\,\text{m}$ is subjected to the temperature change $-50\,\text{K}$, calculate the resulting thermal stress. The coefficient of linear thermal expansion and Young's modulus are $\alpha = 11.2 \times 10^6\,1/\text{K}$ and $E = 206\,\text{GPa}$, respectively.

Solution. The free thermal elongation λ_T is

$$\lambda_T = \alpha\tau l \tag{13.16}$$

The cross-sectional area A_x at the position x is given by

$$A_x = \frac{\pi}{4}d_x^2 = \frac{\pi}{4}\left[d_0 + (d_1 - d_0)\frac{x}{l}\right]^2 \tag{13.17}$$

Thus, the strain ϵ_x of the frustum at x due to an internal force Q becomes

$$\epsilon_x = \frac{\sigma_x}{E} = \frac{Q}{EA_x} = \frac{4Q}{E\pi\left[d_0 + (d_1 - d_0)\frac{x}{l}\right]^2} \tag{13.18}$$

and the elongation λ_s of the frustum due to the internal force Q equals

$$\begin{aligned}\lambda_s &= \int d\lambda_s = \int_0^l \epsilon_x\,dx = \int_0^l \frac{4Q}{E\pi\left[d_0 + (d_1 - d_0)\frac{x}{l}\right]^2}\,dx \\ &= -\frac{4Ql}{E\pi(d_1 - d_0)}\left[\frac{1}{d_0 + (d_1 - d_0)\frac{x}{l}}\right]_0^l = \frac{4Ql}{E\pi d_1 d_0}\end{aligned} \tag{13.19}$$

As the frustum is perfectly constrained in the x direction, the combined elongation of the free thermal elongation λ_T and the elongation λ_s due to the internal force Q must be zero

$$\lambda = \lambda_T + \lambda_s = 0 \tag{13.20}$$

From Eqs. (13.16), (13.19), and (13.20) the internal force Q is

$$Q = -\alpha E\tau\frac{\pi}{4}d_1 d_0 \tag{13.21}$$

Then, the thermal stress is

$$\sigma_x = \frac{Q}{A_x} = -\alpha E\tau \frac{d_1 d_0}{\left[d_0 + (d_1 - d_0)\frac{x}{l}\right]^2} \tag{13.22}$$

If $d_1 > d_0$, the maximum thermal stress $(\sigma_x)_{\max}$ occurs at the minimum cross-sectional area and the minimum thermal stress $(\sigma_x)_{\min}$ occurs at the maximum cross-sectional area

$$(\sigma_x)_{\max} = -\alpha E\tau \frac{d_1}{d_0}, \quad (\sigma_x)_{\min} = -\alpha E\tau \frac{d_0}{d_1} \tag{13.23}$$

The thermal stress σ_x is calculated from Eq. (13.22)

$$\begin{aligned} \sigma_x &= -11.2 \times 10^{-6} \times 206 \times 10^9 \times (-50) \\ &\times \frac{1 \times 10^{-2} \times 2 \times 10^{-2}}{\left[1 \times 10^{-2} + (2 \times 10^{-2} - 1 \times 10^{-2})\frac{x}{2}\right]^2} \\ &= \frac{230.7 \times 10^6}{\left(1 + \frac{x}{2}\right)^2}\,\text{Pa} = \frac{231}{\left(1 + \frac{x}{2}\right)^2}\,\text{MPa} \qquad \text{(Answer)} \end{aligned} \tag{13.24}$$

The maximum and minimum thermal stresses are from Eq. (13.24)

$$(\sigma_x)_{\max} = 231\,\text{MPa}, \quad (\sigma_x)_{\min} = \frac{231}{\left(1 + \frac{2}{2}\right)^2}\,\text{MPa} = 57.8\,\text{MPa} \qquad \text{(Answer)}$$

Problem 13.8. If the temperature of a clamped circular frustum of mild steel with $d_0 = 1\,\text{cm}$, $d_1 = 2\,\text{cm}$, and $l = 2\,\text{m}$ changes linearly from 0 K at one end to −50 K at the other end, calculate the resulting thermal stress. The coefficient of linear thermal expansion and Young's modulus are $\alpha = 11.2 \times 10^6\,1/\text{K}$ and $E = 206\,\text{GPa}$, respectively.

Solution. The distribution of the temperature change $\tau(x)$ is

$$\tau(x) = -50\frac{x}{l} \tag{13.25}$$

The free thermal elongation λ_T is

$$\lambda_T = \int_0^l \alpha\tau(x)dx = \alpha\int_0^l \frac{-50x}{l}dx = -\left[\frac{25\alpha x^2}{l}\right]_0^l = -0.56 \times 10^{-3}\,\text{m} \tag{13.26}$$

The cross-sectional area A_x at the position x is given by

$$A_x = \frac{\pi}{4}d_x^2 = \frac{\pi}{4}\left[d_0 + (d_1 - d_0)\frac{x}{l}\right]^2 \tag{13.27}$$

Thus, the strain ϵ_x of the frustum at x due to an internal force Q becomes

$$\epsilon_x = \frac{\sigma_x}{E} = \frac{Q}{EA_x} = \frac{4Q}{E\pi\left[d_0 + (d_1 - d_0)\frac{x}{l}\right]^2} \tag{13.28}$$

and the elongation λ_s of the frustum due to the internal force Q equals

$$\begin{aligned}\lambda_s &= \int d\lambda_s = \int_0^l \epsilon_x\, dx = \int_0^l \frac{4Q}{E\pi\left[d_0 + (d_1 - d_0)\frac{x}{l}\right]^2} dx \\ &= -\frac{4Ql}{E\pi(d_1 - d_0)}\left[\frac{1}{d_0 + (d_1 - d_0)\frac{x}{l}}\right]_0^l = \frac{4Ql}{E\pi d_1 d_0}\end{aligned} \tag{13.29}$$

As the frustum is perfectly constrained in the x direction, the summation of elongation of the free thermal elongation λ_T and the elongation λ_s due to the internal force Q must be zero

$$\lambda = \lambda_T + \lambda_s = 0 \tag{13.30}$$

From Eqs. (13.29) and (13.30) the internal force Q is

$$Q = -\frac{E\pi d_1 d_0 \lambda_T}{4l} \tag{13.31}$$

and the thermal stress is calculated to be

$$\begin{aligned}\sigma_x &= \frac{Q}{A_x} = -\frac{E\pi d_1 d_0 \lambda_T}{4l\frac{\pi}{4}\left[d_0 + (d_1 - d_0)\frac{x}{l}\right]^2} \\ &= \frac{(0.56 \times 10^{-3}) \times (206 \times 10^9) \times (1 \times 10^{-2}) \times (2 \times 10^{-2})}{2 \times \left[1 \times 10^{-2} + (2 \times 10^{-2} - 1 \times 10^{-2})\frac{x}{2}\right]^2} \\ &= \frac{115.36 \times 10^6}{\left(1 + \frac{x}{2}\right)^2}\ \text{Pa} = \frac{115}{\left(1 + \frac{x}{2}\right)^2}\ \text{MPa}\end{aligned} \quad \text{(Answer)}$$

The maximum and minimum thermal stresses are

$$(\sigma_x)_{\max} = 115\,\text{MPa}, \quad (\sigma_x)_{\min} = 28.8\,\text{MPa} \quad \text{(Answer)}$$

Problem 13.9. If a bar with a small gap e between its free end and a rigid wall is subjected to the positive temperature change $\tau(x)$, and the cross-sectional area of the bar is given by $A(x)$, calculate the thermal stress produced in the bar.

Solution. The small elongation $d\lambda(x)$ of the small element dx is

$$d\lambda(x) = \epsilon(x)dx = \left[\frac{\sigma(x)}{E} + \alpha\tau(x)\right]dx \tag{13.32}$$

The elongation λ of the bar with length l is

$$\lambda = \int d\lambda = \int_0^l \left[\frac{\sigma(x)}{E} + \alpha\tau(x)\right]dx = \int_0^l \left[\frac{Q}{EA(x)} + \alpha\tau(x)\right]dx \tag{13.33}$$

in which Q is an internal force. The free thermal elongation is assumed to be longer than the gap. The summation of elongation due to the free thermal elongation and elongation due to the stress is equal to the small gap e

$$\int_0^l \alpha\tau(x)dx + \frac{1}{E}\int_0^l \frac{Q}{A(x)}dx = e \tag{13.34}$$

Then, we get

$$Q = -\frac{E}{\displaystyle\int_0^l \frac{1}{A(x)}dx}\left[\alpha\int_0^l \tau(x)dx - e\right] \tag{13.35}$$

Thermal stress is

$$\sigma = \frac{Q}{A(x)} = -\frac{E}{A(x)\displaystyle\int_0^l \frac{1}{A(x)}dx}\left[\alpha\int_0^l \tau(x)dx - e\right] \qquad \text{(Answer)}$$

The maximum and minimum thermal stresses are

$$(\sigma)_{\max} = -\frac{E}{A(x)_{\min}\displaystyle\int_0^l \frac{1}{A(x)}dx}\left[\alpha\int_0^l \tau(x)dx - e\right]$$

$$(\sigma)_{\min} = -\frac{E}{A(x)_{\max}\displaystyle\int_0^l \frac{1}{A(x)}dx}\left[\alpha\int_0^l \tau(x)dx - e\right] \qquad \text{(Answer)}$$

Problem 13.10. A hollow cylinder with a bar of the same length l and the same centerline, shown in Fig. 13.3 is subjected to different temperature changes τ_i, $(i = 1, 2)$. The hollow cylinder and the bar are connected to two rigid plates. Calculate the thermal stresses produced in both the hollow cylinder and the bar, and the elongations.

Solution. The elongations λ_i due to both the free thermal elongation and the thermal stress are

$$\lambda_1 = \alpha_1\tau_1 l + \frac{\sigma_1}{E_1}l, \quad \lambda_2 = \alpha_2\tau_2 l + \frac{\sigma_2}{E_2}l \tag{13.36}$$

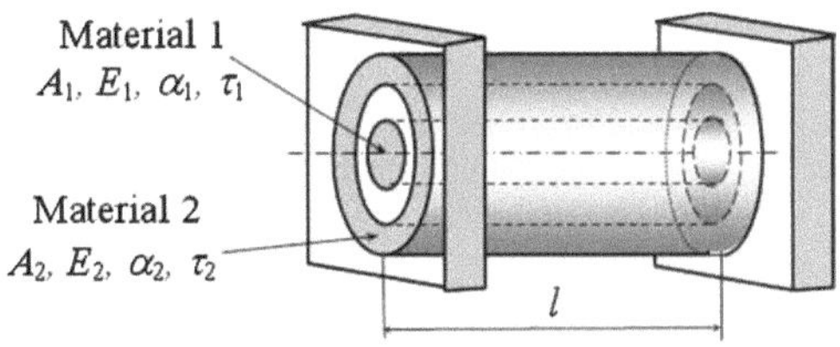

Fig. 13.3 A bar and a hollow cylinder with both ends clamped to rigid plates

where A_i, E_i, and α_i denote cross-sectional area, Young's modulus, and the coefficient of linear thermal expansion of the i-th material, respectively. Since the final length of both the cylinder and the bar after deformation is the same, the following relation holds

$$l + \alpha_1 \tau_1 l + \frac{\sigma_1}{E_1} l = l + \alpha_2 \tau_2 l + \frac{\sigma_2}{E_2} l \tag{13.37}$$

The equilibrium of the internal forces is described by

$$\sigma_1 A_1 + \sigma_2 A_2 = 0 \tag{13.38}$$

Solving Eqs. (13.37) and (13.38) gives the stresses

$$\sigma_1 = -\frac{A_2 E_1 E_2 (\alpha_1 \tau_1 - \alpha_2 \tau_2)}{A_1 E_1 + A_2 E_2}$$
$$\sigma_2 = \frac{A_1 E_1 E_2 (\alpha_1 \tau_1 - \alpha_2 \tau_2)}{A_1 E_1 + A_2 E_2} \qquad \text{(Answer)}$$

Substitution of these stresses into Eq. (13.36) gives the elongations of the cylinder and the bar

$$\lambda_1 = \lambda_2 = \frac{(\alpha_1 \tau_1 E_1 A_1 + \alpha_2 \tau_2 E_2 A_2) l}{A_1 E_1 + A_2 E_2} \qquad \text{(Answer)}$$

Problem 13.11. Two circular bars, one is mild steel of length 50 cm and diameter 1 cm, and the other is aluminum of length 25 cm and diameter 2 cm, are attached to each other in series, placed between rigid walls, and subjected to the temperature change $\tau = T_1 - T_0$, as shown in Fig. 13.4. Calculate the temperature rise needed for the thermal stresses in the bars to reach the compressive strength. The coefficient of linear thermal expansion, Young's modulus and the compressive strength for mild steel are $\alpha_1 = 11.2 \times 10^6$ 1/K, $E_1 = 206$ GPa and 400 MPa, respectively. The coefficient of linear thermal expansion, Young's modulus and the compressive strength for aluminum are $\alpha_2 = 23.1 \times 10^6$ 1/K, $E_2 = 72$ GPa and 70 MPa, respectively.

Solution. The elongations of bar 1 and 2 are, respectively, given by

$$\alpha_1 \tau l_1 + \frac{\sigma_1}{E_1} l_1, \quad \alpha_2 \tau l_2 + \frac{\sigma_2}{E_2} l_2 \tag{13.39}$$

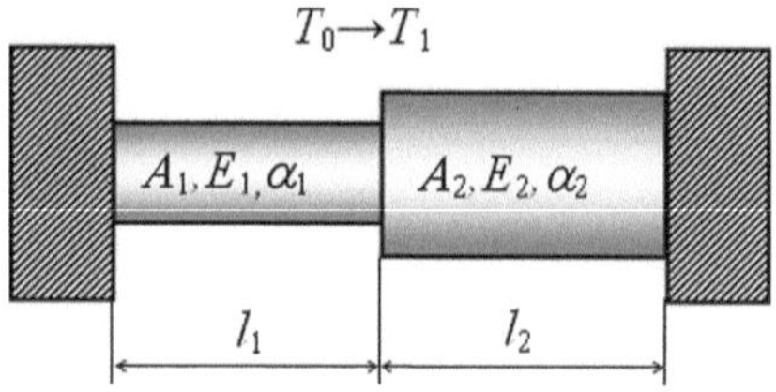

Fig. 13.4 Two bars attached to each other

As two bars are placed between rigid walls, the combined elongation of the bars is zero. Thus,

$$\alpha_1 \tau l_1 + \frac{\sigma_1}{E_1} l_1 + \alpha_2 \tau l_2 + \frac{\sigma_2}{E_2} l_2 = 0 \tag{13.40}$$

From the equilibrium condition of internal forces, the internal force in bar 1 is equal to the internal force in bar 2

$$\sigma_1 A_1 = \sigma_2 A_2 \tag{13.41}$$

From Eqs. (13.40) and (13.41), the thermal stresses σ_1 and σ_2 are given as

$$\sigma_1 = -\frac{\alpha_1 E_1 \tau \left(1 + \dfrac{\alpha_2 l_2}{\alpha_1 l_1}\right)}{1 + \dfrac{A_1 E_1 l_2}{A_2 E_2 l_1}}, \quad \sigma_2 = \sigma_1 \frac{A_1}{A_2} = -\alpha_1 E_1 \tau \frac{A_1}{A_2} \frac{1 + \dfrac{\alpha_2 l_2}{\alpha_1 l_1}}{1 + \dfrac{A_1 E_1 l_2}{A_2 E_2 l_1}} \tag{13.42}$$

Therefore, the necessary temperature rise for bar 1 is

$$\tau = -\frac{\sigma_1}{\alpha_1 E_1} \frac{1 + \dfrac{A_1 E_1 l_2}{A_2 E_2 l_1}}{1 + \dfrac{\alpha_2 l_2}{\alpha_1 l_1}} \tag{13.43}$$

Numerical calculation gives the temperature rise

$$\tau = 115.876\,\mathrm{K} = 116\,\mathrm{K} \tag{13.44}$$

On the other hand, the necessary temperature rise for bar 2 is given by

$$\tau = -\frac{\sigma_2}{\alpha_1 E_1} \frac{A_2}{A_1} \frac{1 + \dfrac{A_1 E_1 l_2}{A_2 E_2 l_1}}{1 + \dfrac{\alpha_2 l_2}{\alpha_1 l_1}} \tag{13.45}$$

Therefore, the necessary temperature rise is

$$\tau = 81.11\,\mathrm{K} = 81\,\mathrm{K} \tag{13.46}$$

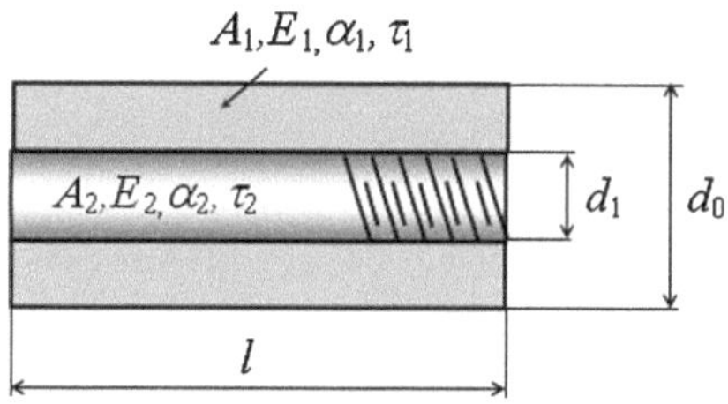

Fig. 13.5 A hollow cylinder with an inserted screw

Then, comparison between Eqs. (13.44) and (13.46) gives the necessary temperature rise 81 K.

Problem 13.12. A hollow cylinder with an inserted screw, shown in Fig. 13.5 is subjected to different temperature changes τ_i, $(i = 1, 2)$. Calculate the thermal stresses produced in both the hollow cylinder and the screw.

Solution. The elongations λ_i due to both the free thermal elongation and the thermal stress are

$$\lambda_1 = \alpha_1 \tau_1 l + \frac{\sigma_1}{E_1} l, \quad \lambda_2 = \alpha_2 \tau_2 l + \frac{\sigma_2}{E_2} l \tag{13.47}$$

where A_i, E_i, and α_i denote cross-sectional area, Young's modulus, and the coefficient of linear thermal expansion of the i-th material, respectively. Since the final length of both the hollow cylinder and the screw after deformation is the same, the following relation holds

$$l + \alpha_1 \tau_1 l + \frac{\sigma_1}{E_1} l = l + \alpha_2 \tau_2 l + \frac{\sigma_2}{E_2} l \tag{13.48}$$

The equilibrium condition of the internal forces is described by

$$\sigma_1 A_1 + \sigma_2 A_2 = 0 \tag{13.49}$$

Solving Eqs. (13.48) and (13.49) gives the thermal stresses

$$\begin{aligned} \sigma_1 &= -\frac{E_1 E_2 A_2 (\alpha_1 \tau_1 - \alpha_2 \tau_2)}{A_1 E_1 + A_2 E_2} \\ \sigma_2 &= \frac{E_1 E_2 A_1 (\alpha_1 \tau_1 - \alpha_2 \tau_2)}{A_1 E_1 + A_2 E_2} \end{aligned} \quad \text{(Answer)}$$

Problem 13.13. A copper tube is fastened by a mild steel bolt, as shown in Fig. 13.6. The length of the tube is 50 cm, and the cross-sectional areas of the bolt and the tube are $A_s = 1\,\text{cm}^2$ and $A_c = 2\,\text{cm}^2$, respectively. Calculate the thermal stresses produced if the system is subjected to the temperature change of 80 K. The coefficient of linear thermal expansion and Young's modulus for mild steel are $\alpha_s = 11.2 \times 10^6$ 1/K and $E_s = 206$ GPa, respectively. The coefficient of linear thermal expansion

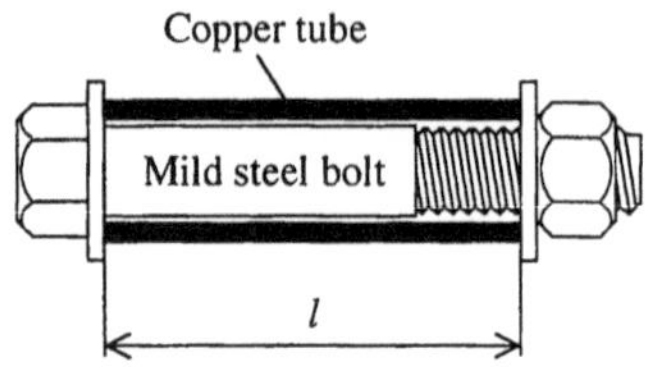

Fig. 13.6 A copper tube fastened by a mild steel bolt

and Young's modulus for copper are $\alpha_c = 16.5 \times 10^6$ 1/K and $E_c = 120\,\text{GPa}$, respectively.

Solution. Since the final length of both the copper tube and the mild steel bolt after deformation is the same, the following relation holds

$$l + \alpha_s \tau l + \frac{\sigma_s}{E_s} l = l + \alpha_c \tau l + \frac{\sigma_c}{E_c} l \tag{13.50}$$

The equilibrium condition of the internal forces is described by

$$\sigma_s A_s + \sigma_c A_c = 0 \tag{13.51}$$

Solving Eqs. (13.50) and (13.51) gives the stresses

$$\sigma_s = -\frac{E_s E_c A_c (\alpha_s - \alpha_c)\tau}{A_s E_s + A_c E_c}, \quad \sigma_c = \frac{E_s E_c A_s (\alpha_s - \alpha_c)\tau}{A_s E_s + A_c E_c} \tag{13.52}$$

The numerical results are

$$\sigma_s = 47.001 \times 10^6\,\text{Pa} = 47\,\text{MPa}, \quad \sigma_c = -23.5\,\text{MPa} \qquad \text{(Answer)}$$

Problem 13.14. In the foregoing problem, calculate the maximum tolerable temperature rise such that stresses in the system do not exceed the compressive or the tensile strength. The tensile strengths of the steel and the copper are $\sigma_{st} = 400\,\text{MPa}$ and $\sigma_{ct} = 300\,\text{MPa}$, respectively. We assume the compressive strength has the same magnitude as the tensile strength. The safety factor (defined by the ratio of yield stress or the tensile strength and the tolerable stress) is $f = 3$.

Solution. The stresses due to the temperature change τ are given by Eq. (13.52), namely

$$\sigma_s = -\frac{E_s E_c A_c (\alpha_s - \alpha_c)\tau}{A_s E_s + A_c E_c}, \quad \sigma_c = \frac{E_s E_c A_s (\alpha_s - \alpha_c)\tau}{A_s E_s + A_c E_c} \tag{13.53}$$

The tolerable stress of a mild steel bolt is

$$\sigma_{sa} = \frac{\sigma_{st}}{f} \tag{13.54}$$

The maximum tolerable temperature rise τ of the mild steel bolt is given by

$$\begin{aligned}\tau &= -\frac{\sigma_{sa}}{\alpha_s E_s\left(1-\frac{\alpha_c}{\alpha_s}\right)\Big/\left(1+\frac{A_s E_s}{A_c E_c}\right)} = -\frac{\sigma_{st}}{f}\frac{1+\frac{A_s E_s}{A_c E_c}}{\alpha_s E_s\left(1-\frac{\alpha_c}{\alpha_s}\right)}\\ &= -\frac{400\times 10^6}{3}\\ &\quad\times\frac{1+1\times 10^{-4}\times 206\times 10^9/(2\times 10^{-4}\times 120\times 10^9)}{11.2\times 10^{-6}\times 206\times 10^9\times[1-16.5\times 10^{-6}/(11.2\times 10^{-6})]}\\ &= 226.944 = 227\,\mathrm{K}\end{aligned} \tag{13.55}$$

Tolerable stress of the copper tube is

$$\sigma_{ca} = \frac{\sigma_{ct}}{f} \tag{13.56}$$

The maximum tolerable temperature rise τ of the copper tube is given by

$$\begin{aligned}\tau &= \frac{\sigma_{ca}}{\alpha_s E_s\frac{A_s}{A_c}\left(1-\frac{\alpha_c}{\alpha_s}\right)\Big/\left(1+\frac{A_s E_s}{A_c E_c}\right)} = \frac{\sigma_{ct}}{f}\frac{A_c}{A_s}\frac{1+\frac{A_s E_s}{A_c E_c}}{\alpha_s E_s\left(1-\frac{\alpha_c}{\alpha_s}\right)}\\ &= -\frac{300\times 10^6}{3}\frac{2\times 10^{-4}}{1\times 10^{-4}}\\ &\quad\times\frac{1+1\times 10^{-4}\times 206\times 10^9/(2\times 10^{-4}\times 120\times 10^9)}{11.2\times 10^{-6}\times 206\times 10^9\times[1-16.5\times 10^{-6}/(11.2\times 10^{-6})]}\\ &= 340.416 = 340\,\mathrm{K}\end{aligned} \tag{13.57}$$

Therefore, from Eqs. (13.55) and (13.57), the maximum tolerable temperature rise is 227 K.

Problem 13.15. A bar of mild steel of cross-sectional area A_s is placed between two parallel bars of copper of cross-sectional area A_c, shown in Fig. 13.7. When the three bars of same length l are bonded together and are subjected to a temperature change of τ_s in the bar of mild steel and τ_c in the bar of copper, calculate the thermal stresses produced in each bar.

Solution. The final lengths of middle steel and two copper bars are same

$$l + \alpha_s \tau_s l + \frac{\sigma_s l}{E_s} = l + \alpha_c \tau_c l + \frac{\sigma_c l}{E_c} \tag{13.58}$$

While the equilibrium condition of internal forces gives

$$\sigma_s A_s + 2\sigma_c A_c = 0 \tag{13.59}$$

From Eqs. (13.58) and (13.59) we get

$$\sigma_s = \frac{E_s(\alpha_c\tau_c - \alpha_s\tau_s)}{1 + \dfrac{A_s E_s}{2A_c E_c}}, \qquad \sigma_c = -\frac{\sigma_s A_s}{2A_c} \qquad \text{(Answer)}$$

Problem 13.16. Calculate the thermal stresses produced in the bars of the truss shown in Fig. 13.8, if the temperature changes of the bars are τ_i.

Solution. The relation between the elongation of bar 1 and bar 2 is

$$\lambda_2 = \lambda_1 \cos\theta \tag{13.60}$$

Therefore,

$$\alpha_2\tau_2 l_2 + \frac{\sigma_2 l_2}{E_2} = \left(\alpha_1\tau_1 l_1 + \frac{\sigma_1 l_1}{E_1}\right)\cos\theta \tag{13.61}$$

The relation between the length of bar 1 and bar 2 gives

$$l_1 = l_2 \cos\theta \tag{13.62}$$

Substitution of Eq. (13.62) into (13.61) reduces to

$$\alpha_2\tau_2 + \frac{\sigma_2}{E_2} = \left(\alpha_1\tau_1 + \frac{\sigma_1}{E_1}\right)\cos^2\theta \tag{13.63}$$

Then

$$\frac{\sigma_1}{E_1}\cos^2\theta - \frac{\sigma_2}{E_2} = -\alpha_1\tau_1\cos^2\theta + \alpha_2\tau_2 \tag{13.64}$$

The equilibrium of internal forces requires

$$\sigma_1 A_1 + 2\sigma_2 A_2 \cos\theta = 0 \tag{13.65}$$

Solution of Eqs. (13.64) and (13.65) gives

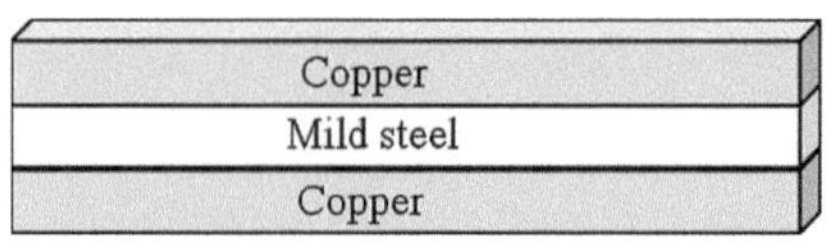

Fig. 13.7 Three bars with rectangular cross section fastened to each other

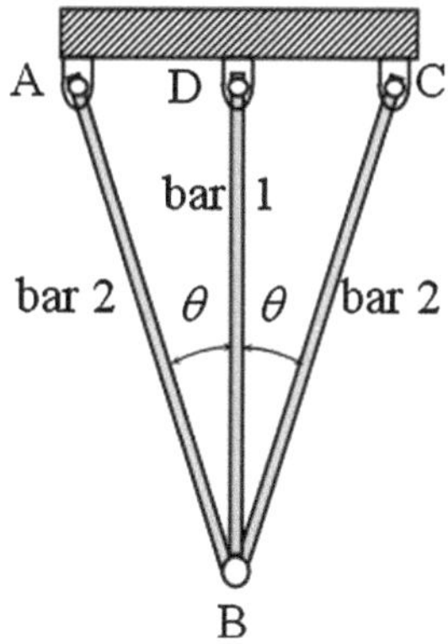

Fig. 13.8 Truss of three bars

$$\sigma_1 = -\alpha_1 E_1 \tau_1 \frac{\cos^2\theta - \dfrac{\alpha_2\tau_2}{\alpha_1\tau_1}}{\cos^2\theta + \dfrac{E_1}{E_2}\dfrac{A_1}{2A_2}\dfrac{1}{\cos\theta}}$$

$$= -\alpha_1 E_1 \tau_1 \frac{1 - \dfrac{\alpha_2\tau_2}{\alpha_1\tau_1\cos^2\theta}}{1 + \dfrac{A_1 E_1}{2A_2 E_2 \cos^3\theta}}$$

$$\sigma_2 = -\frac{A_1}{2A_2\cos\theta}\sigma_1 \quad \text{(Answer)}$$

Problem 13.17. Calculate the thermal stresses produced in the bars which hang from a rigid plate shown in Fig. 13.9, if the temperature changes of the bars are τ_i. The weight of the rigid plate may be neglected.

Solution. The elongations of each bar are

$$\lambda_i = \frac{\sigma_i}{E_i} l + \alpha_i \tau_i l \quad (i = 1, 2, 3) \tag{13.66}$$

The equilibrium condition of the internal forces in each bar requires

$$\sigma_1 A_1 + \sigma_2 A_2 + \sigma_3 A_3 = 0 \tag{13.67}$$

The equilibrium of the moments at the point A is

$$\sigma_2 A_2 a + \sigma_3 A_3 (a + b) = 0 \tag{13.68}$$

The relation between the elongation of each bar is

$$(\lambda_3 - \lambda_1) : (\lambda_2 - \lambda_1) = (a + b) : a \tag{13.69}$$

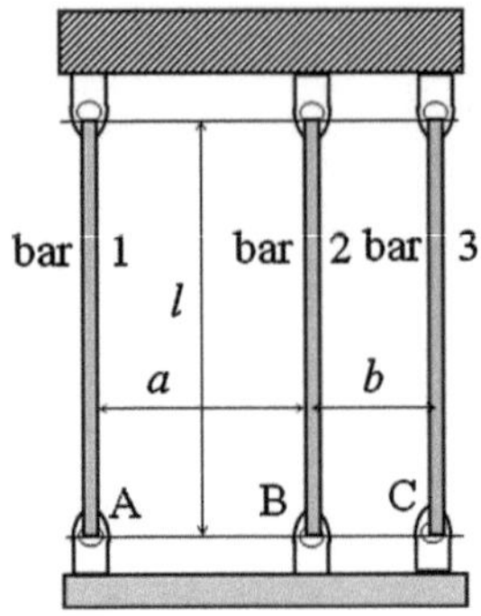

Fig. 13.9 Three bars on whih hangs a rigid plate

Solution of Eqs. (13.67), (13.68) and (13.69) gives

$$\sigma_1 = A_2A_3b\frac{C}{D}, \quad \sigma_2 = -A_1A_3(a+b)\frac{C}{D}, \quad \sigma_3 = A_1A_2a\frac{C}{D} \qquad \text{(Answer)}$$

in which

$$\begin{aligned} C &= -b\alpha_1\tau_1 + (a+b)\alpha_2\tau_2 - a\alpha_3\tau_3 \\ D &= a^2\frac{A_1A_2}{E_3} + b^2\frac{A_2A_3}{E_1} + (a+b)^2\frac{A_1A_3}{E_2} \end{aligned} \qquad (13.70)$$

Chapter 14
Thermal Stresses in Beams

In this chapter, based on the Bernoulli-Euler hypothesis, thermal stresses in beams subjected to thermal and mechanical loads are recalled. Thermal stresses in composite and curved beams, and thermal deflections in beams subjected to a symmetrical thermal load are treated. Furthermore, solutions for stresses in curved beams are included. Problems and solutions for beams subjected to various temperature field or various boundary conditions are presented. [see also Chap. 23.]

14.1 Thermal Stresses in Beams

14.1.1 Thermal Stresses in Beams

We consider the thermal stresses in beams under the Bernoulli-Euler hypothesis.[1] The neutral axis passes through the centroid of the cross section of the beam which is defined by

$$\int_A y\, dA = 0 \tag{14.1}$$

where dA denotes a small element area of the cross section at a distance y from the neutral plane.

The thermal stress is given by

$$\sigma_x = -\alpha E \tau + E\epsilon_0 + E\frac{y}{\rho} \tag{14.2}$$

where ρ denotes the radius of curvature at the neutral plane and ϵ_0 denotes the axial strain at the neutral plane. When the beam is subjected to an axial force N and a

[1] The plane which is perpendicular to the neutral axis before deformation remains plane and perpendicular to the neutral axis after deformation.

M. Reza Eslami et al., *Theory of Elasticity and Thermal Stresses*, Solid Mechanics and Its Applications 197, DOI: 10.1007/978-94-007-6356-2_14,

mechanical bending moment M_M, the axial strain ϵ_0 and the curvature $1/\rho$ at the neutral plane $y = 0$ are

$$\epsilon_0 = \frac{1}{EA}\Big[N + \int_A \alpha E\tau(y)\,dA\Big] \tag{14.3}$$

$$\frac{1}{\rho} = \frac{1}{EI}\Big[M_M + \int_A \alpha E\tau(y)y\,dA\Big] \tag{14.4}$$

where I denotes the moment of inertia of the cross section which is defined by

$$I = \int_A y^2\,dA \tag{14.5}$$

The thermal stress is

$$\begin{aligned}\sigma_x(y) = -\alpha E\tau(y) &+ \frac{1}{A}\Big[N + \int_A \alpha E\tau(y)\,dA\Big]\\ &+ \frac{y}{I}\Big[M_M + \int_A \alpha E\tau(y)y\,dA\Big]\end{aligned} \tag{14.6}$$

The thermal stress in the beam with free boundary conditions under only thermal loads is

$$\sigma_x(y) = -\alpha E\tau(y) + \frac{1}{A}\int_A \alpha E\tau(y)\,dA + \frac{y}{I}\int_A \alpha E\tau(y)y\,dA \tag{14.7}$$

The thermal stress in the beam with rectangular cross section with width b and height h is

$$\begin{aligned}\sigma_x(y) = &- \alpha E\tau(y) + \frac{1}{h}\int_{-h/2}^{h/2} \alpha E\tau(y)\,dy\\ &+ \frac{12y}{h^3}\int_{-h/2}^{h/2} \alpha E\tau(y)y\,dy\end{aligned} \tag{14.8}$$

Next, we consider the thermal stress in the beam subjected to an arbitrary temperature change $\tau(x, y, z)$.

The thermal stress σ_x is

$$\sigma_x = -\alpha E\tau(x, y, z) + E\epsilon_0 + E\frac{y}{\rho_y} + E\frac{z}{\rho_z} \tag{14.9}$$

where ϵ_0 and ρ_y, ρ_z denote the axial strain and the radii of curvature in y and z directions at the centroid of the cross section. When the external force and moments act on the beam, the conditions of the equilibrium of both the force and the moments are

$$\int_A \sigma_x dA = N, \quad \int_A \sigma_x y dA = M_{Mz}, \quad \int_A \sigma_x z dA = M_{My} \tag{14.10}$$

where N denotes the axial force, and M_{My} and M_{Mz} mean the mechanical bending moments with respect to y and z axes, respectively.

The axial strain ϵ_0 and the curvatures $1/\rho_y$ and $1/\rho_z$ at $y = z = 0$ are from Eq. (14.10)

$$\epsilon_0 = \frac{P}{EA} \tag{14.11}$$

$$\frac{1}{\rho_y} = \frac{I_y M_z - I_{yz} M_y}{E(I_y I_z - I_{yz}^2)} \tag{14.12}$$

$$\frac{1}{\rho_z} = \frac{I_z M_y - I_{yz} M_z}{E(I_y I_z - I_{yz}^2)} \tag{14.13}$$

where I_y and I_z are the moments of inertia of the cross section about the y and z axes, respectively, and I_{yz} is the product of inertia about these axes which are defined by

$$I_y = \int_A z^2\, dA, \quad I_z = \int_A y^2\, dA, \quad I_{yz} = \int_A yz\, dA \tag{14.14}$$

and

$$P = N + P_T, \quad M_y = M_{My} + M_{Ty}, \quad M_z = M_{Mz} + M_{Tz} \tag{14.15}$$

$$P_T = \int_A \alpha E \tau(x, y, z)\, dA \tag{14.16}$$

$$M_{Ty} = \int_A \alpha E \tau(x, y, z) z\, dA, \quad M_{Tz} = \int_A \alpha E \tau(x, y, z) y dA \tag{14.17}$$

in which P_T denotes the thermally induced force, and M_{Ty} and M_{Tz} the thermally induced moments about y and z axes, respectively, P the total force, and M_y, M_z the total moments due to both mechanical and thermal loads.

The thermal stress due to both thermal and mechanical loads is given from Eq. (14.9) by

$$\sigma_x(x, y, z) = -\alpha E \tau + \frac{P}{A} + \frac{I_y M_z - I_{yz} M_y}{I_y I_z - I_{yz}^2} y + \frac{I_z M_y - I_{yz} M_z}{I_y I_z - I_{yz}^2} z \tag{14.18}$$

When the cross section of the beam is symmetric about y axis, the thermal stress is simplified from Eq. (14.18)

$$\sigma_x(x, y, z) = -\alpha E\tau + \frac{P}{A} + \frac{M_z}{I_z}y + \frac{M_y}{I_y}z \tag{14.19}$$

Because the product of inertia I_{yz} reduces to zero.

Next, we consider the shearing stress in a beam. The equilibrium of an element of the beam with arbitrary cross section, small length dx and width of the beam $b(y)$ as shown in Fig. 14.1, is

$$-\int_{-b_1}^{b_1} (\sigma_{xy} dx) dz + \int_{-b_1}^{b_1} \left[\int_y^{e_1} (\sigma_x + \frac{\partial \sigma_x}{\partial x} dx) dy \right] dz$$
$$- \int_{-b_1}^{b_1} \int_y^{e_1} \sigma_x dy dz = 0 \tag{14.20}$$

where σ_{xy} is the shearing stress.

The shearing stress is from Eqs. (14.6) and (14.20)

$$\sigma_{xy} = \int_y^{e_1} \frac{\partial}{\partial x} \Bigg\{ - \alpha E\tau(x, y) + \frac{1}{A} \left[N + \int_A \alpha E\tau(x, y)\, dA \right]$$
$$+ \frac{y}{I} \left[M_M + \int_A \alpha E\tau(x, y) y\, dA \right] \Bigg\} dy \tag{14.21}$$

If the bending stress σ_x is independent of the coordinate x, the shearing stress does not occur.

14.1.2 Thermal Stresses in Composite Beams

We consider a multi-layered composite beam, shown in Fig. 14.2, subjected to temperature change $\tau_i(y)$ and mechanical loads. The origin $y = 0$ of the coordinate system (x, y) is taken at the upper surface of the composite beam.

Thermal stresses σ_{xi} are

$$\sigma_{xi} = -\alpha_i E_i \tau_i(y) + E_i \epsilon_0 + E_i \frac{y}{\rho} \quad (i = 1, 2, \ldots, n) \tag{14.22}$$

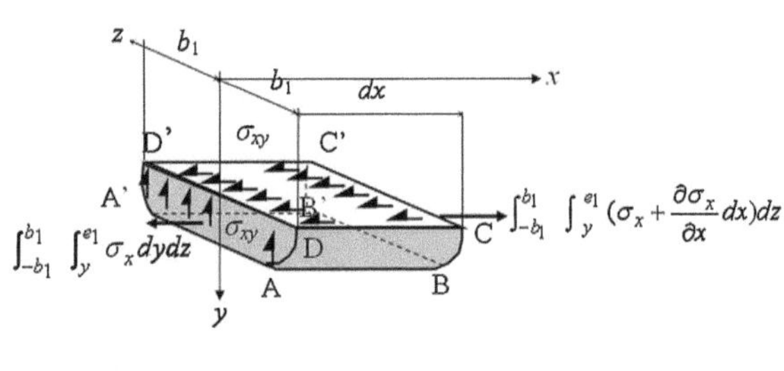

Fig. 14.1 Shearing stress in a beam

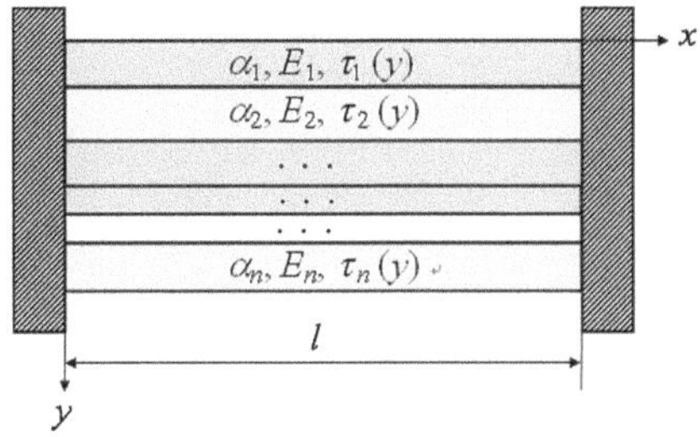

Fig. 14.2 A multi-layered composite beam

where ϵ_0 and $1/\rho$ denote the axial strain at the upper surface $y = 0$ and the curvature at $y = 0$, respectively. When the axial force N and the mechanical bending moment M_M act on the multi-layered composite beam, the axial strain ϵ_0 and the curvature $1/\rho$ at $y = 0$ are

$$\epsilon_0 = \frac{PI_{E2} - MI_{E1}}{I_{E0}I_{E2} - I_{E1}^2}, \quad \frac{1}{\rho} = \frac{MI_{E0} - PI_{E1}}{I_{E0}I_{E2} - I_{E1}^2} \tag{14.23}$$

where

$$\begin{aligned}
I_{E0} &= \sum_{i=1}^{n} E_i b_i (y_i - y_{i-1}) \\
I_{E1} &= \frac{1}{2}\sum_{i=1}^{n} E_i b_i (y_i^2 - y_{i-1}^2), \quad I_{E2} = \frac{1}{3}\sum_{i=1}^{n} E_i b_i (y_i^3 - y_{i-1}^3) \\
P_T &= \sum_{i=1}^{n}\int_{y_{i-1}}^{y_i} \alpha_i E_i \tau_i(y) b_i\, dy, \quad M_T = \sum_{i=1}^{n}\int_{y_{i-1}}^{y_i} \alpha_i E_i \tau_i(y) y b_i\, dy \\
P &= N + P_T, \quad M = M_M + M_T
\end{aligned} \tag{14.24}$$

in which b_i and $y_i - y_{i-1} = h_i$ denote the width and the height of each layer, respectively, and y_i $(i = 1, 2, \ldots, n)$ mean the lower surface of the i-th beam, and $y_0 = 0$.

The thermal stresses in the composite beam may be expressed as

$$\sigma_{xi}(y) = -\alpha_i E_i \tau_i(y) + E_i \frac{PI_{E2} - MI_{E1}}{I_{E0}I_{E2} - I_{E1}^2} + E_i \frac{MI_{E0} - PI_{E1}}{I_{E0}I_{E2} - I_{E1}^2} y \tag{14.25}$$

Let us consider the thermal stress in a nonhomogeneous beam in which the coefficient of linear thermal expansion α and Young's modulus E are functions of position y. Taking the origin $y = 0$ of the coordinate system (x, y) at the centroid of the cross section of the beam, the thermal stress σ_x is

$$\sigma_x = -\alpha(y)E(y)\tau(y) + E(y)\epsilon_0 + E(y)\frac{y}{\rho} \tag{14.26}$$

When the axial force N and the mechanical bending moment M_M act on the beam, we can obtain the axial strain ϵ_0 at $y = 0$ and the curvature $1/\rho$ as

$$\epsilon_0 = \frac{PI_{E2} - MI_{E1}}{I_{E0}I_{E2} - I_{E1}^2}, \quad \frac{1}{\rho} = \frac{MI_{E0} - PI_{E1}}{I_{E0}I_{E2} - I_{E1}^2} \tag{14.27}$$

where

$$\begin{aligned} I_{E0} &= \int_A E(y)\,dA, \quad I_{E1} = \int_A E(y)y\,dA, \quad I_{E2} = \int_A E(y)y^2\,dA \\ P_T &= \int_A \alpha(y)E(y)\tau(y)\,dA, \quad M_T = \int_A \alpha(y)E(y)\tau(y)y\,dA \\ P &= N + P_T, \qquad M = M_M + M_T \end{aligned} \tag{14.28}$$

The thermal stress σ_x in the nonhomogeneous beam may be expressed from Eq. (14.26) as

$$\begin{aligned} \sigma_x(y) = -\alpha(y)E(y)\tau(y) + E(y)\frac{PI_{E2} - MI_{E1}}{I_{E0}I_{E2} - I_{E1}^2} \\ + E(y)\frac{MI_{E0} - PI_{E1}}{I_{E0}I_{E2} - I_{E1}^2}y \end{aligned} \tag{14.29}$$

14.1.3 Thermal Deflection in Beams

When the homogeneous beam is subjected to the symmetrical thermal load $\tau(x, y)$, the axial load N and the mechanical bending moment M_M, the axial displacement u is

$$\begin{aligned} u = u_0 + \int_0^x \left\{ \frac{1}{EA}\left[N + \int_A \alpha E\tau(x, y)\,dA \right] \right. \\ \left. + \frac{y}{EI}\left[M_M + \int_A \alpha E\tau(x, y)y\,dA \right] \right\} dx \end{aligned} \tag{14.30}$$

where u_0 is the axial displacement at $x = 0$. The average axial displacement for the cross section u_{av} is

$$u_{av} = u_0 + \frac{1}{A}\int_A \left\{ \int_0^x \frac{1}{EA}\left[N + \int_A \alpha E\tau(x, y)\,dA \right] dx \right\} dA \tag{14.31}$$

Next, consider the deflection v of the beam subjected to both mechanical and thermal loads. The relationship between the deflection and the curvature is

$$\frac{1}{\rho} = -\frac{d^2 v}{dx^2} \tag{14.32}$$

The governing equation for the deflection v can be obtained by substitution of Eq. (14.4) into Eq. (14.32)

$$\frac{d^2 v}{dx^2} = -\frac{M_M + M_T}{EI} \tag{14.33}$$

Then, the deflection v is given by

$$v = -\int \left(\int \frac{M_M + M_T}{EI} dx \right) dx + C_1 x + C_2 \tag{14.34}$$

The unknown constants C_1, C_2 will be determined from the boundary conditions:

$$\begin{aligned} &\text{for simply supported edge} \quad v = 0, \quad \frac{d^2 v}{dx^2} + \frac{M_T}{EI} = 0 \\ &\text{for built-in edge} \quad v = 0, \quad \frac{dv}{dx} = 0 \end{aligned} \tag{14.35}$$

14.1.4 Curved Beams

We consider a curved beam with the curvature $1/R$ subjected to both mechanical and thermal loads. The origin of the coordinate system is taken at the centroid of the curved beam. When the curvature $1/R$ before deformation deforms to the curvature $1/\rho$ after deformation, the strain ϵ_0 at the center line is given by

$$\epsilon_0 = \frac{\rho - R}{R} + \frac{\rho}{R} \omega_0 \tag{14.36}$$

where ω_0 means the ratio of angle change of the curved beam defined by

$$\omega_0 = \frac{\Delta d\theta}{d\theta} \tag{14.37}$$

The strain ϵ at a distance y from the center line is[2]

$$\epsilon = \alpha\tau + \frac{\sigma_{\theta\theta}}{E} = \frac{1}{R + y} \left(\epsilon_0 R + \omega_0 y + \int_0^y \alpha\tau \, dy \right) \tag{14.38}$$

The hoop stress $\sigma_{\theta\theta}$ can be expressed by

$$\sigma_{\theta\theta} = -\alpha E\tau + \frac{E}{R + y} \left(\epsilon_0 R + \omega_0 y + \int_0^y \alpha\tau \, dy \right) \tag{14.39}$$

---

[2] N. Noda, R. Hetnarski, Y. Tanigawa, *Thermal Stresses* (Taylor & Francis, New York, 2004).

When the curved beam is subjected to both the axial force N and the mechanical bending moment M_M, the axial strain ϵ_0 and ω_0 at the center surface $y = 0$ are

$$\epsilon_0 = \frac{P}{EA} + \frac{M}{EAR}, \quad \omega_0 = \frac{P}{EA} + \frac{M}{EAR}\left(1 + \frac{1}{\kappa}\right) \tag{14.40}$$

where

$$\kappa = -\frac{1}{A}\int_A \frac{y}{R+y}\, dA \tag{14.41}$$

$$P = N + N_T, \qquad M = M_M + M_T \tag{14.42}$$

$$\begin{aligned} N_T &= \int_A \alpha E\tau dA - \int_A \frac{E}{R+y}\left(\int_0^y \alpha\tau dy\right) dA \\ M_T &= \int_A \alpha E\tau y\, dA - \int_A \frac{Ey}{R+y}\left(\int_0^y \alpha\tau dy\right) dA \end{aligned} \tag{14.43}$$

The radius of curvature after deformation at the center surface $y = 0$ is obtained from Eqs. (14.36) and (14.40)

$$\rho = \left(\frac{1+\epsilon_0}{1+\omega_0}\right) R = \left(1 - \frac{\dfrac{M}{EAR}\dfrac{1}{\kappa}}{1 + \dfrac{P}{EA} + \dfrac{M}{EAR}\left(1 + \dfrac{1}{\kappa}\right)}\right) R \tag{14.44}$$

Then, the hoop stress $\sigma_{\theta\theta}$ due to both mechanical and thermal loads is

$$\begin{aligned} \sigma_{\theta\theta} &= -\alpha E\tau + \frac{1}{A}\left[P + \frac{M}{R}\left(1 + \frac{y}{\kappa(R+y)}\right)\right] + \frac{E}{R+y}\int_0^y \alpha\tau\, dy \\ &= -\alpha E\tau + \frac{1}{A}\left[N + \int_A \alpha E\tau\, dA \right. \\ &\quad \left. + \frac{1}{R}\left(1 + \frac{y}{\kappa(R+y)}\right)(M_M + \int_A \alpha E\tau y\, dA)\right] \\ &\quad - \frac{1}{A}\left[\int_A \frac{E}{R+y}\int_0^y \alpha\tau\, dy\, dA \right. \\ &\quad \left. + \frac{1}{R}\left(1 + \frac{y}{\kappa(R+y)}\right)\int_A \frac{Ey}{R+y}\int_0^y \alpha\tau\, dy\, dA\right] + \frac{E}{R+y}\int_0^y \alpha\tau\, dy \end{aligned} \tag{14.45}$$

14.2 Problems and Solutions Related to Thermal Stresses in Beams

Problem 14.1. When the boundary conditions of the beams are given by

[1] perfectly clamped ends
[2] free expansion and restrained bending
[3] restrained expansion and free bending
derive the thermal stress in the beam.

Solution. Thermal stress in the beam is given by Eq. (14.2). The axial strain ϵ_0 at the neutral plane and the curvature $1/\rho$ at the neutral plane can be determined by following boundary conditions.

[1] perfectly clamped ends
From the boundary conditions of perfectly clamped ends, the axial strain ϵ_0 and the curvature $1/\rho$ are zero. Then,

$$\sigma_x(y) = -\alpha E\tau(y) \qquad \text{(Answer)}$$

[2] free expansion and restrained bending
From the boundary conditions of free thermal expansion and restrained bending, the axial strain ϵ_0 is given by Eq. (14.3) and the curvature $1/\rho$ is zero. Then,

$$\sigma_x(y) = -\alpha E\tau(y) + \frac{1}{A}\int_A \alpha E\tau(y)\, dA \qquad \text{(Answer)}$$

[3] restrained expansion and free bending
From the boundary conditions of restrained expansion and free bending, the axial strain ϵ_0 is zero and the curvature $1/\rho$ is given by Eq. (14.4). Then,

$$\sigma_x(y) = -\alpha E\tau(y) + \frac{y}{I}\int_A \alpha E\tau(y) y\, dA \qquad \text{(Answer)}$$

Problem 14.2. A rectangular beam with a cross section $b \times h$ is subjected to the temperature change $\tau(y) = C_1 y + C_0$. Calculate the thermal stresses in the beam for the following cases:

[1] perfectly clamped ends
[2] free expansion and restrained bending
[3] restrained expansion and free bending
[4] free expansion and free bending.

Solution. From the Problem 14.1, we can get the thermal stress for each boundary condition.

[1] perfectly clamped ends

$$\sigma_x = -\alpha E \tau(y) \tag{14.46}$$

Substitution of $\tau(y) = C_1 y + C_0$ into Eq. (14.46) gives

$$\sigma_x = -\alpha E(C_1 y + C_0) \qquad \text{(Answer)}$$

[2] free expansion and restrained bending

$$\sigma_x = -\alpha E \tau(y) + \frac{1}{h}\int_{-h/2}^{h/2} \alpha E \tau(y) dy \tag{14.47}$$

Substitution of $\tau(y) = C_1 y + C_0$ into Eq. (14.47) gives

$$\begin{aligned}\sigma_x &= -\alpha E(C_1 y + C_0) + \frac{1}{h}\int_{-h/2}^{h/2} \alpha E(C_1 y + C_0) dy \\ &= -\alpha E(C_1 y + C_0) + \frac{1}{h}\alpha E C_0 h = -\alpha E C_1 y \qquad \text{(Answer)}\end{aligned}$$

[3] restrained expansion and free bending

$$\sigma_x = -\alpha E \tau(y) + \frac{12y}{h^3}\int_{-h/2}^{h/2} \alpha E \tau(y) y dy \tag{14.48}$$

Substitution of $\tau(y) = C_1 y + C_0$ into Eq. (14.48) gives

$$\begin{aligned}\sigma_x &= -\alpha E(C_1 y + C_0) + \frac{12y}{h^3}\int_{-h/2}^{h/2} \alpha E(C_1 y + C_0) y dy \\ &= -\alpha E(C_1 y + C_0) + \frac{12y}{h^3}\alpha E \frac{2}{3} C_1 \frac{h^3}{8} = -\alpha E C_0 \qquad \text{(Answer)}\end{aligned}$$

[4] free expansion and free bending

$$\sigma_x = -\alpha E \tau(y) + \frac{1}{h}\int_{-h/2}^{h/2} \alpha E \tau(y) dy + \frac{12y}{h^3}\int_{-h/2}^{h/2} \alpha E \tau(y) y dy \tag{14.49}$$

Substitution of $\tau(y) = C_1 y + C_0$ into Eq. (14.49) gives

$$\begin{aligned}\sigma_x = &-\alpha E(C_1 y + C_0) + \frac{1}{h}\int_{-h/2}^{h/2} \alpha E(C_1 y + C_0) dy \\ &+ \frac{12y}{h^3}\int_{-h/2}^{h/2} \alpha E(C_1 y + C_0) y dy\end{aligned}$$

$$= -\alpha E(C_1 y + C_0) + \frac{1}{h}\alpha E C_0 h + \frac{12y}{h^3}\alpha E C_1 \frac{h^3}{12} = 0 \qquad \text{(Answer)}$$

Problem 14.3. When the origin of the coordinate system does not coincide with the centroid of the section, find the thermal stress in the beam with temperature rise τ.

Solution. The variable y denotes the distance from the origin of the coordinate system. If ε_0 and ρ denote the axial strain and the radius of curvature at $y = 0$, respectively, then the stress σ_x is

$$\sigma_x = -\alpha E\tau + E\varepsilon_0 + E\frac{y}{\rho} \tag{14.50}$$

Since external forces do not act on the beam

$$\int_A \sigma_x dA = 0, \quad \int_A \sigma_x y dA = 0 \tag{14.51}$$

Substitution of Eq. (14.50) into Eq. (14.51) gives

$$\begin{aligned} E\varepsilon_0 A + E\frac{1}{\rho}\int_A y dA &= \int_A \alpha E\tau dA \\ E\varepsilon_0 \int_A y dA + E\frac{1}{\rho}\int_A y^2 dA &= \int_A \alpha E\tau y dA \end{aligned} \tag{14.52}$$

The solutions of algebraic equations (14.52) are

$$\begin{aligned} \varepsilon_0 &= \frac{1}{E(AI_2 - I_1^2)}\left(I_2 \int_A \alpha E\tau dA - I_1 \int_A \alpha E\tau y dA\right) \\ \frac{1}{\rho} &= \frac{1}{E(AI_2 - I_1^2)}\left(A \int_A \alpha E\tau y dA - I_1 \int_A \alpha E\tau dA\right) \end{aligned} \tag{14.53}$$

where

$$I_1 = \int_A y dA, \quad I_2 = \int_A y^2 dA \tag{14.54}$$

The stress σ_x is

$$\begin{aligned} \sigma_x = -\alpha E\tau(y) &+ \frac{1}{AI_2 - I_1^2}\left[I_2 \int_A \alpha E\tau(y) dA - I_1 \int_A \alpha E\tau(y) y dA\right] \\ &+ \frac{y}{AI_2 - I_1^2}\left[A \int_A \alpha E\tau(y) y dA - I_1 \int_A \alpha E\tau(y) dA\right] \end{aligned} \qquad \text{(Answer)}$$

Problem 14.4. A rectangular beam with length l, height h, and width b is subjected to the temperature change

$$\tau = \sum_{n=0}^{\infty} T_{2n} y^{2n} + \sum_{n=0}^{\infty} T_{2n+1} y^{2n+1} \tag{14.55}$$

Calculate the thermal stress and curvature produced in the beam assuming a stress free boundary condition. Furthermore, calculate the thermal deflection in the simply supported beam .

Solution. Substitution of the temperature change (14.55) into Eq. (14.8) gives

$$\begin{aligned}
\sigma_x &= -\alpha E\tau(y) + \frac{1}{h}\int_{-h/2}^{h/2} \alpha E\tau(y)dy + \frac{12y}{h^3}\int_{-h/2}^{h/2} \alpha E\tau(y)ydy \\
&= -\alpha E \sum_{n=0}^{\infty} (T_{2n}y^{2n} + T_{2n+1}y^{2n+1}) \\
&\quad + \frac{1}{h}\alpha E \sum_{n=0}^{\infty} \left[\frac{1}{2n+1}T_{2n}y^{2n+1} + \frac{1}{2n+2}T_{2n+1}y^{2n+2}\right]_{-h/2}^{h/2} \\
&\quad + \frac{12y}{h^3}\alpha E \sum_{n=0}^{\infty} \left[\frac{1}{2n+2}T_{2n}y^{2n+2} + \frac{1}{2n+3}T_{2n+1}y^{2n+3}\right]_{-h/2}^{h/2} \\
&= -\alpha E \sum_{n=0}^{\infty} \left\{ T_{2n}\left[y^{2n} - \frac{1}{2n+1}\left(\frac{h}{2}\right)^{2n}\right] \right. \\
&\quad \left. + T_{2n+1}\left[y^{2n+1} - \frac{3}{2n+3}\left(\frac{h}{2}\right)^{2n} y\right]\right\} \qquad \text{(Answer)}
\end{aligned}$$

The curvature is given by Eq. (14.4)

$$\begin{aligned}
\frac{1}{\rho} &= \frac{1}{EI}\int_{-h/2}^{h/2} \alpha E\tau(y)ydA \\
&= \frac{12}{h^3}\alpha \sum_{n=0}^{\infty} \left[\frac{1}{2n+2}T_{2n}y^{2n+2} + \frac{1}{2n+3}T_{2n+1}y^{2n+3}\right]_{-h/2}^{h/2} \\
&= 3\alpha \sum_{n=0}^{\infty} \frac{1}{2n+3}T_{2n+1}\left(\frac{h}{2}\right)^{2n} \qquad \text{(Answer)}
\end{aligned}$$

The deflection v is given by Eq. (14.34)

$$v = -\iint \frac{M_T}{EI}dxdx + C_1x + C_2 \tag{14.56}$$

where

$$M_T = \int_A \alpha E\tau y dA, \quad I = \frac{bh^3}{12} \tag{14.57}$$

Calculation of M_T gives

$$M_T = \int_A \alpha E\tau y dA = 2\alpha Eb \sum_{n=0}^{\infty} \frac{T_{2n+1}}{2n+3}\left(\frac{h}{2}\right)^{2n+3} \tag{14.58}$$

From Eqs. (14.56) and (14.58) we get

$$\begin{aligned} v &= -2 \iint \frac{\alpha Eb}{EI} \sum_{n=0}^{\infty} \frac{T_{2n+1}}{2n+3}\left(\frac{h}{2}\right)^{2n+3} dx dx + C_1 x + C_2 \\ &= -\frac{3\alpha}{2} \sum_{i=0}^{\infty} \frac{T_{2n+1}}{2n+3}\left(\frac{h}{2}\right)^{2n} x^2 + C_1 x + C_2 \end{aligned} \tag{14.59}$$

The boundary conditions of the simply supported beam are

$$v = 0 \quad \text{at} \quad x = 0 \quad \text{and} \quad x = l \tag{14.60}$$

The boundary conditions (14.60) give

$$C_1 = \frac{3\alpha}{2} \sum_{n=0}^{n} \frac{T_{2n+1}}{2n+3}\left(\frac{h}{2}\right)^{2n} l, \quad C_2 = 0 \tag{14.61}$$

From Eqs. (14.59) and (14.61) we get the deflection v

$$v = \frac{3}{2}\alpha \sum_{n=0}^{\infty} \frac{T_{2n+1}}{2n+3}\left(\frac{h}{2}\right)^{2n} x(l - x) \qquad \text{(Answer)}$$

Problem 14.5. When the beam is subjected to both thermal and mechanical loads, derive the axial strain ϵ_0, the curvature $1/\rho$ at the plane $y = 0$ and the thermal stress.

Solution. The thermal stress is given by Eq. (14.2)

$$\sigma_x = -\alpha E\tau + \epsilon_0 E + E\frac{y}{\rho} \tag{14.62}$$

As the beam is subjected to the external force N and bending moment M_M

$$\int_A \sigma_x \, dA = N, \quad \int_A \sigma_x y \, dA = M_M \tag{14.63}$$

Substitution of Eq. (14.62) into Eq. (14.63) gives the axial strain ϵ_0 and the curvature $1/\rho$ at the plane $y = 0$ in the beam

$$\begin{aligned}\epsilon_0 &= \frac{1}{D}\Big[N\int_A Ey^2\,dA - M_M\int_A Ey\,dA \\ &\quad + \int_A Ey^2\,dA\int_A \alpha E\tau(y)\,dA - \int_A Ey\,dA\int_A \alpha E\tau(y)y\,dA\Big] \\ \frac{1}{\rho} &= \frac{1}{D}\Big[M_M\int_A E\,dA - N\int_A Ey\,dA \\ &\quad + \int_A E\,dA\int_A \alpha E\tau(y)y\,dA - \int_A Ey\,dA\int_A \alpha E\tau(y)\,dA\Big] \qquad \text{(Answer)}\end{aligned} \tag{14.64}$$

where

$$D = \int_A E\,dA\int_A Ey^2\,dA - \left(\int_A Ey\,dA\right)^2 \tag{14.65}$$

Then, the substitution of Eq. (14.64) into Eq. (14.62) gives the thermal stress

$$\begin{aligned}\sigma_x(y) &= -\alpha E\tau(y) \\ &\quad + \frac{E}{D}\Big[N\int_A Ey^2\,dA - M_M\int_A Ey\,dA \\ &\quad + \int_A Ey^2\,dA\int_A \alpha E\tau(y)\,dA - \int_A Ey\,dA\int_A \alpha E\tau(y)y\,dA\Big] \\ &\quad + \frac{Ey}{D}\Big[M_M\int_A E\,dA - N\int_A Ey\,dA \\ &\quad + \int_A E\,dA\int_A \alpha E\tau(y)y\,dA - \int_A Ey\,dA\int_A \alpha E\tau(y)\,dA\Big] \qquad \text{(Answer)}\end{aligned} \tag{14.66}$$

If the origin of the coordinate system is selected in the neutral plane, that is $\int_A y\,dA = 0$, and Young's modulus E is independent of y axis, the axial strain ϵ_0 and the curvature $1/\rho$ at the plane $y = 0$ reduce to Eqs. (14.3) and (14.4), respectively

$$\epsilon_0 = \frac{1}{EA}\left[N + \int_A \alpha E\tau(y)\,dA\right], \quad \frac{1}{\rho} = \frac{1}{EI}\left[M_M + \int_A \alpha E\tau(y)y\,dA\right] \qquad \text{(Answer)}$$

and the thermal stress Eq. (14.66) reduces to Eq. (14.6)

$$\sigma_x(y) = -\alpha E\tau(y) + \frac{1}{A}\left[N + \int_A \alpha E\tau(y)\,dA\right] + \frac{y}{I}\left[M_M + \int_A \alpha E\tau(y) y\,dA\right]$$

(Answer)

Problem 14.6. When a two-layered beam with the same dimensional cross sections is subjected to the same temperature change τ, calculate the thermal stress produced in the beam. The upper beam and the lower beam are made of mild steel and of aluminum, respectively, and the temperature change is $\tau = 100\,\text{K}$.

Solution. We take the origin ($y = 0$) at the bonded surface of the two-layered beam. The subscript i ($i = 1, 2$) refer to the mild steel ($i = 1$) and the aluminum ($i = 2$), and b_i and h_i denote the width and the height of each layer, respectively. The stresses are from Eq. (14.22)

$$\sigma_{xi} = -\alpha_i E_i T_i(y) + E_i \epsilon_0 + E_i \frac{y}{\rho} \quad (i = 1, 2) \tag{14.67}$$

From the conditions of no external forces, the strain ϵ_0 and the curvature $1/\rho$ at the bonding surface $y = 0$ are obtained

$$\begin{aligned}
\epsilon_0 = \frac{2}{D}\Bigg\{ & 2\left[\int_{-h_1}^{0} \alpha_1 E_1 \tau_1(y) b_1\,dy + \int_0^{h_2} \alpha_2 E_2 \tau_2(y) b_2\,dy\right] \\
& \times (E_2 h_2^3 b_2 + E_1 h_1^3 b_1) \\
& - 3\left[\int_{-h_1}^{0} \alpha_1 E_1 \tau_1(y) b_1 y\,dy + \int_0^{h_2} \alpha_2 E_2 \tau_2(y) b_2 y\,dy\right] \\
& \times (E_2 h_2^2 b_2 - E_1 h_1^2 b_1)\Bigg\} \\
\frac{1}{\rho} = \frac{6}{D}\Bigg\{ & 2\left[\int_{-h_1}^{0} \alpha_1 E_1 \tau_1(y) b_1 y\,dy \right.\\
& \left. + \int_0^{h_2} \alpha_2 E_2 \tau_2(y) b_2 y\,dy\right](E_2 h_2 b_2 + E_1 h_1 b_1) \\
& - \left[\int_{-h_1}^{0} \alpha_1 E_1 \tau_1(y) b_1\,dy + \int_0^{h_2} \alpha_2 E_2 \tau_2(y) b_2\,dy\right] \\
& \times (E_2 h_2^2 b_2 - E_1 h_1^2 b_1)\Bigg\}
\end{aligned} \tag{14.68}$$

where

$$D = (E_2 h_2^2 b_2 - E_1 h_1^2 b_1)^2 + 4E_1 E_2 h_1 h_2 (h_1 + h_2)^2 b_1 b_2 \tag{14.69}$$

Thus, thermal stresses in the two-layered beam are given by

$$\begin{aligned}
\sigma_{xi}(y) &= -\alpha_i E_i \tau_i(y) \\
&+ \frac{2E_i}{D}\Bigg\{2\Bigg[\int_{-h_1}^{0} \alpha_1 E_1 \tau_1(y) b_1\, dy + \int_{0}^{h_2} \alpha_2 E_2 \tau_2(y) b_2\, dy\Bigg] \\
&\times (E_2 h_2^3 b_2 + E_1 h_1^3 b_1) \\
&- 3\Bigg[\int_{-h_1}^{0} \alpha_1 E_1 \tau_1(y) b_1 y\, dy \\
&+ \int_{0}^{h_2} \alpha_2 E_2 \tau_2(y) b_2 y\, dy\Bigg](E_2 h_2^2 b_2 - E_1 h_1^2 b_1)\Bigg\} \\
&+ \frac{6E_i y}{D}\Bigg\{2\Bigg[\int_{-h_1}^{0} \alpha_1 E_1 \tau_1(y) b_1 y\, dy + \int_{0}^{h_2} \alpha_2 E_2 \tau_2(y) b_2 y\, dy\Bigg] \\
&\times (E_2 h_2 b_2 + E_1 h_1 b_1) \\
&- \Bigg[\int_{-h_1}^{0} \alpha_1 E_1 \tau_1(y) b_1\, dy + \int_{0}^{h_2} \alpha_2 E_2 \tau_2(y) b_2\, dy\Bigg] \\
&\times (E_2 h_2^2 b_2 - E_1 h_1^2 b_1)\Bigg\} \quad (i = 1, 2)
\end{aligned} \tag{14.70}$$

The thermal stresses for this problem given by Eq. (14.70) reduce to

$$\begin{aligned}
\sigma_{x1} &= \frac{E_1 E_2}{D}(\alpha_2 - \alpha_1)(7E_1 + E_2 + 12\frac{y}{h}E_1)\tau \\
\sigma_{x2} &= -\frac{E_1 E_2}{D}(\alpha_2 - \alpha_1)(7E_2 + E_1 - 12\frac{y}{h}E_2)\tau
\end{aligned} \tag{14.71}$$

where

$$D = (E_1 + E_2)^2 + 12E_1E_2 \tag{14.72}$$

When the material properties are given by

$$\begin{aligned}
\alpha_1 &= 11.2 \times 10^{-6}\ 1/\mathrm{K}, \quad E_1 = 206 \times 10^9\ \mathrm{Pa} \\
\alpha_2 &= 23.1 \times 10^{-6}\ 1/\mathrm{K}, \quad E_2 = 72 \times 10^9\ \mathrm{Pa}
\end{aligned} \tag{14.73}$$

numerical results give

$$\begin{aligned}
\sigma_{xs} &= \frac{206 \times 10^9 \times 72 \times 10^9 \times (23.1 \times 10^{-6} - 11.2 \times 10^{-6})}{(206 \times 10^9 + 72 \times 10^9)^2 + 12 \times 206 \times 10^9 \times 72 \times 10^9} \\
&\quad \times (7 \times 206 \times 10^9 + 72 \times 10^9 + 12\frac{y}{h} \times 206 \times 10^9) \times 100 \\
&= (104.683 + 170.922\frac{y}{h}) \times 10^6\ \mathrm{Pa} \\
&= (104.7 + 170.9\frac{y}{h})\ \mathrm{MPa}
\end{aligned}$$

$$\sigma_{xa} = -\frac{206 \times 10^9 \times 72 \times 10^9 \times (23.1 \times 10^{-6} - 11.2 \times 10^{-6})}{(206 \times 10^9 + 72 \times 10^9)^2 + 12 \times 206 \times 10^9 \times 72 \times 10^9}$$
$$\times (7 \times 72 \times 10^9 + 206 \times 10^9 - 12\frac{y}{h} \times 72 \times 10^9) \times 100$$
$$= -(49.0917 - 59.7398\frac{y}{h}) \times 10^6 \text{ Pa}$$
$$= -(49.1 - 59.7\frac{y}{h}) \text{ MPa} \qquad \text{(Answer)}$$

Problem 14.7. When the temperature on the upper and the lower surfaces of the two-layered beam with rectangular cross section is prescribed to T_a and T_b, respectively, the temperature is given by

$$\begin{aligned} T_1 &= T_a - (T_a - T_b)\frac{1 + y/h_1}{1 + \lambda_1 h_2/\lambda_2 h_1} \qquad -h_1 \leq y \leq 0 \\ T_2 &= T_a - (T_a - T_b)\frac{1 + \lambda_1 y/\lambda_2 h_1}{1 + \lambda_1 h_2/\lambda_2 h_1} \qquad 0 \leq y \leq h_2 \end{aligned} \tag{14.74}$$

in which λ_1 and λ_2 denote the thermal conductivities of the upper and lower beams, respectively. Calculate the curvature and the thermal stress, when the two-layered beam is subjected to the temperature (14.74).

Solution. The temperature changes from the initial temperature T_0 are rewritten from Eq. (14.74) as

$$\tau_i = T_i - T_0 = (T_a - T_0) + (T_b - T_a)K(1 + C_i y) \qquad (i = 1, 2) \tag{14.75}$$

where

$$C_1 = \frac{1}{h_1}, \quad C_2 = \frac{\lambda_1}{\lambda_2 h_1}, \quad K = \frac{1}{1 + \dfrac{\lambda_1 h_2}{\lambda_2 h_1}} \tag{14.76}$$

Substitution of Eq. (14.75) into Eq. (14.68) yields

$$\frac{1}{\rho} = \frac{6}{D}\Bigg\{2\Bigg[\alpha_1 E_1\Bigg\{-(T_a - T_0)\frac{h_1^2}{2} + (T_b - T_a)K\left(-\frac{h_1^2}{2} + C_1\frac{h_1^3}{3}\right)\Bigg\}b_1$$
$$+ \alpha_2 E_2\Bigg\{(T_a - T_0)\frac{h_2^2}{2} + (T_b - T_a)K\left(\frac{h_2^2}{2} + C_2\frac{h_2^3}{3}\right)\Bigg\}b_2\Bigg]$$
$$\times (E_2 h_2 b_2 + E_1 h_1 b_1)$$
$$-\Bigg[\alpha_1 E_1\Bigg\{(T_a - T_0)h_1 + (T_b - T_a)K\left(h_1 - C_1\frac{h_1^2}{2}\right)\Bigg\}b_1$$

$$+ \alpha_2 E_2 \left\{ (T_a - T_0) h_2 + (T_b - T_a) K \left(h_2 + C_2 \frac{h_2^2}{2} \right) \right\} b_2 \Bigg]$$

$$\times \, (E_2 h_2^2 b_2 - E_1 h_1^2 b_1) \Bigg\} \qquad \text{(Answer)}$$

where

$$D = (E_2 h_2^2 b_2 - E_1 h_1^2 b_1)^2 + 4 E_1 E_2 h_1 h_2 (h_1 + h_2)^2 b_1 b_2 \tag{14.77}$$

Substitution of Eq. (14.75) into Eq. (14.70) gives

$$\begin{aligned}
\sigma_{xi}(y) = & -\alpha_i E_i (T_a - T_0) \Bigg\{ 1 - \frac{1}{D\alpha_i} \Big[4(\alpha_1 E_1 h_1 b_1 + \alpha_2 E_2 h_2 b_2) \\
& \times (E_2 h_2^3 b_2 + E_1 h_1^3 b_1) \\
& + 3(\alpha_1 E_1 h_1^2 b_1 - \alpha_2 E_2 h_2^2 b_2)(E_2 h_2^2 b_2 - E_1 h_1^2 b_1) \Big] \\
& + \frac{6y}{D\alpha_i} \Big[(\alpha_1 E_1 h_1^2 b_1 - \alpha_2 E_2 h_2^2 b_2)(E_2 h_2 b_2 + E_1 h_1 b_1) \\
& + (\alpha_1 E_1 h_1 b_1 + \alpha_2 E_2 h_2 b_2)(E_2 h_2^2 b_2 - E_1 h_1^2 b_1) \Big] \Bigg\} \\
& - \alpha_i E_i (T_b - T_a) K \Bigg\{ (1 + C_i y) \\
& - \frac{1}{D\alpha_i} \Big[2\{\alpha_1 E_1 (2 - C_1 h_1) h_1 b_1 + \alpha_2 E_2 (2 + C_2 h_2) h_2 b_2\} \\
& \times (E_2 h_2^3 b_2 + E_1 h_1^3 b_1) \\
& + \{\alpha_1 E_1 (3 - 2C_1 h_1) h_1^2 b_1 - \alpha_2 E_2 (3 + 2C_2 h_2) h_2^2 b_2\} \\
& \times (E_2 h_2^2 b_2 - E_1 h_1^2 b_1) \Big] \\
& + \frac{y}{D\alpha_i} \Big[2\{\alpha_1 E_1 (3 - 2C_1 h_1) h_1^2 b_1 - \alpha_2 E_2 (3 + 2C_2 h_2) h_2^2 b_2\} \\
& \times (E_2 h_2 b_2 + E_1 h_1 b_1) \\
& - 3\{\alpha_1 E_1 (2 - C_1 h_1) h_1 b_1 + \alpha_2 E_2 (2 + C_2 h_2) h_2 b_2\} \\
& \times (E_2 h_2^2 b_2 - E_1 h_1^2 b_1) \Big] \Bigg\} \quad (i = 1, 2) \qquad \text{(Answer)}
\end{aligned}$$

Problem 14.8. When two parallel beams with rectangular cross section are clamped to rigid plates, and are subjected to the different constant temperature rises τ_i as in Fig. 14.3, calculate the thermal stresses produced in each beam.

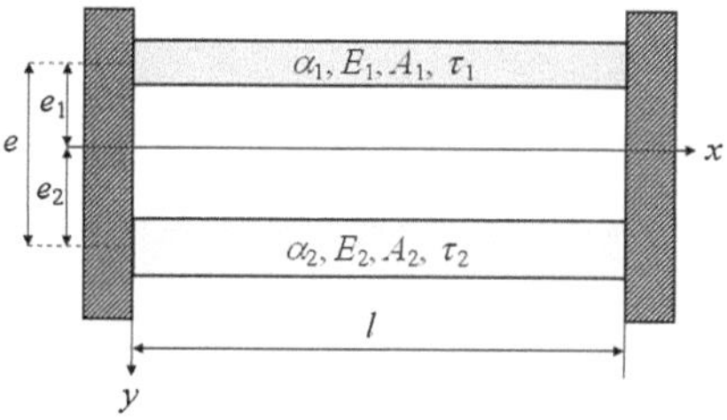

Fig. 14.3 Two parallel beams clamped to rigid plates

Solution. The origin of the coordinate y is taken at an arbitrary position between two beams, and the origin of the local coordinate y_i is taken at the centroid of the cross section of each beam. e denotes the distance between the centroids of cross sections of both beams, and e_i denotes the distance from $y = 0$ to the centroid of the cross section of each bar. The moments of the area A_i of the cross section for each beam are zero

$$\int_{A_i} y_i \, dA_i = 0 \quad (i = 1, 2) \tag{14.78}$$

because the origin of the local coordinate system passes through the centroid of the section of each beam.

When ϵ_0 and ρ denote the axial strain and the radius of curvature at $y = 0$, respectively, the stress σ_x is

$$\sigma_x = -\alpha E \tau + E\epsilon_0 + E\frac{y}{\rho} \tag{14.79}$$

Since external forces are not applied to the beam, we get

$$\int_A \sigma_x \, dA = 0, \quad \int_A \sigma_x y \, dA = 0 \tag{14.80}$$

where the integration extends from the top of the upper beam to the bottom of the lower beam. Using the relationship $y_2 = y - e_2$ and $y_1 = y + e_1$, we obtain the axial strain ϵ_0 and the curvature $1/\rho$ at $y = 0$ from Eqs. (14.79) and (14.80)

$$\epsilon_0 = \frac{P_T I_{E2} - M_T I_{E1}}{I_{E0} I_{E2} - I_{E1}^2}, \quad \frac{1}{\rho} = \frac{M_T I_{E0} - P_T I_{E1}}{I_{E0} I_{E2} - I_{E1}^2} \tag{14.81}$$

where $I_i = \int_{A_i} y_i^2 \, dA_i$ denotes the moment of inertia of the cross section of each beam with respect to its neutral axis and

$$I_{E0} = (E_1 A_1 + E_2 A_2), \quad I_{E1} = (e_2 E_2 A_2 - e_1 E_1 A_1)$$
$$I_{E2} = E_1 (I_1 + A_1 e_1^2) + E_2 (I_2 + A_2 e_2^2)$$
$$P_T = P_{T1} + P_{T2}$$

$$
\begin{aligned}
P_{T1} &= \int_{A_1} \alpha_1 E_1 \tau_1(y_1)\, dA_1, \quad P_{T2} = \int_{A_2} \alpha_2 E_2 \tau_2(y_2)\, dA_2 \\
M_T &= M_{T1} + M_{T2} + e_2 P_{T2} - e_1 P_{T1} \\
M_{T1} &= \int_{A_1} \alpha_1 E_1 \tau(y_1) y_1\, dA_1, \quad M_{T2} = \int_{A_2} \alpha_2 E_2 \tau_2(y_2) y_2\, dA_2
\end{aligned}
\tag{14.82}
$$

Thus, the thermal stresses are expressed by

$$
\begin{aligned}
\sigma_{x1}(y_1) &= -\alpha_1 E_1 \tau_1(y_1) + E_1 \left[\frac{P_T I_{E2} - M_T I_{E1}}{I_{E0} I_{E2} - I_{E1}^2} \right] \\
&\quad + (y_1 - e_1) E_1 \left[\frac{M_T I_{E0} - P_T I_{E1}}{I_{E0} I_{E2} - I_{E1}^2} \right] \\
\sigma_{x2}(y_2) &= -\alpha_2 E_2 \tau_2(y_2) + E_2 \left[\frac{P_T I_{E2} - M_T I_{E1}}{I_{E0} I_{E2} - I_{E1}^2} \right] \\
&\quad + (y_2 + e_2) E_2 \left[\frac{M_T I_{E0} - P_T I_{E1}}{I_{E0} I_{E2} - I_{E1}^2} \right]
\end{aligned}
\tag{14.83}
$$

For the different constant temperature rise τ_i, we have

$$
\begin{aligned}
\sigma_{x1}(y_1) &= -\alpha_1 E_1 \tau_1 + \frac{E_1}{D} \Big[(P_{T1} + P_{T2})(E_1 I_1 + E_2 I_2) \\
&\quad + P_{T1} E_2 A_2 e^2 - (M_{T1} + M_{T2}) E_2 A_2 e \Big] \\
&\quad + \frac{E_1 y_1}{D} \Big[(M_{T1} + M_{T2})(E_1 A_1 + E_2 A_2) \\
&\quad + (P_{T2} E_1 A_1 - P_{T1} E_2 A_2) e \Big] \\
\sigma_{x2}(y_2) &= -\alpha_2 E_2 \tau_2 + \frac{E_2}{D} \Big[(P_{T1} + P_{T2})(E_1 I_1 + E_2 I_2) \\
&\quad + P_{T2} E_1 A_1 e^2 + (M_{T1} + M_{T2}) E_1 A_1 e \Big] \\
&\quad + \frac{E_2 y_2}{D} \Big[(M_{T1} + M_{T2})(E_1 A_1 + E_2 A_2) \\
&\quad + (P_{T2} E_1 A_1 - P_{T1} E_2 A_2) e \Big]
\end{aligned}
\tag{14.84}
$$

where

$$
\begin{aligned}
D &= (E_1A_1 + E_2A_2)(E_1I_1 + E_2I_2) + E_1E_2A_1A_2e^2 \\
P_{T1} &= \int_{A_1} \alpha_1 E_1 \tau_1 dA_1 = \alpha_1 E_1 \tau_1 A_1 \\
P_{T2} &= \int_{A_2} \alpha_2 E_2 \tau_2 dA_2 = \alpha_2 E_2 \tau_2 A_2 \\
M_{T1} &= \int_{A_1} \alpha_1 E_1 \tau_1 y_1 dA_1 = \alpha_1 E_1 \tau_1 \int_{A_1} y_1 dA_1 = 0 \\
M_{T2} &= \int_{A_2} \alpha_2 E_2 \tau_2 y_2 dA_2 = \alpha_2 E_2 \tau_2 \int_{A_2} y_2 dA_2 = 0
\end{aligned} \tag{14.85}
$$

Substitution of Eq. (14.85) into Eq. (14.84) gives

$$
\begin{aligned}
\sigma_{x1}(y_1) &= \frac{E_1}{D}[-\alpha_1\tau_1(E_1A_1 + E_2A_2)(E_1I_1 + E_2I_2) - \alpha_1\tau_1E_1E_2A_1A_2e^2 \\
&+ (\alpha_1E_1\tau_1A_1 + \alpha_2E_2\tau_2A_2)(E_1I_1 + E_2I_2) + \alpha_1E_1\tau_1A_1E_2A_2e^2] \\
&+ \frac{E_1y_1}{D}(\alpha_2E_2\tau_2A_2E_1A_1 - \alpha_1E_1\tau_1A_1E_2A_2)e \\
&= \frac{E_1}{D}(\alpha_2\tau_2 - \alpha_1\tau_1)[E_2A_2(E_1I_1 + E_2I_2) + y_1E_2A_2E_1A_1e] \\
&= \frac{1}{D}(\alpha_2\tau_2 - \alpha_1\tau_1)E_1E_2A_2(E_1I_1 + E_2I_2 + y_1eE_1A_1) \\
\sigma_{x2}(y_2) &= \frac{E_2}{D}[-\alpha_2\tau_2(E_1A_1 + E_2A_2)(E_1I_1 + E_2I_2) - \alpha_2\tau_2E_1E_2A_1A_2e^2 \\
&+ (\alpha_1E_1\tau_1A_1 + \alpha_2E_2\tau_2A_2)(E_1I_1 + E_2I_2) + \alpha_2E_2\tau_2A_2E_1A_1e^2] \\
&+ \frac{E_2y_2}{D}(\alpha_2E_2\tau_2A_2E_1A_1 - \alpha_1E_1\tau_1A_1E_2A_2)e \\
&= \frac{1}{D}(\alpha_1\tau_1 - \alpha_2\tau_2)E_1E_2A_1(E_1I_1 + E_2I_2 - y_2eE_2A_2) \\
&= \frac{1}{D}(\alpha_2\tau_2 - \alpha_1\tau_1)E_1E_2A_1[-(E_1I_1 + E_2I_2) + y_2eE_2A_2] \qquad \text{(Answer)}
\end{aligned}
$$

Problem 14.9. When two parallel beams clamped at each end to rigid plates are subjected to different temperature changes τ_i shown in Fig. 14.3, calculate the thermal stress and the curvature produced in each beam if the elongation at $y = 0$ is restrained to zero.

Solution. Thermal stress is

$$
\sigma_x = -\alpha E\tau + E\varepsilon_0 + E\frac{y}{\rho} \tag{14.86}
$$

From the condition $\varepsilon_0 = 0$, Eq. (14.86) gives

$$\sigma_x = -\alpha E \tau + E\frac{y}{\rho} \tag{14.87}$$

Since an external moment is zero, we get

$$\int_A \sigma_x y dA = 0 \tag{14.88}$$

Substitution of Eq. (14.87) into Eq. (14.88) gives

$$\begin{aligned}
&\int_A E(\frac{y}{\rho} - \alpha\tau) y dA \\
&= \int_{A_1} E_1\left(\frac{y_1 - e_1}{\rho} - \alpha_1\tau_1\right)(y_1 - e_1)dA_1 \\
&\quad + \int_{A_2} E_2\left(\frac{y_2 + e_2}{\rho} - \alpha_2\tau_2\right)(y_2 + e_2)dA_2 \\
&= \frac{1}{\rho}E_1\int_{A_1}(y_1^2 - 2y_1e_1 + e_1^2)dA_1 - \int_{A_1}\alpha_1E_1\tau_1(y_1 - e_1)dA_1 \\
&\quad + \frac{1}{\rho}E_2\int_{A_2}(y_2^2 + 2y_2e_2 + e_2^2)dA_2 - \int_{A_2}\alpha_2E_2\tau_2(y_2 + e_2)dA_2 \\
&= \frac{1}{\rho}I_{E2} - (M_{T1} + M_{T2} - e_1P_{T1} + e_2P_{T2}) = 0
\end{aligned} \tag{14.89}$$

where

$$\begin{aligned}
I_{E2} &= E_1(I_1 + A_1e_1^2) + E_2(I_2 + A_2e_2^2) \\
P_{T1} &= \int_{A_1}\alpha_1E_1\tau_1(y_1)dA_1, \quad P_{T2} = \int_{A_2}\alpha_2E_2\tau_2(y_2)dA_2 \\
M_{T1} &= \int_{A_1}\alpha_1E_1\tau_1(y_1)y_1dA_1, \quad M_{T2} = \int_{A_2}\alpha_2E_2\tau_2(y_2)y_2dA_2
\end{aligned} \tag{14.90}$$

Then

$$\frac{1}{\rho} = \frac{1}{I_{E2}}(M_{T1} + M_{T2} - e_1P_{T1} + e_2P_{T2}) \tag{Answer}$$

Substitution of above equation into Eq. (14.87) gives

$$\begin{aligned}
\sigma_x(y_1) &= -\alpha_1E_1\tau_1(y_1) + E_1(y_1 - e_1)\frac{M_T}{I_{E2}} \\
\sigma_x(y_2) &= -\alpha_2E_2\tau_2(y_2) + E_2(y_2 + e_2)\frac{M_T}{I_{E2}}
\end{aligned} \tag{Answer}$$

where

$$M_T = M_{T1} + M_{T2} + e_2 P_{T2} - e_1 P_{T1} \tag{14.91}$$

Problem 14.10. Prove that the boundary condition of simply supported edge of the beam is given by Eq. (14.35).

Solution. When the beam is subjected to an external moment M_M and the temperature change τ, the axial strain ε_0 and the curvature $1/\rho$ reduce from Eq. (14.3) and Eq. (14.4) to

$$\varepsilon_0 = \frac{1}{EA}\int_A \alpha E\tau(y)dA, \quad \frac{1}{\rho} = \frac{M_M}{EI} + \frac{1}{EI}\int_A \alpha E\tau(y)ydA \tag{14.92}$$

From Eqs. (14.32) and (14.92), we get

$$-\frac{d^2v}{dx^2} = \frac{M_M}{EI} + \frac{1}{EI}\int_A \alpha E\tau(y)ydA \tag{14.93}$$

The deflection and the external moment are zero at the simply supported edge so that

$$v = 0, \quad \frac{d^2v}{dx^2} + \frac{1}{EI}\int_A \alpha E\tau(y)ydA = 0 \tag{14.94}$$

Then, the boundary conditions are

$$v = 0, \quad \frac{d^2v}{dx^2} + \frac{M_T}{EI} = 0 \qquad \text{(Answer)}$$

where

$$M_T = \int_A \alpha E\tau(y)ydA \tag{14.95}$$

Problem 14.11. When the temperature change in a rectangular beam with width b, height h and length l is given by $\tau(y) = A + By$, calculate the deflections of four kinds of beams:

[1] the cantilever beam
[2] the simply supported beam
[3] the clamped-simply supported beam
[4] the perfectly clamped beam

Solution. The deflection v is given by Eq. (14.34)

$$v = -\int\left(\int \frac{M_T}{EI}dx\right)dx + C_1 x + C_2 \tag{14.96}$$

where M_T is given by Eq. (14.28). Substitution of the temperature change $\tau(y) = A + By$ into Eq. (14.28) gives

$$M_T = \int_A \alpha E\tau(y)y dA = \int_{-h/2}^{h/2} \alpha E(A + By)ybdy$$
$$= \alpha E\frac{bh^3}{12}B \tag{14.97}$$

Substitution of Eq. (14.97) into Eq. (14.96) gives

$$v = -\alpha B\frac{x^2}{2} + C_1x + C_2 \tag{14.98}$$

The unknown coefficients C_1 and C_2 in Eq. (14.98) are determined by the boundary conditions.

[1] the cantilever beam
The boundary conditions are

$$v = 0, \quad \frac{dv}{dx} = 0 \quad \text{on} \quad x = 0 \tag{14.99}$$

Substitution of Eq. (14.98) into Eq. (14.99) gives

$$C_1 = 0, \quad C_2 = 0 \tag{14.100}$$

The deflection v is

$$v = -\frac{\alpha B}{2}x^2 \qquad \text{(Answer)}$$

[2] the simply supported beam
The boundary conditions are

$$v = 0 \quad \text{on} \quad x = 0,\ l \tag{14.101}$$

Substitution of Eq. (14.98) into Eq. (14.101) gives

$$C_1 = \frac{\alpha B}{2}l, \quad C_2 = 0 \tag{14.102}$$

Then the deflection is

$$v = \frac{\alpha B}{2}x(l - x) \qquad \text{(Answer)}$$

[3] the clamped-simply supported beam
We put M_A as the supporting bending moment and R_A as the supporting reaction force at the clamped edge.

The equilibrium of the force and the moment gives

$$M_M(x) = -M_A + R_A x \tag{14.103}$$

The deflection v due to both mechanical and thermal loads is given by Eq. (14.34)

$$v = -\int \left(\int \frac{M_M + M_T}{EI} dx \right) dx + C_1 x + C_2 \tag{14.104}$$

Substitution of M_M given by Eq. (14.103) and M_T given by Eq. (14.97) into Eq. (14.104) gives

$$v = -\frac{1}{6EI}(R_A x^3 - 3M_A x^2 + 3M_T x^2) + C_1 x + C_2 \tag{14.105}$$

The unknown coefficients R_A, M_A, C_1 and C_2 are determined by the boundary conditions:

$$\begin{aligned} &v = 0, \quad \frac{dv}{dx} = 0 \quad \text{on} \quad x = 0 \\ &v = 0, \quad \frac{d^2v}{dx^2} + \frac{M_T}{EI} = 0 \quad \text{on} \quad x = l \end{aligned} \tag{14.106}$$

Substitution of Eq. (14.105) into Eq. (14.106) gives

$$C_1 = C_2 = 0, \quad M_A = \frac{3}{2}M_T, \quad R_A = \frac{3}{2l}M_T \tag{14.107}$$

Then, the deflection v is

$$v = \frac{M_T}{4EIl}x^2(l - x) = \frac{\alpha B}{4l}x^2(l - x) \qquad \text{(Answer)}$$

[4] the perfectly clamped beam
We put M_A as the supporting bending moment and R_A as the supporting reaction force at the clamped edges. The equilibrium of the force and the moment gives

$$M_M(x) = -M_A + R_A x \tag{14.108}$$

The deflection v due to both mechanical and thermal loads is given by Eq. (14.105). The unknown coefficients R_A, M_A, C_1 and C_2 are determined by the boundary conditions:

$$v = 0, \quad \frac{dv}{dx} = 0 \quad \text{on} \quad x = 0, \ l \tag{14.109}$$

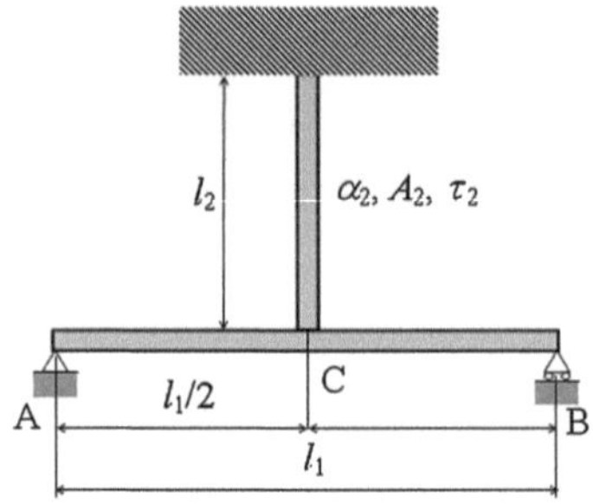

Fig. 14.4 A structure made by the bar and the beam

Substitution of Eq. (14.105) into Eq. (14.109) gives

$$C_1 = C_2 = 0, \quad M_A = M_T, \quad R_A = 0 \tag{14.110}$$

Then, the deflection v is

$$v = 0 \qquad \text{(Answer)}$$

That is, the thermal deflection does not occur.

Problem 14.12. There is a structure made by a simply supported beam and a bar hung from the ceiling shown in Fig. 14.4. When the temperature rise τ_2 occurs in the bar, calculate the thermal stress in both the bar and the beam, and the deflection in the beam.

Solution. We put the internal force produced in the bar and the reaction force from the beam to Q and P, respectively. The equilibrium of the forces is

$$Q + P = 0 \tag{14.111}$$

The elongation of the bar is

$$\lambda_2 = \frac{Ql_2}{E_2A_2} + \alpha_2\tau_2 l_2 \tag{14.112}$$

where A_2, α_2 and τ_2 denote the cross sectional area, the coefficient of linear thermal expansion and the temperature change in the bar, respectively. When the simply supported beam is subjected to the external force P at the center of the beam, the deflection v of the beam is

$$v = \frac{P}{48E_1I_1}x(3l_1^2 - 4x^2) \quad \text{for} \quad 0 \le x \le l_1/2 \tag{14.113}$$

where I_1 means the moment of inertia of the cross section of the beam. The deflection $v_{x=l_1/2}$ of the beam at $x = l_1/2$ is

$$v_{x=l_1/2} = \frac{Pl_1^3}{48E_1I_1} \tag{14.114}$$

As the elongation of the bar is equal to the deflection $v_{x=l_1/2}$ of the beam at $x = l_1/2$, we get

$$\frac{Ql_2}{E_2A_2} + \alpha_2\tau_2 l_2 = \frac{Pl_1^3}{48E_1I_1} \tag{14.115}$$

From Eqs. (14.111) and (14.115), the internal force Q produced in the bar and the reaction force P can be obtained as

$$P = -Q = \frac{48E_1E_2A_2I_1\alpha_2\tau_2l_2}{48E_1I_1l_2 + E_2A_2l_1^3} \tag{14.116}$$

Therefore, the thermal stress produced in the bar is

$$\sigma_2 = -\frac{48E_1E_2I_1\alpha_2\tau_2l_2}{48E_1I_1l_2 + E_2A_2l_1^3} \qquad \text{(Answer)}$$

The thermal stress and the deflection of the beam for $0 \leq x \leq l_1/2$ are

$$\begin{aligned}
\sigma_1 &= \frac{M}{I_1}y = \frac{Pxy}{2I_1} = \frac{24E_1E_2A_2\alpha_2\tau_2l_2}{48E_1I_1l_2 + E_2A_2l_1^3}xy \\
v &= \frac{P}{48E_1I_1}x(3l_1^2 - 4x^2) = \frac{E_2A_2\alpha_2\tau_2l_2}{48E_1I_1l_2 + E_2A_2l_1^3}x(3l_1^2 - 4x^2)
\end{aligned} \qquad \text{(Answer)}$$

The elongation λ_2 and maximum deflection of the beam y_{max} are

$$\lambda_2 = y_{max} = \frac{E_2A_2\alpha_2\tau_2l_2l_1^3}{48E_1I_1l_2 + E_2A_2l_1^3} \qquad \text{(Answer)}$$

Problem 14.13. When a curved beam with uniform rectangular cross section $b \times h$ is subjected to the uniform temperature rise τ, calculate the thermal stress, the radius of curvature, and the change of θ in the curved beam.

Solution. Equation (14.40) give

$$\varepsilon_0 = \frac{N_T}{EA} + \frac{M_T}{EAR}, \quad \omega_0 = \frac{N_T}{EA} + \frac{M_T}{EAR}\left(1 + \frac{1}{\kappa}\right) \tag{14.117}$$

where

$$\kappa = -\frac{1}{A}\int_A \frac{y}{R+y}dA$$

$$
\begin{aligned}
N_T &= \int_A \alpha E\tau dA - \int_A \frac{E}{R+y}\left(\int_0^y \alpha\tau dy\right)dA \\
M_T &= \int_A \alpha E\tau y dA - \int_A \frac{Ey}{R+y}\left(\int_0^y \alpha\tau dy\right)dA
\end{aligned} \tag{14.118}
$$

As τ is constant, we get

$$
\begin{aligned}
\kappa &= -\frac{1}{h}\int_{-h/2}^{h/2} \frac{y}{R+y}dy \\
N_T &= \alpha E\tau b\left(h - \int_{-h/2}^{h/2} \frac{y}{R+y}dy\right) = \alpha E\tau bh(1+\kappa) \\
M_T &= -\alpha E\tau b\int_{-h/2}^{h/2} \frac{y^2}{R+y}dy \\
&= -\alpha E\tau b\int_{-h/2}^{h/2} \left(y - \frac{yR}{R+y}\right)dy = -\alpha E\tau bh\kappa R
\end{aligned} \tag{14.119}
$$

Therefore

$$
\begin{aligned}
\varepsilon_0 &= \frac{N_T}{EA} + \frac{M_T}{EAR} = \frac{1}{EA}\left\{\alpha E\tau bh(1+\kappa) + \frac{1}{R}(-\alpha E\tau bh\kappa R)\right\} = \alpha\tau \\
\omega_0 &= \frac{N_T}{EA} + \frac{M_T}{EAR}\left(1+\frac{1}{\kappa}\right) \\
&= \frac{1}{EA}\left\{\alpha E\tau bh(1+\kappa) + \frac{1}{R}(-\alpha E\tau bh\kappa R)\left(1+\frac{1}{\kappa}\right)\right\} = 0
\end{aligned} \tag{14.120}
$$

Then

$$
\begin{aligned}
\sigma_{\theta\theta} &= -\alpha E\tau + \frac{E}{R+y}\left(\varepsilon_0 R + \omega_0 y + \int_0^y \alpha\tau dy\right) \\
&= -\alpha E\tau + \frac{E}{R+y}(\alpha\tau R + \alpha\tau y) = -\alpha E\tau + \alpha E\tau = 0 \\
\rho &= \frac{1+\varepsilon_0}{1+\omega_0}R = (1+\alpha\tau)R
\end{aligned} \tag{Answer}
$$

Using Eqs. (14.37) and (14.120), we get

$$
\Delta\theta = \omega_0 d\theta = 0 \tag{Answer}
$$

Problem 14.14. Derive the thermal stress in the curved beam for three kinds of boundary conditions:

[1] perfectly clamped ends

[2] free expansion and restrained bending
[3] restrained expansion and free bending

Solution. [1] As perfectly clamped ends are considered, ϵ_0 and ω_0 in Eq. (14.39) become zero. Then thermal stress is given by

$$\sigma_{\theta\theta} = -\alpha E\tau + \frac{E}{R+y}\int_0^y \alpha\tau\, dy \qquad \text{(Answer)}$$

[2] The conditions of free expansion and restrained bending give

$$\int_A \sigma_{\theta\theta}\, dA = 0, \quad \omega_0 = 0 \qquad (14.121)$$

Substitution of Eq. (14.39) into Eq. (14.121) yields

$$\epsilon_0 = \frac{N_T}{EA(1+\kappa)} \qquad (14.122)$$

Then thermal stress is

$$\sigma_{\theta\theta} = -\alpha E\tau + \frac{E}{R+y}\left(\frac{N_T R}{EA(1+\kappa)} + \int_0^y \alpha\tau\, dy\right) \qquad \text{(Answer)}$$

[3] The conditions of restrained expansion and free bending give

$$\epsilon_0 = 0, \quad \int_A \sigma_{\theta\theta} y\, dA = 0 \qquad (14.123)$$

Substitution of Eq. (14.39) into Eq. (14.123) yields to

$$\omega_0 = \frac{M_T}{EAR\kappa} \qquad (14.124)$$

Then thermal stress is

$$\sigma_{\theta\theta} = -\alpha E\tau + \frac{E}{R+y}\left(\frac{M_T}{EAR\kappa} y + \int_0^y \alpha\tau\, dy\right) \qquad \text{(Answer)}$$

Problem 14.15. When a partial circular curved beam with a radius R, shown in Fig. 14.5 is subjected to the temperature change $\tau(y)$ along the radial direction, derive the thermal deflection at a free end B by use of the Castigliano theorem.[3] One end A of the partial circular curved beam is clamped and the other end B is free.

[3] The Castigliano theorem: The displacement δ_{P_i} of the point where the load P_i is applied in the direction of the load is given by the partial derivative of the strain energy $U(P_0, P_1, P_2, ...)$ with respect to the load P_i.

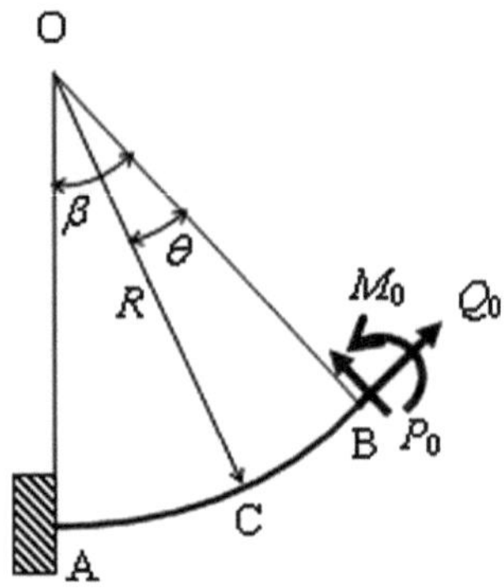

Fig. 14.5 A partial circular curved beam

Solution. The strain energy U due to the bending stress σ is given by

$$U = \int_V \frac{\sigma^2}{2E} dV \tag{14.125}$$

Substitution of Eq. (14.45) into Eq. (14.125) gives

$$\begin{aligned} U = \int_V \frac{1}{2E}\Bigg[-\alpha E\tau + \frac{E}{R+y}\int_0^y \alpha\tau dy \\ + \frac{1}{A}\left(N + N_T + \frac{M_M + M_T}{R} + \frac{M_M + M_T}{\kappa R}\frac{y}{R+y}\right)\Bigg]^2 dV \end{aligned} \tag{14.126}$$

We apply the virtual axial force Q_0, the virtual lateral force P_0, the virtual bending moment M_0 on the end B. Using the Castigliano theorem, the displacements on the end B can be obtained as

$$\begin{aligned} \delta_{P_0} &= \int_s \left\{ \frac{1}{EA}\int_A F(s,y)\left[\frac{\partial N}{\partial P_0} + \frac{\partial M_M}{\partial P_0}\frac{1}{R}(1 + \frac{1}{\kappa}\frac{y}{R+y})\right] dA \right\} ds \\ \delta_{Q_0} &= \int_s \left\{ \frac{1}{EA}\int_A F(s,y)\left[\frac{\partial N}{\partial Q_0} + \frac{\partial M_M}{\partial Q_0}\frac{1}{R}(1 + \frac{1}{\kappa}\frac{y}{R+y})\right] dA \right\} ds \\ \delta_{M_0} &= \int_s \left\{ \frac{1}{EA}\int_A F(s,y)\left[\frac{\partial N}{\partial M_0} + \frac{\partial M_M}{\partial M_0}\frac{1}{R}(1 + \frac{1}{\kappa}\frac{y}{R+y})\right] dA \right\} ds \end{aligned} \tag{14.127}$$

where

$$\begin{aligned} F(s,y) = -\alpha E\tau + \frac{E}{R+y}\int_0^y \alpha\tau dy \\ + \frac{1}{A}\left(N + N_T + \frac{M_M + M_T}{R} + \frac{M_M + M_T}{\kappa R}\frac{y}{R+y}\right) \end{aligned} \tag{14.128}$$

The the axial force N, the lateral force F_M, the bending moment M_M at the position (R, θ) are obtained from the equilibrium of the forces and moment

$$
\begin{aligned}
&N = -P_0\sin\theta + Q_0\cos\theta, \quad F_M = P_0\cos\theta + Q_0\sin\theta \\
&M_M = M_0 + P_0R\sin\theta + Q_0R(1-\cos\theta)
\end{aligned} \tag{14.129}
$$

Substituting of Eq. (14.129) into Eq. (14.127), performing the partial differentiation and taking into consideration of zero virtual forces and $ds = Rd\theta$, we get

$$
\begin{aligned}
\delta_R = \delta_{P_0} &= \frac{R}{EA}\int_0^{\beta}\left\{\int_A\left[-\alpha E\tau + \frac{E}{R+y}\int_0^{y}\alpha\tau dy\right.\right.\\
&\left.\left.+\frac{1}{A}\left(N_T + \frac{M_T}{R} + \frac{M_T}{\kappa R}\frac{y}{R+y}\right)\right]\frac{\sin\theta}{\kappa}\frac{y}{R+y}dA\right\}d\theta \\
\delta_\theta = \delta_{Q_0} &= \frac{R}{EA}\int_0^{\beta}\left\{\int_A\left[-\alpha E\tau + \frac{E}{R+y}\int_0^{y}\alpha\tau dy\right.\right.\\
&\left.\left.+\frac{1}{A}\left(N_T + \frac{M_T}{R} + \frac{M_T}{\kappa R}\frac{y}{R+y}\right)\right]\left[1 + \frac{1-\cos\theta}{\kappa}\frac{y}{R+y}\right]dA\right\}d\theta \\
\delta_M = \delta_{M_0} &= \frac{1}{EA}\int_0^{\beta}\left\{\int_A\left[-\alpha E\tau + \frac{E}{R+y}\int_0^{y}\alpha\tau dy\right.\right.\\
&\left.\left.+\frac{1}{A}\left(N_T + \frac{M_T}{R} + \frac{M_T}{\kappa R}\frac{y}{R+y}\right)\right]\left(1 + \frac{1}{\kappa}\frac{y}{R+y}\right)dA\right\}d\theta \quad \text{(Answer)}
\end{aligned} \tag{14.130}
$$

Problem 14.16. Derive the thermal deflection at a free end B, when the quarter circular curved beam with a radius R and the rectangular section $b \times h$ is subjected to the linear temperature change $\tau(y) = C_1 + C_2y$ along the radial direction. One end A of the quarter circular curved beam is clamped and the other end B is free.

Solution. Equation (14.130) reduce to for this problem

$$
\begin{aligned}
\delta_R = \delta_{P_0} &= \frac{bR}{EA}\int_0^{\pi/2}\left\{\int_{-h/2}^{h/2}\left[-\alpha E(C_1 + C_2y)\right.\right.\\
&+\frac{E}{R+y}\int_0^{y}\alpha(C_1 + C_2y)dy \\
&\left.\left.+\frac{1}{A}\left(N_T + \frac{M_T}{R} + \frac{M_T}{\kappa R}\frac{y}{R+y}\right)\right]\frac{\sin\theta}{\kappa}\frac{y}{R+y}dy\right\}d\theta \\
\delta_\theta = \delta_{Q_0} &= \frac{bR}{EA}\int_0^{\pi/2}\left\{\int_{-h/2}^{h/2}\left[-\alpha E(C_1 + C_2y)\right.\right.\\
&+\frac{E}{R+y}\int_0^{y}\alpha(C_1 + C_2y)dy
\end{aligned}
$$

$$
\begin{aligned}
&+\frac{1}{A}\left(N_T+\frac{M_T}{R}+\frac{M_T}{\kappa R}\frac{y}{R+y}\right)\bigg]\left[1+\frac{1-\cos\theta}{\kappa}\frac{y}{R+y}\right]dy\bigg\}d\theta\\
\delta_M=\delta_{M_0}=&\frac{b}{EA}\int_0^{\pi/2}\bigg\{\int_{-h/2}^{h/2}\bigg[-\alpha E(C_1+C_2y)\\
&+\frac{E}{R+y}\int_0^y\alpha(C_1+C_2y)dy\\
&+\frac{1}{A}\left(N_T+\frac{M_T}{R}+\frac{M_T}{\kappa R}\frac{y}{R+y}\right)\bigg]\left(1+\frac{1}{\kappa}\frac{y}{R+y}\right)dy\bigg\}d\theta
\end{aligned}
\tag{14.131}
$$

We use the integration as follows:

$$
\begin{aligned}
&\int_{-h/2}^{h/2}\frac{y}{R+y}dy=h-R\ln\left(\frac{2R+h}{2R-h}\right)\\
&\int_{-h/2}^{h/2}\frac{y^2}{R+y}dy=-hR+R^2\ln\left(\frac{2R+h}{2R-h}\right)\\
&\int_{-h/2}^{h/2}\frac{y^3}{R+y}dy=\frac{h}{12}(h^2+12R^2)-R^3\ln\left(\frac{2R+h}{2R-h}\right)\\
&\int_{-h/2}^{h/2}\frac{y^2}{(R+y)^2}dy=\frac{h(8R^2-h^2)}{4R^2-h^2}-2R\ln\left(\frac{2R+h}{2R-h}\right)\\
&\int_{-h/2}^{h/2}\frac{y^3}{(R+y)^2}dy=-2hR\frac{6R^2-h^2}{4R^2-h^2}+3R^2\ln\left(\frac{2R+h}{2R-h}\right)
\end{aligned}
\tag{14.132}
$$

N_T and N_T are evaluated as

$$
\begin{aligned}
N_T&=\int_A\alpha E\tau(y)dA-\int_A\frac{E}{R+y}\left[\int_A\alpha E\tau(y)dy\right]dA\\
&=\int_A\alpha E(C_1+C_2y)dA-\int_A\frac{E}{R+y}\left[\int_A\alpha E(C_1+C_2y)dy\right]dA\\
&=\alpha EC_1bR\ln\left(\frac{2R+h}{2R-h}\right)-\frac{1}{2}\alpha EC_2b\left[-hR+R^2\ln\left(\frac{2R+h}{2R-h}\right)\right]\\
M_T&=\int_A\alpha E\tau(y)ydA-\int_A\frac{Ey}{R+y}\left[\int_A\alpha E\tau(y)dy\right]dA\\
&=\int_A\alpha E(C_1+C_2y)ydA-\int_A\frac{Ey}{R+y}\left[\int_A\alpha E(C_1+C_2y)dy\right]dA\\
&=-\alpha EC_1b\left[-hR+R^2\ln\left(\frac{2R+h}{2R-h}\right)\right]\\
&\quad-\frac{1}{2}\alpha EC_2b\left[-\frac{h^3}{12}+hR^2-R^3\ln\left(\frac{2R+h}{2R-h}\right)\right]
\end{aligned}
\tag{14.133}
$$

Substituting Eqs. (14.132) and (14.133) into Eq. (14.131), the thermal deflections at the free end B can be obtained as follows:

$$\begin{aligned}
\delta_R = \delta_{P_0} = \frac{\alpha R}{\kappa h}\Bigg\{ & \frac{1}{\alpha Ebh}\left(N_T + \frac{M_T}{R} - C_1\alpha Ebh\right)\left[h - R\ln\left(\frac{2R+h}{2R-h}\right)\right] \\
& - C_2\left[-hR + R^2\ln\left(\frac{2R+h}{2R-h}\right)\right] \\
& + \left(C_1 + \frac{M_T}{\kappa\alpha ERbh}\right)\left[\frac{h(8R^2-h^2)}{4R^2-h^2} - 2R\ln\left(\frac{2R+h}{2R-h}\right)\right] \\
& + \frac{1}{2}C_2\left[3R^2\ln\left(\frac{2R+h}{2R-h}\right) - 2hR\frac{6R^2-h^2}{4R^2-h^2}\right]\Bigg\}
\end{aligned}$$

$$\begin{aligned}
\delta_\theta = \delta_{Q_0} = \frac{\pi\alpha R}{2h}\Bigg\{ & \frac{1}{\alpha Eb}\left(N_T + \frac{M_T}{R} - C_1\alpha Ebh\right) \\
& + \left(C_1 + \frac{M_T}{\kappa\alpha ERbh}\right)\left[h - R\ln\left(\frac{2R+h}{2R-h}\right)\right] \\
& + \frac{1}{2}C_2\left[-hR + R^2\ln\left(\frac{2R+h}{2R-h}\right)\right] \\
& + (1-\frac{2}{\pi})\frac{1}{\kappa}\Bigg(\frac{1}{\alpha Ebh}\left(N_T + \frac{M_T}{R} - C_1\alpha Ebh\right)\left[h - R\ln\left(\frac{2R+h}{2R-h}\right)\right] \\
& - C_2\left[-hR + R^2\ln\left(\frac{2R+h}{2R-h}\right)\right] \\
& + \left(C_1 + \frac{M_T}{\kappa\alpha ERbh}\right)\left[\frac{h(8R^2-h^2)}{4R^2-h^2} - 2R\ln\left(\frac{2R+h}{2R-h}\right)\right] \\
& + \frac{1}{2}C_2\left[3R^2\ln\left(\frac{2R+h}{2R-h}\right) - 2hR\frac{6R^2-h^2}{4R^2-h^2}\right]\Bigg)\Bigg\}
\end{aligned}$$

$$\begin{aligned}
\delta_M = \delta_{M_0} = \frac{\pi\alpha}{2h}\Bigg\{ & \frac{1}{\alpha Eb}\left(N_T + \frac{M_T}{R} - C_1\alpha Ebh\right) \\
& + \left(C_1 + \frac{M_T}{\kappa\alpha ERbh}\right)\left[h - R\ln\left(\frac{2R+h}{2R-h}\right)\right] \\
& + \frac{1}{2}C_2\left[-hR + R^2\ln\left(\frac{2R+h}{2R-h}\right)\right] \\
& + \frac{1}{\kappa}\Bigg(\frac{1}{\alpha Ebh}\left(N_T + \frac{M_T}{R} - C_1\alpha Ebh\right)\left[h - R\ln\left(\frac{2R+h}{2R-h}\right)\right] \\
& - C_2\left[-hR + R^2\ln\left(\frac{2R+h}{2R-h}\right)\right] \\
& + \left(C_1 + \frac{M_T}{\kappa\alpha ERbh}\right)\left[\frac{h(8R^2-h^2)}{4R^2-h^2} - 2R\ln\left(\frac{2R+h}{2R-h}\right)\right]
\end{aligned}$$

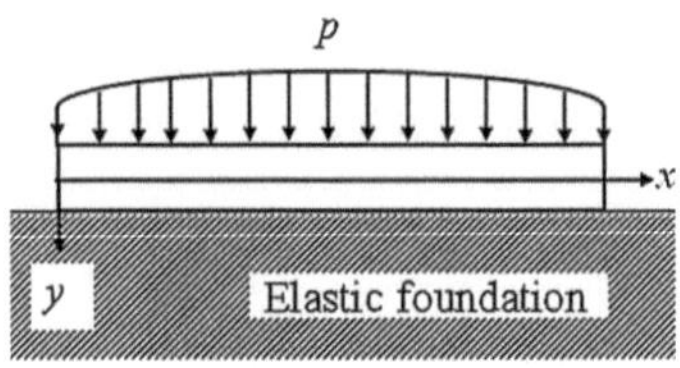

Fig. 14.6 A beam on elastic foundation

$$+\frac{1}{2}C_2\left[3R^2\ln\left(\frac{2R+h}{2R-h}\right)-2hR\frac{6R^2-h^2}{4R^2-h^2}\right]\Big)\Big\} \quad \text{(Answer)}$$

Problem 14.17. Derive the deflection and the thermal stress in a beam with a rectangular cross section $b \times h$ on the elastic foundation with simply supported edges subjected to the temperature change

$$\tau = \sum_{i=0}^{n}(T_{2i}y^{2i} + T_{2i+1}y^{2i+1}) \tag{14.134}$$

Solution. When the beam supported by an elastic foundation along its whole length, shown in Fig. 14.6 is subjected to thermal load $\tau(y)$ and mechanical load $p(x)$, the relation between stress and strain is given by Eq. (14.2)

$$\sigma_x = -\alpha E\tau + \epsilon_0 E + E\frac{y}{\rho} \tag{14.135}$$

where the axial strain ϵ_0 and the curvature $1/\rho$ at the neutral plane $y = 0$ are

$$\epsilon_0 = \frac{\alpha}{A}\int_A \tau(y)\,dA\,, \quad \frac{1}{\rho} = \frac{M_M}{EI} + \frac{\alpha}{I}\int_A \tau(y)y\,dA \tag{14.136}$$

When the beam is deflected, the reaction from the elastic foundation which is proportional to the deflection v supports the beam. The reaction per unit length along the beam can be expressed by kv, where k is a constant. The relation between the bending moment M_M and the load $p - kv$ is given by[4]

$$\frac{d^2M_M}{dx^2} = -p + kv \tag{14.137}$$

Using Eqs. (14.32), (14.136) and (14.137), the differential equation for the deflection v is

$$\frac{d^4v}{dx^4} + \frac{k}{EI}v = \frac{p}{EI} - \frac{\alpha}{I}\frac{d^2}{dx^2}\int_A \tau(y)y\,dA \tag{14.138}$$

[4] See: S. Timoshenko, *Strength of Materials*, Part 1 Elementary, 3rd edn. (Van Nostrand Reinhold, New York, 1995), Eqs. (50) and (51), pp. 77–78.

Since the temperature change (14.134) is independent of a variable x and the mechanical load p does not act on the beam, the differential equation for deflection (14.138) reduces to

$$\frac{d^4 v}{dx^4} + \frac{k}{EI} v = 0 \tag{14.139}$$

The boundary conditions for simply supported edges are

$$v = 0, \quad \frac{d^2 v}{dx^2} + \frac{\alpha}{I} \int_A \tau(y) y \, dA = 0 \quad \text{at} \quad x = 0, \ l \tag{14.140}$$

The general solution of Eq. (14.139) is

$$v(x) = e^{\beta x}(C_1 \cos \beta x + C_2 \sin \beta x) + e^{-\beta x}(C_3 \cos \beta x + C_4 \sin \beta x) \tag{14.141}$$

From the boundary conditions (14.140), we get the unknown constants

$$\begin{aligned}
C_1 &= B \frac{\sin \beta l(\cosh \beta l - \cos \beta l)}{4\beta^2(\sinh^2 \beta l + \sin^2 \beta l)} \\
C_2 &= B \frac{e^{-\beta l} \sinh \beta l - \sin^2 \beta l - \sinh \beta l \cos \beta l}{4\beta^2(\sinh^2 \beta l + \sin^2 \beta l)} \\
C_3 &= -B \frac{\sin \beta l(\cosh \beta l - \cos \beta l)}{4\beta^2(\sinh^2 \beta l + \sin^2 \beta l)} \\
C_4 &= B \frac{e^{\beta l} \sinh \beta l + \sin^2 \beta l - \sinh \beta l \cos \beta l}{4\beta^2(\sinh^2 \beta l + \sin^2 \beta l)}
\end{aligned} \tag{14.142}$$

where

$$B = \alpha \sum_{i=0}^{n} \frac{3}{2i+3} \left(\frac{h}{2}\right)^{2i} T_{2i+1} \tag{14.143}$$

Then, the deflection v is

$$\begin{aligned}
v = \alpha \sum_{i=0}^{n} \frac{3}{2i+3} T_{2i+1} \left(\frac{h}{2}\right)^{2i} \frac{\cosh \beta l - \cos \beta l}{2\beta^2(\sinh^2 \beta l + \sin^2 \beta l)} \\
\times [\sinh \beta x \sin \beta(l-x) + \sinh \beta(l-x) \sin \beta x] \quad \text{(Answer)}
\end{aligned} \tag{14.144}$$

The thermal stress is determined as

$$\sigma_x = -\alpha E \tau(x, y) + \frac{\alpha E}{A} \int_A \tau(x, y) \, dA - Ey \frac{d^2 v}{dx^2} \tag{14.145}$$

Substitution of the deflection (14.144) into Eq. (14.145) gives the thermal stress in the beam

$$\sigma_x = -\alpha E \sum_{i=0}^{n} \left\{ T_{2i} y^{2i} + T_{2i+1} y^{2i+1} - \frac{T_{2i}}{2i+1} \left(\frac{h}{2} \right)^{2i} \right.$$

$$- \frac{3}{2i+3} T_{2i+1} \left(\frac{h}{2} \right)^{2i} y \frac{\cosh \beta l - \cos \beta l}{\sinh^2 \beta l + \sin^2 \beta l}$$

$$\left. \times \left[\cosh \beta x \cos \beta (l - x) + \cosh \beta (l - x) \cos \beta x \right] \right\} \qquad \text{(Answer)}$$

Chapter 15
Heat Conduction

In this chapter the Fourier heat conduction equation along with the boundary conditions and the initial conditions for various coordinate systems are recalled. One-dimensional heat conduction problems in Cartesian coordinates, cylindrical coordinates and spherical coordinates are treated for both the steady and the transient temperature fields. The particular problems and solutions for heat conduction in a strip, a solid cylinder, a hollow circular cylinder and a hollow sphere are presented for various boundary conditions. [See also Chap. 22.]

15.1 Heat Conduction Equation

Heat conduction equation
The Fourier law of heat conduction is

$$q = -\lambda \frac{\partial T}{\partial n} \tag{15.1}$$

where q denotes the heat flux with dimension $[\mathrm{W/m^2}]$ and λ is the thermal conductivity of the solid with dimension $[\mathrm{W/(m \cdot K)}]$. Here, $\partial/\partial n$ denotes differentiation along out-drawn normal n to the isothermal surface.

The Fourier heat conduction equation for the homogeneous isotropic solid based on the Fourier law of heat conduction (15.1) is

$$c\rho \frac{\partial T}{\partial t} = \lambda \nabla^2 T + Q \tag{15.2}$$

An alternative form is

$$\frac{1}{\kappa} \frac{\partial T}{\partial t} = \nabla^2 T + \frac{Q}{\lambda} \tag{15.2$'$}$$

M. Reza Eslami et al., *Theory of Elasticity and Thermal Stresses*, Solid Mechanics and Its Applications 197, DOI: 10.1007/978-94-007-6356-2_15,

where

$$\kappa = \frac{\lambda}{c\rho} \tag{15.3}$$

in which Q is the internal heat generation per unit volume per unit time, c is the specific heat with dimension [J/(kg · K)], ρ is the density with dimension [kg/m^3] of the solid, and κ means the thermal diffusivity with dimension [m^2/s], and the expression for the Laplacian operator ∇^2 is different for each coordinate system:

$$\begin{aligned}
\nabla^2 &= \frac{\partial^2}{\partial x^2} + \frac{\partial^2}{\partial y^2} + \frac{\partial^2}{\partial z^2} && \text{: for Cartesian coordinates} \\
&= \frac{\partial^2}{\partial r^2} + \frac{1}{r}\frac{\partial}{\partial r} + \frac{1}{r^2}\frac{\partial^2}{\partial \theta^2} + \frac{\partial^2}{\partial z^2} && \text{: for cylindrical coordinates} \\
&= \frac{\partial^2}{\partial r^2} + \frac{2}{r}\frac{\partial}{\partial r} + \frac{1}{r^2 \sin\theta}\frac{\partial}{\partial \theta}\left(\sin\theta \frac{\partial}{\partial \theta}\right) + \frac{1}{r^2 \sin^2\theta}\frac{\partial^2}{\partial \phi^2} \\
& && \text{: for spherical coordinates}
\end{aligned} \tag{15.4}$$

The heat conduction equation for a nonhomogeneous anisotropic solid is

$$c\rho\frac{\partial T}{\partial t} = \frac{\partial}{\partial x}\left(\lambda_x \frac{\partial T}{\partial x}\right) + \frac{\partial}{\partial y}\left(\lambda_y \frac{\partial T}{\partial y}\right) + \frac{\partial}{\partial z}\left(\lambda_z \frac{\partial T}{\partial z}\right) + Q \tag{15.5}$$

where λ_x, λ_y, and λ_z denote the thermal conductivities along the x, y, and z directions, respectively, and depend on the position.

The heat conduction equation for a nonhomogeneous isotropic solid is

$$c\rho\frac{\partial T}{\partial t} = \frac{\partial}{\partial x}\left(\lambda \frac{\partial T}{\partial x}\right) + \frac{\partial}{\partial y}\left(\lambda \frac{\partial T}{\partial y}\right) + \frac{\partial}{\partial z}\left(\lambda \frac{\partial T}{\partial z}\right) + Q \tag{15.6}$$

The heat conduction equation for homogeneous anisotropic solid is

$$c\rho\frac{\partial T}{\partial t} = \lambda_x \frac{\partial^2 T}{\partial x^2} + \lambda_y \frac{\partial^2 T}{\partial y^2} + \lambda_z \frac{\partial^2 T}{\partial z^2} + Q \tag{15.7}$$

The heat conduction equation for a homogeneous isotropic solid without internal heat generation is

$$\frac{1}{\kappa}\frac{\partial T}{\partial t} = \nabla^2 T \tag{15.8}$$

The steady state heat conduction equation for the homogeneous isotropic solid with the internal heat generation Q is

$$\nabla^2 T + \frac{Q}{\lambda} = 0 \tag{15.9}$$

The steady state heat conduction equation for the homogeneous isotropic solid without internal heat generation is

$$\nabla^2 T = 0 \tag{15.10}$$

Boundary conditions

When heat transfer between the boundary surface of the solid and the surrounding medium occurs by convection, the boundary condition is

$$-\lambda \frac{\partial T}{\partial n} + q_b = h(T - \Theta) \tag{15.11}$$

where h denotes the heat transfer coefficient with dimension $[\mathrm{W}/(\mathrm{m}^2 \cdot \mathrm{K})]$, q_b means heat generation per unit area per unit time on the boundary surface and Θ is the temperature of the surrounding medium which is a given function of position and time.

When the surfaces of two solids are in perfect thermal contact, the temperature on the contact surface and the heat flow through the contact surface are the same for both solids

$$T_1 = T_2, \quad \lambda_1 \frac{\partial T_1}{\partial n} = \lambda_2 \frac{\partial T_2}{\partial n} \tag{15.12}$$

where subscripts 1 and 2 refer to the solid 1 and 2, respectively, and n is the common normal direction on the contact surface.

Initial condition

When the transient heat conduction Eq. (15.2) is considered, an initial condition which expresses the temperature distribution in the solid at initial time must be specified

$$T = \Phi(P) \tag{15.13}$$

where $\Phi(P)$ is the initial temperature distribution and P is a position in the solid.

15.2 One-Dimensional Heat Conduction Problems

Temperature in a strip

The heat conduction Eq. (15.9) simplifies to the form for one-dimensional steady state heat conduction problems of a homogeneous isotropic solid with the internal heat generation Q

$$\frac{d^2 T}{dx^2} = -\frac{Q}{\lambda} \tag{15.14}$$

If there is no internal heat generation Q, Eq. (15.14) reduces to

$$\frac{d^2T}{dx^2} = 0 \tag{15.15}$$

The steady temperature in a strip of width l with constant internal heat generation Q is given for the heat transfer boundary conditions

$$\begin{aligned} T &= T_a + (T_b - T_a)\frac{h_b(h_a x + \lambda)}{\lambda(h_a + h_b) + h_a h_b l} \\ &+ \frac{Ql^2}{2\lambda}\left[\frac{(2\lambda + h_b l)(h_b x + \lambda)}{\lambda(h_a + h_b)l + h_a h_b l^2} - \frac{x^2}{l^2}\right] \end{aligned} \tag{15.16}$$

where T_a and T_b are the temperatures of the surrounding media, h_a and h_b are the heat transfer coefficients, and subscripts a and b denote boundaries at $x = 0$ and $x = l$, respectively.

The heat conduction Eq. (15.8) simplifies to the form for one-dimensional transient heat conduction problems of a homogeneous isotropic solid

$$\frac{\partial T}{\partial t} = \kappa \frac{\partial^2 T}{\partial x^2} \tag{15.17}$$

The transient temperature in a strip of width l with the initial temperature $T_i(x)$ is given for the heat transfer boundary conditions

$$\begin{aligned} T(x,t) &= T_a + (T_b - T_a)\frac{h_b(h_a x + \lambda)}{\lambda(h_a + h_b) + h_a h_b l} \\ &+ 2\sum_{n=1}^{\infty} \frac{(\lambda^2 s_n^2 + h_b^2)(h_a \sin s_n x + \lambda s_n \cos s_n x)e^{-\kappa s_n^2 t}}{l(\lambda^2 s_n^2 + h_a^2)(\lambda^2 s_n^2 + h_b^2) + \lambda(h_a + h_b)(\lambda^2 s_n^2 + h_a h_b)} \\ &\times \int_0^l \left\{T_i(x) - \left[T_a + (T_b - T_a)\frac{h_b(h_a x + \lambda)}{\lambda(h_a + h_b) + h_a h_b l}\right]\right\} \\ &\times (h_a \sin s_n x + \lambda s_n \cos s_n x)\, dx \end{aligned} \tag{15.18}$$

where s_n are eigenvalues of the transcendental equation

$$\tan s_n l = \frac{\lambda s_n (h_a + h_b)}{\lambda^2 s_n^2 - h_a h_b} \tag{15.19}$$

Temperature in a hollow cylinder

The heat conduction Eq. (15.9) simplifies to the form for one-dimensional steady state heat conduction problems of the homogeneous isotropic cylinder with an internal heat generation Q

$$\frac{d^2T}{dr^2} + \frac{1}{r}\frac{dT}{dr} = -\frac{O}{\lambda} \tag{15.20}$$

Furthermore, if there is no internal heat generation Q, Eq. (15.20) reduces to

$$\frac{d^2T}{dr^2} + \frac{1}{r}\frac{dT}{dr} = 0 \tag{15.21}$$

The steady temperature in a hollow cylinder of inner radius a and outer radius b with constant internal heat generation Q is given for the heat transfer boundary conditions

$$\begin{aligned} T &= T_a + (T_b - T_a)\frac{\ln\frac{r}{a} + \frac{\lambda}{h_a a}}{\ln\frac{b}{a} + \frac{\lambda}{h_a a} + \frac{\lambda}{h_b b}} \\ &\quad - \frac{Q}{4\lambda}r^2 + \frac{Q}{4\lambda}\frac{\begin{pmatrix} a^2\left(1 - 2\frac{\lambda}{a h_a}\right)\ln\frac{b}{r} + b^2\left(1 + 2\frac{\lambda}{b h_b}\right)\ln\frac{r}{a} \\ + \frac{\lambda}{a h_a}b^2 + \frac{\lambda}{b h_b}a^2 + 2(b^2 - a^2)\frac{\lambda}{a h_a}\frac{\lambda}{b h_b} \end{pmatrix}}{\ln\frac{b}{a} + \frac{\lambda}{h_a a} + \frac{\lambda}{h_b b}} \end{aligned} \tag{15.22}$$

where subscripts a and b denote the boundaries at $r = a$ and $r = b$, respectively.

The heat conduction Eq. (15.8) simplifies to the form for one-dimensional transient heat conduction problems of a homogeneous isotropic cylinder

$$\frac{\partial T}{\partial t} = \kappa\left(\frac{\partial^2 T}{\partial r^2} + \frac{1}{r}\frac{\partial T}{\partial r}\right) \tag{15.23}$$

The transient temperature in a hollow cylinder of inner radius a and outer radius b with the initial temperature $T_i(r)$ is given for the heat transfer boundary conditions

$$\begin{aligned} T &= T_a + (T_b - T_a)\frac{\ln(r/a) + \lambda/(h_a a)}{\ln(b/a) + \lambda/(h_a a) + \lambda/(h_b b)} \\ &\quad - \pi\sum_{n=1}^{\infty}\frac{T_a h_a - T_b h_b G_n}{(h_a^2 + \lambda^2 s_n^2) - (h_b^2 + \lambda^2 s_n^2)G_n^2} f(s_n, r)e^{-\kappa s_n^2 t} \\ &\quad - \frac{\pi^2}{2}\sum_{n=1}^{\infty}\frac{s_n^2 f(s_n, r)}{(h_a^2 + \lambda^2 s_n^2) - (h_b^2 + \lambda^2 s_n^2)G_n^2}\int_a^b T_i(\eta) f(s_n, \eta)\eta d\eta\, e^{-\kappa s_n^2 t} \end{aligned} \tag{15.24}$$

where s_n are eigenvalues of the transcendental equation

$$\begin{aligned}&[h_a J_0(s_n a) + \lambda s_n J_1(s_n a)][h_b Y_0(s_n b) - \lambda s_n Y_1(s_n b)]\\&- [h_b J_0(s_n b) - \lambda s_n J_1(s_n b)][h_a Y_0(s_n a) + \lambda s_n Y_1(s_n a)] = 0\end{aligned} \tag{15.25}$$

and

$$G_n = \frac{h_a J_0(s_n a) + \lambda s_n J_1(s_n a)}{h_b J_0(s_n b) - \lambda s_n J_1(s_n b)} = \frac{h_a Y_0(s_n a) + \lambda s_n Y_1(s_n a)}{h_b Y_0(s_n b) - \lambda s_n Y_1(s_n b)} \tag{15.26}$$

$$\begin{aligned}f(s_n, r) = &J_0(s_n r)[h_a Y_0(s_n a) + \lambda s_n Y_1(s_n a)]\\&- Y_0(s_n r)[h_a J_0(s_n a) + \lambda s_n J_1(s_n a)]\end{aligned} \tag{15.27}$$

in which $J_0(sr)$ is the Bessel function of the first kind of order zero, and $Y_0(sr)$ is the Bessel function of the second kind of order zero.

Temperature in a hollow sphere

The heat conduction Eq. (15.9) simplifies to the form for one-dimensional steady state heat conduction problems of the homogeneous isotropic sphere with the internal heat generation Q

$$\frac{d^2T}{dr^2} + \frac{2}{r}\frac{dT}{dr} = -\frac{Q}{\lambda} \tag{15.28}$$

Furthermore, if there is no internal heat generation Q, Eq. (15.28) reduces to

$$\frac{d^2T}{dr^2} + \frac{2}{r}\frac{dT}{dr} = 0 \tag{15.29}$$

The steady temperature in a hollow sphere of inner radius a and outer radius b with constant internal heat generation Q is given for the heat transfer boundary conditions

$$\begin{aligned}T = T_a + (T_b - T_a)&\frac{1 + \dfrac{\lambda}{h_a a} - \dfrac{a}{r}}{1 - \dfrac{a}{b} + \dfrac{\lambda}{h_a a} + \dfrac{a}{b}\dfrac{\lambda}{h_b b}} - \frac{Q}{6\lambda}r^2\\
&- \frac{Q}{6\lambda}b^2 \frac{\left(\begin{aligned}&\left[1 - \left(\frac{a}{b}\right)^2 + 2\frac{\lambda}{a h_a}\left(\frac{a}{b}\right)^2 + 2\frac{\lambda}{b h_b}\right]\frac{a}{r}\\&-\left[\left(1 + 2\frac{\lambda}{b h_b}\right)\left(1 + \frac{\lambda}{a h_a}\right) - \left(\frac{a}{b}\right)^3\left(1 - 2\frac{\lambda}{a h_a}\right)\left(1 - \frac{\lambda}{b h_b}\right)\right]\end{aligned}\right)}{1 - \dfrac{a}{b} + \dfrac{\lambda}{h_a a} + \dfrac{a}{b}\dfrac{\lambda}{h_b b}}\end{aligned} \tag{15.30}$$

The heat conduction Eq. (15.8) simplifies to the form for one-dimensional transient heat conduction problems of a homogeneous isotropic sphere

$$\frac{\partial T}{\partial t} = \kappa \left(\frac{\partial^2 T}{\partial r^2} + \frac{2}{r}\frac{\partial T}{\partial r} \right) \tag{15.31}$$

The transient temperature in a hollow sphere of inner radius a and outer radius b with the initial temperature $T_i(r)$ is given for the heat transfer boundary conditions

$$\begin{aligned} T(r,t) &= T_a + (T_b - T_a)\frac{1 + \dfrac{\lambda}{h_a a} - \dfrac{a}{r}}{1 - \dfrac{a}{b} + \dfrac{\lambda}{h_a a} + \dfrac{a}{b}\dfrac{\lambda}{h_b b}} \\ &+ \sum_{n=1}^{\infty} \frac{C_n}{r}\Big[(h_a a + \lambda)\sin s_n(r-a) + \lambda s_n a \cos s_n(r-a)\Big] e^{-\kappa s_n^2 t} \end{aligned} \tag{15.32}$$

where coefficients C_n are

$$C_n = \frac{\left(\begin{aligned} &2[(h_b b - \lambda)^2 + \lambda^2 s_n^2 b^2] \int_a^b [T_i(\eta) - T_s(\eta)]\eta \\ &\times [(h_a a + \lambda)\sin s_n(\eta - a) + \lambda s_n a \cos s_n(\eta - a)]\, d\eta \end{aligned} \right)}{\left(\begin{aligned} &(b-a)[(h_a a + \lambda)^2 + \lambda^2 s_n^2 a^2][(h_b b - \lambda)^2 + \lambda^2 s_n^2 b^2] \\ &+ \lambda[b(h_a a + \lambda) + a(h_b b - \lambda)][(h_a a + \lambda)(h_b b - \lambda) + \lambda^2 s_n^2 ab] \end{aligned} \right)} \tag{15.33}$$

and s_n are eigenvalues of the transcendental equation

$$\begin{aligned} &\Big[(h_a a + \lambda)(h_b b - \lambda) - \lambda^2 s_n^2 ab\Big] \sin s_n(b-a) \\ &+ \lambda s_n [b(h_a a + \lambda) + a(h_b b - \lambda)] \cos s_n(b-a) = 0 \end{aligned} \tag{15.34}$$

15.3 Problems and Solutions Related to Heat Conduction

Problem 15.1. When the boundary conditions of a strip are given by following three cases (1), (2) and (3), find the steady temperatures in the strip.

[1] Prescribed surface temperatures T_a and T_b at both surfaces $x = 0$ and $x = l$, respectively.
[2] Prescribed surface temperature T_a at the left surface $x = 0$ and constant heat flux q_b $(= -\lambda(dT/dx))$ at the right surface $x = l$.
[3] Constant heat flux q_a $(= \lambda(dT/dx))$ at the left surface $x = 0$ and prescribed surface temperature T_b at the right surface $x = l$.

Solution. The general solution of the governing Eq. (15.15) is

$$T = A + Bx \tag{15.35}$$

where unknown coefficients A and B are determined by each boundary condition.

[1] The boundary conditions are

$$T = T_a \quad \text{on} \quad x = 0, \qquad T = T_b \quad \text{on} \quad x = l \tag{15.36}$$

Substituting Eq. (15.35) into Eq. (15.36), unknown coefficients A and B can be determined as

$$A = T_a, \qquad B = (T_b - T_a)\frac{1}{l} \tag{15.37}$$

The temperature becomes

$$T = T_a + (T_b - T_a)\frac{x}{l} \qquad \text{(Answer)}$$

[2] The boundary conditions are

$$T = T_a \quad \text{on} \quad x = 0, \qquad -\lambda\frac{dT}{dx} = q_b \quad \text{on} \quad x = l \tag{15.38}$$

Substituting Eq. (15.35) into Eq. (15.38), unknown coefficients A and B can be determined as

$$A = T_a, \quad B = -\frac{q_b}{\lambda} \tag{15.39}$$

The temperature becomes

$$T = T_a - \frac{q_b}{\lambda}x \qquad \text{(Answer)}$$

[3] The boundary conditions are

$$\lambda\frac{dT}{dx} = q_a \quad \text{on} \quad x = 0, \quad T = T_b \quad \text{on } x = l \tag{15.40}$$

Substituting Eq. (15.35) into Eq. (15.40), unknown coefficients A and B can be determined as

$$A = T_b - \frac{q_a}{\lambda}l, \quad B = \frac{q_a}{\lambda} \tag{15.41}$$

The temperature becomes

$$T = T_b - \frac{q_a}{\lambda}(l - x) \qquad \text{(Answer)}$$

Problem 15.2. When the boundary conditions of a hollow cylinder are given by following five cases (1)–(5), find the steady temperatures in the hollow cylinder.

[1] Prescribed surface temperatures T_a and T_b at both surfaces $r = a$ and $r = b$, respectively.

[2] Prescribed surface temperature T_a at the inner surface $r = a$, and heat transfer between the outer surface and the surrounding medium with temperature T_b at the outer surface $r = b$.
[3] Prescribed surface temperature T_b at the outer surface $r = b$, and heat transfer between the inner surface and the surrounding medium with temperature T_a at the inner surface $r = a$.
[4] Constant heat flux q_a ($= \lambda(dT/dr)$) at the inner surface $r = a$, and heat transfer between the outer surface and the surrounding medium with temperature T_b at the outer surface $r = b$.
[5] Constant heat flux q_b ($= -\lambda(dT/dr)$) at the outer surface $r = b$, and heat transfer between the inner surface and the surrounding medium with temperature T_a at the inner surface $r = a$.

Solution. The general solution of the governing Eq. (15.21) is

$$T = A + B \ln r \tag{15.42}$$

where unknown coefficients A and B are determined by each boundary condition.

[1] The boundary conditions are

$$T = T_a \quad \text{on} \quad r = a, \qquad T = T_b \quad \text{on} \quad r = b \tag{15.43}$$

Substituting Eq. (15.42) into Eq. (15.43), unknown coefficients A and B can be determined as

$$A = T_a - \frac{T_b - T_a}{\ln b - \ln a} \ln a, \quad B = \frac{T_b - T_a}{\ln b - \ln a} \tag{15.44}$$

The temperature becomes

$$T = T_a + (T_b - T_a) \frac{\ln \dfrac{r}{a}}{\ln \dfrac{b}{a}} \qquad \text{(Answer)}$$

[2] The boundary conditions are

$$T = T_a \quad \text{on} \quad r = a, \qquad -\lambda \frac{\partial T}{\partial r} = h_b (T - T_b) \quad \text{on} \quad r = b \tag{15.45}$$

Substitution of Eq. (15.42) into Eq. (15.45) gives

$$A = T_a - (T_b - T_a) \frac{\ln a}{\ln \dfrac{b}{a} + \dfrac{\lambda}{h_b b}}, \quad B = (T_b - T_a) \frac{1}{\ln \dfrac{b}{a} + \dfrac{\lambda}{h_b b}} \tag{15.46}$$

The temperature becomes

$$T = T_a + (T_b - T_a)\frac{\ln\frac{r}{a}}{\ln\frac{b}{a} + \frac{\lambda}{h_b b}} \qquad \text{(Answer)}$$

[3] The boundary conditions are

$$\lambda\frac{dT}{dr} = h_a(T - T_a) \quad \text{on} \quad r = a, \quad T = T_b \quad \text{on} \quad r = b \tag{15.47}$$

Substitution of Eq. (15.42) into Eq. (15.47) gives

$$A = T_a - (T_b - T_a)\frac{\ln a - \frac{\lambda}{h_a a}}{\ln\frac{b}{a} + \frac{\lambda}{h_a a}}, \quad B = (T_b - T_a)\frac{1}{\ln\frac{b}{a} + \frac{\lambda}{h_a a}} \tag{15.48}$$

The temperature becomes

$$T = T_a + (T_b - T_a)\frac{\ln\frac{r}{a} + \frac{\lambda}{h_a a}}{\ln\frac{b}{a} + \frac{\lambda}{h_a a}} \qquad \text{(Answer)}$$

[4] The boundary conditions are

$$\lambda\frac{dT}{dr} = q_a \quad \text{on} \quad r = a, \quad -\lambda\frac{dT}{dr} = h_b(T - T_b) \quad \text{on} \quad r = b \tag{15.49}$$

Substitution of Eq. (15.42) into Eq. (15.49) gives

$$A = T_b - \frac{q_a a}{\lambda}(\ln b + \frac{\lambda}{h_b b}), \quad B = \frac{q_a a}{\lambda} \tag{15.50}$$

The temperature becomes

$$T = T_b + \frac{q_a a}{\lambda}\left(\ln\frac{r}{b} - \frac{\lambda}{h_b b}\right) \qquad \text{(Answer)}$$

[5] The boundary conditions are

$$\lambda\frac{dT}{dr} = h_a(T - T_a) \quad \text{on} \quad r = a, \quad -\lambda\frac{dT}{dr} = q_b \quad \text{on} \quad r = b \tag{15.51}$$

Substitution of Eq. (15.42) into Eq. (15.51) gives

$$A = T_a + \frac{q_b b}{\lambda}(\ln a - \frac{\lambda}{h_a a}), \quad B = -\frac{q_b b}{\lambda} \tag{15.52}$$

The temperature becomes

$$T = T_a - \frac{q_b b}{\lambda}\left(\ln \frac{r}{a} + \frac{\lambda}{h_a a}\right) \qquad \text{(Answer)}$$

Problem 15.3. When the boundary conditions of a hollow sphere are given by following five cases (1)–(5), find the steady temperatures in the hollow sphere.

[1] Prescribed surface temperatures T_a and T_b at both surfaces $r = a$ and $r = b$, respectively.
[2] Prescribed surface temperature T_a at the inner surface $r = a$, and heat transfer between the outer surface and the surrounding medium with temperature T_b at the outer surface $r = b$.
[3] Prescribed surface temperature T_b at the outer surface $r = b$, and heat transfer between the inner surface and the surrounding medium with temperature T_a at the inner surface $r = a$.
[4] Constant heat flux q_a $(= \lambda(dT/dr))$ at the inner surface $r = a$, and heat transfer between the outer surface and the surrounding medium with temperature T_b at the outer surface $r = b$.
[5] Constant heat flux q_b $(= -\lambda(dT/dr))$ at the outer surface $r = b$, and heat transfer between the inner surface and the surrounding medium with temperature T_a at the inner surface $r = a$.

Solution. The general solution of the governing Eq. (15.29) is

$$T = A + \frac{B}{r} \tag{15.53}$$

where unknown coefficients A and B are determined by each boundary condition.

[1] The boundary conditions are

$$T = T_a \quad \text{on} \quad r = a, \quad T = T_b \quad \text{on} \quad r = b \tag{15.54}$$

Substitution of Eq. (15.53) into Eq. (15.54) gives

$$A + \frac{B}{a} = T_a\,, \quad A + \frac{B}{b} = T_b \tag{15.55}$$

Equation (15.55) gives

$$A = T_a + \frac{T_b - T_a}{1 - \dfrac{a}{b}}, \quad B = -\frac{a(T_b - T_a)}{1 - \dfrac{a}{b}} \tag{15.56}$$

The temperature becomes

$$T = T_a + (T_b - T_a)\frac{1 - \dfrac{a}{r}}{1 - \dfrac{a}{b}} \quad \text{(Answer)}$$

[2] The boundary conditions are

$$T = T_a \quad \text{on} \quad r = a, \quad -\lambda\frac{dT}{dr} = h_b(T - T_b) \quad \text{on} \quad r = b \tag{15.57}$$

Substitution of Eq. (15.53) into Eq. (15.57) gives

$$A = T_a + (T_b - T_a)\frac{1}{1 - \dfrac{a}{b} + \dfrac{a}{b}\dfrac{\lambda}{bh_b}}, \quad B = -\frac{a(T_b - T_a)}{1 - \dfrac{a}{b} + \dfrac{a}{b}\dfrac{\lambda}{bh_b}} \tag{15.58}$$

The temperature becomes

$$T = T_a + (T_b - T_a)\frac{1 - \dfrac{a}{r}}{1 - \dfrac{a}{b} + \dfrac{a}{b}\dfrac{\lambda}{bh_b}} \quad \text{(Answer)}$$

[3] The boundary conditions are

$$\lambda\frac{dT}{dr} = h_a(T - T_a) \quad \text{on} \quad r = a, \quad T = T_b \quad \text{on} \quad r = b \tag{15.59}$$

Substitution of Eq. (15.53) into Eq. (15.59) gives

$$A = T_a + (T_b - T_a)\frac{1 + \dfrac{\lambda}{ah_a}}{1 - \dfrac{a}{b} + \dfrac{\lambda}{ah_a}}, \quad B = -\frac{a(T_b - T_a)}{1 - \dfrac{a}{b} + \dfrac{\lambda}{ah_a}} \tag{15.60}$$

The temperature becomes

$$T = T_a + (T_b - T_a)\frac{1 + \dfrac{\lambda}{ah_a} - \dfrac{a}{r}}{1 - \dfrac{a}{b} + \dfrac{\lambda}{ah_a}} \qquad \text{(Answer)}$$

[4] The boundary conditions are

$$\lambda\frac{dT}{dr} = q_a \quad \text{on} \quad r = a, \quad -\lambda\frac{dT}{dr} = h_b(T - T_b) \quad \text{on} \quad r = b \tag{15.61}$$

Substitution of Eq. (15.53) into Eq. (15.61) gives

$$B = -\frac{a^2 q_a}{\lambda}, \quad A = T_b + \frac{aq_a}{\lambda}\frac{a}{b}\left(1 - \frac{\lambda}{bh_b}\right) \tag{15.62}$$

The temperature becomes

$$T = T_b - \frac{aq_a}{\lambda}\frac{a}{b}\left(\frac{b}{r} - 1 + \frac{\lambda}{bh_b}\right) \qquad \text{(Answer)}$$

[5] The boundary conditions are

$$\lambda\frac{dT}{dr} = h_a(T - T_a) \quad \text{on} \quad r = a, \quad -\lambda\frac{dT}{dr} = q_b \quad \text{on} \quad r = b \tag{15.63}$$

Substitution of Eq. (15.53) into Eq. (15.63) gives

$$B = \frac{b^2 q_b}{\lambda}, \quad A = T_a - \frac{bq_b}{\lambda}\frac{b}{a}\left(1 + \frac{\lambda}{ah_a}\right) \tag{15.64}$$

The temperature becomes

$$T = T_a - \frac{bq_b}{\lambda}\frac{b}{a}\left(1 + \frac{\lambda}{ah_a} - \frac{a}{r}\right) \qquad \text{(Answer)}$$

Problem 15.4. Find the steady temperature in a strip of width l with constant internal heat generation Q under heat transfer boundary conditions.

Solution. The steady state heat conduction equation is given by Eq. (15.14). The boundary conditions are

$$\lambda\frac{dT}{dx} = h_a(T - T_a) \quad \text{on} \quad x = 0, \quad -\lambda\frac{dT}{dx} = h_b(T - T_b) \quad \text{on} \quad x = l \tag{15.65}$$

where T_a and T_b are the temperatures of the surrounding media, h_a and h_b are the heat transfer coefficients, and subscripts a and b denote boundaries at $x = 0$ and $x = l$, respectively. A general solution of Eq. (15.14) is

$$T = A + Bx - \frac{Q}{2\lambda}x^2 \tag{15.66}$$

The coefficients A and B can be determined from the boundary conditions (15.65)

$$\begin{aligned} A &= T_a + \frac{\lambda h_b(T_b - T_a)}{\lambda(h_a + h_b) + h_a h_b l} + \frac{Q}{2\lambda}\frac{\lambda l(2\lambda + h_b l)}{\lambda(h_a + h_b) + h_a h_b l} \\ B &= \frac{h_a h_b(T_b - T_a)}{\lambda(h_a + h_b) + h_a h_b l} + \frac{Q}{2\lambda}\frac{h_a l(2\lambda + h_b l)}{\lambda(h_a + h_b) + h_a h_b l} \end{aligned} \tag{15.67}$$

Substitution of Eq. (15.67) into Eq. (15.66) gives the temperature

$$\begin{aligned} T = T_a &+ (T_b - T_a)\frac{h_b(h_a x + \lambda)}{\lambda(h_a + h_b) + h_a h_b l} \\ &+ \frac{Ql^2}{2\lambda}\left[\frac{(2\lambda + h_b l)(h_b x + \lambda)}{\lambda(h_a + h_b)l + h_a h_b l^2} - \frac{x^2}{l^2}\right] \end{aligned} \tag{Answer}$$

Problem 15.5. Find the steady temperature in a hollow cylinder of inner radius a and outer radius b with constant internal heat generation Q under heat transfer boundary conditions.

Solution. The steady heat conduction equation in the hollow cylinder is given by Eq. (15.20). The boundary conditions are

$$\begin{aligned} \lambda\frac{dT}{dr} &= h_a(T - T_a) \quad \text{on} \quad r = a \\ -\lambda\frac{dT}{dr} &= h_b(T - T_b) \quad \text{on} \quad r = b \end{aligned} \tag{15.68}$$

The general solution of Eq. (15.20) is

$$T = A + B\ln r - \frac{Q}{4\lambda}r^2 \tag{15.69}$$

The coefficients A and B can be determined from the boundary conditions (15.68)

$$A = T_a - (T_b - T_a)\frac{\ln a - \dfrac{\lambda}{h_a a}}{\ln\dfrac{b}{a} + \dfrac{\lambda}{h_a a} + \dfrac{\lambda}{h_b b}}$$

$$+\frac{Q}{4\lambda}\frac{\left(\begin{array}{l} a^2\ln b - b^2\ln a - 2\dfrac{\lambda}{bh_b}b^2\ln a - 2\dfrac{\lambda}{ah_a}a^2\ln b \\ +\dfrac{\lambda}{ah_a}b^2 + \dfrac{\lambda}{bh_b}a^2 + 2(b^2-a^2)\dfrac{\lambda}{ah_a}\dfrac{\lambda}{bh_b} \end{array}\right)}{\ln\dfrac{b}{a} + \dfrac{\lambda}{h_a a} + \dfrac{\lambda}{h_b b}}$$

$$B = \frac{T_b - T_a}{\ln\dfrac{b}{a} + \dfrac{\lambda}{h_a a} + \dfrac{\lambda}{h_b b}} + \frac{Q}{4\lambda}\frac{\left(1 + 2\dfrac{\lambda}{bh_b}\right)b^2 - \left(1 - 2\dfrac{\lambda}{ah_a}\right)a^2}{\ln\dfrac{b}{a} + \dfrac{\lambda}{h_a a} + \dfrac{\lambda}{h_b b}} \tag{15.70}$$

Thus the temperature is

$$T = T_a + (T_b - T_a)\frac{\ln\dfrac{r}{a} + \dfrac{\lambda}{h_a a}}{\ln\dfrac{b}{a} + \dfrac{\lambda}{h_a a} + \dfrac{\lambda}{h_b b}}$$

$$-\frac{Q}{4\lambda}r^2 + \frac{Q}{4\lambda}\frac{\left(\begin{array}{l} a^2\left(1 - 2\dfrac{\lambda}{ah_a}\right)\ln\dfrac{b}{r} + b^2\left(1 + 2\dfrac{\lambda}{bh_b}\right)\ln\dfrac{r}{a} \\ +\dfrac{\lambda}{ah_a}b^2 + \dfrac{\lambda}{bh_b}a^2 + 2(b^2-a^2)\dfrac{\lambda}{ah_a}\dfrac{\lambda}{bh_b} \end{array}\right)}{\ln\dfrac{b}{a} + \dfrac{\lambda}{h_a a} + \dfrac{\lambda}{h_b b}} \quad \text{(Answer)}$$

Problem 15.6. Find the steady temperature in a hollow sphere of inner radius a and outer radius b with constant internal heat generation Q under heat transfer boundary conditions.

Solution. The steady heat conduction equation in the hollow sphere is given by Eq. (15.28). The boundary conditions are

$$\begin{aligned} \lambda\frac{dT}{dr} &= h_a(T - T_a) \quad \text{on} \quad r = a \\ -\lambda\frac{dT}{dr} &= h_b(T - T_b) \quad \text{on} \quad r = b \end{aligned} \tag{15.71}$$

The general solution of Eq. (15.28) is

$$T = A + \frac{B}{r} - \frac{Q}{6\lambda}r^2 \tag{15.72}$$

The coefficients A and B can be determined from the boundary conditions (15.71)

$$A = T_a + (T_b - T_a)\frac{1+\dfrac{\lambda}{h_a a}}{1-\dfrac{a}{b}+\dfrac{\lambda}{h_a a}+\dfrac{a}{b}\dfrac{\lambda}{h_b b}} + \frac{Q}{6\lambda}b^2\frac{\left(1+2\dfrac{\lambda}{bh_b}\right)\left(1+\dfrac{\lambda}{ah_a}\right)-\left(\dfrac{a}{b}\right)^3\left(1-2\dfrac{\lambda}{ah_a}\right)\left(1-\dfrac{\lambda}{bh_b}\right)}{1-\dfrac{a}{b}+\dfrac{\lambda}{h_a a}+\dfrac{a}{b}\dfrac{\lambda}{h_b b}}$$

$$B = -(T_b - T_a)\frac{a}{1-\dfrac{a}{b}+\dfrac{\lambda}{h_a a}+\dfrac{a}{b}\dfrac{\lambda}{h_b b}} - \frac{Q}{6\lambda}ab^2\frac{1-\left(\dfrac{a}{b}\right)^2+2\dfrac{\lambda}{ah_a}\left(\dfrac{a}{b}\right)^2+2\dfrac{\lambda}{bh_b}}{1-\dfrac{a}{b}+\dfrac{\lambda}{h_a a}+\dfrac{a}{b}\dfrac{\lambda}{h_b b}} \tag{15.73}$$

Thus, the temperature is

$$T = T_a + (T_b - T_a)\frac{1+\dfrac{\lambda}{h_a a}-\dfrac{a}{r}}{1-\dfrac{a}{b}+\dfrac{\lambda}{h_a a}+\dfrac{a}{b}\dfrac{\lambda}{h_b b}} - \frac{Q}{6\lambda}r^2 - \frac{Q}{6\lambda}b^2\frac{\left(\begin{array}{l}\left[1-\left(\dfrac{a}{b}\right)^2+2\dfrac{\lambda}{ah_a}\left(\dfrac{a}{b}\right)^2+2\dfrac{\lambda}{bh_b}\right]\dfrac{a}{r}\\ -\left[\left(1+2\dfrac{\lambda}{bh_b}\right)\left(1+\dfrac{\lambda}{ah_a}\right)-\left(\dfrac{a}{b}\right)^3\left(1-2\dfrac{\lambda}{ah_a}\right)\left(1-\dfrac{\lambda}{bh_b}\right)\right]\end{array}\right)}{1-\dfrac{a}{b}+\dfrac{\lambda}{h_a a}+\dfrac{a}{b}\dfrac{\lambda}{h_b b}}$$

(Answer)

Problem 15.7. Determine the one-dimensional steady temperature of a two-layered hollow cylinder for the heat transfer boundary conditions.

Solution. The temperature of each layer is

$$T_i = A_i + B_i \ln r \quad (i = 1, 2) \tag{15.74}$$

The boundary conditions at each boundary are

$$
\begin{aligned}
\lambda_1 \frac{dT_1}{dr} &= h_a(T_1 - T_a) && \text{on} \quad r = a \\
T_1 = T_2 \,, \quad \lambda_1 \frac{dT_1}{dr} &= \lambda_2 \frac{dT_2}{dr} && \text{on} \quad r = c \\
-\lambda_2 \frac{dT_2}{dr} &= h_b(T_2 - T_b) && \text{on} \quad r = b
\end{aligned} \tag{15.75}
$$

We get the coefficients A_i and B_i in Eq. (15.74) from (15.74) and (15.75)

$$
\begin{aligned}
A_1 &= T_a - \frac{T_b - T_a}{D} b h_b \lambda_2 (a h_a \ln a - \lambda_1) \,, \quad B_1 = \frac{T_b - T_a}{D} a b h_a h_b \lambda_2 \\
A_2 &= T_a + \frac{T_b - T_a}{D} b h_b \left(a h_a \lambda_2 \ln \frac{c}{a} - a h_a \lambda_1 \ln c + \lambda_1 \lambda_2 \right) \\
B_2 &= \frac{T_b - T_a}{D} a b h_a h_b \lambda_1
\end{aligned} \tag{15.76}
$$

where

$$
D = \lambda_1 \lambda_2 (a h_a + b h_b) + a b h_a h_b \left(\lambda_1 \ln \frac{b}{c} + \lambda_2 \ln \frac{c}{a} \right) \tag{15.77}
$$

Then, the temperature of each layer is

$$
\begin{aligned}
T_1 &= T_a + \frac{T_b - T_a}{D} b h_b \lambda_2 \left(a h_a \ln \frac{r}{a} + \lambda_1 \right) \\
T_2 &= T_a + \frac{T_b - T_a}{D} b h_b \lambda_2 \left\{ a h_a \left(\ln \frac{c}{a} + \frac{\lambda_1}{\lambda_2} \ln \frac{r}{c} \right) + \lambda_1 \right\}
\end{aligned} \tag{Answer}
$$

Problem 15.8. Find the transient temperature in a strip, when the initial temperature is $T_i(r)$, and the boundary conditions of the strip are given by following two cases:

[1] Prescribed surface temperatures T_a and T_b at both surfaces $x = 0$ and $x = l$, respectively.
[2] Prescribed surface temperature T_a at $x = 0$ and constant heat flux $q_b(= -\lambda(\partial T/\partial x))$ at $x = l$.

Solution. When the heat transfer conditions at both surfaces are

$$
\lambda \frac{\partial T}{\partial x} = h_a(T - T_a) \quad \text{on} \quad x = 0, \quad -\lambda \frac{\partial T}{\partial x} = h_b(T - T_b) \quad \text{on} \quad x = l \tag{15.78}
$$

the temperature is given by Eq. (15.18). Comparing between the heat transfer conditions (15.78) and each boundary condition, the temperature for each boundary condition can easily be obtained.

[1] The boundary conditions on $x = 0$ and $x = l$ for this problem are

$$
T = T_a \quad \text{on} \quad x = 0, \quad T = T_b \quad \text{on} \quad x = l \tag{15.79}
$$

Rewriting the boundary conditions (15.78) gives

$$T = T_a + \frac{\lambda}{h_a}\frac{\partial T}{\partial x} \quad \text{on} \quad x = 0, \quad T = T_b - \frac{\lambda}{h_b}\frac{\partial T}{\partial x} \quad \text{on} \quad x = l \tag{15.80}$$

Putting $h_a \to \infty$ and $h_b \to \infty$, Eq. (15.80) reduces to Eq. (15.79). Therefore, we can obtain the temperature from Eq. (15.18) after putting $h_a \to \infty$ and $h_b \to \infty$.

$$\begin{aligned} T(x,t) &= T_a + (T_b - T_a)\frac{x}{l} \\ &\quad + 2\sum_{n=1}^{\infty}\left\{\int_0^l \left[T_i(x) - T_a - (T_b - T_a)\frac{x}{l}\right]\sin n\pi\frac{x}{l}dx\right\} \\ &\quad \times \sin n\pi\frac{x}{l}e^{-\kappa(n\pi/l)^2 t} \end{aligned} \quad \text{(Answer)}$$

[2] The boundary conditions of this case are

$$T = T_a \quad \text{on} \quad x = 0, \quad -\lambda\frac{\partial T}{\partial x} = q_b \quad \text{on} \quad x = l \tag{15.81}$$

If we rewrite $h_b T_b = -q_b$, and put $h_a \to \infty$ and $h_b = 0$ in Eq. (15.78), Eq. (15.78) reduces to Eq. (15.81). Therefore, we can obtain the temperature from Eq. (15.18), after rewriting $h_b T_b = -q_b$ and putting $h_a \to \infty$ and $h_b = 0$.

$$\begin{aligned} T(x,t) &= T_a - \frac{q_b}{\lambda}x \\ &\quad + 2\sum_{n=1}^{\infty}\left\{\int_0^l \left[T_i(x) - T_a + \frac{q_b}{\lambda}x\right]\sin\left(\frac{2n-1}{2}\pi\frac{x}{l}\right)dx\right\} \\ &\quad \times \sin\left(\frac{2n-1}{2}\pi\frac{x}{l}\right)e^{-\kappa[(2n-1)\pi/2]^2 t} \end{aligned} \quad \text{(Answer)}$$

Problem 15.9. When the boundary condition of the solid cylinder is heat transfer between the surface and the surrounding medium with the temperature T_a, and the initial temperature is $T_i(r)$, find the transient temperature in the solid cylinder.

Solution. When the boundary condition of the solid cylinder is heat transfer between the surface of the cylinder and the surrounding medium with the temperature T_a, and the initial temperature is $T_i(r)$, the equations to be solved are

(1) Governing equation

$$\frac{\partial T}{\partial t} = \kappa\left(\frac{\partial^2 T}{\partial r^2} + \frac{1}{r}\frac{\partial T}{\partial r}\right) \tag{15.82}$$

(2) Boundary condition

$$-\lambda \frac{\partial T}{\partial r} = h_a (T - T_a) \quad \text{on} \quad r = a \tag{15.83}$$

(3) Initial condition

$$T = T_i(r) \quad \text{at} \quad t = 0 \tag{15.84}$$

The solution of Eq. (15.82) can be obtained by use of the method of separation of variables. We put the temperature to

$$T(r, t) = f(r)g(t) \tag{15.85}$$

Substitution of Eq. (15.85) into Eq. (15.82) leads to a pair of ordinary differential equation for $g(t)$ and $f(t)$

$$\frac{dg(t)}{dt} + \kappa s^2 g(t) = 0 \tag{15.86}$$

$$\frac{d^2 f(r)}{dr^2} + \frac{1}{r}\frac{df(r)}{dr} + s^2 f(r) = 0 \tag{15.87}$$

where s is arbitrary constant, and Eq. (15.87) is Bessel's differential equation of order zero.[1]

The linearly independent solutions of Eq. (15.86) are

$$g(t) = 1 \quad \text{for} \quad s = 0, \quad g(t) = \exp(-\kappa s^2 t) \quad \text{for} \quad s \neq 0 \tag{15.88}$$

and the linearly independent solutions of Eq. (15.87) are

$$f(r) = \begin{pmatrix} 1 \\ \ln r \end{pmatrix} \quad \text{for} \quad s = 0, \quad f(r) = \begin{pmatrix} J_0(sr) \\ Y_0(sr) \end{pmatrix} \quad \text{for} \quad s \neq 0 \tag{15.89}$$

where $J_0(sr)$ is the Bessel function of the first kind of order zero, and $Y_0(sr)$ is the Bessel function of the second kind of order zero. As these solutions exist for arbitrary values of s, the general solution of temperature $T(r, t)$ may be given by

$$T(r, t) = A_0 + B_0 r + \sum_{n=1}^{\infty} [A_n J_0(s_n r) + B_n Y_0(s_n r)] e^{-\kappa s_n^2 t} \tag{15.90}$$

As the Bessel function $Y_0(sr)$ and $\ln r$ are infinite at $r = 0$, B_0 and B_n must be zero for the solid cylinder. Therefore, the temperature for this problem reduces to

[1] G. N. Watson, Theory of Bessel Functions (2nd ed.), Cambridge University Press, Cambridge (1944).

$$T(r,t) = A_0 + \sum_{n=1}^{\infty} A_n J_0(s_n r) e^{-\kappa s_n^2 t} \tag{15.91}$$

where A_n are unknown coefficients.

Substitution of Eq. (15.91) into the boundary condition (15.83) gives

$$h_a(A_0 - T_a) + \sum_{n=1}^{\infty} A_n [h_a J_0(s_n a) - \lambda s_n J_1(s_n a)] e^{-\kappa s_n^2 t} = 0 \tag{15.92}$$

If $A_0 = T_a$ and s_n are eigenvalues of the eigenfunction

$$h_a J_0(s_n a) - \lambda s_n J_1(s_n a) = 0 \tag{15.93}$$

then, Eq. (15.92) is identically satisfied. Therefore, the temperature is

$$T = T_a + \sum_{n=1}^{\infty} A_n J_0(s_n r) e^{-\kappa s_n^2 t} \tag{15.94}$$

Using the initial condition (15.84), Eq. (15.94) yields

$$\sum_{n=1}^{\infty} A_n J_0(s_n r) = T_i(r) - T_a \tag{15.95}$$

Multiplying both sides of Eq. (15.95) by $r J_0(s_m r)$, and integrating from 0 to a, we find

$$A_m = \frac{2}{a^2[J_0^2(s_m a) + J_1^2(s_m a)]} \int_0^a [T_i(r) - T_a] J_0(s_m r) r dr \tag{15.96}$$

in which the following relations are used

$$\begin{aligned}
&\int_0^a J_0(s_m r) J_0(s_n r) r dr \\
&= \frac{a}{s_m^2 - s_n^2} [s_m J_1(s_m a) J_0(s_n a) - s_n J_1(s_n a) J_0(s_m a)] \\
&= \frac{a}{\lambda(s_m^2 - s_n^2)} [h_a J_0(s_m a) J_0(s_n a) - h_a J_0(s_n a) J_0(s_m a)] = 0 \\
&\qquad\qquad\qquad \text{for} \quad m \neq n \\
&\int_0^a J_0^2(s_m r) r dr = \frac{a^2}{2} [J_0^2(s_m a) + J_1^2(s_m a)]
\end{aligned} \tag{15.97}$$

Therefore, the temperature is given by

$$T(r,t) = T_a - \frac{2}{a^2}\sum_{n=1}^{\infty}\left\{\frac{1}{J_0^2(s_n a) + J_1^2(s_n a)}\int_0^a [T_a - T_i(r)]J_0(s_n r) r dr\right\} \times J_0(s_n r)e^{-\kappa s_n^2 t} \qquad \text{(Answer)}$$

If the surface temperature on the surface $r = a$ is prescribed, the boundary condition is

$$T = T_a \quad \text{on} \quad r = a \tag{15.98}$$

For this case the temperature becomes

$$T(r,t) = T_a - \frac{2}{a^2}\sum_{n=1}^{\infty}\left\{\frac{1}{J_1^2(s_n a)}\int_0^a [T_a - T_i(r)]J_0(s_n r) r\, dr\right\} J_0(s_n r)e^{-\kappa s_n^2 t} \qquad \text{(Answer)}$$

where s_n are eigenvalues of eigenfunction

$$J_0(s_n a) = 0 \tag{15.99}$$

Problem 15.10. Find the transient temperature in a hollow cylinder of inner radius a and outer radius b with the initial temperature $T_i(r)$ under heat transfer boundary conditions.

Solution. The problem to be solved consists of

(1) Governing equation:

$$\frac{\partial T}{\partial t} = \kappa\left(\frac{\partial^2 T}{\partial r^2} + \frac{1}{r}\frac{\partial T}{\partial r}\right) \tag{15.100}$$

(2) Boundary conditions:

$$\begin{aligned} \lambda\frac{\partial T}{\partial r} &= h_a(T - T_a) \quad \text{on} \quad r = a \\ -\lambda\frac{\partial T}{\partial r} &= h_b(T - T_b) \quad \text{on} \quad r = b \end{aligned} \tag{15.101}$$

(3) Initial condition:

$$T = T_i(r) \quad \text{at} \quad t = 0 \tag{15.102}$$

The general solution of Eq. (15.100) is from (15.90)

$$T(r,t) = A_0 + B_0 \ln r + \sum_{n=1}^{\infty}[A_n J_0(s_n r) + B_n Y_0(s_n r)]e^{-\kappa s_n^2 t} \tag{15.103}$$

Substitution of Eq. (15.103) into Eq. (15.101) gives

$$
\begin{aligned}
&ah_a A_0 + B_0(ah_a \ln a - \lambda) - ah_a T_a + \sum_{n=1}^{\infty}\{A_n[ah_a J_0(s_n a) + \lambda s_n a J_1(s_n a)] \\
&\quad + B_n[ah_a Y_0(s_n a) + \lambda s_n a Y_1(s_n a)]\}e^{-\kappa s_n^2 t} = 0 \\
&bh_b A_0 + B_0(bh_b \ln b + \lambda) - bh_b T_b + \sum_{n=1}^{\infty}\{A_n[bh_b J_0(s_n b) - \lambda s_n b J_1(s_n b)] \\
&\quad + B_n[bh_b Y_0(s_n b) - \lambda s_n b Y_1(s_n b)]\}e^{-\kappa s_n^2 t} = 0
\end{aligned}
\tag{15.104}
$$

Equation (15.104) gives

$$
\begin{aligned}
ah_a A_0 + B_0(ah_a \ln a - \lambda) &= ah_a T_a \\
bh_b A_0 + B_0(bh_b \ln b + \lambda) &= bh_b T_b
\end{aligned}
\tag{15.105}
$$

and

$$
\begin{aligned}
A_n[ah_a J_0(s_n a) + \lambda s_n a J_1(s_n a)] + B_n[ah_a Y_0(s_n a) + \lambda s_n a Y_1(s_n a)] &= 0 \\
A_n[bh_b J_0(s_n b) - \lambda s_n b J_1(s_n b)] + B_n[bh_b Y_0(s_n b) - \lambda s_n b Y_1(s_n b)] &= 0
\end{aligned}
\tag{15.106}
$$

Solving Eq. (15.105) for A_0 and B_0, we get

$$
A_0 = T_a - (T_b - T_a)\frac{\ln a - \dfrac{\lambda}{ah_a}}{\ln\dfrac{b}{a} + \dfrac{\lambda}{ah_a} + \dfrac{\lambda}{bh_b}}, \quad B_0 = \frac{T_b - T_a}{\ln\dfrac{b}{a} + \dfrac{\lambda}{ah_a} + \dfrac{\lambda}{bh_b}} \tag{15.107}
$$

Equation (15.106) is satisfied, if s_n are eigenvalues of the transcendental equation

$$
\begin{aligned}
&[h_a J_0(s_n a) + \lambda s_n J_1(s_n a)][h_b Y_0(s_n b) - \lambda s_n Y_1(s_n b)] \\
&- [h_b J_0(s_n b) - \lambda s_n J_1(s_n b)][h_a Y_0(s_n a) + \lambda s_n Y_1(s_n a)] = 0
\end{aligned}
\tag{15.108}
$$

Referring to Eq. (15.108) we may put

$$
G_n = \frac{h_a J_0(s_n a) + \lambda s_n J_1(s_n a)}{h_b J_0(s_n b) - \lambda s_n J_1(s_n b)} = \frac{h_a Y_0(s_n a) + \lambda s_n Y_1(s_n a)}{h_b Y_0(s_n b) - \lambda s_n Y_1(s_n b)} \tag{15.109}
$$

Equation (15.108) can be written as

$$
h_b f_0(s_n, b) - \lambda s_n f_1(s_n, b) = 0 \tag{15.110}
$$

where

$$
\begin{aligned}
f_i(s_n, r) &= J_i(s_n r)[h_a Y_0(s_n a) + \lambda s_n Y_1(s_n a)] \\
&\quad - Y_i(s_n r)[h_a J_0(s_n a) + \lambda s_n J_1(s_n a)] \quad (i = 0, 1) \qquad (15.111)
\end{aligned}
$$

The temperature (15.103) reduces to

$$
T = A_0 + B_0 \ln r + \sum_{n=1}^{\infty} A_n f_0(s_n, r) e^{-\kappa s_n^2 t} \qquad (15.112)
$$

where $A_n/[ah_a Y_0(s_n a) + \lambda s_n Y_1(s_n a)]$ is rewritten as A_n. From the initial condition (15.102), Eq. (15.112) reduces to

$$
\sum_{n=1}^{\infty} A_n f_0(s_n, r) = T_i(r) - (A_0 + B_0 \ln r) \qquad (15.113)
$$

Multiplying $r f_0(s_m r)$ to both sides of Eq. (15.113) and integrating form a to b, we get[2]

$$
\begin{aligned}
A_m &= \frac{\pi^2 s_m^2}{2[(h_b^2 + \lambda^2 s_m^2)G_m^2 - (h_a^2 + \lambda^2 s_m^2)]} \\
&\quad \times \int_a^b [T_i(r) - (A_0 + B_0 \ln r)] f_0(s_m, r) r dr \qquad (15.114)
\end{aligned}
$$

Substitution of Eq. (15.107) into Eq. (15.114), we get

$$
\begin{aligned}
A_m = & \frac{\pi^2 s_m^2}{2[(h_b^2 + \lambda^2 s_m^2)G_m^2 - (h_a^2 + \lambda^2 s_m^2)]} \int_a^b T_i(r) f_0(s_m, r) r dr \\
& - \frac{\pi(T_b h_b G_m - T_a h_a)}{(h_b^2 + \lambda^2 s_m^2)G_m^2 - (h_a^2 + \lambda^2 s_m^2)} \qquad (15.115)
\end{aligned}
$$

Rewriting $f_0(s_m, r)$ as $f(s_m, r)$, the temperature can be expressed as

$$
\begin{aligned}
T = & T_a + (T_b - T_a) \frac{\ln(r/a) + \lambda/(h_a a)}{\ln(b/a) + \lambda/(h_a a) + \lambda/(h_b b)} \\
& - \pi \sum_{n=1}^{\infty} \frac{T_a h_a - T_b h_b G_n}{(h_a^2 + \lambda^2 s_n^2) - (h_b^2 + \lambda^2 s_n^2) G_n^2} f(s_n, r) e^{-\kappa s_n^2 t} \\
& - \frac{\pi^2}{2} \sum_{n=1}^{\infty} \frac{s_n^2 f(s_n, r)}{(h_a^2 + \lambda^2 s_n^2) - (h_b^2 + \lambda^2 s_n^2) G_n^2} \int_a^b T_i(\eta) f(s_n, \eta) \eta d\eta \, e^{-\kappa s_n^2 t}
\end{aligned}
$$

(Answer)

[2] see: Problem 15.11.

Problem 15.11. Derive Eqs. (15.114) and (15.115) from Eq. (15.113) in Problem 15.10.

Solution. From Eq. (15.113), we have

$$\sum_{n=1}^{\infty} A_n f_0(s_n, r) = T_i(r) - (A_0 + B_0 \ln r) \tag{15.116}$$

Multiplying $r f_0(s_m, r)$ to Eq. (15.116) and integrating form a to b, we get

$$\int_a^b f_0(s_m, r) \sum_{n=1}^{\infty} A_n f_0(s_n, r) r dr = \int_a^b f_0(s_m, r)[T_i(r) - (A_0 + B_0 \ln r)] r dr \tag{15.117}$$

where

$$\begin{aligned} f_i(s_n, r) &= J_i(s_n r)[h_a Y_0(s_n a) + \lambda s_n Y_1(s_n a)] \\ &\quad - Y_i(s_n r)[h_a J_0(s_n a) + \lambda s_n J_1(s_n a)] \qquad (i = 0, 1) \end{aligned} \tag{15.118}$$

Before performing integration of Eq. (15.117), we calculate the following functions:

$$\begin{aligned} f_0(s_n, a) &= -\lambda s_n \frac{2}{\pi a s_n} = -\frac{2\lambda}{\pi a} \\ f_1(s_n, a) &= \frac{2h_a}{\pi s_n a} \\ f_0(s_n, b) &= J_0(s_n b)[h_a Y_0(s_n a) + \lambda s_n Y_1(s_n a)] \\ &\quad - Y_0(s_n b)[h_a J_0(s_n a) + \lambda s_n J_1(s_n a)] \\ &= J_0(s_n b) G_n [h_b Y_0(s_n b) - \lambda s_n Y_1(s_n b)] \\ &\quad - Y_0(s_n b) G_n [h_b J_0(s_n b) - \lambda s_n J_1(s_n b)] \\ &= G_n \lambda s_n [J_1(s_n b) Y_0(s_n b) - Y_1(s_n b) J_0(s_n b)] \\ &= G_n \lambda s_n \frac{2}{\pi s_n b} = G_n \frac{2\lambda}{\pi b} \\ f_1(s_n, b) &= J_1(s_n b)[h_a Y_0(s_n a) + \lambda s_n Y_1(s_n a)] \\ &\quad - Y_1(s_n b)[h_a J_0(s_n a) + \lambda s_n J_1(s_n a)] \\ &= J_1(s_n b) G_n [h_b Y_0(s_n b) - \lambda s_n Y_1(s_n b)] \\ &\quad - Y_1(s_n b) G_n [h_b J_0(s_n b) - \lambda s_n J_1(s_n b)] \\ &= G_n h_b [J_1(s_n b) Y_0(s_n b) - Y_1(s_n b) J_0(s_n b)] = G_n \frac{2h_b}{\pi s_n b} \end{aligned} \tag{15.119}$$

in which the following formula of Bessel functions is used:

$$J_{n+1}(z)Y_n(z) - J_n(z)Y_{n+1}(z) = \frac{2}{\pi z} \tag{15.120}$$

and

$$G_n = \frac{h_a J_0(s_n a) + \lambda s_n J_1(s_n a)}{h_b J_0(s_n b) - \lambda s_n J_1(s_n b)} = \frac{h_a Y_0(s_n a) + \lambda s_n Y_1(s_n a)}{h_b Y_0(s_n b) - \lambda s_n Y_1(s_n b)} \tag{15.121}$$

By use of Eq. (15.119), the integration of the left side in Eq. (15.117) becomes

$$\begin{aligned}
&\int_a^b f_0(s_m, r) f_0(s_n, r) r dr \\
&\quad = \frac{b}{s_m^2 - s_n^2}\Big[s_m f_1(s_m, b) f_0(s_n, b) - s_n f_1(s_n, b) f_0(s_m, b)\Big] \\
&\quad\quad - \frac{a}{s_m^2 - s_n^2}\Big[s_m f_1(s_m, a) f_0(s_n, a) - s_n f_1(s_n, a) f_0(s_m, a)\Big] \\
&\quad = \frac{b}{s_m^2 - s_n^2}\Big[\frac{h_b}{\lambda s_m} s_m f_0(s_m, b) f_0(s_n, b) - \frac{h_b}{\lambda s_n} s_n f_0(s_n, b) f_0(s_m, b)\Big] \\
&\quad\quad - \frac{a}{s_m^2 - s_n^2}\Big[s_m \frac{2h_a}{\pi s_m a}\Big(-\frac{2\lambda}{\pi a}\Big) - s_n \frac{2h_a}{\pi s_n a}\Big(-\frac{2\lambda}{\pi a}\Big)\Big] = 0 \\
&\int_a^b f_0^2(s_m, r) r dr = \frac{b^2}{2}\Big[f_0^2(s_m b) + f_1^2(s_m b)\Big] - \frac{a^2}{2}\Big[f_0^2(s_m a) + f_1^2(s_m a)\Big] \\
&\quad = \frac{b^2}{2}\frac{h_b^2 + \lambda^2 s_m^2}{h_b^2} f_1^2(s_m b) - \frac{a^2}{2}[f_0^2(s_m a) + f_1^2(s_m a)] \\
&\quad = \frac{b^2}{2}\frac{h_b^2 + \lambda^2 s_m^2}{h_b^2}\Big(G_m \frac{2h_b}{\pi s_m b}\Big)^2 - \frac{a^2}{2}\Big[\Big(\frac{2h_a}{\pi s_m a}\Big)^2 + \Big(-\frac{2\lambda}{\pi a}\Big)^2\Big] \\
&\quad = \frac{2}{\pi^2 s_m^2}\Big[(h_b^2 + \lambda^2 s_m^2) G_m^2 - (h_a^2 + \lambda^2 s_m^2)\Big]
\end{aligned} \tag{15.122}$$

From Eqs. (15.117) and (15.122), A_m is determined as

$$\begin{aligned}
A_m = &\frac{\pi^2 s_m^2}{2[(h_b^2 + \lambda^2 s_m^2) G_m^2 - (h_a^2 + \lambda^2 s_m^2)]} \\
&\times \int_a^b [T_i(r) - (A_0 + B_0 \ln r)] f_0(s_m, r) r dr
\end{aligned} \tag{Answer}$$

Calculating the following integral:

$$\begin{aligned}
&\int_a^b (A_0 + B_0 \ln r) f_0(s_m, r) r dr \\
&\quad = \Big[(A_0 + B_0 \ln r)\frac{r}{s_m} f_1(s_m, r) - \int \frac{B_0}{s_m} f_1(s_m, r) dr\Big]_a^b
\end{aligned}$$

$$
\begin{aligned}
&= \left[(A_0 + B_0 \ln r)\frac{r}{s_m} f_1(s_m, r) + \frac{B_0}{s_m^2} f_0(s_m, r)\right]_a^b \\
&= \frac{2}{\pi s_m^2}\left\{(h_b G_m - h_a)A_0 + B_0\left[\left(\ln b + \frac{\lambda}{b h_b}\right)h_b G_m\right.\right. \\
&\quad \left.\left. - \left(\ln a - \frac{\lambda}{a h_a}\right)h_a\right]\right\} \\
&= \frac{2}{\pi s_m^2}(h_b G_m - h_a)T_a + \frac{2}{\pi s_m^2}(T_b - T_a)h_b G_m \\
&= \frac{2}{\pi s_m^2}(T_b h_b G_m - T_a h_a)
\end{aligned} \tag{15.123}
$$

we get

$$
\begin{aligned}
A_m = &\frac{\pi^2 s_m^2}{2[(h_b^2 + \lambda^2 s_m^2)G_m^2 - (h_a^2 + \lambda^2 s_m^2)]}\int_a^b T_i(r) f_0(s_m, r) r dr \\
&- \frac{\pi(T_b h_b G_m - T_a h_a)}{(h_b^2 + \lambda^2 s_m^2)G_m^2 - (h_a^2 + \lambda^2 s_m^2)}
\end{aligned} \tag{Answer}
$$

Problem 15.12. Find the solution of the differential equation

$$
\frac{d^2 f(r)}{d^2 r} + \frac{1}{r}\frac{df(r)}{dr} - q^2 f(r) = g(r) \tag{15.124}
$$

Equation (15.124) is appeared in the derivation of the transient temperature in the cylinder when Laplace transform technique is used.

Solution. The solution of homogeneous differential equation of Eq. (15.124) is

$$
f(r) = A I_0(qr) + B K_0(qr) \tag{15.125}
$$

where $I_0(qr)$ and $K_0(qr)$ are modified Bessel functions.

We introduce the method of variation of parameters to obtain the particular solution of Eq. (15.124). We put $f(r)$ instead of Eq. (15.125)

$$
f(r) = A(r) I_0(qr) + B(r) K_0(qr) \tag{15.126}
$$

Differentiation of Eq. (15.126) with respect to r gives

$$
\begin{aligned}
\frac{df(r)}{dr} &= \frac{dA(r)}{dr} I_0(qr) + \frac{dB(r)}{dr} K_0(qr) + A(r)\frac{dI_0(qr)}{dr} + B(r)\frac{dK_0(qr)}{dr} \\
\frac{d^2 f(r)}{dr^2} &= \frac{d}{dr}\left[\frac{dA(r)}{dr} I_0(qr) + \frac{dB(r)}{dr} K_0(qr)\right] + \frac{dA(r)}{dr}\frac{dI_0(qr)}{dr} \\
&\quad + \frac{dB(r)}{dr}\frac{dK_0(qr)}{dr} + A(r)\frac{d^2 I_0(qr)}{dr} + B(r)\frac{d^2 K_0(qr)}{dr}
\end{aligned} \tag{15.127}
$$

Substitution of Eqs. (15.127) into Eq. (15.124) yields

$$\left(\frac{d}{dr}+\frac{1}{r}\right)\left[\frac{dA(r)}{dr}I_0(qr)+\frac{dB(r)}{dr}K_0(qr)\right] + \frac{dA(r)}{dr}\frac{dI_0(qr)}{dr}+\frac{dB(r)}{dr}\frac{dK_0(qr)}{dr}=g(r) \tag{15.128}$$

Equation (15.128) can be satisfied when we take

$$\begin{aligned}&\frac{dA(r)}{dr}I_0(qr)+\frac{dB(r)}{dr}K_0(qr)=0\\&\frac{dA(r)}{dr}\frac{dI_0(qr)}{dr}+\frac{dB(r)}{dr}\frac{dK_0(qr)}{dr}=g(r)\end{aligned} \tag{15.129}$$

Solving Eq. (15.129), we get

$$\begin{aligned}\frac{dA(r)}{dr}&=g(r)\frac{K_0(qr)}{q[I_0(qr)K_1(qr)+I_1(qr)K_0(qr)]}\\\frac{dB(r)}{dr}&=-g(r)\frac{I_0(qr)}{q[I_0(qr)K_1(qr)+I_1(qr)K_0(qr)]}\end{aligned} \tag{15.130}$$

where

$$\frac{dI_0(qr)}{dr}=qI_1(qr),\quad \frac{dK_0}{dr}=-qK_1(qr) \tag{15.131}$$

Using the following formula

$$I_0(qr)K_1(qr)+I_1(qr)K_0(qr)=\frac{1}{qr} \tag{15.132}$$

Equation (15.130) reduce to

$$\frac{dA(r)}{dr}=rg(r)K_0(qr),\quad \frac{dB(r)}{dr}=-rg(r)I_0(qr) \tag{15.133}$$

Then, we obtain

$$A(r)=\int_r rg(r)K_0(qr)dr,\quad B(r)=-\int_r rg(r)I_0(qr)dr \tag{15.134}$$

Therefore the general solution is given by

$$f(r)=AI_0(qr)+BK_0(qr)+\int_r \eta g(\eta)[I_0(qr)K_0(q\eta)-K_0(qr)I_0(q\eta)]d\eta \text{ (Answer)} \tag{15.135}$$

Problem 15.13. Find the transient temperature in a hollow cylinder, when the initial condition is $T_i(r)$, and the boundary conditions of the hollow cylinder are given by following five cases (1)–(5):

[1] Prescribed surface temperatures T_a and T_b at both surfaces $r = a$ and $r = b$, respectively.
[2] Prescribed surface temperature T_a at the inner surface $r = a$, and heat transfer between the outer surface and the surrounding medium with temperature T_b at the outer surface $r = b$.
[3] Prescribed surface temperature T_b at the outer surface $r = b$, and heat transfer between the inner surface and the surrounding medium with temperature T_a at the inner surface $r = a$.
[4] Constant heat flux $q_a (= \lambda(\partial T/\partial r))$ at the inner surface $r = a$, and heat transfer between the outer surface and the surrounding medium with temperature T_b at the outer surface $r = b$.
[5] Constant heat flux $q_b (= -\lambda(\partial T/\partial r))$ at the outer surface $r = b$, and heat transfer between the inner surface and the surrounding medium with temperature T_a at the inner surface $r = a$.

Solution. When the both surfaces are heat transfer conditions

$$\lambda \frac{\partial T}{\partial r} = h_a(T - T_a) \quad \text{on} \quad r = a, \quad -\lambda \frac{\partial T}{\partial r} = h_b(T - T_b) \quad \text{on} \quad r = b \tag{15.136}$$

the temperature is given by Eq. (15.24). Comparing between the heat transfer conditions (15.136) and each boundary condition, the temperature for each boundary condition can be obtained.

[1] The boundary conditions on $r = a$ and $r = b$ for this problem are

$$T = T_a \quad \text{on} \quad r = a, \quad T = T_b \quad \text{on} \quad r = b \tag{15.137}$$

Rewriting the boundary conditions (15.136) gives

$$T = T_a + \frac{\lambda}{h_a} \frac{\partial T}{\partial r} \quad \text{on} \quad r = a, \quad T = T_b - \frac{\lambda}{h_b} \frac{\partial T}{\partial r} \quad \text{on} \quad r = b \tag{15.138}$$

Putting $h_a \to \infty$ and $h_b \to \infty$ in Eq. (15.138), Eq. (15.138) reduces to Eq. (15.137). Therefore, we can obtain the temperature from Eq. (15.24) after putting $h_a \to \infty$ and $h_b \to \infty$.

$$
\begin{aligned}
T &= T_a + (T_b - T_a)\frac{\ln\dfrac{r}{a}}{\ln\dfrac{b}{a}} \\
&\quad - \pi \sum_{n=1}^{\infty} \frac{T_a J_0(s_n b) - T_b J_0(s_n a)}{J_0^2(s_n b) - J_0^2(s_n a)} J_0(s_n b) f(s_n, r)\, e^{-\kappa s_n^2 t} \\
&\quad - \frac{\pi^2}{2} \sum_{n=1}^{\infty} \frac{s_n^2 J_0^2(s_n b) f(s_n, r)}{J_0^2(s_n b) - J_0^2(s_n a)} \int_a^b T_i(\eta) f(s_n, \eta)\eta\, d\eta\, e^{-\kappa s_n^2 t} \qquad \text{(Answer)}
\end{aligned}
$$

where

$$
f(s_n, r) = Y_0(s_n a) J_0(s_n r) - J_0(s_n a) Y_0(s_n r) \tag{15.139}
$$

and s_n are eigenvalues of the eigenfunction $f(s_n, b) = 0$.

[2] The boundary conditions of this case are

$$
T = T_a \quad \text{on} \quad r = a, \quad -\lambda \frac{\partial T}{\partial r} = h_b(T - T_b) \quad \text{on} \quad r = b \tag{15.140}
$$

If we put $h_a \to \infty$ in Eq. (15.138), Eq. (15.138) reduces to Eq. (15.140). Therefore, we can obtain the temperature from Eq. (15.24) after putting $h_a \to \infty$.

$$
\begin{aligned}
T &= T_a + (T_b - T_a)\frac{\ln\dfrac{r}{a}}{\ln\dfrac{b}{a} + \dfrac{\lambda}{h_b b}} \\
&\quad - \pi \sum_{n=1}^{\infty} \frac{(T_a - T_b G_n h_b) f(s_n, r)}{1 - G_n^2(h_b^2 + \lambda^2 s_n^2)} e^{-\kappa s_n^2 t} \\
&\quad - \frac{\pi^2}{2} \sum_{n=1}^{\infty} \frac{s_n^2 f(s_n, r) e^{-\kappa s_n^2 t}}{1 - G_n^2(h_b^2 + \lambda^2 s_n^2)} \int_a^b T_i(\eta) f(s_n, \eta)\eta\, d\eta \qquad \text{(Answer)}
\end{aligned}
$$

where $f(s_n, r)$ and G_n are given by

$$
\begin{aligned}
f(s_n, r) &= Y_0(s_n a) J_0(s_n r) - J_0(s_n a) Y_0(s_n r) \\
G_n &= \frac{J_0(s_n a)}{h_b J_0(s_n b) - \lambda s_n J_1(s_n b)} = \frac{Y_0(s_n a)}{h_b Y_0(s_n b) - \lambda s_n Y_1(s_n b)}
\end{aligned} \tag{15.141}
$$

and s_n are eigenvalues of the eigenfunction

$$
[h_b Y_0(s_n b) - \lambda s_n Y_1(s_n b)] J_0(s_n a) - [h_b J_0(s_n b) - \lambda s_n J_1(s_n b)] Y_0(s_n a) = 0 \tag{15.142}
$$

[3] The boundary conditions of this case are

$$\lambda\frac{\partial T}{\partial r} = h_a(T - T_a) \quad \text{on} \quad r = a, \quad T = T_b \quad \text{on} \quad r = b \tag{15.143}$$

By comparison between Eqs. (15.138) and (15.143), Eq. (15.138) reduces to Eq. (15.143) if we put $h_b \to \infty$. Therefore, we can obtain the temperature from Eq. (15.24) after putting $h_b \to \infty$.

$$\begin{aligned} T = {} & T_a + (T_b - T_a)\frac{\ln\dfrac{r}{a} + \dfrac{\lambda}{h_a a}}{\ln\dfrac{b}{a} + \dfrac{\lambda}{h_a a}} \\ & - \pi\sum_{n=1}^{\infty}\frac{(T_a h_a - T_b G_n) f(s_n, r)}{h_a^2 + \lambda^2 s_n^2 - G_n^2} e^{-\kappa s_n^2 t} \\ & - \frac{\pi^2}{2}\sum_{n=1}^{\infty}\frac{s_n^2 f(s_n, r) e^{-\kappa s_n^2 t}}{h_a^2 + \lambda^2 s_n^2 - G_n^2}\int_a^b T_i(\eta) f(s_n, \eta)\eta\, d\eta \end{aligned} \qquad \text{(Answer)}$$

where $f(s_n, r)$ and G_n are given by

$$\begin{aligned} f(s_n, r) &= [h_a Y_0(s_n a) + \lambda s_n Y_1(s_n a)] J_0(s_n r) \\ &\quad - [h_a J_0(s_n a) + \lambda s_n J_1(s_n a)] Y_0(s_n r) \\ G_n &= \frac{h_a J_0(s_n a) + \lambda s_n J_1(s_n a)}{J_0(s_n b)} = \frac{h_a Y_0(s_n a) + \lambda s_n Y_1(s_n a)}{Y_0(s_n b)} \end{aligned} \qquad \text{(Answer)}$$

and s_n are eigenvalues of the eigenfunction $f(s_n, b) = 0$.

[4] The boundary conditions of this case are

$$\lambda\frac{\partial T}{\partial r} = q_a \quad \text{on} \quad r = a, \quad -\lambda\frac{\partial T}{\partial r} = h_b(T - T_b) \quad \text{on} \quad r = b \tag{15.144}$$

By comparison between Eqs. (15.136) and (15.144), after rewriting $h_a T_a = -q_a$ and putting $h_a \to 0$, Eq. (15.136) reduces to Eq. (15.144). Therefore, we can obtain the temperature from Eq. (15.24), after rewriting $h_a T_a = -q_a$ and putting $h_a \to 0$.

$$\begin{aligned} T = {} & T_b + \frac{q_a a}{\lambda}\left(\ln\frac{r}{b} - \frac{\lambda}{h_b b}\right) \\ & + \pi\sum_{n=1}^{\infty}\frac{(q_a + T_b G_n h_b \lambda s_n) f(s_n, r)}{\lambda s_n[1 - G_n^2(h_b^2 + \lambda^2 s_n^2)]} e^{-\kappa s_n^2 t} \end{aligned}$$

$$- \frac{\pi^2}{2} \sum_{n=1}^{\infty} \frac{s_n^2 f(s_n, r) e^{-\kappa s_n^2 t}}{1 - G_n^2 (h_b^2 + \lambda^2 s_n^2)} \int_a^b T_i(\eta) f(s_n, \eta) \eta \, d\eta \qquad \text{(Answer)}$$

where $f(s_n, r)$ and G_n are given by

$$f(s_n, r) = Y_1(s_n a) J_0(s_n r) - J_1(s_n a) Y_0(s_n r)$$
$$G_n = \frac{J_1(s_n a)}{h_b J_0(s_n b) - \lambda s_n J_1(s_n b)} = \frac{Y_1(s_n a)}{h_b Y_0(s_n b) - \lambda s_n Y_1(s_n b)} \qquad (15.145)$$

and s_n are eigenvalues of the eigenfunction

$$[h_b Y_0(s_n b) - \lambda s_n Y_1(s_n b)] J_1(s_n a) - [h_b J_0(s_n b) - \lambda s_n J_1(s_n b)] Y_1(s_n a) = 0 \qquad (15.146)$$

[5] The boundary conditions of this case are

$$\lambda \frac{\partial T}{\partial r} = h_a (T - T_a) \quad \text{on} \quad r = a, \quad -\lambda \frac{\partial T}{\partial r} = q_b \quad \text{on} \quad r = b \qquad (15.147)$$

By comparison between Eqs. (15.136) and (15.147), after rewriting $h_b T_b = -q_b$ and putting $h_b \to 0$, Eq. (15.136) reduces to Eq. (15.147). Therefore, we can obtain the temperature from the temperature Eq. (15.24), after rewriting $h_b T_b = -q_b$ and putting $h_b \to 0$.

$$T = T_a - \frac{q_b b}{\lambda} \left(\ln \frac{r}{a} + \frac{\lambda}{h_a a} \right)$$
$$- \pi \sum_{n=1}^{\infty} \frac{(T_a h_a \lambda s_n - q_b G_n) f(s_n, r)}{\lambda s_n (h_a^2 + \lambda^2 s_n^2 - G_n^2)} e^{-\kappa s_n^2 t}$$
$$- \frac{\pi^2}{2} \sum_{n=1}^{\infty} \frac{s_n^2 f(s_n, r) e^{-\kappa s_n^2 t}}{h_a^2 + \lambda^2 s_n^2 - G_n^2} \int_a^b T_i(\eta) f(s_n, \eta) \eta \, d\eta \qquad \text{(Answer)}$$

where $f(s_n, r)$ and G_n are given by

$$f(s_n, r) = [h_a Y_0(s_n a) + \lambda s_n Y_1(s_n a)] J_0(s_n r)$$
$$- [h_a J_0(s_n a) + \lambda s_n J_1(s_n a)] Y_0(s_n r)$$
$$G_n = \frac{h_a J_0(s_n a) + \lambda s_n J_1(s_n a)}{J_1(s_n b)} = \frac{h_a Y_0(s_n a) + \lambda s_n Y_1(s_n a)}{Y_1(s_n b)} \qquad (15.148)$$

and s_n are eigenvalues of the eigenfunction

$$[h_a Y_0(s_n a) + \lambda s_n Y_1(s_n a)] J_1(s_n b) - [h_a J_0(s_n a) + \lambda s_n J_1(s_n a)] Y_1(s_n b) = 0 \qquad (15.149)$$

Problem 15.14. Find the transient temperature in the hollow sphere, when the initial condition is $T_i(r)$, and the boundary conditions of a hollow sphere are given by following five cases (1)–(5):

[1] Prescribed surface temperatures T_a and T_b at both surfaces $r = a$ and $r = b$, respectively.
[2] Prescribed surface temperature T_a at the inner surface $r = a$, and heat transfer between the outer surface and the surrounding medium with temperature T_b at the outer surface $r = b$.
[3] Prescribed surface temperature T_b at the outer surface $r = b$, and heat transfer between the inner surface and the surrounding medium with temperature T_a at the inner surface $r = a$.
[4] Constant heat flux $q_a (= \lambda(\partial T/\partial r))$ at the inner surface $r = a$, and heat transfer between the outer surface and the surrounding medium with temperature T_b at the outer surface $r = b$.
[5] Constant heat flux $q_b (= -\lambda(\partial T/\partial r))$ at the outer surface $r = b$, and heat transfer between the inner surface and the surrounding medium with temperature T_a at the inner surface $r = a$.

Solution. When at both surfaces the heat transfer conditions are

$$\lambda \frac{\partial T}{\partial r} = h_a (T - T_a) \quad \text{on} \quad r = a, \quad -\lambda \frac{\partial T}{\partial r} = h_b (T - T_b) \quad \text{on} \quad r = b \tag{15.150}$$

the temperature is given by Eq. (15.32). By comparing between the heat transfer conditions (15.150) and each boundary condition, the temperature for each boundary condition can be obtained.

[1] The boundary conditions on $r = a$ and $r = b$ for this problem are

$$T = T_a \quad \text{on} \quad r = a, \quad T = T_b \quad \text{on} \quad r = b \tag{15.151}$$

Rewriting the boundary conditions (15.150) gives

$$T = T_a + \frac{\lambda}{h_a} \frac{\partial T}{\partial r} \quad \text{on} \quad r = a, \quad T = T_b - \frac{\lambda}{h_b} \frac{\partial T}{\partial r} \quad \text{on} \quad r = b \tag{15.152}$$

If we put $h_a \to \infty$ and $h_b \to \infty$ in Eq. (15.152), Eq. (15.152) reduces to Eq. (15.151). Therefore, we can obtain the temperature from Eq. (15.32) after putting $h_a \to \infty$ and $h_b \to \infty$

$$T = T_a + (T_b - T_a) \frac{1 - \dfrac{a}{r}}{1 - \dfrac{a}{b}} + \frac{2}{(b-a)r} \sum_{n=1}^{\infty} \sin s_n (r-a) e^{-\kappa s_n^2 t}$$

$$\times \int_a^b \left[T_i(\eta) - T_a - (T_b - T_a) \frac{1 - \dfrac{a}{\eta}}{1 - \dfrac{a}{b}} \right] \eta \sin s_n (\eta - a) d\eta \qquad \text{(Answer)}$$

where $s_n = n\pi/(b-a)$.

[2] The boundary conditions of this case are

$$T = T_a \quad \text{on} \quad r = a, \quad -\lambda \frac{\partial T}{\partial r} = h_b(T - T_b) \quad \text{on} \quad r = b \tag{15.153}$$

If we put $h_a \to \infty$ in Eq. (15.152), Eq. (15.152) reduces to Eq. (15.153). Therefore, we can obtain the temperature from Eq. (15.32), after putting $h_a \to \infty$.

$$\begin{aligned} T = {} & T_a + (T_b - T_a) \frac{1 - \dfrac{a}{r}}{1 - \dfrac{a}{b} + \dfrac{a}{b}\dfrac{\lambda}{h_b b}} + \frac{2}{r} \sum_{n=1}^{\infty} \sin s_n(r-a) e^{-\kappa s_n^2 t} \\ & \times \frac{\lambda^2 s_n^2 b^2 + (h_b b - \lambda)^2}{(b-a)[\lambda^2 s_n^2 b^2 + (h_b b - \lambda)^2] + \lambda b(h_b b - \lambda)} \\ & \times \int_a^b \left[T_i(\eta) - T_a - \frac{(T_b - T_a)\left(1 - \dfrac{a}{\eta}\right)}{1 - \dfrac{a}{b} + \dfrac{a}{b}\dfrac{\lambda}{h_b b}} \right] \eta \sin s_n(\eta - a)\, d\eta \end{aligned}$$

(Answer)

where s_n are eigenvalues of the eigenfunction

$$(h_b b - \lambda) \sin s_n\,(b-a) + \lambda s_n b \cos s_n\,(b-a) = 0 \tag{15.154}$$

[3] The boundary conditions of this case are

$$\lambda \frac{\partial T}{\partial r} = h_a(T - T_a) \quad \text{on} \quad r = a, \quad T = T_b \quad \text{on} \quad r = b \tag{15.155}$$

If we put $h_b \to \infty$ in Eq. (15.152), Eq. (15.152) reduces to Eq. (15.155). Therefore, we can obtain the temperature from Eq. (15.32) after putting $h_b \to \infty$.

$$\begin{aligned} T = {} & T_a + (T_b - T_a) \frac{1 + \dfrac{\lambda}{h_a a} - \dfrac{a}{r}}{1 - \dfrac{a}{b} + \dfrac{\lambda}{h_a a}} \\ & + \frac{2}{r} \sum_{n=1}^{\infty} \frac{(h_a a + \lambda) \sin s_n(r-a) + \lambda s_n a \cos s_n(r-a)}{(b-a)[\lambda^2 s_n^2 a^2 + (h_a a + \lambda)^2] + \lambda a(h_a a + \lambda)} e^{-\kappa s_n^2 t} \end{aligned}$$

$$\times \int_a^b \left[T_i(\eta) - T_a - (T_b - T_a) \left(\frac{1 + \dfrac{\lambda}{h_a a} - \dfrac{a}{\eta}}{1 - \dfrac{a}{b} + \dfrac{\lambda}{h_a a}} \right) \right]$$
$$\times\ \eta[(h_a a + \lambda) \sin s_n(\eta - a) + \lambda s_n a \cos s_n(\eta - a)]\, d\eta \qquad \text{(Answer)}$$

where s_n are eigenvalues of the eigenfunction

$$(h_a a + \lambda) \sin s_n\,(b - a) + \lambda s_n a \cos s_n\,(b - a) = 0 \tag{15.156}$$

[4] The boundary conditions of this case are

$$\lambda \frac{\partial T}{\partial r} = q_a \quad \text{on} \quad r = a, \quad -\lambda \frac{\partial T}{\partial r} = h_b(T - T_b) \quad \text{on} \quad r = b \tag{15.157}$$

If we rewrite $h_a T_a = -q_a$ and put $h_a \to 0$ in Eq. (15.150), Eq. (15.150) reduces to Eq. (15.157). Therefore, we can obtain the temperature from Eq. (15.32) after rewriting $h_a T_a = -q_a$ and putting $h_a \to 0$.

$$T = T_b - \frac{q_a a}{\lambda} \frac{a}{b} \left(\frac{b}{r} - 1 + \frac{\lambda}{h_b b} \right)$$
$$+ \frac{2}{r} \sum_{n=1}^{\infty} \frac{[\lambda^2 s_n^2 b^2 + (h_b b - \lambda)^2][\sin s_n(r - a) + s_n a \cos s_n(r - a)]}{\left(\begin{array}{l} (b - a)(1 + s_n^2 a^2)[\lambda^2 s_n^2 b^2 + (h_b b - \lambda)^2] \\ + [b\lambda + a(h_b b - \lambda)][\lambda s_n^2 ab + (h_b b - \lambda)] \end{array} \right)}$$
$$\times\ e^{-\kappa s_n^2 t} \int_a^b \left[T_i(\eta) - T_b + \frac{q_a a}{\lambda} \frac{a}{b} \left(\frac{b}{\eta} - 1 + \frac{\lambda}{h_b b} \right) \right]$$
$$\times\ \eta[\sin s_n(\eta - a) + s_n a \cos s_n(\eta - a)]\, d\eta \qquad \text{(Answer)}$$

where s_n are eigenvalues of the eigenfunction

$$\left(h_b b - \lambda - \lambda s_n^2 ab \right) \sin s_n\,(b - a) + s_n\,[a\,(h_b b - \lambda) + b\lambda] \cos s_n\,(b - a) = 0 \tag{15.158}$$

[5] The boundary conditions of this case are

$$\lambda \frac{\partial T}{\partial r} = h_a(T - T_a) \quad \text{on} \quad r = a, \quad -\lambda \frac{\partial T}{\partial r} = q_b \quad \text{on} \quad r = b \tag{15.159}$$

If we rewrite $h_b T_b = -q_b$ and put $h_b \to 0$ in Eq. (15.150), Eq. (15.150) reduces to Eq. (15.159). Therefore, we can obtain the temperature from Eq. (15.32) after rewriting $h_b T_b = -q_b$ and putting $h_b \to 0$.

$$T = T_a - \frac{q_b b}{\lambda} \frac{b}{a} \left(1 - \frac{a}{r} + \frac{\lambda}{h_a a} \right)$$

$$+\frac{2}{r}\sum_{n=1}^{\infty}\frac{(s_n^2b^2+1)[(h_aa+\lambda)\sin s_n(r-a)+\lambda s_na\cos s_n(r-a)]}{\begin{pmatrix}(b-a)(1+s_n^2b^2)[\lambda^2s_n^2a^2+(h_aa+\lambda)^2]\\+[b(h_aa+\lambda)-a\lambda][\lambda s_n^2ab-(h_aa+\lambda)]\end{pmatrix}}$$
$$\times e^{-\kappa s_n^2t}\int_a^b\left[T_i(\eta)-T_a+\frac{q_bb}{\lambda}\frac{b}{a}\left(1-\frac{a}{\eta}+\frac{\lambda}{h_aa}\right)\right]$$
$$\times\eta[(h_aa+\lambda)\sin s_n(\eta-a)+\lambda s_na\cos s_n(\eta-a)]\,d\eta \qquad \text{(Answer)}$$

where s_n are eigenvalues of the eigenfunction

$$\left(h_aa+\lambda+\lambda s_n^2ab\right)\sin s_n\,(b-a)-s_n\,[b\,(h_aa+\lambda)-a\lambda]\cos s_n\,(b-a)=0 \tag{15.160}$$

Problem 15.15. When a solid cylinder with the initial temperature $T_i(r)$ is exposed to heat transfer between the surface of radius a and the surrounding medium with time dependent temperature $T_a(t)$, find the transient temperature in the solid cylinder.

Solution. The equations to be solved are

(1) Governing equation

$$\frac{1}{\kappa}\frac{\partial T}{\partial t}=\frac{\partial^2T}{\partial r^2}+\frac{1}{r}\frac{\partial T}{\partial r} \tag{15.161}$$

(2) Boundary condition

$$-\lambda\frac{\partial T}{\partial r}=h_a[T-T_a(t)]\quad\text{on}\quad r=a \tag{15.162}$$

(3) Initial condition

$$T=T_i(r)\quad\text{at}\quad t=0 \tag{15.163}$$

Applying the Laplace transform with respect to the time t and taking the initial condition into consideration, we obtain
Governing equation:

$$\frac{d^2\bar{T}}{dr^2}+\frac{1}{r}\frac{d\bar{T}}{dr}-q^2\bar{T}=-\frac{1}{\kappa}T_i(r) \tag{15.161$'$}$$

Boundary condition:

$$-\lambda\frac{d\bar{T}}{dr}=h_a(\bar{T}-\bar{T}_a)\quad\text{on}\quad r=a \tag{15.162$'$}$$

where $q^2=p/\kappa$.

The general solution of Eq. (15.161′) for a solid cylinder is obtained from Eq. (15.135) of Problem 15.12 by putting $B=0$ and $g(\eta)=-T_i(\eta)/\kappa$

$$\begin{aligned}\bar{T} &= AI_0(qr) - \frac{1}{\kappa}\int_0^r T_i(\eta)\eta[I_0(qr)K_0(q\eta) - I_0(q\eta)K_0(qr)]d\eta \\ &= AI_0(qr) + G(qr)\end{aligned} \tag{15.164}$$

where

$$G(qr) = -\frac{1}{\kappa}\int_0^r T_i(\eta)\eta[I_0(qr)K_0(q\eta) - I_0(q\eta)K_0(qr)]d\eta \tag{15.165}$$

Differentiation of Eq. (15.164) with respect to r gives

$$\frac{d\bar{T}}{dr} = AqI_1(qr) + G_1(qr) \tag{15.166}$$

where

$$G_1(qr) \equiv \frac{dG(qr)}{dr} = -\frac{1}{\kappa}\int_0^r T_i(\eta)\eta q[I_1(qr)K_0(q\eta) + I_0(q\eta)K_1(qr)]d\eta \tag{15.167}$$

The boundary condition (15.162′) gives

$$A\lambda qI_1(qa) + \lambda G_1(qa) + Ah_aI_0(qa) + h_aG(qa) = h_a\bar{T}_a \tag{15.168}$$

Then, A is given by

$$\begin{aligned}A &= \frac{h_a\bar{T}_a}{\lambda qI_1(qa) + h_aI_0(qa)} - \frac{\lambda G_1(qa) + h_aG(qa)}{\lambda qI_1(qa) + h_aI_0(qa)} \\ &= \frac{h_a\bar{T}_a}{\lambda qI_1(qa) + h_aI_0(qa)} + \frac{\lambda qK_1(qa) - h_aK_0(qa)}{\lambda qI_1(qa) + h_aI_0(qa)} \\ &\quad \times \frac{1}{\kappa}\int_0^a T_i(\eta)\eta I_0(q\eta)d\eta + \frac{1}{\kappa}\int_0^a T_i(\eta)\eta K_0(q\eta)d\eta\end{aligned} \tag{15.169}$$

Hence, the temperature in the Laplace transformed domain is

$$\begin{aligned}\bar{T} &= \frac{h_a\bar{T}_aI_0(qr)}{\lambda qI_1(qa) + h_aI_0(qa)} \\ &\quad + \frac{\lambda qK_1(qa) - h_aK_0(qa)}{\lambda qI_1(qa) + h_aI_0(qa)}\frac{I_0(qr)}{\kappa}\int_0^a T_i(\eta)\eta I_0(q\eta)d\eta \\ &\quad + \frac{I_0(qr)}{\kappa}\int_0^a T_i(\eta)\eta K_0(q\eta)d\eta + G(qr)\end{aligned} \tag{15.170}$$

or an alternative form

$$\bar{T} = \bar{T}_a\bar{T}_1 + \bar{T}_2 + \bar{T}_3 \tag{15.171}$$

where

$$\bar{T}_1 = \frac{h_a I_0(qr)}{\lambda q I_1(qa) + h_a I_0(qa)} \tag{15.171'}$$

$$\bar{T}_2 = \frac{\lambda q K_1(qa) - h_a K_0(qa)}{\lambda q I_1(qa) + h_a I_0(qa)} \frac{I_0(qr)}{\kappa} \int_0^a T_i(\eta)\eta I_0(q\eta)d\eta \tag{15.171''}$$

$$\bar{T}_3 = \frac{I_0(qr)}{\kappa} \int_0^a T_i(\eta)\eta K_0(q\eta)d\eta + G(qr) \tag{15.171'''}$$

The inverse Laplace transform of Eq. (15.170) reduces to calculation of the sum of the residues at the poles in the inner region with the contour. Since Eq. (15.171‴) has no pole, the inverse Laplace transform of Eq. (15.171‴) reduces to zero. Equations (15.171′) and (15.171″) have poles at $p = -\kappa s_n^2$, and s_n are eigenvalues of the eigenfunction

$$\lambda s_n J_1(s_n a) - h_a J_0(s_n a) = 0 \tag{15.172}$$

The residue of $\bar{T}_1$ is

$$\begin{aligned}
\left.\frac{h_a I_0(qr)e^{pt}}{\frac{d}{dp}[\lambda q I_1(qa) + h_a I_0(qa)]}\right|_{p=-\kappa s_n^2} &= \left.\frac{h_a I_0(qr)e^{-\kappa s_n^2 t}}{\frac{1}{2q\kappa}\frac{d}{dq}[\lambda q I_1(qa) + h_a I_0(qa)]}\right|_{q=is_n} \\
= \frac{2is_n \kappa h_a I_0(is_n r)e^{-\kappa s_n^2 t}}{a[\lambda i s_n I_0(is_n a) + h_a I_1(is_n a)]} &= \frac{2is_n \kappa h_a J_0(s_n r)e^{-\kappa s_n^2 t}}{a[\lambda i s_n J_0(s_n a) + i h_a J_1(s_n a)]} \\
= \frac{2s_n \kappa h_a J_0(s_n r)e^{-\kappa s_n^2 t}}{a[\lambda s_n J_0(s_n a) + h_a J_1(s_n a)]} &= \frac{2\kappa\lambda s_n^2 h_a J_0(s_n r)e^{-\kappa s_n^2 t}}{a(\lambda^2 s_n^2 + h_a^2)J_0(s_n a)}
\end{aligned} \tag{15.173}$$

where $i^2 = -1$. On the other hand,

$$\begin{aligned}
&\lambda q K_1(qa) - h_a K_0(qa)|_{q=is_n} \\
&= \lambda i s_n(-\frac{\pi}{2})[J_1(s_n a) - iY_1(s_n a)] - h_a(-\frac{\pi}{2}i)[J_0(s_n a) - iY_0(s_n a)] \\
&= \frac{\pi}{2}i[h_a J_0(s_n a) - \lambda s_n J_1(s_n a)] + \frac{\pi}{2}[h_a Y_0(s_n a) - \lambda s_n Y_1(s_n a)]
\end{aligned} \tag{15.174}$$

Using Eq. (15.172), Eq. (15.174) reduces to

$$\begin{aligned}
&\lambda q K_1(qa) - h_a K_0(qa)|_{q=is_n} \\
&= \frac{\pi}{2}\left\{h_a Y_0(s_n a) - \frac{\lambda s_n}{J_0(s_n a)}\left[J_1(s_n a)Y_0(s_n a) - \frac{2}{\pi s_n a}\right]\right\}
\end{aligned}$$

$$= \frac{\pi}{2}[h_a J_0(s_n a) - \lambda s_n J_1(s_n a)]\frac{Y_0(s_n a)}{J_0(s_n a)} + \frac{\lambda}{a J_0(s_n a)}$$
$$= \frac{\lambda}{a J_0(s_n a)} \tag{15.175}$$

Residue of $\bar{T}_2$ is

$$\frac{\lambda q K_1(qa) - h_a K_0(qa)}{\frac{d}{dp}[\lambda q I_1(qa) + h_a I_0(qa)]} \frac{I_0(qr)}{\kappa} \int_0^a T_i(\eta)\eta I_0(q\eta) d\eta e^{pt}\Big|_{p=-\kappa s_n^2}$$
$$= \frac{2\lambda^2 s_n^2 J_0(s_n r) e^{-\kappa s_n^2 t}}{a^2(\lambda^2 s_n^2 + h_a^2) J_0^2(s_n a)} \int_0^a T_i(\eta)\eta J_0(s_n \eta) d\eta \tag{15.176}$$

We get

$$L^{-1}[\bar{T}_a \bar{T}_1] = \int_0^t T_a(\tau) T_1(t-\tau) d\tau$$
$$= \frac{2\kappa\lambda h_a}{a} \int_0^t T_a(\tau) \sum_{n=1}^{\infty} \frac{s_n^2 J_0(s_n r) e^{-\kappa s_n^2 (t-\tau)}}{(\lambda^2 s_n^2 + h_a^2) J_0(s_n a)} d\tau \tag{15.177}$$

Therefore, the temperature is given by

$$T = \frac{2\kappa\lambda h_a}{a} \int_0^t T_a(\tau) \sum_{n=1}^{\infty} \frac{s_n^2 J_0(s_n r) e^{-\kappa s_n^2 (t-\tau)}}{(\lambda^2 s_n^2 + h_a^2) J_0(s_n a)} d\tau$$
$$+ \frac{2\lambda^2}{a^2} \sum_{n=1}^{\infty} \frac{s_n^2 J_0(s_n r) e^{-\kappa s_n^2 t}}{(\lambda^2 s_n^2 + h_a^2) J_0^2(s_n a)} \int_0^a T_i(\eta)\eta J_0(s_n \eta) d\eta \tag{Answer}$$

Chapter 16
Basic Equations of Thermoelasticity

In this chapter the basic governing equations of thermoelasticity for three-dimensional bodies are recalled. The equilibrium equations of stresses, Cauchy's relations between the tractions and stresses, and the compatibility equations of strains in Cartesian coordinates are presented. The formulae for coordinate transformation of stress, strain and displacement components are included. A solution of Navier's equations is carried out wherein Goodier's thermoelastic potential is used in conjunction with harmonic functions of various types. The equilibrium equations, stress, strain, the compatibility equations, Navier's equations in cylindrical and spherical coordinates are also presented. [see also Chaps. 2, 3, 6, and 7.]

16.1 Governing Equations of Thermoelasticity

16.1.1 Stress and Strain in a Cartesian Coordinate System

Stress
The equilibrium equations of the elastic body from Eq. (2.21) are

$$\sigma_{ji,j} + F_i = 0 \quad (i, j = 1, 2, 3) \tag{16.1}$$

where σ_{ji} denote the components of stress, F_i mean the components of body force per unit volume. The components of stress satisfy symmetry relations

$$\sigma_{ij} = \sigma_{ji} \quad (i, j = 1, 2, 3) \tag{16.2}$$

Cauchy's fundamental relations are

$$\sigma_{ji} n_j = p_{ni} \quad (i, j = 1, 2, 3) \tag{16.3}$$

M. Reza Eslami et al., *Theory of Elasticity and Thermal Stresses*, Solid Mechanics and Its Applications 197, DOI: 10.1007/978-94-007-6356-2_16,

where n_j denote the direction cosines between the external normal of the surface and each axis.

The formulae for coordinate transformation of stress components between the components of stress $(\sigma_{xx}, \sigma_{xy}, \ldots)$ referred to an old Cartesian coordinate system (x, y, z) and the components of stress $(\sigma_{x'x'}, \sigma_{x'y'}, \ldots)$ referred to a new Cartesian coordinate system (x', y', z') are

$$\sigma_{i'j'} = l_{i'k} l_{j'l} \sigma_{kl} \qquad (i', j' = 1, 2, 3) \tag{16.4}$$

where $l_{i'k}$ denote the direction cosines between the axis $x_{i'}$ of the new Cartesian coordinate system (x'_1, x'_2, x'_3) and the axis x_k of the old (x_1, x_2, x_3).

Strain
The strains are from Eq. (2.5)

$$\epsilon_{ij} = \frac{1}{2}(u_{i,j} + u_{j,i}) \qquad (i, j = 1, 2, 3) \tag{16.5}$$

where u_i are the components of displacement. The components of strain are symmetric

$$\epsilon_{ij} = \epsilon_{ji} \qquad (i, j = 1, 2, 3) \tag{16.6}$$

The transformation of coordinates between the the new Cartesian coordinate system $(x_{i'})$ and the old system (x_k) are

$$x_{i'} = l_{i'j} x_j, \quad x_i = l_{j'i} x_{j'} \qquad (i, i' = 1, 2, 3) \tag{16.7}$$

The relationship between the components of the displacement in each coordinate system are

$$u_{i'} = l_{i'j} u_j, \qquad u_i = l_{j'i} u_{j'} \qquad (i, i' = 1, 2, 3) \tag{16.8}$$

The coordinate transformation of strain components is

$$\epsilon_{i'j'} = l_{i'k} l_{j'l} \epsilon_{kl} \qquad (i', j' = 1, 2, 3) \tag{16.9}$$

16.1.2 Navier's Equations, Compatibility Equations and Boundary Conditions

Navier's equations
The constitutive equations for a homogeneous, isotropic body which are known as the generalized Hooke's law are

$$\epsilon_{ij} = \frac{1}{2G}\left(\sigma_{ij} - \frac{\nu}{1+\nu}\Theta\delta_{ij}\right) + \alpha\tau\delta_{ij} \quad (i, j = 1, 2, 3) \tag{16.10}$$

an alternative form

$$\sigma_{ij} = 2\mu\epsilon_{ij} + (\lambda e - \beta\tau)\delta_{ij} \quad (i, j = 1, 2, 3) \tag{16.11}$$

where τ is the temperature change from the reference temperature T_0

$$\tau = T - T_0 \tag{16.12}$$

and G is the shear modulus, ν is Poisson's ratio, α is the coefficient of linear thermal expansion, λ and μ are the Lamé elastic constants, and β is the thermoelastic constant, and Θ denotes the sum of the normal stresses

$$\Theta = \sigma_{kk} = \sigma_{xx} + \sigma_{yy} + \sigma_{zz} \tag{16.13}$$

e is the dilatation

$$e = \epsilon_{kk} = \epsilon_{xx} + \epsilon_{yy} + \epsilon_{zz} \tag{16.14}$$

and δ_{ij} is Kronecker's symbol

$$\delta_{ij} = \begin{cases} 1 & \text{for} \quad i = j \\ 0 & \text{for} \quad i \neq j \end{cases} \tag{16.15}$$

The relationship between the elastic constants (E, G, ν, λ, μ) and the thermoelastic constant β is given by

$$\begin{aligned} 2G &= \frac{E}{1+\nu}, \quad \lambda = \frac{\nu E}{(1+\nu)(1-2\nu)} = \frac{2\nu G}{1-2\nu} \\ \mu &= G, \quad \beta = \frac{\alpha E}{1-2\nu} = \alpha(3\lambda + 2\mu) \end{aligned} \tag{16.16}$$

Navier's equations of thermoelasticity, or the displacement equations of thermoelasticity expressed in terms of the components of displacement are from Eq. (3.22)

$$\mu\nabla^2 u_i + (\lambda + \mu)u_{k,ki} - \beta\tau_{,i} + F_i = 0 \qquad (i = 1, 2, 3) \tag{16.17}$$

an alternative form

$$(\lambda + 2\mu)u_{k,ki} - 2\mu\varepsilon_{ijk}\omega_{k,j} - \beta\tau_{,i} + F_i = 0 \qquad (i = 1, 2, 3) \tag{16.18}$$

where ∇^2 is the Laplacian operator defined as

$$\nabla^2 = \frac{\partial^2}{\partial x^2} + \frac{\partial^2}{\partial y^2} + \frac{\partial^2}{\partial z^2} \tag{16.19}$$

and ω_k denote the rotations

$$\omega_x = \frac{1}{2}\left(\frac{\partial u_z}{\partial y} - \frac{\partial u_y}{\partial z}\right), \quad \omega_y = \frac{1}{2}\left(\frac{\partial u_x}{\partial z} - \frac{\partial u_z}{\partial x}\right), \quad \omega_z = \frac{1}{2}\left(\frac{\partial u_y}{\partial x} - \frac{\partial u_x}{\partial y}\right) \tag{16.20}$$

and ε_{ijk} is the alternating tensor, also known as permutation symbol, and is defined by

$$\varepsilon_{ijk} = \begin{cases} 1 & \text{if } ijk \text{ represents an even permutation of 123} \\ 0 & \text{if any two of the } ijk \text{ indices are equal} \\ -1 & \text{if } ijk \text{ represents an odd permutation of 123} \end{cases} \tag{16.21}$$

Compatibility equations
The compatibility equations are from Eq. (2.18)

$$\epsilon_{ij,kl} + \epsilon_{kl,ij} - \epsilon_{ik,jl} - \epsilon_{jl,ik} = 0 \tag{16.22}$$

The compatibility equations (16.22) can be expressed in terms of the components of stress

$$\begin{aligned} \nabla^2\sigma_{ij} + \frac{1}{1+\nu}\sigma_{kk,ij} + \alpha E\left(\frac{1}{1-\nu}\nabla^2\tau\delta_{ij} + \frac{1}{1+\nu}\tau_{,ij}\right) \\ = -\left(\frac{\nu}{1-\nu}F_{k,k}\delta_{ij} + F_{i,j} + F_{j,i}\right) \end{aligned} \tag{16.23}$$

These equations are called the Beltrami-Michell compatibility equations for thermoelasticity.

Boundary conditions
The boundary conditions have been explained in Chap. 3. The three kinds of boundary conditions are

(1) Traction boundary condition

$$\sigma_{ji}n_j = p_{ni} \quad (i = 1, 2, 3) \tag{16.24}$$

where p_{ni} denote the prescribed surface tractions, and n_j denote the direction cosines between the external normal and each axis.

(2) Displacement boundary condition

$$u_i = \overline{u}_i \quad (i = 1, 2, 3) \tag{16.25}$$

where the displacements with overbar denote the prescribed boundary displacements.

(3) Mixed boundary condition

$$\begin{aligned} &\sigma_{ji} n_j = p_{ni} \quad (i = 1, 2, 3) \text{ on the part of the boundary } B_1 \\ &u_i = \bar{u}_i \qquad (i = 1, 2, 3) \text{ on the part of the boundary } B_2 \end{aligned} \tag{16.26}$$

16.1.3 General Solution of Navier's Equations

When the body forces are absent, the boundary-value problem for thermoelasticity may be written in the form

$$\mu \nabla^2 u_i + (\lambda + \mu) u_{k,ki} = \beta \tau_{,i} \quad \text{in the body } D \tag{16.27}$$

$$\sigma_{ji} n_j = p_{ni} \quad \text{on the part of the boundary } B_1 \tag{16.28}$$

$$u_i = \overline{u}_i \quad \text{on the rest of the boundary } B_2 \tag{16.29}$$

The general solution of Eq. (16.27) may be expressed as the sum of a complementary solution u_i^c and a particular solution u_i^p.

$$u_i = u_i^c + u_i^p \tag{16.30}$$

The particular solution u_i^p may be expressed in terms of a scalar potential function as follows

$$u_i^p = \Phi_{,i} \tag{16.31}$$

where Φ is called Goodier's thermoelastic potential, and Φ should satisfy the equation

$$\nabla^2 \Phi = K \tau \tag{16.32}$$

where

$$K = \frac{\beta}{\lambda + 2\mu} = \frac{1+\nu}{1-\nu} \alpha \tag{16.33}$$

Goodier's thermoelastic potential for transient thermoelasticity can be calculated from

$$\Phi = \kappa K \int_{t_r}^{t} \tau \, dt + \Phi_r + (t - t_r) \Phi_0 \tag{16.34}$$

where t_r denotes the reference time, and Φ_r and Φ_0 denote solutions of the following Poisson's equation and Laplace's equation, respectively:

$$\nabla^2 \Phi_r = K \tau_r \tag{16.35}$$

$$\nabla^2 \Phi_0 = 0 \tag{16.36}$$

where τ_r denotes the temperature change at the reference time.

The complementary solutions u_i^c for Navier's equations (16.27) are discussed in Chap. 6. The Boussinesq solution is

$$\boldsymbol{u}^c = \text{grad}\,\varphi + 2\,\text{curl}\,[0, 0, \vartheta] + \text{grad}\,(z\psi) - 4(1-\nu)[0, 0, \psi] \tag{16.37}$$

where φ, ϑ, and ψ are harmonic functions.

Hence, a typical example of the general solution of Navier's equations (16.27) is given as

$$\boldsymbol{u} = \text{grad}\,\Phi + \text{grad}\,\varphi + 2\,\text{curl}\,[0, 0, \vartheta] + \text{grad}\,(z\psi) - 4(1-\nu)[0, 0, \psi] \tag{16.38}$$

where

$$\nabla^2 \Phi = K\tau, \quad \nabla^2 \varphi = 0, \quad \nabla^2 \vartheta = 0, \quad \nabla^2 \psi = 0 \tag{16.39}$$

16.1.4 Thermal Stresses in a Cylindrical Coordinate System

The equilibrium equations in a cylindrical coordinate system (r, θ, z) are

$$\begin{aligned}
&\frac{\partial \sigma_{rr}}{\partial r} + \frac{1}{r}\frac{\partial \sigma_{\theta r}}{\partial \theta} + \frac{\partial \sigma_{zr}}{\partial z} + \frac{\sigma_{rr} - \sigma_{\theta\theta}}{r} + F_r = 0 \\
&\frac{\partial \sigma_{r\theta}}{\partial r} + \frac{1}{r}\frac{\partial \sigma_{\theta\theta}}{\partial \theta} + \frac{\partial \sigma_{z\theta}}{\partial z} + 2\frac{\sigma_{r\theta}}{r} + F_\theta = 0 \\
&\frac{\partial \sigma_{rz}}{\partial r} + \frac{1}{r}\frac{\partial \sigma_{\theta z}}{\partial \theta} + \frac{\partial \sigma_{zz}}{\partial z} + \frac{\sigma_{rz}}{r} + F_z = 0
\end{aligned} \tag{16.40}$$

The components of stress in a cylindrical coordinate system expressed in terms of those of a Cartesian coordinate system are

$$\begin{aligned}
\sigma_{rr} &= \sigma_{xx}\cos^2\theta + \sigma_{yy}\sin^2\theta + \sigma_{xy}\sin 2\theta \\
\sigma_{\theta\theta} &= \sigma_{xx}\sin^2\theta + \sigma_{yy}\cos^2\theta - \sigma_{xy}\sin 2\theta \\
\sigma_{zz} &= \sigma_{zz} \\
\sigma_{r\theta} &= -\frac{1}{2}(\sigma_{xx} - \sigma_{yy})\sin 2\theta + \sigma_{xy}\cos 2\theta \\
\sigma_{\theta z} &= \sigma_{yz}\cos\theta - \sigma_{zx}\sin\theta \\
\sigma_{zr} &= \sigma_{zx}\cos\theta + \sigma_{yz}\sin\theta
\end{aligned} \tag{16.41}$$

The components of strain and dilatation e in a cylindrical coordinate system in terms of the components of displacement are

$$\begin{aligned}
\epsilon_{rr} &= \frac{\partial u_r}{\partial r}, \quad \epsilon_{\theta\theta} = \frac{u_r}{r} + \frac{1}{r}\frac{\partial u_\theta}{\partial \theta}, \quad \epsilon_{zz} = \frac{\partial u_z}{\partial z} \\
\epsilon_{r\theta} &= \frac{1}{2}\left(\frac{1}{r}\frac{\partial u_r}{\partial \theta} + \frac{\partial u_\theta}{\partial r} - \frac{u_\theta}{r}\right), \quad \epsilon_{\theta z} = \frac{1}{2}\left(\frac{\partial u_\theta}{\partial z} + \frac{1}{r}\frac{\partial u_z}{\partial \theta}\right) \\
\epsilon_{zr} &= \frac{1}{2}\left(\frac{\partial u_r}{\partial z} + \frac{\partial u_z}{\partial r}\right) \\
e &= \frac{\partial u_r}{\partial r} + \frac{u_r}{r} + \frac{1}{r}\frac{\partial u_\theta}{\partial \theta} + \frac{\partial u_z}{\partial z}
\end{aligned} \tag{16.42}$$

where u_r, u_θ, u_z are the components of displacement in the r, θ, z directions, respectively.

The compatibility equations of strain in a cylindrical coordinate system are

$$\begin{aligned}
&2\frac{\partial^2(r\epsilon_{r\theta})}{\partial r \partial \theta} = \frac{\partial^2 \epsilon_{rr}}{\partial \theta^2} + r\frac{\partial^2(r\epsilon_{\theta\theta})}{\partial r^2} - r\frac{\partial \epsilon_{rr}}{\partial r} \\
&2\frac{\partial^2 \epsilon_{zr}}{\partial r \partial z} = \frac{\partial^2 \epsilon_{zz}}{\partial r^2} + \frac{\partial^2 \epsilon_{rr}}{\partial z^2} \\
&2\frac{\partial^2(r\epsilon_{\theta z})}{\partial \theta \partial z} = r^2\frac{\partial^2 \epsilon_{\theta\theta}}{\partial z^2} - 2r\frac{\partial \epsilon_{zr}}{\partial z} + r\frac{\partial \epsilon_{zz}}{\partial r} + \frac{\partial^2 \epsilon_{zz}}{\partial \theta^2} \\
&\frac{\partial}{\partial z}\left(-\frac{\partial \epsilon_{r\theta}}{\partial z} + \frac{1}{r}\frac{\partial \epsilon_{zr}}{\partial \theta} + \frac{\partial \epsilon_{\theta z}}{\partial r}\right) = \frac{\partial^2}{\partial r \partial \theta}\left(\frac{\epsilon_{zz}}{r}\right) + \frac{1}{r}\frac{\partial \epsilon_{\theta z}}{\partial z} \\
&\frac{\partial}{\partial \theta}\left(\frac{\partial \epsilon_{r\theta}}{\partial z} - \frac{1}{r}\frac{\partial \epsilon_{zr}}{\partial \theta} + \frac{\partial \epsilon_{\theta z}}{\partial r}\right) = \frac{\partial^2(r\epsilon_{\theta\theta})}{\partial r \partial z} - \frac{\partial \epsilon_{rr}}{\partial z} - \frac{1}{r}\frac{\partial \epsilon_{\theta z}}{\partial \theta} \\
&\frac{\partial}{\partial r}\left(\frac{\partial \epsilon_{r\theta}}{\partial z} + \frac{1}{r}\frac{\partial \epsilon_{zr}}{\partial \theta} - \frac{\partial \epsilon_{\theta z}}{\partial r}\right) = \frac{1}{r}\frac{\partial^2 \epsilon_{rr}}{\partial \theta \partial z} - \frac{2}{r}\frac{\partial \epsilon_{r\theta}}{\partial z} + \frac{\partial}{\partial r}\left(\frac{\epsilon_{\theta z}}{r}\right)
\end{aligned} \tag{16.43}$$

The coordinate transformations of strain components between a cylindrical coordinate system and a Cartesian coordinate system are

$$\begin{aligned}
\epsilon_{rr} &= \epsilon_{xx}\cos^2\theta + \epsilon_{yy}\sin^2\theta + \epsilon_{xy}\sin 2\theta \\
\epsilon_{\theta\theta} &= \epsilon_{xx}\sin^2\theta + \epsilon_{yy}\cos^2\theta - \epsilon_{xy}\sin 2\theta \\
\epsilon_{zz} &= \epsilon_{zz} \\
\epsilon_{r\theta} &= -\frac{1}{2}(\epsilon_{xx} - \epsilon_{yy})\sin 2\theta + \epsilon_{xy}\cos 2\theta \\
\epsilon_{\theta z} &= \epsilon_{yz}\cos\theta - \epsilon_{zx}\sin\theta \\
\epsilon_{zr} &= \epsilon_{zx}\cos\theta + \epsilon_{yz}\sin\theta
\end{aligned} \tag{16.44}$$

The constitutive equations, or the generalized Hooke's law, for a homogeneous, isotropic body in a cylindrical coordinate system are

$$
\begin{aligned}
\epsilon_{rr} &= \frac{1}{E}[\sigma_{rr} - \nu(\sigma_{\theta\theta} + \sigma_{zz})] + \alpha\tau = \frac{1}{2G}\left(\sigma_{rr} - \frac{\nu}{1+\nu}\Theta\right) + \alpha\tau \\
\epsilon_{\theta\theta} &= \frac{1}{E}[\sigma_{\theta\theta} - \nu(\sigma_{zz} + \sigma_{rr})] + \alpha\tau = \frac{1}{2G}\left(\sigma_{\theta\theta} - \frac{\nu}{1+\nu}\Theta\right) + \alpha\tau \\
\epsilon_{zz} &= \frac{1}{E}[\sigma_{zz} - \nu(\sigma_{rr} + \sigma_{\theta\theta})] + \alpha\tau = \frac{1}{2G}\left(\sigma_{zz} - \frac{\nu}{1+\nu}\Theta\right) + \alpha\tau \\
\epsilon_{r\theta} &= \frac{\sigma_{r\theta}}{2G}, \quad \epsilon_{\theta z} = \frac{\sigma_{\theta z}}{2G}, \quad \epsilon_{zr} = \frac{\sigma_{zr}}{2G}
\end{aligned}
\tag{16.45}
$$

An alternative form

$$
\begin{aligned}
\sigma_{rr} &= 2\mu\epsilon_{rr} + \lambda e - \beta\tau, \quad & \sigma_{r\theta} &= 2\mu\epsilon_{r\theta} \\
\sigma_{\theta\theta} &= 2\mu\epsilon_{\theta\theta} + \lambda e - \beta\tau, \quad & \sigma_{\theta z} &= 2\mu\epsilon_{\theta z} \\
\sigma_{zz} &= 2\mu\epsilon_{zz} + \lambda e - \beta\tau, \quad & \sigma_{zr} &= 2\mu\epsilon_{zr}
\end{aligned}
\tag{16.46}
$$

where $\Theta = \sigma_{rr} + \sigma_{\theta\theta} + \sigma_{zz}$ and $e = \epsilon_{rr} + \epsilon_{\theta\theta} + \epsilon_{zz}$.

Navier's equations (16.17) for thermoelasticity can be expressed in a cylindrical coordinate system as

$$
\begin{aligned}
&(\lambda + 2\mu)\frac{\partial e}{\partial r} - 2\mu\left(\frac{1}{r}\frac{\partial \omega_z}{\partial \theta} - \frac{\partial \omega_\theta}{\partial z}\right) - \beta\frac{\partial \tau}{\partial r} + F_r = 0 \\
&(\lambda + 2\mu)\frac{1}{r}\frac{\partial e}{\partial \theta} - 2\mu\left(\frac{\partial \omega_r}{\partial z} - \frac{\partial \omega_z}{\partial r}\right) - \beta\frac{1}{r}\frac{\partial \tau}{\partial \theta} + F_\theta = 0 \\
&(\lambda + 2\mu)\frac{\partial e}{\partial z} - \frac{2\mu}{r}\left[\frac{\partial (r\omega_\theta)}{\partial r} - \frac{\partial \omega_r}{\partial \theta}\right] - \beta\frac{\partial \tau}{\partial z} + F_z = 0
\end{aligned}
\tag{16.47}
$$

where

$$
\begin{aligned}
\omega_r &= \frac{1}{2}\left(\frac{1}{r}\frac{\partial u_z}{\partial \theta} - \frac{\partial u_\theta}{\partial z}\right), \quad \omega_\theta = \frac{1}{2}\left(\frac{\partial u_r}{\partial z} - \frac{\partial u_z}{\partial r}\right) \\
\omega_z &= \frac{1}{2r}\left(\frac{\partial (r u_\theta)}{\partial r} - \frac{\partial u_r}{\partial \theta}\right)
\end{aligned}
\tag{16.48}
$$

The solution of Navier's equations (16.47) without the body force can be expressed, for example, by the thermoelastic potential Φ and the Boussinesq harmonic functions as follows:

$$
u_r = \frac{\partial \Phi}{\partial r} + \frac{\partial \varphi}{\partial r} + \frac{2}{r}\frac{\partial \vartheta}{\partial \theta} + z\frac{\partial \psi}{\partial r}
$$

$$u_\theta = \frac{1}{r}\frac{\partial \Phi}{\partial \theta} + \frac{1}{r}\frac{\partial \varphi}{\partial \theta} - 2\frac{\partial \vartheta}{\partial r} + \frac{z}{r}\frac{\partial \psi}{\partial \theta} \tag{16.49}$$
$$u_z = \frac{\partial \Phi}{\partial z} + \frac{\partial \varphi}{\partial z} + z\frac{\partial \psi}{\partial z} - (3-4\nu)\psi$$

where the four functions Φ, φ, ϑ, and ψ must satisfy

$$\nabla^2\Phi = K\tau, \quad \nabla^2\varphi = 0, \quad \nabla^2\vartheta = 0, \quad \nabla^2\psi = 0 \tag{16.50}$$

and

$$\nabla^2 = \frac{\partial^2}{\partial r^2} + \frac{1}{r}\frac{\partial}{\partial r} + \frac{1}{r^2}\frac{\partial^2}{\partial \theta^2} + \frac{\partial^2}{\partial z^2} \tag{16.51}$$

16.1.5 Thermal Stresses in a Spherical Coordinate System

The equilibrium equations in a spherical coordinate system (r, θ, ϕ) are

$$\frac{\partial \sigma_{rr}}{\partial r} + \frac{1}{r}\frac{\partial \sigma_{\theta r}}{\partial \theta} + \frac{1}{r\sin\theta}\frac{\partial \sigma_{\phi r}}{\partial \phi} + \frac{1}{r}(2\sigma_{rr} - \sigma_{\theta\theta} - \sigma_{\phi\phi} + \sigma_{\theta r}\cot\theta) + F_r = 0$$
$$\frac{\partial \sigma_{r\theta}}{\partial r} + \frac{1}{r}\frac{\partial \sigma_{\theta\theta}}{\partial \theta} + \frac{1}{r\sin\theta}\frac{\partial \sigma_{\phi\theta}}{\partial \phi} + \frac{1}{r}\Big[(\sigma_{\theta\theta} - \sigma_{\phi\phi})\cot\theta + 3\sigma_{r\theta}\Big] + F_\theta = 0$$
$$\frac{\partial \sigma_{r\phi}}{\partial r} + \frac{1}{r}\frac{\partial \sigma_{\theta\phi}}{\partial \theta} + \frac{1}{r\sin\theta}\frac{\partial \sigma_{\phi\phi}}{\partial \phi} + \frac{1}{r}(3\sigma_{r\phi} + 2\sigma_{\theta\phi}\cot\theta) + F_\phi = 0 \tag{16.52}$$

The components of stress in a spherical coordinate system expressed in terms of those of a Cartesian coordinate system are

$$\sigma_{rr} = \sigma_{xx}\sin^2\theta\cos^2\phi + \sigma_{yy}\sin^2\theta\sin^2\phi + \sigma_{zz}\cos^2\theta$$
$$+ \sigma_{xy}\sin^2\theta\sin 2\phi + \sigma_{yz}\sin 2\theta\sin\phi + \sigma_{zx}\sin 2\theta\cos\phi$$
$$\sigma_{\theta\theta} = \sigma_{xx}\cos^2\theta\cos^2\phi + \sigma_{yy}\cos^2\theta\sin^2\phi + \sigma_{zz}\sin^2\theta$$
$$+ \sigma_{xy}\cos^2\theta\sin 2\phi - \sigma_{yz}\sin 2\theta\sin\phi - \sigma_{zx}\sin 2\theta\cos\phi$$
$$\sigma_{\phi\phi} = \sigma_{xx}\sin^2\phi + \sigma_{yy}\cos^2\phi - \sigma_{xy}\sin 2\phi$$
$$\sigma_{r\theta} = \frac{1}{2}\sin 2\theta\,(\sigma_{xx}\cos^2\phi + \sigma_{yy}\sin^2\phi - \sigma_{zz}) \tag{16.53}$$
$$+ \frac{1}{2}\sigma_{xy}\sin 2\theta\sin 2\phi + \sigma_{yz}\cos 2\theta\sin\phi + \sigma_{zx}\cos 2\theta\cos\phi$$
$$\sigma_{\theta\phi} = -\frac{1}{2}\cos\theta\sin 2\phi\,(\sigma_{xx} - \sigma_{yy})$$
$$+ \sigma_{xy}\cos\theta\cos 2\phi - \sigma_{yz}\sin\theta\cos\phi + \sigma_{zx}\sin\theta\sin\phi$$
$$\sigma_{\phi r} = -\frac{1}{2}\sin\theta\sin 2\phi\,(\sigma_{xx} - \sigma_{yy})$$

$$+ \sigma_{xy} \sin\theta \cos 2\phi + \sigma_{yz} \cos\theta \cos\phi - \sigma_{zx} \cos\theta \sin\phi$$

The components of strain and dilatation e in terms of the components of displacement are

$$\begin{aligned}
\epsilon_{rr} &= \frac{\partial u_r}{\partial r}, \quad \epsilon_{\theta\theta} = \frac{u_r}{r} + \frac{1}{r}\frac{\partial u_\theta}{\partial \theta} \\
\epsilon_{\phi\phi} &= \frac{u_r}{r} + \cot\theta \frac{u_\theta}{r} + \frac{1}{r\sin\theta}\frac{\partial u_\phi}{\partial \phi}, \quad \epsilon_{r\theta} = \frac{1}{2}\left(\frac{1}{r}\frac{\partial u_r}{\partial \theta} + \frac{\partial u_\theta}{\partial r} - \frac{u_\theta}{r}\right) \\
\epsilon_{\theta\phi} &= \frac{1}{2}\left(\frac{1}{r}\frac{\partial u_\phi}{\partial \theta} - \cot\theta \frac{u_\phi}{r} + \frac{1}{r\sin\theta}\frac{\partial u_\theta}{\partial \phi}\right) \\
\epsilon_{\phi r} &= \frac{1}{2}\left(\frac{1}{r\sin\theta}\frac{\partial u_r}{\partial \phi} + \frac{\partial u_\phi}{\partial r} - \frac{u_\phi}{r}\right) \\
e &= \frac{\partial u_r}{\partial r} + 2\frac{u_r}{r} + \frac{1}{r}\frac{\partial u_\theta}{\partial \theta} + \cot\theta \frac{u_\theta}{r} + \frac{1}{r\sin\theta}\frac{\partial u_\phi}{\partial \phi}
\end{aligned} \tag{16.54}$$

where u_r, u_θ, u_ϕ are the components of displacement in the r, θ, ϕ directions, respectively.

The compatibility equations of strain in a spherical coordinate system are

$$\begin{aligned}
&-\frac{\epsilon_{\theta\theta,\phi\phi}}{r^2\sin^2\theta} + \frac{\epsilon_{\theta\theta,\theta}}{r^2\tan\theta} - \frac{\epsilon_{\theta\theta,r}}{r} + \frac{2\epsilon_{r\theta,\theta}}{r^2} + \frac{2\epsilon_{r\theta}}{r^2\tan\theta} - \frac{2\epsilon_{\theta\theta}}{r^2} - \frac{\epsilon_{\phi\phi,\theta\theta}}{r^2} \\
&\quad - \frac{\epsilon_{\phi\phi,r}}{r} - \frac{2\epsilon_{\phi\phi,\theta}}{r^2\tan\theta} + \frac{\epsilon_{\phi r}}{r^2\sin\theta} + \frac{2\cos\theta\, \epsilon_{\theta\phi,\phi}}{r^2\sin^2\theta} + \frac{2\epsilon_{\theta\phi,\theta\phi}}{r^2\sin\theta} + \frac{2\epsilon_{rr}}{r^2} = 0 \\
&\frac{2\epsilon_{\phi r,\phi r}}{r\sin\theta} + \frac{\epsilon_{\phi r,\phi}}{r^2\sin\theta} - \frac{\epsilon_{rr,\phi\phi}}{r^2\sin^2\theta} - \frac{2\epsilon_{\phi\phi,r}}{r} - \epsilon_{\phi\phi,rr} \\
&\quad + \frac{2\epsilon_{r\theta,r}}{r\tan\theta} + \frac{2\epsilon_{r\theta}}{r^2\tan\theta} - \frac{\epsilon_{rr,\theta}}{r^2\tan\theta} + \frac{\epsilon_{rr,r}}{r} = 0 \\
&\frac{2\epsilon_{r\theta,r\theta}}{r} - \frac{2\epsilon_{\theta\theta,r}}{r} - \epsilon_{\theta\theta,rr} - \frac{\epsilon_{rr,r}}{r} - \frac{\epsilon_{rr,\theta\theta}}{r^2} + \frac{2\epsilon_{r\theta,\theta}}{r^2} = 0 \\
&\frac{\epsilon_{rr,\theta\phi}}{r^2\sin\theta} - \frac{\epsilon_{r\theta,\phi r}}{r\sin\theta} - \frac{\epsilon_{r\theta,\phi}}{r^2\sin\theta} + \frac{\epsilon_{\phi r}}{r^2\tan\theta} - \frac{\epsilon_{\phi r,\theta}}{r^2} - \frac{\epsilon_{\phi r,r\theta}}{r} \\
&\quad + \epsilon_{\theta\phi,rr} + \frac{2\epsilon_{\theta\phi,r}}{r} + \frac{\epsilon_{\phi r,r}}{r\tan\theta} - \frac{\cos\theta\, \epsilon_{rr,\phi}}{r^2\sin^2\theta} = 0 \\
&\frac{\epsilon_{\theta\theta,\phi r}}{r\sin\theta} - \frac{\epsilon_{\theta\phi,r\theta}}{r} - \frac{2\epsilon_{\theta\phi,r}}{r\tan\theta} + \frac{\cos\theta \epsilon_{r\theta,\phi}}{r^2\sin^2\theta} + \frac{\epsilon_{\phi r,\theta}}{r^2\tan\theta} \\
&\quad - \frac{\epsilon_{r\theta,\theta\phi}}{r^2\sin\theta} + \frac{\epsilon_{\phi r,\theta\theta}}{r^2} - \frac{\cos 2\theta\, \epsilon_{\phi r}}{r^2\sin^2\theta} - \frac{\epsilon_{rr,\phi}}{r^2\sin\theta} = 0 \\
&\frac{\epsilon_{\phi\phi,r\theta}}{r} + \frac{\epsilon_{\phi\phi,r}}{r\tan\theta} - \frac{\epsilon_{\phi r,\theta\phi}}{r^2\sin\theta} + \frac{2\epsilon_{r\theta}}{r^2} - \frac{\epsilon_{rr,\theta}}{r^2} \\
&\quad - \frac{\epsilon_{\theta\phi,\phi r}}{r\sin\theta} + \frac{\epsilon_{r\theta,\phi\phi}}{r^2\sin^2\theta} - \frac{\cos\theta\, \epsilon_{\phi r,\phi}}{r^2\sin^2\theta} - \frac{\epsilon_{\theta\theta,r}}{r\tan\theta} = 0
\end{aligned} \tag{16.55}$$

The coordinate transformations of the displacement between a spherical coordinate system and a Cartesian coordinate system are

$$\begin{aligned} u_x &= u_r \sin\theta\cos\phi + u_\theta\cos\theta\cos\phi - u_\phi\sin\phi \\ u_y &= u_r \sin\theta\sin\phi + u_\theta\cos\theta\sin\phi + u_\phi\cos\phi \\ u_z &= u_r\cos\theta - u_\theta\sin\theta \end{aligned} \tag{16.56}$$

The coordinate transformations of the strain components between a spherical coordinate system and a Cartesian coordinate system are

$$\begin{aligned} \epsilon_{rr} &= \epsilon_{xx}\sin^2\theta\cos^2\phi + \epsilon_{yy}\sin^2\theta\sin^2\phi + \epsilon_{zz}\cos^2\theta \\ &\quad + \epsilon_{xy}\sin^2\theta\sin 2\phi + \epsilon_{yz}\sin 2\theta\sin\phi + \epsilon_{zx}\sin 2\theta\cos\phi \\ \epsilon_{\theta\theta} &= \epsilon_{xx}\cos^2\theta\cos^2\phi + \epsilon_{yy}\cos^2\theta\sin^2\phi + \epsilon_{zz}\sin^2\theta \\ &\quad + \epsilon_{xy}\cos^2\theta\sin 2\phi - \epsilon_{yz}\sin 2\theta\sin\phi - \epsilon_{zx}\sin 2\theta\cos\phi \\ \epsilon_{\phi\phi} &= \epsilon_{xx}\sin^2\phi + \epsilon_{yy}\cos^2\phi - \epsilon_{xy}\sin 2\phi \\ \epsilon_{r\theta} &= \frac{1}{2}\sin 2\theta\,(\epsilon_{xx}\cos^2\phi + \epsilon_{yy}\sin^2\phi - \epsilon_{zz}) \\ &\quad + \frac{1}{2}\epsilon_{xy}\sin 2\theta\sin 2\phi + \epsilon_{yz}\cos 2\theta\sin\phi + \epsilon_{zx}\cos 2\theta\cos\phi \\ \epsilon_{\theta\phi} &= -\frac{1}{2}\cos\theta\sin 2\phi\,(\epsilon_{xx} - \epsilon_{yy}) \\ &\quad + \epsilon_{xy}\cos\theta\sin 2\phi - \epsilon_{yz}\sin\theta\cos\phi + \epsilon_{zx}\sin\theta\sin\phi \\ \epsilon_{\phi r} &= -\frac{1}{2}\sin\theta\sin 2\phi\,(\epsilon_{xx} - \epsilon_{yy}) \\ &\quad + \epsilon_{xy}\sin\theta\cos 2\phi + \epsilon_{yz}\cos\theta\cos\phi - \epsilon_{zx}\cos\theta\sin\phi \end{aligned} \tag{16.57}$$

The constitutive equations for a homogeneous, isotropic body in a spherical coordinate system are

$$\begin{aligned} \epsilon_{rr} &= \frac{1}{E}[\sigma_{rr} - \nu(\sigma_{\theta\theta} + \sigma_{\phi\phi})] + \alpha\tau = \frac{1}{2G}\left(\sigma_{rr} - \frac{\nu}{1+\nu}\Theta\right) + \alpha\tau \\ \epsilon_{\theta\theta} &= \frac{1}{E}[\sigma_{\theta\theta} - \nu(\sigma_{\phi\phi} + \sigma_{rr})] + \alpha\tau = \frac{1}{2G}\left(\sigma_{\theta\theta} - \frac{\nu}{1+\nu}\Theta\right) + \alpha\tau \\ \epsilon_{\phi\phi} &= \frac{1}{E}[\sigma_{\phi\phi} - \nu(\sigma_{rr} + \sigma_{\theta\theta})] + \alpha\tau = \frac{1}{2G}\left(\sigma_{\phi\phi} - \frac{\nu}{1+\nu}\Theta\right) + \alpha\tau \\ \epsilon_{r\theta} &= \frac{\sigma_{r\theta}}{2G}, \quad \epsilon_{\theta\phi} = \frac{\sigma_{\theta\phi}}{2G}, \quad \epsilon_{\phi r} = \frac{\sigma_{\phi r}}{2G} \end{aligned} \tag{16.58}$$

An alternative form

$$\begin{aligned}\sigma_{rr} &= 2\mu\epsilon_{rr} + \lambda e - \beta\tau, \quad \sigma_{r\theta} = 2\mu\epsilon_{r\theta}\\ \sigma_{\theta\theta} &= 2\mu\epsilon_{\theta\theta} + \lambda e - \beta\tau, \quad \sigma_{\theta\phi} = 2\mu\epsilon_{\theta\phi}\\ \sigma_{\phi\phi} &= 2\mu\epsilon_{\phi\phi} + \lambda e - \beta\tau, \quad \sigma_{\phi r} = 2\mu\epsilon_{\phi r}\end{aligned} \tag{16.59}$$

where $\Theta = \sigma_{rr} + \sigma_{\theta\theta} + \sigma_{\phi\phi}$ and $e = \epsilon_{rr} + \epsilon_{\theta\theta} + \epsilon_{\phi\phi}$.

Navier's equations (16.17) of thermoelasticity can be expressed in a spherical coordinate system as

$$\begin{aligned}&(\lambda + 2\mu)\frac{\partial e}{\partial r} - \frac{2\mu}{r\sin\theta}\left[\frac{\partial(\omega_\phi \sin\theta)}{\partial\theta} - \frac{\partial\omega_\theta}{\partial\phi}\right] - \beta\frac{\partial\tau}{\partial r} + F_r = 0\\ &(\lambda + 2\mu)\frac{1}{r}\frac{\partial e}{\partial\theta} - \frac{2\mu}{r\sin\theta}\left[\frac{\partial\omega_r}{\partial\phi} - \sin\theta\frac{\partial(r\omega_\phi)}{\partial r}\right] - \beta\frac{1}{r}\frac{\partial\tau}{\partial\theta} + F_\theta = 0\\ &(\lambda + 2\mu)\frac{1}{r\sin\theta}\frac{\partial e}{\partial\phi} - \frac{2\mu}{r}\left[\frac{\partial(r\omega_\theta)}{\partial r} - \frac{\partial\omega_r}{\partial\theta}\right] - \beta\frac{1}{r\sin\theta}\frac{\partial\tau}{\partial\phi} + F_\phi = 0\end{aligned} \tag{16.60}$$

where

$$\begin{aligned}\omega_r &= \frac{1}{2r\sin\theta}\left[\frac{\partial(u_\phi\sin\theta)}{\partial\theta} - \frac{\partial u_\theta}{\partial\phi}\right], \quad \omega_\theta = \frac{1}{2r\sin\theta}\left[\frac{\partial u_r}{\partial\phi} - \sin\theta\frac{\partial(ru_\phi)}{\partial r}\right]\\ \omega_\phi &= \frac{1}{2r}\left[\frac{\partial(ru_\theta)}{\partial r} - \frac{\partial u_r}{\partial\theta}\right]\end{aligned} \tag{16.61}$$

The solution of Navier's equations (16.60) without the body force in a spherical coordinate system can be expressed, for example, by the thermoelastic potential Φ and the Boussinesq harmonic functions ϕ, ϑ, ψ:

$$\begin{aligned}u_r &= \frac{\partial\Phi}{\partial r} + \frac{\partial\varphi}{\partial r} + \frac{2}{r}\frac{\partial\vartheta}{\partial\phi} + r\cos\theta\frac{\partial\psi}{\partial r} - (3 - 4\nu)\psi\cos\theta\\ u_\theta &= \frac{1}{r}\frac{\partial\Phi}{\partial\theta} + \frac{1}{r}\frac{\partial\varphi}{\partial\theta} + \frac{2}{r\tan\theta}\frac{\partial\vartheta}{\partial\phi} + \cos\theta\frac{\partial\psi}{\partial\theta} + (3 - 4\nu)\psi\sin\theta\\ u_\phi &= \frac{1}{r\sin\theta}\frac{\partial\Phi}{\partial\phi} + \frac{1}{r\sin\theta}\frac{\partial\varphi}{\partial\phi} - 2\sin\theta\frac{\partial\vartheta}{\partial r} - 2\frac{\cos\theta}{r}\frac{\partial\vartheta}{\partial\theta}\\ &\quad + \frac{1}{\tan\theta}\frac{\partial\psi}{\partial\phi}\end{aligned} \tag{16.62}$$

where the four functions must satisfy

$$\nabla^2\Phi = K\tau, \quad \nabla^2\varphi = 0, \quad \nabla^2\vartheta = 0, \quad \nabla^2\psi = 0 \tag{16.63}$$

and

$$\nabla^2 = \frac{\partial^2}{\partial r^2} + \frac{2}{r}\frac{\partial}{\partial r} + \frac{1}{r^2}\frac{\partial^2}{\partial \theta^2} + \frac{1}{r^2 \tan\theta}\frac{\partial}{\partial \theta} + \frac{1}{r^2 \sin^2\theta}\frac{\partial^2}{\partial \phi^2} \tag{16.64}$$

16.2 Problems and Solutions Related to Basic Equations of Thermoelasticity

Problem 16.1. When the elastic body moves under the mechanical and thermal loads, find the equations of motion.

Solution. The motion of the element in the x direction is

$$\left(\sigma_{xx} + \frac{\partial \sigma_{xx}}{\partial x}dx\right)dydz - \sigma_{xx}dydz + \left(\sigma_{yx} + \frac{\partial \sigma_{yx}}{\partial y}dy\right)dzdx - \sigma_{yx}dzdx$$
$$+ \left(\sigma_{zx} + \frac{\partial \sigma_{zx}}{\partial z}dz\right)dxdy - \sigma_{zx}dxdy + F_x dxdydz = \rho\frac{\partial^2 u_x}{\partial t^2}dxdydz \tag{16.65}$$

where ρ means the density. Simplification of Eq. (16.65) gives

$$\frac{\partial \sigma_{xx}}{\partial x} + \frac{\partial \sigma_{yx}}{\partial y} + \frac{\partial \sigma_{zx}}{\partial z} + F_x = \rho\frac{\partial^2 u_x}{\partial t^2} \tag{16.66}$$

The two other equations of motion in the y and z directions can be obtained in the same way. Then, we get the equations of motion

$$\sigma_{ji,j} + F_i = \rho\ddot{u}_i \qquad (i = 1, 2, 3) \qquad \text{(Answer)} \tag{16.67}$$

where u_i denote the components of displacement and the dot denotes partial differentiation with respect to the time.

Problem 16.2. Derive the compatibility equations (16.22).

Solution. Using the definition of strain

$$2\varepsilon_{ij} = u_{i,j} + u_{j,i} \tag{16.68}$$

we get

$$\begin{aligned} 2\varepsilon_{ij,kl} &= u_{i,jkl} + u_{j,ikl}, \quad 2\varepsilon_{kl,ij} = u_{k,lij} + u_{l,kij} \\ 2\varepsilon_{ik,jl} &= u_{i,kjl} + u_{k,ijl}, \quad 2\varepsilon_{jl,ik} = u_{j,lik} + u_{l,jik} \end{aligned} \tag{16.69}$$

Therefore,

$$\begin{aligned}
&\varepsilon_{ij,kl}+\varepsilon_{kl,ij}-\varepsilon_{ik,jl}-\varepsilon_{jl,ik}\\
&\quad=\frac{1}{2}(u_{i,jkl}+u_{j,ikl}+u_{k,lij}+u_{l,kij}\\
&\quad-u_{i,kjl}-u_{k,ijl}-u_{j,lik}-u_{l,jik})=0 \qquad \text{(Answer)}
\end{aligned}$$

Problem 16.3. Show that the strain components can be transformed by the following equations

$$\begin{aligned}
\epsilon_{x'x'}&=\epsilon_{xx}l_1^2+\epsilon_{yy}m_1^2+\epsilon_{zz}n_1^2+2(\epsilon_{xy}l_1m_1+\epsilon_{yz}m_1n_1+\epsilon_{zx}n_1l_1)\\
\epsilon_{y'y'}&=\epsilon_{xx}l_2^2+\epsilon_{yy}m_2^2+\epsilon_{zz}n_2^2+2(\epsilon_{xy}l_2m_2+\epsilon_{yz}m_2n_2+\epsilon_{zx}n_2l_2)\\
\epsilon_{z'z'}&=\epsilon_{xx}l_3^2+\epsilon_{yy}m_3^2+\epsilon_{zz}n_3^2+2(\epsilon_{xy}l_3m_3+\epsilon_{yz}m_3n_3+\epsilon_{zx}n_3l_3)\\
\epsilon_{x'y'}&=\epsilon_{xx}l_1l_2+\epsilon_{yy}m_1m_2+\epsilon_{zz}n_1n_2+\epsilon_{xy}(l_1m_2+l_2m_1)\\
&\quad+\epsilon_{yz}(m_1n_2+m_2n_1)+\epsilon_{zx}(n_1l_2+n_2l_1)\\
\epsilon_{y'z'}&=\epsilon_{xx}l_2l_3+\epsilon_{yy}m_2m_3+\epsilon_{zz}n_2n_3+\epsilon_{xy}(l_2m_3+l_3m_2)\\
&\quad+\epsilon_{yz}(m_2n_3+m_3n_2)+\epsilon_{zx}(n_2l_3+n_3l_2)\\
\epsilon_{z'x'}&=\epsilon_{xx}l_3l_1+\epsilon_{yy}m_3m_1+\epsilon_{zz}n_3n_1+\epsilon_{xy}(l_3m_1+l_1m_3)\\
&\quad+\epsilon_{yz}(m_3n_1+m_1n_3)+\epsilon_{zx}(n_3l_1+n_1l_3)
\end{aligned} \tag{16.70}$$

Solution. Equations (16.9) are written as

$$\begin{aligned}
\varepsilon_{i'j'}&=l_{i'k}l_{j'l}\varepsilon_{kl}\\
&=l_{i'1}l_{j'1}\varepsilon_{11}+l_{i'1}l_{j'2}\varepsilon_{12}+l_{i'1}l_{j'3}\varepsilon_{13}+l_{i'2}l_{j'1}\varepsilon_{21}+l_{i'2}l_{j'2}\varepsilon_{22}\\
&\quad+l_{i'2}l_{j'3}\varepsilon_{23}+l_{i'3}l_{j'1}\varepsilon_{31}+l_{i'3}l_{j'2}\varepsilon_{32}+l_{i'3}l_{j'3}\varepsilon_{33}
\end{aligned} \tag{16.71}$$

Using the notation

$$\begin{aligned}
&l_{1'1}=l_1 \quad l_{1'2}=m_1 \quad l_{1'3}=n_1\\
&l_{2'1}=l_2 \quad l_{2'2}=m_2 \quad l_{2'3}=n_2\\
&l_{3'1}=l_3 \quad l_{3'2}=m_3 \quad l_{3'3}=n_3
\end{aligned} \tag{16.72}$$

and rewriting subscript $(1,2,3)$ as (x,y,z), we get

$$\begin{aligned}
\varepsilon_{x'x'}&=\varepsilon_{xx}l_1^2+\varepsilon_{yy}m_1^2+\varepsilon_{zz}n_1^2+2(\varepsilon_{xy}l_1m_1+\varepsilon_{yz}m_1n_1+\varepsilon_{zx}n_1l_1)\\
\varepsilon_{y'y'}&=\varepsilon_{xx}l_2^2+\varepsilon_{yy}m_2^2+\varepsilon_{zz}n_2^2+2(\varepsilon_{xy}l_2m_2+\varepsilon_{yz}m_2n_2+\varepsilon_{zx}n_2l_2)\\
\varepsilon_{z'z'}&=\varepsilon_{xx}l_3^2+\varepsilon_{yy}m_3^2+\varepsilon_{zz}n_3^2+2(\varepsilon_{xy}l_3m_3+\varepsilon_{yz}m_3n_3+\varepsilon_{zx}n_3l_3)\\
\varepsilon_{x'y'}&=\varepsilon_{xx}l_1l_2+\varepsilon_{yy}m_1m_2+\varepsilon_{zz}n_1n_2\\
&\quad+\varepsilon_{xy}(l_1m_2+l_2m_1)+\varepsilon_{yz}(m_1n_2+m_2n_1)+\varepsilon_{zx}(n_1l_2+n_2l_1)
\end{aligned}$$

$$
\begin{aligned}
\varepsilon_{y'z'} &= \varepsilon_{xx} l_2 l_3 + \varepsilon_{yy} m_2 m_3 + \varepsilon_{zz} n_2 n_3 \\
&\quad + \varepsilon_{xy}(l_2 m_3 + l_3 m_2) + \varepsilon_{yz}(m_2 n_3 + m_3 n_2) + \varepsilon_{zx}(n_2 l_3 + n_3 l_2) \\
\varepsilon_{z'x'} &= \varepsilon_{xx} l_3 l_1 + \varepsilon_{yy} m_3 m_1 + \varepsilon_{zz} n_3 n_1 \\
&\quad + \varepsilon_{xy}(l_3 m_1 + l_1 m_3) + \varepsilon_{yz}(m_3 n_1 + m_1 n_3) + \varepsilon_{zx}(n_3 l_1 + n_1 l_3) \qquad \text{(Answer)}
\end{aligned}
$$

Problem 16.4. Derive Navier's equations of thermoelasticity with motion of the body.

Solution. The equations of motion are given from Eq. (16.67)

$$\sigma_{ji,j} + F_i = \rho \ddot{u}_i \quad (i = 1, 2, 3) \tag{16.73}$$

Substitution of Eq. (16.11) into Eq. (16.73) gives

$$
\begin{aligned}
&\sigma_{ji,j} + F_i - \rho \ddot{u}_i \\
&\quad = \mu(u_{i,jj} + u_{j,ji}) + \lambda u_{j,ji} - \beta \tau_{,i} + F_i - \rho \ddot{u}_i \\
&\quad = \mu \nabla^2 u_i + (\lambda + \mu) u_{k,ki} - \beta \tau_{,i} + F_i - \rho \ddot{u}_i = 0
\end{aligned} \tag{16.74}
$$

Then, we get

$$\mu \nabla^2 u_i + (\lambda + \mu) u_{k,ki} - \beta \tau_{,i} + F_i = \rho \ddot{u}_i \quad (i = 1, 2, 3) \qquad \text{(Answer)}$$

Problem 16.5. Derive the Beltrami-Michell compatibility equations (16.23).

Solution. The compatibility equations are from Eq. (16.22)

$$
\begin{aligned}
&\frac{\partial^2 \varepsilon_{xx}}{\partial y^2} + \frac{\partial^2 \varepsilon_{yy}}{\partial x^2} = 2\frac{\partial^2 \varepsilon_{xy}}{\partial x \partial y} \\
&\frac{\partial^2 \varepsilon_{yy}}{\partial z^2} + \frac{\partial^2 \varepsilon_{zz}}{\partial y^2} = 2\frac{\partial^2 \varepsilon_{yz}}{\partial y \partial z} \\
&\frac{\partial^2 \varepsilon_{zz}}{\partial x^2} + \frac{\partial^2 \varepsilon_{xx}}{\partial z^2} = 2\frac{\partial^2 \varepsilon_{zx}}{\partial z \partial x} \\
&\frac{\partial^2 \varepsilon_{xx}}{\partial y \partial z} = \frac{\partial}{\partial x}\left(-\frac{\partial \varepsilon_{yz}}{\partial x} + \frac{\partial \varepsilon_{zx}}{\partial y} + \frac{\partial \varepsilon_{xy}}{\partial z} \right) \\
&\frac{\partial^2 \varepsilon_{yy}}{\partial z \partial x} = \frac{\partial}{\partial y}\left(\frac{\partial \varepsilon_{yz}}{\partial x} - \frac{\partial \varepsilon_{zx}}{\partial y} + \frac{\partial \varepsilon_{xy}}{\partial z} \right) \\
&\frac{\partial^2 \varepsilon_{zz}}{\partial x \partial y} = \frac{\partial}{\partial z}\left(\frac{\partial \varepsilon_{yz}}{\partial x} + \frac{\partial \varepsilon_{zx}}{\partial y} - \frac{\partial \varepsilon_{xy}}{\partial z} \right)
\end{aligned} \tag{16.75}
$$

The constitutive equations are from Eq. (16.10)

$$\varepsilon_{ij} = \frac{1}{2G}\left(\sigma_{ij} - \frac{\nu}{1+\nu}\Theta\delta_{ij}\right) + \alpha\tau\delta_{ij} \tag{16.76}$$

Substitution of Eq. (16.76) into the first three equations in Eq. (16.75) gives

$$\begin{aligned}
&\frac{\partial^2\sigma_{xx}}{\partial y^2} + \frac{\partial^2\sigma_{yy}}{\partial x^2} - \frac{\nu}{1+\nu}\left(\frac{\partial^2\Theta}{\partial y^2} + \frac{\partial^2\Theta}{\partial x^2}\right) + 2G\alpha\left(\frac{\partial^2\tau}{\partial y^2} + \frac{\partial^2\tau}{\partial x^2}\right) = 2\frac{\partial^2\sigma_{xy}}{\partial x\partial y}\\
&\frac{\partial^2\sigma_{yy}}{\partial z^2} + \frac{\partial^2\sigma_{zz}}{\partial y^2} - \frac{\nu}{1+\nu}\left(\frac{\partial^2\Theta}{\partial z^2} + \frac{\partial^2\Theta}{\partial y^2}\right) + 2G\alpha\left(\frac{\partial^2\tau}{\partial z^2} + \frac{\partial^2\tau}{\partial y^2}\right) = 2\frac{\partial^2\sigma_{yz}}{\partial y\partial z}\\
&\frac{\partial^2\sigma_{zz}}{\partial x^2} + \frac{\partial^2\sigma_{xx}}{\partial z^2} - \frac{\nu}{1+\nu}\left(\frac{\partial^2\Theta}{\partial x^2} + \frac{\partial^2\Theta}{\partial z^2}\right) + 2G\alpha\left(\frac{\partial^2\tau}{\partial x^2} + \frac{\partial^2\tau}{\partial z^2}\right) = 2\frac{\partial^2\sigma_{zx}}{\partial z\partial x}
\end{aligned} \tag{16.77}$$

The summation of Eq. (16.77) gives

$$\begin{aligned}
&\nabla^2\Theta - \left(\frac{\partial^2\sigma_{xx}}{\partial x^2} + \frac{\partial^2\sigma_{yy}}{\partial y^2} + \frac{\partial^2\sigma_{zz}}{\partial z^2}\right) - \frac{2\nu}{1+\nu}\nabla^2\Theta + 4G\alpha\nabla^2\tau\\
&\quad = 2\left(\frac{\partial^2\sigma_{xy}}{\partial x\partial y} + \frac{\partial^2\sigma_{yz}}{\partial y\partial z} + \frac{\partial^2\sigma_{zx}}{\partial z\partial x}\right)
\end{aligned} \tag{16.78}$$

Therefore,

$$\begin{aligned}
&\frac{1-\nu}{1+\nu}\nabla^2\Theta + 4G\alpha\nabla^2\tau = \left(\frac{\partial^2\sigma_{xx}}{\partial x^2} + \frac{\partial^2\sigma_{xy}}{\partial x\partial y} + \frac{\partial^2\sigma_{zx}}{\partial z\partial x}\right)\\
&\quad + \left(\frac{\partial^2\sigma_{xy}}{\partial x\partial y} + \frac{\partial^2\sigma_{yy}}{\partial y^2} + \frac{\partial^2\sigma_{yz}}{\partial y\partial z}\right) + \left(\frac{\partial^2\sigma_{zx}}{\partial z\partial x} + \frac{\partial^2\sigma_{yz}}{\partial y\partial z} + \frac{\partial^2\sigma_{zz}}{\partial z^2}\right)
\end{aligned} \tag{16.79}$$

Taking Eq. (16.1) into consideration, Eq. (16.79) becomes

$$\nabla^2\Theta = -\frac{2E}{1-\nu}\alpha\nabla^2\tau - \frac{1+\nu}{1-\nu}\left(\frac{\partial F_x}{\partial x} + \frac{\partial F_y}{\partial y} + \frac{\partial F_z}{\partial z}\right) \tag{16.80}$$

The summation of the second and the third equation in Eq. (16.77) gives

$$\begin{aligned}
&\frac{\partial^2\sigma_{zz}}{\partial y^2} + \frac{\partial^2\sigma_{zz}}{\partial x^2} + \frac{\partial^2(\sigma_{xx}+\sigma_{yy})}{\partial z^2} - \frac{\nu}{1+\nu}\left(\nabla^2\Theta + \frac{\partial^2\Theta}{\partial z^2}\right)\\
&\quad + 2G\alpha\left(\nabla^2\tau + \frac{\partial^2\tau}{\partial z^2}\right) - 2\frac{\partial}{\partial z}\left(\frac{\partial\sigma_{xz}}{\partial x} + \frac{\partial\sigma_{yz}}{\partial y}\right)\\
&= \nabla^2\sigma_{zz} + \frac{\partial^2\Theta}{\partial z^2} - 2\frac{\partial}{\partial z}\left(\frac{\partial\sigma_{xz}}{\partial x} + \frac{\partial\sigma_{yz}}{\partial y} + \frac{\partial\sigma_{zz}}{\partial z}\right) + 2G\alpha\left(\nabla^2\tau + \frac{\partial^2\tau}{\partial z^2}\right)
\end{aligned}$$

$$
\begin{aligned}
&-\frac{\nu}{1+\nu}\frac{\partial^2\Theta}{\partial z^2}+\frac{\nu}{1+\nu}\left[\frac{2E}{1-\nu}\alpha\nabla^2\tau+\frac{1+\nu}{1-\nu}\left(\frac{\partial F_x}{\partial x}+\frac{\partial F_y}{\partial y}+\frac{\partial F_z}{\partial z}\right)\right]\\
&=\nabla^2\sigma_{zz}+\frac{1}{1+\nu}\frac{\partial^2\Theta}{\partial z^2}+\frac{\alpha E}{1+\nu}\frac{\partial^2\tau}{\partial z^2}+\frac{\alpha E}{1-\nu}\nabla^2\tau\\
&+\frac{\nu}{1-\nu}\left(\frac{\partial F_x}{\partial x}+\frac{\partial F_y}{\partial y}+\frac{\partial F_z}{\partial z}\right)+2\frac{\partial F_z}{\partial z}=0
\end{aligned}
\tag{16.81}
$$

From Eq. (16.81), we get

$$
\begin{aligned}
&\nabla^2\sigma_{zz}+\frac{1}{1+\nu}\frac{\partial^2\Theta}{\partial z^2}+\alpha E\left(\frac{1}{1+\nu}\frac{\partial^2\tau}{\partial z^2}+\frac{1}{1-\nu}\nabla^2\tau\right)\\
&=-\frac{\nu}{1-\nu}\left(\frac{\partial F_x}{\partial x}+\frac{\partial F_y}{\partial y}+\frac{\partial F_z}{\partial z}\right)-2\frac{\partial F_z}{\partial z}
\end{aligned}
\qquad\text{(Answer)}
$$

The two other equations can be obtained in the same way. Then, we get the first, second and third equation in Eq. (16.23).

Substitution of Eq. (16.76) into the fourth equation in Eq. (16.75) gives

$$
\begin{aligned}
&2G\left[\frac{\partial^2\varepsilon_{xx}}{\partial y\partial z}-\frac{\partial}{\partial x}\left(-\frac{\partial\varepsilon_{yz}}{\partial x}+\frac{\partial\varepsilon_{zx}}{\partial y}+\frac{\partial\varepsilon_{xy}}{\partial z}\right)\right]\\
&=\frac{\partial^2}{\partial y\partial z}\left(\sigma_{xx}-\frac{\nu}{1+\nu}\Theta+2G\alpha\tau\right)-\frac{\partial}{\partial x}\left(-\frac{\partial\sigma_{yz}}{\partial x}+\frac{\partial\sigma_{zx}}{\partial y}+\frac{\partial\sigma_{xy}}{\partial z}\right)\\
&=\frac{1}{1+\nu}\frac{\partial^2\Theta}{\partial y\partial z}+2G\alpha\frac{\partial^2\tau}{\partial y\partial z}\\
&\quad-\left[\frac{\partial}{\partial z}\left(\frac{\partial\sigma_{yy}}{\partial y}+\frac{\partial\sigma_{xy}}{\partial x}\right)+\frac{\partial}{\partial y}\left(\frac{\partial\sigma_{zz}}{\partial z}+\frac{\partial\sigma_{zx}}{\partial x}\right)-\frac{\partial^2\sigma_{yz}}{\partial x^2}\right]\\
&=\frac{1}{1+\nu}\frac{\partial^2\Theta}{\partial y\partial z}+2G\alpha\frac{\partial^2\tau}{\partial y\partial z}\\
&\quad+\left[\frac{\partial}{\partial z}\left(\frac{\partial\sigma_{yz}}{\partial z}+F_y\right)+\frac{\partial}{\partial y}\left(\frac{\partial\sigma_{yz}}{\partial y}+F_z\right)+\frac{\partial^2\sigma_{yz}}{\partial x^2}\right]\\
&=\nabla^2\sigma_{yz}+\frac{1}{1+\nu}\frac{\partial^2\Theta}{\partial y\partial z}+2G\alpha\frac{\partial^2\tau}{\partial y\partial z}+\left(\frac{\partial F_y}{\partial z}+\frac{\partial F_z}{\partial y}\right)=0
\end{aligned}
\tag{16.82}
$$

From Eq. (16.82), we get the fifth equation in Eq. (16.23)

$$
\nabla^2\sigma_{yz}+\frac{1}{1+\nu}\frac{\partial^2\Theta}{\partial y\partial z}+\frac{\alpha E}{1+\nu}\frac{\partial^2\tau}{\partial y\partial z}=-\left(\frac{\partial F_y}{\partial z}+\frac{\partial F_z}{\partial y}\right)\qquad\text{(Answer)}
$$

The fourth and sixth equations can be obtained in the same way.

Problem 16.6. Derive the equilibrium equations (16.40) in a cylindrical coordinate system.

Solution. The equilibrium equation of the forces in the r direction acting on the element is

$$\begin{aligned}
&\left(\sigma_{rr} + \frac{\partial \sigma_{rr}}{\partial r} dr\right)(r+dr)\, d\theta dz - \sigma_{rr} r \, d\theta dz \\
&\quad + \left(\sigma_{zr} + \frac{\partial \sigma_{zr}}{\partial z} dz\right)\left[\pi(r+dr)^2 - \pi r^2\right]\frac{d\theta}{2\pi} - \sigma_{zr}\left[\pi(r+dr)^2 - \pi r^2\right]\frac{d\theta}{2\pi} \\
&\quad + \left(\sigma_{\theta r} + \frac{\partial \sigma_{\theta r}}{\partial \theta} d\theta\right) drdz \cos\frac{d\theta}{2} - \sigma_{\theta r} drdz \cos\frac{d\theta}{2} \\
&\quad - \left(\sigma_{\theta\theta} + \frac{\partial \sigma_{\theta\theta}}{\partial \theta} d\theta\right) drdz \sin\frac{d\theta}{2} - \sigma_{\theta\theta} drdz \sin\frac{d\theta}{2} \\
&\quad + F_r\left[\pi(r+dr)^2 - \pi r^2\right]\frac{d\theta}{2\pi} dz = 0
\end{aligned} \tag{16.83}$$

After dividing Eq. (16.83) by $rdrd\theta dz$ and omitting higher infinitesimal terms, we obtain

$$\frac{\partial \sigma_{rr}}{\partial r} + \frac{1}{r}\frac{\partial \sigma_{\theta r}}{\partial \theta} + \frac{\sigma_{zr}}{\partial z} + \frac{\sigma_{rr} - \sigma_{\theta\theta}}{r} + F_r = 0 \qquad \text{(Answer)}$$

The equilibrium equation of the forces in the $\theta + d\theta/2$ direction acting on the element is

$$\begin{aligned}
&\left(\sigma_{\theta\theta} + \frac{\partial \sigma_{\theta\theta}}{\partial \theta} d\theta\right) drdz \cos\frac{d\theta}{2} + \left(\sigma_{\theta r} + \frac{\partial \sigma_{\theta r}}{\partial \theta} d\theta\right) drdz \sin\frac{d\theta}{2} \\
&\quad - \sigma_{\theta\theta} drdz \cos\frac{d\theta}{2} + \sigma_{\theta r} drdz \sin\frac{d\theta}{2} \\
&\quad + \left(\sigma_{z\theta} + \frac{\partial \sigma_{z\theta}}{\partial z} dz\right)[\pi(r+dr)^2 - \pi r^2]\frac{d\theta}{2\pi} - \sigma_{z\theta}[\pi(r+dr)^2 - \pi r^2]\frac{d\theta}{2\pi} \\
&\quad + \left(\sigma_{r\theta} + \frac{\partial \sigma_{r\theta}}{\partial r} dr\right)(r+dr) d\theta dz - \sigma_{r\theta} r d\theta dz \\
&\quad + F_\theta\,[\pi(r+dr)^2 - \pi r^2]\frac{d\theta}{2\pi} dz = 0
\end{aligned} \tag{16.84}$$

Letting $\cos\frac{d\theta}{2} \to 1$, $\sin\frac{d\theta}{2} \to \frac{d\theta}{2}$, we get

$$\frac{\partial \sigma_{\theta\theta}}{\partial \theta} dr d\theta dz + \sigma_{\theta r} dr d\theta dz + \frac{\partial \sigma_{\theta r}}{\partial \theta} dr d\theta dz \frac{d\theta}{2} + \frac{\partial \sigma_{z\theta}}{\partial z}\left(r + \frac{dr}{2}\right) dr d\theta dz$$
$$+ \sigma_{r\theta} dr d\theta dz + \frac{\partial \sigma_{r\theta}}{\partial r}(r + dr) dr d\theta dz + F_\theta \left(r + \frac{dr}{2}\right) dr d\theta dz$$
$$= \left(\frac{\partial \sigma_{r\theta}}{\partial r} + \frac{1}{r}\frac{\partial \sigma_{\theta\theta}}{\partial \theta} + \frac{\partial \sigma_{z\theta}}{\partial z} + 2\frac{\sigma_{r\theta}}{r} + F_\theta\right) r dr d\theta dz$$
$$+ \left(\frac{\partial \sigma_{\theta r}}{\partial \theta}\frac{d\theta}{2} + \frac{\partial \sigma_{z\theta}}{\partial z}\frac{dr}{2} + \frac{\partial \sigma_{r\theta}}{\partial r} dr + F_\theta \frac{dr}{2}\right) dr d\theta dz = 0 \tag{16.85}$$

After dividing Eq. (16.85) by $r dr d\theta dz$ and omitting higher infinitesimal terms, we obtain

$$\frac{\partial \sigma_{r\theta}}{\partial r} + \frac{1}{r}\frac{\partial \sigma_{\theta\theta}}{\partial \theta} + \frac{\partial \sigma_{z\theta}}{\partial z} + 2\frac{\sigma_{r\theta}}{r} + F_\theta = 0 \qquad \text{(Answer)}$$

The equilibrium equation of the forces in the z direction acting on the element is

$$\left(\sigma_{zz} + \frac{\partial \sigma_{zz}}{\partial z} dz\right)[\pi(r+dr)^2 - \pi r^2]\frac{d\theta}{2\pi} - \sigma_{zz}[\pi(r+dr)^2 - \pi r^2]\frac{d\theta}{2\pi}$$
$$+ \left(\sigma_{\theta z} + \frac{\partial \sigma_{z\theta}}{\partial \theta} d\theta\right) dr dz - \sigma_{\theta z} dr dz + \left(\sigma_{rz} + \frac{\partial \sigma_{rz}}{\partial r} dr\right)(r + dr) d\theta dz$$
$$- \sigma_{rz} r d\theta dz + F_z\,[\pi(r+dr)^2 - \pi r^2]\frac{d\theta}{2\pi} dz$$
$$= \frac{\partial \sigma_{zz}}{\partial z}\left(r + \frac{dr}{2}\right) dr d\theta dz + \frac{\partial \sigma_{z\theta}}{\partial \theta} d\theta dr dz + \sigma_{rz} dr d\theta dz$$
$$+ \frac{\partial \sigma_{rz}}{\partial r}(r + dr) dr d\theta dz + F_z\left(r + \frac{dr}{2}\right) dr d\theta dz = 0 \tag{16.86}$$

After dividing Eq. (16.86) by $r dr d\theta dz$ and omitting higher infinitesimal terms, we obtain

$$\frac{\partial \sigma_{rz}}{\partial r} + \frac{1}{r}\frac{\partial \sigma_{\theta z}}{\partial \theta} + \frac{\partial \sigma_{zz}}{\partial z} + \frac{\sigma_{rz}}{r} + F_z = 0 \qquad \text{(Answer)}$$

Problem 16.7. Derive Eq. (16.47) in a cylindrical coordinate system.

Solution. Substitution of Eq. (16.46) into the first equation of Eq. (16.40) gives

$$\frac{\partial}{\partial r}(2\mu\epsilon_{rr} + \lambda e - \beta\tau) + \frac{2\mu}{r}\frac{\partial \epsilon_{\theta r}}{\partial \theta} + 2\mu\frac{\partial \epsilon_{zr}}{\partial z} + 2\mu\frac{\epsilon_{rr} - \epsilon_{\theta\theta}}{r} + F_r = 0 \tag{16.87}$$

Using the strain-displacement relation (16.42), Eq. (16.87) reduces to

$$(\lambda+2\mu)\frac{\partial e}{\partial r}+2\mu\left\{\frac{1}{2}\frac{\partial}{\partial z}\left(\frac{\partial u_r}{\partial z}-\frac{\partial u_z}{\partial r}\right)-\frac{1}{r}\frac{1}{2r}\frac{\partial}{\partial \theta}\left[\frac{\partial (ru_\theta)}{\partial r}-\frac{\partial u_r}{\partial \theta}\right]\right\}$$
$$-\beta\frac{\partial \tau}{\partial r}+F_r=0 \tag{16.88}$$

Then, we can get

$$(\lambda+2\mu)\frac{\partial e}{\partial r}-2\mu\left(\frac{1}{r}\frac{\partial \omega_z}{\partial \theta}-\frac{\partial \omega_\theta}{\partial z}\right)-\beta\frac{\partial \tau}{\partial r}+F_r=0 \qquad \text{(Answer)}$$

where

$$\omega_\theta=\frac{1}{2}\left(\frac{\partial u_r}{\partial z}-\frac{\partial u_z}{\partial r}\right),\quad \omega_z=\frac{1}{2r}\left(\frac{\partial (ru_\theta)}{\partial r}-\frac{\partial u_r}{\partial \theta}\right)$$

We can obtain the second and third equations of Navier's equations (16.47) by the same technique.

Problem 16.8. Derive the solutions of Laplace's equation in a cylindrical coordinate system.

Solution. We consider the solutions of Laplace's equation in a cylindrical coordinate system by use of the method of separation of variables. Laplace's equation is

$$\left(\frac{\partial^2}{\partial r^2}+\frac{1}{r}\frac{\partial}{\partial r}+\frac{1}{r^2}\frac{\partial^2}{\partial \theta^2}+\frac{\partial^2}{\partial z^2}\right)\varphi=0 \tag{16.89}$$

We assume that the harmonic function can be expressed by the product of three unknown functions, each of only one variable

$$\varphi(r,\theta,z)=f(r)g(\theta)h(z) \tag{16.90}$$

Substitution of Eq. (16.90) into Eq. (16.89) gives

$$\frac{1}{f(r)}\left(\frac{d^2}{dr^2}+\frac{1}{r}\frac{d}{dr}\right)f(r)+\frac{1}{r^2}\frac{1}{g(\theta)}\frac{d^2g(\theta)}{d\theta^2}+\frac{1}{h(z)}\frac{d^2h(z)}{dz^2}=0 \tag{16.91}$$

Equation (16.91) will be satisfied if the functions are selected as

$$\begin{aligned}&\frac{d^2f(r)}{dr^2}+\frac{1}{r}\frac{df(r)}{dr}+\left(a^2-\frac{b^2}{r^2}\right)f(r)=0\\&\frac{d^2g(\theta)}{d\theta^2}+b^2g(\theta)=0\\&\frac{d^2h(z)}{dz^2}-a^2h(z)=0\end{aligned} \tag{16.92}$$

where a and b are constants. The first equation of Eq. (16.92) is called the Bessel's differential equation, and has two independent solutions $f(r) = J_b(ar)$ and $Y_b(ar)$ for $a \neq 0,\ |b| < \infty$. $J_b(ar)$ and $Y_b(ar)$ are the Bessel function of first kind of order b and of second kind of order b, respectively. Similarly, the first equation of Eq. (16.92) has two independent solutions $f(r) = 1$ and ln r for $a = b = 0$, and $f(r) = r^b$ and r^{-b} for $a = 0, b \neq 0$.

The linearly independent solutions of Eq. (16.92) are

$$\begin{aligned}
&f(r) = \begin{pmatrix} 1 \\ \ln r \end{pmatrix} \quad \text{for} \quad a = b = 0, \quad f(r) = \begin{pmatrix} r^b \\ r^{-b} \end{pmatrix} \quad \text{for} \quad a = 0, \quad b \neq 0 \\
&f(r) = \begin{pmatrix} J_b(ar) \\ Y_b(ar) \end{pmatrix} \quad \text{for} \quad a \neq 0, \quad |b| < \infty \\
&g(\theta) = \begin{pmatrix} 1 \\ \theta \end{pmatrix} \quad \text{for} \quad b = 0, \quad g(\theta) = \begin{pmatrix} \sin b\theta \\ \cos b\theta \end{pmatrix} \quad \text{for} \quad b \neq 0 \\
&h(z) = \begin{pmatrix} 1 \\ z \end{pmatrix} \quad \text{for} \quad a = 0, \quad h(z) = \begin{pmatrix} e^{az} \\ e^{-az} \end{pmatrix} \quad \text{for} \quad a \neq 0
\end{aligned} \tag{16.93}$$

In another case, Eq. (16.91) will be satisfied if the functions are selected as

$$\begin{aligned}
&\frac{d^2 f(r)}{dr^2} + \frac{1}{r}\frac{df(r)}{dr} - \left(a^2 + \frac{b^2}{r^2}\right) f(r) = 0 \\
&\frac{d^2 g(\theta)}{d\theta^2} + b^2 g(\theta) = 0 \\
&\frac{d^2 h(z)}{dz^2} + a^2 h(z) = 0
\end{aligned} \tag{16.94}$$

The first equation of Eq. (16.94) is called the modified Bessel's differential equation, and has two independent solutions $f(r) = I_b(ar)$ and $K_b(ar)$ for $a \neq 0,\ |b| < \infty$. $I_b(ar)$ and $K_b(ar)$ are the Bessel function of first kind of order b and of second kind of order b, respectively.

The linearly independent solutions of Eq. (16.94) are

$$\begin{aligned}
&f(r) = \begin{pmatrix} 1 \\ \ln r \end{pmatrix} \quad \text{for} \quad a = b = 0, \quad f(r) = \begin{pmatrix} r^b \\ r^{-b} \end{pmatrix} \quad \text{for} \quad a = 0, \quad b \neq 0 \\
&f(r) = \begin{pmatrix} I_b(ar) \\ K_b(ar) \end{pmatrix} \quad \text{for} \quad a \neq 0, \quad |b| < \infty \\
&g(\theta) = \begin{pmatrix} 1 \\ \theta \end{pmatrix} \quad \text{for} \quad b = 0, \quad g(\theta) = \begin{pmatrix} \sin b\theta \\ \cos b\theta \end{pmatrix} \quad \text{for} \quad b \neq 0 \\
&h(z) = \begin{pmatrix} 1 \\ z \end{pmatrix} \quad \text{for} \quad a = 0, \quad h(z) = \begin{pmatrix} \sin az \\ \cos az \end{pmatrix} \quad \text{for} \quad a \neq 0
\end{aligned} \tag{16.95}$$

Therefore, the particular solutions of Laplace's equation in a cylindrical coordinate system may be expressed as follows:

$$\begin{aligned}
&\begin{pmatrix}1\\ \ln r\end{pmatrix}\begin{pmatrix}1\\ \theta\end{pmatrix}\begin{pmatrix}1\\ z\end{pmatrix}, \quad \begin{pmatrix}r^b\\ r^{-b}\end{pmatrix}\begin{pmatrix}\sin b\theta\\ \cos b\theta\end{pmatrix}\begin{pmatrix}1\\ z\end{pmatrix}\\
&\begin{pmatrix}J_0(ar)\\ Y_0(ar)\end{pmatrix}\begin{pmatrix}1\\ \theta\end{pmatrix}\begin{pmatrix}e^{az}\\ e^{-az}\end{pmatrix}, \quad \begin{pmatrix}I_0(ar)\\ K_0(ar)\end{pmatrix}\begin{pmatrix}1\\ \theta\end{pmatrix}\begin{pmatrix}\sin az\\ \cos az\end{pmatrix}\\
&\begin{pmatrix}J_b(ar)\\ Y_b(ar)\end{pmatrix}\begin{pmatrix}\sin b\theta\\ \cos b\theta\end{pmatrix}\begin{pmatrix}e^{az}\\ e^{-az}\end{pmatrix}\\
&\begin{pmatrix}I_b(ar)\\ K_b(ar)\end{pmatrix}\begin{pmatrix}\sin b\theta\\ \cos b\theta\end{pmatrix}\begin{pmatrix}\sin az\\ \cos az\end{pmatrix}
\end{aligned} \qquad \text{(Answer)} \quad (16.96)$$

where $\begin{pmatrix}\sinh az\\ \cosh az\end{pmatrix}$ can be used instead of $\begin{pmatrix}e^{az}\\ e^{-az}\end{pmatrix}$.

In Eq. (16.96), we used the following notation for the product of three one-column matrices

$$\begin{pmatrix}f_1\\ f_2\end{pmatrix}\begin{pmatrix}g_1\\ g_2\end{pmatrix}\begin{pmatrix}h_1\\ h_2\end{pmatrix} = \begin{pmatrix}f_1g_1h_1\\ f_2g_1h_1\\ f_1g_2h_1\\ f_2g_2h_1\\ f_1g_1h_2\\ f_2g_1h_2\\ f_1g_2h_2\\ f_2g_2h_2\end{pmatrix} \qquad (16.97)$$

Then, $\begin{pmatrix}I_b(ar)\\ K_b(ar)\end{pmatrix}\begin{pmatrix}\sin b\theta\\ \cos b\theta\end{pmatrix}\begin{pmatrix}\sin az\\ \cos az\end{pmatrix}$, for example means

$$\begin{pmatrix}I_b(ar)\\ K_b(ar)\end{pmatrix}\begin{pmatrix}\sin b\theta\\ \cos b\theta\end{pmatrix}\begin{pmatrix}\sin az\\ \cos az\end{pmatrix} = \begin{pmatrix}I_b(ar)\sin b\theta\sin az\\ K_b(ar)\sin b\theta\sin az\\ I_b(ar)\cos b\theta\sin az\\ K_b(ar)\cos b\theta\sin az\\ I_b(ar)\sin b\theta\cos az\\ K_b(ar)\sin b\theta\cos az\\ I_b(ar)\cos b\theta\cos az\\ K_b(ar)\cos b\theta\cos az\end{pmatrix} \qquad (16.98)$$

Therefore, a product of the three one-column matrices in Eq. (16.96) produces 8 particular solutions of Laplace's equation, and Eq. (16.96) represent an ordered array of 8×6 particular harmonic solutions in cylindrical coordinates.

Problem 16.9. Derive the equilibrium equations (16.52) in a spherical coordinate system.

Solution. We consider the infinitesimal element in a spherical coordinate system.
The area of the infinitesimal element of ϕ plane is $rdrd\theta$
The area of the infinitesimal element of $(\phi + d\phi)$ plane is $rdrd\theta$
The area of the infinitesimal element of θ plane is $drr\sin\theta d\phi$
The area of the infinitesimal element of $(\theta + d\theta)$ plane is

$$\begin{aligned} drr\sin(\theta + d\theta)d\phi &= drr\sin\theta\cos d\theta d\phi + drr\cos\theta\sin d\theta d\phi \\ &\cong drr\sin\theta d\phi + drr\cos\theta d\theta d\phi \end{aligned}$$

The area of the infinitesimal element of r plane is $r^2\sin\theta d\theta d\phi$
The area of the infinitesimal element of $(r + dr)$ plane is

$$(r + dr)^2 d\theta\sin\theta d\phi \cong (r^2 + 2rdr)d\theta\sin\theta d\phi$$

The volume of the infinitesimal element is $r^2\sin\theta drd\theta d\phi$

The equilibrium equation of the forces in the r direction acting on the element is

$$\begin{aligned} &\left(\sigma_{rr} + \frac{\partial\sigma_{rr}}{\partial r}dr\right)(r^2 + 2rdr)\sin\theta d\theta d\phi - \sigma_{rr}r^2\sin\theta d\theta d\phi \\ &+ \left(\sigma_{\theta r} + \frac{\partial\sigma_{\theta r}}{\partial\theta}d\theta\right)(\sin\theta + \cos\theta d\theta)rdrd\phi\cos\frac{d\theta}{2} - \sigma_{\theta r}\sin\theta rdrd\phi\cos\frac{d\theta}{2} \\ &- \left(\sigma_{\theta\theta} + \frac{\partial\sigma_{\theta\theta}}{\partial\theta}d\theta\right)(\sin\theta + \cos\theta d\theta)rdrd\phi\sin\frac{d\theta}{2} - \sigma_{\theta\theta}\sin\theta rdrd\phi\sin\frac{d\theta}{2} \\ &+ \left(\sigma_{\phi r} + \frac{\partial\sigma_{\phi r}}{\partial\phi}d\phi\right)rdrd\theta\cos\frac{\sin\theta d\phi}{2} - \sigma_{\phi r}rdrd\theta\cos\frac{\sin\theta d\phi}{2} \\ &- \left(\sigma_{\phi\phi} + \frac{\partial\sigma_{\phi\phi}}{\partial\phi}d\phi\right)rdrd\theta\sin\frac{\sin\theta d\phi}{2} - \sigma_{\phi\phi}rdrd\theta\sin\frac{\sin\theta d\phi}{2} \\ &+ F_r r^2\sin\theta drd\theta d\phi = 0 \end{aligned} \tag{16.99}$$

Letting $\cos\frac{d\theta}{2} \to 1$, $\sin\frac{d\theta}{2} \to \frac{d\theta}{2}$, $\cos\frac{\sin\theta d\phi}{2} \to 1$, $\sin\frac{\sin\theta d\phi}{2} \to \frac{\sin\theta d\phi}{2}$, we get

$$\begin{aligned} &\frac{\partial\sigma_{rr}}{\partial r}r^2\sin\theta drd\theta d\phi + 2\sigma_{rr}r\sin\theta drd\theta d\phi + 2\frac{\partial\sigma_{rr}}{\partial r}r\sin\theta drd\theta d\phi dr \\ &+ \sigma_{\theta r}r\cos\theta drd\theta d\phi + \frac{\partial\sigma_{\theta r}}{\partial\theta}r\sin\theta drd\theta d\phi + \frac{\partial\sigma_{\theta r}}{\partial\theta}r\cos\theta drd\theta d\phi d\theta \\ &- \sigma_{\theta\theta}r\sin\theta drd\theta d\phi \\ &- \frac{1}{2}\left(\sigma_{\theta\theta}\cos\theta d\theta + \frac{\partial\sigma_{\theta\theta}}{\partial\theta}\sin\theta d\theta + \frac{\partial\sigma_{\theta\theta}}{\partial\theta}\cos\theta d\theta d\theta\right)rdrd\theta d\phi \end{aligned}$$

$$+ \frac{\partial \sigma_{\phi r}}{\partial \phi} r dr d\theta d\phi - \sigma_{\phi\phi} r \sin\theta dr d\theta d\phi - \frac{1}{2}\frac{\partial \sigma_{\phi r}}{\partial \phi} r \sin\theta dr d\theta d\phi d\phi$$
$$+ F_r r^2 \sin\theta dr d\theta d\phi = 0 \tag{16.100}$$

After dividing Eq. (16.100) by $r^2 \sin\theta dr d\theta d\phi$ and omitting higher infinitesimal terms, we obtain

$$\frac{\partial \sigma_{rr}}{\partial r} + \frac{1}{r}\frac{\partial \sigma_{\theta r}}{\partial \theta} + \frac{1}{r\sin\theta}\frac{\partial \sigma_{\phi r}}{\partial \phi} + \frac{1}{r}(2\sigma_{rr} - \sigma_{\theta\theta} - \sigma_{\phi\phi} + \sigma_{\theta r}\cot\theta) + F_r = 0$$

(Answer)

The equilibrium equation of the forces in the $(\theta + d\theta/2)$ direction acting on the element is

$$\left(\sigma_{r\theta} + \frac{\partial \sigma_{r\theta}}{\partial r} dr\right)(r^2 + 2r dr)\sin\theta d\theta d\phi - \sigma_{r\theta} r^2 \sin\theta d\theta d\phi$$
$$+ \left(\sigma_{\theta r} + \frac{\partial \sigma_{\theta r}}{\partial \theta} d\theta\right)(\sin\theta + \cos\theta d\theta) r dr d\phi \sin\frac{d\theta}{2} + \sigma_{\theta r}\sin\theta r dr d\phi \sin\frac{d\theta}{2}$$
$$+ \left(\sigma_{\theta\theta} + \frac{\partial \sigma_{\theta\theta}}{\partial \theta} d\theta\right)(\sin\theta + \cos\theta d\theta) r dr d\phi \cos\frac{d\theta}{2} - \sigma_{\theta\theta}\sin\theta r dr d\phi \cos\frac{d\theta}{2}$$
$$+ \left(\sigma_{\phi\theta} + \frac{\partial \sigma_{\phi\theta}}{\partial \phi} d\phi\right) r dr d\theta \cos\frac{\cos\theta d\phi}{2} - \sigma_{\phi\theta} r dr d\theta \cos\frac{\cos\theta d\phi}{2}$$
$$- \left(\sigma_{\phi\phi} + \frac{\partial \sigma_{\phi\phi}}{\partial \phi} d\phi\right) r dr d\theta \sin\frac{\cos\theta d\phi}{2} - \sigma_{\phi\phi} r dr d\theta \sin\frac{\cos\theta d\phi}{2}$$
$$+ F_\theta r^2 \sin\theta dr d\theta d\phi = 0 \tag{16.101}$$

Letting $\cos\frac{d\theta}{2} \to 1$, $\sin\frac{d\theta}{2} \to \frac{d\theta}{2}$, $\cos\frac{\cos\theta d\phi}{2} \to 1$, $\sin\frac{\cos\theta d\phi}{2} \to \frac{\cos\theta d\phi}{2}$, we get

$$2\sigma_{r\theta} r \sin\theta dr d\theta d\phi + \frac{\partial \sigma_{r\theta}}{\partial r} r^2 \sin\theta dr d\theta d\phi + 2\frac{\partial \sigma_{r\theta}}{\partial r} r \sin\theta dr d\theta d\phi dr$$
$$+ \sigma_{\theta r} r \sin\theta dr d\theta d\phi$$
$$+ \frac{1}{2}\left(\sigma_{\theta r}\cot\theta + \frac{\partial \sigma_{\theta r}}{\partial \theta} + \frac{\partial \sigma_{\theta r}}{\partial \theta}\cot\theta d\theta\right) r \sin\theta dr d\theta d\phi d\theta$$
$$+ \left(\sigma_{\theta\theta}\cot\theta + \frac{\partial \sigma_{\theta\theta}}{\partial \theta} + \frac{\partial \sigma_{\theta\theta}}{\partial \theta}\cot\theta d\theta\right) r \sin\theta dr d\theta d\phi$$
$$+ \frac{1}{\sin\theta}\frac{\partial \sigma_{\phi\theta}}{\partial \phi} r \sin\theta dr d\theta d\phi - \left(\sigma_{\phi\phi} + \frac{1}{2}\frac{\partial \sigma_{\phi\phi}}{\partial \phi} d\phi\right)\cot\theta r \sin\theta dr d\theta d\phi$$
$$+ F_\theta r^2 \sin\theta dr d\theta d\phi = 0 \tag{16.102}$$

After dividing Eq. (16.102) by $r^2 \sin\theta dr d\theta d\phi$ and omitting higher infinitesimal terms, we obtain

$$\frac{\partial \sigma_{r\theta}}{\partial r} + \frac{1}{r}\frac{\partial \sigma_{\theta\theta}}{\partial \theta} + \frac{1}{r\sin\theta}\frac{\partial \sigma_{\phi\theta}}{\partial \phi} + \frac{1}{r}(3\sigma_{r\theta} + \sigma_{\theta\theta}\cot\theta - \sigma_{\phi\phi}\cot\theta) + F_\theta = 0$$

The equilibrium equation of the forces in $(\phi + d\phi/2)$ direction acting on the element is

$$\begin{aligned}
&\left(\sigma_{r\phi} + \frac{\partial \sigma_{r\phi}}{\partial r}dr\right)(r^2 + 2rdr)\sin\theta d\theta d\phi - \sigma_{r\phi}r^2\sin\theta d\theta d\phi \\
&+ \left(\sigma_{\theta\phi} + \frac{\partial \sigma_{\theta\phi}}{\partial \theta}d\theta\right)(\sin\theta + \cos\theta d\theta)rdrd\phi - \sigma_{\theta\phi}\sin\theta rdrd\phi \\
&+ \left(\sigma_{\phi r} + \frac{\partial \sigma_{\phi r}}{\partial \phi}d\phi\right)rdrd\theta \sin\frac{\sin\theta d\phi}{2} + \sigma_{\phi r}rdrd\theta\sin\frac{\sin\theta d\phi}{2} \\
&+ \left(\sigma_{\phi\phi} + \frac{\partial \sigma_{\phi\phi}}{\partial \phi}d\phi\right)rdrd\theta\cos\frac{\cos\theta d\phi}{2}\cos\frac{\sin\theta d\phi}{2} \\
&- \sigma_{\phi\phi}rdrd\theta\cos\frac{\cos\theta d\phi}{2}\cos\frac{\sin\theta d\phi}{2} + \left(\sigma_{\phi\theta} + \frac{\partial \sigma_{\phi\theta}}{\partial \phi}d\phi\right)rdrd\theta\sin\frac{\cos\theta d\phi}{2} \\
&+ \sigma_{\phi\theta}rdrd\theta\sin\frac{\cos\theta d\phi}{2} + F_\phi r^2\sin\theta drd\theta d\phi = 0
\end{aligned} \tag{16.103}$$

Letting $\cos\frac{\sin\theta d\phi}{2} \to 1$, $\cos\frac{\cos\theta d\phi}{2} \to 1$, $\sin\frac{\sin\theta d\phi}{2} \to \frac{\sin\theta d\phi}{2}$, $\sin\frac{\cos\theta d\phi}{2} \to \frac{\cos\theta d\phi}{2}$, we get

$$\begin{aligned}
&2\sigma_{r\phi}r\sin\theta drd\theta d\phi + \frac{\partial \sigma_{r\phi}}{\partial r}r^2\sin\theta drd\theta d\phi + 2\frac{\partial \sigma_{r\phi}}{\partial r}rdr\sin\theta drd\theta d\phi \\
&+ \sigma_{\theta\phi}r\cos\theta drd\theta d\phi + \frac{\partial \sigma_{\theta\phi}}{\partial \theta}r\sin\theta drd\theta d\phi + \frac{\partial \sigma_{\theta\phi}}{\partial \theta}d\theta r\cos\theta drd\theta d\phi \\
&+ \left(\sigma_{\phi r} + \frac{1}{2}\frac{\partial \sigma_{\phi r}}{\partial \phi}d\phi\right)r\sin\theta drd\theta d\phi + \frac{\partial \sigma_{\phi\phi}}{\partial \phi}rdrd\theta d\phi \\
&+ \sigma_{\phi\theta}r\cos\theta drd\theta d\phi + \frac{1}{2}\frac{\partial \sigma_{\phi\theta}}{\partial \phi}d\phi r\cos\theta drd\theta d\phi \\
&+ F_\phi r^2\sin\theta drd\theta d\phi = 0
\end{aligned} \tag{16.104}$$

After dividing Eq. (16.104) by $r^2\sin\theta drd\theta d\phi$ and omitting higher infinitesimal terms, we obtain

$$\frac{\partial \sigma_{r\phi}}{\partial r} + \frac{1}{r}\frac{\partial \sigma_{\theta\phi}}{\partial \theta} + \frac{1}{r\sin\theta}\frac{\partial \sigma_{\phi\phi}}{\partial \phi} + \frac{1}{r}(3\sigma_{r\phi} + 2\cot\theta\sigma_{\theta\phi}) + F_\phi = 0 \quad \text{(Answer)}$$

Problem 16.10. Derive Navier's equations (16.60) in a spherical coordinate system.

Solution. Substitution of Hooke's law (16.59) into the first equation of the equilibrium equations (16.52) yields

$$\begin{aligned}
&\frac{\partial}{\partial r}(2\mu\epsilon_{rr}+\lambda e-\beta\tau)+\frac{2\mu}{r}\frac{\partial\epsilon_{r\theta}}{\partial\theta}+\frac{2\mu}{r\sin\theta}\frac{\partial\epsilon_{\phi r}}{\partial\phi}\\
&+\frac{1}{r}[2(2\mu\epsilon_{rr}+\lambda e-\beta\tau)-(2\mu\epsilon_{\theta\theta}+\lambda e-\beta\tau)\\
&-(2\mu\epsilon_{\phi\phi}+\lambda e-\beta\tau)+(2\mu\epsilon_{r\theta})\cot\theta]+F_r=0
\end{aligned}\tag{16.105}$$

Using Eq. (16.54), Eq. (16.105) reduces to

$$\begin{aligned}
&(\lambda+2\mu)\frac{\partial e}{\partial r}-2\mu\left[\left(\frac{1}{r}\frac{\partial u_r}{\partial r}-\frac{u_r}{r^2}-\frac{1}{r^2}\frac{\partial u_\theta}{\partial\theta}+\frac{1}{r}\frac{\partial^2 u_\theta}{\partial r\partial\theta}\right)+\frac{1}{r}\frac{\partial u_r}{\partial r}-\frac{u_r}{r^2}\right.\\
&\left.+\cot\theta\left(\frac{1}{r}\frac{\partial u_\theta}{\partial r}-\frac{u_\theta}{r^2}\right)-\frac{1}{r^2\sin\theta}\frac{\partial u_\phi}{\partial\phi}+\frac{1}{r\sin\theta}\frac{\partial^2 u_\phi}{\partial r\partial\phi}\right)\Bigg]\\
&+\mu\left(\frac{1}{r^2}\frac{\partial^2 u_r}{\partial\theta^2}+\frac{1}{r}\frac{\partial^2 u_\theta}{\partial r\partial\theta}-\frac{1}{r^2}\frac{\partial u_\theta}{\partial\theta}\right)\\
&+\frac{\mu}{r\sin\theta}\left(\frac{1}{r\sin\theta}\frac{\partial^2 u_r}{\partial\phi^2}+\frac{\partial^2 u_\phi}{\partial r\partial\phi}-\frac{1}{r}\frac{\partial u_\phi}{\partial\phi}\right)\\
&+\frac{\mu}{r}\left[4\frac{\partial u_r}{\partial r}-4\frac{u_r}{r}-\frac{2}{r}\frac{\partial u_\theta}{\partial\theta}-3\cot\theta\frac{u_\theta}{r}-\frac{2}{r\sin\theta}\frac{\partial u_\phi}{\partial\phi}+\cot\theta\frac{1}{r}\frac{\partial u_r}{\partial\theta}\right.\\
&\left.+\cot\theta\frac{\partial u_\theta}{\partial r}\right]-\beta\frac{\partial\tau}{\partial r}+F_r=0
\end{aligned}\tag{16.106}$$

From Eq. (16.106), we get

$$\begin{aligned}
&(\lambda+2\mu)\frac{\partial e}{\partial r}-\frac{\mu}{r\sin\theta}\left[\frac{\sin\theta}{r}\left(\frac{\partial u_\theta}{\partial\theta}+r\frac{\partial^2 u_\theta}{\partial r\partial\theta}-\frac{\partial^2 u_r}{\partial^2\theta}\right)\right.\\
&\left.+\frac{\cos\theta}{r}\left(u_\theta+r\frac{\partial u_\theta}{\partial r}-\frac{\partial u_r}{\partial\theta}\right)-\frac{1}{r}\left(\frac{1}{\sin\theta}\frac{\partial^2 u_r}{\partial^2\phi}-\frac{\partial u_\phi}{\partial\phi}-r\frac{\partial^2 u_\phi}{\partial r\partial\phi}\right)\right]\\
&-\beta\frac{\partial\tau}{\partial r}+F_r=0
\end{aligned}\tag{16.107}$$

Taking into the consideration of

$$\begin{aligned}
2\frac{\partial(\omega_\phi\sin\theta)}{\partial\theta}&=2\sin\theta\frac{\partial\omega_\phi}{\partial\theta}+2\cos\theta\omega_\phi\\
&=\frac{\sin\theta}{r}\left(\frac{\partial u_\theta}{\partial\theta}+r\frac{\partial^2 u_\theta}{\partial r\partial\theta}-\frac{\partial^2 u_r}{\partial^2\theta}\right)+\frac{\cos\theta}{r}\left(u_\theta+r\frac{\partial u_\theta}{\partial r}-\frac{\partial u_r}{\partial\theta}\right)\\
2\frac{\partial\omega_\theta}{\partial\phi}&=\frac{1}{r}\left(\frac{1}{\sin\theta}\frac{\partial^2 u_r}{\partial^2\phi}-\frac{\partial u_\phi}{\partial\phi}-r\frac{\partial^2 u_\phi}{\partial r\partial\phi}\right)
\end{aligned}\tag{16.108}$$

equation (16.107) reduces to

$$(\lambda+2\mu)\frac{\partial e}{\partial r}-\frac{2\mu}{r\sin\theta}\left[\sin\theta\frac{\partial\omega_\phi}{\partial\theta}+\cos\theta\omega_\phi-\frac{\partial\omega_\theta}{\partial\phi}\right]-\beta\frac{\partial\tau}{\partial r}+F_r=0 \quad (16.109)$$

We finally obtain

$$(\lambda+2\mu)\frac{\partial e}{\partial r}-\frac{2\mu}{r\sin\theta}\left[\frac{\partial(\omega_\phi\sin\theta)}{\partial\theta}-\frac{\partial\omega_\theta}{\partial\phi}\right]-\beta\frac{\partial\tau}{\partial r}+F_r=0 \qquad \text{(Answer)}$$

We can obtain the second and third equations of Navier's equations (16.60) by same technique.

Problem 16.11. Derive the solutions of Laplace's equation in a spherical coordinate system.

Solution. We consider the solution of Laplace's equation in a spherical coordinate system by use of the method of separation of variables. Laplace's equation in a spherical coordinate system is

$$\left(\frac{\partial^2}{\partial r^2}+\frac{2}{r}\frac{\partial}{\partial r}+\frac{1}{r^2}\frac{\partial^2}{\partial\theta^2}+\frac{1}{r^2\tan\theta}\frac{\partial}{\partial\theta}+\frac{1}{r^2\sin^2\theta}\frac{\partial^2}{\partial\phi^2}\right)\varphi=0 \qquad (16.110)$$

We assume that the harmonic function can be expressed by the product of three unknown functions, each of only one variable

$$\varphi(r,\theta,\phi)=f(r)g(\theta)h(\phi) \qquad (16.111)$$

Substitution of Eq. (16.111) into Eq. (16.110) gives

$$\frac{r^2}{f(r)}\left(\frac{d^2}{dr^2}+\frac{2}{r}\frac{d}{dr}\right)f(r)+\frac{1}{g(\theta)}\left(\frac{d^2}{d\theta^2}+\frac{1}{\tan\theta}\frac{d}{d\theta}\right)g(\theta)$$
$$+\frac{1}{\sin^2\theta h(\phi)}\frac{d^2h(\phi)}{d\phi^2}=0 \qquad (16.112)$$

Equation (16.112) will be satisfied if the functions are selected as follows:

$$\frac{d^2f(r)}{dr^2}+\frac{2}{r}\frac{df(r)}{dr}-\frac{\nu(\nu+1)}{r^2}f(r)=0$$
$$\frac{d^2g(\theta)}{d\theta^2}+\frac{1}{\tan\theta}\frac{dg(\theta)}{d\theta}+\left[\nu(\nu+1)-\frac{\mu^2}{\sin^2\theta}\right]g(\theta)=0$$
$$\frac{d^2h(\phi)}{d\phi^2}+\mu^2h(\phi)=0 \qquad (16.113)$$

The linearly independent solutions of the first and the third equations of Eq. (16.113) are

$$f(r) = \begin{pmatrix} r^{\nu} \\ r^{-(\nu+1)} \end{pmatrix}$$

$$h(\phi) = \begin{pmatrix} 1 \\ \phi \end{pmatrix} \quad \text{for} \quad \mu = 0, \quad h(\phi) = \begin{pmatrix} \sin \mu\phi \\ \cos \mu\phi \end{pmatrix} \quad \text{for} \quad \mu \neq 0 \tag{16.114}$$

Application of the transformation of a variable $x = \cos\theta$ to the second equation in Eq. (16.113) gives

$$(1 - x^2)\frac{d^2 g(x)}{dx^2} - 2x\frac{dg(x)}{dx} + \left[\nu(\nu + 1) - \frac{\mu^2}{1 - x^2}\right] g(x) = 0 \tag{16.115}$$

The Eq. (16.115) is called the associated Legendre's differential equation, and the linearly independent solutions are given by

$$g(x) = \begin{pmatrix} P_{\nu}^{\mu}(x) \\ Q_{\nu}^{\mu}(x) \end{pmatrix} \tag{16.116}$$

Therefore, the particular solutions of the harmonic equation in a spherical coordinate system may be expressed as follows:

$$\begin{aligned} &\begin{pmatrix} r^{\nu} \\ r^{-(\nu+1)} \end{pmatrix} \begin{pmatrix} P_{\nu}(\cos\theta) \\ Q_{\nu}(\cos\theta) \end{pmatrix} \begin{pmatrix} 1 \\ \phi \end{pmatrix} \\ &\begin{pmatrix} r^{\nu} \\ r^{-(\nu+1)} \end{pmatrix} \begin{pmatrix} P_{\nu}^{\mu}(\cos\theta) \\ Q_{\nu}^{\mu}(\cos\theta) \end{pmatrix} \begin{pmatrix} \sin \mu\phi \\ \cos \mu\phi \end{pmatrix} \end{aligned} \tag{Answer}$$

where μ and ν are constants, $P_{\nu}(\cos\theta)$ is the Legendre function of the first kind, $Q_{\nu}(\cos\theta)$ is the Legendre function of the second kind, $P_{\nu}^{\mu}(\cos\theta)$ is the associated Legendre function of the first kind, and $Q_{\nu}^{\mu}(\cos\theta)$ is the associated Legendre function of the second kind. The notation for the product of three one-column matrices is explained by Eqs.(16.97) and (16.98).

Problem 16.12. Express the displacements and strains in a spherical coordinate system by use of the thermoelastic potential Φ and the Boussinesq functions φ, ϑ, ψ.

Solution. The relationship between the Cartesian and the spherical coordinate systems is

$$\begin{aligned} &x = r\sin\theta\cos\phi, \quad y = r\sin\theta\sin\phi, \quad z = r\cos\theta \\ &r = \sqrt{x^2 + y^2 + z^2}, \quad \tan\theta = \sqrt{x^2 + y^2}/z, \quad \tan\phi = \frac{y}{x} \end{aligned} \tag{16.117}$$

The direction cosines in both coordinate systems are

$$
\begin{aligned}
&l_1 = \sin\theta\cos\phi, \quad m_1 = \sin\theta\sin\phi, \quad n_1 = \cos\theta \\
&l_2 = \cos\theta\cos\phi, \quad m_2 = \cos\theta\sin\phi, \quad n_2 = -\sin\theta \\
&l_3 = -\sin\phi, \quad m_3 = \cos\phi, \quad n_3 = 0
\end{aligned} \tag{16.118}
$$

The partial derivatives are

$$
\begin{aligned}
\frac{\partial}{\partial x} &= \sin\theta\cos\phi\frac{\partial}{\partial r} + \frac{\cos\theta\cos\phi}{r}\frac{\partial}{\partial\theta} - \frac{\sin\phi}{r\sin\theta}\frac{\partial}{\partial\phi} \\
\frac{\partial}{\partial y} &= \sin\theta\sin\phi\frac{\partial}{\partial r} + \frac{\cos\theta\sin\phi}{r}\frac{\partial}{\partial\theta} + \frac{\cos\phi}{r\sin\theta}\frac{\partial}{\partial\phi} \\
\frac{\partial}{\partial z} &= \cos\theta\frac{\partial}{\partial r} - \frac{\sin\theta}{r}\frac{\partial}{\partial\theta}
\end{aligned} \tag{16.119}
$$

Displacements u_x, u_y, u_z in a Cartesian coordinate system are expressed by the thermoelastic potential Φ and the Boussinesq functions φ, ϑ, ψ from Eq. (16.38)

$$
\begin{aligned}
u_x &= \frac{\partial\Phi}{\partial x} + \frac{\partial\varphi}{\partial x} + 2\frac{\partial\vartheta}{\partial y} + z\frac{\partial\psi}{\partial x} \\
u_y &= \frac{\partial\Phi}{\partial y} + \frac{\partial\varphi}{\partial y} - 2\frac{\partial\vartheta}{\partial x} + z\frac{\partial\psi}{\partial y} \\
u_z &= \frac{\partial\Phi}{\partial z} + \frac{\partial\varphi}{\partial z} + z\frac{\partial\psi}{\partial z} - (3-4\nu)\psi
\end{aligned} \tag{16.120}
$$

Displacements u_r, u_θ, u_ϕ in a spherical coordinate system are expressed by the displacements u_x, u_y, u_z in a Cartesian coordinate system

$$
\begin{aligned}
u_r &= u_x\sin\theta\cos\phi + u_y\sin\theta\sin\phi + u_z\cos\theta \\
u_\theta &= u_x\cos\theta\cos\phi + u_y\cos\theta\sin\phi - u_z\sin\theta \\
u_\phi &= -u_x\sin\phi + u_y\cos\phi
\end{aligned} \tag{16.121}
$$

Substituting Eq. (16.120) into Eq. (16.121) and translating the partial differentials from the Cartesian to the spherical coordinate systems, we get

$$
\begin{aligned}
u_r &= u_x\sin\theta\cos\phi + u_y\sin\theta\sin\phi + u_z\cos\theta \\
&= \left[\left(\sin\theta\cos\phi\frac{\partial\Phi}{\partial r} + \frac{\cos\theta\cos\phi}{r}\frac{\partial\Phi}{\partial\theta} - \frac{\sin\phi}{r\sin\theta}\frac{\partial\Phi}{\partial\phi}\right)\right. \\
&\quad + \left(\sin\theta\cos\phi\frac{\partial\varphi}{\partial r} + \frac{\cos\theta\cos\phi}{r}\frac{\partial\varphi}{\partial\theta} - \frac{\sin\phi}{r\sin\theta}\frac{\partial\varphi}{\partial\phi}\right) \\
&\quad + 2\left(\sin\theta\sin\phi\frac{\partial\vartheta}{\partial r} + \frac{\cos\theta\sin\phi}{r}\frac{\partial\vartheta}{\partial\theta} + \frac{\cos\phi}{r\sin\theta}\frac{\partial\vartheta}{\partial\phi}\right) \\
&\quad \left. + r\cos\theta\left(\sin\theta\cos\phi\frac{\partial\psi}{\partial r} + \frac{\cos\theta\cos\phi}{r}\frac{\partial\psi}{\partial\theta} - \frac{\sin\phi}{r\sin\theta}\frac{\partial\psi}{\partial\phi}\right)\right]\sin\theta\cos\phi
\end{aligned}
$$

$$
\begin{aligned}
&+\Bigg[\left(\sin\theta\sin\phi\frac{\partial\Phi}{\partial r}+\frac{\cos\theta\sin\phi}{r}\frac{\partial\Phi}{\partial\theta}+\frac{\cos\phi}{r\sin\theta}\frac{\partial\Phi}{\partial\phi}\right)\\
&+\left(\sin\theta\sin\phi\frac{\partial\varphi}{\partial r}+\frac{\cos\theta\sin\phi}{r}\frac{\partial\varphi}{\partial\theta}+\frac{\cos\phi}{r\sin\theta}\frac{\partial\varphi}{\partial\phi}\right)\\
&-2\left(\sin\theta\cos\phi\frac{\partial\vartheta}{\partial r}+\frac{\cos\theta\cos\phi}{r}\frac{\partial\vartheta}{\partial\theta}-\frac{\sin\phi}{r\sin\theta}\frac{\partial\vartheta}{\partial\phi}\right)\\
&+r\cos\theta\left(\sin\theta\sin\phi\frac{\partial\psi}{\partial r}+\frac{\cos\theta\sin\phi}{r}\frac{\partial\psi}{\partial\theta}+\frac{\cos\phi}{r\sin\theta}\frac{\partial\psi}{\partial\phi}\right)\Bigg]\sin\theta\sin\phi\\
&+\Bigg[\left(\cos\theta\frac{\partial\Phi}{\partial r}-\frac{\sin\theta}{r}\frac{\partial\Phi}{\partial\theta}\right)+\left(\cos\theta\frac{\partial\varphi}{\partial r}-\frac{\sin\theta}{r}\frac{\partial\varphi}{\partial\theta}\right)\\
&+r\cos\theta\left(\cos\theta\frac{\partial\psi}{\partial r}-\frac{\sin\theta}{r}\frac{\partial\psi}{\partial\theta}\right)-(3-4\nu)\phi\Bigg]\cos\theta\\
=&\ \frac{\partial\Phi}{\partial r}+\frac{\partial\varphi}{\partial r}+\frac{2}{r}\frac{\partial\vartheta}{\partial\theta}+r\cos\theta\frac{\partial\psi}{\partial r}-(3-4\nu)\psi\cos\theta
\end{aligned}
\tag{16.122}
$$

$$
\begin{aligned}
u_\theta =&\ u_x\cos\theta\cos\phi+u_y\cos\theta\sin\phi-u_z\sin\theta\\
=&\Bigg[\left(\sin\theta\cos\phi\frac{\partial\Phi}{\partial r}+\frac{\cos\theta\cos\phi}{r}\frac{\partial\Phi}{\partial\theta}-\frac{\sin\phi}{r\sin\theta}\frac{\partial\Phi}{\partial\phi}\right)\\
&+\left(\sin\theta\cos\phi\frac{\partial\varphi}{\partial r}+\frac{\cos\theta\cos\phi}{r}\frac{\partial\varphi}{\partial\theta}-\frac{\sin\phi}{r\sin\theta}\frac{\partial\varphi}{\partial\phi}\right)\\
&+2\left(\sin\theta\sin\phi\frac{\partial\vartheta}{\partial r}+\frac{\cos\theta\sin\phi}{r}\frac{\partial\vartheta}{\partial\theta}+\frac{\cos\phi}{r\sin\theta}\frac{\partial\vartheta}{\partial\phi}\right)\\
&+r\cos\theta\left(\sin\theta\cos\phi\frac{\partial\psi}{\partial r}+\frac{\cos\theta\cos\phi}{r}\frac{\partial\psi}{\partial\theta}-\frac{\sin\phi}{r\sin\theta}\frac{\partial\psi}{\partial\phi}\right)\Bigg]\cos\theta\cos\phi\\
&+\Bigg[\left(\sin\theta\sin\phi\frac{\partial\Phi}{\partial r}+\frac{\cos\theta\sin\phi}{r}\frac{\partial\Phi}{\partial\theta}+\frac{\cos\phi}{r\sin\theta}\frac{\partial\Phi}{\partial\phi}\right)\\
&+\left(\sin\theta\sin\phi\frac{\partial\varphi}{\partial r}+\frac{\cos\theta\sin\phi}{r}\frac{\partial\varphi}{\partial\theta}+\frac{\cos\phi}{r\sin\theta}\frac{\partial\varphi}{\partial\phi}\right)\\
&-2\left(\sin\theta\cos\phi\frac{\partial\vartheta}{\partial r}+\frac{\cos\theta\cos\phi}{r}\frac{\partial\vartheta}{\partial\theta}-\frac{\sin\phi}{r\sin\theta}\frac{\partial\vartheta}{\partial\phi}\right)\\
&+r\cos\theta\left(\sin\theta\sin\phi\frac{\partial\psi}{\partial r}+\frac{\cos\theta\sin\phi}{r}\frac{\partial\psi}{\partial\theta}+\frac{\cos\phi}{r\sin\theta}\frac{\partial\psi}{\partial\phi}\right)\Bigg]\cos\theta\sin\phi\\
&-\Bigg[\left(\cos\theta\frac{\partial\Phi}{\partial r}-\frac{\sin\theta}{r}\frac{\partial\Phi}{\partial\theta}\right)+\left(\cos\theta\frac{\partial\varphi}{\partial r}-\frac{\sin\theta}{r}\frac{\partial\varphi}{\partial\theta}\right)\\
&+r\cos\theta\left(\cos\theta\frac{\partial\psi}{\partial r}-\frac{\sin\theta}{r}\frac{\partial\psi}{\partial\theta}\right)-(3-4\nu)\phi\Bigg]\sin\theta\\
=&\ \frac{1}{r}\frac{\partial\Phi}{\partial\phi}+\frac{1}{r}\frac{\partial\varphi}{\partial\phi}+\frac{2}{r\tan\theta}\frac{\partial\vartheta}{\partial\phi}+\cos\theta\frac{\partial\psi}{\partial\phi}+(3-4\nu)\psi\sin\theta
\end{aligned}
\tag{16.123}
$$

$$
\begin{aligned}
u_\phi &= -u_x \sin\phi + u_y \cos\phi \\
&= -\left[\left(\sin\theta\cos\phi\frac{\partial\Phi}{\partial r} + \frac{\cos\theta\cos\phi}{r}\frac{\partial\Phi}{\partial\theta} - \frac{\sin\phi}{r\sin\theta}\frac{\partial\Phi}{\partial\phi}\right)\right. \\
&\quad + \left(\sin\theta\cos\phi\frac{\partial\varphi}{\partial r} + \frac{\cos\theta\cos\phi}{r}\frac{\partial\varphi}{\partial\theta} - \frac{\sin\phi}{r\sin\theta}\frac{\partial\varphi}{\partial\phi}\right) \\
&\quad + 2\left(\sin\theta\sin\phi\frac{\partial\vartheta}{\partial r} + \frac{\cos\theta\sin\phi}{r}\frac{\partial\vartheta}{\partial\theta} + \frac{\cos\phi}{r\sin\theta}\frac{\partial\vartheta}{\partial\phi}\right) \\
&\quad \left. + r\cos\theta\left(\sin\theta\cos\phi\frac{\partial\psi}{\partial r} + \frac{\cos\theta\cos\phi}{r}\frac{\partial\psi}{\partial\theta} - \frac{\sin\phi}{r\sin\theta}\frac{\partial\psi}{\partial\phi}\right)\right]\sin\phi \\
&\quad + \left[\left(\sin\theta\sin\phi\frac{\partial\Phi}{\partial r} + \frac{\cos\theta\sin\phi}{r}\frac{\partial\Phi}{\partial\theta} + \frac{\cos\phi}{r\sin\theta}\frac{\partial\Phi}{\partial\phi}\right)\right. \\
&\quad + \left(\sin\theta\sin\phi\frac{\partial\varphi}{\partial r} + \frac{\cos\theta\sin\phi}{r}\frac{\partial\varphi}{\partial\theta} + \frac{\cos\phi}{r\sin\theta}\frac{\partial\varphi}{\partial\phi}\right) \\
&\quad - 2\left(\sin\theta\cos\phi\frac{\partial\vartheta}{\partial r} + \frac{\cos\theta\cos\phi}{r}\frac{\partial\vartheta}{\partial\theta} - \frac{\sin\phi}{r\sin\theta}\frac{\partial\vartheta}{\partial\phi}\right) \\
&\quad \left. + r\cos\theta\left(\sin\theta\sin\phi\frac{\partial\psi}{\partial r} + \frac{\cos\theta\sin\phi}{r}\frac{\partial\psi}{\partial\theta} + \frac{\cos\phi}{r\sin\theta}\frac{\partial\psi}{\partial\phi}\right)\right]\cos\phi \\
&= \frac{1}{r\sin\theta}\frac{\partial\Phi}{\partial\phi} + \frac{1}{r\sin\theta}\frac{\partial\varphi}{\partial\phi} - 2\sin\theta\frac{\partial\vartheta}{\partial r} - \frac{2\cos\theta}{r}\frac{\partial\psi}{\partial\theta} + \frac{1}{\tan\theta}\frac{\partial\psi}{\partial\phi}
\end{aligned}
\tag{16.124}
$$

The displacement-strain relations (16.54) are

$$
\begin{aligned}
\epsilon_{rr} &= \frac{\partial u_r}{\partial r}, \quad \epsilon_{\theta\theta} = \frac{u_r}{r} + \frac{1}{r}\frac{\partial u_\theta}{\partial\theta} \\
\epsilon_{\phi\phi} &= \frac{u_r}{r} + \cot\theta\frac{u_\theta}{r} + \frac{1}{r\sin\theta}\frac{\partial u_\phi}{\partial\phi}, \quad \epsilon_{r\theta} = \frac{1}{2}\left(\frac{1}{r}\frac{\partial u_r}{\partial\theta} + \frac{\partial u_\theta}{\partial r} - \frac{u_\theta}{r}\right) \\
\epsilon_{\theta\phi} &= \frac{1}{2}\left(\frac{1}{r}\frac{\partial u_\phi}{\partial\theta} - \cot\theta\frac{u_\phi}{r} + \frac{1}{r\sin\theta}\frac{\partial u_\theta}{\partial\phi}\right) \\
\epsilon_{\phi r} &= \frac{1}{2}\left(\frac{1}{r\sin\theta}\frac{\partial u_r}{\partial\phi} + \frac{\partial u_\phi}{\partial r} - \frac{u_\phi}{r}\right)
\end{aligned}
\tag{16.125}
$$

Substitution of Eqs. (16.122)–(16.124) into Eq. (16.125) gives

$$
\begin{aligned}
\epsilon_{rr} &= \frac{\partial^2\Phi}{\partial r^2} + \frac{\partial^2\varphi}{\partial r^2} - \frac{2}{r^2}\frac{\partial\vartheta}{\partial\phi} + \frac{2}{r}\frac{\partial^2\vartheta}{\partial r\partial\phi} + r\cos\theta\frac{\partial^2\psi}{\partial r^2} \\
&\quad - 2(1-2\nu)\cos\theta\frac{\partial\psi}{\partial r} \\
\epsilon_{\theta\theta} &= \frac{1}{r}\frac{\partial\Phi}{\partial r} + \frac{1}{r^2}\frac{\partial^2\Phi}{\partial\theta^2} + \frac{1}{r}\frac{\partial\varphi}{\partial r} + \frac{1}{r^2}\frac{\partial^2\varphi}{\partial\theta^2} + \frac{2}{r^2}\frac{\partial\vartheta}{\partial\phi} - \frac{2}{r^2\sin^2\theta}\frac{\partial\vartheta}{\partial\phi}
\end{aligned}
$$

$$
\begin{aligned}
&+\frac{2}{r^2\tan\theta}\frac{\partial^2\vartheta}{\partial\phi\partial\theta}+\cos\theta\frac{\partial\psi}{\partial r}+\frac{\cos\theta}{r}\frac{\partial^2\psi}{\partial\theta^2}+2(1-2\nu)\frac{\sin\theta}{r}\frac{\partial\psi}{\partial\theta}\\
\epsilon_{\phi\phi}=&\frac{1}{r}\frac{\partial\Phi}{\partial r}+\frac{\cot\theta}{r^2}\frac{\partial\Phi}{\partial\theta}+\frac{1}{r^2\sin^2\theta}\frac{\partial^2\Phi}{\partial\phi^2}+\frac{1}{r}\frac{\partial\varphi}{\partial r}+\frac{\cot\theta}{r^2}\frac{\partial\varphi}{\partial\theta}\\
&+\frac{1}{r^2\sin^2\theta}\frac{\partial^2\varphi}{\partial\phi^2}+\frac{2}{r^2\sin^2\theta}\frac{\partial\vartheta}{\partial\phi}-\frac{2}{r}\frac{\partial^2\vartheta}{\partial r\partial\phi}-\frac{2\cos\theta}{r^2\sin\theta}\frac{\partial^2\vartheta}{\partial\theta\partial\phi}\\
&+\cos\theta\frac{\partial\psi}{\partial r}+\frac{\cos^2\theta}{r\sin\theta}\frac{\partial\psi}{\partial\theta}+\frac{\cos\theta}{r\sin^2\theta}\frac{\partial^2\psi}{\partial\phi^2}\\
\epsilon_{r\theta}=&\frac{1}{r}\frac{\partial^2\Phi}{\partial r\partial\theta}-\frac{1}{r^2}\frac{\partial\Phi}{\partial\theta}+\frac{1}{r}\frac{\partial^2\varphi}{\partial r\partial\theta}-\frac{1}{r^2}\frac{\partial\varphi}{\partial\theta}\\
&+\frac{1}{r^2}\frac{\partial^2\vartheta}{\partial\theta\partial\phi}-\frac{2}{r^2\tan\theta}\frac{\partial\vartheta}{\partial\phi}+\frac{1}{r\tan\theta}\frac{\partial^2\vartheta}{\partial r\partial\phi}\\
&-2(1-\nu)\sin\theta\frac{\partial\psi}{\partial r}-2(1-\nu)\frac{\cos\theta}{r}\frac{\partial\psi}{\partial\theta}+\cos\theta\frac{\partial^2\psi}{\partial r\partial\theta}\\
\epsilon_{\theta\phi}=&-\frac{\cos\theta}{r^2\sin^2\theta}\frac{\partial\Phi}{\partial\phi}+\frac{1}{r^2\sin\theta}\frac{\partial^2\Phi}{\partial\theta\partial\phi}-\frac{\cos\theta}{r^2\sin^2\theta}\frac{\partial\varphi}{\partial\phi}+\frac{1}{r^2\sin\theta}\frac{\partial^2\varphi}{\partial\theta\partial\phi}\\
&+\frac{1}{r^2\sin\theta}\frac{\partial\vartheta}{\partial\theta}-\frac{\sin\theta}{r}\frac{\partial^2\vartheta}{\partial r\partial\theta}-\frac{\cos\theta}{r^2}\frac{\partial^2\vartheta}{\partial\theta^2}+\frac{\cos\theta}{r^2\sin^2\theta}\frac{\partial^2\vartheta}{\partial\phi^2}\\
&+\frac{1}{r\tan\theta}\frac{\partial^2\psi}{\partial\theta\partial\phi}+\left(1-2\nu-\frac{1}{\tan^2\theta}\right)\frac{1}{r}\frac{\partial\psi}{\partial\phi}\\
\epsilon_{r\phi}=&\frac{1}{r\sin\theta}\frac{\partial^2\Phi}{\partial r\partial\phi}-\frac{1}{r^2\sin\theta}\frac{\partial\Phi}{\partial\phi}+\frac{1}{r\sin\theta}\frac{\partial^2\varphi}{\partial r\partial\phi}-\frac{1}{r^2\sin\theta}\frac{\partial\varphi}{\partial\phi}\\
&+\frac{1}{r^2\sin\theta}\frac{\partial^2\vartheta}{\partial\phi^2}-\sin\theta\frac{\partial^2\vartheta}{\partial r^2}+\frac{\cos\theta}{r^2}\frac{\partial\vartheta}{\partial\theta}-\frac{\cos\theta}{r}\frac{\partial^2\vartheta}{\partial r\partial\theta}\\
&+\frac{1}{\tan\theta}\frac{\partial^2\psi}{\partial r\partial\phi}-2(1-\nu)\frac{1}{r\tan\theta}\frac{\partial\psi}{\partial\phi}
\end{aligned}
\qquad\text{(Answer)}
$$

Chapter 17
Plane Thermoelastic Problems

In this chapter the basic treatment of plane thermoelastic problems in a state of plane strain and a plane stress are recalled. Typical three methods for the solution of plane problems are presented: the thermal stress function method for both simply connected and multiply connected bodies, the complex variable method with use of the conformal mapping technique, and potential method for Navier's equations [See also Chap. 7].

17.1 Plane Strain and Plane Stress

The unified systems of the governing equations for both plane strain and plane stress are as follows:

The generalized Hooke's law is

$$\begin{aligned} \epsilon_{xx} &= \frac{1}{E^*}\left(\sigma_{xx} - \nu^*\sigma_{yy}\right) + \alpha^*\tau - c^* \\ \epsilon_{yy} &= \frac{1}{E^*}\left(\sigma_{yy} - \nu^*\sigma_{xx}\right) + \alpha^*\tau - c^* \\ \epsilon_{xy} &= \frac{1}{2G}\sigma_{xy} \end{aligned} \tag{17.1}$$

An alternative form

$$\begin{aligned} \sigma_{xx} &= (\lambda^* + 2\mu)\epsilon_{xx} + \lambda^*\epsilon_{yy} - \beta^*\tau \\ \sigma_{yy} &= (\lambda^* + 2\mu)\epsilon_{yy} + \lambda^*\epsilon_{xx} - \beta^*\tau \\ \sigma_{xy} &= 2\mu\epsilon_{xy} \end{aligned} \tag{17.1$'$}$$

where

$$E^* = \begin{cases} E' = \dfrac{E}{1-\nu^2} & \text{for plane strain} \\ E & \text{for plane stress} \end{cases}$$

M. Reza Eslami et al., *Theory of Elasticity and Thermal Stresses*, Solid Mechanics and Its Applications 197, DOI: 10.1007/978-94-007-6356-2_17,

$$
\nu^* = \begin{cases} \nu' = \dfrac{\nu}{1-\nu} & \text{for plane strain} \\ \nu & \text{for plane stress} \end{cases}
$$
$$
\alpha^* = \begin{cases} \alpha' = (1+\nu)\alpha & \text{for plane strain} \\ \alpha & \text{for plane stress} \end{cases}
$$
$$
\lambda^* = \begin{cases} \lambda & \text{for plane strain} \\ \lambda' = \dfrac{2\mu\lambda}{\lambda+2\mu} & \text{for plane stress} \end{cases} \tag{17.2}
$$
$$
\beta^* = \begin{cases} \beta & \text{for plane strain} \\ \beta' = \dfrac{2\mu\beta}{\lambda+2\mu} & \text{for plane stress} \end{cases}
$$
$$
c^* = \begin{cases} \nu\epsilon_0 & \text{for plane strain} \\ 0 & \text{for plane stress} \end{cases}
$$

The equilibrium equations in the absence of body forces are

$$
\frac{\partial \sigma_{xx}}{\partial x} + \frac{\partial \sigma_{yx}}{\partial y} = 0, \quad \frac{\partial \sigma_{xy}}{\partial x} + \frac{\partial \sigma_{yy}}{\partial y} = 0 \tag{17.3}
$$

The compatibility equation is

$$
\frac{\partial^2 \epsilon_{xx}}{\partial y^2} + \frac{\partial^2 \epsilon_{yy}}{\partial x^2} = 2\frac{\partial^2 \epsilon_{xy}}{\partial x \partial y} \tag{17.4}
$$

Navier's equations are from Eqs. (7.25) and (7.35)

$$
\begin{aligned}
\mu\nabla^2 u_x + (\lambda^* + \mu)\frac{\partial e}{\partial x} - \beta^* \frac{\partial \tau}{\partial x} = 0 \\
\mu\nabla^2 u_y + (\lambda^* + \mu)\frac{\partial e}{\partial y} - \beta^* \frac{\partial \tau}{\partial y} = 0
\end{aligned} \tag{17.5}
$$

where $e = \epsilon_{xx} + \epsilon_{yy} + c^*$.

The boundary conditions are

$$
\sigma_{xx} l + \sigma_{yx} m = p_{nx}, \quad \sigma_{xy} l + \sigma_{yy} m = p_{ny} \tag{17.6}
$$

Next, we show typical three analytical methods for the plane problem.

Thermal stress function method

We introduce a thermal stress function χ related to the components of stress as follows

$$
\sigma_{xx} = \frac{\partial^2 \chi}{\partial y^2}, \quad \sigma_{yy} = \frac{\partial^2 \chi}{\partial x^2}, \quad \sigma_{xy} = -\frac{\partial^2 \chi}{\partial x \partial y} \tag{17.7}
$$

The governing equation for the thermal stress function χ is

$$\nabla^4\chi = -\alpha^* E^* \nabla^2 \tau \tag{17.8}$$

where

$$\nabla^4 = \nabla^2\nabla^2 = \left(\frac{\partial^2}{\partial x^2} + \frac{\partial^2}{\partial y^2}\right)\left(\frac{\partial^2}{\partial x^2} + \frac{\partial^2}{\partial y^2}\right) = \frac{\partial^4}{\partial x^4} + 2\frac{\partial^4}{\partial x^2 \partial y^2} + \frac{\partial^4}{\partial y^4} \tag{17.9}$$

$$\alpha^* E^* = \begin{cases} \dfrac{\alpha E}{1-\nu} & \text{for plane strain} \\ \alpha E & \text{for plane stress} \end{cases} \tag{17.10}$$

The components of displacement can be expressed in the form

$$\begin{aligned} u_x &= \frac{1}{2G}\left[-\frac{\partial \chi}{\partial x} + \frac{1}{1+\nu^*}\frac{\partial \psi}{\partial y}\right] - c^* x \\ u_y &= \frac{1}{2G}\left[-\frac{\partial \chi}{\partial y} + \frac{1}{1+\nu^*}\frac{\partial \psi}{\partial x}\right] - c^* y \end{aligned} \tag{17.11}$$

where c^* is a constant and the function ψ satisfies the equation

$$\sigma_{xx} + \sigma_{yy} + \alpha^* E^* \tau = \nabla^2\chi + \alpha^* E^* \tau \equiv \frac{\partial^2 \psi}{\partial x \partial y} \tag{17.12}$$

in which

$$\frac{\partial^2}{\partial x \partial y}\nabla^2\psi = 0 \tag{17.13}$$

When the external force does not apply to the body, the boundary conditions of pure thermal stress problems are

$$\begin{aligned} \chi(P) &= C_1 x + C_2 y + C_3 \\ \frac{\partial \chi(P)}{\partial n'} &= C_1 \cos(n', x) + C_2 \cos(n', y) \end{aligned} \tag{17.14}$$

where n' denotes some direction which does not coincide with the direction of the contour, and C_1, C_2, and C_3 are arbitrary integration constants. The arbitrary integration constants C_1, C_2, and C_3 can be taken zero for a simply connected body. On the other hand, for a multiply connected body whose boundary consists of $m+1$ simply closed contours L_i $(i = 0, 1, \ldots, m)$, Eq. (17.14) can be rewritten as

$$\left.\begin{aligned} \chi(P_i) &= C_{1i} x + C_{2i} y + C_{3i} \\ \frac{\partial \chi(P_i)}{\partial n'} &= C_{1i} \cos(n', x) + C_{2i} \cos(n', y) \end{aligned}\right\} \quad \text{on } L_i \ (i = 0, 1, \cdots, m) \tag{17.15}$$

where P_i is an arbitrary point on the i-th boundary contour L_i $(i = 0, 1, \dots, m)$, C_{1i}, C_{2i}, and C_{3i} are the integration constants on the boundary contour L_i $(i = 0, 1, \dots, m)$, and the integration constants on only one contour can be zero.

The conditions of single-valuedness of rotation and displacements in $(m+1)$-tuply connected body with traction free surfaces are

$$\oint_{L_i} \frac{\partial}{\partial n}(\nabla^2\chi + \alpha^* E^* \tau)\, ds = 0 \qquad (i = 1, \dots, m) \tag{17.16}$$

$$\oint_{L_i} \left(x_1\frac{\partial}{\partial s} - x_2\frac{\partial}{\partial n}\right)(\nabla^2\chi + \alpha^* E^* \tau)\, ds = 0 \qquad (i = 1, \dots, m) \tag{17.17}$$

$$\oint_{L_i} \left(x_1\frac{\partial}{\partial n} + x_2\frac{\partial}{\partial s}\right)(\nabla^2\chi + \alpha^* E^* \tau)\, ds = 0 \qquad (i = 1, \dots, m) \tag{17.18}$$

The general solution of Eq. (17.8) for the thermal stress function χ may be expressed as the sum of the complementary solution χ_c and the particular solution χ_p

$$\chi = \chi_c + \chi_p \tag{17.19}$$

where the complementary solution χ_c and the particular solution χ_p are governed by

$$\nabla^4\chi_c = 0 \tag{17.20}$$

$$\nabla^2\chi_p = -\alpha^* E^* \tau \tag{17.21}$$

When the transient heat conduction equation with no heat generation is discussed, the particular solution χ_p is

$$\chi_p = -\alpha^* E^* \kappa \int_{t_r}^{t} \tau(x, y, t')\, dt' + \chi_{pr} + (t - t_r)\chi_{p0} \tag{17.22}$$

where t_r denotes the reference time, and χ_{pr} and χ_{p0} denote solutions of the following Poisson's and Laplace's equations, respectively

$$\nabla^2\chi_{pr} = -\alpha^* E^* \tau_r, \quad \nabla^2\chi_{p0} = 0 \tag{17.23}$$

in which τ_r denotes the temperature at the reference time t_r.

Complex variable method

The biharmonic function χ_c governed by Eq. (17.20) can be represented by two complex functions $\varphi(z)$ and $\psi_1(z)$ as follows

$$\chi_c = \frac{1}{2}\left[\bar{z}\varphi(z) + z\overline{\varphi(z)} + \psi_1(z) + \overline{\psi_1(z)}\right] \tag{17.24}$$

where the upper bar denotes its conjugate complex function

$$\bar{z} = x - iy, \quad \overline{\varphi(z)} = p - iq \tag{17.25}$$

where $i^2 = -1$. Hence, the thermal stress function χ can be represented by two complex functions and the particular solution χ_p

$$\chi = \chi_c + \chi_p = \frac{1}{2}\left[\bar{z}\varphi(z) + z\overline{\varphi(z)} + \psi_1(z) + \overline{\psi_1(z)}\right] + \chi_p \tag{17.26}$$

The plane thermal stresses are given by

$$\begin{aligned} \sigma_{xx} + \sigma_{yy} &= 4\mathrm{Re}\left[\varphi'(z)\right] - \alpha^* E^* \tau \\ \sigma_{yy} - \sigma_{xx} + 2i\sigma_{xy} &= 2\left[\bar{z}\varphi''(z) + \psi'(z)\right] + \left(\frac{\partial}{\partial x} - i\frac{\partial}{\partial y}\right)^2 \chi_p \end{aligned} \tag{17.27}$$

where $\psi(z) \equiv \psi_1'(z)$, and the complex functions $\varphi(z)$ and $\psi(z)$ are called the complex stress functions.

The components of displacement are

$$u_x + iu_y = \frac{1}{2G}\left[\frac{3-\nu^*}{1+\nu^*}\varphi(z) - z\overline{\varphi'(z)} - \overline{\psi(z)} - \left(\frac{\partial \chi_p}{\partial x} + i\frac{\partial \chi_p}{\partial y}\right)\right] - c^*(x+iy) \tag{17.28}$$

The boundary condition for the pure thermoelastic problem without traction is

$$\varphi(z) + z\overline{\varphi'(z)} + \overline{\psi(z)} = -\left(\frac{\partial \chi_p}{\partial x} + i\frac{\partial \chi_p}{\partial y}\right) + C \tag{17.29}$$

The resultant moment M about the origin of the coordinate system is

$$M = \mathrm{Re}\Big[\psi_1(z) - z\psi_1'(z) - z\bar{z}\varphi'(z)\Big]_A^P - \left[x\frac{\partial \chi_p}{\partial x} + y\frac{\partial \chi_p}{\partial y} - \chi_p\right]_A^P \tag{17.30}$$

Let us translate a given region S in the complex z-plane into a region Σ in the complex ζ-plane by use of the conformal mapping function $\omega(\zeta)$

$$z = x + iy = \omega(\zeta), \quad \zeta = \xi + i\eta = \rho e^{i\theta} \tag{17.31}$$

A curvilinear coordinate system (ρ, θ) consists of curves ρ = constant and radii θ = constant. The components (u_ρ, u_θ) of displacement vector $\boldsymbol{u}$ in the ζ-plane referred to a curvilinear coordinate system (ρ, θ) can be expressed by the components (u_x, u_y) of displacement vector $\boldsymbol{u}$ in the z-plane referred to a Cartesian coordinate system (x, y)

$$u_\rho + iu_\theta = e^{-i\alpha}(u_x + iu_y) \tag{17.32}$$

where α denotes an angle between x axis and ρ axis. The components of stress in plane problems referred to a curvilinear coordinate system (ρ, θ) can be expressed by the components referred to a Cartesian coordinate system (x, y) as follows

$$\begin{aligned}\sigma_{\rho\rho} + \sigma_{\theta\theta} &= \sigma_{xx} + \sigma_{yy} \\ \sigma_{\theta\theta} - \sigma_{\rho\rho} + 2i\sigma_{\rho\theta} &= e^{2i\alpha}(\sigma_{yy} - \sigma_{xx} + 2i\sigma_{xy})\end{aligned} \tag{17.33}$$

With the conformal mapping function $\omega(\zeta)$, Eq. (17.32) with $c^* = 0$ becomes

$$\begin{aligned}u_\rho + iu_\theta = \frac{1}{2G}\frac{\bar{\zeta}}{\rho}\frac{\overline{\omega'(\zeta)}}{|\omega'(\zeta)|}\Big[&\frac{3-\nu^*}{1+\nu^*}\phi(\zeta) - \frac{\omega(\zeta)}{\overline{\omega'(\zeta)}}\overline{\phi'(\zeta)} - \overline{\Psi(\zeta)} \\ &- \frac{\zeta}{\rho}\frac{1}{\overline{\omega'(\zeta)}}\left(\frac{\partial \chi_p}{\partial \rho} + i\frac{1}{\rho}\frac{\partial \chi_p}{\partial \theta}\right)\Big]\end{aligned} \tag{17.34}$$

The stress fields (17.33) are expressed by

$$\begin{aligned}\sigma_{\rho\rho} + \sigma_{\theta\theta} &= 4\,\mathrm{Re}\left[\frac{\phi'(\zeta)}{\omega'(\zeta)}\right] - \alpha^* E^* \tau \\ \sigma_{\theta\theta} - \sigma_{\rho\rho} + 2i\sigma_{\rho\theta} &= \frac{2\zeta^2}{\rho^2\overline{\omega'(\zeta)}}\left\{\overline{\omega(\zeta)}\phi(\zeta)\left[\frac{\phi'(\zeta)}{\omega'(\zeta)}\right]' + \Psi'(\zeta)\right\} \\ &\quad + \frac{4\zeta^2}{\rho^2\overline{\omega'(\zeta)}}\left\{\frac{\partial^2 \chi_p}{\partial \zeta^2}\frac{1}{\omega'(\zeta)} - \frac{\partial \chi_p}{\partial \zeta}\frac{\omega''(\zeta)}{[\omega'(\zeta)]^2}\right\}\end{aligned} \tag{17.35}$$

Potential method

Navier's equations (17.5) can be rewritten as

$$\begin{aligned}\mu\nabla^2 u_x + (\lambda^* + \mu)\left(\frac{\partial^2 u_x}{\partial x^2} + \frac{\partial^2 u_y}{\partial x \partial y}\right) - \beta^*\frac{\partial \tau}{\partial x} = 0 \\ \mu\nabla^2 u_y + (\lambda^* + \mu)\left(\frac{\partial^2 u_x}{\partial x \partial y} + \frac{\partial^2 u_y}{\partial y^2}\right) - \beta^*\frac{\partial \tau}{\partial y} = 0\end{aligned} \tag{17.36}$$

The general solutions of Navier's equations (17.36) for the plane problem can be expressed as the sum of the complementary solutions u_x^c and u_y^c, and the particular solutions u_x^p and u_y^p

$$u_x = u_x^c + u_x^p, \quad u_y = u_y^c + u_y^p \tag{17.37}$$

The particular solutions u_x^p and u_y^p can be expressed in terms of Goodier's thermoelastic potential Φ as follows:

$$u_x^p = \Phi_{,x}, \quad u_y^p = \Phi_{,y} \tag{17.38}$$

Φ must satisfy the equation as

$$\nabla^2 \Phi = K\tau \tag{17.39}$$

where

$$K = \frac{\beta^*}{\lambda^* + 2\mu} = (1 + \nu^*)\alpha^* \tag{17.40}$$

The complementary solutions u_x^c and u_y^c of Navier's equations (17.36) are expressed by two plane harmonic functions.

$$u_x^c = \frac{3 - \nu^*}{1 + \nu^*}\phi_1 - x\frac{\partial \phi_1}{\partial x} - y\frac{\partial \phi_2}{\partial x}, \quad u_y^c = \frac{3 - \nu^*}{1 + \nu^*}\phi_2 - x\frac{\partial \phi_1}{\partial y} - y\frac{\partial \phi_2}{\partial y} \tag{17.41}$$

where two functions ϕ_1 and ϕ_2 are harmonic

$$\nabla^2 \phi_1 = 0, \quad \nabla^2 \phi_2 = 0 \tag{17.42}$$

17.2 Problems and Solutions Related to Plane Thermoelastic Problems

Problem 17.1. Derive the governing equation for χ to be expressed by Eq. (17.8).

Solution. From Eq. (17.7) the thermal stress function χ is defined by

$$\sigma_{xx} = \frac{\partial^2 \chi}{\partial y^2}, \quad \sigma_{yy} = \frac{\partial^2 \chi}{\partial x^2}, \quad \sigma_{xy} = -\frac{\partial^2 \chi}{\partial x \partial y} \tag{17.43}$$

The equilibrium equations (17.3) are automatically satisfied by use of the thermal stress function χ. The compatibility equation is from Eq. (17.4)

$$\frac{\partial^2 \epsilon_{xx}}{\partial y^2} + \frac{\partial^2 \epsilon_{yy}}{\partial x^2} = 2\frac{\partial^2 \epsilon_{xy}}{\partial x \partial y} \tag{17.44}$$

Using Hooke's law, and substituting the thermal stress function χ into Eq. (17.44), we obtain

$$\begin{aligned}
&\frac{\partial^2 \epsilon_{xx}}{\partial y^2} + \frac{\partial^2 \epsilon_{yy}}{\partial x^2} - 2\frac{\partial^2 \epsilon_{xy}}{\partial x \partial y} \\
&= \frac{\partial^2}{\partial y^2}\left[\frac{1}{E^*}\left(\sigma_{xx} - \nu^* \sigma_{yy}\right) + \alpha^* \tau - c^*\right] \\
&\quad + \frac{\partial^2}{\partial x^2}\left[\frac{1}{E^*}\left(\sigma_{yy} - \nu^* \sigma_{xx}\right) + \alpha^* \tau - c^*\right] - 2\frac{\partial^2}{\partial x \partial y}\left(\frac{1}{2G}\sigma_{xy}\right)
\end{aligned}$$

$$
\begin{aligned}
&= \frac{\partial^2}{\partial y^2}\Big[\frac{1}{E^*}\Big(\frac{\partial^2 \chi}{\partial y^2} - \nu^* \frac{\partial^2 \chi}{\partial x^2}\Big) + \alpha^* \tau - c^*\Big] \\
&\quad + \frac{\partial^2}{\partial x^2}\Big[\frac{1}{E^*}\Big(\frac{\partial^2 \chi}{\partial x^2} - \nu^* \frac{\partial^2 \chi}{\partial y^2}\Big) + \alpha^* \tau - c^*\Big] \\
&\quad + 2\frac{\partial^2}{\partial x \partial y}\Big(\frac{1+\nu^*}{E^*}\frac{\partial^2 \chi}{\partial x \partial y}\Big) \\
&= \frac{1}{E^*}\Big[\frac{\partial^4 \chi}{\partial x^4} + 2\frac{\partial^4 \chi}{\partial x^2 \partial y^2} + \frac{\partial^4 \chi}{\partial y^4} - \alpha^* E^* \Big(\frac{\partial^2}{\partial x^2} + \frac{\partial^2}{\partial y^2}\Big)\tau\Big] = 0
\end{aligned} \tag{17.45}
$$

Therefore, the governing equation for thermal stress function χ is

$$
\nabla^4 \chi = -\alpha^* E^* \nabla^2 \tau \qquad \text{(Answer)}
$$

where

$$
\alpha^* E^* = \begin{cases} (1+\nu)\alpha \dfrac{E}{1-\nu^2} = \dfrac{\alpha E}{1-\nu} & \text{for plane strain} \\ \alpha E & \text{for plane stress} \end{cases} \tag{17.46}
$$

Problem 17.2. Prove that the arbitrary integration constants C_1, C_2, and C_3 in Eq. (17.14) may be taken as zero for a simply connected body.

Solution. We take

$$
\chi = \chi^* + C_1 x + C_2 y + C_3 \tag{17.47}
$$

Substitution of Eq. (17.47) into Eqs. (17.8), (17.7) and (17.14) gives the governing equation

$$
\nabla^4 \chi^* = -\alpha^* E^* \nabla^2 \tau \tag{17.48}
$$

the stresses

$$
\sigma_{xx} = \frac{\partial^2 \chi^*}{\partial y^2}, \quad \sigma_{yy} = \frac{\partial^2 \chi^*}{\partial x^2}, \quad \sigma_{xy} = -\frac{\partial^2 \chi^*}{\partial x \partial y} \tag{17.49}
$$

and the boundary conditions

$$
\chi^*(P) = 0, \quad \frac{\partial \chi^*(P)}{\partial n'} = 0 \quad \text{on } L \tag{17.50}
$$

Since the function $C_1 x + C_2 y + C_3$ does not appear in the governing Eq. (17.48), the stresses (17.49) and the boundary conditions (17.50), the integration constants can be taken zero for the simply connected body.

Problem 17.3. Prove that the integration constants C_{1i}, C_{2i}, and C_{3i} on the boundary contour L_i $(i = 0, 1, \ldots, m)$ in Eq. (17.15) can be taken zero on only one contour.

Solution. We take

$$
\chi = \chi^* + C_{10} x + C_{20} y + C_{30} \tag{17.51}
$$

Substitution of Eq. (17.51) into Eqs. (17.8), (17.7) and (17.15) gives the governing equation

$$\nabla^4 \chi^* = -\alpha^* E^* \nabla^2 \tau \tag{17.52}$$

the stresses

$$\sigma_{xx} = \frac{\partial^2 \chi^*}{\partial y^2}, \quad \sigma_{yy} = \frac{\partial^2 \chi^*}{\partial x^2}, \quad \sigma_{xy} = -\frac{\partial^2 \chi^*}{\partial x \partial y} \tag{17.53}$$

and the boundary conditions

$$\begin{aligned} &\chi^*(P_0) = 0, \quad \frac{\partial \chi^*(P_0)}{\partial n'} = 0 \quad \text{on } L_0 \\ &\chi^*(P_i) = (C_{1i} - C_{10})x + (C_{2i} - C_{20})y + (C_{3i} - C_{30}) \\ &\frac{\partial \chi^*(P_i)}{\partial n'} = (C_{1i} - C_{10})\cos(n', x) + (C_{2i} - C_{20})\cos(n', y) \\ &\qquad\qquad\qquad\qquad\qquad\qquad \text{on } L_i (i = 1, 2, \cdots, n) \end{aligned} \tag{17.54}$$

If we put

$$C^*_{1i} = C_{1i} - C_{10}, \quad C^*_{2i} = C_{2i} - C_{20}, \quad C^*_{3i} = C_{3i} - C_{30} \tag{17.55}$$

equations (17.54) reduce to

$$\begin{aligned} &\chi^*(P_0) = 0, \quad \frac{\partial \chi^*(P_0)}{\partial n'} = 0 \quad \text{on } L_0 \\ &\chi^*(P_i) = C^*_{1i} x + C^*_{2i} y + C^*_{3i} \\ &\frac{\partial \chi^*(P_i)}{\partial n'} = C^*_{1i} \cos(n', x) + C^*_{2i} \cos(n', y) \quad \text{on} \quad L_i (i = 1, 2, \cdots, n) \end{aligned} \tag{17.56}$$

Taking Eqs. (17.52), (17.53) and (17.56) into consideration, the integration constants on only one contour can be taken zero.

Problem 17.4. Prove that the thermal stress is not produced in a strip with thickness l, when the steady temperature distribution without the internal heat generation is treated.

Solution. The heat conduction equation without internal heat generation is

$$\nabla^2 T = 0 \tag{17.57}$$

The thermal stress function χ satisfies the equation

$$\nabla^4 \chi = -\alpha^* E^* \nabla^2 \tau \tag{17.58}$$

where $\tau = T - T_0$. From Eqs. (17.57) and (17.58) we get

$$\nabla^4 \chi = 0 \tag{17.59}$$

The general solution of Eq. (17.59) is

$$\chi = A_0 + A_1 x + B_1 y + A_2 x^2 + B_2 y^2 + C_2 xy + A_3 x^3 + B_3 y^3 + C_3 x^2 y + D_3 xy^2 \tag{17.60}$$

Thermal stresses are

$$\begin{aligned}\sigma_{xx} &= \frac{\partial^2 \chi}{\partial^2 y} = 2B_2 + 6B_3 y + 2D_3 x\\ \sigma_{yy} &= \frac{\partial^2 \chi}{\partial x^2} = 2A_2 + 6A_3 x + 2C_3 y\\ \sigma_{xy} &= -\frac{\partial^2 \chi}{\partial x \partial y} = -C_2 - 2C_3 x - 2D_3 y\end{aligned} \tag{17.61}$$

The boundary conditions are

$$\sigma_{xx} = 0\,, \quad \sigma_{xy} = 0 \quad \text{on } x = 0,\ l \tag{17.62}$$

The unknown coefficients are determined from Eq. (17.62) as

$$B_2 = 0, \quad B_3 = 0, \quad D_3 = 0\,, \quad C_2 = 0, \quad C_3 = 0 \tag{17.63}$$

From the condition of $\lim_{y \to \infty} \sigma_{yy} = 0$, we get

$$A_2 = 0, \quad A_3 = 0 \tag{17.64}$$

Then, the thermal stress is not produced in the strip.

Problem 17.5. Find the displacements in a strip when a steady temperature is given by

$$T = T_a + (T_b - T_a)\frac{x}{l} \tag{17.65}$$

Solution. Thermal stress is not produced in a strip, since the temperature given by Eq. (17.65) is the steady temperature without internal heat generation. As the thermal stress is not produced in a strip, a harmonic function ψ expressed by Eq. (17.12) reduces to

$$\frac{\partial^2 \psi}{\partial x \partial y} = \nabla^2 \chi + \alpha^* E^* \tau = \alpha^* E^* \tau = \alpha^* E^* \left[T_a - T_0 + (T_b - T_a)\frac{x}{l} \right] \tag{17.66}$$

where T_0 denotes the initial temperature. The integration of Eq. (17.66) gives

$$\psi = A + Bx + Cy + \alpha^* E^* \left[(T_a - T_0) xy + (T_b - T_a)\frac{x^2 y}{2l} \right] \tag{17.67}$$

The displacements (17.11) with no thermal stress reduce to

$$u_x = \frac{1}{2G(1+\nu^*)}\frac{\partial\psi}{\partial y} - c^* x, \quad u_y = \frac{1}{2G(1+\nu^*)}\frac{\partial\psi}{\partial x} - c^* y \tag{17.68}$$

Substitution of Eq. (17.67) into Eq. (17.68) gives

$$\begin{aligned} u_x &= \frac{C}{E^*} - c^* x + \alpha^*\Big[(T_a - T_0)x + (T_b - T_a)\frac{x^2}{2l}\Big] \\ u_y &= \frac{B}{E^*} - c^* y + \alpha^*\Big[(T_a - T_0)y + (T_b - T_a)\frac{xy}{l}\Big] \end{aligned} \qquad \text{(Answer)}$$

Problem 17.6. Derive the solutions of Eq. (17.20) in a Cartesian coordinate system.

Solution. First, we consider the solutions of Laplace's equation in a Cartesian coordinate system by use of the method of separation of variables. Laplace's equation is

$$\Big(\frac{\partial^2}{\partial x^2} + \frac{\partial^2}{\partial y^2}\Big)h(x, y) = 0 \tag{17.69}$$

We assume that the harmonic function can be expressed by the product of two unknown functions, each of which has only one variable

$$h(x, y) = f(r)g(y) \tag{17.70}$$

Substitution of Eq. (17.70) into Eq. (17.69) gives

$$\frac{d^2 f(x)}{dx^2} + a^2 f(x) = 0, \quad \frac{d^2 g(y)}{dy^2} - a^2 g(y) = 0 \tag{17.71}$$

or

$$\frac{d^2 f(x)}{dx^2} - a^2 f(x) = 0, \quad \frac{d^2 g(y)}{dy^2} + a^2 g(y) = 0 \tag{17.72}$$

where a is a constant. The linearly independent solutions of Eq. (17.71) are

$$\begin{aligned} f(x) &= \begin{pmatrix} 1 \\ x \end{pmatrix} \text{ for } a = 0, \quad f(x) = \begin{pmatrix} \cos ax \\ \sin ax \end{pmatrix} \text{ for } a \neq 0 \\ g(y) &= \begin{pmatrix} 1 \\ y \end{pmatrix} \text{ for } a = 0, \quad g(y) = \begin{pmatrix} \cosh ay \\ \sinh ay \end{pmatrix} \text{ for } a \neq 0 \end{aligned} \tag{17.73}$$

and the linearly independent solutions of Eq. (17.72) are

$$f(x) = \begin{pmatrix} 1 \\ x \end{pmatrix} \quad \text{for } a = 0, \quad f(x) = \begin{pmatrix} \cosh ax \\ \sinh ax \end{pmatrix} \quad \text{for } a \neq 0$$
$$g(y) = \begin{pmatrix} 1 \\ y \end{pmatrix} \quad \text{for } a = 0, \quad g(y) = \begin{pmatrix} \cos ay \\ \sin ay \end{pmatrix} \quad \text{for } a \neq 0 \tag{17.74}$$

Now, we show that a function

$$p(x, y) = [Ax + By + C(x^2 + y^2)]h(x, y) \tag{17.75}$$

is a biharmonic function, where a function $h(x, y)$ is harmonic, and A, B, C are arbitrary constants. Differentiation of Eq. (17.75) gives

$$\frac{\partial^2 p(x, y)}{\partial x^2} = 2Ch(x, y) + 2(A + 2Cx)\frac{\partial h(x, y)}{\partial x} + [Ax + By + C(x^2 + y^2)]\frac{\partial^2 h(x, y)}{\partial x^2}$$
$$\frac{\partial^2 p(x, y)}{\partial y^2} = 2Ch(x, y) + 2(B + 2Cy)\frac{\partial h(x, y)}{\partial y} + [Ax + By + C(x^2 + y^2)]\frac{\partial^2 h(x, y)}{\partial y^2} \tag{17.76}$$

As the function $h(x, y)$ is harmonic, we get

$$\nabla^2 p(x, y) = 4Ch(x, y) + 2(A + 2Cx)\frac{\partial h(x, y)}{\partial x} + 2(B + 2Cy)\frac{\partial h(x, y)}{\partial y} \tag{17.77}$$

Differentiation of Eq. (17.77) gives

$$\frac{\partial^2 \nabla^2 p(x, y)}{\partial x^2} = 12C\frac{\partial^2 h(x, y)}{\partial x^2} + 2(A + 2Cx)\frac{\partial^3 h(x, y)}{\partial x^3} + 2(B + 2Cy)\frac{\partial^3 h(x, y)}{\partial x^2 \partial y}$$
$$\frac{\partial^2 \nabla^2 p(x, y)}{\partial y^2} = 12C\frac{\partial^2 h(x, y)}{\partial y^2} + 2(B + 2Cy)\frac{\partial^3 h(x, y)}{\partial y^3} + 2(A + 2Cx)\frac{\partial^3 h(x, y)}{\partial y^2 \partial x} \tag{17.78}$$

Therefore, we obtain

$$\nabla^4 p(x, y) = 12C\nabla^2 h(x, y) + 2(A + 2Cx)\frac{\partial}{\partial x}\nabla^2 h(x, y) + 2(B + 2Cy)\frac{\partial}{\partial y}\nabla^2 h(x, y) = 0 \tag{17.79}$$

From Eqs. (17.73), (17.74) and (17.75), the particular solutions of a biharmonic equation (17.20) in a Cartesian coordinate system may be expressed as follows:

$$\begin{pmatrix} 1 \\ x \\ y \\ x^2+y^2 \end{pmatrix} \begin{pmatrix} 1 \\ x \end{pmatrix} \begin{pmatrix} 1 \\ y \end{pmatrix}$$
$$\begin{pmatrix} 1 \\ x \\ y \\ x^2+y^2 \end{pmatrix} \begin{pmatrix} \sin ax \\ \cos ax \end{pmatrix} \begin{pmatrix} \sinh ay \\ \cosh ay \end{pmatrix}$$
$$\begin{pmatrix} 1 \\ x \\ y \\ x^2+y^2 \end{pmatrix} \begin{pmatrix} \sinh ax \\ \cosh ax \end{pmatrix} \begin{pmatrix} \sin ay \\ \cos ay \end{pmatrix} \qquad \text{(Answer)} \quad (17.80)$$

The notation for the product of three one-column matrices is explained by Eqs. (16.97) and (16.98).

Next, we show another type of the solutions of the biharmonic equation. A complex function $\varphi(z)$ is introduced.

$$z = x + iy, \quad \varphi(z) = p + iq \tag{17.81}$$

where $i^2 = -1$, and p and q are harmonic functions. Therefore

$$2p = \varphi(z) + \overline{\varphi(z)}, \quad 2iq = \varphi(z) - \overline{\varphi(z)} \tag{17.82}$$

We assume that $\varphi(z)$ is expressed by

$$\varphi(z) = 1 + \sum_{n=1}^{\infty}(A_n z^n + B_n z^{-n}) \tag{17.83}$$

where A_n, B_n are real constants. Then, the harmonic function p of the real part of $\varphi(z)$ is written as

$$\begin{aligned} p &= \frac{1}{2}[\varphi(z) + \overline{\varphi(z)}] \\ &= 1 + \frac{1}{2}\sum_{n=1}^{\infty}[A_n(z^n + \overline{z}^n) + B_n(z^{-n} + \overline{z}^{-n}) \\ &= 1 + \frac{1}{2}\sum_{n=1}^{\infty}(z^n + \overline{z}^n)\Big[A_n + B_n\frac{1}{(x^2+y^2)^n}\Big] \end{aligned} \tag{17.84}$$

We obtain the following harmonic functions from Eq. (17.84)

$$\begin{aligned} z+\bar{z} &= (x+iy)+(x-iy) = 2x \\ z^2+\bar{z}^2 &= (x+iy)^2+(x-iy)^2 = 2(x^2-y^2) \\ z^3+\bar{z}^3 &= (x+iy)^3+(x-iy)^3 = 2x(x^2-3y^2) \\ z^4+\bar{z}^4 &= (x+iy)^4+(x-iy)^4 = 2(x^4-6x^2y^2+y^4) \end{aligned} \tag{17.85}$$

In the similar way, we obtain the harmonic function q of the imaginary part of $\varphi(z)$

$$\begin{aligned} q &= \frac{1}{2i}[\varphi(z)-\overline{\varphi(z)}] \\ &= \frac{1}{2i}\sum_{n=1}^{\infty}[A_n(z^n-\bar{z}^n)+B_n(z^{-n}-\bar{z}^{-n})] \\ &= \frac{1}{2i}\sum_{n=1}^{\infty}(z^n-\bar{z}^n)\left[A_n-B_n\frac{1}{(x^2+y^2)^n}\right] \end{aligned} \tag{17.86}$$

From Eq. (17.86) we obtain the following imaginary parts

$$\begin{aligned} z-\bar{z} &= (x+iy)-(x-iy) = 2iy \\ z^2-\bar{z}^2 &= (x+iy)^2-(x-iy)^2 = 4ixy \\ z^3-\bar{z}^3 &= (x+iy)^3-(x-iy)^3 = -2iy(y^2-3x^2) \\ z^4-\bar{z}^4 &= (x+iy)^4-(x-iy)^4 = 8izy(x^2-y^2) \end{aligned} \tag{17.87}$$

Therefore, taking into the consideration of Eqs. (17.84)–(17.87), we obtain the alternative forms of the particular solutions of the biharmonic equation

$$\begin{pmatrix} 1 \\ x \\ y \\ x^2+y^2 \end{pmatrix}, \begin{pmatrix} 1 \\ x \\ y \\ x^2+y^2 \end{pmatrix}\begin{pmatrix} 1 \\ r^{-2} \end{pmatrix}\begin{pmatrix} x \\ y \end{pmatrix}$$

$$\begin{pmatrix} 1 \\ x \\ y \\ x^2+y^2 \end{pmatrix}\begin{pmatrix} 1 \\ r^{-4} \end{pmatrix}\begin{pmatrix} x^2-y^2 \\ xy \end{pmatrix}$$

$$\begin{pmatrix} 1 \\ x \\ y \\ x^2+y^2 \end{pmatrix}\begin{pmatrix} 1 \\ r^{-6} \end{pmatrix}\begin{pmatrix} x(x^2-3y^2) \\ y(y^2-3x^2) \end{pmatrix}$$

$$\begin{pmatrix} 1 \\ x \\ y \\ x^2+y^2 \end{pmatrix}\begin{pmatrix} 1 \\ r^{-8} \end{pmatrix}\begin{pmatrix} x^4-6x^2y^2+y^4 \\ xy(x^2-y^2) \end{pmatrix}, \ldots \qquad \text{(Answer)}$$

in which $r = \sqrt{x^2 + y^2}$. The notation for the product of three one-column matrices is explained by Eqs. (16.97) and (16.98).

Problem 17.7. Derive the solutions of Eq. (17.20) in the polar coordinate system.

Solution. First, we consider the solutions of Laplace's equation in the polar coordinate system by use of the method of separation of variables. Laplace's equation is

$$\left(\frac{\partial^2}{\partial r^2} + \frac{1}{r}\frac{\partial}{\partial r} + \frac{1}{r^2}\frac{\partial^2}{\partial \theta^2}\right)h(r,\theta) = 0 \tag{17.88}$$

We assume that the harmonic function can be expressed by the product of two unknown functions, each of which has only one variable

$$h(r,\theta) = f(r)g(\theta) \tag{17.89}$$

Substitution of Eq. (17.89) into Eq. (17.88) gives

$$\begin{aligned}
&\frac{d^2 f(r)}{dr^2} + \frac{1}{r}\frac{df(r)}{dr} - \frac{n^2}{r^2}f(r) = 0\\
&\frac{d^2 g(\theta)}{d\theta^2} + n^2 g(\theta) = 0
\end{aligned} \tag{17.90}$$

where n is the integer. The linearly independent solutions of Eq. (17.90) are

$$\begin{aligned}
f(r) &= \begin{pmatrix} 1 \\ \ln r \end{pmatrix} \quad \text{for } n = 0, \quad f(r) = \begin{pmatrix} r^n \\ r^{-n} \end{pmatrix} \quad \text{for } n \neq 0\\
g(\theta) &= \begin{pmatrix} 1 \\ \theta \end{pmatrix} \quad \text{for } n = 0, \quad g(\theta) = \begin{pmatrix} \sin n\theta \\ \cos n\theta \end{pmatrix} \quad \text{for } n \neq 0
\end{aligned} \tag{17.91}$$

Next, we consider the particular solution $p(r,\theta)$ which satisfies the equation

$$\nabla^2 p(r,\theta) = f(r)g(\theta) \tag{17.92}$$

The particular solution p is assumed to be expressed by the product of two functions, each of which has only one variable

$$p(r,\theta) = F(r)g(\theta) \tag{17.93}$$

Substitution of Eq. (17.93) into Eq. (17.92) gives

$$\frac{d^2 F(r)}{dr^2} + \frac{1}{r}\frac{dF(r)}{dr} - \frac{n^2}{r^2}F(r) = f(r) \tag{17.94}$$

and a particular solution $F(r)$ of Eq. (17.94) takes the form

$$
\begin{aligned}
F(r) &= \begin{pmatrix} r^2/4 \\ r^2(\ln r - 1)/4 \end{pmatrix} \quad \text{when} \, f(r) = \begin{pmatrix} 1 \\ \ln r \end{pmatrix} \\
F(r) &= \begin{pmatrix} r^3/8 \\ r \ln r/2 \end{pmatrix} \quad \text{when} \, f(r) = \begin{pmatrix} r \\ r^{-1} \end{pmatrix} \\
F(r) &= \begin{pmatrix} r^{n+2}/(4n+4) \\ -r^{-n+2}/(4n-4) \end{pmatrix} \quad \text{when} \, f(r) = \begin{pmatrix} r^n \\ r^{-n} \end{pmatrix}
\end{aligned} \tag{17.95}
$$

From Eqs. (17.91) and (17.95), the particular solutions of Eq. (17.20) in the polar coordinate system are

$$
\begin{pmatrix} 1 \\ r^2 \\ \ln r \\ r^2 \ln r \end{pmatrix} \begin{pmatrix} 1 \\ \theta \end{pmatrix}, \quad \begin{pmatrix} r \\ r^{-1} \\ r^3 \\ r \ln r \end{pmatrix} \begin{pmatrix} \cos\theta \\ \sin\theta \end{pmatrix}
$$

$$
\begin{pmatrix} r^n \\ r^{-n} \\ r^{n+2} \\ r^{-n+2} \end{pmatrix} \begin{pmatrix} \cos n\theta \\ \sin n\theta \end{pmatrix} \qquad \text{(Answer)} \tag{17.96}
$$

In Eq. (17.96), we used the following notation for the product of two one-column matrices

$$
\begin{pmatrix} f_1 \\ f_2 \\ f_3 \\ f_4 \end{pmatrix} \begin{pmatrix} g_1 \\ g_2 \end{pmatrix} = \begin{pmatrix} f_1 g_1 \\ f_2 g_1 \\ f_3 g_1 \\ f_4 g_1 \\ f_1 g_2 \\ f_2 g_2 \\ f_3 g_2 \\ f_4 g_2 \end{pmatrix} \tag{17.97}
$$

Then, for example

$$
\begin{pmatrix} r^n \\ r^{-n} \\ r^{n+2} \\ r^{-n+2} \end{pmatrix} \begin{pmatrix} \cos n\theta \\ \sin n\theta \end{pmatrix} = \begin{pmatrix} r^n \cos n\theta \\ r^{-n} \cos n\theta \\ r^{n+2} \cos n\theta \\ r^{-n+2} \cos n\theta \\ r^n \sin n\theta \\ r^{-n} \sin n\theta \\ r^{n+2} \sin n\theta \\ r^{-n+2} \sin n\theta \end{pmatrix} \tag{17.98}
$$

Problem 17.8. Derive Eq. (17.34).

Solution. The relationship for the displacement between a curvilinear coordinate system and a Cartesian coordinate system is given by Eq. (17.32)

$$u_\rho + iu_\theta = e^{-i\alpha}(u_x + iu_y) \tag{17.99}$$

When a small displacement dz is produced, a corresponding point ζ undergoes a small displacement $d\zeta$

$$dz = |dz|e^{i\alpha}, \quad d\zeta = |d\zeta|e^{i\theta} \tag{17.100}$$

From Eq. (17.100), we get

$$\begin{aligned}
e^{i\alpha} &= \frac{dz}{|dz|} = \frac{\omega'(\zeta)d\zeta}{|\omega'(\zeta)| \cdot |d\zeta|} = e^{i\theta}\frac{\omega'(\zeta)}{|\omega'(\zeta)|} = \frac{\zeta}{\rho}\frac{\omega'(\zeta)}{|\omega'(\zeta)|} \\
e^{-i\alpha} &= e^{-i\theta}\frac{\overline{\omega'(\zeta)}}{|\omega'(\zeta)|} = \frac{\bar{\zeta}}{\rho}\frac{\overline{\omega'(\zeta)}}{|\omega'(\zeta)|} \\
e^{2i\alpha} &= \left[\frac{\zeta}{\rho}\frac{\omega'(\zeta)}{|\omega'(\zeta)|}\right]^2 = \frac{\zeta^2}{\rho^2}\frac{\omega'(\zeta)\omega'(\zeta)}{\omega'(\zeta)\overline{\omega'(\zeta)}} = \frac{\zeta^2}{\rho^2}\frac{\omega'(\zeta)}{\overline{\omega'(\zeta)}}
\end{aligned} \tag{17.101}$$

Next, we introduce the new notation

$$\begin{aligned}
&\varphi(z) = \varphi(\omega(\zeta)) \equiv \phi(\zeta), \quad \psi(z) = \psi(\omega(\zeta)) \equiv \Psi(\zeta) \\
&\varphi'(z) = \frac{d\varphi(z)}{dz} = \frac{d\phi(\zeta)}{d\zeta}\frac{d\zeta}{dz} = \frac{1}{\omega'(\zeta)}\frac{d\phi(\zeta)}{d\zeta} = \frac{\phi'(\zeta)}{\omega'(\zeta)}
\end{aligned} \tag{17.102}$$

Substitution of Eqs. (17.28) with $c^* = 0$, (17.101) and (17.102) into Eq. (17.99) yields

$$\begin{aligned}
u_\rho + iu_\theta &= e^{-i\alpha}\frac{1}{2G}\Big[\frac{3-\nu^*}{1+\nu^*}\varphi(z) - z\overline{\varphi'(z)} - \overline{\psi(z)} - \Big(\frac{\partial \chi_p}{\partial x} + i\frac{\partial \chi_p}{\partial y}\Big)\Big] \\
&= \frac{1}{2G}\frac{\bar{\zeta}}{\rho}\frac{\overline{\omega'(\zeta)}}{|\omega'(\zeta)|}\Big[\frac{3-\nu^*}{1+\nu^*}\phi(\zeta) - \frac{\omega(\zeta)}{\overline{\omega'(\zeta)}}\overline{\phi'(\zeta)} - \overline{\Psi(\zeta)} \\
&\quad - \Big(\frac{\partial \chi_p}{\partial x} + i\frac{\partial \chi_p}{\partial y}\Big)\Big]
\end{aligned} \tag{17.103}$$

Taking into the consideration the following relationship

$$\begin{aligned}
\frac{\partial}{\partial x} + i\frac{\partial}{\partial y} &= \Big(\frac{\partial}{\partial z}\frac{\partial z}{\partial x} + \frac{\partial}{\partial \bar{z}}\frac{\partial \bar{z}}{\partial x}\Big) + i\Big(\frac{\partial}{\partial z}\frac{\partial z}{\partial y} + \frac{\partial}{\partial \bar{z}}\frac{\partial \bar{z}}{\partial y}\Big) \\
&= \Big(\frac{\partial}{\partial z} + \frac{\partial}{\partial \bar{z}}\Big) + i^2\Big(\frac{\partial}{\partial z} - \frac{\partial}{\partial \bar{z}}\Big) \\
&= 2\frac{\partial}{\partial \bar{z}} = 2\frac{\partial}{\partial \bar{\zeta}}\frac{d\bar{\zeta}}{d\bar{z}} = 2\frac{1}{\overline{\omega'(\zeta)}}\frac{\partial}{\partial \bar{\zeta}} = 2\frac{1}{\overline{\omega'(\zeta)}}\Big(\frac{\partial}{\partial \rho}\frac{\partial \rho}{\partial \bar{\zeta}} + \frac{\partial}{\partial \theta}\frac{\partial \theta}{\partial \bar{\zeta}}\Big)
\end{aligned}$$

$$
\begin{aligned}
&= 2\frac{1}{\overline{\omega'(\zeta)}}\left\{\frac{\partial}{\partial\rho}\frac{\partial\sqrt{\zeta\bar{\zeta}}}{\partial\bar{\zeta}} + \frac{\partial}{\partial\theta}\frac{\partial}{\partial\bar{\zeta}}\left[-\frac{i}{2}(\ln\zeta - \ln\bar{\zeta})\right]\right\} \\
&= \frac{1}{\overline{\omega'(\zeta)}}\left(\frac{\partial}{\partial\rho}\sqrt{\frac{\zeta}{\bar{\zeta}}} + i\frac{\partial}{\partial\theta}\frac{1}{\bar{\zeta}}\right) = \frac{e^{i\theta}}{\overline{\omega'(\zeta)}}\left(\frac{\partial}{\partial\rho} + i\frac{1}{\rho}\frac{\partial}{\partial\theta}\right) \\
&= \frac{\zeta}{\rho}\frac{1}{\overline{\omega'(\zeta)}}\left(\frac{\partial}{\partial\rho} + i\frac{1}{\rho}\frac{\partial}{\partial\theta}\right)
\end{aligned}
\tag{17.104}
$$

we obtain the displacement

$$
\begin{aligned}
u_\rho + iu_\theta = &\frac{1}{2G}\frac{\bar{\zeta}}{\rho}\frac{\overline{\omega'(\zeta)}}{|\omega'(\zeta)|}\left[\frac{3-\nu^*}{1+\nu^*}\phi(\zeta) - \frac{\omega(\zeta)}{\overline{\omega'(\zeta)}}\overline{\phi'(\zeta)} - \overline{\Psi(\zeta)}\right. \\
&\left. - \frac{\zeta}{\rho}\frac{1}{\overline{\omega'(\zeta)}}\left(\frac{\partial}{\partial\rho} + i\frac{1}{\rho}\frac{\partial}{\partial\theta}\right)\chi_p\right]
\end{aligned}
\tag{Answer}
$$

Problem 17.9. Derive Eq. (17.35).

Solution. The relationship for the stress between a curvilinear coordinate system and a Cartesian coordinate system is given by Eq. (17.33):

$$
\begin{aligned}
\sigma_{\rho\rho} + \sigma_{\theta\theta} &= \sigma_{xx} + \sigma_{yy} \\
\sigma_{\theta\theta} - \sigma_{\rho\rho} + 2i\sigma_{\theta\rho} &= e^{2i\alpha}(\sigma_{yy} - \sigma_{xx} + 2i\sigma_{xy})
\end{aligned}
\tag{17.105}
$$

Substitution of Eq. (17.27) into Eq. (17.105) yields

$$
\begin{aligned}
\sigma_{\rho\rho} + \sigma_{\theta\theta} &= 4Re[\varphi'(z)] - \alpha^* E^* \tau \\
\sigma_{\theta\theta} - \sigma_{\rho\rho} + 2i\sigma_{\theta\rho} &= e^{2i\alpha}\left\{2[\bar{z}\varphi''(z) + \psi'(z)] + \left(\frac{\partial}{\partial x} - i\frac{\partial}{\partial y}\right)^2\chi_p\right\}
\end{aligned}
\tag{17.106}
$$

By transforming the variable from z to ζ, and using Eqs. (17.101) and (17.102) in Problem 17.8, Eq. (17.106) reduce to

$$
\begin{aligned}
\sigma_{\rho\rho} + \sigma_{\theta\theta} &= 4Re\left[\frac{\phi'(\zeta)}{\omega'(\zeta)}\right] - \alpha^* E^* \tau \\
\sigma_{\theta\theta} - \sigma_{\rho\rho} + 2i\sigma_{\theta\rho} &= \frac{\zeta^2}{\rho^2}\frac{\omega'(\zeta)}{\overline{\omega'(\zeta)}}\left\{2\left[\frac{\overline{\omega(\zeta)}}{\omega'(\zeta)}\left(\frac{\phi'(\zeta)}{\omega'(\zeta)}\right)' + \frac{\Psi'(\zeta)}{\omega'(\zeta)}\right]\right. \\
&\quad \left. + \left(\frac{\partial}{\partial x} - i\frac{\partial}{\partial y}\right)^2\chi_p\right\}
\end{aligned}
\tag{17.107}
$$

Taking into the consideration of the relationship

$$\frac{\partial}{\partial x} - i\frac{\partial}{\partial y} = \left(\frac{\partial}{\partial z}\frac{\partial z}{\partial x} + \frac{\partial}{\partial \bar{z}}\frac{\partial \bar{z}}{\partial x}\right) - i\left(\frac{\partial}{\partial z}\frac{\partial z}{\partial y} + \frac{\partial}{\partial \bar{z}}\frac{\partial \bar{z}}{\partial y}\right)$$
$$= \left(\frac{\partial}{\partial z} + \frac{\partial}{\partial \bar{z}}\right) - i^2\left(\frac{\partial}{\partial z} - \frac{\partial}{\partial \bar{z}}\right) = 2\frac{\partial}{\partial z} = 2\frac{\partial}{\partial \zeta}\frac{d\zeta}{dz} = 2\frac{1}{\omega'(\zeta)}\frac{\partial}{\partial \zeta}$$
$$\left(\frac{\partial}{\partial x} - i\frac{\partial}{\partial y}\right)^2 = 4\frac{1}{\omega'(\zeta)}\frac{\partial}{\partial \zeta}\left[\frac{1}{\omega'(\zeta)}\frac{\partial}{\partial \zeta}\right]$$
$$= 4\frac{1}{\omega'(\zeta)}\left[\frac{1}{\omega'(\zeta)}\frac{\partial^2}{\partial \zeta^2} - \frac{\omega''(\zeta)}{[\omega'(\zeta)]^2}\frac{\partial}{\partial \zeta}\right] \tag{17.108}$$

we obtain the stress

$$\sigma_{\rho\rho} + \sigma_{\theta\theta} = 4Re\left[\frac{\phi'(\zeta)}{\omega'(\zeta)}\right] - \alpha^* E^* \tau$$
$$\sigma_{\theta\theta} - \sigma_{\rho\rho} + 2i\sigma_{\theta\rho} = 2\frac{\zeta^2}{\rho^2}\frac{1}{\overline{\omega'(\zeta)}}\left\{\overline{\omega(\zeta)}\left[\frac{\phi'(\zeta)}{\omega'(\zeta)}\right]' + \Psi'(\zeta)\right\}$$
$$+ 4\frac{\zeta^2}{\rho^2}\frac{1}{\overline{\omega''(\zeta)}}\left[\frac{1}{\omega'(\zeta)}\frac{\partial^2 \chi_p}{\partial \zeta^2} - \frac{\omega'(\zeta)}{[\omega'(\zeta)]^2}\frac{\partial \chi_p}{\partial \zeta}\right] \quad \text{(Answer)}$$

Problem 17.10. Airy's stress function F related to the components of stress

$$\sigma_{xx} = \frac{\partial^2 F}{\partial y^2}, \quad \sigma_{yy} = \frac{\partial^2 F}{\partial x^2}, \quad \sigma_{xy} = -\frac{\partial^2 F}{\partial x \partial y} \tag{17.109}$$

is usually used in isothermal plane problems, where a governing equation of F is $\nabla^4 F = 0$. Prove that the thermal stress in plane problems can be expressed by

$$\sigma_{xx} = \frac{\partial^2}{\partial y^2}(F - 2\mu\Phi), \quad \sigma_{yy} = \frac{\partial^2}{\partial x^2}(F - 2\mu\Phi)$$
$$\sigma_{xy} = -\frac{\partial^2}{\partial x \partial y}(F - 2\mu\Phi) \tag{17.110}$$

where Φ is Goodier's thermoelastic potential and F is Airy's stress function.

Solution. Using Eqs. (17.37) and (17.38), the strains are expressed by

$$\varepsilon_{xx} = \varepsilon_{xx}^c + \Phi_{,xx}, \quad \varepsilon_{yy} = \varepsilon_{yy}^c + \Phi_{,yy}, \quad \varepsilon_{yx} = \varepsilon_{xy}^c + \Phi_{,xy} \tag{17.111}$$

From Eqs. (17.1′) and (17.111), we get

$$\sigma_{xx} = (\lambda^* + 2\mu)\varepsilon_{xx}^c + \lambda^*\varepsilon_{yy}^c - 2\mu\Phi_{,yy} + (\lambda^* + 2\mu)\nabla^2\Phi - \beta^*\tau$$
$$\sigma_{yy} = (\lambda^* + 2\mu)\varepsilon_{yy}^c + \lambda^*\varepsilon_{xx}^c - 2\mu\Phi_{,xx} + (\lambda^* + 2\mu)\nabla^2\Phi - \beta^*\tau$$
$$\sigma_{xy} = 2\mu\varepsilon_{xy}^c + 2\mu\Phi_{,xy} \tag{17.112}$$

Since the governing equation for Goodier's thermoelastic potential function Φ is given by Eq. (17.39), Eq. (17.112) reduces to

$$\begin{aligned}
\sigma_{xx} &= (\lambda^* + 2\mu)\varepsilon^c_{xx} + \lambda^*\varepsilon^c_{yy} - 2\mu\Phi,_{yy} = \sigma^c_{xx} - 2\mu\Phi,_{yy} = F,_{yy} - 2\mu\Phi,_{yy} \\
\sigma_{yy} &= (\lambda^* + 2\mu)\varepsilon^c_{yy} + \lambda^*\varepsilon^c_{xx} - 2\mu\Phi,_{xx} = \sigma^c_{yy} - 2\mu\Phi,_{xx} = F,_{xx} - 2\mu\Phi,_{xx} \\
\sigma_{xy} &= 2\mu\varepsilon^c_{xy} + 2\mu\Phi,_{xy} = \sigma^c_{xy} + 2\mu\Phi,_{xy} = -(F,_{xy} - 2\mu\Phi,_{xy}) \qquad \text{(Answer)}
\end{aligned}$$

Next, we derive the governing equation of Airy's stress function F. Substitution of Eq. (17.1) into Eq. (17.4) gives

$$\begin{aligned}
&\frac{1}{E^*}\frac{\partial^2\sigma_{xx}}{\partial y^2} - \frac{\nu^*}{E^*}\frac{\partial^2\sigma_{yy}}{\partial y^2} + \alpha^*\frac{\partial^2\tau}{\partial y^2} + \frac{1}{E^*}\frac{\partial^2\sigma_{yy}}{\partial x^2} - \frac{\nu^*}{E^*}\frac{\partial^2\sigma_{xx}}{\partial x^2} + \alpha^*\frac{\partial^2\tau}{\partial x^2} \\
&\quad = 2\frac{1+\nu^*}{E^*}\frac{\partial^2\sigma_{xy}}{\partial x\partial y}
\end{aligned} \tag{17.113}$$

Simplification of Eq. (17.113) reduces to

$$\begin{aligned}
&\frac{\partial^2\sigma_{xx}}{\partial y^2} - 2\frac{\partial^2\sigma_{xy}}{\partial x\partial y} + \frac{\partial^2\sigma_{yy}}{\partial x^2} - \nu^*\left(\frac{\partial^2\sigma_{xx}}{\partial x^2} + 2\frac{\partial^2\sigma_{xy}}{\partial x\partial y} + \frac{\partial^2\sigma_{yy}}{\partial y^2}\right) \\
&\quad = -\alpha^* E^*\nabla^2\tau
\end{aligned} \tag{17.114}$$

Substitution of Eq. (17.110) into Eq. (17.114) gives

$$\begin{aligned}
&\frac{\partial^2(F,_{yy} - 2\mu\Phi,_{yy})}{\partial y^2} + 2\frac{\partial^2(F,_{xy} - 2\mu\Phi,_{xy})}{\partial x\partial y} + \frac{\partial^2(F,_{xx} - 2\mu\Phi,_{xx})}{\partial x^2} \\
&\quad - \nu^*\left[\frac{\partial^2(F,_{yy} - 2\mu\Phi,_{yy})}{\partial x^2} - 2\frac{\partial^2(F,_{xy} - 2\mu\Phi,_{xy})}{\partial x\partial y} + \frac{\partial^2(F,_{xx} - 2\mu\Phi,_{xx})}{\partial y^2}\right] \\
&= -\alpha^* E^*\nabla^2\tau
\end{aligned} \tag{17.115}$$

Simplification of above equation reduces to

$$\nabla^4 F - 2\mu\nabla^2\left(\nabla^2\Phi - \frac{\alpha^* E^*}{2\mu}\tau\right) = 0 \tag{17.116}$$

By the use of Eq. (17.39), Eq. (17.116) reduces to

$$\nabla^4 F = 0 \tag{17.117}$$

Problem 17.11. Prove that the components of thermal stress in plane problems can be expressed by

$$\sigma_{xx} = 2\mu\left[-\Phi_{,yy} + x\phi_{1,yy} + y\phi_{2,yy} + \frac{2}{1+\nu^*}(\phi_{1,x} + \nu^*\phi_{2,y})\right]$$

$$\sigma_{yy} = 2\mu\Big[-\Phi_{,xx} + x\phi_{1,xx} + y\phi_{2,xx} + \frac{2}{1+\nu^*}(\phi_{2,y} + \nu^*\phi_{1,x})\Big]$$
$$\sigma_{xy} = 2\mu\Big[\Phi_{,xy} - (x\phi_1 + y\phi_2)_{,xy} + \frac{1-\nu^*}{1+\nu^*}(\phi_{1,y} + \phi_{2,x})\Big] \tag{17.118}$$

where Φ denotes Goodier's thermoelastic potential given by Eq. (17.38) and two harmonic functions ϕ_1, ϕ_2 are given by Eq. (17.41).

Solution. The displacement may be expressed from Eqs. (17.38) and (17.41)

$$u_x = \Phi,_x + \frac{3-\nu^*}{1+\nu^*}\phi_1 - x\phi_1,_x - y\phi_2,_x$$
$$u_y = \Phi,_y + \frac{3-\nu^*}{1+\nu^*}\phi_2 - x\phi_1,_y - y\phi_2,_y \tag{17.119}$$

Equation (17.119) give the strains

$$\varepsilon_{xx} = \Phi,_{xx} + 2\frac{1-\nu^*}{1+\nu^*}\phi_1,_x - x\phi_1,_{xx} - y\phi_2,_{xx}$$
$$\varepsilon_{yy} = \Phi,_{yy} + 2\frac{1-\nu^*}{1+\nu^*}\phi_2,_y - x\phi_1,_{yy} - y\phi_2,_{yy}$$
$$\varepsilon_{xy} = \Phi,_{xy} + \frac{1-\nu^*}{1+\nu^*}(\phi_1,_y + \phi_2,_x) - x\phi_1,_{xy} - y\phi_2,_{xy} \tag{17.120}$$

From Eqs. (17.1′) and (17.120), we get

$$\sigma_{xx} = (\lambda^* + 2\mu)(\Phi,_{xx} + 2\frac{1-\nu^*}{1+\nu^*}\phi_1,_x - x\phi_1,_{xx} - y\phi_2,_{xx})$$
$$+ \lambda^*(\Phi,_{yy} + 2\frac{1-\nu^*}{1+\nu^*}\phi_2,_y - x\phi_1,_{yy} - y\phi_2,_{yy}) - \beta^*\tau$$
$$\sigma_{yy} = (\lambda^* + 2\mu)(\Phi,_{yy} + 2\frac{1-\nu^*}{1+\nu^*}\phi_2,_y - x\phi_1,_{yy} - y\phi_2,_{yy})$$
$$+ \lambda^*(\Phi,_{xx} + 2\frac{1-\nu^*}{1+\nu^*}\phi_1,_x - x\phi_1,_{xx} - y\phi_2,_{xx}) - \beta^*\tau$$
$$\sigma_{xy} = 2\mu[\Phi,_{xy} + \frac{1-\nu^*}{1+\nu^*}(\phi_1,_y + \phi_2,_x) - x\phi_1,_{xy} - y\phi_2,_{xy}] \tag{17.121}$$

By the use of Eqs. (17.39) and (17.42), we can obtain

$$\sigma_{xx} = 2\mu(-\Phi,_{yy} - x\phi_1,_{xx} - y\phi_2,_{xx}) + 2\frac{1-\nu^*}{1+\nu^*}[(\lambda^* + 2\mu)\phi_1,_x + \lambda^*\phi_2,_y]$$
$$\sigma_{yy} = 2\mu(-\Phi,_{xx} - x\phi_1,_{yy} - y\phi_2,_{yy}) + 2\frac{1-\nu^*}{1+\nu^*}[(\lambda^* + 2\mu)\phi_2,_y + \lambda^*\phi_1,_x]$$
$$\sigma_{xy} = 2\mu[\Phi,_{xy} + \frac{1-\nu^*}{1+\nu^*}(\phi_1,_y + \phi_2,_x) - x\phi_1,_{xy} - y\phi_2,_{xy}] \tag{17.122}$$

Material constants are rewritten as for plane strain

$$
\begin{aligned}
&\nu^* = \frac{\nu}{1-\nu} \quad \leftrightarrow \nu = \frac{\nu^*}{1+\nu^*} \\
&\lambda^* = \lambda = \frac{2\nu\mu}{1-2\nu} = \frac{2\mu\nu^*/(1+\nu^*)}{1-2\nu^*/(1+\nu^*)} = \frac{2\mu\nu^*}{1-\nu^*} \\
&\frac{1-\nu^*}{1+\nu^*}(\lambda^* + 2\mu) = 2\mu\frac{1-\nu^*}{1+\nu^*}\left(\frac{\nu^*}{1-\nu^*} + 1\right) = 2\mu\frac{1}{1+\nu^*} \\
&\frac{1-\nu^*}{1+\nu^*}\lambda^* = 2\mu\frac{1-\nu^*}{1+\nu^*}\frac{\nu^*}{1-\nu^*} = 2\mu\frac{\nu^*}{1+\nu^*}
\end{aligned}
\tag{17.123}
$$

and for plane stress

$$
\begin{aligned}
&\nu^* = \nu, \quad \lambda^* = \frac{2\mu\lambda}{\lambda+2\mu} = \frac{2\mu 2\nu\mu/(1-2\nu)}{2\nu\mu/(1-2\nu)+2\mu} = \frac{2\mu\nu}{1-\nu} = \frac{2\mu\nu^*}{1-\nu^*} \\
&\frac{1-\nu^* *}{1+\nu^*}(\lambda^* + 2\mu) = 2\mu\frac{1-\nu^*}{1+\nu^*}\left(\frac{\nu^*}{1-\nu^*} + 1\right) = 2\mu\frac{1}{1+\nu^*} \\
&\frac{1-\nu^*}{1+\nu^*}\lambda^* = 2\mu\frac{1-\nu^*}{1+\nu^*}\frac{\nu^*}{1-\nu^*} = 2\mu\frac{\nu^*}{1+\nu^*}
\end{aligned}
\tag{17.124}
$$

By the use of Eqs. (17.123), (17.124) and (17.42), Eq. (17.122) reduce to

$$
\begin{aligned}
\sigma_{xx} &= 2\mu\Big[-\Phi,_{yy} + x\phi_1,_{yy} + y\phi_2,_{yy} + \frac{2}{1+\nu^*}(\phi_1,_x + \nu^*\phi_2,_y)\Big] \\
\sigma_{yy} &= 2\mu\Big[-\Phi,_{xx} + x\phi_1,_{xx} + y\phi_2,_{xx} + \frac{2}{1+\nu^*}(\phi_2,_y + \nu^*\phi_1,_x)\Big] \\
\sigma_{xy} &= 2\mu\Big[\Phi,_{xy} - (x\phi_1 + y\phi_2),_{xy} + \frac{1-\nu^*}{1+\nu^*}(\phi_1,_y + \phi_2,_x)\Big]
\end{aligned}
\qquad \text{(Answer)}
$$

Chapter 18
Thermal Stresses in Circular Cylinders

In this chapter various techniques are presented to determine the thermal stresses in solid and hollow cylinders. The one-dimensional problems of cylindrical bodies are treated by the displacement method. Plane problems for infinitely long cylinders and for circular plates are treated by the thermal stress function method. Two-dimensional axisymmetric problems and three-dimensional problems are treated with Goodier's thermoelastic potential and the Boussinesq harmonic functions or Michell's biharmonic function. The derivation and the general solution of the basic equations related to thermal stresses in circular cylinders are treated in a number of problems. [See also Chap. 24.]

18.1 One-Dimensional Problems

The one-dimensional equilibrium equation of a cylindrical body due to axi-symmetric temperature field is

$$\frac{d\sigma_{rr}}{dr} + \frac{\sigma_{rr} - \sigma_{\theta\theta}}{r} = 0 \tag{18.1}$$

The generalized Hooke's law for plane problems is

$$\begin{aligned}
\epsilon_{rr} &= \frac{1}{E^*}(\sigma_{rr} - \nu^* \sigma_{\theta\theta}) + \alpha^* \tau \\
\epsilon_{\theta\theta} &= \frac{1}{E^*}(\sigma_{\theta\theta} - \nu^* \sigma_{rr}) + \alpha^* \tau \\
\epsilon_{r\theta} &= \frac{1}{2G}\sigma_{r\theta}
\end{aligned} \tag{18.2}$$

M. Reza Eslami et al., *Theory of Elasticity and Thermal Stresses*, Solid Mechanics and Its Applications 197, DOI: 10.1007/978-94-007-6356-2_18,

where

$$
\begin{aligned}
E^* &= \begin{cases} \dfrac{E}{1-\nu^2} & \text{for plane strain} \\ E & \text{for plane stress} \end{cases} \\
\nu^* &= \begin{cases} \dfrac{\nu}{1-\nu} & \text{for plane strain} \\ \nu & \text{for plane stress} \end{cases} \\
\alpha^* &= \begin{cases} (1+\nu)\alpha & \text{for plane strain} \\ \alpha & \text{for plane stress} \end{cases}
\end{aligned}
\tag{18.3}
$$

The strain-displacement relations are

$$
\epsilon_{rr} = \frac{du}{dr}, \quad \epsilon_{\theta\theta} = \frac{u}{r}, \quad \epsilon_{r\theta} = 0 \tag{18.4}
$$

where u is the radial displacement.

The components of stress are

$$
\begin{aligned}
\sigma_{rr} &= \frac{E^*}{1-\nu^{*2}}\left[\frac{du}{dr} + \nu^*\frac{u}{r} - (1+\nu^*)\alpha^*\tau\right] \\
\sigma_{\theta\theta} &= \frac{E^*}{1-\nu^{*2}}\left[\nu^*\frac{du}{dr} + \frac{u}{r} - (1+\nu^*)\alpha^*\tau\right] \\
\sigma_{r\theta} &= 0
\end{aligned}
\tag{18.5}
$$

The equilibrium equation with respect to the displacement is

$$
\frac{d}{dr}\left[\frac{1}{r}\frac{d(ru)}{dr}\right] = (1+\nu^*)\alpha^*\frac{d\tau}{dr} \tag{18.6}
$$

The general solution of Eq. (18.6) is

$$
u = (1+\nu^*)\alpha^*\frac{1}{r}\int \tau r\, dr + C_1 r + \frac{C_2}{r} \tag{18.7}
$$

where C_1 and C_2 are constants which may be determined from the boundary conditions.

The stresses are

$$
\begin{aligned}
\sigma_{rr} &= -\frac{\alpha^* E^*}{r^2}\int \tau r\, dr + \frac{E^*}{1-\nu^*}C_1 - \frac{E^*}{1+\nu^*}\frac{C_2}{r^2} \\
\sigma_{\theta\theta} &= \frac{\alpha^* E^*}{r^2}\int \tau r\, dr - \alpha^* E^* \tau + \frac{E^*}{1-\nu^*}C_1 + \frac{E^*}{1+\nu^*}\frac{C_2}{r^2} \\
\sigma_{r\theta} &= 0
\end{aligned}
\tag{18.8}
$$

The displacement and stresses in a solid cylinder of radius a with free traction are

$$u = (1+\nu^*)\alpha^*\Big(\frac{1}{r}\int_0^r \tau r\,dr + \frac{1-\nu^*}{1+\nu^*}\frac{r}{a^2}\int_0^a \tau r\,dr\Big) \tag{18.9}$$

$$\sigma_{rr} = \alpha^* E^*\Big(-\frac{1}{r^2}\int_0^r \tau r\,dr + \frac{1}{a^2}\int_0^a \tau r\,dr\Big)$$

$$\sigma_{\theta\theta} = \alpha^* E^*\Big(\frac{1}{r^2}\int_0^r \tau r\,dr + \frac{1}{a^2}\int_0^a \tau r\,dr - \tau\Big) \tag{18.10}$$

$$\sigma_{zz} = \begin{cases} 0 & \text{for plane stress} \\ \dfrac{\alpha E}{1-\nu}\Big(\dfrac{2\nu}{a^2}\displaystyle\int_0^a \tau r\,dr - \tau\Big) & \text{for plane strain} \end{cases}$$

The displacement and stresses in a hollow cylinder with inner radius a and outer radius b with free traction are

$$u = (1+\nu^*)\alpha^*\Big[\frac{1}{r}\int_a^r \tau r\,dr + \Big(\frac{1-\nu^*}{1+\nu^*}r + \frac{a^2}{r}\Big)\frac{1}{b^2-a^2}\int_a^b \tau r\,dr\Big] \tag{18.11}$$

$$\sigma_{rr} = \alpha^* E^*\Big[-\frac{1}{r^2}\int_a^r \tau r\,dr + \frac{r^2-a^2}{r^2(b^2-a^2)}\int_a^b \tau r\,dr\Big]$$

$$\sigma_{\theta\theta} = \alpha^* E^*\Big[\frac{1}{r^2}\int_a^r \tau r\,dr + \frac{r^2+a^2}{r^2(b^2-a^2)}\int_a^b \tau r\,dr - \tau\Big] \tag{18.12}$$

$$\sigma_{zz} = \begin{cases} 0 & \text{for plane stress} \\ \dfrac{\alpha E}{1-\nu}\Big(\dfrac{2\nu}{b^2-a^2}\displaystyle\int_a^b \tau r\,dr - \tau\Big) & \text{for plane strain} \end{cases}$$

Since there is no restriction on the temperature field in above solutions, the formulae for displacement and stresses are valid for both steady and transient temperature fields in the cylindrical body.

18.2 Plane Problems

We introduce the thermal stress function method which is discussed in Chap. 17 for the plane thermal stress problems. The governing equation for the thermal stress function χ is from Eq. (17.8)

$$\nabla^4\chi = -\alpha^* E^* \nabla^2 \tau \tag{18.13}$$

where

$$\nabla^4 = \nabla^2\nabla^2 = \Big(\frac{\partial^2}{\partial r^2} + \frac{1}{r}\frac{\partial}{\partial r} + \frac{1}{r^2}\frac{\partial^2}{\partial \theta^2}\Big)^2 \tag{18.14}$$

The components of stress in a cylindrical coordinate system are expressed by χ as

$$\sigma_{rr} = \frac{1}{r}\frac{\partial \chi}{\partial r} + \frac{1}{r^2}\frac{\partial^2 \chi}{\partial \theta^2}, \quad \sigma_{\theta\theta} = \frac{\partial^2 \chi}{\partial r^2}, \quad \sigma_{r\theta} = -\frac{\partial}{\partial r}\left(\frac{1}{r}\frac{\partial \chi}{\partial \theta}\right) \tag{18.15}$$

The conditions of single-valuedness of rotation and displacements (17.16), (17.17) and (17.18) for a cylindrical coordinate system reduce to

$$\begin{aligned}
&\int_0^{2\pi} r\frac{\partial}{\partial r}\left(\nabla^2\chi + \alpha^* E^* \tau\right) d\theta = 0 \\
&\int_0^{2\pi} r\left(\cos\theta\frac{\partial}{\partial \theta} - r\sin\theta\frac{\partial}{\partial r}\right)(\nabla^2\chi + \alpha^* E^* \tau)\, d\theta = 0 \\
&\int_0^{2\pi} r\left(r\cos\theta\frac{\partial}{\partial r} + \sin\theta\frac{\partial}{\partial \theta}\right)(\nabla^2\chi + \alpha^* E^* \tau)\, d\theta = 0
\end{aligned} \tag{18.16}$$

18.3 Two-Dimensional Axisymmetric Problems (*r*, *z*)

We now consider axisymmetric problems (r, z) of a homogeneous, isotropic cylindrical body.

The equilibrium equations are from Eq. (16.40)

$$\begin{aligned}
&\frac{\partial \sigma_{rr}}{\partial r} + \frac{\partial \sigma_{zr}}{\partial z} + \frac{\sigma_{rr} - \sigma_{\theta\theta}}{r} + F_r = 0 \\
&\frac{\partial \sigma_{rz}}{\partial r} + \frac{\partial \sigma_{zz}}{\partial z} + \frac{\sigma_{rz}}{r} + F_z = 0
\end{aligned} \tag{18.17}$$

The constitutive equations are from Eq. (16.45)

$$\begin{aligned}
\epsilon_{rr} &= \frac{1}{E}[\sigma_{rr} - \nu(\sigma_{\theta\theta} + \sigma_{zz})] + \alpha\tau = \frac{1}{2G}\left(\sigma_{rr} - \frac{\nu}{1+\nu}\Theta\right) + \alpha\tau \\
\epsilon_{\theta\theta} &= \frac{1}{E}[\sigma_{\theta\theta} - \nu(\sigma_{zz} + \sigma_{rr})] + \alpha\tau = \frac{1}{2G}\left(\sigma_{\theta\theta} - \frac{\nu}{1+\nu}\Theta\right) + \alpha\tau \\
\epsilon_{zz} &= \frac{1}{E}[\sigma_{zz} - \nu(\sigma_{rr} + \sigma_{\theta\theta})] + \alpha\tau = \frac{1}{2G}\left(\sigma_{zz} - \frac{\nu}{1+\nu}\Theta\right) + \alpha\tau \\
\epsilon_{zr} &= \frac{\sigma_{zr}}{2G}
\end{aligned} \tag{18.18}$$

where $\Theta = \sigma_{rr} + \sigma_{\theta\theta} + \sigma_{zz}$.

The strain-displacement relations are from Eq. (16.42)

$$\epsilon_{rr} = \frac{\partial u_r}{\partial r}, \quad \epsilon_{\theta\theta} = \frac{u_r}{r}, \quad \epsilon_{zz} = \frac{\partial u_z}{\partial z}$$
$$\epsilon_{zr} = \frac{1}{2}\Big(\frac{\partial u_r}{\partial z} + \frac{\partial u_z}{\partial r}\Big), \quad e = \frac{\partial u_r}{\partial r} + \frac{u_r}{r} + \frac{\partial u_z}{\partial z} \tag{18.19}$$

where u_r and u_z are the components of displacement in the direction of r and z, respectively.

Navier's equations are from Eq. (16.47)

$$\nabla^2 u_r - \frac{u_r}{r^2} + \frac{1}{1-2\nu}\frac{\partial e}{\partial r} - 2\frac{1+\nu}{1-2\nu}\alpha\frac{\partial \tau}{\partial r} + \frac{2(1+\nu)}{E}F_r = 0$$
$$\nabla^2 u_z + \frac{1}{1-2\nu}\frac{\partial e}{\partial z} - 2\frac{1+\nu}{1-2\nu}\alpha\frac{\partial \tau}{\partial z} + \frac{2(1+\nu)}{E}F_z = 0 \tag{18.20}$$

in which

$$\nabla^2 = \frac{\partial^2}{\partial r^2} + \frac{1}{r}\frac{\partial}{\partial r} + \frac{\partial^2}{\partial z^2} \tag{18.21}$$

The solution of Navier's Eq. (18.20) without the body forces can be expressed, for example, by Goodier's thermoelastic potential Φ and the Boussinesq harmonic functions φ and ψ. Referring to Eq. (16.49) under the axisymmetric condition, the displacements and the stresses can be expressed by

$$u_r = \frac{\partial \Phi}{\partial r} + \frac{\partial \varphi}{\partial r} + z\frac{\partial \psi}{\partial r}$$
$$u_z = \frac{\partial \Phi}{\partial z} + \frac{\partial \varphi}{\partial z} + z\frac{\partial \psi}{\partial z} - (3-4\nu)\psi \tag{18.22}$$

$$\sigma_{rr} = 2G\Big(\frac{\partial^2 \Phi}{\partial r^2} - K\tau + \frac{\partial^2 \varphi}{\partial r^2} + z\frac{\partial^2 \psi}{\partial r^2} - 2\nu\frac{\partial \psi}{\partial z}\Big)$$
$$\sigma_{\theta\theta} = 2G\Big(\frac{1}{r}\frac{\partial \Phi}{\partial r} - K\tau + \frac{1}{r}\frac{\partial \varphi}{\partial r} + \frac{z}{r}\frac{\partial \psi}{\partial r} - 2\nu\frac{\partial \psi}{\partial z}\Big)$$
$$\sigma_{zz} = 2G\Big[\frac{\partial^2 \Phi}{\partial z^2} - K\tau + \frac{\partial^2 \varphi}{\partial z^2} + z\frac{\partial^2 \psi}{\partial z^2} - 2(1-\nu)\frac{\partial \psi}{\partial z}\Big]$$
$$\sigma_{rz} = 2G\Big[\frac{\partial^2 \Phi}{\partial r \partial z} + \frac{\partial^2 \varphi}{\partial r \partial z} + z\frac{\partial^2 \psi}{\partial r \partial z} - (1-2\nu)\frac{\partial \psi}{\partial r}\Big] \tag{18.23}$$

in which Goodier's thermoelastic potential Φ and the Boussinesq harmonic functions φ, ψ must satisfy the following governing equations

$$\nabla^2 \Phi = K\tau, \quad \nabla^2 \varphi = 0, \quad \nabla^2 \psi = 0 \tag{18.24}$$

On the other hand, Michell's function M may be used instead of the Boussinesq harmonic functions φ and ψ. If we take

$$M = -\int (\varphi + z\psi) dz \tag{18.25}$$

the displacements and the stresses are represented by Michell's function M as follows:

$$\begin{aligned} u_r &= \frac{\partial \Phi}{\partial r} - \frac{\partial^2 M}{\partial r \partial z} \\ u_z &= \frac{\partial \Phi}{\partial z} + 2(1-\nu)\nabla^2 M - \frac{\partial^2 M}{\partial z^2} \end{aligned} \tag{18.26}$$

$$\begin{aligned} \sigma_{rr} &= 2G\Big[\frac{\partial^2 \Phi}{\partial r^2} - K\tau + \frac{\partial}{\partial z}\Big(\nu\nabla^2 M - \frac{\partial^2 M}{\partial r^2}\Big)\Big] \\ \sigma_{\theta\theta} &= 2G\Big[\frac{1}{r}\frac{\partial \Phi}{\partial r} - K\tau + \frac{\partial}{\partial z}\Big(\nu\nabla^2 M - \frac{1}{r}\frac{\partial M}{\partial r}\Big)\Big] \\ \sigma_{zz} &= 2G\Big\{\frac{\partial^2 \Phi}{\partial z^2} - K\tau + \frac{\partial}{\partial z}\Big[(2-\nu)\nabla^2 M - \frac{\partial^2 M}{\partial z^2}\Big]\Big\} \\ \sigma_{rz} &= 2G\Big\{\frac{\partial^2 \Phi}{\partial r \partial z} + \frac{\partial}{\partial r}\Big[(1-\nu)\nabla^2 M - \frac{\partial^2 M}{\partial z^2}\Big]\Big\} \end{aligned} \tag{18.27}$$

where Michell's function M must satisfy the equation

$$\nabla^2 \nabla^2 M = 0 \tag{18.28}$$

18.4 Three-Dimensional Problems

We now consider three-dimensional problems in a cylindrical coordinate system (r, θ, z). The equilibrium equations are from Eq. (16.40)

$$\begin{aligned} &\frac{\partial \sigma_{rr}}{\partial r} + \frac{1}{r}\frac{\partial \sigma_{\theta r}}{\partial \theta} + \frac{\partial \sigma_{zr}}{\partial z} + \frac{\sigma_{rr} - \sigma_{\theta\theta}}{r} + F_r = 0 \\ &\frac{\partial \sigma_{r\theta}}{\partial r} + \frac{1}{r}\frac{\partial \sigma_{\theta\theta}}{\partial \theta} + \frac{\partial \sigma_{z\theta}}{\partial z} + 2\frac{\sigma_{r\theta}}{r} + F_\theta = 0 \\ &\frac{\partial \sigma_{rz}}{\partial r} + \frac{1}{r}\frac{\partial \sigma_{\theta z}}{\partial \theta} + \frac{\partial \sigma_{zz}}{\partial z} + \frac{\sigma_{rz}}{r} + F_z = 0 \end{aligned} \tag{18.29}$$

The constitutive equations for a homogeneous, isotropic body are from Eq. (16.45)

$$\begin{aligned}
\epsilon_{rr} &= \frac{1}{E}[\sigma_{rr} - \nu(\sigma_{\theta\theta} + \sigma_{zz})] + \alpha\tau = \frac{1}{2G}\left(\sigma_{rr} - \frac{\nu}{1+\nu}\Theta\right) + \alpha\tau \\
\epsilon_{\theta\theta} &= \frac{1}{E}[\sigma_{\theta\theta} - \nu(\sigma_{zz} + \sigma_{rr})] + \alpha\tau = \frac{1}{2G}\left(\sigma_{\theta\theta} - \frac{\nu}{1+\nu}\Theta\right) + \alpha\tau \\
\epsilon_{zz} &= \frac{1}{E}[\sigma_{zz} - \nu(\sigma_{rr} + \sigma_{\theta\theta})] + \alpha\tau = \frac{1}{2G}\left(\sigma_{zz} - \frac{\nu}{1+\nu}\Theta\right) + \alpha\tau \\
\epsilon_{r\theta} &= \frac{\sigma_{r\theta}}{2G}, \quad \epsilon_{\theta z} = \frac{\sigma_{\theta z}}{2G}, \quad \epsilon_{zr} = \frac{\sigma_{zr}}{2G}
\end{aligned} \tag{18.30}$$

where $\Theta = \sigma_{rr} + \sigma_{\theta\theta} + \sigma_{zz}$.

The strain-displacement relations are from Eq. (16.42)

$$\begin{aligned}
\epsilon_{rr} &= \frac{\partial u_r}{\partial r}, \quad \epsilon_{\theta\theta} = \frac{u_r}{r} + \frac{1}{r}\frac{\partial u_\theta}{\partial \theta}, \quad \epsilon_{zz} = \frac{\partial u_z}{\partial z} \\
\epsilon_{r\theta} &= \frac{1}{2}\left(\frac{1}{r}\frac{\partial u_r}{\partial \theta} + \frac{\partial u_\theta}{\partial r} - \frac{u_\theta}{r}\right), \quad \epsilon_{\theta z} = \frac{1}{2}\left(\frac{\partial u_\theta}{\partial z} + \frac{1}{r}\frac{\partial u_z}{\partial \theta}\right) \\
\epsilon_{zr} &= \frac{1}{2}\left(\frac{\partial u_r}{\partial z} + \frac{\partial u_z}{\partial r}\right), \quad e = \frac{\partial u_r}{\partial r} + \frac{u_r}{r} + \frac{1}{r}\frac{\partial u_\theta}{\partial \theta} + \frac{\partial u_z}{\partial z}
\end{aligned} \tag{18.31}$$

where u_r, u_θ, u_z are the components of displacement in the r, θ, z directions, respectively.

Navier's equations for three-dimensional thermoelastic problems are expressed from Eq. (16.47) as

$$\begin{aligned}
&(\lambda + 2\mu)\frac{\partial e}{\partial r} - \mu\left[\frac{1}{r^2}\frac{\partial^2(ru_\theta)}{\partial r\partial\theta} - \frac{1}{r^2}\frac{\partial^2 u_r}{\partial\theta^2} - \frac{\partial^2 u_r}{\partial z^2} + \frac{\partial^2 u_z}{\partial r\partial z}\right] \\
&\quad - \beta\frac{\partial\tau}{\partial r} + F_r = 0 \\
&(\lambda + 2\mu)\frac{1}{r}\frac{\partial e}{\partial\theta} \\
&\quad - \mu\left[\frac{1}{r}\frac{\partial^2 u_z}{\partial z\partial\theta} - \frac{\partial^2 u_\theta}{\partial z^2} + \frac{1}{r^2}\frac{\partial(ru_\theta)}{\partial r} - \frac{1}{r}\frac{\partial^2(ru_\theta)}{\partial r^2} - \frac{1}{r^2}\frac{\partial u_r}{\partial\theta} + \frac{1}{r}\frac{\partial^2 u_r}{\partial r\partial\theta}\right] \\
&\quad - \beta\frac{1}{r}\frac{\partial\tau}{\partial\theta} + F_\theta = 0 \\
&(\lambda + 2\mu)\frac{\partial e}{\partial z} - \mu\left[\frac{1}{r}\frac{\partial u_r}{\partial z} + \frac{\partial^2 u_r}{\partial r\partial z} - \frac{\partial^2 u_z}{\partial r^2} - \frac{1}{r}\frac{\partial u_z}{\partial r} - \frac{1}{r^2}\frac{\partial^2 u_z}{\partial\theta^2} + \frac{1}{r}\frac{\partial^2 u_\theta}{\partial z\partial\theta}\right] \\
&\quad - \beta\frac{\partial\tau}{\partial z} + F_z = 0
\end{aligned} \tag{18.32}$$

The solution of Navier's equations (18.32) without body forces in a cylindrical coordinate system can be expressed by Goodier's thermoelastic potential Φ and the Boussinesq harmonic functions φ, ϑ, and ψ. The displacements and the stresses can be expressed by

$$
\begin{aligned}
u_r &= \frac{\partial \Phi}{\partial r} + \frac{\partial \varphi}{\partial r} + \frac{2}{r}\frac{\partial \vartheta}{\partial \theta} + z\frac{\partial \psi}{\partial r} \\
u_\theta &= \frac{1}{r}\frac{\partial \Phi}{\partial \theta} + \frac{1}{r}\frac{\partial \varphi}{\partial \theta} - 2\frac{\partial \vartheta}{\partial r} + \frac{z}{r}\frac{\partial \psi}{\partial \theta} \\
u_z &= \frac{\partial \Phi}{\partial z} + \frac{\partial \varphi}{\partial z} + z\frac{\partial \psi}{\partial z} - (3-4\nu)\psi
\end{aligned}
\tag{18.33}
$$

$$
\begin{aligned}
\sigma_{rr} &= 2G\Big[\frac{\partial^2 \Phi}{\partial r^2} - K\tau + \frac{\partial^2 \varphi}{\partial r^2} + 2\frac{\partial}{\partial r}\Big(\frac{1}{r}\frac{\partial \vartheta}{\partial \theta}\Big) + z\frac{\partial^2 \psi}{\partial r^2} - 2\nu\frac{\partial \psi}{\partial z}\Big] \\
\sigma_{\theta\theta} &= 2G\Big[\frac{1}{r}\frac{\partial \Phi}{\partial r} + \frac{1}{r^2}\frac{\partial^2 \Phi}{\partial \theta^2} - K\tau + \frac{1}{r}\frac{\partial \varphi}{\partial r} + \frac{1}{r^2}\frac{\partial^2 \varphi}{\partial \theta^2} - 2\frac{\partial}{\partial r}\Big(\frac{1}{r}\frac{\partial \vartheta}{\partial \theta}\Big) \\
&\quad + \frac{z}{r}\frac{\partial \psi}{\partial r} + \frac{z}{r^2}\frac{\partial^2 \psi}{\partial \theta^2} - 2\nu\frac{\partial \psi}{\partial z}\Big] \\
\sigma_{zz} &= 2G\Big[\frac{\partial^2 \Phi}{\partial z^2} - K\tau + \frac{\partial^2 \varphi}{\partial z^2} + z\frac{\partial^2 \psi}{\partial z^2} - 2(1-\nu)\frac{\partial \psi}{\partial z}\Big] \\
\sigma_{rz} &= 2G\Big[\frac{\partial^2 \Phi}{\partial r \partial z} + \frac{\partial^2 \varphi}{\partial r \partial z} + \frac{1}{r}\frac{\partial^2 \vartheta}{\partial \theta \partial z} + z\frac{\partial^2 \psi}{\partial r \partial z} - (1-2\nu)\frac{\partial \psi}{\partial r}\Big] \\
\sigma_{r\theta} &= 2G\Big[\frac{\partial}{\partial r}\Big(\frac{1}{r}\frac{\partial \Phi}{\partial \theta}\Big) + \frac{\partial}{\partial r}\Big(\frac{1}{r}\frac{\partial \varphi}{\partial \theta}\Big) - \frac{\partial^2 \vartheta}{\partial r^2} + \frac{1}{r}\frac{\partial \vartheta}{\partial r} + \frac{1}{r^2}\frac{\partial^2 \vartheta}{\partial \theta^2} \\
&\quad + z\frac{\partial}{\partial r}\Big(\frac{1}{r}\frac{\partial \psi}{\partial \theta}\Big)\Big] \\
\sigma_{\theta z} &= 2G\Big(\frac{1}{r}\frac{\partial^2 \Phi}{\partial \theta \partial z} + \frac{1}{r}\frac{\partial^2 \varphi}{\partial \theta \partial z} - \frac{\partial^2 \vartheta}{\partial r \partial z} + \frac{z}{r}\frac{\partial^2 \psi}{\partial \theta \partial z} - \frac{1-2\nu}{r}\frac{\partial \psi}{\partial \theta}\Big)
\end{aligned}
\tag{18.34}
$$

in which the four functions must satisfy

$$
\nabla^2 \Phi = K\tau, \quad \nabla^2 \varphi = \nabla^2 \vartheta = \nabla^2 \psi = 0 \tag{18.35}
$$

where

$$
\nabla^2 = \frac{\partial^2}{\partial r^2} + \frac{1}{r}\frac{\partial}{\partial r} + \frac{1}{r^2}\frac{\partial^2}{\partial \theta^2} + \frac{\partial^2}{\partial z^2} \tag{18.36}
$$

On the other hand, Michell's function M may be used instead of the Boussinesq harmonic functions φ and ψ. If we take

$$
M = -\int (\varphi + z\psi)dz \tag{18.37}
$$

the displacements and the stress components are represented by Goodier's thermoelastic potential Φ, Michell's function M, and the Boussinesq harmonic function ϑ

$$
\begin{aligned}
u_r &= \frac{\partial \Phi}{\partial r} - \frac{\partial^2 M}{\partial r \partial z} + \frac{2}{r}\frac{\partial \vartheta}{\partial \theta} \\
u_\theta &= \frac{1}{r}\frac{\partial \Phi}{\partial \theta} - \frac{1}{r}\frac{\partial^2 M}{\partial z \partial \theta} - 2\frac{\partial \vartheta}{\partial r} \\
u_z &= \frac{\partial \Phi}{\partial z} + 2(1-\nu)\nabla^2 M - \frac{\partial^2 M}{\partial z^2}
\end{aligned}
\tag{18.38}
$$

$$
\begin{aligned}
\sigma_{rr} &= 2G\Big[\frac{\partial^2 \Phi}{\partial r^2} - K\tau + \frac{\partial}{\partial z}\Big(\nu\nabla^2 M - \frac{\partial^2 M}{\partial r^2}\Big) + \frac{2}{r}\frac{\partial^2 \vartheta}{\partial r \partial \theta} - \frac{2}{r^2}\frac{\partial \vartheta}{\partial \theta}\Big] \\
\sigma_{\theta\theta} &= 2G\Big[\frac{1}{r}\frac{\partial \Phi}{\partial r} + \frac{1}{r^2}\frac{\partial^2 \Phi}{\partial \theta^2} - K\tau + \frac{\partial}{\partial z}\Big(\nu\nabla^2 M - \frac{1}{r}\frac{\partial M}{\partial r} - \frac{1}{r^2}\frac{\partial^2 M}{\partial \theta^2}\Big) \\
&\quad + \frac{2}{r^2}\frac{\partial \vartheta}{\partial \theta} - \frac{2}{r}\frac{\partial^2 \vartheta}{\partial r \partial \theta}\Big] \\
\sigma_{zz} &= 2G\Big\{\frac{\partial^2 \Phi}{\partial z^2} - K\tau + \frac{\partial}{\partial z}\Big[(2-\nu)\nabla^2 M - \frac{\partial^2 M}{\partial z^2}\Big]\Big\} \\
\sigma_{rz} &= 2G\Big\{\frac{\partial^2 \Phi}{\partial r \partial z} + \frac{\partial}{\partial r}\Big[(1-\nu)\nabla^2 M - \frac{\partial^2 M}{\partial z^2}\Big] + \frac{1}{r}\frac{\partial^2 \vartheta}{\partial \theta \partial z}\Big\} \\
\sigma_{r\theta} &= 2G\Big[\frac{\partial}{\partial r}\Big(\frac{1}{r}\frac{\partial \Phi}{\partial \theta}\Big) - \frac{\partial}{\partial r}\Big(\frac{1}{r}\frac{\partial^2 M}{\partial \theta \partial z}\Big) - \frac{\partial^2 \vartheta}{\partial r^2} + \frac{1}{r}\frac{\partial \vartheta}{\partial r} + \frac{1}{r^2}\frac{\partial^2 \vartheta}{\partial \theta^2}\Big] \\
\sigma_{\theta z} &= 2G\Big\{\frac{1}{r}\frac{\partial^2 \Phi}{\partial \theta \partial z} + \frac{1}{r}\frac{\partial}{\partial \theta}\Big[(1-\nu)\nabla^2 M - \frac{\partial^2 M}{\partial z^2}\Big] - \frac{\partial^2 \vartheta}{\partial r \partial z}\Big\}
\end{aligned}
\tag{18.39}
$$

in which Michell's function M must satisfy the equation

$$
\nabla^2\nabla^2 M = 0 \tag{18.40}
$$

18.5 Problems and Solutions Related to Thermal Stresses in Circular Cylinders

Problem 18.1. Derive Eq. (18.13).

Solution. The equations of equilibrium in a cylindrical coordinate system for the plane problem are from Eq. (16.40)

$$
\begin{aligned}
&\frac{\partial \sigma_{rr}}{\partial r} + \frac{1}{r}\frac{\partial \sigma_{\theta r}}{\partial \theta} + \frac{\sigma_{rr} - \sigma_{\theta\theta}}{r} = 0 \\
&\frac{\partial \sigma_{r\theta}}{\partial r} + \frac{1}{r}\frac{\partial \sigma_{\theta\theta}}{\partial \theta} + 2\frac{\sigma_{r\theta}}{r} = 0
\end{aligned}
\tag{18.41}
$$

Substitution of Eq. (18.15) into Eq. (18.41) gives

$$
\begin{aligned}
&\frac{\partial}{\partial r}\Big(\frac{1}{r}\frac{\partial \chi}{\partial r} + \frac{1}{r^2}\frac{\partial^2 \chi}{\partial \theta^2}\Big) - \frac{1}{r}\frac{\partial^2}{\partial r \partial \theta}\Big(\frac{1}{r}\frac{\partial \chi}{\partial \theta}\Big) + \frac{1}{r}\Big(\frac{1}{r}\frac{\partial \chi}{\partial r} + \frac{1}{r^2}\frac{\partial^2 \chi}{\partial \theta^2} - \frac{\partial^2 \chi}{\partial r^2}\Big) \\
&= -\frac{1}{r^2}\frac{\partial \chi}{\partial r} + \frac{1}{r}\frac{\partial^2 \chi}{\partial r^2} - \frac{2}{r^3}\frac{\partial^2 \chi}{\partial \theta^2} + \frac{1}{r^2}\frac{\partial^3 \chi}{\partial r \partial \theta^2} - \frac{1}{r}\Big(-\frac{1}{r^2}\frac{\partial^2 \chi}{\partial \theta^2} + \frac{1}{r}\frac{\partial^3 \chi}{\partial r \partial \theta^2}\Big) \\
&\quad + \frac{1}{r^2}\frac{\partial \chi}{\partial r} + \frac{1}{r^3}\frac{\partial^2 \chi}{\partial \theta^2} - \frac{1}{r}\frac{\partial^2 \chi}{\partial r^2} = 0 \\
&- \frac{\partial^2}{\partial r^2}\Big(\frac{1}{r}\frac{\partial \chi}{\partial \theta}\Big) + \frac{1}{r}\frac{\partial^3 \chi}{\partial r^2 \partial \theta} - \frac{2}{r}\frac{\partial}{\partial r}\Big(\frac{1}{r}\frac{\partial \chi}{\partial \theta}\Big) \\
&= -\Big(\frac{2}{r^3}\frac{\partial \chi}{\partial \theta} - \frac{2}{r^2}\frac{\partial^2 \chi}{\partial r \partial \theta} + \frac{1}{r}\frac{\partial^3 \chi}{\partial r^2 \partial \theta}\Big) \\
&\quad + \frac{1}{r}\frac{\partial^3 \chi}{\partial r^2 \partial \theta} - \frac{2}{r}\Big(-\frac{1}{r^2}\frac{\partial \chi}{\partial \theta} + \frac{1}{r}\frac{\partial^2 \chi}{\partial r \partial \theta}\Big) = 0
\end{aligned}
\tag{18.42}
$$

The thermal stress function χ automatically satisfies the equations of equilibrium (18.41).

Next, the compatibility equation is from Eq. (16.43)

$$
2\frac{1}{r^2}\frac{\partial^2 (r\varepsilon_{r\theta})}{\partial r \partial \theta} - \frac{1}{r^2}\frac{\partial^2 \varepsilon_{rr}}{\partial \theta^2} - \frac{1}{r}\frac{\partial^2 (r\varepsilon_{\theta\theta})}{\partial r^2} + \frac{1}{r}\frac{\partial \varepsilon_{rr}}{\partial r} = 0 \tag{18.43}
$$

The constitutive equations are

$$
\begin{aligned}
\varepsilon_{rr} &= \frac{1}{2G}[\sigma_{rr} - \frac{\nu^*}{1+\nu^*}(\sigma_{rr} + \sigma_{\theta\theta}) + 2G\alpha^* \tau] \\
\varepsilon_{\theta\theta} &= \frac{1}{2G}[\sigma_{\theta\theta} - \frac{\nu^*}{1+\nu^*}(\sigma_{rr} + \sigma_{\theta\theta}) + 2G\alpha^* \tau] \\
\varepsilon_{r\theta} &= \frac{1}{2G}\sigma_{r\theta}
\end{aligned}
\tag{18.44}
$$

Substitution of Eq. (18.15) into Eq. (18.44) gives

$$
\begin{aligned}
\varepsilon_{rr} &= \frac{1}{2G}\Big[\frac{1}{r}\frac{\partial \chi}{\partial r} + \frac{1}{r^2}\frac{\partial^2 \chi}{\partial \theta^2} - \frac{\nu^*}{1+\nu^*}\nabla^2 \chi + 2G\alpha^* \tau\Big] \\
\varepsilon_{\theta\theta} &= \frac{1}{2G}\Big[\frac{\partial^2 \chi}{\partial r^2} - \frac{\nu^*}{1+\nu^*}\nabla^2 \chi + 2G\alpha^* \tau\Big] \\
\varepsilon_{r\theta} &= -\frac{1}{2G}\frac{\partial}{\partial r}\Big(\frac{1}{r}\frac{\partial \chi}{\partial \theta}\Big)
\end{aligned}
\tag{18.45}
$$

where

$$
\nabla^2 = \frac{\partial^2}{\partial r^2} + \frac{1}{r}\frac{\partial}{\partial r} + \frac{1}{r^2}\frac{\partial^2}{\partial \theta^2} \tag{18.46}
$$

Multiplying $2G$ to the left hand side of Eq. (18.43) and representing it to CE, we obtain

$$2G\Big[2\frac{1}{r^2}\frac{\partial^2(r\varepsilon_{r\theta})}{\partial r\partial\theta}-\frac{1}{r^2}\frac{\partial^2\varepsilon_{rr}}{\partial\theta^2}-\frac{1}{r}\frac{\partial^2(r\varepsilon_{\theta\theta})}{\partial r^2}+\frac{1}{r}\frac{\partial\varepsilon_{rr}}{\partial r}\Big]=CE \tag{18.47}$$

From Eqs. (18.43) and (18.47) we have

$$\begin{aligned}
CE &= 2G\left\{2\frac{1}{r^2}\frac{\partial^2(r\varepsilon_{r\theta})}{\partial r\partial\theta}-\Big[\frac{1}{r^2}\frac{\partial^2\varepsilon_{rr}}{\partial\theta^2}+\frac{1}{r}\frac{\partial^2(r\varepsilon_{\theta\theta})}{\partial r^2}-\frac{1}{r}\frac{\partial\varepsilon_{rr}}{\partial r}\Big]\right\}\\
&= 2\frac{1}{r^2}\frac{\partial^2}{\partial r\partial\theta}\Big[-r\frac{\partial}{\partial r}\Big(\frac{1}{r}\frac{\partial\chi}{\partial\theta}\Big)\Big]-\frac{1}{r^2}\frac{\partial^2}{\partial\theta^2}\Big[\Big(\nabla^2\chi-\frac{\partial^2\chi}{\partial r^2}\Big)\\
&\quad -\frac{\nu^*}{1+\nu^*}\nabla^2\chi+2G\alpha^*\tau\Big]\\
&\quad -\frac{1}{r}\frac{\partial^2}{\partial r^2}\Big[r\frac{\partial^2\chi}{\partial r^2}-\frac{\nu^*}{1+\nu^*}r\nabla^2\chi+2G\alpha^* r\tau\Big]\\
&\quad +\frac{1}{r}\frac{\partial}{\partial r}\Big[\Big(\nabla^2\chi-\frac{\partial^2\chi}{\partial r^2}\Big)-\frac{\nu^*}{1+\nu^*}\nabla^2\chi+2G\alpha^*\tau\Big]\\
&= -\frac{2}{r^4}\frac{\partial^2\chi}{\partial\theta^2}+\frac{2}{r^3}\frac{\partial^3\chi}{\partial r\partial\theta^2}-\frac{2}{r^2}\frac{\partial^4\chi}{\partial r^2\partial\theta^2}-\Big(\frac{1}{r^2}\frac{\partial^2}{\partial\theta^2}-\frac{1}{r}\frac{\partial}{\partial r}\Big)\nabla^2\chi\\
&\quad +\frac{1}{r^2}\frac{\partial^4\chi}{\partial r^2\partial\theta^2}-\frac{1}{r}\frac{\partial^2}{\partial r^2}\Big(r\frac{\partial^2\chi}{\partial r^2}\Big)\\
&\quad -\frac{1}{r}\frac{\partial^3\chi}{\partial r^3}+\frac{\nu^*}{1+\nu^*}\nabla^2\nabla^2\chi-2G\alpha^*\nabla^2\tau\\
&= -\Big(\frac{\partial^4}{\partial r^4}+\frac{3}{r}\frac{\partial^3}{\partial r^3}+\frac{2}{r^4}\frac{\partial^2}{\partial\theta^2}-\frac{2}{r^3}\frac{\partial^3}{\partial r\partial\theta^2}+\frac{1}{r^2}\frac{\partial^4}{\partial r^2\partial\theta^2}\Big)\chi\\
&\quad -\Big(\frac{1}{r^2}\frac{\partial^2}{\partial\theta^2}-\frac{1}{r}\frac{\partial}{\partial r}\Big)\nabla^2\chi+\frac{\nu^*}{1+\nu^*}\nabla^2\nabla^2\chi-2G\alpha^*\nabla^2\tau
\end{aligned} \tag{18.48}$$

On the other hand,

$$\begin{aligned}
\nabla^2\nabla^2\chi &= \Big(\frac{\partial^2}{\partial r^2}+\frac{1}{r}\frac{\partial}{\partial r}+\frac{1}{r^2}\frac{\partial^2}{\partial\theta^2}\Big)\nabla^2\chi\\
&= \Big(\frac{\partial^2}{\partial r^2}+\frac{2}{r}\frac{\partial}{\partial r}\Big)\nabla^2\chi+\Big(\frac{1}{r^2}\frac{\partial^2}{\partial\theta^2}-\frac{1}{r}\frac{\partial}{\partial r}\Big)\nabla^2\chi\\
&= \Big(\frac{\partial^2}{\partial r^2}+\frac{2}{r}\frac{\partial}{\partial r}\Big)\Big(\frac{\partial^2}{\partial r^2}+\frac{1}{r}\frac{\partial}{\partial r}+\frac{1}{r^2}\frac{\partial^2}{\partial\theta^2}\Big)\chi+\Big(\frac{1}{r^2}\frac{\partial^2}{\partial\theta^2}-\frac{1}{r}\frac{\partial}{\partial r}\Big)\nabla^2\chi\\
&= \Big[\frac{\partial^4}{\partial r^4}+\frac{2}{r^3}\frac{\partial}{\partial r}-\frac{2}{r^2}\frac{\partial^2}{\partial r^2}+\frac{1}{r}\frac{\partial^3}{\partial r^3}\\
&\quad +\frac{6}{r^4}\frac{\partial^2}{\partial\theta^2}-\frac{4}{r^3}\frac{\partial^3}{\partial r\partial\theta^2}+\frac{1}{r^2}\frac{\partial^4}{\partial r^2\partial\theta^2}
\end{aligned}$$

$$
+ \frac{2}{r}\frac{\partial^3}{\partial r^3} - \frac{2}{r^3}\frac{\partial}{\partial r} + \frac{2}{r^2}\frac{\partial^2}{\partial r^2} - \frac{4}{r^4}\frac{\partial^2}{\partial \theta^2} + \frac{2}{r^3}\frac{\partial^3}{\partial r \partial \theta^2}\Big]\chi
$$

$$
+ \Big(\frac{1}{r^2}\frac{\partial^2}{\partial \theta^2} - \frac{1}{r}\frac{\partial}{\partial r}\Big)\nabla^2\chi
$$

$$
= \Big(\frac{\partial^4}{\partial r^4} + \frac{3}{r}\frac{\partial^3}{\partial r^3} + \frac{2}{r^4}\frac{\partial^2}{\partial \theta^2} - \frac{2}{r^3}\frac{\partial^3}{\partial r \partial \theta^2} + \frac{1}{r^2}\frac{\partial^4}{\partial r^2 \partial \theta^2}\Big)\chi
$$

$$
+ \Big(\frac{1}{r^2}\frac{\partial^2}{\partial \theta^2} - \frac{1}{r}\frac{\partial}{\partial r}\Big)\nabla^2\chi \tag{18.49}
$$

From Eqs. (18.48) and (18.49), we get

$$
\begin{aligned}
CE &= -\nabla^2\nabla^2\chi + \frac{\nu^*}{1+\nu^*}\nabla^2\nabla^2\chi - \frac{\alpha^* E^*}{1+\nu^*}\nabla^2\tau \\
&= -\frac{1}{1+\nu^*}\nabla^2\nabla^2\chi - \frac{\alpha^* E^*}{1+\nu^*}\nabla^2\tau = 0
\end{aligned} \tag{18.50}
$$

Therefore, the governing equation of the thermal stress function χ becomes

$$
\nabla^2\nabla^2\chi = -\alpha^* E^* \nabla^2\tau \qquad \text{(Answer)}
$$

Problem 18.2. Derive Eq. (18.16).

Solution. The conditions of single-valuedness of rotation and displacements are given by Eqs. (17.16), (17.17) and (17.18). When variables $(n,\ ds,\ x_1,\ x_2)$ are rewritten with variables $(r,\ r d\theta,\ r\cos\theta,\ r\sin\theta)$ in a cylindrical coordinate system, Eqs. (17.16)–(17.18) reduce to Eq. (18.16) in a cylindrical coordinate system

$$
\int_0^{2\pi} r\frac{\partial}{\partial r}\Big(\nabla^2\chi + \alpha^* E^*\tau\Big)\, d\theta = 0
$$

$$
\int_0^{2\pi} r\Big(\cos\theta\frac{\partial}{\partial\theta} - r\sin\theta\frac{\partial}{\partial r}\Big)(\nabla^2\chi + \alpha^* E^*\tau)\, d\theta = 0
$$

$$
\int_0^{2\pi} r\Big(r\cos\theta\frac{\partial}{\partial r} + \sin\theta\frac{\partial}{\partial\theta}\Big)(\nabla^2\chi + \alpha^* E^*\tau)\, d\theta = 0 \qquad \text{(Answer)}
$$

Problem 18.3. Find the thermal stresses in a hollow cylinder subjected to asymmetric temperature change by use of the thermal stress function χ.

Solution. We divide the thermal stress function χ into two functions:

$$
\chi = \chi_c + \chi_p \tag{18.51}
$$

where the functions χ_c and χ_p are complementary and particular solutions of Eq. (18.13), respectively, and are governed by

$$\nabla^4 \chi_c = 0, \quad \nabla^4 \chi_p = -\alpha^* E^* \nabla^2 \tau \tag{18.52}$$

General solutions of the thermal stress function χ are from Eq. (17.96)

$$\begin{aligned}\chi &= \chi_c + \chi_p \\ &= E_0 + F_0 \ln r + G_0 r^2 + H_0 r^2 \ln r \\ &\quad + (E_1 r + F_1 r^{-1} + G_1 r^3 + H_1 r \ln r) \cos\theta \\ &\quad + (E_1' + F_1' r^{-1} + G_1' r^3 + H_1' r \ln r) \sin\theta \\ &\quad + \sum_{n=2}^{\infty} \big[(E_n r^n + F_n r^{-n} + G_n r^{n+2} + H_n r^{-n+2}) \cos n\theta \\ &\quad + (E_n' r^n + F_n' r^{-n} + G_n' r^{n+2} + H_n' r^{-n+2}) \sin n\theta \big] + \chi_p \end{aligned} \tag{18.53}$$

The differentiation of Eq. (18.53) gives

$$\begin{aligned}\frac{\partial \chi}{\partial r} &= F_0 r^{-1} + 2G_0 r + H_0 r (2\ln r + 1) \\ &\quad + [E_1 - F_1 r^{-2} + 3G_1 r^2 + H_1 (\ln r + 1)] \cos\theta \\ &\quad + [E_1' - F_1' r^{-2} + 3G_1' r^2 + H_1' (\ln r + 1)] \sin\theta \\ &\quad + \sum_{n=2}^{\infty} \big\{ [n E_n r^{n-1} - n F_n r^{-n-1} + (n+2) G_n r^{n+1} \\ &\quad - (n-2) H_n r^{-n+1} \big] \cos n\theta + \big[n E_n' r^{n-1} - n F_n' r^{-n-1} \\ &\quad + (n+2) G_n' r^{n+1} - (n-2) H_n' r^{-n+1} \big] \sin n\theta \big\} + \frac{\partial \chi_p}{\partial r} \end{aligned} \tag{18.54}$$

$$\begin{aligned}\frac{1}{r}\frac{\partial \chi}{\partial r} &= F_0 r^{-2} + 2G_0 + H_0 (2\ln r + 1) \\ &\quad + [E_1 r^{-1} - F_1 r^{-3} + 3G_1 r + H_1 r^{-1} (\ln r + 1)] \cos\theta \\ &\quad + [E_1' r^{-1} - F_1' r^{-3} + 3G_1' r + H_1' r^{-1} (\ln r + 1)] \sin\theta \\ &\quad + \sum_{n=2}^{\infty} \big\{ [n E_n r^{n-2} - n F_n r^{-n-2} + (n+2) G_n r^n \\ &\quad - (n-2) H_n r^{-n}] \cos n\theta + [n E_n' r^{n-2} - n F_n' r^{-n-2} \\ &\quad + (n+2) G_n' r^n - (n-2) H_n' r^{-n}] \sin n\theta \big\} + \frac{1}{r}\frac{\partial \chi_p}{\partial r} \end{aligned} \tag{18.55}$$

$$
\begin{aligned}
\frac{\partial^2 \chi}{\partial r^2} &= -F_0 r^{-2} + 2G_0 + H_0(2\ln r + 3) \\
&+ (2F_1 r^{-3} + 6G_1 r + H_1 r^{-1})\cos\theta \\
&+ (2F_1' r^{-3} + 6G_1' r + H_1' \ln r^{-1})\sin\theta \\
&+ \sum_{n=2}^{\infty} \{[n(n-1)E_n r^{n-2} + n(n+1)F_n r^{-n-2} \\
&+ (n+1)(n+2)G_n r^n + (n-1)(n-2)H_n r^{-n}]\cos n\theta \\
&+ [n(n-1)E_n' r^{n-2} + n(n+1)F_n' r^{-n-2} \\
&+ (n+1)(n+2)G_n' r^n + (n-1)(n-2)H_n' r^{-n}]\sin n\theta\} + \frac{\partial^2 \chi_p}{\partial r^2}
\end{aligned} \tag{18.56}
$$

$$
\begin{aligned}
\frac{1}{r^2}\frac{\partial^2 \chi}{\partial \theta^2} &= -(E_1 r^{-1} + F_1 r^{-3} + G_1 r + H_1 r^{-1}\ln r)\cos\theta \\
&- (E_1' r^{-1} + F_1' r^{-3} + G_1' r + H_1' r^{-1}\ln r)\sin\theta \\
&- \sum_{n=2}^{\infty} n^2\big[(E_n r^{n-2} + F_n r^{-n-2} + G_n r^n + H_n r^{-n})\cos n\theta \\
&+ (E_n' r^{n-2} + F_n' r^{-n-2} + G_n' r^{n2} + H_n' r^{-n})\sin n\theta\big] + \frac{1}{r^2}\frac{\partial^2 \chi_p}{\partial \theta^2}
\end{aligned} \tag{18.57}
$$

$$
\begin{aligned}
\frac{\partial}{\partial r}\Big(\frac{1}{r}\frac{\partial \chi}{\partial \theta}\Big) &= -(-2F_1 r^{-3} + 2G_1 r + H_1 r^{-1})\sin\theta \\
&+ (-2F_1' r^{-3} + 2G_1' r + H_1' r^{-1})\cos\theta \\
&+ \sum_{n=2}^{\infty} n\big\{-[(n-1)E_n r^{n-2} - (n+1)F_n r^{-n-2} \\
&+ (n+1)G_n r^n - (n-1)H_n r^{-n}]\sin n\theta \\
&+ [(n-1)E_n' r^{n-2} - (n+1)F_n' r^{-n-2} \\
&+ (n+1)G_n' r^n - (n-1)H_n' r^{-n}]\cos n\theta\big\} + \frac{\partial}{\partial r}\Big(\frac{1}{r}\frac{\partial \chi_p}{\partial \theta}\Big)
\end{aligned} \tag{18.58}
$$

Substitution of Eqs. (18.55)–(18.58) into Eq. (18.15) gives the thermal stress components

$$
\begin{aligned}
\sigma_{rr} = {} & F_0 r^{-2} + 2G_0 + H_0(2\ln r + 1) \\
& + (-2F_1 r^{-3} + 2G_1 r + H_1 r^{-1}) \cos\theta \\
& + (-2F_1' r^{-3} + 2G_1' r + H_1' r^{-1}) \sin\theta \\
& - \sum_{n=2}^{\infty} \Big\{ \big[n(n-1)E_n r^{n-2} + n(n+1)F_n r^{-n-2} \\
& + (n-2)(n+1)G_n r^n + (n-1)(n+2)H_n r^{-n} \big] \cos n\theta \\
& + \big[n(n-1)E_n' r^{n-2} + n(n+1)F_n' r^{-n-2} \\
& + (n-2)(n+1)G_n' r^n + (n-1)(n+2)H_n' r^{-n} \big] \sin n\theta \Big\} + \sigma_{rr}^p \\
\sigma_{\theta\theta} = {} & -F_0 r^{-2} + 2G_0 + H_0(2\ln r + 3) \\
& + (2F_1 r^{-3} + 6G_1 r + H_1 r^{-1}) \cos\theta \\
& + (2F_1' r^{-3} + 6G_1' r + H'_1 r^{-1}) \sin\theta \\
& + \sum_{n=2}^{\infty} \Big\{ \big[n(n-1)E_n r^{n-2} + n(n+1)F_n r^{-n-2} \\
& + (n+1)(n+2)G_n r^n + (n-1)(n-2)H_n r^{-n} \big] \cos n\theta \\
& + \big[n(n-1)E_n' r^{n-2} + n(n+1)F_n' r^{-n-2} \\
& + (n+1)(n+2)G_n' r^n + (n-1)(n-2)H_n' r^{-n} \big] \sin n\theta \Big\} \\
& + \sigma_{\theta\theta}^p \\
\sigma_{r\theta} = {} & (-2F_1 r^{-3} + 2G_1 r + H_1 r^{-1}) \sin\theta \\
& - (-2F_1' r^{-3} + 2G_1' r + H_1' r^{-1}) \cos\theta \\
& + \sum_{n=2}^{\infty} n \Big\{ \big[(n-1)E_n r^{n-2} - (n+1)F_n r^{-n-2} \\
& + (n+1)G_n r^n - (n-1)H_n r^{-n} \big] \sin n\theta \\
& - \big[(n-1)E_n' r^{n-2} - (n+1)F_n' r^{-n-2} \\
& + (n+1)G_n' r^n - (n-1)H_n' r^{-n} \big] \cos n\theta \Big\} + \sigma_{r\theta}^p \qquad \text{(Answer)}
\end{aligned}
$$

in which

$$
\sigma_{rr}^p = \frac{1}{r}\frac{\partial \chi_p}{\partial r} + \frac{1}{r^2}\frac{\partial^2 \chi_p}{\partial \theta^2}, \quad \sigma_{\theta\theta}^p = \frac{\partial^2 \chi_p}{\partial r^2}, \quad \sigma_{r\theta}^p = -\frac{\partial}{\partial r}\left(\frac{1}{r}\frac{\partial \chi_p}{\partial \theta}\right) \tag{18.59}
$$

The unknown coefficients can be determined from the boundary conditions and the conditions of single-valuedness of rotation and displacements (18.16).

Problem 18.4. Find the steady thermal stresses in a long hollow cylinder of inner radius a and outer radius b with the constant initial temperature T_i, when the hollow cylinder is subjected to asymmetric temperature $T(r, \theta)$.

Solution. The governing equation of the heat conduction problem without the internal heat generation is

$$\frac{\partial^2 T}{\partial r^2} + \frac{1}{r}\frac{\partial T}{\partial r} + \frac{1}{r^2}\frac{\partial^2 T}{\partial \theta^2} = 0 \tag{18.60}$$

The general solution of Eq. (18.60) is

$$T(r, \theta) = A_0 + B_0 \ln r + \sum_{n=1}^{\infty} \left[(A_n r^n + B_n r^{-n}) \cos n\theta + (A'_n r^n + B'_n r^{-n}) \sin n\theta \right] \tag{18.61}$$

The temperature change from the initial temperature T_i is given by

$$\begin{aligned} \tau(r, \theta) = {} & A_0 - T_i + B_0 \ln r \\ & + (A_1 r + B_1 r^{-1}) \cos \theta + (C_1 r + D_1 r^{-1}) \sin \theta \\ & + \sum_{n=2}^{\infty} \left[(A_n r^n + B_n r^{-n}) \cos n\theta + (C_n r^n + D_n r^{-n}) \sin n\theta \right] \end{aligned} \tag{18.62}$$

From Eq. (18.62) and the second equation in Eq. (18.52), the particular solution χ_p becomes zero. Using the boundary conditions for traction free surfaces and the conditions of single-valuedness of rotation and displacements (18.16), unknown coefficients can be determined. Therefore, thermal stress components can be obtained as

$$\begin{aligned} \sigma_{rr} = {} & F_0 r^{-2} + 2G_0 + H_0(2 \ln r + 1) \\ & + (-2F_1 r^{-3} + 2G_1 r + H_1 r^{-1}) \cos \theta \\ & + (-2F'_1 r^{-3} + 2G'_1 r + H'_1 r^{-1}) \sin \theta \\ \sigma_{\theta\theta} = {} & -F_0 r^{-2} + 2G_0 + H_0(2 \ln r + 3) \\ & + (2F_1 r^{-3} + 6G_1 r + H_1 r^{-1}) \cos \theta \\ & + (2F'_1 r^{-3} + 6G'_1 r + H'_1 r^{-1}) \sin \theta \\ \sigma_{r\theta} = {} & (-2F_1 r^{-3} + 2G_1 r + H_1 r^{-1}) \sin \theta \\ & - (-2F'_1 r^{-3} + 2G'_1 r + H'_1 r^{-1}) \cos \theta \end{aligned} \tag{Answer}$$

From the conditions of single-valuedness of rotation and displacements (18.16), we get

$$H_0 = -\frac{1}{4}\alpha^* E^* B_0, \quad H_1 = -\frac{1}{2}\alpha^* E^* B_1, \quad H'_1 = -\frac{1}{2}\alpha^* E^* B'_1$$

The components of $\cos n\theta$ and $\sin n\theta$ for $n \geq 2$ in the temperature do not affect the thermal stress.

Problem 18.5. Prove that Navier's equation in a cylindrical coordinate system in plane problems can be expressed by

$$(\lambda^* + 2\mu)\frac{\partial e}{\partial r} - \frac{2\mu}{r}\frac{\partial \omega}{\partial \theta} = \beta^* \frac{\partial \tau}{\partial r}$$
$$(\lambda^* + 2\mu)\frac{1}{r}\frac{\partial e}{\partial \theta} + 2\mu\frac{\partial \omega}{\partial r} = \beta^* \frac{1}{r}\frac{\partial \tau}{\partial \theta} \tag{18.63}$$

where

$$e = \frac{\partial u_r}{\partial r} + \frac{u_r}{r} + \frac{1}{r}\frac{\partial u_\theta}{\partial \theta}, \quad \omega = \frac{1}{2r}\left[\frac{\partial (r u_\theta)}{\partial r} - \frac{\partial u_r}{\partial \theta}\right] \tag{18.64}$$

Solution. From the equilibrium equations (16.40) in a cylindrical coordinate system, the equilibrium equations without the body forces reduce for the plane problem to

$$\frac{\partial \sigma_{rr}}{\partial r} + \frac{1}{r}\frac{\partial \sigma_{\theta r}}{\partial \theta} + \frac{\sigma_{rr} - \sigma_{\theta\theta}}{r} = 0$$
$$\frac{\partial \sigma_{r\theta}}{\partial r} + \frac{1}{r}\frac{\partial \sigma_{\theta\theta}}{\partial \theta} + 2\frac{\sigma_{r\theta}}{r} = 0 \tag{18.65}$$

Hooke's law is

$$\begin{aligned}
\sigma_{rr} &= (\lambda^* + 2\mu)\varepsilon_{rr} + \lambda^* \varepsilon_{\theta\theta} - \beta^* \tau \\
&= (\lambda^* + 2\mu)(\varepsilon_{rr} + \varepsilon_{\theta\theta}) - 2\mu\varepsilon_{\theta\theta} - \beta^* \tau \\
\sigma_{\theta\theta} &= (\lambda^* + 2\mu)\varepsilon_{\theta\theta} + \lambda^* \varepsilon_{rr} - \beta^* \tau \\
&= (\lambda^* + 2\mu)(\varepsilon_{rr} + \varepsilon_{\theta\theta}) - 2\mu\varepsilon_{rr} - \beta^* \tau \\
\sigma_{r\theta} &= 2\mu\varepsilon_{\theta r}
\end{aligned} \tag{18.66}$$

The components of strain and rotation are

$$\begin{aligned}
\varepsilon_{rr} &= \frac{\partial u_r}{\partial r}, \quad \varepsilon_{\theta\theta} = \frac{u_r}{r} + \frac{1}{r}\frac{\partial u_\theta}{\partial \theta}, \quad \varepsilon_{r\theta} = \frac{1}{2}\left(\frac{1}{r}\frac{\partial u_r}{\partial \theta} + \frac{\partial u_\theta}{\partial r} - \frac{u_\theta}{r}\right) \\
e &= \frac{\partial u_r}{\partial r} + \frac{u_r}{r} + \frac{1}{r}\frac{\partial u_\theta}{\partial \theta} \\
\omega &= \frac{1}{2r}\left[\frac{\partial (r u_\theta)}{\partial r} - \frac{\partial u_r}{\partial \theta}\right] = \frac{1}{2}\left(\frac{u_\theta}{r} + \frac{\partial u_\theta}{\partial r} - \frac{1}{r}\frac{\partial u_r}{\partial \theta}\right)
\end{aligned} \tag{18.67}$$

The shear strain may be written as

$$\varepsilon_{r\theta} = -\omega + \frac{\partial u_\theta}{\partial r} \tag{18.68}$$

Substitution of Eq. (18.66) into the first equation of Eq. (18.65) gives

$$
\begin{aligned}
&(\lambda^* + 2\mu)\frac{\partial e}{\partial r} - 2\mu\frac{\partial \varepsilon_{\theta\theta}}{\partial r} - \beta^*\frac{\partial \tau}{\partial r} - 2\mu\frac{1}{r}\frac{\partial \omega}{\partial \theta} \\
&+ 2\mu\frac{1}{r}\frac{\partial^2 u_\theta}{\partial r\partial \theta} + 2\mu\frac{\varepsilon_{rr} - \varepsilon_{\theta\theta}}{r} \\
&= (\lambda^* + 2\mu)\frac{\partial e}{\partial r} - \frac{2\mu}{r}\frac{\partial \omega}{\partial \theta} - \beta^*\frac{\partial \tau}{\partial r} \\
&- 2\mu\left(\frac{1}{r}\frac{\partial u_r}{\partial r} - \frac{u_r}{r^2} + \frac{1}{r}\frac{\partial^2 u_\theta}{\partial r\partial \theta} - \frac{1}{r^2}\frac{\partial u_\theta}{\partial \theta} - \frac{1}{r}\frac{\partial^2 u_\theta}{\partial r\partial \theta}\right. \\
&\left. - \frac{1}{r}\frac{\partial u_r}{\partial r} + \frac{u_r}{r^2} + \frac{1}{r^2}\frac{\partial u_\theta}{\partial \theta}\right) = 0
\end{aligned}
\tag{18.69}
$$

Therefore, we get

$$
(\lambda^* + 2\mu)\frac{\partial e}{\partial r} - \frac{2\mu}{r}\frac{\partial \omega}{\partial \theta} = \beta^*\frac{\partial \tau}{\partial r} \qquad \text{(Answer)}
$$

Substitution of Eq. (18.66) into the second equation of Eq. (18.65) gives

$$
\begin{aligned}
&(\lambda^* + 2\mu)\frac{1}{r}\frac{\partial e}{\partial \theta} - 2\mu\frac{1}{r}\frac{\partial \varepsilon_{rr}}{\partial \theta} - \beta^*\frac{1}{r}\frac{\partial \tau}{\partial \theta} + 2\mu\frac{\partial \omega}{\partial r} - 2\mu\frac{\partial^2 u_\theta}{\partial r^2} \\
&+ 4\mu\left(\frac{\partial \varepsilon_{r\theta}}{\partial r} + \frac{\varepsilon_{r\theta}}{r}\right) \\
&= (\lambda^* + 2\mu)\frac{1}{r}\frac{\partial e}{\partial \theta} + 2\mu\frac{\partial \omega}{\partial r} - \beta^*\frac{1}{r}\frac{\partial \tau}{\partial \theta} - 2\mu\left[\frac{1}{r}\frac{\partial^2 u_r}{\partial r\partial \theta}\right. \\
&+ \frac{\partial^2 u_\theta}{\partial r^2} - \left(-\frac{1}{r^2}\frac{\partial u_r}{\partial \theta} + \frac{1}{r}\frac{\partial^2 u_r}{\partial r\partial \theta} + \frac{\partial^2 u_\theta}{\partial r^2} + \frac{u_\theta}{r^2} - \frac{1}{r}\frac{\partial u_\theta}{\partial r}\right. \\
&\left.\left. + \frac{1}{r^2}\frac{\partial u_r}{\partial \theta} + \frac{1}{r}\frac{\partial u_\theta}{\partial r} - \frac{u_\theta}{r^2}\right)\right] = 0
\end{aligned}
\tag{18.70}
$$

Therefore, we get

$$
(\lambda^* + 2\mu)\frac{1}{r}\frac{\partial e}{\partial \theta} + 2\mu\frac{\partial \omega}{\partial r} = \frac{\beta^*}{r}\frac{\partial \tau}{\partial \theta} \qquad \text{(Answer)}
$$

Problem 18.6. Prove that the governing equation (18.63) can be solved by use of Goodier's thermoelastic potential Φ and the Boussinesq harmonic functions φ, ϑ which are related to the components of displacement as follows

$$
\begin{aligned}
u_r &= \frac{\partial \Phi}{\partial r} + \frac{\partial \varphi}{\partial r} + \frac{2}{r}\frac{\partial \vartheta}{\partial \theta} \\
u_\theta &= \frac{1}{r}\frac{\partial \Phi}{\partial \theta} + \frac{1}{r}\frac{\partial \varphi}{\partial \theta} - 2\frac{\partial \vartheta}{\partial r}
\end{aligned}
\tag{18.71}
$$

where Goodier's thermoelastic potential Φ and the Boussinesq harmonic functions φ, ϑ must satisfy the equations

$$\nabla^2\Phi = K\tau, \quad \nabla^2\varphi = 0, \quad \nabla^2\vartheta = 0 \tag{18.72}$$

in which

$$K = \frac{\beta^*}{\lambda^* + 2\mu}, \quad \nabla^2 = \frac{\partial^2}{\partial r^2} + \frac{1}{r}\frac{\partial}{\partial r} + \frac{1}{r^2}\frac{\partial^2}{\partial \theta^2} \tag{18.73}$$

Solution. The strain, the dilatation, and the rotation are derived from Eq. (18.71)

$$\begin{aligned}
\varepsilon_{rr} &= \frac{\partial^2\Phi}{\partial r^2} + \frac{\partial^2\varphi}{\partial r^2} + \frac{2}{r}\frac{\partial^2\vartheta}{\partial r\partial\theta} - \frac{2}{r^2}\frac{\partial\vartheta}{\partial\theta}\\
\varepsilon_{\theta\theta} &= \frac{1}{r}\frac{\partial\Phi}{\partial r} + \frac{1}{r^2}\frac{\partial^2\Phi}{\partial\theta^2} + \frac{1}{r}\frac{\partial\varphi}{\partial r} + \frac{1}{r^2}\frac{\partial^2\varphi}{\partial\theta^2} + \frac{2}{r^2}\frac{\partial\vartheta}{\partial\theta} - \frac{2}{r}\frac{\partial^2\vartheta}{\partial r\partial\theta}\\
e &= \nabla^2\Phi + \nabla^2\varphi\\
\omega &= \frac{1}{2r}\left[\frac{\partial(ru_\theta)}{\partial r} - \frac{\partial u_r}{\partial\theta}\right]\\
&= \frac{1}{2r}\left[\frac{\partial^2\Phi}{\partial r\partial\theta} + \frac{\partial^2\varphi}{\partial r\partial\theta} - 2\frac{\partial\vartheta}{\partial r} - 2r\frac{\partial^2\vartheta}{\partial r^2} - \frac{\partial^2\Phi}{\partial r\partial\theta} - \frac{\partial^2\varphi}{\partial r\partial\theta} - \frac{2}{r}\frac{\partial^2\vartheta}{\partial\theta^2}\right]\\
&= -\left(\frac{\partial^2\vartheta}{\partial r^2} + \frac{1}{r}\frac{\partial\vartheta}{\partial r} + \frac{1}{r^2}\frac{\partial^2\vartheta}{\partial\theta^2}\right) = -\nabla^2\vartheta
\end{aligned} \tag{18.74}$$

Navier's equations are

$$(\lambda^* + 2\mu)\frac{\partial e}{\partial r} - \frac{2\mu}{r}\frac{\partial\omega}{\partial\theta} = \beta^*\frac{\partial\tau}{\partial r} \tag{18.75}$$

$$(\lambda^* + 2\mu)\frac{1}{r}\frac{\partial e}{\partial\theta} + 2\mu\frac{\partial\omega}{\partial r} = \frac{\beta^*}{r}\frac{\partial\tau}{\partial\theta} \tag{18.76}$$

Eliminating ω from Eqs. (18.75) and (18.76) by $\frac{\partial}{\partial r}[(18.75)\times r] + \frac{\partial}{\partial\theta}[(18.76)]$, we get

$$(\lambda^* + 2\mu)\left[\frac{\partial}{\partial r}\left(r\frac{\partial e}{\partial r}\right) + \frac{1}{r}\frac{\partial^2 e}{\partial\theta^2}\right] = \beta^*\left[\frac{\partial}{\partial r}\left(r\frac{\partial\tau}{\partial r}\right) + \frac{1}{r}\frac{\partial^2\tau}{\partial\theta^2}\right] \tag{18.77}$$

Simplification of Eq. (18.77) yields

$$\nabla^2\left(e - \frac{\beta^*}{\lambda^* + 2\mu}\tau\right) = 0 \;\rightarrow\; \nabla^2(e - K\tau) = 0 \tag{18.78}$$

Calculating the operation of $\frac{\partial}{\partial r}[(18.76)\times r] - \frac{\partial}{\partial\theta}[(18.75)]$, we get

$$2\mu\Big[\frac{\partial}{\partial r}\Big(r\frac{\partial \omega}{\partial r}\Big) + \frac{1}{r}\frac{\partial^2 \omega}{\partial \theta^2}\Big] = 0 \ \rightarrow\ \nabla^2\omega = 0 \tag{18.79}$$

Equations (18.78) and (18.79) are satisfied from Eqs. (18.72) and (18.74).

Problem 18.7. Derive Eq. (18.26).

Solution. The definition of Michell's function M is from Eq. (18.25)

$$M = -\int(\varphi + z\psi)dz \tag{18.80}$$

Application of the Laplacian operator to Eq. (18.80) gives

$$\begin{aligned}\nabla^2 M &= -\Big(\frac{\partial^2}{\partial r^2} + \frac{1}{r}\frac{\partial}{\partial r} + \frac{\partial^2}{\partial z^2}\Big)\int(\varphi + z\psi)dz \\ &= -\int\Big[\Big(\frac{\partial^2}{\partial r^2} + \frac{1}{r}\frac{\partial}{\partial r}\Big)\varphi + z\Big(\frac{\partial^2}{\partial r^2} + \frac{1}{r}\frac{\partial}{\partial r}\Big)\psi\Big]dz \\ &\quad -\Big(\frac{\partial\varphi}{\partial z} + z\frac{\partial\psi}{\partial z} + \psi\Big)\end{aligned} \tag{18.81}$$

Since the functions φ, ψ are harmonic functions, Eq. (18.81) reduces to

$$\begin{aligned}\nabla^2 M &= \int\Big(\frac{\partial^2\varphi}{\partial z^2} + z\frac{\partial^2\psi}{\partial z^2}\Big)dz - \Big(\frac{\partial\varphi}{\partial z} + z\frac{\partial\psi}{\partial z} + \psi\Big) \\ &= \Big(\frac{\partial\varphi}{\partial z} + z\frac{\partial\psi}{\partial z} - \psi\Big) - \Big(\frac{\partial\varphi}{\partial z} + z\frac{\partial\psi}{\partial z} + \psi\Big) = -2\psi\end{aligned} \tag{18.82}$$

Therefore, we get from Eq. (18.82)

$$\nabla^2\nabla^2 M = -2\nabla^2\psi = 0 \tag{18.83}$$

Substitution of Eq. (18.80) into Eq. (18.22) yields

$$\begin{aligned}u_r &= \frac{\partial\Phi}{\partial r} + \frac{\partial\varphi}{\partial r} + z\frac{\partial\psi}{\partial r} = \frac{\partial\Phi}{\partial r} + \frac{\partial^2}{\partial r\partial z}\int(\varphi + z\psi)dz = \frac{\partial\Phi}{\partial r} - \frac{\partial^2 M}{\partial r\partial z} \\ u_z &= \frac{\partial\Phi}{\partial z} + \frac{\partial\varphi}{\partial z} + z\frac{\partial\psi}{\partial z} - (3 - 4\nu)\psi \\ &= \frac{\partial\Phi}{\partial z} - 4(1-\nu)\psi + \Big(\frac{\partial\varphi}{\partial z} + z\frac{\partial\psi}{\partial z} + \psi\Big) \\ &= \frac{\partial\Phi}{\partial z} + 2(1-\nu)\nabla^2 M - \frac{\partial^2 M}{\partial z^2}\end{aligned} \qquad \text{(Answer)}$$

Problem 18.8. Derive Eq. (18.27).

Solution. The strains are expressed by Goodier's thermoelastic potential Φ and Michell's function M

$$\begin{aligned}
\varepsilon_{rr} &= \frac{\partial u_r}{\partial r} = \frac{\partial^2 \Phi}{\partial r^2} - \frac{\partial^3 M}{\partial r^2 \partial z} \\
\varepsilon_{\theta\theta} &= \frac{u_r}{r} = \frac{1}{r}\frac{\partial \Phi}{\partial r} - \frac{1}{r}\frac{\partial^2 M}{\partial r \partial z} \\
\varepsilon_{zz} &= \frac{\partial u_z}{\partial z} = \frac{\partial^2 \Phi}{\partial z^2} + 2(1-\nu)\frac{\partial \nabla^2 M}{\partial z} - \frac{\partial^3 M}{\partial z^3} \\
\varepsilon_{rz} &= \frac{\partial^2 \Phi}{\partial r \partial z} + (1-\nu)\frac{\partial \nabla^2 M}{\partial r} - \frac{\partial^3 M}{\partial r \partial z^2} \\
e &= \varepsilon_{rr} + \varepsilon_{\theta\theta} + \varepsilon_{zz} = \nabla^2 \Phi + (1-2\nu)\frac{\partial}{\partial z}\nabla^2 M \\
&= K\tau + (1-2\nu)\frac{\partial}{\partial z}\nabla^2 M
\end{aligned} \tag{18.84}$$

Therefore, Hooke's law gives

$$\begin{aligned}
\sigma_{rr} &= 2\mu\varepsilon_{rr} + \lambda e - \beta\tau \\
&= 2\mu\Big(\frac{\partial^2 \Phi}{\partial r^2} - \frac{\partial^3 M}{\partial r^2 \partial z}\Big) + \lambda\Big[K\tau + (1-2\nu)\frac{\partial}{\partial z}\nabla^2 M\Big] - \beta\tau \\
&= 2G\Big[\frac{\partial^2 \Phi}{\partial r^2} - K\tau + \frac{\partial}{\partial z}\Big(\nu\nabla^2 M - \frac{\partial^2 M}{\partial r^2}\Big)\Big] \\
\sigma_{\theta\theta} &= 2\mu\varepsilon_{\theta\theta} + \lambda e - \beta\tau \\
&= 2\mu\Big(\frac{1}{r}\frac{\partial \Phi}{\partial r} - \frac{1}{r}\frac{\partial^2 M}{\partial r \partial z}\Big) + \lambda\Big[K\tau + (1-2\nu)\frac{\partial}{\partial z}\nabla^2 M\Big] - \beta\tau \\
&= 2G\Big[\frac{1}{r}\frac{\partial \Phi}{\partial r} - K\tau + \frac{\partial}{\partial z}\Big(\nu\nabla^2 M - \frac{1}{r}\frac{\partial M}{\partial r}\Big)\Big] \\
\sigma_{zz} &= 2\mu\varepsilon_{zz} + \lambda e - \beta\tau \\
&= 2\mu\Big[\frac{\partial^2 \Phi}{\partial z^2} + 2(1-\nu)\frac{\partial \nabla^2 M}{\partial z} - \frac{\partial^3 M}{\partial z^3}\Big] \\
&\quad + \lambda\Big[K\tau + (1-2\nu)\frac{\partial}{\partial z}\nabla^2 M\Big] - \beta\tau \\
&= 2G\Big\{\frac{\partial^2 \Phi}{\partial z^2} - K\tau + \frac{\partial}{\partial z}\Big[(2-\nu)\nabla^2 M - \frac{\partial^2 M}{\partial z^2}\Big]\Big\} \\
\sigma_{rz} &= 2G\varepsilon_{rz} = 2G\Big\{\frac{\partial^2 \Phi}{\partial r \partial z} + \frac{\partial}{\partial r}\Big[(1-\nu)\nabla^2 M - \frac{\partial^2 M}{\partial z^2}\Big]\Big\}
\end{aligned} \tag{Answer}$$

where

$$\lambda K - \beta = \frac{\lambda \beta}{\lambda + 2\mu} - \beta = -2\mu \frac{\beta}{\lambda + 2\mu} = -2\mu K = 2GK$$
$$\lambda(1-2\nu) = \frac{2\nu\mu}{1-2\nu}(1-2\nu) = 2\nu\mu = 2G\nu \tag{18.85}$$

Problem 18.9. We assume that the steady temperature change τ in a long circular cylinder is given by

$$\tau = T - T_i = \int_0^\infty A(s) I_0(sr) \cos sz\, ds \tag{18.86}$$

where

$$A(s) = \frac{2}{\pi} \frac{h \int_0^\infty [T_a(z) - T_i] \cos sz\, dz}{h I_0(sa) + \lambda s I_1(sa)} \tag{18.87}$$

$T_a(z)$ means the temperature of the surrounding medium, T_i is the constant initial temperature, h means the heat transfer coefficient, and λ denotes the heat conductivity. Then, find the thermal stresses by use of Goodier's thermoelastic potential Φ and Michell's function M.

Solution. Michell's function M and Goodier's thermoelastic potential Φ are

$$M = \int_0^\infty [B(s) I_0(sr) + C(s) r I_1(sr)] \sin szds$$
$$\Phi = \frac{K}{2} \int_0^\infty A(s) \frac{r}{s} I_1(sr) \cos szds \tag{18.88}$$

Since the displacements and the stresses are expressed by Eqs. (18.26) and (18.27), we first calculate following integrals

$$\frac{\partial M}{\partial r} = \int_0^\infty [B(s) s I_1(sr) + C(s) sr I_0(sr)] \sin szds$$
$$\frac{1}{r}\frac{\partial M}{\partial r} = \int_0^\infty [B(s) \frac{s}{r} I_1(sr) + C(s) s I_0(sr)] \sin szds$$
$$\frac{\partial^2 M}{\partial r^2} = \int_0^\infty \{B(s) s [s I_0(sr) - \frac{1}{r} I_1(sr)]$$
$$+ C(s) s [I_0(sr) + sr I_1(sr)]\} \sin szds$$
$$\frac{\partial^2 M}{\partial z^2} = -\int_0^\infty s^2 [B(s) I_0(sr) + C(s) r I_1(sr)] \sin szds$$
$$\nabla^2 M = 2 \int_0^\infty C(s) s I_0(sr) \sin szds$$
$$\frac{\partial \Phi}{\partial r} = \frac{K}{2} \int_0^\infty A(s) r I_0(sr) \cos szds$$

$$\frac{\partial^2\Phi}{\partial r^2} = \frac{K}{2}\int_0^\infty A(s)[I_0(sr) + rsI_1(sr)]\cos szds \tag{18.89}$$

Then, the displacements and the stresses are

$$\begin{aligned}
u_r &= \frac{\partial\Phi}{\partial r} - \frac{\partial^2 M}{\partial r\partial z}\\
&= \int_0^\infty \Big[\frac{K}{2}A(s)rI_0(sr) - s^2B(s)I_1(sr) - s^2C(s)rI_0(sr)\Big]\cos szds\\
u_z &= \frac{\partial\Phi}{\partial z} + 2(1-\nu)\nabla^2 M - \frac{\partial^2 M}{\partial z^2}\\
&= \int_0^\infty \Big\{-\frac{K}{2}A(s)rI_1(sr) + s^2B(s)I_0(sr)\\
&\quad + sC(s)[4(1-\nu)I_0(sr) + srI_1(sr)]\Big\}\sin szds\\
\sigma_{rr} &= 2G\Big[\frac{\partial^2\Phi}{\partial r^2} - K\tau + \frac{\partial}{\partial z}\Big(\nu\nabla^2 M - \frac{\partial^2 M}{\partial r^2}\Big)\Big]\\
&= 2G\int_0^\infty \Big\{-\frac{K}{2}A(s)[I_0(sr) - rI_1(sr)] + s^3B(s)\Big[\frac{1}{sr}I_1(sr) - I_0(sr)\Big]\\
&\quad - s^2C(s)[(1-2\nu)I_0(sr) + srI_1(sr)]\Big\}\cos szds\\
\sigma_{\theta\theta} &= 2G\Big[\frac{1}{r}\frac{\partial^2\Phi}{\partial r} - K\tau + \frac{\partial}{\partial z}\Big(\nu\nabla^2 M - \frac{1}{r}\frac{\partial M}{\partial r}\Big)\Big]\\
&= -2G\int_0^\infty \Big[\frac{K}{2}A(s)I_0(sr)\\
&\quad + s^3B(s)\frac{I_1(sr)}{sr} + s^2C(s)(1-2\nu)I_0(sr)\Big]\cos szds\\
\sigma_{zz} &= 2G\Big\{\frac{\partial^2\Phi}{\partial z^2} - K\tau + \frac{\partial}{\partial z}\Big[(2-\nu)\nabla^2 M - \frac{\partial^2 M}{\partial z^2}\Big]\Big\}\\
&= 2G\int_0^\infty \Big\{-KA(s)[I_0(sr) + \frac{1}{2}srI_1(sr)] + s^3B(s)I_0(sr)\\
&\quad + s^2C(s)[2(2-\nu)I_0(sr) + srI_1(sr)]\Big\}\cos szds\\
\sigma_{rz} &= 2G\Big\{\frac{\partial^2\Phi}{\partial r\partial z} + \frac{\partial}{\partial r}\Big[(1-\nu)\nabla^2 M - \frac{\partial^2 M}{\partial z^2}\Big]\Big\}\\
&= 2G\int_0^\infty \Big\{-\frac{K}{2}A(s)srI_0(sr) + s^3B(s)I_1(sr)\\
&\quad + s^2C(s)[2(1-\nu)I_1(sr) + srI_0(sr)]\Big\}\sin szds
\end{aligned} \tag{18.90}$$

The boundary conditions of the traction free surface are

$$\sigma_{rr} = 0, \quad \sigma_{rz} = 0 \quad \text{on} \quad r = a \tag{18.91}$$

The boundary conditions (18.91) give $B(s)$ and $C(s)$ as follows:

$$\begin{aligned} B(s) &= (1-\nu)KA(s)\frac{a}{s^2}\frac{saI_0^2(sa) - saI_1^2(sa) + I_0(sa)I_1(sa)}{(2-2\nu+s^2a^2)I_1^2(sa) - s^2a^2I_0^2(sa)} \\ C(s) &= KA(s)\frac{a^2}{2}\frac{I_1^2(sa) - I_0^2(sa)}{(2-2\nu+s^2a^2)I_1^2(sa) - s^2a^2I_0^2(sa)} \end{aligned} \tag{18.92}$$

Substitution of Eq. (18.92) into Eq. (18.90) gives the displacements and the thermal stresses.

Problem 18.10. Derive a biharmonic function M in a cylindrical coordinate system (r, θ, z).

Solution. The biharmonic function M in a cylindrical coordinate system may be expressed as $M = M_c + M_p$, where M_c and M_p satisfy the equations

$$\nabla^2 M_c = 0, \quad \nabla^2 M_p = L, \quad \nabla^2 L = 0 \tag{18.93}$$

Since two functions M_c and L are harmonic functions, we first obtain the harmonic functions by use of the method of separation of variables. We assume that the harmonic function may be expressed by the product of three functions, each of only one variable

$$L(r, \theta, z) = f(r)g(\theta)h(z) \tag{18.94}$$

Substitution of Eq. (18.94) into the third equation in Eq. (18.93) gives the governing equations for functions $f(r)$, $g(\theta)$ and $h(z)$

$$\begin{aligned} &\frac{d^2 f(r)}{dr^2} + \frac{1}{r}\frac{df(r)}{dr} - \left(a^2 + \frac{n^2}{r^2}\right) f(r) = 0 \\ &\frac{d^2 g(\theta)}{d\theta^2} + n^2 g(\theta) = 0 \\ &\frac{d^2 h(z)}{dz^2} + a^2 h(z) = 0 \end{aligned} \tag{18.95}$$

in which a is an arbitrary constant, and n is the integer. Therefore, the linearly independent solutions of Eq. (18.95) are

$$\begin{aligned} f(r) &= \begin{pmatrix} 1 \\ \ln r \end{pmatrix} \qquad \text{for} \quad n = a = 0 \\ f(r) &= \begin{pmatrix} r^n \\ r^{-n} \end{pmatrix} \qquad \text{for} \quad n \neq 0, \ a = 0 \end{aligned}$$

$$f(r) = \begin{pmatrix} I_n(ar) \\ K_n(ar) \end{pmatrix} \quad \text{for} \quad a \neq 0$$

$$g(\theta) = \begin{pmatrix} 1 \\ \theta \end{pmatrix} \quad \text{for} \quad n = 0, \quad g(\theta) = \begin{pmatrix} \sin n\theta \\ \cos n\theta \end{pmatrix} \quad \text{for} \quad n \neq 0$$

$$h(z) = \begin{pmatrix} 1 \\ z \end{pmatrix} \quad \text{for} \quad a = 0, \quad h(z) = \begin{pmatrix} \sin az \\ \cos az \end{pmatrix} \quad \text{for} \quad a \neq 0 \tag{18.96}$$

Next, we consider a particular solution M_p which satisfies the equation

$$\nabla^2 M_p = f(r)g(\theta)h(z) \tag{18.97}$$

The particular solution M_p is assumed to be expressed by the product of three functions, each of only one variable

$$[\text{Case 1}] \quad M_p(r, \theta, z) = F(r)g(\theta)h(z) \tag{18.98}$$

$$[\text{Case 2}] \quad M_p(r, \theta, z) = f(r)G(\theta)h(z) \tag{18.99}$$

$$[\text{Case 3}] \quad M_p(r, \theta, z) = f(r)g(\theta)H(z) \tag{18.100}$$

[Case 1] Substitution of Eq. (18.98) into Eq. (18.97) gives

$$\frac{d^2F(r)}{dr^2} + \frac{1}{r}\frac{dF(r)}{dr} - \left(a^2 + \frac{n^2}{r^2}\right)F(r) = f(r) \tag{18.101}$$

We get a particular solution $F(r)$:

$$F(r) = \begin{pmatrix} r^2/4 \\ r^2(\ln r - 1)/4 \end{pmatrix}$$

$$\text{when} \quad f(r) = \begin{pmatrix} 1 \\ \ln r \end{pmatrix} \quad \text{for} \quad n = a = 0$$

$$F(r) = \begin{pmatrix} r^3/8 \\ r \ln r/2 \end{pmatrix}$$

$$\text{when} \quad f(r) = \begin{pmatrix} r \\ r^{-1} \end{pmatrix} \quad \text{for} \quad n = 1, \ a = 0$$

$$F(r) = \begin{pmatrix} r^{n+2}/(4n+4) \\ -r^{-n+2}/(4n-4) \end{pmatrix}$$

$$\text{when} \quad f(r) = \begin{pmatrix} r^n \\ r^{-n} \end{pmatrix} \quad \text{for} \quad n \geq 2, \ a = 0$$

$$F(r) = \begin{pmatrix} [arI_{n+1}(ar) + nI_n(ar)]/(2a^2) \\ [-arK_{n+1}(ar) + nK_n(ar)]/(2a^2) \end{pmatrix}$$

$$\text{when} \quad f(r) = \begin{pmatrix} I_n(ar) \\ K_n(ar) \end{pmatrix} \quad \text{for} \quad a \neq 0 \tag{18.102}$$

[Case 2] Substitution of Eq. (18.99) into Eq. (18.97) gives

$$\frac{d^2G(\theta)}{d\theta^2} + n^2G(\theta) = r^2g(\theta) \tag{18.103}$$

The expression on the left-hand side in Eq. (18.103) is a function of θ. However, the expression on the right-hand side in Eq. (18.103) is a function of θ and r. Because of this reason, the assumption of [Case 2] is not acceptable.

[Case 3] Substitution of Eq. (18.100) into Eq. (18.97) gives

$$\frac{d^2H(z)}{dz^2} + a^2H(z) = h(z) \tag{18.104}$$

and we get a particular solution $H(z)$:

$$H(z) = \begin{pmatrix} z^2/2 \\ z^3/6 \end{pmatrix}$$

$$\text{when} \quad h(z) = \begin{pmatrix} 1 \\ z \end{pmatrix} \quad \text{for} \quad a = 0$$

$$H(z) = \begin{pmatrix} z \sin az/(2a) \\ -z \cos az/(2a) \end{pmatrix}$$

$$\text{when} \quad h(z) = \begin{pmatrix} \cos az \\ \sin az \end{pmatrix} \quad \text{for} \quad a \neq 0 \tag{18.105}$$

Next, we consider the second case in which the harmonic function can be expressed by

$$L(r, \theta, z) = f(r)g(\theta)h(z) \tag{18.106}$$

where the governing equations of functions $f(r)$, $g(\theta)$ and $h(z)$ are

$$\begin{aligned} &\frac{d^2f(r)}{dr^2} + \frac{1}{r}\frac{df(r)}{dr} + \left(a^2 - \frac{n^2}{r^2}\right)f(r) = 0 \\ &\frac{d^2g(\theta)}{d\theta^2} + n^2g(\theta) = 0 \\ &\frac{d^2h(z)}{dz^2} - a^2h(z) = 0 \end{aligned} \tag{18.107}$$

The linearly independent solutions of Eq. (18.107) are

$$f(r) = \begin{pmatrix} 1 \\ \ln r \end{pmatrix} \quad \text{for} \quad n = a = 0$$

$$f(r) = \begin{pmatrix} r^n \\ r^{-n} \end{pmatrix} \quad \text{for} \quad n \neq 0, \ a = 0$$

$$f(r) = \begin{pmatrix} J_n(ar) \\ Y_n(ar) \end{pmatrix} \quad \text{for} \quad a \neq 0$$

$$g(\theta) = \begin{pmatrix} 1 \\ \theta \end{pmatrix} \quad \text{for} \quad n = 0, \quad g(\theta) = \begin{pmatrix} \sin n\theta \\ \cos n\theta \end{pmatrix} \quad \text{for} \quad n \neq 0$$

$$h(z) = \begin{pmatrix} 1 \\ z \end{pmatrix} \quad \text{for} \quad a = 0, \quad h(z) = \begin{pmatrix} e^{az} \\ e^{-az} \end{pmatrix} \quad \text{for} \quad a \neq 0 \quad (18.108)$$

Next, we consider a particular solution M_p of Eq. (18.97). A particular solution M_p is assumed to be expressed by the product of three functions, each of only one variable

$$\text{[Case 4]} \quad M_p(r, \theta, z) = F(r)g(\theta)h(z) \quad (18.109)$$

$$\text{[Case 5]} \quad M_p(r, \theta, z) = f(r)G(\theta)h(z) \quad (18.110)$$

$$\text{[Case 6]} \quad M_p(r, \theta, z) = f(r)g(\theta)H(z) \quad (18.111)$$

[Case 4] Substitution of Eq. (18.18.109) into Eq. (18.97) gives

$$\frac{d^2F(r)}{dr^2} + \frac{1}{r}\frac{dF(r)}{dr} + \left(a^2 - \frac{n^2}{r^2}\right)F(r) = f(r) \quad (18.112)$$

We get a particular solution $F(r)$:

$$F(r) = \begin{pmatrix} r^2/4 \\ r^2(\ln r - 1)/4 \end{pmatrix}$$

$$\text{when} \quad f(r) = \begin{pmatrix} 1 \\ \ln r \end{pmatrix} \quad \text{for} \quad n = a = 0$$

$$F(r) = \begin{pmatrix} r^3/8 \\ r \ln r/2 \end{pmatrix}$$

$$\text{when} \quad f(r) = \begin{pmatrix} r \\ r^{-1} \end{pmatrix} \quad \text{for} \quad n = 1, \ a = 0$$

$$F(r) = \begin{pmatrix} r^{n+2}/(4n+4) \\ -r^{-n+2}/(4n-4) \end{pmatrix}$$

$$\text{when} \quad f(r) = \begin{pmatrix} r^n \\ r^{-n} \end{pmatrix} \quad \text{for} \quad n \geq 2, \ a = 0$$

$$F(r) = \begin{pmatrix} [arJ_{n+1}(ar) - nJ_n(ar)]/(2a^2) \\ [arY_{n+1}(ar) - nY_n(ar)]/(2a^2) \end{pmatrix}$$

$$\text{when} \quad f(r) = \begin{pmatrix} J_n(ar) \\ Y_n(ar) \end{pmatrix} \quad \text{for} \quad a \neq 0 \tag{18.113}$$

[Case 5] Substitution of Eq. (18.110) into Eq. (18.97) gives

$$\frac{d^2G(\theta)}{d\theta^2} + n^2G(\theta) = r^2g(\theta) \tag{18.114}$$

The expression on the left-hand side in Eq. (18.114) is a function of θ. However, the expression on the right-hand side in Eq. (18.114) is a function of θ and r. Because of this, the assumption of [Case 5] is not acceptable.

[Case 6] Substitution of Eq. (18.111) into Eq. (18.97) gives

$$\frac{d^2H(z)}{dz^2} - a^2H(z) = h(z) \tag{18.115}$$

and we get a particular solution $H(z)$:

$$H(z) = \begin{pmatrix} z^2/2 \\ z^3/6 \end{pmatrix}$$

$$\text{when} \quad h(z) = \begin{pmatrix} 1 \\ z \end{pmatrix} \quad \text{for} \quad a = 0$$

$$H(z) = \begin{pmatrix} ze^{az}/(2a) \\ -ze^{-az}/(2a) \end{pmatrix}$$

$$\text{when} \quad h(z) = \begin{pmatrix} e^{az} \\ e^{-az} \end{pmatrix} \quad \text{for} \quad a \neq 0 \tag{18.116}$$

Finally, the particular solutions of the Laplace's equations in Eq. (18.35) in a cylindrical coordinate system are

$$\begin{pmatrix} 1 \\ \ln r \end{pmatrix} \begin{pmatrix} 1 \\ \theta \end{pmatrix} \begin{pmatrix} 1 \\ z \end{pmatrix}, \qquad \begin{pmatrix} J_0(ar) \\ Y_0(ar) \end{pmatrix} \begin{pmatrix} 1 \\ \theta \end{pmatrix} \begin{pmatrix} e^{az} \\ e^{-az} \end{pmatrix}$$

$$\begin{pmatrix} I_0(ar) \\ K_0(ar) \end{pmatrix} \begin{pmatrix} 1 \\ \theta \end{pmatrix} \begin{pmatrix} \sin az \\ \cos az \end{pmatrix}, \quad \begin{pmatrix} r^n \\ r^{-n} \end{pmatrix} \begin{pmatrix} \cos n\theta \\ \sin n\theta \end{pmatrix} \begin{pmatrix} 1 \\ z \end{pmatrix}$$

$$\begin{pmatrix} J_n(ar) \\ Y_n(ar) \end{pmatrix} \begin{pmatrix} \cos n\theta \\ \sin n\theta \end{pmatrix} \begin{pmatrix} e^{az} \\ e^{-az} \end{pmatrix}, \begin{pmatrix} I_n(ar) \\ K_n(ar) \end{pmatrix} \begin{pmatrix} \cos n\theta \\ \sin n\theta \end{pmatrix} \begin{pmatrix} \cos az \\ \sin az \end{pmatrix} \quad \text{(Answer)} \tag{18.117}$$

where a is an arbitrary constant, $J_n(ar)$ and $Y_n(ar)$ are Bessel functions, $I_n(ar)$ and $K_n(ar)$ are the modified Bessel functions. Moreover, functions $\begin{pmatrix} \sinh az \\ \cosh az \end{pmatrix}$ can be used instead of $\begin{pmatrix} e^{az} \\ e^{-az} \end{pmatrix}$.

The particular solutions of Eq. (18.40) in a cylindrical coordinate system are given

$$
\begin{aligned}
&\begin{pmatrix} 1 \\ \ln r \\ r^2 \\ r^2 \ln r \end{pmatrix} \begin{pmatrix} 1 \\ \theta \end{pmatrix} \begin{pmatrix} 1 \\ z \end{pmatrix}, \qquad \begin{pmatrix} 1 \\ \ln r \end{pmatrix} \begin{pmatrix} 1 \\ \theta \end{pmatrix} \begin{pmatrix} z^2 \\ z^3 \end{pmatrix} \\
&\begin{pmatrix} J_0(ar) \\ Y_0(ar) \\ rJ_1(ar) \\ rY_1(ar) \end{pmatrix} \begin{pmatrix} 1 \\ \theta \end{pmatrix} \begin{pmatrix} e^{az} \\ e^{-az} \end{pmatrix}, \qquad \begin{pmatrix} J_0(ar) \\ Y_0(ar) \end{pmatrix} \begin{pmatrix} 1 \\ \theta \end{pmatrix} \begin{pmatrix} ze^{az} \\ ze^{-az} \end{pmatrix} \\
&\begin{pmatrix} I_0(ar) \\ K_0(ar) \\ rI_1(ar) \\ rK_1(ar) \end{pmatrix} \begin{pmatrix} 1 \\ \theta \end{pmatrix} \begin{pmatrix} \sin az \\ \cos az \end{pmatrix}, \qquad \begin{pmatrix} I_0(ar) \\ K_0(ar) \end{pmatrix} \begin{pmatrix} 1 \\ \theta \end{pmatrix} \begin{pmatrix} z \sin az \\ z \cos az \end{pmatrix} \\
&\begin{pmatrix} 0 \\ r \end{pmatrix} \begin{pmatrix} \theta \cos\theta \\ \theta \sin\theta \end{pmatrix} \begin{pmatrix} 1 \\ z \end{pmatrix}, \qquad \begin{pmatrix} r^3 \\ r \ln r \end{pmatrix} \begin{pmatrix} \cos\theta \\ \sin\theta \end{pmatrix} \begin{pmatrix} 1 \\ z \end{pmatrix} \\
&\begin{pmatrix} r^n \\ r^{-n} \\ r^{n+2} \\ r^{-n+2} \end{pmatrix} \begin{pmatrix} \cos n\theta \\ \sin n\theta \end{pmatrix} \begin{pmatrix} 1 \\ z \end{pmatrix}, \qquad \begin{pmatrix} r^n \\ r^{-n} \end{pmatrix} \begin{pmatrix} \cos n\theta \\ \sin n\theta \end{pmatrix} \begin{pmatrix} z^2 \\ z^3 \end{pmatrix} \\
&\begin{pmatrix} J_n(ar) \\ Y_n(ar) \\ rJ_{n+1}(ar) \\ rY_{n+1}(ar) \end{pmatrix} \begin{pmatrix} \cos n\theta \\ \sin n\theta \end{pmatrix} \begin{pmatrix} e^{az} \\ e^{-az} \end{pmatrix} \\
&\begin{pmatrix} J_n(ar) \\ Y_n(ar) \end{pmatrix} \begin{pmatrix} \cos n\theta \\ \sin n\theta \end{pmatrix} \begin{pmatrix} ze^{az} \\ ze^{-az} \end{pmatrix} \\
&\begin{pmatrix} I_n(ar) \\ K_n(ar) \\ rI_{n+1}(ar) \\ rK_{n+1}(ar) \end{pmatrix} \begin{pmatrix} \cos n\theta \\ \sin n\theta \end{pmatrix} \begin{pmatrix} \cos az \\ \sin az \end{pmatrix} \\
&\begin{pmatrix} I_n(ar) \\ K_n(ar) \end{pmatrix} \begin{pmatrix} \cos n\theta \\ \sin n\theta \end{pmatrix} \begin{pmatrix} z \cos az \\ z \sin az \end{pmatrix} \qquad \text{(Answer)}
\end{aligned}
$$

In these equations, the notation for the product of three one-column matrices is explained by Eqs. (16.97) and (16.98).

Chapter 19
Thermal Stresses in Spherical Bodies

In this chapter the thermal stresses in spherical bodies are presented. First, one-dimensional problems for a solid and a hollow sphere are discussed. Next, two-dimensional axisymmetric problems are treated by Goodier's thermoelastic potential and the Boussinesq harmonic functions. Problems and solutions for thermal stresses in a solid and a hollow cylinder subjected to the steady and the transient temperature field are presented. [See also Chap. 24.]

19.1 One-Dimensional Problems in Spherical Bodies

The equilibrium equation without body force for a one-dimensional problem in a spherical coordinate system is obtained from Eq. (16.52)

$$\frac{d\sigma_{rr}}{dr} + \frac{2}{r}(\sigma_{rr} - \sigma_{\theta\theta}) = 0 \tag{19.1}$$

Hooke's law is from Eq. (16.59)

$$\begin{aligned} \sigma_{rr} &= 2\mu\epsilon_{rr} + \lambda e - \beta\tau \\ \sigma_{\theta\theta} &= \sigma_{\phi\phi} = 2\mu\epsilon_{\theta\theta} + \lambda e - \beta\tau \end{aligned} \tag{19.2}$$

where $e = \epsilon_{rr} + 2\epsilon_{\theta\theta}$.

The strain-displacement relations are

$$\epsilon_{rr} = \frac{du_r}{dr}, \qquad \epsilon_{\theta\theta} = \epsilon_{\phi\phi} = \frac{u_r}{r} \tag{19.3}$$

M. Reza Eslami et al., *Theory of Elasticity and Thermal Stresses*, Solid Mechanics and Its Applications 197, DOI: 10.1007/978-94-007-6356-2_19,

The stress components in terms of the displacement component u_r are

$$\begin{aligned}\sigma_{rr} &= \frac{E}{(1+\nu)(1-2\nu)}\left[(1-\nu)\frac{du_r}{dr} + 2\nu\frac{u_r}{r} - (1+\nu)\alpha\tau\right]\\ \sigma_{\theta\theta} = \sigma_{\phi\phi} &= \frac{E}{(1+\nu)(1-2\nu)}\left[\nu\frac{du_r}{dr} + \frac{u_r}{r} - (1+\nu)\alpha\tau\right]\end{aligned} \tag{19.4}$$

The equilibrium equation in terms of the displacement component u_r is

$$\frac{d}{dr}\left[\frac{1}{r^2}\frac{d}{dr}\left(r^2 u_r\right)\right] = \frac{1+\nu}{1-\nu}\alpha\frac{d\tau}{dr} \tag{19.5}$$

The general solution of Eq. (19.5) is

$$u_r = \frac{1+\nu}{1-\nu}\alpha\frac{1}{r^2}\int \tau r^2\,dr + C_1 r + C_2\frac{1}{r^2} \tag{19.6}$$

where C_1 and C_2 are constants.

The stresses are expressed by

$$\begin{aligned}\sigma_{rr} &= \frac{E}{(1+\nu)(1-2\nu)}\left[-2\frac{(1+\nu)(1-2\nu)}{1-\nu}\alpha\frac{1}{r^3}\int \tau r^2\,dr\right.\\ &\qquad \left. + (1+\nu)C_1 - 2(1-2\nu)C_2\frac{1}{r^3}\right]\\ \sigma_{\theta\theta} &= \sigma_{\phi\phi}\\ &= \frac{E}{(1+\nu)(1-2\nu)}\left[\frac{(1+\nu)(1-2\nu)}{1-\nu}\alpha\left(\frac{1}{r^3}\int \tau r^2\,dr - \tau\right)\right.\\ &\qquad \left. + (1+\nu)C_1 + (1-2\nu)C_2\frac{1}{r^3}\right]\end{aligned} \tag{19.7}$$

The displacement and the thermal stresses in a solid sphere of radius a with free traction are

$$\begin{aligned}u_r &= \frac{\alpha}{1-\nu}\left[(1+\nu)\frac{1}{r^2}\int_0^r \tau r^2\,dr + 2(1-2\nu)\frac{r}{a^3}\int_0^a \tau r^2\,dr\right]\\ \sigma_{rr} &= \frac{\alpha E}{1-\nu}\left[\frac{2}{a^3}\int_0^a \tau r^2\,dr - \frac{2}{r^3}\int_0^r \tau r^2\,dr\right]\\ \sigma_{\theta\theta} = \sigma_{\phi\phi} &= \frac{\alpha E}{1-\nu}\left[\frac{2}{a^3}\int_0^a \tau r^2\,dr + \frac{1}{r^3}\int_0^r \tau r^2\,dr - \tau\right]\end{aligned} \tag{19.8}$$

The displacement and the thermal stresses in a hollow sphere of inner radius a and outer radius b with free traction are

$$\begin{aligned} u_r &= \frac{1+\nu}{1-\nu}\alpha\Big[\frac{1}{r^2}\int_a^r \tau r^2\,dr \\ &\quad + \frac{2(1-2\nu)}{1+\nu}\frac{r}{b^3-a^3}\int_a^b \tau r^2\,dr + \frac{a^3}{b^3-a^3}\frac{1}{r^2}\int_a^b \tau r^2\,dr\Big] \\ \sigma_{rr} &= \frac{\alpha E}{1-\nu}\Big[\frac{2(r^3-a^3)}{r^3(b^3-a^3)}\int_a^b \tau r^2\,dr - \frac{2}{r^3}\int_a^r \tau r^2\,dr\Big] \\ \sigma_{\theta\theta} = \sigma_{\phi\phi} &= \frac{\alpha E}{1-\nu}\Big[\frac{2r^3+a^3}{r^3(b^3-a^3)}\int_a^b \tau r^2\,dr + \frac{1}{r^3}\int_a^r \tau r^2\,dr - \tau\Big] \end{aligned} \tag{19.9}$$

The displacement and the thermal stresses in an infinite space with a spherical cavity of radius a with free traction are

$$\begin{aligned} u_r &= \frac{1+\nu}{1-\nu}\alpha\frac{1}{r^2}\int_a^r \tau r^2\,dr \\ \sigma_{rr} &= -\frac{2\alpha E}{1-\nu}\frac{1}{r^3}\int_a^r \tau r^2\,dr \\ \sigma_{\theta\theta} &= \sigma_{\phi\phi} = \frac{\alpha E}{1-\nu}\Big(\frac{1}{r^3}\int_a^r \tau r^2\,dr - \tau\Big) \end{aligned} \tag{19.10}$$

19.2 Two-Dimensional Axisymmetric Problems

We now consider two-dimensional axisynetric problems of a spherical body. The equilibrium equations in the directions of r and θ are obtained form Eq. (16.52)

$$\begin{aligned} &\frac{\partial \sigma_{rr}}{\partial r} + \frac{1}{r}\frac{\partial \sigma_{\theta r}}{\partial \theta} + \frac{1}{r}\big(2\sigma_{rr} - \sigma_{\theta\theta} - \sigma_{\phi\phi} + \sigma_{\theta r}\cot\theta\big) + F_r = 0 \\ &\frac{\partial \sigma_{r\theta}}{\partial r} + \frac{1}{r}\frac{\partial \sigma_{\theta\theta}}{\partial \theta} + \frac{1}{r}\big[(\sigma_{\theta\theta} - \sigma_{\phi\phi})\cot\theta + 3\sigma_{r\theta}\big] + F_\theta = 0 \end{aligned} \tag{19.11}$$

The constitutive equations in a spherical coordinate system are from Eq. (16.58)

$$\begin{aligned} \epsilon_{rr} &= \frac{1}{E}\big[\sigma_{rr} - \nu\left(\sigma_{\theta\theta} + \sigma_{\phi\phi}\right)\big] + \alpha\tau = \frac{1}{2G}\Big(\sigma_{rr} - \frac{\nu}{1+\nu}\Theta\Big) + \alpha\tau \\ \epsilon_{\theta\theta} &= \frac{1}{E}\big[\sigma_{\theta\theta} - \nu\left(\sigma_{\phi\phi} + \sigma_{rr}\right)\big] + \alpha\tau = \frac{1}{2G}\Big(\sigma_{\theta\theta} - \frac{\nu}{1+\nu}\Theta\Big) + \alpha\tau \\ \epsilon_{\phi\phi} &= \frac{1}{E}\big[\sigma_{\phi\phi} - \nu\left(\sigma_{rr} + \sigma_{\theta\theta}\right)\big] + \alpha\tau = \frac{1}{2G}\Big(\sigma_{\phi\phi} - \frac{\nu}{1+\nu}\Theta\Big) + \alpha\tau \\ \epsilon_{r\theta} &= \frac{1}{2G}\sigma_{r\theta} \end{aligned} \tag{19.12}$$

where $\Theta = \sigma_{rr} + \sigma_{\theta\theta} + \sigma_{\phi\phi}$. Alternative forms are

$$\begin{aligned} \sigma_{rr} &= 2\mu\epsilon_{rr} + \lambda e - \beta\tau, \qquad \sigma_{\theta\theta} = 2\mu\epsilon_{\theta\theta} + \lambda e - \beta\tau \\ \sigma_{\phi\phi} &= 2\mu\epsilon_{\phi\phi} + \lambda e - \beta\tau, \qquad \sigma_{r\theta} = 2\mu\epsilon_{r\theta} \end{aligned} \tag{19.13}$$

where $e = \epsilon_{rr} + \epsilon_{\theta\theta} + \epsilon_{\phi\phi}$.

The components of strain for an axisymmetric deformation are from Eq. (16.54)

$$\begin{aligned} \epsilon_{rr} &= \frac{\partial u_r}{\partial r}, \qquad \epsilon_{\theta\theta} = \frac{u_r}{r} + \frac{1}{r}\frac{\partial u_\theta}{\partial \theta} \\ \epsilon_{\phi\phi} &= \frac{u_r}{r} + \cot\theta\frac{u_\theta}{r}, \qquad \epsilon_{r\theta} = \frac{1}{2}\left(\frac{1}{r}\frac{\partial u_r}{\partial \theta} + \frac{\partial u_\theta}{\partial r} - \frac{u_\theta}{r}\right) \\ e &= \frac{\partial u_r}{\partial r} + 2\frac{u_r}{r} + \cot\theta\frac{u_\theta}{r} + \frac{1}{r}\frac{\partial u_\theta}{\partial \theta} \end{aligned} \tag{19.14}$$

Substituting Eqs. (19.13) and (19.14) into Eq. (19.11), Navier's equations of thermoelasticity for axisymmetric problems may be expressed as

$$\begin{aligned} (\lambda + 2\mu)\frac{\partial e}{\partial r} - \frac{2\mu}{r\sin\theta}\frac{\partial(\omega_\phi \sin\theta)}{\partial \theta} - \beta\frac{\partial \tau}{\partial r} + F_r = 0 \\ (\lambda + 2\mu)\frac{1}{r}\frac{\partial e}{\partial \theta} + \frac{2\mu}{r}\frac{\partial(r\omega_\phi)}{\partial r} - \beta\frac{1}{r}\frac{\partial \tau}{\partial \theta} + F_\theta = 0 \end{aligned} \tag{19.15}$$

where

$$\omega_\phi = \frac{1}{2r}\left[\frac{\partial(ru_\theta)}{\partial r} - \frac{\partial u_r}{\partial \theta}\right] \tag{19.16}$$

The solution of Navier's equations (19.15) without body force for axisymmetric problems in a spherical coordinate system can be expressed, for example, by Goodier's thermoelastic potential Φ and the Boussinesq harmonic functions φ and ψ from Eq. (16.62)

$$\begin{aligned} u_r &= \frac{\partial \Phi}{\partial r} + \frac{\partial \varphi}{\partial r} + r\cos\theta\frac{\partial \psi}{\partial r} - (3 - 4\nu)\,\psi\cos\theta \\ u_\theta &= \frac{1}{r}\frac{\partial \Phi}{\partial \theta} + \frac{1}{r}\frac{\partial \varphi}{\partial \theta} + \cos\theta\frac{\partial \psi}{\partial \theta} + (3 - 4\nu)\,\psi\sin\theta \end{aligned} \tag{19.17}$$

where the three functions must satisfy the equations

$$\nabla^2\Phi = K\tau, \quad \nabla^2\varphi = 0, \quad \nabla^2\psi = 0 \tag{19.18}$$

in which

$$\nabla^2 = \frac{\partial^2}{\partial r^2} + \frac{2}{r}\frac{\partial}{\partial r} + \frac{1}{r^2}\frac{\partial^2}{\partial \theta^2} + \frac{1}{r^2}\cot\theta\frac{\partial}{\partial \theta}, \qquad K = \frac{1+\nu}{1-\nu}\alpha \tag{19.19}$$

Making use of Eq. (19.17), the strain components are represented as

$$\begin{aligned}
\epsilon_{rr} &= \frac{\partial^2 \Phi}{\partial r^2} + \frac{\partial^2 \varphi}{\partial r^2} + r\cos\theta \frac{\partial^2 \psi}{\partial r^2} - 2(1-2\nu)\cos\theta \frac{\partial \psi}{\partial r} \\
\epsilon_{\theta\theta} &= \frac{1}{r}\frac{\partial \Phi}{\partial r} + \frac{1}{r^2}\frac{\partial^2 \Phi}{\partial \theta^2} + \frac{1}{r}\frac{\partial \varphi}{\partial r} + \frac{1}{r^2}\frac{\partial^2 \varphi}{\partial \theta^2} \\
&\quad + \cos\theta \frac{\partial \psi}{\partial r} + \frac{1}{r}\cos\theta \frac{\partial^2 \psi}{\partial \theta^2} + 2(1-2\nu)\frac{1}{r}\sin\theta \frac{\partial \psi}{\partial \theta} \\
\epsilon_{\phi\phi} &= \frac{1}{r}\frac{\partial \Phi}{\partial r} + \cot\theta \frac{1}{r^2}\frac{\partial \Phi}{\partial \theta} + \frac{1}{r}\frac{\partial \varphi}{\partial r} + \cot\theta \frac{1}{r^2}\frac{\partial \varphi}{\partial \theta} \\
&\quad + \cos\theta \frac{\partial \psi}{\partial r} + \frac{\cos^2\theta}{\sin\theta}\frac{1}{r}\frac{\partial \psi}{\partial \theta} \\
\epsilon_{r\theta} &= \frac{\partial^2}{\partial r \partial \theta}\left(\frac{\Phi}{r}\right) + \frac{\partial^2}{\partial r \partial \theta}\left(\frac{\varphi}{r}\right) \\
&\quad + (1-2\nu)\sin\theta \frac{\partial \psi}{\partial r} + \cos\theta \frac{\partial^2 \psi}{\partial r \partial \theta} - 2(1-\nu)\frac{1}{r}\cos\theta \frac{\partial \psi}{\partial \theta} \\
e &= K\tau - 2(1-2\nu)\left(\cos\theta \frac{\partial \psi}{\partial r} - \frac{1}{r}\sin\theta \frac{\partial \psi}{\partial \theta}\right)
\end{aligned} \tag{19.20}$$

Substitution of Eq. (19.20) into Eq. (19.13) gives the stress components

$$\begin{aligned}
\sigma_{rr} &= 2G\Big[\frac{\partial^2 \Phi}{\partial r^2} + \frac{\partial^2 \varphi}{\partial r^2} + r\cos\theta \frac{\partial^2 \psi}{\partial r^2} - 2(1-\nu)\cos\theta \frac{\partial \psi}{\partial r} \\
&\quad + 2\nu \frac{1}{r}\sin\theta \frac{\partial \psi}{\partial \theta} - K\tau\Big] \\
\sigma_{\theta\theta} &= 2G\Big[\frac{1}{r}\frac{\partial \Phi}{\partial r} + \frac{1}{r^2}\frac{\partial^2 \Phi}{\partial \theta^2} + \frac{1}{r}\frac{\partial \varphi}{\partial r} + \frac{1}{r^2}\frac{\partial^2 \varphi}{\partial \theta^2} \\
&\quad + (1-2\nu)\cos\theta \frac{\partial \psi}{\partial r} + \frac{1}{r}\cos\theta \frac{\partial^2 \psi}{\partial \theta^2} + 2(1-\nu)\frac{1}{r}\sin\theta \frac{\partial \psi}{\partial \theta} - K\tau\Big] \\
\sigma_{\phi\phi} &= 2G\Big[\frac{1}{r}\frac{\partial \Phi}{\partial r} + \cot\theta \frac{1}{r^2}\frac{\partial \Phi}{\partial \theta} + \frac{1}{r}\frac{\partial \varphi}{\partial r} + \cot\theta \frac{1}{r^2}\frac{\partial \varphi}{\partial \theta} \\
&\quad + (1-2\nu)\cos\theta \frac{\partial \psi}{\partial r} + (\cos\theta\cot\theta + 2\nu\sin\theta)\frac{1}{r}\frac{\partial \psi}{\partial \theta} - K\tau\Big] \\
\sigma_{r\theta} &= 2G\Big[\frac{\partial^2}{\partial r \partial \theta}\left(\frac{\Phi}{r}\right) + \frac{\partial^2}{\partial r \partial \theta}\left(\frac{\varphi}{r}\right) \\
&\quad + (1-2\nu)\sin\theta \frac{\partial \psi}{\partial r} + \cos\theta \frac{\partial^2 \psi}{\partial r \partial \theta} - 2(1-\nu)\frac{1}{r}\cos\theta \frac{\partial \psi}{\partial \theta}\Big]
\end{aligned} \tag{19.21}$$

The particular solutions φ and ψ for the axisymmetric bodies can be expressed as

$$\begin{pmatrix} \varphi \\ \psi \end{pmatrix} = \begin{pmatrix} r^n P_n(\cos\theta) \\ r^{-n-1} P_n(\cos\theta) \end{pmatrix} \qquad (n = 0, 1, 2, \ldots) \tag{19.22}$$

where $P_n(\cos\theta)$ is the Legendre function of the first kind of order n.

19.3 Problems and Solutions Related to Thermal Stresses in Spherical Bodies

Problem 19.1. Find the displacement and the thermal stresses in a solid sphere of radius a with constant heat generation Q under a steady temperature field. Furthermore derive the displacement and the thermal stresses in a solid sphere with constant temperature T_a at $r = a$.

Solution. The one-dimensional steady state heat conduction equation with constant heat generation Q is

$$\frac{d^2T}{dr^2} + \frac{2}{r}\frac{dT}{dr} = -\frac{Q}{\lambda} \tag{19.23}$$

The general solution of Eq. (19.23) is

$$T = A + Br^{-1} + \left(-\frac{Q}{6\lambda}\right)r^2 \tag{19.24}$$

Since T is finite at $r = 0$, the temperature change from the initial temperature T_i reduces to

$$\tau = (A - T_i) + \left(-\frac{Q}{6\lambda}\right)r^2 \tag{19.25}$$

The following integral relation is obtained

$$\int \tau r^2\, dr = \frac{1}{3}(A - T_i)r^3 + \left(-\frac{Q}{30\lambda}\right)r^5 \tag{19.26}$$

Substitution of Eq. (19.26) into Eq. (19.8) gives the radial displacement and the thermal stresses as

$$\begin{aligned} u_r &= \alpha(A - T_i)r + \frac{\alpha}{1-\nu}\left(-\frac{Q}{30\lambda}\right)r[(1+\nu)r^2 + 2(1-2\nu)a^2] \\ \sigma_{rr} &= \frac{\alpha E}{1-\nu}\left(-\frac{Q}{15\lambda}\right)(a^2 - r^2) \\ \sigma_{\theta\theta} = \sigma_{\phi\phi} &= \frac{\alpha E}{1-\nu}\left(-\frac{Q}{15\lambda}\right)(a^2 - 2r^2) \end{aligned} \qquad \text{(Answer)} \tag{19.27}$$

The boundary condition of the temperature is

$$T = T_a \quad \text{at} \quad r = a \tag{19.28}$$

From Eq. (19.28), the unknown constant A is determined as

$$A = T_a - \left(-\frac{Q}{6\lambda}\right)a^2 \tag{19.29}$$

Then, the temperature change is

$$\tau = T_a - T_i + \left(-\frac{Q}{6\lambda}\right)(r^2 - a^2) \tag{19.30}$$

Substitution of Eq. (19.29) into Eq. (19.27) gives the displacement and the thermal stresses

$$\begin{aligned} u_r &= \alpha(T_a - T_i)r + \frac{\alpha}{1-\nu}\left(-\frac{Q}{30\lambda}\right)r[(1+\nu)r^2 - (3-\nu)a^2] \\ \sigma_{rr} &= \frac{\alpha E}{1-\nu}\left(-\frac{Q}{15\lambda}\right)(a^2 - r^2) \\ \sigma_{\theta\theta} = \sigma_{\phi\phi} &= \frac{\alpha E}{1-\nu}\left(-\frac{Q}{15\lambda}\right)(a^2 - 2r^2) \end{aligned} \quad \text{(Answer)}$$

Problem 19.2. Find the displacement and the thermal stresses in a hollow sphere of inner radius a and outer radius b with constant heat generation Q under a steady temperature field. Furthermore derive the displacement and the thermal stresses in a hollow sphere with constant temperature T_a at $r = a$ and with constant temperature T_b at $r = b$.

Solution. The one-dimensional steady state heat conduction equation with constant heat generation Q is

$$\frac{d^2T}{dr^2} + \frac{2}{r}\frac{dT}{dr} = -\frac{Q}{\lambda} \tag{19.31}$$

The general solution of Eq. (19.31) is

$$T = A + Br^{-1} + \left(-\frac{Q}{6\lambda}\right)r^2 \tag{19.32}$$

The temperature change from the initial temperature T_i reduces to

$$\tau = (A - T_i) + Br^{-1} + \left(-\frac{Q}{6\lambda}\right)r^2 \tag{19.33}$$

and the following integral relation is obtained

$$\int \tau r^2\, dr = \frac{1}{3}(A - T_i)r^3 + \frac{1}{2}Br^2 + \left(-\frac{Q}{30\lambda}\right)r^5 \tag{19.34}$$

Substitution of Eq. (19.34) into Eq. (19.9) gives the displacement and the thermal stresses as

$$u_r = \alpha(A - T_i)r + \frac{1+\nu}{1-\nu}\alpha\frac{B}{2}\left[1 + \frac{2(1-2\nu)}{1+\nu}\frac{b^2-a^2}{b^3-a^3}r - \frac{a^2b^2(b-a)}{b^3-a^3}\frac{1}{r^2}\right]$$
$$+ \frac{1+\nu}{1-\nu}\alpha\Big(-\frac{Q}{30\lambda}\Big)\left[r^3 + \frac{b^2-a^2}{b^3-a^3}\frac{a^3b^3}{r^2} + \frac{2(1-2\nu)}{1+\nu}\frac{b^5-a^5}{b^3-a^3}r\right]$$
$$\sigma_{rr} = \frac{\alpha E}{1-\nu}B\frac{b-a}{b^3-a^3}\Big(1-\frac{a}{r}\Big)\Big(1-\frac{b}{r}\Big)\Big(b+a+\frac{ab}{r}\Big)$$
$$+ \frac{\alpha E}{1-\nu}\Big(-\frac{Q}{15\lambda}\Big)\frac{1}{(b^3-a^3)r^3}[(r^3-a^3)(b^5-a^5)$$
$$-(b^3-a^3)(r^5-a^5)]$$
$$\sigma_{\theta\theta} = \sigma_{\phi\phi}$$
$$= \frac{\alpha E}{1-\nu}B\frac{b-a}{2(b^3-a^3)}\left[2(b+a) - (b^2+ab+a^2)\frac{1}{r} - \frac{a^2b^2}{r^3}\right]$$
$$+ \frac{\alpha E}{1-\nu}\Big(-\frac{Q}{30\lambda}\Big)\frac{1}{r^3}\left[\frac{(2r^3+a^3)(b^5-a^5)}{b^3-a^3} - (4r^5+a^5)\right] \qquad \text{(Answer)}$$
$$(19.35)$$

The boundary conditions of the temperature are

$$T = T_a \quad \text{at} \quad r = a, \qquad T = T_b \quad \text{at} \quad r = b \tag{19.36}$$

From Eq. (19.36), the unknown constants A and B are determined as

$$A = \frac{1}{b-a}\left[(bT_b - aT_a) - \Big(-\frac{Q}{6\lambda}\Big)(b^3-a^3)\right]$$
$$B = \frac{1}{b-a}\left[-ab(T_b - T_a) + \Big(-\frac{Q}{6\lambda}\Big)ab(b^2-a^2)\right] \tag{19.37}$$

The temperature change is

$$\tau = T_a - T_i + (T_b - T_a)\frac{1-\dfrac{a}{r}}{1-\dfrac{a}{b}}$$
$$+ \Big(-\frac{Q}{6\lambda}\Big)\left[-(b^2+ab+a^2) + \frac{ab(b+b)}{r} + r^2\right] \tag{19.38}$$

Substitution of Eq. (19.37) into Eq. (19.35) gives the displacement and the thermal stresses

$$
\begin{aligned}
u_r &= \alpha(T_a - T_i)r + \frac{1+\nu}{2(1-\nu)}\alpha(T_b - T_a)\frac{b}{b^3 - a^3} \\
&\quad \times \left\{ \frac{2r}{1+\nu}\left[(1-\nu)b^2 + \nu(ab + a^2)\right] - a(b^2 + ab + a^2) + \frac{a^3 b^2}{r^2} \right\} \\
&\quad + \frac{1+\nu}{1-\nu}\alpha\Big(-\frac{Q}{60\lambda}\Big)\Big\{ -\frac{1-\nu}{1+\nu}10(b^2 + ab + a^2)r + 5ab(a+b) + 2r^3 \\
&\quad + \frac{2(1-2\nu)}{1+\nu}\frac{r}{b^3 - a^3}\left[5(b^2 - a^2)(a+b)ab + 2(b^5 - a^5)\right] \\
&\quad - \frac{3(b^2 - a^2)}{b^3 - a^3}\frac{a^3 b^3}{r^2} \Big\} \\
\sigma_{rr} &= -\frac{\alpha E}{1-\nu}(T_b - T_a)\frac{ab}{b^3 - a^3}\Big(1 - \frac{a}{r}\Big)\Big(1 - \frac{b}{r}\Big)\Big(b + a + \frac{ab}{r}\Big) \\
&\quad + \frac{\alpha E}{1-\nu}\Big(-\frac{Q}{30\lambda}\Big)\frac{1}{r^3(b^3 - a^3)} \\
&\quad \times \Big\{ (r^3 - a^3)[5(a+b)ab(b^2 - a^2)r + 2(b^5 - a^5)] \\
&\quad - 5(a+b)ab(r^2 - a^2)(b^3 - a^3) - 2(r^5 - a^5)(b^3 - a^3) \Big\} \\
\sigma_{\theta\theta} = \sigma_{\phi\phi} &= -\frac{\alpha E}{1-\nu}(T_b - T_a)\frac{ab}{b^3 - a^3}\Big[b + a - (b^2 + ab + a^2)\frac{1}{2r} - \frac{a^2 b^2}{2r^3} \Big] \\
&\quad + \frac{\alpha E}{1-\nu}\Big(-\frac{Q}{60\lambda}\Big)\frac{1}{r^3(b^3 - a^3)}\{5(a+b)ab[(2r^3 + a^3)(b^2 - a^2) \\
&\quad - (r^2 + a^2)(b^3 - a^3)] + 2(2r^3 + a^3)(b^5 - a^5) - 2(4r^5 + a^5)(b^3 - a^3)\}
\end{aligned}
$$

(Answer) (19.39)

Problem 19.3. Find the displacement and the thermal stresses in an infinite body with a spherical cavity of radius a with heat generation $Q = Q_0 r^{-m}$, $(m \geq 4)$ under a steady temperature field. Furthermore derive the displacement and the thermal stresses in the infinite body with a spherical cavity with constant temperature T_a at $r = a$.

Solution. The one-dimensional steady state heat conduction equation with a heat generation $Q = Q_0 r^{-m}$ $(m \geq 4)$, is

$$
\frac{d^2T}{dr^2} + \frac{2}{r}\frac{dT}{dr} = -\frac{Q_0}{\lambda}r^{-m} \tag{19.40}
$$

The general solution of Eq. (19.40) is

$$T = A + Br^{-1} + \left(-\frac{Q_0}{\lambda}\right)\frac{r^{2-m}}{(m-2)(m-3)} \tag{19.41}$$

The temperature change from the initial temperature T_i reduces to

$$\tau = (A - T_i) + Br^{-1} + \left(-\frac{Q_0}{\lambda}\right)\frac{r^{2-m}}{(m-2)(m-3)} \tag{19.42}$$

and the following integral relation is obtained

$$\int \tau r^2\,dr = (A - T_i)\frac{r^3}{3} + B\frac{r^2}{2} - \left(-\frac{Q_0}{\lambda}\right)\frac{r^{5-m}}{(m-2)(m-3)(m-5)} \tag{19.43}$$

Substitution of Eq. (19.43) into Eq. (19.10) gives the displacement and the thermal stresses as

$$\begin{aligned}
u_r &= \frac{1+\nu}{1-\nu}\alpha\frac{1}{r^2}\Big[(A - T_i)\frac{r^3 - a^3}{3} + B\frac{r^2 - a^2}{2} \\
&\quad - \left(-\frac{Q}{\lambda}\right)\frac{r^{5-m} - a^{5-m}}{(m-2)(m-3)(m-5)}\Big] \\
\sigma_{rr} &= -\frac{2\alpha E}{1-\nu}\frac{1}{r^3}\Big[(A - T_i)\frac{r^3 - a^3}{3} + B\frac{r^2 - a^2}{2} \\
&\quad - \left(-\frac{Q}{\lambda}\right)\frac{r^{5-m} - a^{5-m}}{(m-2)(m-3)(m-5)}\Big] \\
\sigma_{\theta\theta} = \sigma_{\phi\phi} &= -\frac{\alpha E}{1-\nu}\frac{1}{r^3}\Big[(A - T_i)\frac{2r^3 + a^3}{3} + B\frac{r^2 + a^2}{2} \\
&\quad + \left(-\frac{Q}{\lambda}\right)\frac{(m-4)r^{5-m} - a^{5-m}}{(m-2)(m-3)(m-5)}\Big]
\end{aligned} \tag{Answer) (19.44}$$

Since the temperature at cavity is given by the constant temperature T_a, and the temperature at infinite point equals the initial temperature T_i, the boundary conditions are

$$T = T_a \quad \text{at} \quad r = a, \quad T = T_i \quad \text{at} \quad r = \infty \tag{19.45}$$

From Eq. (19.45), the unknown constants A and B are determined as

$$A = T_i, \quad B = (T_a - T_i)a - \left(-\frac{Q}{\lambda}\right)\frac{a^{3-m}}{(m-2)(m-3)} \tag{19.46}$$

The temperature change is

$$\tau = (T_a - T_i)\frac{a}{r} + \left(-\frac{Q}{\lambda}\right)\frac{r^{3-m} - a^{3-m}}{(m-2)(m-3)r} \tag{19.47}$$

Substitution of Eq. (19.46) into Eq. (19.44) gives the displacement and the thermal stresses

$$\begin{aligned}
u_r &= \frac{1+\nu}{1-\nu}\alpha(T_a - T_i)\frac{a(r^2-a^2)}{2r^2} \\
&\quad - \frac{1+\nu}{1-\nu}\alpha\left(-\frac{Q_0}{\lambda}\right)\frac{2r^{5-m} + (m-5)r^2a^{3-m} - (m-3)a^{5-m}}{2(m-2)(m-3)(m-5)r^2} \\
\sigma_{rr} &= -\frac{\alpha E}{1-\nu}(T_a - T_i)\frac{a(r^2-a^2)}{r^3} \\
&\quad + \frac{\alpha E}{1-\nu}\left(-\frac{Q_0}{\lambda}\right)\frac{2r^{5-m} + (m-5)r^2a^{3-m} - (m-3)a^{5-m}}{(m-2)(m-3)(m-5)r^3} \\
\sigma_{\theta\theta} &= \sigma_{\phi\phi} \\
&= -\frac{\alpha E}{1-\nu}(T_a - T_i)\frac{a(r^2+a^2)}{2r^3} \\
&\quad + \frac{\alpha E}{1-\nu}\left(-\frac{Q_0}{\lambda}\right)\frac{-2(m-4)r^{5-m} + (m-5)r^2a^{3-m} + (m-3)a^{5-m}}{2(m-2)(m-3)(m-5)r^3}
\end{aligned}$$

(Answer) (19.48)

Problem 19.4. Both the inner surface ($r = a$) and the outer surface ($r = b$) of a hollow sphere without heat generation are kept at the constant temperature T_a and T_b, respectively. When $b = 2a$ and $\nu = 0.3$, calculate the steady temperature change and extremum values of u_r, σ_{rr} and $\sigma_{\theta\theta}$ for the following two cases:
[Case 1] $T_a - T_i = 50$ K, $T_b - T_i = 0$ K
[Case 2] $T_a - T_i = 0$ K, $T_b - T_i = 50$ K

Solution. The steady temperature change and thermal stresses can be obtained from Eqs. (19.38) and (19.39) as

$$\tau = T_a - T_i + (T_b - T_a)\frac{1 - \dfrac{a}{r}}{1 - \dfrac{a}{b}} \tag{19.49}$$

$$\begin{aligned}
u_r &= \alpha(T_a - T_i)r + \frac{1+\nu}{1-\nu}\alpha(T_b - T_a)\frac{b}{2(b^3-a^3)} \\
&\quad \times \left\{\frac{2r}{1+\nu}[(1-\nu)b^2 + \nu(ab+a^2)] - a(b^2+ab+a^2) + \frac{a^3b^2}{r^2}\right\}
\end{aligned} \tag{19.50}$$

$$\sigma_{rr} = -\frac{\alpha E(T_b - T_a)}{1-\nu}\frac{ab}{b^3-a^3}\left(1-\frac{a}{r}\right)\left(1-\frac{b}{r}\right)\left(b+a+\frac{ab}{r}\right) \tag{19.51}$$

$$\begin{aligned}
\sigma_{\theta\theta} &= \sigma_{\phi\phi} \\
&= -\frac{\alpha E(T_b - T_a)}{1-\nu}\frac{ab}{b^3-a^3}\left[b + a - (b^2+ab+a^2)\frac{1}{2r} - \frac{a^2b^2}{2r^3}\right]
\end{aligned} \tag{19.52}$$

Making use of Eq. (19.51), it can be seen that the extremum value of σ_{rr} appears at the position $r = \left[\dfrac{3a^2b^2}{b^2+ab+a^2}\right]^{1/2}$, which is

$$(\sigma_{rr})_{\text{extremum}} = -\frac{\alpha E}{1-\nu}(T_b - T_a)\frac{ab}{b^3-a^3} \times \left[a + b - \frac{2\sqrt{3}}{9ab}\left(a^2+ab+b^2\right)^{3/2}\right] \tag{19.53}$$

Furthermore, the extremum values of $\sigma_{\theta\theta}(= \sigma_{\phi\phi})$ appear at the inner and outer surfaces, and they are

$$\sigma_{\theta\theta}|_{r=a} = \frac{\alpha E}{2(1-\nu)}(T_b - T_a)\frac{b(2b+a)}{b^2+ab+a^2}$$
$$\sigma_{\theta\theta}|_{r=b} = -\frac{\alpha E}{2(1-\nu)}(T_b - T_a)\frac{a(b+2a)}{b^2+ab+a^2} \tag{19.54}$$

Numerical assumptions are

$$b = 2a, \quad \nu = 0.3$$
$$\text{[Case 1]} \quad T_a - T_i = 50\ \text{K}, \quad T_b - T_i = 0\ \text{K}, \quad T_b - T_a = -50\ \text{K}$$
$$\text{[Case 2]} \quad T_a - T_i = 0\ \text{K}, \quad T_b - T_i = 50\ \text{K}, \quad T_b - T_a = 50\ \text{K} \tag{19.55}$$

[Temperature change]
Substituting Eq. (19.55) into Eq. (19.49), the temperature change τ is determined as

$$\text{[Case 1]} \quad \tau = 50 - 50 \times 2l\left(1 - \frac{a}{r}\right) = 50\left(2\frac{a}{r} - 1\right)\ \text{K}$$
$$\text{[Case 2]} \quad \tau = 0 + 50 \times 2\left(1 - \frac{a}{r}\right) = 100\left(1 - \frac{a}{r}\right)\ \text{K} \qquad \text{(Answer)}$$

[Thermal displacement]
Substituting the relation $b = 2a$ into Eq. (19.50), the displacement u_r is represented as

$$u_r = (T_a - T_i)\alpha r + (T_b - T_a)\alpha\frac{1+\nu}{1-\nu}\frac{1}{7}\left[\frac{2(4-\nu)}{1+\nu}r - 7a + 4\frac{a^3}{r^2}\right] \tag{19.56}$$

Substituting Eq. (19.55) into Eq. (19.56), we have

$$\begin{aligned}
\text{[Case 1]}\quad u_r &= 50\alpha r - 50\alpha\frac{1+\nu}{1-\nu}\frac{1}{7}\Big[\frac{2(4-\nu)}{1+\nu}r - 7a + 4\frac{a^3}{r^2}\Big]\\
u_r\,|_{r=a} &= 50\alpha a - 50\alpha\frac{1+\nu}{1-\nu}\frac{1}{7}\Big[\frac{2(4-\nu)}{1+\nu}a - 7a + 4a\Big]\\
&= 50\alpha a\Big(1 - \frac{5}{7}\Big) = 14.3\alpha a\\
u_r\,|_{r=b} &= 100\alpha a - 50\alpha\frac{1+\nu}{1-\nu}\frac{1}{7}\Big[\frac{4(4-\nu)}{1+\nu}a - 7a + a\Big]\\
&= 100\alpha a\Big(1 - \frac{5}{7}\Big) = 28.6\alpha a\\
\text{[Case 2]}\quad u_r &= 50\alpha\frac{1+\nu}{1-\nu}\frac{1}{7}\Big[\frac{2(4-\nu)}{1+\nu}r - 7a + 4\frac{a^3}{r^2}\Big]\\
u_r\,|_{r=a} &= 50\alpha\frac{1+\nu}{1-\nu}\frac{1}{7}\Big[\frac{2(4-\nu)}{1+\nu}a - 7a + 4a\Big]\\
&= 50\alpha a\frac{5}{7} = 35.7\alpha a\\
u_r\,|_{r=b} &= 50\alpha\frac{1+\nu}{1-\nu}\frac{1}{7}\Big[\frac{4(4-\nu)}{1+\nu}a - 7a + a\Big]\\
&= 50\alpha a\frac{10}{7} = 71.4\alpha a \qquad \text{(Answer)}
\end{aligned}$$

[Stress component σ_{rr}]
Substituting the relations $b = 2a$ and $\nu = 0.3$ into Eq. (19.53), the extremum value of σ_{rr} is

$$\begin{aligned}
(\sigma_{rr})_{\text{extremum}} &= -\alpha E(T_b - T_a)\frac{1}{0.7}\frac{2}{7}\Big[3 - \frac{7\sqrt{21}}{9}\Big]\\
&= -\alpha E(T_b - T_a)\times(-0.230) \qquad (19.57)
\end{aligned}$$

Substituting Eq. (19.55) into Eq. (19.57), we have

$$\begin{aligned}
&\text{[Case 1]}\quad (\sigma_{rr})_{\text{extremum}} = -\alpha E\times(-50)\times(-0.230) = -11.5\alpha E\\
&\text{[Case 2]}\quad (\sigma_{rr})_{\text{extremum}} = -\alpha E\times(+50)\times(-0.230) = 11.5\alpha E \qquad \text{(Answer)}
\end{aligned}$$

[Stress component $\sigma_{\theta\theta}$]
Substituting the relation $b = 2a$, $\nu = 0.3$ into Eq. (19.52), $\sigma_{\theta\theta}$ is

$$\sigma_{\theta\theta} = -\alpha E(T_b - T_a)\frac{1}{0.7}\times\frac{2}{7}\Big[3 - \frac{7}{2}\frac{a}{r} - 2\frac{a^3}{r^3}\Big] \qquad (19.58)$$

Substituting of Eq. (19.55) into Eq. (19.58) gives

$$
\begin{aligned}
\text{[Case 1]}\quad \sigma_{\theta\theta} &= +50\alpha E\frac{1}{0.7}\times\frac{2}{7}\left[3-\frac{7}{2}\frac{a}{r}-2\frac{a^3}{r^3}\right]\\
\sigma_{\theta\theta}\left.\right|_{r=a} &= +50\alpha E\frac{1}{0.7}\times\frac{2}{7}\left[3-\frac{7}{2}-2\right]=-51.0\alpha E\\
\sigma_{\theta\theta}\left.\right|_{r=b} &= +50\alpha E\frac{1}{0.7}\times\frac{2}{7}\left[3-\frac{7}{4}-\frac{1}{4}\right]=20.4\alpha E\\
\text{[Case 2]}\quad \sigma_{\theta\theta} &= -50\alpha E\frac{1}{0.7}\times\frac{2}{7}\left[3-\frac{7}{2}\frac{a}{r}-2\frac{a^3}{r^3}\right]\\
\sigma_{\theta\theta}\left.\right|_{r=a} &= -50\alpha E\frac{1}{0.7}\times\frac{2}{7}\left[3-\frac{7}{2}-2\right]=51.0\alpha E\\
\sigma_{\theta\theta}\left.\right|_{r=b} &= -50\alpha E\frac{1}{0.7}\times\frac{2}{7}\left[3-\frac{7}{4}-\frac{1}{4}\right]=-20.4\alpha E \qquad \text{(Answer)}
\end{aligned}
$$

Problem 19.5. When the temperature on the surface of the cavity with radius a of an infinite body without heat generation is kept at the constant temperature T_a, calculate the steady temperature change and extremum values of σ_{rr} and $\sigma_{\theta\theta}$, where $T_a - T_i = 50$ K and $\nu = 0.3$.

Solution. The steady temperature change τ, stress components σ_{rr} and $\sigma_{\theta\theta}$ for an infinite body with the spherical cavity are given by Eqs. (19.47) and (19.48), namely

$$
\begin{aligned}
\tau &= (T_a - T_i)\frac{a}{r}\\
\sigma_{rr} &= -\frac{\alpha E}{1-\nu}(T_a - T_i)\frac{a}{r}\left(1-\frac{a^2}{r^2}\right)\\
\sigma_{\theta\theta} &= -\frac{\alpha E}{1-\nu}(T_a - T_i)\frac{1}{2}\frac{a}{r}\left(1+\frac{a^2}{r^2}\right)
\end{aligned}
\tag{19.59}
$$

When $T_a - T_i = 50$ K, $\nu = 0.3$, we obtain τ, σ_{rr} and $\sigma_{\theta\theta}$ from Eq. (19.59).

$$
\tau = 50\frac{a}{r}\ \text{K}
$$

$$
\sigma_{rr} = -\frac{500}{7}\alpha E\frac{a}{r}\left(1-\frac{a^2}{r^2}\right),\qquad \sigma_{\theta\theta} = -\frac{250}{7}\alpha E\frac{a}{r}\left(1+\frac{a^2}{r^2}\right) \qquad \text{(Answer)}
$$

Now, the extremum value of σ_{rr} is obtained from the condition $d\sigma_{rr}/dr = 0$. The condition gives

$$
-\frac{1}{r^2}+3\frac{a^2}{r^4}=0 \tag{19.60}
$$

The solution of Eq. (19.60) is

$$
r=\sqrt{3}a \tag{19.61}
$$

Therefore, the extremum value of σ_{rr} is

$$(\sigma_{rr})_{\text{extremum}} = (\sigma_{rr})\Big|_{r=\sqrt{3}a} = -\frac{500}{7}\alpha E\frac{1}{\sqrt{3}}\left(1-\frac{1}{3}\right) = -27.5\alpha E \qquad (19.62)$$

On the other hand, the extremum value of $\sigma_{\theta\theta}$ occurs at the boundary $r = a$, namely

$$(\sigma_{\theta\theta})_{\text{extremum}} = (\sigma_{\theta\theta})\Big|_{r=a} = -\frac{250}{7}\alpha E \times 2 = -71.4\alpha \qquad (19.63)$$

From Eqs. (19.62) and (19.63), it can be seen that

$$(\sigma)_{\text{extremum}} = (\sigma_{\theta\theta})\,|_{r=a} = -\,71.4\alpha E \qquad \text{(Answer)}$$

Problem 19.6. When the surface $r = a$ of a solid sphere with the initial temperature T_i is heated to the temperature T_a, find the radial displacement and the thermal stresses in the unsteady state.

Solution. Consider a solid sphere of radius a with the initial temperature $T_i(r)$ and the boundary surface temperature T_a. One-dimensional heat conduction equation is

$$\frac{\partial T}{\partial t} = \kappa\left(\frac{\partial^2 T}{\partial r^2} + \frac{2}{r}\frac{\partial T}{\partial r}\right) \qquad (19.64)$$

Boundary condition is

$$T = T_a \quad \text{on} \quad r = a \qquad (19.65)$$

Initial condition is

$$T = T_i(r) \quad \text{at} \quad t = 0 \qquad (19.66)$$

Applying the Laplace transform with respect to the time t to the above equations, the heat conduction equation reduces to

$$\frac{d^2\bar{T}}{dr^2} + \frac{2}{r}\frac{d\bar{T}}{dr} - \frac{p}{\kappa}\bar{T} = -\frac{1}{\kappa}T_i(r) \qquad (19.65')$$

and the boundary condition reduces to

$$\bar{T} = \frac{T_a}{p} \quad \text{on} \quad r = a \qquad (19.66')$$

Then, we obtain the solution of Eq. (19.65′) with the method of variation of parameters

$$\bar{T} = A\frac{1}{r}\sinh qr + B\frac{1}{r}\cosh qr - \frac{1}{\kappa q r}\int_0^r T_i(\eta)\eta\sinh q(r-\eta)d\eta \qquad (19.67)$$

where $q = \sqrt{p/\kappa}$.

Since the temperature must be bounded at $r = 0$, Eq. (19.67) reduces to

$$\bar{T} = A\frac{1}{r}\sinh qr - \frac{1}{\kappa qr}\int_0^r T_i(\eta)\eta \sinh q(r-\eta)d\eta \tag{19.68}$$

Substitution of Eq. (19.68) into Eq. (19.66′) gives the coefficient A

$$A = \frac{T_a a}{p\sinh qa} + \frac{a}{\kappa qa \sinh qa}\int_0^a T_i(\eta)\eta \sinh q(a-\eta)d\eta \tag{19.69}$$

Substitution of Eq. (19.69) into Eq. (19.68) gives the temperature $\bar{T}$

$$\bar{T} = \left[\frac{T_a a}{p\sinh qa} + \frac{a}{\kappa qa \sinh qa}\int_0^a T_i(\eta)\eta \sinh q(a-\eta)d\eta\right]\frac{1}{r}\sinh qr - \frac{1}{\kappa qr}\int_0^r T_i(\eta)\eta \sinh q(r-\eta)d\eta \tag{19.70}$$

Performing the inverse Laplace transform on Eq. (19.70), we can obtain the temperature.

When the initial temperature is constant, the inverse Laplace transform on Eq. (19.70) gives the temperature in the form

$$T(r,t) = T_a + 2(T_i - T_a)\sum_{n=1}^{\infty}\int_0^a \eta \sin s_n\eta\, d\eta \frac{\sin s_n r}{ar}e^{-\kappa s_n^2 t} \tag{19.71}$$

where s_n are eigenvalues of $\sin s_n a = 0$, that is

$$s_n = \frac{n\pi}{a} \quad (n = 1, 2, 3, \ldots) \tag{19.72}$$

Performing the integration in Eq. (19.71), the temperature is obtained as

$$T = T_a + 2(T_a - T_i)\sum_{n=1}^{\infty}(-1)^n \frac{\sin n\pi\frac{r}{a}}{n\pi\frac{r}{a}}e^{-\kappa n^2\pi^2 t/a^2} \tag{19.73}$$

Then the temperature change τ is

$$\tau = (T_a - T_i)\left[1 + 2\sum_{n=1}^{\infty}(-1)^n \frac{\sin n\pi\frac{r}{a}}{n\pi\frac{r}{a}}e^{-\kappa n^2\pi^2 t/a^2}\right] \tag{19.74}$$

The thermal stresses are from Eq. (19.8)

$$
\begin{aligned}
u_r &= \frac{\alpha}{1-\nu}\Big[(1+\nu)\frac{1}{r^2}\int_0^r \tau r^2\,dr + 2(1-2\nu)\frac{r}{a^3}\int_0^a \tau r^2\,dr\Big] \\
\sigma_{rr} &= \frac{\alpha E}{1-\nu}\Big[\frac{2}{a^3}\int_0^a \tau r^2\,dr - \frac{2}{r^3}\int_0^r \tau r^2\,dr\Big] \\
\sigma_{\theta\theta} &= \sigma_{\phi\phi} = \frac{\alpha E}{1-\nu}\Big[\frac{2}{a^3}\int_0^a \tau r^2\,dr + \frac{1}{r^3}\int_0^r \tau r^2\,dr - \tau\Big] \qquad (19.75)
\end{aligned}
$$

We first calculate the following integral:

$$
\int_0^r \tau r^2\,dr = (T_a - T_i)\left\{\frac{r^3}{3} + 2\sum_{n=1}^{\infty}\frac{(-1)^n}{s_n^3}[\sin(s_n r) - s_n r\cos(s_n r)]e^{-\kappa s_n^2 t}\right\} \qquad (19.76)
$$

Substitution of Eq. (19.76) into Eq. (19.75) gives the radial displacement and the stresses

$$
\begin{aligned}
u_r &= \alpha(T_a - T_i)r - 4\frac{1-2\nu}{1-\nu}\alpha(T_a - T_i)r\sum_{n=1}^{\infty}\frac{1}{n^2\pi^2}e^{-\kappa n^2\pi^2 t/a^2} \\
&\quad + 2\frac{1+\nu}{1-\nu}\alpha(T_a - T_i)a\sum_{n=1}^{\infty}\frac{(-1)^n}{n^3\pi^3(r/a)^2} \\
&\quad \times\left(\sin n\pi\frac{r}{a} - n\pi\frac{r}{a}\cos n\pi\frac{r}{a}\right)e^{-\kappa n^2\pi^2 t/a^2} \\
\sigma_{rr} &= -4\frac{\alpha E}{1-\nu}(T_a - T_i)\sum_{n=1}^{\infty}\Big[\frac{1}{n^2\pi^2} + \frac{(-1)^n}{n^3\pi^3(r/a)^3} \\
&\quad \times\left(\sin n\pi\frac{r}{a} - n\pi\frac{r}{a}\cos n\pi\frac{r}{a}\right)\Big]e^{-\kappa n^2\pi^2 t/a^2} \\
\sigma_{\theta\theta} &= \sigma_{\phi\phi} = 2\frac{\alpha E}{1-\nu}(T_a - T_i)\sum_{n=1}^{\infty}\Bigg\{\frac{(-1)^n}{n^3\pi^3(r/a)^3} \\
&\quad \times\Big[\left(1 - n^2\pi^2\frac{r^2}{a^2}\right)\sin n\pi\frac{r}{a} - n\pi\frac{r}{a}\cos n\pi\frac{r}{a}\Big] - \frac{2}{n^2\pi^2}\Bigg\}e^{-\kappa n^2\pi^2 t/a^2}
\end{aligned}
$$

(Answer)

Problem 19.7. Let us consider a hollow sphere with the initial temperature T_i and with prescribed surface temperatures T_a and T_b at both surfaces at $r = a$ and $r = b$, respectively. Find the thermal displacement and thermal stresses in a hollow sphere subjected to the transient temperature field.

Solution. Consider the transient temperature in a hollow sphere of inner radius a and outer radius b.

One-dimensional heat conduction equation is

$$\frac{\partial T}{\partial t} = \kappa\left(\frac{\partial^2 T}{\partial r^2} + \frac{2}{r}\frac{\partial T}{\partial r}\right) \tag{19.77}$$

Boundary conditions are

$$T = T_a \quad \text{on} \quad r = a, \qquad T = T_b \quad \text{on} \quad r = b \tag{19.78}$$

Initial condition is

$$T = T_i \quad \text{at} \quad t = 0 \tag{19.79}$$

We introduce the temperature change τ from the initial temperature T_i as

$$\tau = T - T_i \tag{19.80}$$

Then, Eqs. (19.77)–(19.79) can be rewritten by τ.

One-dimensional heat conduction equation is

$$\frac{\partial \tau}{\partial t} = \kappa\left(\frac{\partial^2 \tau}{\partial r^2} + \frac{2}{r}\frac{\partial \tau}{\partial r}\right) \tag{19.77$'$}$$

Boundary conditions are

$$\tau = T_a - T_i \quad \text{on} \quad r = a, \qquad \tau = T_b - T_i \quad \text{on} \quad r = b \tag{19.78$'$}$$

Initial condition is

$$\tau = 0 \quad \text{at} \quad t = 0 \tag{19.79$'$}$$

By use of the method of separation of variables, the general solution of the heat conduction equation (19.77′) may be expressed by

$$\tau = A_0 + B_0 r^{-1} + \sum_{n=1}^{\infty}\left(A_n \frac{\sin s_n r}{r} + B_n \frac{\cos s_n r}{r}\right)e^{-\kappa s_n^2 t} \tag{19.81}$$

From the boundary conditions (19.78′), the coefficients A_0, B_0 and B_n are

$$\begin{aligned} A_0 &= T_a - T_i + (T_b - T_a)\frac{b}{b-a}, \qquad B_0 = -(T_b - T_a)\frac{ab}{b-a} \\ B_n &= -\frac{\sin s_n a}{\cos s_n a} A_n \end{aligned} \tag{19.82}$$

and s_n are eigenvalues of eigenfunction $\sin s_n(b-a) = 0$, that is

$$s_n = \frac{n\pi}{b-a} \qquad (n = 1, 2, 3, \ldots) \tag{19.83}$$

Therefore, the temperature change τ is

$$\begin{aligned} \tau &= T_a - T_i + (T_b - T_a)\frac{b}{b-a}\left(1 - \frac{a}{r}\right) \\ &\quad + \sum_{n=1}^{\infty} A_n \frac{\sin s_n(r-a)}{r \cos s_n a} e^{-\kappa s_n^2 t} \end{aligned} \tag{19.84}$$

The initial condition (19.79′) gives

$$\sum_{n=1}^{\infty} A_n \frac{\sin s_n(r-a)}{r \cos s_n a} = -\left[T_a - T_i + (T_b - T_a)\frac{b}{b-a}\left(1 - \frac{a}{r}\right)\right] \tag{19.85}$$

Multiplying $r \sin s_m(r-a)$ to both sides of Eq. (19.85) and integrating from a to b, the coefficient A_m can be determined as

$$A_m = 2\frac{\cos s_m a}{s_m(b-a)}[(-1)^m(T_b - T_i)b - (T_a - T_i)a] \tag{19.86}$$

Then, the temperature change τ is

$$\begin{aligned} \tau &= (T_a - T_i) + (T_b - T_a)\frac{1 - \frac{a}{r}}{1 - \frac{a}{b}} \\ &\quad + \frac{2}{\pi r}\sum_{n=1}^{\infty} \frac{1}{n} \sin\left[n\pi\left(\frac{r-a}{b-a}\right)\right] e^{-\kappa n^2 \pi^2 t/(b-a)^2} \\ &\quad \times \left[(T_b - T_i)(-1)^n b - (T_a - T_i)a\right] \end{aligned} \tag{19.87}$$

Substituting Eq. (19.87) into Eq. (19.9), the radial displacement and the stresses are given as

$$\begin{aligned} u_r &= \alpha(T_a - T_i)r \\ &\quad + \frac{1+\nu}{1-\nu}\alpha(T_b - T_a)\frac{b}{2(b^3 - a^3)} \\ &\quad \times \left\{\frac{2r}{1+\nu}\left[(1-\nu)b^2 + \nu(ab + a^2)\right] - a(b^2 + ab + a^2) + \frac{a^3 b^2}{r^2}\right\} \\ &\quad - \frac{2(b-a)}{\pi^3}\sum_{n=1}^{\infty} \frac{1}{n^3} e^{-\kappa n^2 \pi^2 t/(b-a)^2} [(T_b - T_i)(-1)^n b - (T_a - T_i)a] \end{aligned}$$

$$
\begin{aligned}
&\times \left(\frac{2(1-2\nu)}{1+\nu} r \frac{n\pi}{b^3-a^3}[(-1)^n b - a] + \frac{n\pi}{r}\cos\left[n\pi\left(\frac{r-a}{b-a}\right)\right] \right.\\
&\left. + \frac{1}{r^2}\left\{ -(b-a)\sin\left[n\pi\left(\frac{r-a}{b-a}\right)\right] + \frac{n\pi ab}{b^3-a^3}[a^2(-1)^n - b^2] \right\} \right)\\
\sigma_{rr} &= \frac{\alpha E}{1-\nu}\left\{ -(T_b - T_a)\frac{ab}{b^3-a^3}\left(1-\frac{a}{r}\right)\left(1-\frac{b}{r}\right)\left(b+a+\frac{ab}{r}\right) \right.\\
&- \frac{4(b-a)}{\pi^3}\sum_{n=1}^{\infty}\frac{1}{n^3}e^{-\kappa n^2\pi^2 t/(b-a)^2}[(T_b - T_i)(-1)^n b - (T_a - T_i)a]\\
&\times \left(\frac{(r^3-a^3)n\pi}{r^3(b^3-a^3)}[(-1)^n b - a] - \frac{1}{r^3}\left\{ n\pi r\cos\left[n\pi\left(\frac{r-a}{b-a}\right)\right] \right.\right.\\
&\left.\left.\left. - (b-a)\sin\left[n\pi\left(\frac{r-a}{b-a}\right)\right] - n\pi a \right\}\right)\right\}\\
\sigma_{\theta\theta} &= \sigma_{\phi\phi}\\
&= \frac{\alpha E}{1-\nu}\left\{ -(T_b - T_a)\frac{ab}{b^3-a^3}\left[b+a-(b^2+ab+a^2)\frac{1}{2r} - \frac{a^2b^2}{2r^3}\right] \right.\\
&- \frac{2(b-a)}{\pi^3}\sum_{n=1}^{\infty}\frac{1}{n^3}e^{-\kappa n^2\pi^2 t/(b-a)^2}[(T_b - T_i)(-1)^n b - (T_a - T_i)a]\\
&\times \left(\frac{(2r^3+a^3)n\pi}{r^3(b^3-a^3)}[(-1)^n b - a] \right.\\
&+ \frac{1}{r^3}\left\{ n\pi r\cos\left[n\pi\left(\frac{r-a}{b-a}\right)\right] - (b-a)\sin\left[n\pi\left(\frac{r-a}{b-a}\right)\right] - n\pi a \right\}\\
&\left.\left. + \frac{n^2\pi^2}{b-a}\frac{1}{r}\sin\left[n\pi\left(\frac{r-a}{b-a}\right)\right]\right)\right\} \qquad \text{(Answer)}
\end{aligned}
$$

Problem 19.8. Find solutions of the steady axisymmetric temperature field in the spherical body without internal heat generation.

Solution. Referring to Eq. (15.10), the steady state axisymmetric heat conduction equation without internal heat generation is given by

$$
\frac{\partial^2 T}{\partial r^2} + \frac{2}{r}\frac{\partial T}{\partial r} + \frac{1}{r^2\sin\theta}\frac{\partial}{\partial\theta}\left(\sin\theta\frac{\partial T}{\partial\theta}\right) = 0 \tag{19.88}
$$

We put the new variable as follows:

$$
\mu = \cos\theta \tag{19.89}
$$

By use of Eq. (19.89), Eq. (19.88) can be rewritten as

$$\frac{\partial^2 T}{\partial r^2} + \frac{2}{r}\frac{\partial T}{\partial r} + \frac{1}{r^2}\frac{\partial}{\partial \mu}\left[(1-\mu^2)\frac{\partial T}{\partial \mu}\right] = 0 \tag{19.90}$$

Now, we use the method of separation of variables to solve Eq. (19.90). If the steady temperature can be expressed by

$$T = f(r)g(\mu) \tag{19.91}$$

equation (19.90) reduces to

$$\frac{r^2}{f(r)}\left[\frac{d^2 f(r)}{dr^2} + \frac{2}{r}\frac{df(r)}{dr}\right] = -\frac{1}{g(\mu)}\frac{d}{d\mu}\left[(1-\mu^2)\frac{dg(\mu)}{d\mu}\right] \equiv \nu(1+\nu) \tag{19.92}$$

where $\nu(1+\nu)$ means an arbitrary constant. Equation (19.92) gives two separate equations

$$\frac{d^2 f(r)}{dr^2} + \frac{2}{r}\frac{df(r)}{dr} - \frac{\nu(1+\nu)}{r^2}f(r) = 0 \tag{19.93}$$

and

$$\frac{d}{d\mu}\left[(1-\mu^2)\frac{dg(\mu)}{d\mu}\right] + \nu(1+\nu)g(\mu) = 0 \tag{19.94}$$

The linearly independent solutions of Eq. (19.93) are

$$f(r) = \begin{pmatrix} r^{\nu} \\ r^{-(1+\nu)} \end{pmatrix} \tag{19.95}$$

We recall that Eq. (19.94) is Legendre's differential equation. The linearly independent solutions of Legendre's differential equation (19.94) are

$$g(\mu) = \begin{pmatrix} P_\nu(\mu) \\ Q_\nu(\mu) \end{pmatrix} \tag{19.96}$$

where $P_\nu(\mu)$ is the Legendre function of the first kind of order ν and $Q_\nu(\mu)$ is the Legendre function of the second kind of order ν.

Then, the particular solutions for the steady temperature equation in the axisymmetrical spherical coordinate system are

$$\begin{pmatrix} r^{\nu} P_\nu(\cos\theta) \\ r^{-(1+\nu)} P_\nu(\cos\theta) \\ r^{\nu} Q_\nu(\cos\theta) \\ r^{-(1+\nu)} Q_\nu(\cos\theta) \end{pmatrix} \quad \text{(Answer)} \tag{19.97}$$

When an arbitrary constant ν is an integer n, the Legendre functions satisfy the following relations:

$$P_{-n-1}(\cos\theta) = P_n(\cos\theta)$$
$$Q_{-n-1}(\cos\theta) = Q_n(\cos\theta) \qquad (n = 0, \pm 1, \pm 2, \ldots) \tag{19.98}$$

Since the Legendre function of the second kind $Q_n(\cos\theta)$ has a singular value when $\cos\theta = 1$ ($\theta = 0$), the particular solutions of the steady state temperature equation for the axisymmetrical spherical coordinate system are

$$\begin{pmatrix} r^n P_n(\cos\theta) \\ r^{-n-1} P_n(\cos\theta) \end{pmatrix} \qquad (n = 0, 1, 2, \ldots) \qquad \text{(Anwer)} \tag{19.99}$$

Since the Boussinesq harmonic functions φ and ψ satisfy Eq. (19.88), the particular Boussinesq harmonic functions φ and ψ are also expressed by Eqs. (19.97) or (19.99).

Problem 19.9. Find the solutions of the unsteady axisymmetric temperature field in the spherical body without internal heat generation.

Solution. Referring to Eq. (15.8), the unsteady state axisymmetric heat conduction equation without internal heat generation is given as

$$\frac{\partial^2 T}{\partial r^2} + \frac{2}{r}\frac{\partial T}{\partial r} + \frac{1}{r^2 \sin\theta}\frac{\partial}{\partial\theta}\left(\sin\theta \frac{\partial T}{\partial\theta}\right) = \frac{1}{\kappa}\frac{\partial T}{\partial t} \tag{19.100}$$

We put the new variable as follows:

$$\mu = \cos\theta \tag{19.101}$$

By use of Eq. (19.101), Eq. (19.100) can be rewritten as

$$\frac{\partial^2 T}{\partial r^2} + \frac{2}{r}\frac{\partial T}{\partial r} + \frac{1}{r^2}\frac{\partial}{\partial\mu}\left[(1-\mu^2)\frac{\partial T}{\partial\mu}\right] = \frac{1}{\kappa}\frac{\partial T}{\partial t} \tag{19.102}$$

We use the method of separation of variables to solve Eq. (19.102). If the unsteady temperature can be expressed by

$$T = f(r)g(\mu)h(t) \tag{19.103}$$

equation (19.102) reduces to

$$\frac{1}{f(r)}\left[\frac{d^2 f(r)}{dr^2} + \frac{2}{r}\frac{df(r)}{dr}\right] + \frac{1}{r^2 g(\mu)}\frac{d}{d\mu}\left[(1-\mu^2)\frac{dg(\mu)}{d\mu}\right] = \frac{1}{\kappa h(t)}\frac{dh(t)}{dt} \tag{19.104}$$

Equation (19.104) gives three separate equations

$$\frac{dh(t)}{dt} + \kappa s^2 h(t) = 0 \tag{19.105}$$

$$\frac{d}{d\mu}\left[(1-\mu^2)\frac{dg(\mu)}{d\mu}\right]+n(1+n)g(\mu)=0 \tag{19.106}$$

$$\frac{d^2 f(r)}{dr^2}+\frac{2}{r}\frac{df(r)}{dr}+\left(s^2-\frac{n(1+n)}{r^2}\right)f(r)=0 \tag{19.107}$$

where s is an arbitrary constant and n means the positive integer.

The linearly independent solutions of Eq. (19.105) are

$$h(t)=1 \ \text{ for } \ s=0, \qquad h(t)=e^{-\kappa s^2 t} \ \text{ for } \ s\neq 0 \tag{19.108}$$

That is, $s=0$ and $s\neq 0$ correspond to the steady and the unsteady temperature change, respectively.

Since Eq. (19.106) is Legendre's differential equation, the linearly independent solutions of Eq. (19.106) are

$$g(\mu)=\begin{pmatrix} P_n(\mu) \\ Q_n(\mu) \end{pmatrix}=\begin{pmatrix} P_n(\cos\theta) \\ Q_n(\cos\theta) \end{pmatrix} \tag{19.109}$$

where $P_n(\mu)$ is the Legendre function of the first kind of order n and $Q_n(\mu)$ is the Legendre function of the second kind of order n.

When $s\neq 0$, Eq. (19.107) is called the spherical Bessel's differential equation. Then the linearly independent solutions of Eq. (19.107) is

$$f(r)=\begin{pmatrix} j_n(sr) \\ y_n(sr) \end{pmatrix} \tag{19.110}$$

where $j_n(sr)$ is the spherical Bessel function of the first kind of order n and $y_n(sr)$ is the spherical Bessel function of the second kind of order n.

The linearly independent solutions of Eq. (19.107) for $s=0$ are

$$f(r)=\begin{pmatrix} r^n \\ r^{-(1+n)} \end{pmatrix} \tag{19.111}$$

Then, the particular solutions for the unsteady temperature equation in the axisymmetrical spherical coordinate system are

$$T=\begin{pmatrix} r^n P_n(\cos\theta) \\ r^{-(1+n)} P_n(\cos\theta) \\ r^n Q_n(\cos\theta) \\ r^{-(1+n)} Q_n(\cos\theta) \end{pmatrix}, \quad \begin{pmatrix} j_n(sr)P_n(\cos\theta)e^{-\kappa s^2 t} \\ y_n(sr)P_n(\cos\theta)e^{-\kappa s^2 t} \\ j_n(sr)Q_n(\cos\theta)e^{-\kappa s^2 t} \\ y_n(sr)Q_n(\cos\theta)e^{-\kappa s^2 t} \end{pmatrix} \qquad \text{(Answer)(19.112)}$$

Since the Legendre function of the second kind $Q_n(\cos\theta)$ has a singular value when $\cos\theta=1$ $(\theta=0)$, the particular solutions for the unsteady temperature equation for the axisymmetrical spherical coordinate system are

$$T = \begin{pmatrix} r^n P_n(\cos\theta) \\ r^{-(1+n)} P_n(\cos\theta) \end{pmatrix}, \quad \begin{pmatrix} j_n(sr) P_n(\cos\theta) e^{-\kappa s^2 t} \\ y_n(sr) P_n(\cos\theta) e^{-\kappa s^2 t} \end{pmatrix} \quad \text{(Answer)} \tag{19.113}$$

Problem 19.10. Derive a particular thermoelastic displacement potential Φ, when the spherical body is subjected to the unsteady axisymmetric heating.

Solution. The temperature change τ is given from Eq. (19.113) as

$$\tau = \tau_s + \tau_u \tag{19.114}$$

where τ_s is the steady temperature change given by

$$\tau_s = (A_0 - T_i) + B_0 r^{-1} + \sum_{n=1}^{\infty} (A_n r^n + B_n r^{-(n+1)}) P_n(\mu) \tag{19.115}$$

and τ_u is the unsteady temperature change given by

$$\tau_u = \sum_{n=0}^{\infty} [C_n j_n(sr) + D_n y_n(sr)] P_n(\mu) e^{-\kappa s^2 t} \tag{19.116}$$

Corresponding to the temperature change (19.114), the thermoelastic displacement potential Φ is divided into two parts

$$\Phi = \Phi_s + \Phi_u \tag{19.117}$$

where Φ_s and Φ_u satisfy

$$\nabla^2 \Phi_s = K\tau_s, \quad \nabla^2 \Phi_u = K\tau_u \tag{19.118}$$

The fundamental equation for the thermoelastic displacement potential Φ_s given by Eq. (19.18) reduces to

$$\frac{\partial^2 \Phi_s}{\partial r^2} + \frac{2}{r}\frac{\partial \Phi_s}{\partial r} + \frac{1}{r^2}\frac{\partial}{\partial \mu}\left[(1-\mu^2)\frac{\partial \Phi_s}{\partial \mu}\right] = K\left[(A_0 - T_i) + B_0 r^{-1} + \sum_{n=1}^{\infty} (A_n r^n + B_n r^{-(n+1)}) P_n(\mu)\right] \tag{19.119}$$

To solve Eq. (19.119), we now assume that Φ_s is given in the form

$$\Phi_s = K\left[(A_0 - T_i)\frac{r^2}{6} + B_0 \frac{r}{2}\right] + \Phi_{s1} \tag{19.120}$$

and

$$\Phi_{s1} = \sum_{n=1}^{\infty} F_n(r) P_n(\mu) \tag{19.121}$$

Substituting Eq. (19.120) and (19.121) into Eq. (19.119), we have

$$\sum_{n=1}^{\infty} \left\{ \left(\frac{d^2 F_n}{dr^2} + \frac{2}{r} \frac{dF_n}{dr} \right) P_n(\mu) + F_n \frac{1}{r^2} \frac{d}{d\mu} \left[(1-\mu^2) \frac{dP_n}{d\mu} \right] \right\}$$
$$= K \sum_{n=1}^{\infty} \left(A_n r^n + B_n r^{-(n+1)} \right) P_n(\mu) \tag{19.122}$$

We first calculate

$$\frac{d}{d\mu}\left[(1-\mu^2)\frac{dP_n}{d\mu}\right] = \frac{d}{d\mu}\{(n+1)[\mu P_n(\mu) - P_{n+1}(\mu)]\} = -n(n+1)P_n(\mu) \tag{19.123}$$

Then, Eq. (19.122) is rewritten as

$$\sum_{n=1}^{\infty} \left[\frac{d^2 F_n}{dr^2} + \frac{2}{r} \frac{dF_n}{dr} - \frac{n(n+1)}{r^2} F_n \right] P_n(\mu)$$
$$= K \sum_{n=1}^{\infty} (A_n r^n + B_n r^{-(n+1)}) P_n(\mu) \tag{19.124}$$

From Eq. (19.124), the following equations are derived

$$\frac{d^2 F_n}{dr^2} + \frac{2}{r} \frac{dF_n}{dr} - \frac{n(n+1)}{r^2} F_n = K(A_n r^n + B_n r^{-(n+1)}) \tag{19.125}$$

Here we put

$$F_n = A'_n r^{n+2} + B'_n r^{1-n} \tag{19.126}$$

Substitution of Eq. (19.126) into Eq. (19.125) gives

$$2(2n+3)A'_n r^n + 2(1-2n)B'_n r^{-(n+1)} = K\left(A_n r^n + B_n r^{-(n+1)}\right) \tag{19.127}$$

From Eq. (19.127), A_n and B_n are

$$A'_n = K\frac{A_n}{2(2n+3)}, \qquad B'_n = K\frac{B_n}{2(1-2n)} \tag{19.128}$$

Then, Φ_{s1} is

$$\Phi_{s1} = K\sum_{n=1}^{\infty}\Big[\frac{1}{2(2n+3)}A_n r^{n+2} + B_n\frac{1}{2(1-2n)}r^{1-n}\Big]P_n(\mu) \tag{19.129}$$

From Eqs. (19.120) and (19.129), the thermoelastic displacement potential Φ_s is given by

$$\begin{aligned}\Phi_s &= K\Big\{(A_0 - T_i)\frac{r^2}{6} + B_0\frac{r}{2}\\ &+\sum_{n=1}^{\infty}\Big[\frac{1}{2(2n+3)}A_n r^{n+2} + B_n\frac{1}{2(1-2n)}r^{1-n}\Big]P_n(\mu)\Big\}\end{aligned} \tag{19.130}$$

Next, we consider the thermoelastic displacement potential Φ_u. The thermoelastic displacement potential Φ_u for the unsteady temperature can be obtained from Eq. (16.34)

$$\Phi_u = \kappa K\int \tau_u\, dt \tag{19.131}$$

Substitution of Eq. (19.116) into Eq. (19.131) gives

$$\Phi_u = -\frac{K}{s^2}\tau_u \tag{19.132}$$

Then

$$\Phi_u = -K\sum_{n=0}^{\infty}\frac{1}{s^2}[C_n j_n(sr) + D_n y_n(sr)]P_n(\mu)e^{-\kappa s^2 t} \tag{19.133}$$

From Eqs. (19.117), (19.130) and (19.133), the thermoelastic displacement potential Φ for the axisymmetric temperature distribution is

$$\begin{aligned}\Phi = K\Big\{&(A_0 - T_i)\frac{r^2}{6} + B_0\frac{r}{2}\\ &+\sum_{n=1}^{\infty}\Big[\frac{1}{2(2n+3)}A_n r^{n+2} - B_n\frac{1}{2(2n-1)}r^{1-n}\Big]P_n(\mu)\\ &-\sum_{n=0}^{\infty}\frac{1}{s^2}\Big[C_n j_n(sr) + D_n y_n(sr)\Big]P_n(\mu)e^{-\kappa s^2 t}\Big\}\end{aligned} \qquad \text{(Answer)} \tag{19.134}$$

Problem 19.11. When the temperature at boundary surface ($r = a$) of a solid sphere is given by Fig. 19.1, find the temperature in a solid sphere with the initial temperature T_i under a steady temperature field.

Fig. 19.1 Band temperature

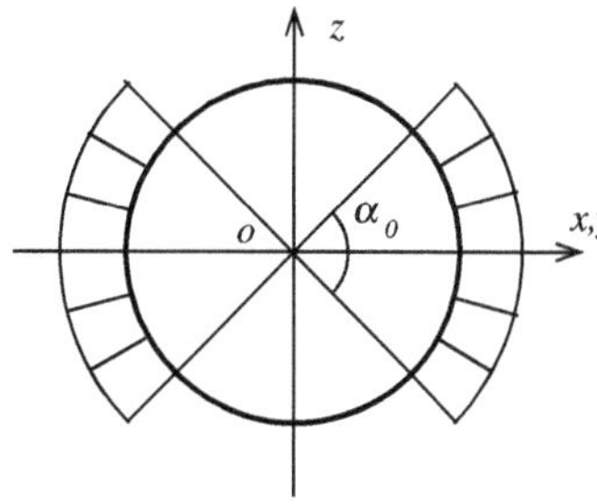

Solution. The heat conduction equation is

$$\frac{\partial^2 T}{\partial r^2} + \frac{2}{r}\frac{\partial T}{\partial r} + \frac{1}{r^2 \sin\theta}\frac{\partial}{\partial \theta}\left(\sin\theta \frac{\partial T}{\partial \theta}\right) = 0 \tag{19.135}$$

The boundary condition is

$$T = T_a F(\theta) \quad \text{on} \quad r = a \tag{19.136}$$

where

$$F(\theta) = \sum_{n=0}^{\infty} f_n P_n(\mu), \qquad \mu = \cos\theta \tag{19.137}$$

The initial temperature is

$$T = T_i \quad \text{at} \quad t = 0 \tag{19.138}$$

Multiplying both sides of Eq. (19.137) by $P_m(\mu)$, integrating from -1 to 1, and using the integration as follows:

$$\begin{aligned} \int_{-1}^{1} P_n(\mu) P_m(\mu)\, d\mu &= 0 \qquad &\text{for} \quad m \neq n \\ \int_{-1}^{1} P_n(\mu) P_m(\mu)\, d\mu &= \frac{2}{2n+1} \qquad &\text{for} \quad m = n \end{aligned} \tag{19.139}$$

we obtain the coefficients f_n as

$$f_n = \frac{2n+1}{2}\int_{-1}^{1} F(\theta) P_n(\mu) d\mu \tag{19.140}$$

When boundary temperature $T_a F(\theta)$ is shown in Fig. 19.1, $F(\theta)$ is represented as

$$F(\theta) = H\left[\frac{1}{2}(\pi + \alpha_0) - \theta\right] - H\left[\frac{1}{2}(\pi - \alpha_0) - \theta\right] \tag{19.141}$$

where $H(x)$ means the Heaviside step function.

If $F(\theta)$ is given by

$$F(\theta) = H(\theta_0 - \theta), \quad \mu_0 \equiv \cos\theta_0 \tag{19.142}$$

substitution of Eq. (19.142) into Eq. (19.140) gives

$$\begin{aligned} f_0 &= \frac{1}{2}(1-\mu_0) && \text{for} \quad n=0 \\ f_n &= \frac{2n+1}{2n}\{\mu_0 P_n(\mu_0) - P_{n+1}(\mu_0)\} && \text{for} \quad n \geq 1 \end{aligned} \tag{19.143}$$

Now putting

$$\theta_1 = \frac{1}{2}(\pi + \alpha_0), \quad \mu_1 = \cos\theta_1 \quad ; \quad \theta_2 = \frac{1}{2}(\pi - \alpha_0), \quad \mu_2 = \cos\theta_2 \tag{19.144}$$

we have for Eq. (19.141)

$$\begin{aligned} f_0 &= \frac{1}{2}(1-\mu_1) - \frac{1}{2}(1-\mu_2) = \frac{1}{2}(\mu_2 - \mu_1) \quad \text{for} \quad n=0 \\ f_n &= \frac{2n+1}{2n}\{\mu_1 P_n(\mu_1) - P_{n+1}(\mu_1)\} - \frac{2n+1}{2n}\{\mu_2 P_n(\mu_2) - P_{n+1}(\mu_2)\} \\ &= \frac{2n+1}{2n}[\{\mu_1 P_n(\mu_1) - P_{n+1}(\mu_1)\} - \{\mu_2 P_n(\mu_2) - P_{n+1}(\mu_2)\}] \\ &\qquad \text{for} \quad n \geq 1 \end{aligned} \tag{19.145}$$

where

$$\begin{aligned} \mu_1 &= \cos\theta_1 = \cos\left\{\frac{1}{2}(\pi + \alpha_0)\right\} = -\sin\frac{\alpha_0}{2} \\ \mu_2 &= \cos\theta_2 = \cos\left\{\frac{1}{2}(\pi - \alpha_0)\right\} = \sin\frac{\alpha_0}{2} \end{aligned} \tag{19.146}$$

Referring to the linearly independent solutions of the temperature distribution given by Eq. (19.113), the temperature can be expressed by

$$T = \sum_{n=0}^{\infty} A_n r^n P_n(\mu) \tag{19.147}$$

Substitution of Eq. (19.147) into the boundary condition (19.136) gives the unknown constants A_n as

$$A_n = \frac{f_n}{a^n} \tag{19.148}$$

Therefore, the steady state temperature distribution is

$$T = T_a \sum_{n=0}^{\infty} \frac{f_n}{a^n} r^n P_n(\mu) \qquad \text{(Answer)} \, (19.149)$$

Problem 19.12. When the temperature at boundary surface ($r = a$) of a solid sphere is given by Fig. 19.1, find the displacements and stresses in the solid sphere with the initial temperature T_i under a steady temperature field.

Solution. Since the temperature is given by Eq. (19.149), the temperature change is

$$\tau = T - T_i = -T_i + T_a \sum_{n=0}^{\infty} \frac{f_n}{a^n} r^n P_n(\mu) \qquad (19.150)$$

The displacements and stresses are given by Eq. (19.17) and (19.21) by use of Goodier's thermoelastic displacement potential Φ and the Boussinesq harmonic functions φ and ψ. Referring Eqs. (19.22) and (19.129), the Boussinesq harmonic functions φ, ψ and Goodier's thermoelastic displacement potential Φ are

$$\varphi = \sum_{n=0}^{\infty} C'_n r^n P_n(\mu), \quad \psi = \sum_{n=0}^{\infty} D'_n r^n P_n(\mu) \qquad (19.151)$$

$$\Phi = K \left[-\frac{1}{6} T_i r^2 + \sum_{n=0}^{\infty} \frac{1}{2(2n+3)} A_n r^{n+2} P_n(\mu) \right] \qquad (19.152)$$

Furthermore, we introduce new constants

$$C'_n = C_n - (n - 4 + 4\nu) D_{n-2}, \qquad D'_n = (2n+1) D_{n-1} \qquad (19.153)$$

Using Eq. (19.153), φ and ψ are rewritten as

$$\varphi = \sum_{n=0}^{\infty} \left[C_n - (n - 4 + 4\nu) D_{n-2} \right] r^n P_n(\mu)$$

$$\psi = \sum_{n=0}^{\infty} (2n+1) D_{n-1} r^n P_n(\mu) \qquad (19.154)$$

Substitution of Eqs. (19.152) and (19.154) into Eqs. (19.17) and (19.21) gives the displacements and stresses

$$u_r = -\frac{1}{3}KT_i r + \sum_{n=0}^{\infty}\Big[nC_n r^{n-1} + (n+1)(n-2+4\nu)D_n r^{n+1} + \frac{n+2}{2(2n+3)}KA_n r^{n+1}\Big]P_n(\mu) \tag{19.155}$$

$$u_\theta = -(1-\mu^2)^{1/2}\sum_{n=1}^{\infty}\Big[C_n r^{n-1} + (n+5-4\nu)D_n r^{n+1} + \frac{1}{2(2n+3)}KA_n r^{n+1}\Big]\frac{n+1}{1-\mu^2}\left[\mu P_n(\mu) - P_{n+1}(\mu)\right] \tag{19.156}$$

$$\sigma_{rr} = \frac{4}{3}GKT_i + 2G\sum_{n=0}^{\infty}\Big[n(n-1)C_n r^{n-2} + (n+1)(n^2-n-2-2\nu)D_n r^n + \frac{n^2-n-4}{2(2n+3)}KA_n r^n\Big]P_n(\mu) \tag{19.157}$$

$$\begin{aligned}\sigma_{\theta\theta} = \frac{4}{3}GKT_i + 2G\Big\{&-\sum_{n=0}^{\infty}\Big[n^2 C_n r^{n-2} + (n+1)(n^2+4n+2+2\nu)D_n r^n + \frac{(n+2)^2}{2(2n+3)}KA_n r^n\Big]P_n(\mu)\\ &+\sum_{n=1}^{\infty}\Big[C_n r^{n-2} + (n+5-4\nu)D_n r^n + \frac{1}{2(2n+3)}KA_n r^n\Big]\\ &\times (n+1)\frac{\mu}{1-\mu^2}\left[\mu P_n(\mu) - P_{n+1}(\mu)\right]\Big\}\end{aligned} \tag{19.158}$$

$$\begin{aligned}\sigma_{\phi\phi} = \frac{4}{3}GKT_i + 2G\Big\{&\sum_{n=0}^{\infty}\Big\{nC_n r^{n-2} + (n+1)\big[n-2-2\nu(2n+1)\big]D_n r^n - \frac{3n+4}{2(2n+3)}KA_n r^n\Big\}P_n(\mu)\\ &-\sum_{n=1}^{\infty}\Big[C_n r^{n-2} + (n+5-4\nu)D_n r^n + \frac{1}{2(2n+3)}KA_n r^n\Big]\\ &\times (n+1)\frac{\mu}{1-\mu^2}\left[\mu P_n(\mu) - P_{n+1}(\mu)\right]\Big\}\end{aligned} \tag{19.159}$$

$$\sigma_{r\theta} = -2G(1-\mu^2)^{1/2}\sum_{n=1}^{\infty}\Big[(n-1)C_n r^{n-2} + (n^2+2n-1+2\nu)D_n r^n + \frac{n+1}{2(2n+3)}KA_n r^n\Big]\frac{n+1}{1-\mu^2}\big[\mu P_n(\mu) - P_{n+1}(\mu)\big] \tag{19.160}$$

The mechanical boundary conditions on the traction free surface $r = a$ are

$$\sigma_{rr} = 0, \quad \sigma_{r\theta} = 0 \quad \text{on } r = a \tag{19.161}$$

From the boundary conditions (19.161), the unknown constants C_n and D_n can be determined as

$$\begin{aligned} C_n &= \frac{1-\nu}{2[(n^2+n+1)+\nu(2n+1)]}KA_n a^2 \quad (n = 2, 3, \ldots) \\ D_0 &= -\frac{K}{3(1+\nu)}(A_0 - T_i) \\ D_n &= -\frac{n+2}{2(2n+3)[(n^2+n+1)+\nu(2n+1)]}KA_n \quad (n = 1, 2, \ldots) \end{aligned} \tag{19.162}$$

Then, the displacements and thermal stresses are obtained by substitution of Eq. (19.162) into Eqs. (19.155) to (19.160).

Problem 19.13. Find the displacements and stresses in a spherical coordinate system using the displacement functions φ and ψ for the axisymmetric thermoelastic deformation.

Solution. Using Eq. (19.22), the displacement functions φ and ψ are represented by series forms

$$\begin{aligned} \varphi &= \sum_{n=0}^{\infty}\left(C'_{1,n}r^n + C'_{2,n}r^{-n-1}\right)P_n(\cos\theta) \\ \psi &= \sum_{n=0}^{\infty}\left(D'_{1,n}r^n + D'_{2,n}r^{-n-1}\right)P_n(\cos\theta) \end{aligned} \tag{19.163}$$

Substituting Eq. (19.163) into Eq. (19.17), the radial displacement $\bar{u}_r$ in terms of φ and ψ is obtained

$$\begin{aligned} \bar{u}_r &= \frac{\partial\varphi}{\partial r} + \mu\Big[r\frac{\partial\psi}{\partial r} - (3-4\nu)\psi\Big] \\ &= \sum_{n=0}^{\infty}\big[nC'_{1,n}r^{n-1} - (n+1)C'_{2,n}r^{-n-2}\big]P_n(\mu) \end{aligned}$$

$$
\begin{aligned}
&+\sum_{n=0}^{\infty}\left[(n-3+4\nu)D'_{1,n}r^{n}-(n+4-4\nu)D'_{2,n}r^{-n-1}\right]\\
&\times\frac{1}{2n+1}\left[(n+1)P_{n+1}(\mu)+nP_{n-1}(\mu)\right]
\end{aligned}
\tag{19.164}
$$

in which $\mu = \cos\theta$.

Since Eq. (19.164) contains three kinds of Legendre functions with different orders $n-1$, n, $n+1$ under the summation signs, we introduce new unknown constants as follows:

$$
\begin{aligned}
C'_{1,n} &= C_{1,n}-(n-4+4\nu)D_{1,n-2}\\
C'_{2,n} &= C_{2,n}-(n+5-4\nu)D_{2,n+2}\\
D'_{1,n} &= (2n+1)D_{1,n-1}, \quad D'_{2,n} = (2n+1)D_{2,n+1}
\end{aligned}
\tag{19.165}
$$

Using new unknown constants (19.165), the displacement functions φ and ψ are represented by

$$
\begin{aligned}
\varphi &= \sum_{n=0}^{\infty}\Big\{[C_{1,n}-(n-4+4\nu)D_{1,n-2}]r^{n}\\
&\quad+[C_{2,n}-(n+5-4\nu)D_{2,n+2}]r^{-n-1}\Big\}P_n(\cos\theta)\\
\psi &= \sum_{n=0}^{\infty}\left[(2n+1)D_{1,n-1}r^{n}+(2n+1)D_{2,n+1}r^{-n-1}\right]P_n(\cos\theta)
\end{aligned}
\tag{19.166}
$$

The radial displacement $\bar{u}_r$ is then reduced to

$$
\begin{aligned}
\bar{u}_r &= \sum_{n=0}^{\infty}\{n[C_{1,n}-(n-4+4\nu)D_{1,n-2}]r^{n-1}\\
&\quad-(n+1)[C_{2,n}-(n+5-4\nu)D_{2,n+2}]r^{-n-2}\}P_n(\mu)\\
&\quad+\sum_{n=0}^{\infty}[(n-3+4\nu)(2n+1)D_{1,n-1}r^{n}\\
&\quad-(n+4-4\nu)(2n+1)D_{2,n+1}r^{-n-1}]\\
&\quad\times\frac{1}{2n+1}[(n+1)P_{n+1}(\mu)+nP_{n-1}(\mu)]\\
&= \sum_{n=0}^{\infty}[nC_{1,n}r^{n-1}-(n+1)C_{2,n}r^{-n-2}]P_n(\mu)\\
&\quad+\sum_{n=0}^{\infty}[-n(n-4+4\nu)D_{1,n-2}r^{n-1}
\end{aligned}
$$

$$
\begin{aligned}
&+ (n+1)(n+5-4\nu)D_{2,n+2}r^{-n-2}]P_n(\mu) \\
&+ \sum_{n=0}^{\infty}[(n-3+4\nu)(2n+1)D_{1,n-1}r^n \\
&- (n+4-4\nu)(2n+1)D_{2,n+1}r^{-n-1}] \\
&\times \frac{1}{2n+1}[(n+1)P_{n+1}(\mu) + nP_{n-1}(\mu)]
\end{aligned} \tag{19.167}
$$

We now change the index in Eq. (19.167) as follows:

$$
\begin{aligned}
&-\sum_{n=0}^{\infty} n(n-4+4\nu)D_{1,n-2}r^{n-1}P_n(\mu) \qquad [n-2 \to m] \\
&+\sum_{n=0}^{\infty}(n-3+4\nu)(2n+1)D_{1,n-1}r^n \\
&\qquad \times \frac{1}{2n+1}[(n+1)P_{n+1}(\mu) + nP_{n-1}(\mu)] \qquad [n-1 \to m] \\
&\quad = -\sum_{m=-2}^{\infty}(m+2)(m-2+4\nu)D_{1,m}r^{m+1}P_{m+2}(\mu) \\
&\qquad + \sum_{m=-1}^{\infty}(m-2+4\nu)D_{1,m}r^{m+1}[(m+2)P_{m+2}(\mu) + (m+1)P_m(\mu)] \\
&\quad = \sum_{m=-1}^{\infty}(m+1)(m-2+4\nu)D_{1,m}r^{m+1}P_m(\mu)] \\
&\quad = \sum_{n=0}^{\infty}(n+1)(n-2+4\nu)D_{1,n}r^{n+1}P_n(\mu)]
\end{aligned} \tag{19.168}
$$

Similarly,

$$
\begin{aligned}
&\sum_{n=0}^{\infty}(n+1)(n+5-4\nu)D_{2,n+2}r^{-n-2}P_n(\mu) \qquad [n+2 \to m] \\
&\qquad -\sum_{n=0}^{\infty}(n+4-4\nu)(2n+1)D_{2,n+1}r^{-n-1} \\
&\qquad \times \frac{1}{2n+1}[(n+1)P_{n+1}(\mu) + nP_{n-1}(\mu)] \qquad [n+1 \to m] \\
&\quad = \sum_{m=2}^{\infty}(m-1)(m+3-4\nu)D_{2,m}r^{-m}P_{m-2}(\mu)
\end{aligned}
$$

$$
\begin{aligned}
&- \sum_{m=+1}^{\infty} (m+3-4\nu) D_{2,m} r^{-m} [m P_m(\mu) + (m-1) P_{m-2}(\mu)] \\
&= - \sum_{m=+1}^{\infty} m(m+3-4\nu) D_{2,m} r^{-m} P_m(\mu) \\
&= - \sum_{n=0}^{\infty} n(n+3-4\nu) D_{2,n} r^{-n} P_n(\mu)
\end{aligned}
\tag{19.169}
$$

Substituting Eqs. (19.168) and (19.169) into Eq. (19.167), we have

$$
\begin{aligned}
\bar{u}_r = \sum_{n=0}^{\infty} \big[& n C_{1,n} r^{n-1} - (n+1) C_{2,n} r^{-n-2} \\
& + (n+1)(n-2+4\nu) D_{1,n} r^{n+1} \\
& - n(n+3-4\nu) D_{2,n} r^{-n} \big] P_n(\mu)
\end{aligned}
\qquad \text{(Answer)}
$$

In a similar way, the hoop displacement is

$$
\begin{aligned}
\bar{u}_\theta &= -(1-\mu^2)^{1/2} \left[\frac{1}{r} \frac{\partial \varphi}{\partial \mu} + \mu \frac{\partial \psi}{\partial \mu} - (3-4\nu)\psi \right] \\
&= -(1-\mu^2)^{1/2} \sum_{n=1}^{\infty} \Big[C_{1,n} r^{n-1} + C_{2,n} r^{-n-2} \\
&\quad + (n+5-4\nu) D_{1,n} r^{n+1} + (n-4+4\nu) D_{2,n} r^{-n} \Big] \\
&\quad \times \frac{n+1}{1-\mu^2} \left[\mu P_n(\mu) - P_{n+1}(\mu) \right]
\end{aligned}
\qquad \text{(Answer)}
$$

From Eq. (19.21), the stress component $\bar{\sigma}_{rr}$ in terms of φ, ψ is given by

$$
\bar{\sigma}_{rr} = 2G \Big[\frac{\partial^2 \varphi}{\partial r^2} + \mu r \frac{\partial^2 \psi}{\partial r^2} - 2(1-\nu)\mu \frac{\partial \psi}{\partial r} - 2\nu \frac{1}{r} (1-\mu^2) \frac{\partial \psi}{\partial \mu} \Big]
\tag{19.170}
$$

Using the formula

$$
(1-\mu^2) \frac{\partial P_n}{\partial \mu} = (n+1)[\mu P_n(\mu) - P_{n+1}(\mu)]
\tag{19.171}
$$

the radial stress $\bar{\sigma}_{rr}$ is

$$
\begin{aligned}
\bar{\sigma}_{rr} = 2G \Big[\sum_{n=0}^{\infty} \{ & n(n-1)[C_{1,n} - (n-4+4\nu) D_{1,n-2}] r^{n-2} \\
& + (n+1)(n+2)[C_{2,n} - (n+5-4\nu) D_{2,n+2}] r^{-n-3} \} P_n(\mu)
\end{aligned}
$$

$$
\begin{aligned}
&+\sum_{n=0}^{\infty}[n(n-1)(2n+1)D_{1,n-1}r^{n-1}\\
&+(n+1)(n+2)(2n+1)D_{2,n+1}r^{-n-2}]\mu P_n(\mu)\\
&-2(1-\nu)\sum_{n=0}^{\infty}[n(2n+1)D_{1,n-1}r^{n-1}\\
&-(n+1)(2n+1)D_{2,n+1}r^{-n-2}]\mu P_n(\mu)\\
&-2\nu\sum_{n=0}^{\infty}[(2n+1)D_{1,n-1}r^{n-1}+(2n+1)D_{2,n+1}r^{-n-2}]\\
&\times(n+1)[\mu P_n(\mu)-P_{n+1}(\mu)]\Big]\\
=\,&2G\Big\{\sum_{n=0}^{\infty}[n(n-1)C_{1,n}r^{n-2}+(n+1)(n+2)C_{2,n}r^{-n-3}]P_n(\mu)\\
&+\sum_{n=0}^{\infty}[-n(n-1)(n-4+4\nu)D_{1,n-2}r^{n-2}\\
&-(n+1)(n+2)(n+5-4\nu)D_{2,n+2}r^{-n-3}]P_n(\mu)\\
&+\sum_{n=0}^{\infty}[n(2n+1)(n-3+2\nu)D_{1,n-1}r^{n-1}\\
&+(n+1)(2n+1)(n+4-2\nu)D_{2,n+1}r^{-n-2}]\mu P_n(\mu)\\
&-2\nu\sum_{n=0}^{\infty}[(2n+1)D_{1,n-1}r^{n-1}\\
&+(2n+1)D_{2,n+1}r^{-n-2}](n+1)[\mu P_n(\mu)-P_{n+1}(\mu)]\Big\}\\
\equiv\,&2G\Big\{\sum_{n=0}^{\infty}[n(n-1)C_{1,n}r^{n-2}+(n+1)(n+2)C_{2,n}r^{-n-3}]P_n(\mu)\\
&+S_1+S_2\Big\}
\end{aligned}
\tag{19.172}
$$

Now, we change the index in Eq. (19.172) as follows:

$$
\begin{aligned}
S_1=&-\sum_{n=0}^{\infty}n(n-1)(n-4+4\nu)D_{1,n-2}r^{n-2}P_n(\mu) \qquad [n-2\to m]\\
&+\sum_{n=0}^{\infty}n(2n+1)(n-3+2\nu)D_{1,n-1}r^{n-1}\mu P_n(\mu) \qquad [n-1\to m]
\end{aligned}
$$

$$
\begin{aligned}
&- \sum_{n=0}^{\infty} 2\nu(n+1)(2n+1)D_{1,n-1}r^{n-1}[\mu P_n(\mu) - P_{n+1}(\mu)] \\
&= - \sum_{m=-2}^{\infty} (m+2)(m+1)(m-2+4\nu)D_{1,m}r^m P_{m+2}(\mu) \\
&\quad + \sum_{m=-1}^{\infty} (m+1)(2m+3)(m-2+2\nu)D_{1,m}r^m \mu P_{m+1}(\mu) \\
&\quad - \sum_{m=-1}^{\infty} 2\nu(m+2)(2m+3)D_{1,m}r^m[\mu P_{m+1}(\mu) - P_{m+2}(\mu)] \\
&= - \sum_{m=-1}^{\infty} [(m+1)(m-2) - 2\nu]D_{1,m}r^m[(m+2)P_{m+2}(\mu) \\
&\quad - (2m+3)\mu P_{m+1}(\mu)] \\
&= \sum_{m=-1}^{\infty} (m+1)[(m+1)(m-2) - 2\nu]D_{1,m}r^m P_m(\mu) \\
&= \sum_{n=0}^{\infty} (n+1)[(n+1)(n-2) - 2\nu]D_{1,n}r^n P_n(\mu)
\end{aligned}
\tag{19.173}
$$

Similarly,

$$
\begin{aligned}
S_2 &= - \sum_{n=0}^{\infty} (n+1)(n+2)(n+5-4\nu)D_{2,n+2}r^{-n-3} P_n(\mu) \qquad [n+2 \to m] \\
&\quad + \sum_{n=0}^{\infty} (n+1)(2n+1)(n+4-2\nu)D_{2,n+1}r^{-n-2} \mu P_n(\mu) \qquad [n+1 \to m] \\
&\quad - \sum_{n=0}^{\infty} 2\nu(n+1)(2n+1)D_{2,n+1}r^{-n-2}[\mu P_n(\mu) - P_{n+1}(\mu)] \\
&= - \sum_{m=2}^{\infty} m(m-1)(m+3-4\nu)D_{2,m}r^{-m-1} P_{m-2}(\mu) \\
&\quad + \sum_{m=1}^{\infty} m(2m-1)(m+3-2\nu)D_{2,m}r^{-m-1} \mu P_{m-1}(\mu) \\
&\quad - \sum_{m=1}^{\infty} 2\nu m(2m-1)D_{2,m}r^{-m-1}[\mu P_{m-1}(\mu) - P_m(\mu)]
\end{aligned}
\tag{19.174}
$$

Using formula

$$(m-1)P_{m-2}(\mu) = (2m-1)\mu P_{m-1}(\mu) - mP_m(\mu) \tag{19.175}$$

substitution of Eq. (19.175) into Eq. (19.174) gives

$$\begin{aligned}
S_2 = & -\sum_{m=2}^{\infty} m(m+3-4\nu)D_{2,m}r^{-m-1} \\
& \quad \times [(2m-1)\mu P_{m-1}(\mu) - mP_m(\mu)] \\
& + \sum_{m=1}^{\infty} m(2m-1)(m+3-2\nu)D_{2,m}r^{-m-1}\mu P_{m-1}(\mu) \\
& - \sum_{m=1}^{\infty} 2\nu m(2m-1)D_{2,m}r^{-m-1}[\mu P_{m-1}(\mu) - P_m(\mu)] \\
= & \sum_{m=1}^{\infty} m[m(m+3) - 2\nu]D_{2,m}r^{-m-1}P_m(\mu) \\
= & \sum_{n=0}^{\infty} n[n(n+3) - 2\nu]D_{2,n}r^{-n-1}P_n(\mu)
\end{aligned} \tag{19.176}$$

Substituting Eqs. (19.173) and (19.176) into Eq. (19.172), the radial stress can be obtained

$$\begin{aligned}
\bar{\sigma}_{rr} = 2G\sum_{n=0}^{\infty} \Big[& n(n-1)C_{1,n}r^{n-2} + (n+1)(n+2)C_{2,n}r^{-n-3} \\
& + (n+1)(n^2 - n - 2 - 2\nu)D_{1,n}r^n \\
& + n(n^2 + 3n - 2\nu)D_{2,n}r^{-n-1}\Big] P_n(\mu)
\end{aligned} \qquad \text{(Answer)}$$

In a similar way, the stresses are

$$\begin{aligned}
\bar{\sigma}_{\theta\theta} = & 2G\Big\{\frac{1}{r}\frac{\partial\varphi}{\partial r} + \frac{1}{r^2}\Big[-\mu\frac{\partial\varphi}{\partial\mu} + (1-\mu^2)\frac{\partial^2\varphi}{\partial\mu^2}\Big] + (1-2\nu)\mu\frac{\partial\psi}{\partial r} \\
& + \frac{\mu}{r}\Big[-\mu\frac{\partial\psi}{\partial\mu} + (1-\mu^2)\frac{\partial^2\psi}{\partial\mu^2}\Big] - 2(1-\nu)\frac{1}{r}(1-\mu^2)\frac{\partial\psi}{\partial\mu}\Big\} \\
= & 2G\Big\{-\sum_{n=0}^{\infty}\Big[n^2C_{1,n}r^{n-2} + (n+1)^2C_{2,n}r^{-n-3} \\
& + (n+1)(n^2 + 4n + 2 + 2\nu)D_{1,n}r^n \\
& + n(n^2 - 2n - 1 + 2\nu)D_{2,n}r^{-n-1}\Big] P_n(\mu)
\end{aligned}$$

$$
\begin{aligned}
&+\sum_{n=1}^{\infty}\Big[C_{1,n}r^{n-2}+C_{2,n}r^{-n-3}+(n+5-4\nu)D_{1,n}r^{n}\\
&+(n-4+4\nu)D_{2,n}r^{-n-1}\Big]\\
&\times(n+1)\frac{\mu}{1-\mu^2}\left[\mu P_n(\mu)-P_{n+1}(\mu)\right]\Big\} \qquad \text{(Answer)}
\end{aligned}
$$

$$
\begin{aligned}
\bar{\sigma}_{\phi\phi}&=2G\Big\{\frac{1}{r}\frac{\partial\varphi}{\partial r}-\mu\frac{1}{r^2}\frac{\partial\varphi}{\partial\mu}+(1-2\nu)\mu\frac{\partial\psi}{\partial r}-\Big[2\nu+(1-2\nu)\mu^2\Big]\frac{1}{r}\frac{\partial\psi}{\partial\mu}\Big\}\\
&=2G\Big\{\sum_{n=0}^{\infty}\Big\{nC_{1,n}r^{n-2}-(n+1)C_{2,n}r^{-n-3}\\
&+(n+1)[n-2-2\nu(2n+1)]D_{1,n}r^{n}\\
&-n[n+3-2\nu(2n+1)]D_{2,n}r^{-n-1}\Big\}P_n(\mu)\\
&-\sum_{n=1}^{\infty}\Big[C_{1,n}r^{n-2}+C_{2,n}r^{-n-3}+(n+5-4\nu)D_{1,n}r^{n}\\
&+(n-4+4\nu)D_{2,n}r^{-n-1}\Big]\\
&\times(n+1)\frac{\mu}{1-\mu^2}\left[\mu P_n(\mu)-P_{n+1}(\mu)\right]\Big\} \qquad \text{(Answer)}
\end{aligned}
$$

$$
\begin{aligned}
\bar{\sigma}_{r\theta}&=2G(1-\mu^2)^{1/2}\Big[-\frac{\partial^2}{\partial r\partial\mu}\left(\frac{\varphi}{r}\right)+(1-2\nu)\frac{\partial\psi}{\partial r}\\
&-\mu\frac{\partial^2\psi}{\partial r\partial\mu}+2(1-\nu)\frac{1}{r}\mu\frac{\partial\psi}{\partial\mu}\Big]\\
&=-2G(1-\mu^2)^{1/2}\sum_{n=1}^{\infty}\Big[(n-1)C_{1,n}r^{n-2}-(n+2)C_{2,n}r^{-n-3}\\
&+(n^2+2n-1+2\nu)D_{1,n}r^{n}\\
&-(n^2-2+2\nu)D_{2,n}r^{-n-1}\Big]\\
&\times\frac{n+1}{1-\mu^2}\left[\mu P_n(\mu)-P_{n+1}(\mu)\right] \qquad \text{(Answer)}
\end{aligned}
$$

Chapter 20
Thermal Stresses in Plates

In this chapter the thermal stresses and the deflection in thin rectangular plates subjected to the temperature change in the thickness direction only are recalled. The basic equations are developed with the Kirchhoff-Love hypothesis. Next, the basic equations for the thermal bending of circular plates with various boundary conditions are summarized. A number of problems for rectangular and circular plates are presented.

20.1 Basic Equations for a Rectangular Plate

We consider a thermal stress in a plate shown in Fig. 20.1, due to uniform temperature in flat surface as a simple case of thermal stresses in plates. When the temperature change τ varies in the thickness direction only, and the plate is subjected to the same bending along both x and y axes, the strain components ϵ_x and ϵ_y are

$$\epsilon_x = \epsilon_y = \epsilon_0 + \frac{z}{\rho} \tag{20.1}$$

where ϵ_0 and ρ denote the in-plane strain and the radius of curvature at the neutral plane of $z = 0$, respectively. When the plate is subjected to the in-plane force P per unit length and the bending moment M_M per unit length in the x and y directions, the thermal stress component $\sigma_x (= \sigma_y)$ is given by

$$\begin{aligned}\sigma_x(=\sigma_y) = \frac{P}{h} + \frac{12M_M}{h^3}z + \frac{\alpha E}{1-\nu}\Big[-\tau(z) + \frac{1}{h}\int_{-h/2}^{h/2}\tau(z)\,dz \\ + \frac{12z}{h^3}\int_{-h/2}^{h/2}\tau(z)z\,dz\Big]\end{aligned} \tag{20.2}$$

M. Reza Eslami et al., *Theory of Elasticity and Thermal Stresses*, Solid Mechanics and Its Applications 197, DOI: 10.1007/978-94-007-6356-2_20,

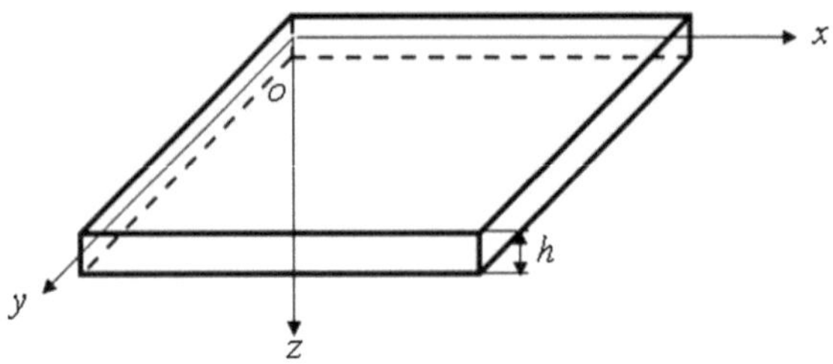

Fig. 20.1 A plate

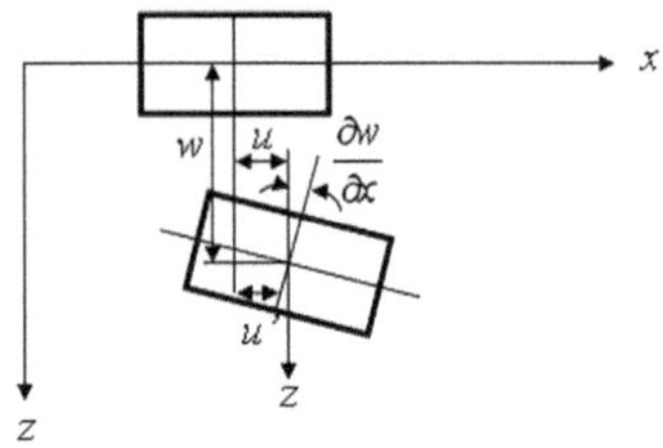

Fig. 20.2 Displacement

For the pure thermal stress problems without external forces, Eq. (20.2) reduces to

$$\sigma_x(=\sigma_y) = \frac{\alpha E}{1-\nu}\Big[-\tau(z) + \frac{1}{h}\int_{-h/2}^{h/2}\tau(z)\,dz + \frac{12z}{h^3}\int_{-h/2}^{h/2}\tau(z)z\,dz\Big] \tag{20.3}$$

It is seen that the thermal stress given by Eq. (20.3) for the plate are $1/(1-\nu)$ times the values for the beam given by Eq. (14.8).

Next, we discuss the general treatment of the thermal bending problems of an isotropic thin plate with thickness h under Kirchhoff-Love hypothesis that the plane initially perpendicular to the neutral plane of the plate remains a plane after deformation and is perpendicular to the deformed neutral plane.

Referring to Fig. 20.2, the displacement components u' and v' in the in-plane direction x and y at the arbitrary point z of the plate are

$$u' = u - z\frac{\partial w}{\partial x}, \qquad v' = v - z\frac{\partial w}{\partial y} \tag{20.4}$$

where u, v, and w are displacement components in the x, y, and z direction at the neutral plane ($z = 0$).

The strain components in the in-plane direction are

$$\begin{aligned}
\epsilon_{xx} &= \frac{\partial u'}{\partial x} = \frac{\partial u}{\partial x} - z\frac{\partial^2 w}{\partial x^2}\\
\epsilon_{yy} &= \frac{\partial v'}{\partial y} = \frac{\partial v}{\partial y} - z\frac{\partial^2 w}{\partial y^2}\\
\epsilon_{xy} &= \frac{1}{2}\Big(\frac{\partial u'}{\partial y} + \frac{\partial v'}{\partial x}\Big) = \frac{1}{2}\Big(\frac{\partial u}{\partial y} + \frac{\partial v}{\partial x}\Big) - z\frac{\partial^2 w}{\partial x\partial y}
\end{aligned} \tag{20.5}$$

Hooke's law is

$$\begin{aligned}\epsilon_{xx} &= \frac{1}{E}\left(\sigma_{xx} - \nu\sigma_{yy}\right) + \alpha\tau \\ \epsilon_{yy} &= \frac{1}{E}\left(\sigma_{yy} - \nu\sigma_{xx}\right) + \alpha\tau \\ \epsilon_{xy} &= \frac{1}{2G}\sigma_{xy} = \frac{1+\nu}{E}\sigma_{xy}\end{aligned} \tag{20.6}$$

The stress components are

$$\begin{aligned}\sigma_{xx} &= \frac{E}{1-\nu^2}\left[\frac{\partial u}{\partial x} + \nu\frac{\partial v}{\partial y} - z\left(\frac{\partial^2 w}{\partial x^2} + \nu\frac{\partial^2 w}{\partial y^2}\right) - (1+\nu)\alpha\tau\right] \\ \sigma_{yy} &= \frac{E}{1-\nu^2}\left[\frac{\partial v}{\partial y} + \nu\frac{\partial u}{\partial x} - z\left(\frac{\partial^2 w}{\partial y^2} + \nu\frac{\partial^2 w}{\partial x^2}\right) - (1+\nu)\alpha\tau\right] \\ \sigma_{xy} &= \frac{E}{2(1+\nu)}\left(\frac{\partial u}{\partial y} + \frac{\partial v}{\partial x} - 2z\frac{\partial^2 w}{\partial x \partial y}\right)\end{aligned} \tag{20.7}$$

Let us introduce the resultant forces N_x, N_y, N_{xy}, and the resultant moments M_x, M_y, M_{xy} per unit length of the plate

$$\begin{aligned}N_x &= \int_{-h/2}^{h/2} \sigma_{xx} dz, \qquad N_y = \int_{-h/2}^{h/2} \sigma_{yy} dz, \qquad N_{xy} = \int_{-h/2}^{h/2} \sigma_{xy} dz \\ M_x &= \int_{-h/2}^{h/2} \sigma_{xx} z dz, \qquad M_y = \int_{-h/2}^{h/2} \sigma_{yy} z dz, \qquad M_{xy} = -\int_{-h/2}^{h/2} \sigma_{xy} z dz\end{aligned} \tag{20.8}$$

Moreover, we introduce N_T and M_T which are the so-called thermally induced resultant force and resultant moment

$$N_T = \alpha E \int_{-h/2}^{h/2} \tau dz, \qquad M_T = \alpha E \int_{-h/2}^{h/2} \tau z dz \tag{20.9}$$

The resultant forces and resultant moments expressed by the displacement components u, v, w are

$$\begin{aligned}N_x &= \frac{Eh}{1-\nu^2}\left(\frac{\partial u}{\partial x} + \nu\frac{\partial v}{\partial y}\right) - \frac{1}{1-\nu}N_T \\ N_y &= \frac{Eh}{1-\nu^2}\left(\frac{\partial v}{\partial y} + \nu\frac{\partial u}{\partial x}\right) - \frac{1}{1-\nu}N_T \\ N_{xy} &= \frac{Eh}{2(1+\nu)}\left(\frac{\partial u}{\partial y} + \frac{\partial v}{\partial x}\right)\end{aligned} \tag{20.10}$$

$$
\begin{aligned}
M_x &= -D\Big(\frac{\partial^2 w}{\partial x^2} + \nu\frac{\partial^2 w}{\partial y^2}\Big) - \frac{1}{1-\nu}M_T \\
M_y &= -D\Big(\frac{\partial^2 w}{\partial y^2} + \nu\frac{\partial^2 w}{\partial x^2}\Big) - \frac{1}{1-\nu}M_T \\
M_{xy} &= (1-\nu)D\frac{\partial^2 w}{\partial x \partial y}
\end{aligned}
\tag{20.11}
$$

where D is the bending rigidity of the plate defined by

$$
D = \frac{Eh^3}{12(1-\nu^2)} \tag{20.12}
$$

The thermal stress components are represented in terms of the resultant forces and the resultant moments

$$
\begin{aligned}
\sigma_{xx} &= \frac{1}{h}N_x + \frac{12z}{h^3}M_x + \frac{1}{1-\nu}\Big(\frac{1}{h}N_T + \frac{12z}{h^3}M_T - \alpha E\tau\Big) \\
\sigma_{yy} &= \frac{1}{h}N_y + \frac{12z}{h^3}M_y + \frac{1}{1-\nu}\Big(\frac{1}{h}N_T + \frac{12z}{h^3}M_T - \alpha E\tau\Big) \\
\sigma_{xy} &= \frac{1}{h}N_{xy} - \frac{12z}{h^3}M_{xy}
\end{aligned}
\tag{20.13}
$$

The equilibrium equations of the forces in the in-plane directions of x and y as shown in Fig. 20.3 are

$$
\frac{\partial N_x}{\partial x} + \frac{\partial N_{xy}}{\partial y} = 0, \quad \frac{\partial N_{xy}}{\partial x} + \frac{\partial N_y}{\partial y} = 0 \tag{20.14}
$$

When the thermal stress function F is defined by

$$
N_x = \frac{\partial^2 F}{\partial y^2}, \qquad N_y = \frac{\partial^2 F}{\partial x^2}, \qquad N_{xy} = -\frac{\partial^2 F}{\partial x \partial y} \tag{20.15}
$$

the equilibrium equations (20.14) are satisfied automatically. The governing equation for F is

$$
\nabla^2\nabla^2 F = -\nabla^2 N_T \tag{20.16}
$$

Consider the equilibrium in the out-of-plane direction (z axis). We define the resultant twisting moment M_{yx}, the resultant shearing forces Q_x and Q_y per unit length parallel to axes x and y in Fig. 20.3

$$
Q_x = \int_{-h/2}^{h/2} \sigma_{xz} dz, \qquad Q_y = \int_{-h/2}^{h/2} \sigma_{yz} dz, \qquad M_{yx} = \int_{-h/2}^{h/2} \sigma_{yx} z dz \tag{20.17}
$$

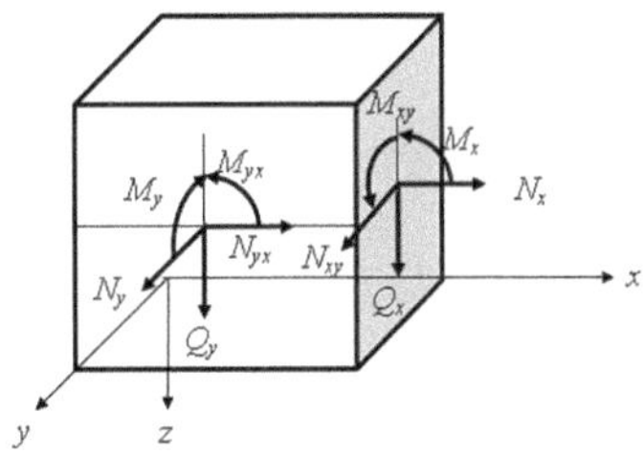

Fig. 20.3 Resultant forces, resultant moments and resultant shearing forces

Comparison between the definition of M_{xy} given by Eq. (20.8) and M_{yx} in Eq. (20.17) gives

$$M_{yx} = -M_{xy} \tag{20.18}$$

The governing equation of deflection w for the thermal bending problems is

$$\nabla^2\nabla^2 w = -\frac{1}{(1-\nu)D}\nabla^2 M_T \tag{20.19}$$

The coordinate transformations of the moments and the shearing forces between a Cartesian coordinate system (x, y) and a Cartesian coordinate system (n, s) are

$$\begin{aligned}
M_n &= M_x \cos^2\alpha + M_y \sin^2\alpha - 2M_{xy}\sin\alpha\cos\alpha \\
M_s &= M_x \sin^2\alpha + M_y \cos^2\alpha + 2M_{xy}\sin\alpha\cos\alpha \\
M_{ns} &= (M_x - M_y)\sin\alpha\cos\alpha + M_{xy}(\cos^2\alpha - \sin^2\alpha) \\
Q_n &= Q_x\cos\alpha + Q_y\sin\alpha \\
Q_s &= -Q_x\sin\alpha + Q_y\cos\alpha
\end{aligned} \tag{20.20}$$

where α means an angle between x axis and n axis.

The moments M_n, M_s, M_{ns}, and the shearing forces Q_n, Q_s in terms of w in the coordinate system (n, s) are

$$\begin{aligned}
M_n &= -D\Big(\frac{\partial^2 w}{\partial n^2} + \nu\frac{\partial^2 w}{\partial s^2}\Big) - \frac{1}{1-\nu}M_T \\
M_s &= -D\Big(\frac{\partial^2 w}{\partial s^2} + \nu\frac{\partial^2 w}{\partial n^2}\Big) - \frac{1}{1-\nu}M_T \\
M_{ns} &= (1-\nu)D\frac{\partial^2 w}{\partial n\partial s} \\
Q_n &= -\frac{\partial}{\partial n}\Big(D\nabla^2 w + \frac{1}{1-\nu}M_T\Big) \\
Q_s &= -\frac{\partial}{\partial s}\Big(D\nabla^2 w + \frac{1}{1-\nu}M_T\Big)
\end{aligned} \tag{20.21}$$

Three kinds of boundary conditions for plate bending problems due to thermal loads are

(1) A built-in edge or fixed end

$$w = 0, \quad \frac{\partial w}{\partial n} = 0 \tag{20.22}$$

(2) A simply supported edge

$$\begin{aligned} & w = 0, \quad M_n = 0 \\ & \text{alternative form} \\ & w = 0, \quad \frac{\partial^2 w}{\partial n^2} + \nu \frac{\partial^2 w}{\partial s^2} = -\frac{1}{(1-\nu)D} M_T \end{aligned} \tag{20.23}$$

(3) A free edge

$$\begin{aligned} & M_n = 0, \quad V_n = Q_n - \frac{\partial M_{ns}}{\partial s} = 0 \\ & \text{alternative form} \\ & \frac{\partial^2 w}{\partial n^2} + \nu \frac{\partial^2 w}{\partial s^2} = -\frac{1}{(1-\nu)D} M_T \\ & \frac{\partial}{\partial n}\left[\frac{\partial^2 w}{\partial n^2} + (2-\nu)\frac{\partial^2 w}{\partial s^2}\right] = -\frac{1}{(1-\nu)D}\frac{\partial M_T}{\partial n} \end{aligned} \tag{20.24}$$

20.2 Basic Equations for a Circular Plate

Let us consider the thermal bending problems of a circular plate of thickness h. We introduce the resultant forces N_r, N_θ, $N_{r\theta}$, the resultant moments M_{rr}, $M_{\theta\theta}$, $M_{r\theta}$, and the shearing forces Q_r, Q_θ per unit length of the circular plate

$$\begin{aligned} & N_r = \int_{-h/2}^{h/2} \sigma_{rr} dz, \quad N_\theta = \int_{-h/2}^{h/2} \sigma_{\theta\theta} dz, \quad N_{r\theta} = \int_{-h/2}^{h/2} \sigma_{r\theta} dz \\ & M_{rr} = \int_{-h/2}^{h/2} \sigma_{rr} z dz, \quad M_{\theta\theta} = \int_{-h/2}^{h/2} \sigma_{\theta\theta} z dz, \quad M_{r\theta} = -\int_{-h/2}^{h/2} \sigma_{r\theta} z dz \\ & Q_r = \int_{-h/2}^{h/2} \sigma_{rz} dz, \quad Q_\theta = \int_{-h/2}^{h/2} \sigma_{\theta z} dz \end{aligned} \tag{20.25}$$

The thermal stress components σ_{rr}, $\sigma_{\theta\theta}$, and $\sigma_{r\theta}$ in terms of the resultant forces and the resultant moments are

$$\sigma_{rr} = \frac{1}{h}N_r + \frac{12z}{h^3}M_{rr} + \frac{1}{1-\nu}\Big(\frac{1}{h}N_T + \frac{12z}{h^3}M_T - \alpha E\tau\Big)$$

$$\sigma_{\theta\theta} = \frac{1}{h}N_\theta + \frac{12z}{h^3}M_{\theta\theta} + \frac{1}{1-\nu}\Big(\frac{1}{h}N_T + \frac{12z}{h^3}M_T - \alpha E\tau\Big) \tag{20.26}$$

$$\sigma_{r\theta} = \frac{1}{h}N_{r\theta} - \frac{12z}{h^3}M_{r\theta}$$

The equilibrium equation of the force in the in-plane direction is

$$\nabla^2\nabla^2 F = -\nabla^2 N_T \tag{20.27}$$

where F denotes the thermal stress function defined by

$$N_r = \frac{1}{r}\frac{\partial F}{\partial \theta} + \frac{1}{r^2}\frac{\partial^2 F}{\partial \theta^2}, \qquad N_\theta = \frac{\partial^2 F}{\partial r^2}, \qquad N_{r\theta} = -\frac{\partial}{r}\Big(\frac{1}{r}\frac{\partial F}{\partial \theta}\Big) \tag{20.28}$$

and

$$\nabla^2 = \frac{\partial^2}{\partial r^2} + \frac{1}{r}\frac{\partial}{\partial \theta} + \frac{1}{r^2}\frac{\partial^2}{\partial \theta^2} \tag{20.29}$$

The governing equation of the deflection w for a circular plate is

$$\nabla^2\nabla^2 w(r,\theta) = -\frac{1}{(1-\nu)D}\nabla^2 M_T(r,\theta) \tag{20.30}$$

The components of the resultant moments and shearing forces are

$$M_{rr} = -D\Big[\frac{\partial^2 w}{\partial r^2} + \nu\Big(\frac{1}{r}\frac{\partial w}{\partial r} + \frac{1}{r^2}\frac{\partial^2 w}{\partial \theta^2}\Big)\Big] - \frac{1}{1-\nu}M_T$$

$$M_{\theta\theta} = -D\Big[\nu\frac{\partial^2 w}{\partial r^2} + \Big(\frac{1}{r}\frac{\partial w}{\partial r} + \frac{1}{r^2}\frac{\partial^2 w}{\partial \theta^2}\Big)\Big] - \frac{1}{1-\nu}M_T$$

$$M_{r\theta} = (1-\nu)D\frac{\partial}{\partial r}\Big(\frac{1}{r}\frac{\partial w}{\partial \theta}\Big) \tag{20.31}$$

$$Q_r = -D\frac{\partial}{\partial r}\big(\nabla^2 w\big) - \frac{1}{1-\nu}\frac{\partial M_T}{\partial r}$$

$$Q_\theta = -D\frac{1}{r}\frac{\partial}{\partial \theta}\big(\nabla^2 w\big) - \frac{1}{1-\nu}\frac{1}{r}\frac{\partial M_T}{\partial \theta}$$

The boundary conditions for a circular plate are
(1) A built-in edge

$$w = 0, \quad \frac{\partial w}{\partial r} = 0 \tag{20.32}$$

(2) A simply supported edge

$$w = 0, \quad M_{rr} = 0$$

alternative form (20.33)

$$w = 0, \qquad \frac{\partial^2 w}{\partial r^2} + \nu\Big(\frac{1}{r}\frac{\partial w}{\partial r} + \frac{1}{r^2}\frac{\partial^2 w}{\partial \theta^2}\Big) = -\frac{M_T}{(1-\nu)D}$$

(3) A free edge

$$M_{rr} = 0, \qquad Q_r - \frac{1}{r}\frac{\partial M_{r\theta}}{\partial \theta} = 0$$

alternative form

$$\frac{\partial^2 w}{\partial r^2} + \nu\Big(\frac{1}{r}\frac{\partial w}{\partial r} + \frac{1}{r^2}\frac{\partial^2 w}{\partial \theta^2}\Big) = -\frac{M_T}{(1-\nu)D}$$
$$\frac{\partial \nabla^2 w}{\partial r} + (1-\nu)\frac{1}{r}\frac{\partial}{\partial r}\Big(\frac{1}{r}\frac{\partial^2 w}{\partial \theta^2}\Big) = -\frac{1}{(1-\nu)D}\frac{\partial M_T}{\partial r} \tag{20.34}$$

20.3 Problems and Solutions Related to Thermal Stresses in Plates

Problem 20.1. Derive Eq. (20.2).

Solution. Since the plate is subjected to the same bending along both x and y axes, the strain components ϵ_x and ϵ_y are

$$\epsilon_x = \epsilon_y = \epsilon_0 + \frac{z}{\rho} \tag{20.35}$$

where ϵ_0 and ρ denote the in-plane strain and the radius of curvature at the neutral plane of $z = 0$, respectively. Using Hooke's law, the strain components ϵ_x and ϵ_y at a distance z from the neutral plane are expressed as

$$\epsilon_x = \frac{1}{E}(\sigma_x - \nu\sigma_y) + \alpha\tau = \epsilon_0 + \frac{z}{\rho}$$
$$\epsilon_y = \frac{1}{E}(\sigma_y - \nu\sigma_x) + \alpha\tau = \epsilon_0 + \frac{z}{\rho} \tag{20.36}$$

Therefore, the stress components σ_x and σ_y are

$$\sigma_x(= \sigma_y) = \frac{E}{1-\nu}\Big(\epsilon_0 + \frac{z}{\rho} - \alpha\tau\Big) \tag{20.37}$$

Since the plate is subjected to the in-plane force P per unit length and the bending moment M_M per unit length in the x and y directions, both the balances of force and moment are

$$\int_{-h/2}^{h/2} \sigma_x \, dz = P, \qquad \int_{-h/2}^{h/2} \sigma_x z \, dz = M_M \tag{20.38}$$

Substitution of Eq. (20.37) into Eq. (20.38) gives

$$\epsilon_0 = \frac{(1-\nu)P}{Eh} + \frac{\alpha}{h}\int_{-h/2}^{h/2} \tau(z) dz, \qquad \frac{1}{\rho} = \frac{12(1-\nu)M_M}{Eh^3} + \frac{12\alpha}{h^3}\int_{-h/2}^{h/2} \tau(z) z dz \tag{20.39}$$

From Eqs. (20.37) and (20.39), the thermal stress component $\sigma_x (= \sigma_y)$ for the plate is

$$\sigma_x (= \sigma_y) = \frac{P}{h} + \frac{12M_M}{h^3} z + \frac{\alpha E}{1-\nu}\left[-\tau(z) + \frac{1}{h}\int_{-h/2}^{h/2} \tau(z)\, dz + \frac{12z}{h^3}\int_{-h/2}^{h/2} \tau(z) z \, dz\right] \quad \text{(Answer)}$$

Problem 20.2. When a thin plate is subjected to the temperature change $\tau(z) = A + Bz$, find the thermal stress.

Solution. Thermal stress is given by Eq. (20.3)

$$\sigma_x = \frac{\alpha E}{1-\nu}\left[-\tau(z) + \frac{1}{h}\int_{-h/2}^{h/2} \tau(z)\, dz + \frac{12z}{h^3}\int_{-h/2}^{h/2} \tau(z) z \, dz\right] \tag{20.40}$$

We calculate each integral

$$\begin{aligned} \int_{-h/2}^{h/2} \tau(z)\, dz &= \int_{-h/2}^{h/2} (A + Bz)\, dz = Ah \\ \int_{-h/2}^{h/2} \tau(z) z \, dz &= \int_{-h/2}^{h/2} (Az + Bz^2)\, dz = \frac{B}{12} h^3 \end{aligned} \tag{20.41}$$

Substitution of Eq. (20.41) into Eq. (20.40) gives thermal stress

$$\sigma_x = \frac{\alpha E}{1-\nu}\left[-(A + Bz) + \frac{1}{h} Ah + \frac{12z}{h^3}\frac{B}{12} h^3\right] = 0 \quad \text{(Answer)}$$

Thus, thermal stress does not occur.

Problem 20.3. When a two-layered thin plate is subjected to the temperature change $\tau(z)$, find the thermal stress components σ_{xi} and σ_{yi}.

Solution. Let us denote the material constants and thickness for each layer by $\alpha_i, E_i, \nu_i, h_i \quad (i = 1, 2)$. Since the components of strain and stress in the x, y directions are identical, they are

$$\begin{aligned} \varepsilon_{xi} &= \varepsilon_0 + \frac{z}{\rho} = \frac{1}{E_i}(\sigma_{xi} - \nu_i \sigma_{yi}) + \alpha_i \tau(z) \\ \varepsilon_{yi} &= \varepsilon_0 + \frac{z}{\rho} = \frac{1}{E_i}(\sigma_{yi} - \nu_i \sigma_{xi}) + \alpha_i \tau(z) \end{aligned} \tag{20.42}$$

where ε_0 and ρ are the uniform strain and the uniform radius of curvature at the bonded surface ($z = 0$) in the x, y directions, respectively.

From Eq. (20.42), the stress components are represented by

$$\sigma_{xi}(= \sigma_{yi}) = \frac{E_i}{1 - \nu_i}\big[\varepsilon_0 + \frac{z}{\rho} - \alpha_i \tau(z)\big] \qquad (i = 1.2) \tag{20.43}$$

The equilibrium conditions of resultant forces and resultant moments are

$$\begin{aligned} \int_{-h_1}^{0} \sigma_{x1} dz + \int_0^{h_2} \sigma_{x2} dz = 0 \\ \int_{-h_1}^{0} \sigma_{x1} z dz + \int_0^{h_2} \sigma_{x2} z dz = 0 \end{aligned} \tag{20.44}$$

Substitution of Eq. (20.43) into Eq. (20.44) gives

$$\begin{aligned} &\int_{-h_1}^{0} \frac{E_1}{1 - \nu_1}\Big[\varepsilon_0 + \frac{z}{\rho} - \alpha_1 \tau(z)\Big] dz \\ &\quad + \int_0^{h_2} \frac{E_2}{1 - \nu_2}\Big[\varepsilon_0 + \frac{z}{\rho} - \alpha_2 \tau(z)\Big] dz = 0 \\ &\int_{-h_1}^{0} \frac{E_1}{1 - \nu_1}\Big[\varepsilon_0 + \frac{z}{\rho} - \alpha_1 \tau(z)\Big] z dz \\ &\quad + \int_0^{h_2} \frac{E_2}{1 - \nu_2}\Big[\varepsilon_0 + \frac{z}{\rho} - \alpha_2 \tau(z)\Big] z dz = 0 \end{aligned} \tag{20.45}$$

Performing the calculation of integrals for Eq. (20.45), we have

$$\begin{aligned} &\varepsilon_0[(1 - \nu_2)E_1 h_1 + (1 - \nu_1)E_2 h_2] + \frac{1}{2\rho}[-(1 - \nu_2)E_1 h_1^2 + (1 - \nu_1)E_2 h_2^2] \\ &\quad = (1 - \nu_2)\alpha_1 E_1 \int_{-h_1}^{0} \tau(z) dz + (1 - \nu_1)\alpha_2 E_2 \int_0^{h_2} \tau(z) dz \\ &\frac{1}{2}\varepsilon_0[-(1 - \nu_2)E_1 h_1^2 + (1 - \nu_1)E_2 h_2^2] + \frac{1}{3}\frac{1}{\rho}[(1 - \nu_2)E_1 h_1^3 + (1 - \nu_1)E_2 h_2^3] \\ &\quad = (1 - \nu_2)\alpha_1 E_1 \int_{-h_1}^{0} \tau(z) z dz + (1 - \nu_1)\alpha_2 E_2 \int_0^{h_2} \tau(z) z dz \end{aligned} \tag{20.46}$$

From Eq. (20.46), ε_0 and $1/\rho$ can be determined as

$$\begin{aligned}\epsilon_0 = \frac{1}{D}\bigg\{&\frac{1}{3}\Big[(1-\nu_2)E_1h_1^3+(1-\nu_1)E_2h_2^3\Big]\\ &\times\Big[(1-\nu_2)\alpha_1E_1\int_{-h_1}^{0}\tau\,dz+(1-\nu_1)\alpha_2E_2\int_0^{h_2}\tau\,dz\Big]\\ &-\frac{1}{2}\Big[-(1-\nu_2)E_1h_1^2+(1-\nu_1)E_2h_2^2\Big]\\ &\times\Big[(1-\nu_2)\alpha_1E_1\int_{-h_1}^{0}\tau z\,dz+(1-\nu_1)\alpha_2E_2\int_0^{h_2}\tau z\,dz\Big]\bigg\}\end{aligned} \tag{20.47}$$

$$\begin{aligned}\frac{1}{\rho} = \frac{1}{D}\bigg\{&\Big[(1-\nu_2)E_1h_1+(1-\nu_1)E_2h_2\Big]\\ &\times\Big[(1-\nu_2)\alpha_1E_1\int_{-h_1}^{0}\tau z\,dz+(1-\nu_1)\alpha_2E_2\int_0^{h_2}\tau z\,dz\Big]\\ &-\frac{1}{2}\Big[-(1-\nu_2)E_1h_1^2+(1-\nu_1)E_2h_2^2\Big]\\ &\times\Big[(1-\nu_2)\alpha_1E_1\int_{-h_1}^{0}\tau\,dz+(1-\nu_1)\alpha_2E_2\int_0^{h_2}\tau\,dz\Big]\bigg\}\end{aligned} \tag{20.48}$$

in which

$$\begin{aligned}D = \frac{1}{12}\bigg\{&\Big[(1-\nu_2)E_1h_1^2+(1-\nu_1)E_2h_2^2\Big]^2\\ &+4(1-\nu_1)(1-\nu_2)E_1E_2h_1h_2(h_1^2+h_1h_2+h_2^2)\bigg\}\end{aligned} \tag{20.49}$$

Substituting Eqs. (20.47) and (20.48) into Eq. (20.43), we can obtain thermal stress components σ_{xi} and σ_{yi} in the two-layered thin plate.

Problem 20.4. When a multi-layered plate is subjected to the temperature change $\tau(z)$, find the thermal stress components σ_{xi} and σ_{yi}.

Solution. The origin of the coordinate system is taken on the upper surface of the multi-layered plate with thickness h. The material constants for each layer are defined by $\alpha_i, E_i, \nu_i \quad (i = 1, \cdots, n)$. The thickness of each layer is given by

$$h_i = z_i - z_{i-1} \quad (i = 1, \ldots, n), \quad z_0 = 0, \quad z_n = h \tag{20.50}$$

Since the components of strain and stress in the x, y directions are identical, the strain components can be expressed by

$$\begin{aligned}\varepsilon_{xi} &= \varepsilon_0 + \frac{z}{\rho} = \frac{1}{E_i}(\sigma_{xi} - \nu_i\sigma_{yi}) + \alpha_i\tau(z)\\ \varepsilon_{yi} &= \varepsilon_0 + \frac{z}{\rho} = \frac{1}{E_i}(\sigma_{yi} - \nu_i\sigma_{xi}) + \alpha_i\tau(z)\end{aligned} \tag{20.51}$$

where ε_0 and ρ are the uniform strain and the uniform radius of curvature at the position $z = 0$ in the x, y directions, respectively.

From Eq. (20.51), the stress components are

$$\sigma_{xi}(=\sigma_{yi}) = \frac{E_i}{1-\nu_i}\left[\varepsilon_0 + \frac{z}{\rho} - \alpha_i\tau(z)\right] \tag{20.52}$$

The equilibrium conditions of the resultant force and the resultant moment are

$$\int_0^h \sigma_x dz = 0, \quad \int_0^h \sigma_x z dz = 0 \tag{20.53}$$

Substitution of Eq. (20.52) into Eq. (20.53) gives

$$\begin{aligned}
&\sum_{i=1}^{n}\int_{z_{i-1}}^{z_i} \frac{E_i}{1-\nu_i}\left[\varepsilon_0 + \frac{z}{\rho} - \alpha_i\tau(z)\right]dz = 0\\
&\sum_{i=1}^{n}\int_{z_{i-1}}^{z_i} \frac{E_i}{1-\nu_i}\left[\varepsilon_0 + \frac{z}{\rho} - \alpha_i\tau(z)\right]zdz = 0
\end{aligned} \tag{20.54}$$

Performing the integration for Eq. (20.54), we have

$$\begin{aligned}
&\varepsilon_0\sum_{i=1}^{n}\frac{E_ih_i}{1-\nu_i} + \frac{1}{\rho}\sum_{i=1}^{n}\frac{E_ih_i}{1-\nu_i}\frac{1}{2}(z_i + z_{i-1})\\
&\quad = \sum_{i=1}^{n}\frac{\alpha_iE_i}{1-\nu_i}\int_{z_{i-1}}^{z_i}\tau(z)dz\\
&\varepsilon_0\sum_{i=1}^{n}\frac{E_ih_i}{1-\nu_i}\frac{1}{2}(z_i + z_{i-1}) + \frac{1}{\rho}\sum_{i=1}^{n}\frac{E_ih_i}{1-\nu_i}\frac{1}{3}(z_i^2 + z_iz_{i-1} + z_{i-1}^2)\\
&\quad = \sum_{i=1}^{n}\frac{\alpha_iE_i}{1-\nu_i}\int_{z_{i-1}}^{z_i}\tau(z)zdz
\end{aligned} \tag{20.55}$$

From Eq. (20.55), ε_0 and $1/\rho$ can be determined as

$$\begin{aligned}
\epsilon_0 = \frac{1}{D}\Bigg[&\sum_{i=1}^{n}\frac{E_ih_i}{1-\nu_i}\frac{1}{3}(z_i^2 + z_iz_{i-1} + z_{i-1}^2)\sum_{j=1}^{n}\frac{\alpha_jE_j}{1-\nu_j}\int_{z_{j-1}}^{z_j}\tau(z)\,dz\\
&-\sum_{i=1}^{n}\frac{E_ih_i}{1-\nu_i}\frac{1}{2}(z_i + z_{i-1})\sum_{j=1}^{n}\frac{\alpha_jE_j}{1-\nu_j}\int_{z_{j-1}}^{z_j}\tau(z)z\,dz\Bigg]
\end{aligned} \tag{20.56}$$

$$\frac{1}{\rho} = \frac{1}{D}\Big[\sum_{i=1}^{n}\frac{E_i h_i}{1-\nu_i}\sum_{j=1}^{n}\frac{\alpha_j E_j}{1-\nu_j}\int_{z_{j-1}}^{z_j}\tau(z)z\,dz - \sum_{i=1}^{n}\frac{E_i h_i}{1-\nu_i}\frac{1}{2}(z_i+z_{i-1})\sum_{j=1}^{n}\frac{\alpha_j E_j}{1-\nu_j}\int_{z_{j-1}}^{z_j}\tau(z)\,dz\Big] \tag{20.57}$$

where

$$D = \sum_{i=1}^{n}\frac{E_i h_i}{1-\nu_i}\sum_{i=1}^{n}\frac{E_i h_i}{1-\nu_i}\frac{1}{3}(z_i^2 + z_i z_{i-1} + z_{i-1}^2) - \Big[\sum_{i=1}^{n}\frac{E_i h_i}{1-\nu_i}\frac{1}{2}(z_i+z_{i-1})\Big]^2 \tag{20.58}$$

Substituting ϵ_0 and $1/\rho$ into Eq. (20.52), we can obtain the thermal stress components σ_{xi} and σ_{yi} in the multi-layered thin plate.

Problem 20.5. When a nonhomogeneous plate is subjected to the temperature change $\tau(z)$, find the thermal stress components σ_x and σ_y.

Solution. Since the components of strain and stress in the x, y directions are identical, they are defined by

$$\begin{aligned}
\varepsilon_x &= \varepsilon_0 + \frac{z}{\rho} = \frac{1}{E(z)}[\sigma_x - \nu(z)\sigma_y] + \alpha(z)\tau(z)\\
\varepsilon_y &= \varepsilon_0 + \frac{z}{\rho} = \frac{1}{E(z)}[\sigma_y - \nu(z)\sigma_x] + \alpha(z)\tau(z)
\end{aligned} \tag{20.59}$$

where ε_0 and ρ are the uniform strain and the uniform radius of curvature at the position $z = 0$ in the x, y directions, respectively.

From Eq. (20.59), the stress component is

$$\sigma_x(=\sigma_y) = \frac{E(z)}{1-\nu(z)}\Big[\varepsilon_0 + \frac{z}{\rho} - \alpha(z)\tau(z)\Big] \tag{20.60}$$

The equilibrium conditions of the resultant force and the resultant moment are

$$\int_{-h/2}^{h/2}\sigma_x dz = 0, \quad \int_{-h/2}^{h/2}\sigma_x z dz = 0 \tag{20.61}$$

Substitution of Eq. (20.60) into Eq. (20.61) gives

$$\varepsilon_0 \int_{-h/2}^{h/2} \frac{E(z)}{1-\nu(z)} dz + \frac{1}{\rho} \int_{-h/2}^{h/2} \frac{E(z)}{1-\nu(z)} z dz = \int_{-h/2}^{h/2} \frac{\alpha(z)E(z)}{1-\nu(z)} \tau(z) dz$$
$$\varepsilon_0 \int_{-h/2}^{h/2} \frac{E(z)}{1-\nu(z)} z dz + \frac{1}{\rho} \int_{-h/2}^{h/2} \frac{E(z)}{1-\nu(z)} z^2 dz = \int_{-h/2}^{h/2} \frac{\alpha(z)E(z)}{1-\nu(z)} \tau(z) z dz \tag{20.62}$$

From Eq. (20.62), ε_0 and $1/\rho$ can be determined as

$$\begin{aligned} \epsilon_0 = \frac{1}{D}\Big[&\int_{-h/2}^{h/2} \frac{E(z)}{1-\nu(z)} z^2\, dz \int_{-h/2}^{h/2} \frac{\alpha(z)E(z)}{1-\nu(z)} \tau(z)\, dz \\ &- \int_{-h/2}^{h/2} \frac{E(z)}{1-\nu(z)} z\, dz \int_{-h/2}^{h/2} \frac{\alpha(z)E(z)}{1-\nu(z)} \tau(z) z\, dz\Big] \end{aligned} \tag{20.63}$$

$$\begin{aligned} \frac{1}{\rho} = \frac{1}{D}\Big[&\int_{-h/2}^{h/2} \frac{E(z)}{1-\nu(z)}\, dz \int_{-h/2}^{h/2} \frac{\alpha(z)E(z)}{1-\nu(z)} \tau(z) z\, dz \\ &- \int_{-h/2}^{h/2} \frac{E(z)}{1-\nu(z)} z\, dz \int_{-h/2}^{h/2} \frac{\alpha(z)E(z)}{1-\nu(z)} \tau(z)\, dz\Big] \end{aligned} \tag{20.64}$$

where

$$D = \int_{-h/2}^{h/2} \frac{E(z)}{1-\nu(z)}\, dz \int_{-h/2}^{h/2} \frac{E(z)}{1-\nu(z)} z^2\, dz - \Big[\int_{-h/2}^{h/2} \frac{E(z)}{1-\nu(z)} z\, dz\Big]^2 \tag{20.65}$$

Substituting ϵ_0 and $1/\rho$ into Eq. (20.60), we can obtain thermal stress components σ_x and σ_y in the nonhomogeneous thin plate.

Problem 20.6. When an orthotropic plate is subjected to the temperature change $\tau(z)$, find the thermal stress components σ_x and σ_y.

Solution. When the principal axes of orthotropy coincide with the coordinate axes x and y, the strain components ε_x and ε_y are

$$\varepsilon_x = \varepsilon_{0x} + \frac{z}{\rho_x}, \qquad \varepsilon_y = \varepsilon_{0y} + \frac{z}{\rho_y} \tag{20.66}$$

From Hooke's law, the stress components σ_x and σ_y are obtained

$$\begin{aligned} \sigma_x - \nu_{xy}\sigma_y &= E_x\big[\varepsilon_x - \alpha_x \tau(z)\big] = E_x\big[\varepsilon_{0x} + \frac{z}{\rho_x} - \alpha_x \tau(z)\big] \\ \sigma_y - \nu_{yx}\sigma_x &= E_y\big[\varepsilon_y - \alpha_y \tau(z)\big] = E_y[\varepsilon_{0y} + \frac{z}{\rho_y} - \alpha_y \tau(z)] \end{aligned} \tag{20.67}$$

From Eq. (20.67), the stress components σ_x and σ_y may be solved as follows:

$$\sigma_x = \frac{1}{1-\nu_{xy}\nu_{yx}}\{E_x[\varepsilon_{0x} + \frac{z}{\rho_x} - \alpha_x\tau(z)] + \nu_{xy}E_y[\varepsilon_{0y} + \frac{z}{\rho_y} - \alpha_y\tau(z)]\}$$
$$\sigma_y = \frac{1}{1-\nu_{xy}\nu_{yx}}\{E_y[\varepsilon_{0y} + \frac{z}{\rho_y} - \alpha_y\tau(z)] + \nu_{yx}E_x[\varepsilon_{0x} + \frac{z}{\rho_x} - \alpha_x\tau(z)]\} \tag{20.68}$$

Now, the unknown constants $\varepsilon_{0x}, \varepsilon_{0y}, 1/\rho_x, 1/\rho_y$ are determined from the equilibrium conditions of the resultant forces and the resultant moments in the x, y directions. Namely,

$$\int_{-h/2}^{h/2}\sigma_x dz = 0,\quad \int_{-h/2}^{h/2}\sigma_y dz = 0,\quad \int_{-h/2}^{h/2}\sigma_x z dz = 0,\quad \int_{-h/2}^{h/2}\sigma_y z dz = 0 \tag{20.69}$$

By thc substitution of Eq. (20.68) into Eq. (20.69), we have

$$\int_{-h/2}^{h/2}\Big\{[E_x\varepsilon_{0x} + \nu_{xy}E_y\varepsilon_{0y}] + \Big[E_x\frac{1}{\rho_x} + \nu_{xy}E_y\frac{1}{\rho_y}\Big]z - [\alpha_x E_x + \nu_{xy}\alpha_y E_y]\tau(z)\Big\}dz = 0$$
$$\int_{-h/2}^{h/2}\Big\{[E_y\varepsilon_{0y} + \nu_{yx}E_x\varepsilon_{0x}] + \Big[E_y\frac{1}{\rho_y} + \nu_{yx}E_x\frac{1}{\rho_x}\Big]z - [\alpha_y E_y + \nu_{yx}\alpha_x E_x]\tau(z)\Big\}dz = 0$$
$$\int_{-h/2}^{h/2}\Big\{[E_x\varepsilon_{0x} + \nu_{xy}E_y\varepsilon_{0y}]z + \Big[E_x\frac{1}{\rho_x} + \nu_{xy}E_y\frac{1}{\rho_y}\Big]z^2 - [\alpha_x E_x + \nu_{xy}\alpha_y E_y]\tau(z)z\Big\}dz = 0$$
$$\int_{-h/2}^{h/2}\Big\{[E_y\varepsilon_{0y} + \nu_{yx}E_x\varepsilon_{0x}]z + \Big[E_y\frac{1}{\rho_y} + \nu_{yx}E_x\frac{1}{\rho_x}\Big]z^2 - [\alpha_y E_y + \nu_{yx}\alpha_x E_x]\tau(z)z\Big\}dz = 0 \tag{20.70}$$

Calculation of integrals in Eq. (20.70) gives

$$(E_x\varepsilon_{0x} + \nu_{xy}E_y\varepsilon_{0y}) = (\alpha_x E_x + \nu_{xy}\alpha_y E_y)\frac{1}{h}\int_{-h/2}^{h/2}\tau(z)dz$$
$$(E_y\varepsilon_{0y} + \nu_{yx}E_x\varepsilon_{0x}) = (\alpha_y E_y + \nu_{yx}\alpha_x E_x)\frac{1}{h}\int_{-h/2}^{h/2}\tau(z)dz$$

$$
\begin{aligned}
(E_x \frac{1}{\rho_x} + \nu_{xy} E_y \frac{1}{\rho_y}) &= (\alpha_x E_x + \nu_{xy} \alpha_y E_y) \frac{12}{h^3} \int_{-h/2}^{h/2} \tau(z) z dz \\
(E_y \frac{1}{\rho_y} + \nu_{yx} E_x \frac{1}{\rho_x}) &= (\alpha_y E_y + \nu_{yx} \alpha_x E_x) \frac{12}{h^3} \int_{-h/2}^{h/2} \tau(z) z dz
\end{aligned} \tag{20.71}
$$

Solving Eq. (20.71), we get

$$
\begin{aligned}
\varepsilon_{0x} &= \alpha_x \frac{1}{h} \int_{-h/2}^{h/2} \tau(z) dz, \qquad \varepsilon_{0y} = \alpha_y \frac{1}{h} \int_{-h/2}^{h/2} \tau(z) dz \\
\frac{1}{\rho_x} &= \alpha_x \frac{12}{h^3} \int_{-h/2}^{h/2} \tau(z) a dz, \qquad \frac{1}{\rho_y} = \alpha_y \frac{12}{h^3} \int_{-h/2}^{h/2} \tau(z) a dz
\end{aligned} \tag{20.72}
$$

By the substitution of Eq. (20.72) into Eq. (20.68), we obtain thermal stresses

$$
\begin{aligned}
\sigma_x &= \frac{\alpha_x E_x + \nu_{xy} \alpha_y E_y}{1 - \nu_{xy} \nu_{yx}} \left\{ -\tau(z) + \frac{1}{h} \int_{-h/2}^{h/2} \tau(z) dz + \frac{12z}{h^3} \int_{-h/2}^{h/2} \tau(z) z dz \right\} \\
\sigma_y &= \frac{\alpha_y E_y + \nu_{yx} \alpha_x E_x}{1 - \nu_{xy} \nu_{yx}} \left\{ -\tau(z) + \frac{1}{h} \int_{-h/2}^{h/2} \tau(z) dz + \frac{12z}{h^3} \int_{-h/2}^{h/2} \tau(z) z dz \right\} \\
&= \frac{\alpha_y E_y + \nu_{yx} \alpha_x E_x}{\alpha_x E_x + \nu_{xy} \alpha_y E_y} \sigma_x
\end{aligned} \tag{Answer}
$$

Problem 20.7. Derive the equilibrium equations for the resultant shearing forces Q_x and Q_y in the plate.

Solution. Considering the equilibrium condition of the resultant moment with respect to the y axis, we have

$$
\left[\left(M_x + \frac{\partial M_x}{\partial x} dx \right) - M_x \right] dy + \left[\left(M_{yx} + \frac{\partial M_{yx}}{\partial y} dy \right) - M_{yx} \right] dx - \left(Q_x + \frac{\partial Q_x}{\partial x} dx \right) dy dx = 0 \tag{20.73}
$$

Equation (20.73) reduces to

$$
\frac{\partial M_x}{\partial x} + \frac{\partial M_{yx}}{\partial y} - Q_x = 0 \tag{Answer}
$$

Similarly, the following relation is obtained from the equilibrium condition of the resultant moment with respect to the x axis

$$
\left[\left(M_y + \frac{\partial M_y}{\partial y} dy \right) - M_y \right] dx - \left[\left(M_{xy} + \frac{\partial M_{xy}}{\partial x} dx \right) - M_{xy} \right] dy - \left(Q_y + \frac{\partial Q_y}{\partial y} dy \right) dx dy = 0 \tag{20.74}
$$

Equation (20.74) reduces to

$$\frac{\partial M_y}{\partial y} - \frac{\partial M_{xy}}{\partial x} - Q_y = 0 \qquad \text{(Answer)}$$

Now, considering the equilibrium condition of the resultant force in the z direction, we have

$$\left[\left(Q_x + \frac{\partial Q_x}{\partial x}dx\right) - Q_x\right]dy + \left[\left(Q_y + \frac{\partial Q_y}{\partial y}dy\right) - Q_y\right]dx + pdxdy = 0 \quad (20.75)$$

where p means an exteral load acting on the flat surface of the palte. Equation (20.75) reduces to

$$\frac{\partial Q_x}{\partial x} + \frac{\partial Q_y}{\partial y} + p = 0 \qquad \text{(Answer)}$$

Problem 20.8. Derive the complementary solution w_c of Eq. (20.19) for the simply supported rectangular plate ($a \times b$).

Solution. The complementary solution of Eq. (20.19) is given by Eq. (17.80). Here, we consider the complementary solution which is suitable for the simply supported rectangular plate. The fundamental equation for the complementary solution w_c of Eq. (20.19) is given by

$$\frac{\partial^4 w_c}{\partial x^4} + 2\frac{\partial^4 w_c}{\partial x^2 \partial y^2} + \frac{\partial^4 w_c}{\partial y^4} = 0 \tag{20.76}$$

Now, we assume that

$$w_c = \sum_{m=1,3,5,}^{\infty} f_m(y) \cos s_m x \tag{20.77}$$

where

$$s_m = \frac{m\pi}{a} \tag{20.78}$$

Then, the deflection w_c is automatically satisfied by the following boundary condition of the simply supported plate

$$w_c = \frac{\partial^2 w_c}{\partial x^2} = 0 \quad \text{on} \quad x = \pm\frac{a}{2} \tag{20.79}$$

By the substitution of Eq. (20.77) into Eq. (20.76), we have

$$\left(\frac{d^4}{dy^4} - 2s_m^2 \frac{d^2}{dy^2} + s_m^4\right) f_m(y) = 0 \tag{20.80}$$

Now, we put $f_m(y)$ into the form

$$f_m(y) = \exp(cy) \tag{20.81}$$

By the substitution of Eq. (20.81) into Eq. (20.80), we have

$$(c - s_m)^2 (c + s_m)^2 = 0 \tag{20.82}$$

From Eqs. (20.81) and (20.82), the linearly independent solutions of Eq. (20.80) are given as

$$f_m(y) = \begin{pmatrix} \exp(s_m y) \\ \exp(-s_m y) \\ y\exp(s_m y) \\ y\exp(-s_m y) \end{pmatrix} \quad \text{or} \quad f_m(y) = \begin{pmatrix} \sinh(s_m y) \\ \cosh(s_m y) \\ y\sinh(s_m y) \\ y\cosh(s_m y) \end{pmatrix} \tag{20.83}$$

Now, we choose as $f_m(y)$ a symmetrical solution with respect to y

$$f_m(y) = C_{1m} \cosh s_m y + C_{2m} s_m y \sinh s_m y \tag{20.84}$$

By the substitution of Eq. (20.84) into Eq. (20.77), the following result is obtained

$$w_c = \sum_{m=1,3,5,}^{\infty} (C_{1m} \cosh s_m y + C_{2m} \alpha_m y \sinh s_m y) \cos s_m x \tag{Answer}$$

Problem 20.9. Derive the thermal deflection w, and bending moments M_{rr} and $M_{\theta\theta}$ for a simply supported solid circular plate when the temperature distribution T is given by

$$T = \frac{1}{2}(T_b + T_a) + (T_b - T_a)\frac{z}{h} \tag{20.85}$$

and the initial temperature is $T = T_i$.

Solution. The governing equation of deflection w is given by Eq. (20.30)

$$\frac{1}{r}\frac{\partial}{\partial r}\left\{r\frac{\partial}{\partial r}\left[\frac{1}{r}\frac{\partial}{\partial r}\left(r\frac{\partial w}{\partial r}\right) + \frac{1}{(1-\nu)D}M_T\right]\right\} = 0 \tag{20.86}$$

A general solution of Eq. (20.86) is

$$\begin{aligned} w(r) = \frac{1}{D}\Big[& C_1 r^2 + C_2 + C_3 r^2 \ln r + C_4 \ln r \\ & - \frac{1}{1-\nu}\int_r \frac{1}{r}\int_r r M_T(r) dr dr \Big] \end{aligned} \tag{20.87}$$

in which C_1 to C_4 are the unknown constants. Substituting Eq. (20.87) into Eq. (20.31), the resultant moments M_{rr}, $M_{\theta\theta}$, and shearing force Q_r are represented as

$$\begin{aligned}
M_{rr} &= -\Big\{2(1+\nu)C_1 + [2(1+\nu)\ln r + 3 + \nu]\,C_3 \\
&\quad - (1-\nu)C_4\frac{1}{r^2} + \frac{1}{r^2}\int_r rM_T(r)dr\Big\} \\
M_{\theta\theta} &= -\Big\{2(1+\nu)C_1 + [2(1+\nu)\ln r + 1 + 3\nu]\,C_3 \\
&\quad + (1-\nu)C_4\frac{1}{r^2} + M_T - \frac{1}{r^2}\int_r rM_T(r)dr\Big\} \\
Q_r &= -4C_3\frac{1}{r}
\end{aligned} \tag{20.88}$$

Since the solid circular plate with radius a is treated, it can be found from Eqs. (20.87) and (20.88) that the solutions for the solid plate are easily derived by neglecting the terms containing C_3 and C_4. Then w, M_{rr}, $M_{\theta\theta}$ and Q_r are

$$\begin{aligned}
w &= \frac{1}{D}\Big[C_1r^2 + C_2 - \frac{1}{1-\nu}\int_0^r\frac{1}{r}\int_0^r rM_T(r)drdr\Big] \\
M_{rr} &= -\Big[2(1+\nu)C_1 + \frac{1}{r^2}\int_0^r rM_T(r)dr\Big] \\
M_{\theta\theta} &= -\Big[2(1+\nu)C_1 + M_T - \frac{1}{r^2}\int_0^r rM_T(r)dr\Big] \\
Q_r &= 0
\end{aligned} \tag{20.89}$$

The simply supported boundary condition is from Eq. (20.33)

$$w = 0, \qquad \frac{\partial^2 w}{\partial r^2} + \nu\frac{1}{r}\frac{\partial w}{\partial r} = -\frac{M_T}{(1-\nu)D} \quad \text{at} \quad r = a \tag{20.90}$$

The unknown constants C_1 and C_2 are determined from Eq. (20.90)

$$\begin{aligned}
C_1 &= -\frac{1}{2(1+\nu)}\frac{1}{a^2}\int_0^a rM_T(r)dr \\
C_2 &= \frac{1}{1-\nu}\int_0^a\frac{1}{r}\int_0^r rM_T(r)drdr + \frac{1}{2(1+\nu)}\int_0^a rM_T(r)dr
\end{aligned} \tag{20.91}$$

Then, w, M_{rr}, $M_{\theta\theta}$, and Q_r are

$$\begin{aligned}
w &= \frac{1}{D}\Big[\frac{1}{1-\nu}\int_r^a\frac{1}{r}\int_0^r rM_T(r)drdr + \frac{1}{2(1+\nu)}\Big(1-\frac{r^2}{a^2}\Big)\int_0^a rM_T(r)dr\Big] \\
M_{rr} &= \frac{1}{a^2}\int_0^a rM_T(r)dr - \frac{1}{r^2}\int_0^r rM_T(r)dr
\end{aligned}$$

$$
\begin{aligned}
M_{\theta\theta} &= \frac{1}{a^2}\int_0^a rM_T(r)dr + \frac{1}{r^2}\int_0^r rM_T(r)dr - M_T(r) \\
Q_r &= 0
\end{aligned}
\tag{20.92}
$$

The temperature change τ is

$$
\tau = \frac{1}{2}(T_b + T_a) - T_i + (T_b - T_a)\frac{z}{h}
\tag{20.93}
$$

The thermally induced resultant moment M_T is given by

$$
\begin{aligned}
M_T &= \alpha E \int_{-h/2}^{h/2}\Big[\frac{1}{2}(T_b + T_a) - T_i + (T_b - T_a)\frac{z}{h}\Big]zdz \\
&= \frac{h^2}{12}\alpha E(T_b - T_a) = \text{const.}
\end{aligned}
\tag{20.94}
$$

Therefore, the following relations are obtained

$$
\begin{aligned}
&\int_0^r rM_T(r)dr = M_T\frac{1}{2}r^2 \\
&\int_r^a \frac{1}{r}\int_0^r rM_T(r)drdr = M_T\int_r^a \frac{1}{2}rdr = \frac{1}{4}M_T(a^2 - r^2)
\end{aligned}
\tag{20.95}
$$

We have w, M_{rr} and $M_{\theta\theta}$

$$
\begin{aligned}
w &= \frac{1}{D}\Big[\frac{1}{1-\nu}\int_r^a \frac{1}{r}\int_0^r rM_T(r)drdr \\
&\quad + \frac{1}{2(1+\nu)}\Big(1 - \frac{r^2}{a^2}\Big)\int_0^a rM_T(r)dr\Big] \\
&= \frac{1}{D}\Big[\frac{1}{1-\nu}\frac{1}{4}M_T(a^2 - r^2) + \frac{1}{2(1+\nu)}\Big(1 - \frac{r^2}{a^2}\Big)\frac{1}{2}M_T a^2\Big] \\
&= \frac{M_T}{2(1-\nu^2)D}(a^2 - r^2) \\
M_{rr} &= \frac{1}{a^2}\int_0^a rM_T(r)dr - \frac{1}{r^2}\int_0^r rM_T(r)dr = \frac{1}{a^2}M_T\frac{a^2}{2} - \frac{1}{r^2}M_T\frac{r^2}{2} \\
&= 0 \\
M_{\theta\theta} &= \frac{1}{a^2}\int_0^a rM_T(r)dr - M_T + \frac{1}{r^2}\int_0^r rM_T(r)dr \\
&= \frac{1}{a^2}M_T\frac{a^2}{2} - M_T + \frac{1}{r^2}M_T\frac{r^2}{2} = 0
\end{aligned}
\tag{Answer}
$$

Problem 20.10. Derive the thermal deflection w and bending moments M_{rr} and $M_{\theta\theta}$ for a solid circular plate with built-in edge when the temperature distribution T

is given by

$$T = \frac{1}{2}(T_b + T_a) + (T_b - T_a)\frac{z}{h} \tag{20.96}$$

and the initial temperature is $T = T_i$.

Solution. From Eq. (20.89), w, M_{rr}, $M_{\theta\theta}$ and Q_r are

$$\begin{aligned}
w &= \frac{1}{D}\Big[C_1 r^2 + C_2 - \frac{1}{1-\nu}\int_0^r \frac{1}{r}\int_0^r r M_T(r)drdr\Big]\\
M_{rr} &= -\Big[2(1+\nu)C_1 + \frac{1}{r^2}\int_0^r r M_T(r)dr\Big]\\
M_{\theta\theta} &= -\Big[2(1+\nu)C_1 + M_T - \frac{1}{r^2}\int_0^r r M_T(r)dr\Big]\\
Q_r &= 0
\end{aligned} \tag{20.97}$$

The boundary conditions are from Eq. (20.32)

$$w = 0, \qquad \frac{dw}{dr} = 0 \quad \text{at} \quad r = a \tag{20.98}$$

The unknown coefficients C_1 and C_2 can be determined from the boundary conditions Eq. (20.98)

$$\begin{aligned}
C_1 &= \frac{1}{1-\nu}\frac{1}{2a^2}\int_0^a r M_T(r)dr\\
C_2 &= \frac{1}{1-\nu}\Big[\int_0^a \frac{1}{r}\int_0^r r M_T(r)drdr - \frac{1}{2}\int_0^a r M_T(r)dr\Big]
\end{aligned} \tag{20.99}$$

Since M_T is constant from Eq. (20.94), w, M_{rr}, $M_{\theta\theta}$, and Q_r are

$$\begin{aligned}
w &= \frac{1}{(1-\nu)D}\Big[\int_r^a \frac{1}{r}\int_0^r r M_T(r)drdr\\
&\quad - \frac{1}{2}\Big(1 - \frac{r^2}{a^2}\Big)\int_0^a r M_T(r)dr\Big] = 0\\
M_{rr} &= -\frac{1+\nu}{1-\nu}\frac{1}{a^2}\int_0^a r M_T(r)dr - \frac{1}{r^2}\int_0^r r M_T(r)dr\\
&= -\frac{M_T}{1-\nu}\\
M_{\theta\theta} &= -\frac{1+\nu}{1-\nu}\frac{1}{a^2}\int_0^a r M_T(r)dr - M_T(r) + \frac{1}{r^2}\int_0^r r M_T(r)dr\\
&= -\frac{M_T}{1-\nu}\\
Q_r &= 0
\end{aligned} \tag{Answer}$$

Chapter 21
Thermally Induced Instability

In this chapter the thermoelastic buckling of beam-columns subjected to both in-plane and lateral loads is recalled for a built-in edge, a simply supported edge and a free edge. Furthermore, the thermoelastic buckling of rectangular and circular plates is also recalled. The problems and solutions for the buckling behavior of beam-columns with various boundary conditions are given. The stress-displacement relations, the relations between the resultant forces and the stress function are treated in illustrative problems.

21.1 Instability of Beam-Column

When the deflection w of the beam-column is produced in the z axial direction, the fundamental equation for thermal bending problems is

$$\frac{d^2w}{dx^2} = -\frac{M_y}{EI_y} \tag{21.1}$$

where

$$M_y = M_{My} + M_{Ty} \tag{21.2}$$

$$M_{Ty} = \int_A \alpha E\tau(x, z)z\, dA \tag{21.3}$$

$$I_y = \int_A z^2 dA \tag{21.4}$$

in which M_y is the total bending moment about y axis, M_{My} is the bending moment about y axis due to the force, M_{Ty} is the thermally induced bending moment about y axis, and I_y is the moment of inertia about the y axis. The boundary conditions are expressed as follows:

M. Reza Eslami et al., *Theory of Elasticity and Thermal Stresses*, Solid Mechanics and Its Applications 197, DOI: 10.1007/978-94-007-6356-2_21,

[1] A built-in edge

$$w = 0, \quad \frac{dw}{dx} = 0 \tag{21.5}$$

[2] A simply supported edge

$$w = 0, \quad EI_y \frac{d^2 w}{dx^2} + M_{Ty} = 0 \tag{21.6}$$

[3] A free edge

$$EI_y \frac{d^2 w}{dx^2} + M_{Ty} = 0, \quad EI_y \frac{d^3 w}{dx^3} + P \frac{dw}{dx} + \frac{dM_{Ty}}{dx} = 0 \tag{21.7}$$

where P means the axial compressive force.

The bukling load P_{cr} for each boundary condition is

$$P_{cr} = \begin{cases} \dfrac{\pi^2 EI_y}{l^2} & \text{for simply supported edges} \\ \dfrac{\pi^2 EI_y}{4l^2} & \text{for a cantilever beam-column} \\ \dfrac{2.046\pi^2 EI_y}{l^2} & \text{for simply supported edge and built-in edge} \\ \dfrac{4\pi^2 EI_y}{l^2} & \text{for built-in edges} \end{cases} \tag{21.8}$$

21.2 Instability of Plate

The governing equation of the thermal stress function F, which is available to the in-plane deformation is

$$\nabla^2 \nabla^2 F = -\nabla^2 N_T \tag{21.9}$$

where N_T means the thermally induced resultant force defined by Eq. (20.9), and

$$\nabla^2 = \frac{\partial^2}{\partial x^2} + \frac{\partial^2}{\partial y^2} \tag{21.10}$$

Here, the thermal stress function F is defined by

$$N_x = \frac{\partial^2 F}{\partial y^2}, \quad N_y = \frac{\partial^2 F}{\partial x^2}, \quad N_{xy} = -\frac{\partial^2 F}{\partial x \partial y} \tag{21.11}$$

and N_x, N_y, N_{xy} denote the resultant forces per unit length of the plate defined by Eq. (20.8).

The boundary conditions for the in-plane deformation are

[1] A built-in edge

$$u_n = 0, \quad u_s = 0 \tag{21.12}$$

[2] A free edge

$$N_n = 0, \quad N_{ns} = 0 \tag{21.13}$$

where u_i $(i = n, s)$ mean in-plane displacement components, n and s are the normal and tangential directions of the boundary surface of the plate.

On the other hand, the governing equation of deflection w is

$$\nabla^2\nabla^2 w = \frac{1}{D}\left(p - \frac{1}{1-\nu}\nabla^2 M_T + N_x\frac{\partial^2 w}{\partial x^2} + N_y\frac{\partial^2 w}{\partial y^2} + 2N_{xy}\frac{\partial^2 w}{\partial x \partial y}\right) \tag{21.14}$$

where p is the lateral load, D means the bending rigidity defined by Eq. (20.12) and M_T denotes thermally induced resultant moment defined by Eq. (20.9).

The boundary conditions for the out-plane deformation are

[1] A built-in edge

$$w = 0, \quad \frac{\partial w}{\partial n} = 0 \tag{21.15}$$

[2] A simply supported edge

$$w = 0, \quad \frac{\partial^2 w}{\partial n^2} + \nu\frac{\partial^2 w}{\partial s^2} = -\frac{1}{(1-\nu)D}M_T \tag{21.16}$$

[3] A free edge

$$\begin{aligned} &\frac{\partial^2 w}{\partial n^2} + \nu\frac{\partial^2 w}{\partial s^2} = -\frac{1}{(1-\nu)D}M_T \\ &\frac{\partial}{\partial n}\left[\frac{\partial^2 w}{\partial n^2} + (2-\nu)\frac{\partial^2 w}{\partial s^2}\right] = -\frac{1}{(1-\nu)D}\frac{\partial M_T}{\partial n} \end{aligned} \tag{21.17}$$

Next, we consider buckling problems in the circular plate. The governing equation of the thermal stress function F for the in-plane deformation is

$$\nabla^2\nabla^2 F = -\nabla^2 N_T \tag{21.18}$$

where

$$\nabla^2 = \frac{\partial^2}{\partial r^2} + \frac{1}{r}\frac{\partial}{\partial r} + \frac{1}{r^2}\frac{\partial^2}{\partial \theta^2} \tag{21.19}$$

Here, the thermal stress function F is defined by

$$N_r = \frac{1}{r}\frac{\partial F}{\partial r} + \frac{1}{r^2}\frac{\partial^2 F}{\partial \theta^2}, \quad N_\theta = \frac{\partial^2 F}{\partial r^2}, \quad N_{r\theta} = -\frac{\partial}{\partial r}\Big(\frac{1}{r}\frac{\partial F}{\partial \theta}\Big) \tag{21.20}$$

and $N_r, N_\theta, N_{r\theta}$ denote the resultant forces per unit length of the circular plate which are expressed by the displacement components as follows

$$\begin{aligned}
N_r &= \int_{-h/2}^{h/2} \sigma_{rr} dz = \frac{Eh}{1-\nu^2}\Big[\frac{\partial u_r}{\partial r} + \nu\Big(\frac{u_r}{r} + \frac{1}{r}\frac{\partial u_\theta}{\partial \theta}\Big)\Big] - \frac{1}{1-\nu}N_T \\
N_\theta &= \int_{-h/2}^{h/2} \sigma_{\theta\theta} dz = \frac{Eh}{1-\nu^2}\Big[\nu\frac{\partial u_r}{\partial r} + \Big(\frac{u_r}{r} + \frac{1}{r}\frac{\partial u_\theta}{\partial \theta}\Big)\Big] - \frac{1}{1-\nu}N_T \\
N_{r\theta} &= \int_{-h/2}^{h/2} \sigma_{r\theta} dz = \frac{Eh}{2(1+\nu)}\Big[\frac{1}{r}\frac{\partial u_r}{\partial \theta} + r\frac{\partial}{\partial r}\Big(\frac{u_\theta}{r}\Big)\Big]
\end{aligned} \tag{21.21}$$

The thermal stress function F are determined under the appropriate mechanical boundary conditions

[1] A built-in edge

$$u_r = 0, \quad u_\theta = 0 \tag{21.22}$$

[2] A free edge

$$N_r = 0, \quad N_{r\theta} = 0 \tag{21.23}$$

On the other hand, the governing equation of deflection w of the circular plate for the out-plane deformation is

$$\begin{aligned}
\nabla^2\nabla^2 w = \frac{1}{D}\Big[p - \frac{1}{1-\nu}\nabla^2 M_T + N_r\frac{\partial^2 w}{\partial r^2} \\
+ N_\theta\Big(\frac{1}{r}\frac{\partial w}{\partial r} + \frac{1}{r^2}\frac{\partial^2 w}{\partial \theta^2}\Big) + 2N_{r\theta}\frac{\partial}{\partial r}\Big(\frac{1}{r}\frac{\partial w}{\partial \theta}\Big)\Big]
\end{aligned} \tag{21.24}$$

The boundary conditions for the thermally induced bending problems of the circular plate are

[1] A built-in edge

$$w = 0, \quad \frac{\partial w}{\partial r} = 0 \tag{21.25}$$

[2] A simply supported edge

$$\begin{aligned}
&w = 0 \\
&\frac{\partial^2 w}{\partial r^2} + \nu\Big(\frac{1}{r}\frac{\partial w}{\partial r} + \frac{1}{r^2}\frac{\partial^2 w}{\partial \theta^2}\Big) = -\frac{1}{(1-\nu)D}M_T
\end{aligned} \tag{21.26}$$

[3] A free edge

$$\begin{aligned}
&\frac{\partial^2 w}{\partial r^2} + \nu\left(\frac{1}{r}\frac{\partial w}{\partial r} + \frac{1}{r^2}\frac{\partial^2 w}{\partial \theta^2}\right) = -\frac{1}{(1-\nu)D}M_T \\
&\frac{\partial}{\partial r}\left(\nabla^2 w\right) + (1-\nu)\frac{1}{r}\frac{\partial}{\partial r}\left(\frac{1}{r}\frac{\partial^2 w}{\partial \theta^2}\right) = -\frac{1}{(1-\nu)D}\frac{\partial M_T}{\partial r}
\end{aligned} \tag{21.27}$$

21.3 Problems and Solutions Related to Thermally Induced Instability

Problem 21.1. Derive the thermal stress distribution σ_x in a beam-column when the thermally induced bending moment M_{Ty} is given by an arbitrary function $f(x)$, and the temperature change $\tau(x, y, z)$ is given by $\tau_0(x)g(y, z)$.

Solution. Making use of Eqs. (14.9), (14.12) and (14.13), we have the following relations when the x and y axes are the principal axes

$$\sigma_x = -\alpha E\tau(x, y, z) + \frac{P_T}{A} + E\frac{y}{\rho_y} + E\frac{z}{\rho_z} \tag{21.28}$$

$$\frac{1}{\rho_y} = \frac{M_{Tz}}{EI_z}, \quad \frac{1}{\rho_z} = \frac{M_{Ty}}{EI_y} \tag{21.29}$$

Therefore,

$$\sigma_x = -\alpha E\tau(x, y, z) + \frac{P_T}{A} + \frac{M_{Tz}}{I_z}y + \frac{M_{Ty}}{I_y}z \tag{21.30}$$

From Eqs. (14.16) and (14.17) we have

$$\begin{aligned}
&P_T = \int_A \alpha E\tau(x, y, z)dA \\
&M_{Tz} = \int_A \alpha E\tau(x, y, z)ydA, \quad M_{Ty} = \int_A \alpha E\tau(x, y, z)zdA
\end{aligned} \tag{21.31}$$

The assumptions for the thermally induced bending moment M_{Ty} and the temperature change $\tau(x, y, z)$ are

$$M_{Ty} = f(x), \quad \tau(x, y, z) = \tau_0(x)g(y, z) \tag{21.32}$$

The substitution of Eq. (21.32) into Eq. (21.31) leads to

$$\alpha E \tau_0(x) = \frac{f(x)}{\displaystyle\int_A g(y,z) z dA} \tag{21.33}$$

By the substitution of Eq. (21.33) into Eq. (21.31), we get

$$\begin{aligned} P_T &= \alpha E \tau_0(x) \int_A g(y,z) dA = f(x) \frac{\displaystyle\int_A g(y,z) dA}{\displaystyle\int_A g(y,z) z dA} \\ M_{Tz} &= \alpha E \tau_0(x) \int_A g(y,z) y dA = f(x) \frac{\displaystyle\int_A g(y,z) y dA}{\displaystyle\int_A g(y,z) z dA} \end{aligned} \tag{21.34}$$

Therefore, thermal stress σ_x can be obtained as

$$\begin{aligned} \sigma_x = \frac{f(x)}{\displaystyle\int_A g(y,z) z dA} \Big[&-g(y,z) + \frac{1}{A} \int_A g(y,z) dA \\ &+ \frac{y}{I_z} \int_A g(y,z) y dA + \frac{z}{I_y} \int_A g(y,z) z dA \Big] \end{aligned} \quad \text{(Answer)}$$

Problem 21.2. Derive the boundary conditions Eq. (21.7) for the free edge.

Solution. The total bending moment M_y is

$$M_y = M_{My} + M_{Ty} \tag{21.35}$$

From Eqs. (21.35) and (21.1), we get

$$M_{My} = M_y - M_{Ty} = -EI_y \frac{d^2 w}{dx^2} - M_{Ty} \tag{21.36}$$

Since the bending moment M_{My} on the free edge is zero, the free edge boundary condition is rewritten by

$$EI_y \frac{d^2 w}{dx^2} + M_{Ty} = 0 \quad \text{(Answer)}$$

Next we consider a small element with length dx at x plane in the beam-column. The bending moment M_{My}, the shearing force F_z, the axial compressive force P and the deflection w are produced at x plane, and on the other hand, the bending moment $M_{My} + dM_{My}$, the shearing force $F_z + dF_z$, the axial compressive force $P + dP$ and the deflection $w + dw$ are produced at $x + dx$ plane. From the moment balance at the neutral axis on x plane, we have

$$M_{My} - (M_{My} + dM_{My}) + (F_z + dF_z)dx + (P + dP)dw = 0 \tag{21.37}$$

Omitting the infinitesimal terms, Eq. (21.37) reduces to

$$F_z - \frac{dM_{My}}{dx} + P\frac{dw}{dx} = 0 \tag{21.38}$$

Then we get

$$F_z = \frac{dM_{My}}{dx} - P\frac{dw}{dx} \tag{21.39}$$

From Eq. (21.35) and (21.1), Eq. (21.39) becomes

$$\begin{aligned} F_z &= \frac{dM_y}{dx} - \frac{dM_{Ty}}{dx} - P\frac{dw}{dx} \\ &= -EI_y\frac{d^3w}{dx^3} - \frac{dM_{Ty}}{dx} - P\frac{dw}{dx} \end{aligned} \tag{21.40}$$

Since $F_z = 0$ for a free edge boundary condition, Eq. (21.40) reduces to

$$EI_y\frac{d^3w}{dx^3} + P\frac{dw}{dx} + \frac{dM_{Ty}}{dx} = 0 \qquad \text{(Answer)}$$

Problem 21.3. Find the deflection w of a beam-column with length l for both simply supported edges shown in Fig. 21.1, and derive the buckling load P_{cr}, when the thermally induced bending moment $M_{Ty}(x)$ is represented by a linear function of x

$$M_{Ty}(x) = D_1 + D_2x \tag{21.41}$$

Solution. The bending moment M_y is

$$M_y = M_{My} + M_{Ty} = Pw + D_1 + D_2x \tag{21.42}$$

where P is the axial compressive force and w means the deflection at x.

Substituting Eq. (21.42) into Eq. (21.1), the fundamental equation for w is rewritten as

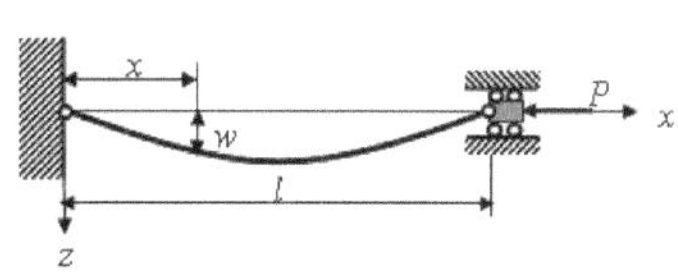

Fig. 21.1 Simply supported beam-column

$$EI_y \frac{d^2 w}{dx^2} + Pw = -(D_1 + D_2 x) \tag{21.43}$$

A solution of Eq. (21.43) is

$$w = C_1 \cos kx + C_2 \sin kx - \frac{1}{P}(D_1 + D_2 x) \tag{21.44}$$

where

$$k^2 = \frac{P}{EI_y} \tag{21.45}$$

and C_1 and C_2 are unknown constants.

When both edges are simply supported, the boundary conditions are

$$w = 0, \quad EI_y \frac{d^2 w}{dx^2} + M_{Ty}(x) = 0 \qquad \text{at } x = 0,\ l \tag{21.46}$$

The two unknown constants C_1 and C_2 can be determined from the boundary conditions (21.46)

$$C_1 = \frac{D_1}{P}, \quad C_2 = \frac{D_1}{P}\frac{1 - \cos kl}{\sin kl} + \frac{D_2}{P}\frac{l}{\sin kl} \tag{21.47}$$

Then the deflection w is

$$w = \frac{1}{P}\left[D_1\left(\cos kx - 1 + \frac{1 - \cos kl}{\sin kl}\sin kx\right) + \frac{D_2}{k}\left(\frac{kl \sin kx}{\sin kl} - kx\right)\right] \qquad \text{(Answer)}$$

The deflection w tends to infinity when kl satisfies the condition

$$\sin kl = 0 \tag{21.48}$$

Solutions of Eq. (21.48) are $kl = n\pi$ $(n = 1, 2, 3, \ldots)$. Since smallest value of kl is π, the buckling load P_{cr} is from Eq. (21.45)

$$P_{cr} = \frac{\pi^2 EI_y}{l^2} \qquad \text{(Answer)}$$

Problem 21.4. Find the deflection w of a cantilever beam-column with length l shown in Fig. 21.2, and derive the buckling load P_{cr}, when the thermally induced bending moment $M_{Ty}(x)$ is represented by a linear function of x

$$M_{Ty}(x) = D_1 + D_2 x \tag{21.49}$$

Solution. When the deflection at the free edge is defined by w_0, the bending moment M_y is

$$M_y = M_{My} + M_{Ty} = D_1 + D_2 x - P(w_0 - w) \tag{21.50}$$

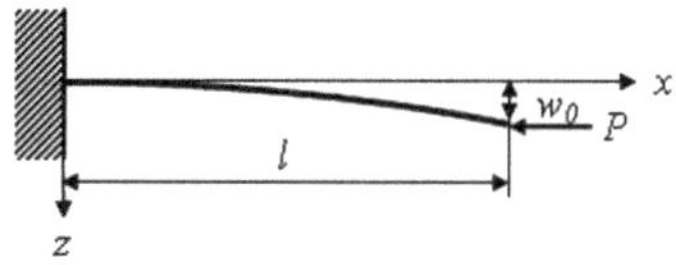

Fig. 21.2 Cantilever beam-column

where P is the axial compressive force. Substituting Eq. (21.50) into Eq. (21.1), the fundamental equation for w is rewritten as

$$EI_y \frac{d^2 w}{dx^2} + Pw = Pw_0 - (D_1 + D_2 x) \tag{21.51}$$

A solution w of Eq. (21.51) is

$$w = C_1 \cos kx + C_2 \sin kx + w_0 - \frac{1}{P}(D_1 + D_2 x) \tag{21.52}$$

where

$$k^2 = \frac{P}{EI_y} \tag{21.53}$$

The boundary conditions for the built-in edge at $x = 0$ and the condition of the deflection $w = w_0$ at $x = l$ are

$$\begin{aligned} &w = 0, \quad \frac{dw}{dx} = 0 && \text{at} \quad x = 0 \\ &w = w_0 && \text{at} \quad x = l \end{aligned} \tag{21.54}$$

The three unknown constants C_1, C_2 and w_0 can be determined from the boundary conditions (21.54)

$$\begin{aligned} C_1 &= \frac{D_1}{P \cos kl} + \frac{D_2}{Pk \cos kl}(kl - \sin kl), \quad C_2 = \frac{D_2}{kP} \\ w_0 &= \frac{D_1}{P} - \frac{D_1}{P \cos kl} - \frac{D_2}{Pk \cos kl}(kl - \sin kl) \end{aligned} \tag{21.55}$$

Then, the deflection w is

$$\begin{aligned} w = &\frac{1}{P \cos kl}\Big\{D_1(\cos kx - 1) \\ &+ \frac{D_2}{k}\big[(kl - \sin kl)(\cos kx - 1) + \cos kl(\sin kx - kx)\big]\Big\} \end{aligned} \tag{Answer}$$

Therefore, the buckling load P_{cr} is derived from the condition

$$\cos kl = 0 \tag{21.56}$$

Then, the buckling load P_{cr} is given by

$$P_{cr} = \frac{\pi^2 EI_y}{4l^2} \qquad \text{(Answer)}$$

Problem 21.5. Find the deflection w of beam-column with length l for one simply supported edge and the other one built-in edge, shown in Fig. 21.3, and derive the buckling load P_{cr}, when the thermally induced bending moment $M_{Ty}(x)$ is represented by a linear function of x

$$M_{Ty}(x) = D_1 + D_2 x \tag{21.57}$$

Solution. Let P, R_0 and w denote the axial compressive force, the reaction force at simply supported edge $x = 0$, and the deflection at x, respectively. Then, the bending moment M_y is given by

$$M_y = M_{My} + M_{Ty} = Pw + R_0 x + D_1 + D_2 x \tag{21.58}$$

Substituting Eq. (21.58) into Eq. (21.1), the fundamental equation for w is rewritten as

$$EI_y \frac{d^2w}{dx^2} + Pw = -R_0 x - (D_1 + D_2 x) \tag{21.59}$$

A solution w of Eq. (21.59) is

$$w = C_1 \cos kx + C_2 \sin kx - \frac{R_0}{P}x - \frac{1}{P}(D_1 + D_2 x) \tag{21.60}$$

where

$$k^2 = \frac{P}{EI_y} \tag{21.61}$$

When the edge $x = 0$ is simply supported and the edge $x = l$ is built-in, the boundary conditions are

$$w = 0, \quad EI_y \frac{d^2w}{dx^2} + (D_1 + D_2 x) = 0 \quad \text{at} \quad x = 0$$

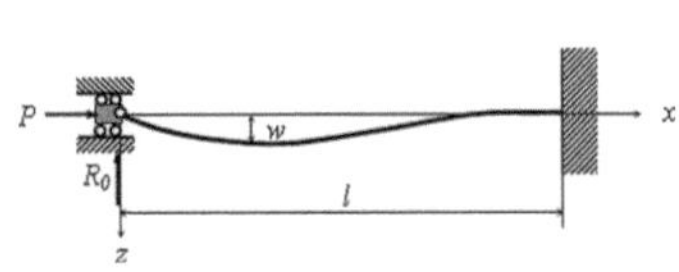

Fig. 21.3 Beam-column with a simply supported edge and a built-in edge

$$w = 0, \quad \frac{dw}{dx} = 0 \qquad \text{at} \quad x = l \tag{21.62}$$

The unknown constants C_1, C_2 and R_0 in Eq. (21.60) can be determined from the boundary conditions (21.62)

$$\begin{aligned} C_1 &= \frac{D_1}{P}, \quad C_2 = \frac{D_1}{P}\frac{kl\sin kl + \cos kl - 1}{kl\cos kl - \sin kl} \\ R_0 &= -D_2 + kD_1\frac{1 - \cos kl}{kl\cos kl - \sin kl} \end{aligned} \tag{21.63}$$

Then, the deflection w is

$$\begin{aligned} w = \frac{D_1}{P}\Big\{&\cos kx - 1 + \frac{1}{kl\cos kl - \sin kl} \\ &\times \big[(\cos kl - 1)(kx + \sin kx) + kl\sin kl\sin kx\big]\Big\} \end{aligned} \qquad \text{(Answer)}$$

Therefore, the buckling load P_{cr} is derived from the condition

$$kl\cos kl - \sin kl = 0 \quad \text{i.e.,} \quad \tan kl = kl \tag{21.64}$$

Then, the buckling load P_{cr} is

$$P_{cr} = \frac{4.4934^2 EI_y}{l^2} \simeq \frac{2.046\pi^2 EI_y}{l^2} \qquad \text{(Answer)}$$

Problem 21.6. Find the deflection w of a beam-column with length l for both built-in edges, shown in Fig. 21.4, and derive the buckling load P_{cr}, when the thermally induced bending moment $M_{Ty}(x)$ is represented by a linear function of x

$$M_{Ty}(x) = D_1 + D_2 x \tag{21.65}$$

Solution. Let P, R_0, M_0 and w denote the axial compressive force, the reaction force at the edge $x = 0$, the bending moment at the edge $x = 0$, and the deflection at x, respectively. The bending moment M_y is given by

$$M_y = M_{My} + M_{Ty} = Pw + R_0 x + M_0 + D_1 + D_2 x \tag{21.66}$$

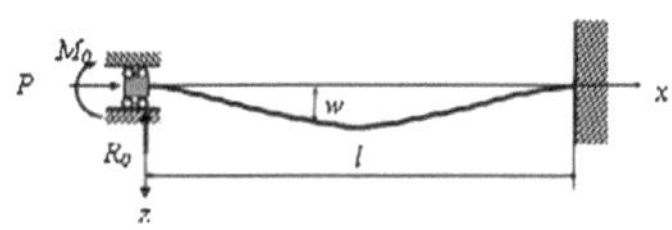

Fig. 21.4 Beam-column with built-in edges

Substituting Eq. (21.66) into Eq. (21.1), the fundamental equation for w is rewritten as

$$EI_y \frac{d^2w}{dx^2} + Pw = -M_0 - R_0 x - (D_1 + D_2 x) \tag{21.67}$$

A solution w of Eq. (21.67) is

$$w = C_1 \cos kx + C_2 \sin kx - \frac{1}{P}[D_1 + M_0 + (D_2 + R_0)x] \tag{21.68}$$

where

$$k^2 = \frac{P}{EI_y} \tag{21.69}$$

The boundary conditions are

$$w = 0, \quad \frac{dw}{dx} = 0 \quad \text{at } x = 0,\ l \tag{21.70}$$

From the boundary conditions (21.70), we have

$$\begin{aligned}
&D_1 + M_0 = C_1 P, \quad D_2 + R_0 = PkC_2 \\
&C_1 \cos kl + C_2 \sin kl = \frac{1}{P}[D_1 + M_0 + (D_2 + R_0)l] \\
&-kC_1 \sin kl + kC_2 \cos kl = \frac{1}{P}(D_2 + R_0)
\end{aligned} \tag{21.71}$$

Elimination of $(D_1 + M_0)$ and $(D_2 + R_0)$ from Eq. (21.71) gives

$$\begin{bmatrix} \cos kl - 1 & \sin kl - kl \\ -\sin kl & \cos kl - 1 \end{bmatrix} \begin{bmatrix} C_1 \\ C_2 \end{bmatrix} = \begin{bmatrix} 0 \\ 0 \end{bmatrix} \tag{21.72}$$

From Eqs. (21.71) and (21.72), the deflection w becomes

$$w = C_2 \left[\sin kx - kx - \frac{\sin kl - kl}{\cos kl - 1}(\cos kx - 1) \right] \tag{Answer}$$

In order to have a non-trivial solution for Eq. (21.72), we find

$$\begin{vmatrix} \cos kl - 1 & \sin kl - kl \\ -\sin kl & \cos kl - 1 \end{vmatrix} = 0 \tag{21.73}$$

From Eq. (21.73), we have

$$\sin \frac{kl}{2} \left(\sin \frac{kl}{2} - \frac{kl}{2} \cos \frac{kl}{2} \right) = 0 \tag{21.74}$$

The smallest value of kl is $kl = 2\pi$. Therefore, the buckling load P_{cr} is

$$P_{cr} = \frac{4\pi^2 EI_y}{l^2} \qquad \text{(Answer)}$$

Problem 21.7. Show that the deflection w equals zero in a pre-buckling state for a beam-column with length l for both built-in edges when the thermally induced bending moment M_{Ty} is given by a linear function of x

$$M_{Ty}(x) = D_1 + D_2 x \tag{21.75}$$

Solution. Let P, R_0, M_0 and w denote the axial compressive force, the reaction force at the edge $x = 0$, the bending moment at the edge $x = 0$, and the deflection at x, respectively. The bending moment M_y is given by

$$M_y = M_{My} + M_{Ty} = Pw + R_0 x + M_0 + D_1 + D_2 x \tag{21.76}$$

Substituting Eq. (21.76) into Eq. (21.1), the fundamental equation for w is rewritten as

$$EI_y \frac{d^2 w}{dx^2} + Pw = -M_0 - R_0 x - (D_1 + D_2 x) \tag{21.77}$$

A solution w of Eq. (21.77) is

$$w = C_1 \cos kx + C_2 \sin kx - \frac{1}{P}[D_1 + M_0 + (D_2 + R_0)x] \tag{21.78}$$

where

$$k^2 = \frac{P}{EI_y} \tag{21.79}$$

The boundary conditions are

$$w = 0, \quad \frac{dw}{dx} = 0 \quad \text{at } x = 0 \text{ and } x = l \tag{21.80}$$

From the boundary conditions (21.80), we have

$$\begin{aligned} &D_1 + M_0 = C_1 P, \qquad D_2 + R_0 = PkC_2 \\ &C_1 \cos kl + C_2 \sin kl = \frac{1}{P}[D_1 + M_0 + (D_2 + R_0)l] \\ &-kC_1 \sin kl + kC_2 \cos kl = \frac{1}{P}(D_2 + R_0) \end{aligned} \tag{21.81}$$

The substitution of Eq. (21.81) into Eq. (21.78) gives

$$w = C_1(\cos kx - 1) + C_2(\sin kx - kx) \tag{21.82}$$

Elimination of $(D_1 + M_0)$ and $(D_2 + R_0)$ from Eq. (21.81) gives

$$\begin{bmatrix} \cos kl - 1 & \sin kl - kl \\ -\sin kl & \cos kl - 1 \end{bmatrix} \begin{bmatrix} C_1 \\ C_2 \end{bmatrix} = \begin{bmatrix} 0 \\ 0 \end{bmatrix} \tag{21.83}$$

From Eq. (21.83), it can be seen that

$$C_1 = 0, \; C_2 = 0 \qquad \text{for prebuckling state} \tag{21.84}$$

$$\begin{vmatrix} \cos kl - 1 & \sin kl - kl \\ -\sin kl & \cos kl - 1 \end{vmatrix} = 0 \qquad \text{for postbuckling state} \tag{21.85}$$

Substitution of Eq. (21.84) into Eq. (21.82) gives the deflection for pre-buckling state

$$w = 0 \qquad \text{(Answer)}$$

Problem 21.8. When the thermally induced bending moment M_{Ty} is given by the parabolic function

$$M_{T_y}(x) = D_0 x(l - x) \tag{21.86}$$

find the deflection w of a beam-column with length l for both simply supported edges, and derive the buckling load P_{cr}.

Solution. The bending moment M_y is

$$M_y = M_{My} + M_{Ty} = Pw + D_0 x(l - x) \tag{21.87}$$

where P is the axial compressive force and w means the deflection at x.

Substituting Eq. (21.87) into Eq. (21.1), the fundamental equation for w is rewritten as

$$EI_y \frac{d^2w}{dx^2} + Pw = -D_0 x(l - x) \tag{21.88}$$

A solution w of Eq. (21.88) is

$$w = C_1 \cos kx + C_2 \sin kx - \frac{D_0}{Pk^2}[2 + k^2 x(l - x)] \tag{21.89}$$

where

$$k^2 = \frac{P}{EI_y} \tag{21.90}$$

and C_1 and C_2 are unknown constants.

When both edges are simply supported, the boundary conditions are

$$w = 0, \quad EI_y \frac{d^2 w}{dx^2} + D_0 x(l - x) = 0 \qquad \text{at } x = 0,\ l \tag{21.91}$$

The two unknown constants C_1 and C_2 can be determined from the boundary conditions (21.91)

$$C_1 = 2\frac{D_0}{Pk^2}, \quad C_2 = 2\frac{D_0}{Pk^2}\frac{1 - \cos kl}{\sin kl} \tag{21.92}$$

Substitution of Eq. (21.92) into Eq. (21.89) gives

$$w = \frac{D_0}{Pk^2}\left[2(\cos kx - 1) + 2\left(\frac{1 - \cos kl}{\sin kl}\right)\sin kx - k^2 x(l - x)\right] \qquad \text{(Answer)}$$

The buckling load P_{cr} is derived from $\sin kl = 0$

$$P_{cr} = \frac{\pi^2 EI_y}{l^2} \qquad \text{(Answer)}$$

Problem 21.9. When the thermally induced bending moment M_{Ty} is given by the parabolic function

$$M_{T_y}(x) = D_0 x(l - x) \tag{21.93}$$

find the deflection w of a cantilever beam-column with length l, and derive the buckling load P_{cr}.

Solution. When the deflection at the free edge is defined by w_0, the bending moment M_y is

$$M_y = M_{My} + M_{Ty} = D_0 x(l - x) - P(w_0 - w) \tag{21.94}$$

where P is the axial compressive force. Substituting Eq. (21.94) into Eq. (21.1), the fundamental equation for w is rewritten as

$$EI_y \frac{d^2 w}{dx^2} + Pw = Pw_0 - D_0 x(l - x) \tag{21.95}$$

A solution w of Eq. (21.95) is taken in the form

$$w = C_1 \cos kx + C_2 \sin kx + w_0 - \frac{D_0}{Pk^2}\left[2 + k^2 x(l - x)\right] \tag{21.96}$$

where

$$k^2 = \frac{P}{EI_y} \tag{21.97}$$

The boundary conditions for the built-in edge at $x = 0$ and the condition of the deflection $w = w_0$ at $x = l$ are

$$\begin{aligned} &w = 0, \quad \frac{dw}{dx} = 0 && \text{at} \quad x = 0 \\ &w = w_0 && \text{at} \quad x = l \end{aligned} \tag{21.98}$$

The three unknown constants can be determined from the boundary conditions (21.98)

$$\begin{aligned} C_1 &= \frac{D_0}{Pk^2} \frac{2 - kl \sin kl}{\cos kl}, \quad C_2 = \frac{D_0 kl}{Pk^2} \\ w_0 &= 2\frac{D_0}{Pk^2} - \frac{D_0}{Pk^2} \frac{2 - kl \sin kl}{\cos kl} \end{aligned} \tag{21.99}$$

Then, the deflection w is

$$w = \frac{D_0}{Pk^2}\left[kl \sin kx + (\cos kx - 1)\frac{2 - kl \sin kl}{\cos kl} - k^2 x(l - x)\right] \qquad \text{(Answer)}$$

The buckling load P_{cr} is determined from $\cos kl = 0$

$$P_{cr} = \frac{\pi^2 EI_y}{4l^2} \qquad \text{(Answer)}$$

Problem 21.10. When the thermally induced bending moment M_{Ty} is given by the parabolic function

$$M_{T_y}(x) = D_0 x(l - x) \tag{21.100}$$

find the deflection w of a beam-column with length l for one simply supported edge and the other one built-in edge, and derive the buckling load P_{cr}.

Solution. Let P, R_0 and w denote the axial compressive force, the reaction force at simply supported edge $x = 0$, and the deflection at x, respectively. The bending moment M_y is given by

$$M_y = M_{My} + M_{Ty} = Pw + R_0 x + D_0 x(l - x) \tag{21.101}$$

Substituting Eq. (21.101) into Eq. (21.1), the fundamental equation for w is rewritten as

$$EI_y \frac{d^2 w}{dx^2} + Pw = -R_0 x - D_0 x(l - x) \tag{21.102}$$

A solution w of Eq. (21.102) is taken in the form

$$w = C_1 \cos kx + C_2 \sin kx - \frac{R_0}{P}x - \frac{D_0}{Pk^2}\left[2 + k^2 x(l - x)\right] \tag{21.103}$$

where

$$k^2 = \frac{P}{EI_y} \tag{21.104}$$

When the edge $x = 0$ is simply supported and the edge $x = l$ is built-in, the boundary conditions are

$$\begin{aligned} w &= 0, \quad EI_y \frac{d^2w}{dx^2} + D_0 x(l - x) = 0 \quad &\text{at} \quad x = 0 \\ w &= 0, \quad \frac{dw}{dx} = 0 \quad &\text{at} \quad x = l \end{aligned} \tag{21.105}$$

The two unknown constants C_1, C_2 and the reaction force R_0 can be determined from the boundary conditions (21.105)

$$\begin{aligned} C_1 &= 2\frac{D_0}{Pk^2}, \quad C_2 = \frac{D_0}{Pk^2} \frac{2kl \sin kl + 2\cos kl - 2 - k^2 l^2}{kl \cos kl - \sin kl} \\ R_0 &= -\frac{D_0}{k} \frac{2\cos kl - 2 + kl \sin kl}{kl \cos kl - \sin kl} \end{aligned} \tag{21.106}$$

Then, the deflection w is

$$\begin{aligned} w = \frac{D_0}{Pk^2} \Big[& 2(\cos kx - 1) - k^2 x(l - x) + kx \frac{2\cos kl - 2 + kl \sin kl}{kl \cos kl - \sin kl} \\ & + \sin kx \frac{2\cos kl - 2 + 2kl \sin kl - k^2 l^2}{kl \cos kl - \sin kl} \Big] \end{aligned} \tag{Answer}$$

The buckling load P_{cr} is given from $kl \cos kl = \sin kl$

$$P_{cr} \simeq \frac{2.046\pi^2 EI_y}{l^2} \tag{Answer}$$

Problem 21.11. When the thermally induced bending moment M_{Ty} is given by the parabolic function

$$M_{T_y}(x) = D_0 x(l - x) \tag{21.107}$$

find the deflection w of a beam-column with length l for both built-in edges, and derive the buckling load P_{cr}.

Solution. Let P, R_0, M_0 and w denote the axial compressive force, the reaction force at the edge $x = 0$, the bending moment at the edge $x = 0$, and the deflection at x, respectively. The bending moment M_y is given by

$$M_y = M_{My} + M_{Ty} = Pw + R_0 x + M_0 + D_0 x(l - x) \tag{21.108}$$

Substituting Eq. (21.108) into Eq. (21.1), the fundamental equation for w is rewritten as

$$EI_y \frac{d^2 w}{dx^2} + Pw = -M_0 - R_0 x - D_0 x(l - x) \tag{21.109}$$

A solution w of Eq. (21.109) is taken in the form

$$w = C_1 \cos kx + C_2 \sin kx - \frac{1}{P}(R_0 x + M_0) - \frac{D_0}{Pk^2}\left[2 + k^2 x(l - x)\right] \tag{21.110}$$

where

$$k^2 = \frac{P}{EI_y} \tag{21.111}$$

The boundary conditions are

$$w = 0, \quad \frac{dw}{dx} = 0 \qquad \text{at } x = 0 \text{ and } x = l \tag{21.112}$$

From the boundary conditions (21.112), we have

$$\begin{aligned} C_1 &= \frac{D_0 kl}{Pk^2} \frac{2 \sin kl - kl \cos kl - kl}{2 - 2 \cos kl - kl \sin kl} \\ C_2 &= \frac{D_0}{Pk^2} kl, \qquad R_0 = 0 \\ M_0 &= \frac{D_0}{k^2}\left[kl \frac{2 \sin kl - kl \cos kl - kl}{2 - 2 \cos kl - kl \sin kl} - 2\right] \end{aligned} \tag{21.113}$$

Then, the deflection w is

$$\begin{aligned} w = \frac{D_0}{Pk^2}\Big\{ & kl \sin kx - k^2 x(l - x) \\ & + \frac{1 - \cos kx}{2 - 2 \cos kl - kl \sin kl}\left[k^2 l^2 (1 + \cos kl) - 2kl \sin kl\right]\Big\} \end{aligned} \qquad \text{(Answer)}$$

The deflection w tends to infinity, when kl satisfies the condition

$$2 - 2 \cos kl - kl \sin kl = 0 \tag{21.114}$$

Equation (21.114) is rewritten by

$$\sin \frac{kl}{2}\left(\sin \frac{kl}{2} - \frac{kl}{2} \cos \frac{kl}{2}\right) = 0 \tag{21.115}$$

Since the smallest value of kl in Eq. (21.115) is $kl = 2\pi$, the buckling load P_{cr} is

$$P_{cr} = \frac{4\pi^2 EI_y}{l^2} \qquad \text{(Answer)}$$

Problem 21.12. When the thermally induced bending moment $M_{Ty}(x)$ is given as an arbitrary function of x, find the deflection w and the buckling load P_{cr} of a beam-column with length l for both simply supported edges.

Solution. The fundamental equation for deflection w is given by Eq. (21.1)

$$EI_y \frac{d^2 w}{dx^2} + Pw = -M_{Ty}(x) \tag{21.116}$$

A solution of Eq. (21.116) can be obtained by use of the method of variation of parameters. A general solution of the homogeneous Eq. (21.116) is given by

$$w_c = C_1 \cos kx + C_2 \sin kx, \qquad k^2 = \frac{P}{EI_y} \tag{21.117}$$

To get a particular solution of Eq. (21.116), we take

$$w_p = A_1(x) \cos kx + A_2(x) \sin kx \tag{21.118}$$

Substitution of Eq. (21.118) into Eq. (21.116) gives

$$\begin{aligned} &\frac{d}{dx}\left[\frac{dA_1(x)}{dx}\cos kx + \frac{dA_2(x)}{dx}\sin kx\right] \\ &\quad -k\frac{dA_1(x)}{dx}\sin kx + k\frac{dA_2(x)}{dx}\cos kx = -\frac{M_{Ty}(x)}{EI_y} \end{aligned} \tag{21.119}$$

Equation (21.119) can be satisfied when we take

$$\begin{aligned} &\frac{dA_1(x)}{dx}\cos kx + \frac{dA_2(x)}{dx}\sin kx = 0 \\ &-k\frac{dA_1(x)}{dx}\sin kx + k\frac{dA_2(x)}{dx}\cos kx = -\frac{M_{Ty}(x)}{EI_y} \end{aligned} \tag{21.120}$$

Solving Eq. (21.120), we get

$$\begin{aligned} \frac{dA_1(x)}{dx} &= \frac{1}{EI_y k} M_{Ty}(x) \sin kx \\ \frac{dA_2(x)}{dx} &= -\frac{1}{EI_y k} M_{Ty}(x) \cos kx \end{aligned} \tag{21.121}$$

Then, $A_1(x)$ and $A_2(x)$ are determined as

$$A_1(x) = \frac{1}{EI_y k}\int M_{Ty}(x)\sin kx dx$$
$$A_2(x) = -\frac{1}{EI_y k}\int M_{Ty}(x)\cos kx dx \tag{21.122}$$

Therefore, a particular solution w_p is given by

$$w_p = \frac{1}{EI_y k}\Big[\cos kx\int M_{Ty}(x)\sin kx dx - \sin kx\int M_{Ty}(x)\cos kx dx\Big] \tag{21.123}$$

From Eqs. (21.117) and (21.123), a solution of Eq. (21.116) is represented by

$$w = C_1\cos kx + C_2\sin kx + \frac{k}{P}\Big[\cos kx\int_0^x M_{Ty}(x')\sin kx' dx'$$
$$- \sin kx\int_0^x M_{Ty}(x')\cos kx' dx'\Big] \tag{21.124}$$

The boundary conditions are given by Eq. (21.6)

$$w = 0, \quad EI_y\frac{d^2w}{dx^2} + M_{Ty}(x) = 0 \quad \text{on} \quad x = 0,\ x = l \tag{21.125}$$

From Eq. (21.124) we have

$$\frac{d^2w}{dx^2} = -k^2(C_1\cos kx + C_2\sin kx)$$
$$-\frac{1}{EI_y}\Big[k\cos kx\int_0^x M_{Ty}(x')\sin kx' dx'$$
$$- k\sin kx\int_0^x M_{Ty}(x')\cos kx' dx' + M_{Ty}(x)\Big] \tag{21.126}$$

The boundary conditions on $x = 0$ give

$$C_1 = 0, \quad EI_y\Big[-k^2C_1 - \frac{1}{EI_y}M_{Ty}(0)\Big] + M_{Ty}(0) = 0 \tag{21.127}$$

The boundary conditions on $x = l$ give

$$C_2\sin kl + \frac{k}{P}\Big[\cos kl\int_0^l M_{Ty}(x')\sin kx' dx'$$
$$- \sin kl\int_0^l M_{Ty}(x')\cos kx' dx'\Big] = 0$$

$$EI_y[-k^2C_2 \sin kl] - \Big[k \cos kl \int_0^l M_{Ty}(x') \sin kx' dx' - k \sin kl \int_0^l M_{Ty}(x') \cos kx' dx' + M_{Ty}(l)\Big] + M_{Ty}(l) = 0 \tag{21.128}$$

From Eq. (21.128), we have

$$C_2 = -\frac{k}{P}\Big[\cot kl \int_0^l M_{Ty}(x') \sin kx' dx' - \int_0^l M_{Ty}(x') \cos kx' dx'\Big] \tag{21.129}$$

Substitution of Eqs. (21.127) and (21.129) into Eq. (21.124) leads the deflection w

$$w = \frac{k}{P}\Big[\sin kx \int_x^l M_{T_y}(x') \cos kx' dx' + \cos kx \int_0^x M_{T_y}(x') \sin kx' dx' - \cot kl \sin kx \int_0^l M_{T_y}(x') \sin kx' dx'\Big] \quad \text{(Answer)}$$

The deflection w tends to infinity, when kl satisfies the condition

$$\sin kl = 0 \tag{21.130}$$

Then, the buckling load P_{cr} is

$$P_{cr} = \frac{\pi^2 EI_y}{l^2} \quad \text{(Answer)}$$

Problem 21.13. When the thermally induced bending moment $M_{Ty}(x)$ is given as an arbitrary function of x, find the deflection w and the buckling load P_{cr} of a cantilever beam-column with length l.

Solution. The fundamental equation for deflection w is given by Eq. (21.1)

$$EI_y \frac{d^2w}{dx^2} + Pw = Pw_0 - M_{Ty}(x) \tag{21.131}$$

where w_0 is the deflection at the free edge.

Using derivation of a solution for deflection w as in Problem 21.12, the fundamental solution w of Eq. (21.131) is

$$w = C_1 \cos kx + C_2 \sin kx + w_0 + \frac{k}{P}\Big[\cos kx \int_0^x M_{T_y}(x') \sin kx' dx' - \sin kx \int_0^x M_{T_y}(x') \cos kx' dx'\Big] \tag{21.132}$$

The boundary conditions are from Eqs. (21.5) and (21.7)

$$\begin{aligned} &w = 0, \quad \frac{dw}{dx} = 0 && \text{on } x = 0 \\ &EI_y \frac{d^2 w}{dx^2} + M_{T_y}(x) = 0 && \\ &EI_y \frac{d^3 w}{dx^3} + P\frac{dw}{dx} + \frac{dM_{T_y}}{dx} = 0 && \text{on } x = l \end{aligned} \tag{21.133}$$

The differentiation of w with respect to x gives

$$\begin{aligned} \frac{dw}{dx} &= -C_1 k \sin kx + C_2 k \cos kx \\ &\quad - \frac{k^2}{P}\Big[\sin kx \int_0^x M_{Ty}(x') \sin kx' dx' + \cos kx \int_0^x M_{Ty}(x') \cos kx' dx'\Big] \\ \frac{d^2 w}{dx^2} &= -k^2 (C_1 \cos kx + C_2 \sin kx) \\ &\quad - \frac{k^2}{P}\Big[k \cos kx \int_0^x M_{Ty}(x') \sin kx' dx' \\ &\quad - k \sin kx \int_0^x M_{Ty}(x') \cos kx' dx' + M_{Ty}(x)\Big] \\ \frac{d^3 w}{dx^3} &= k^3 (C_1 \sin kx - C_2 \cos kx) \\ &\quad - \frac{k^2}{P}\Big[-k^2 \sin kx \int_0^x M_{Ty}(x') \sin kx' dx' \\ &\quad - k^2 \cos kx \int_0^x M_{Ty}(x') \cos kx' dx' + \frac{dM_{Ty}}{dx}\Big] \end{aligned} \tag{21.134}$$

The boundary conditions on $x = 0$ give

$$w_0 = -C_1, \qquad C_2 = 0 \tag{21.135}$$

The boundary conditions on $x = l$ give

$$\begin{aligned} &EI_y \Big\{-k^2 C_1 \cos kl - \frac{k^2}{P}\Big[k \cos kl \int_0^l M_{Ty}(x') \sin kx' dx' \\ &\quad -k \sin kl \int_0^l M_{Ty}(x') \cos kx' dx' + M_{Ty}(l)\Big]\Big\} + M_{Ty}(l) = 0 \\ &EI_y \Big\{k^3 C_1 \sin kl - \frac{k^2}{P}\Big[-k^2 \sin kl \int_0^l M_{Ty}(x') \sin kx' dx' \end{aligned}$$

$$
\begin{aligned}
&-k^2 \cos kl \int_0^l M_{Ty}(x') \cos kx'dx' + \left(\frac{dM_{Ty}}{dx}\right)_{x=l}\Big]\Big\} + P\Big\{-C_1 k \sin kl \\
&-\frac{k^2}{P}\Big[\sin kl \int_0^l M_{Ty}(x') \sin kx'dx' + \cos kl \int_0^l M_{Ty}(x') \cos kx'dx'\Big]\Big\} \\
&+\left(\frac{dM_{Ty}}{dx}\right)_{x=l} = 0
\end{aligned}
\tag{21.136}
$$

From Eq. (21.136) we obtain

$$
C_1 = \frac{k}{P}\Big[-\int_0^l M_{Ty}(x') \sin kx'dx' + \tan kl \int_0^l M_{Ty}(x') \cos kx'dx'\Big] \tag{21.137}
$$

Substitution of Eqs. (21.135) and (21.137) into Eq. (21.132) leads to the deflection w

$$
\begin{aligned}
w = \frac{k}{P}\Big\{&(\cos kx - 1)\Big[-\int_0^l M_{T_y}(x') \sin kx'dx' \\
&+ \tan kl \int_0^l M_{T_y}(x') \cos kx'dx'\Big] + \cos kx \int_0^x M_{T_y}(x') \sin kx'dx' \\
&- \sin kx \int_0^x M_{T_y}(x') \cos kx'dx'\Big\}
\end{aligned}
\qquad \text{(Answer)}
$$

The deflection w tends to infinity, when kl satisfies the condition

$$
\cos kl = 0 \tag{21.138}
$$

Then, the buckling load is

$$
P_{cr} = \frac{\pi^2 EI_y}{4l^2} \qquad \text{(Answer)}
$$

Problem 21.14. When the thermally induced bending moment $M_{Ty}(x)$ is given as an arbitrary function of x, find the deflection w and the buckling load P_{cr} of a beam-column with length l for one simply supported edge and the other one built-in edge.

Solution. The fundamental equation for deflection w is given by Eq. (21.1)

$$
EI_y \frac{d^2w}{dx^2} + Pw = -R_0 x - M_{Ty}(x) \tag{21.139}
$$

where R_0 denotes the reaction force at the edge $x = 0$.

Using derivation of a solution for deflection w as in Problem 21.12, the fundamental solution w of Eq. (21.139) is

$$w = C_1 \cos kx + C_2 \sin kx - \frac{R_0}{P} x + \frac{k}{P} \Big[\cos kx \int_0^x M_{T_y}(x') \sin kx' dx' - \sin kx \int_0^x M_{T_y}(x') \cos kx' dx' \Big] \tag{21.140}$$

The boundary conditions are from Eqs. (21.6) and (21.5)

$$\begin{aligned} w &= 0, \quad EI_y \frac{d^2 w}{dx^2} + M_{Ty}(x) = 0 \quad \text{on } x = 0 \\ w &= 0, \quad \frac{dw}{dx} = 0 \quad \text{on } x = l \end{aligned} \tag{21.141}$$

The differentiation of w with respect to x gives

$$\begin{aligned} \frac{dw}{dx} =& - C_1 k \sin kx + C_2 k \cos kx - \frac{R_0}{P} \\ & - \frac{k^2}{P} \Big[\sin kx \int_0^x M_{Ty}(x') \sin kx' dx' + \cos kx \int_0^x M_{Ty}(x') \cos kx' dx' \Big] \\ \frac{d^2 w}{dx^2} =& - k^2 (C_1 \cos kx + C_2 \sin kx) \\ & - \frac{k^2}{P} \Big[k \cos kx \int_0^x M_{Ty}(x') \sin kx' dx' \\ & - k \sin kx \int_0^x M_{Ty}(x') \cos kx' dx' + M_{Ty}(x) \Big] \end{aligned} \tag{21.142}$$

The boundary conditions on $x = 0$ gives

$$C_1 = 0 \tag{21.143}$$

The boundary conditions on $x = l$ give

$$\begin{aligned} & C_2 \sin kl - \frac{R_0}{P} l + \frac{k}{P} \Big[\cos kl \int_0^l M_{Ty}(x') \sin kx' dx' \\ & - \sin kl \int_0^l M_{Ty}(x') \cos kx' dx' \Big] = 0 \\ & C_2 k \cos kl - \frac{R_0}{P} - \frac{k^2}{P} \Big[\sin kl \int_0^l M_{Ty}(x') \sin kx' dx' \\ & + \cos kl \int_0^l M_{Ty}(x') \cos kx' dx' \Big] = 0 \end{aligned} \tag{21.144}$$

By solving the simultaneous Eq. (21.144) with respect to C_2 and R_0, we get

$$C_2 = \frac{k}{P}\Big[\frac{kl\sin kl + \cos kl}{kl\cos kl - \sin kl}\int_0^l M_{Ty}(x')\sin kx'dx'$$
$$+ \int_0^l M_{Ty}(x')\cos kx'dx'\Big]$$
$$R_0 = \frac{k^2}{kl\cos kl - \sin kl}\int_0^l M_{Ty}(x')\sin kx'dx' \tag{21.145}$$

Therefore, the deflection w is

$$w = \frac{k}{P}\Big[\frac{\sin kx(kl\sin kl + \cos kl) - kx}{kl\cos kl - \sin kl}\int_0^l M_{T_y}(x')\sin kx'dx'$$
$$+ \sin kx\int_x^l M_{T_y}(x')\cos kx'dx' + \cos kx\int_0^x M_{T_y}(x')\sin kx'dx'\Big] \quad \text{(Answer)}$$

The deflection w tends to infinity, when kl satisfies the condition

$$\tan kl = kl \tag{21.146}$$

Then, the buckling load P_{cr} is

$$P_{cr} = \frac{2.046\pi^2 EI_y}{l^2} \quad \text{(Answer)}$$

Problem 21.15. When the thermally induced bending moment $M_{Ty}(x)$ is given as an arbitrary function of x, find the deflection w and the buckling load P_{cr} of a beam-column with length l for both built-in edges.

Solution. The fundamental equation for deflection w is given by Eq. (21.1)

$$EI_y\frac{d^2w}{dx^2} + Pw = -R_0x - M_0 - M_{Ty}(x) \tag{21.147}$$

where R_0 and M_0 denote the reaction force and moment at the edge $x = 0$.

Using derivation of a solution for deflection w as in Problem 21.12, the fundamental solution w of Eq. (21.147) is

$$w = C_1\cos kx + C_2\sin kx - \frac{1}{P}(R_0x + M_0)$$
$$+ \frac{k}{P}\Big[\cos kx\int_0^x M_{T_y}(x')\sin kx'dx' - \sin kx\int_0^x M_{T_y}(x')\cos kx'dx'\Big] \tag{21.148}$$

The boundary conditions are given by Eq. (21.5)

$$w = 0, \qquad \frac{dw}{dx} = 0 \quad \text{on } x = 0,\ l \tag{21.149}$$

The differentiation of w with respect to x gives

$$\begin{aligned}\frac{dw}{dx} =& -C_1 k \sin kx + C_2 k \cos kx - \frac{R_0}{P}\\ &- \frac{k^2}{P}\left[\sin kx \int_0^x M_{Ty}(x')\sin kx' dx' + \cos kx \int_0^x M_{Ty}(x')\cos kx' dx'\right]\end{aligned} \tag{21.150}$$

The boundary conditions on $x = 0$ give

$$M_0 = C_1 P, \qquad R_0 = C_2 k P \tag{21.151}$$

The boundary conditions on $x = l$ give

$$\begin{aligned}&C_1(\cos kl - 1) + C_2(\sin kl - kl)\\ &\quad + \frac{k}{P}\left[\cos kl \int_0^l M_{Ty}(x')\sin kx' dx' - \sin kl \int_0^l M_{Ty}(x')\cos kx' dx'\right] = 0\\ &- C_1 k \sin kl + C_2 k(\cos kl - 1)\\ &\quad - \frac{k^2}{P}\left[\sin kl \int_0^l M_{Ty}(x')\sin kx' dx' + \cos kl \int_0^l M_{Ty}(x')\cos kx' dx'\right] = 0\end{aligned} \tag{21.152}$$

By solving the simultaneous Eq. (21.152) with respect to C_1 and C_2, we get

$$\begin{aligned}C_1 =& -\frac{k}{P}\frac{1}{2(1-\cos kl) - kl\sin kl}\\ &\times\left[(1 - \cos kl - kl\sin kl)\int_0^l M_{Ty}(x')\sin kx' dx'\right.\\ &\left. + (\sin kl - kl\cos kl)\int_0^l M_{Ty}(x')\cos kx' dx'\right]\\ C_2 =& -\frac{k}{P}\frac{1}{2(1-\cos kl) - kl\sin kl}\left[\sin kl \int_0^l M_{Ty}(x')\sin kx' dx'\right.\\ &\left. + (\cos kl - 1)\int_0^l M_{Ty}(x')\cos kx' dx'\right]\end{aligned} \tag{21.153}$$

Then, the deflection w is

$$
\begin{aligned}
w = \frac{k}{P}\Big\{ & \frac{1-\cos kx}{2(1-\cos kl)-kl\sin kl} \\
& \times \Big[(1-\cos kl - kl\sin kl)\int_0^l M_{T_y}(x')\sin kx' dx' \\
& + (\sin kl - kl\cos kl)\int_0^l M_{T_y}(x')\cos kx' dx'\Big] \\
& - \Big(\frac{\sin kx - kx}{2(1-\cos kl)-kl\sin kl}\Big)\Big[\sin kl \int_0^l M_{T_y}(x)\sin kx' dx' \\
& + (\cos kl - 1)\int_0^l M_{T_y}(x')\cos kx' dx'\Big] \\
& + \cos kx \int_0^x M_{T_y}(x')\sin kx' dx' - \sin kx \int_0^x M_{T_y}(x')\cos kx' dx'\Big\}
\end{aligned}
\quad \text{(Answer)}
$$

The deflection w tends to infinity, when kl satisfies the condition

$$2 - 2\cos kl - kl\sin kl = 0 \tag{21.154}$$

That is

$$\sin\frac{kl}{2}\Big(\sin\frac{kl}{2} - \frac{kl}{2}\cos\frac{kl}{2}\Big) = 0 \tag{21.155}$$

Then, the buckling load P_{cr} is

$$P_{cr} = \frac{4\pi^2 EI_y}{l^2} \quad \text{(Answer)}$$

Problem 21.16. Derive the stress-displacement relations for a circular plate.

Solution. The stress-displacement relations in the Cartesian coordinates are given by Eq. (20.7), namely

$$
\begin{aligned}
\sigma_{xx} &= \frac{E}{1-\nu^2}\Big[\frac{\partial u}{\partial x} + \nu\frac{\partial v}{\partial y} - z(\frac{\partial^2 w}{\partial x^2} + \nu\frac{\partial^2 w}{\partial y^2}) - (1+\nu)\alpha\tau\Big] \\
\sigma_{yy} &= \frac{E}{1-\nu^2}\Big[\frac{\partial v}{\partial y} + \nu\frac{\partial u}{\partial x} - z(\frac{\partial^2 w}{\partial y^2} + \nu\frac{\partial^2 w}{\partial x^2}) - (1+\nu)\alpha\tau\Big] \\
\sigma_{xy} &= \frac{E}{2(1+\nu)}\Big[\frac{\partial u}{\partial y} + \frac{\partial v}{\partial x} - 2z\frac{\partial^2 w}{\partial x\partial y}\Big]
\end{aligned}
\tag{21.156}
$$

By use of the coordinate transform between the Cartesian coordinate system and the cylindrical system, we have the following relations from Eqs. (16.8) and (16.4)

$$u_i = l_{j'i}\bar{u}_{j'} \tag{21.157}$$

$$\overline{\sigma}_{i'j'} = l_{i'm} l_{j'n} \sigma_{mn} \tag{21.158}$$

in which, $u_i = u, v, \sigma_{ij} = \sigma_{xx}, \sigma_{yy}, \sigma_{xy}$ are the displacement components and stress components in the Cartesian coordinate system, $\overline{u}_i = u_r, u_\theta, \overline{\sigma}_{i'j'} = \sigma_{rr}, \sigma_{\theta\theta}, \sigma_{r\theta}$ are the displacement components and stress components in the polar coordinate system, $l_{i'j}$ is direction cosines defined by

$$l_{i'j} = \begin{pmatrix} \cos\theta & \sin\theta \\ -\sin\theta & \cos\theta \end{pmatrix} \tag{21.159}$$

The relations between the coordinate variables are given by

$$\begin{pmatrix} x = r\cos\theta \\ y = r\sin\theta \end{pmatrix}, \quad \begin{pmatrix} r^2 = x^2 + y^2 \\ \theta = \arctan\dfrac{y}{x} \end{pmatrix} \tag{21.160}$$

Making use of the relations (21.157), displacement components are transformed by the relations

$$u_x = u_r \cos\theta - u_\theta \sin\theta, \quad u_y = u_r \sin\theta + u_\theta \cos\theta \tag{21.161}$$

Similarly, stress components are transformed by the relations

$$\begin{aligned}
\sigma_{rr} &= \cos^2\theta\sigma_{xx} + \sin^2\theta\sigma_{yy} + 2\sin\theta\cos\theta\sigma_{xy} \\
\sigma_{\theta\theta} &= \sin^2\theta\sigma_{xx} + \cos^2\theta\sigma_{yy} - 2\sin\theta\cos\theta\sigma_{xy} \\
\sigma_{r\theta} &= \sin\theta\cos\theta(\sigma_{yy} - \sigma_{xx}) + (\cos^2\theta - \sin^2\theta)\sigma_{xy}
\end{aligned} \tag{21.162}$$

Making use of relations (21.160), we have

$$\begin{aligned}
\frac{\partial}{\partial x} &= \cos\theta\frac{\partial}{\partial r} - \sin\theta\frac{1}{r}\frac{\partial}{\partial\theta}, \qquad \frac{\partial}{\partial y} = \sin\theta\frac{\partial}{\partial r} + \cos\theta\frac{1}{r}\frac{\partial}{\partial\theta} \\
\frac{\partial^2}{\partial x^2} &= \cos^2\theta\frac{\partial^2}{\partial r^2} + \sin^2\theta\Big(\frac{1}{r}\frac{\partial}{\partial r} + \frac{1}{r^2}\frac{\partial^2}{\partial\theta^2}\Big) \\
&\quad - 2\sin\theta\cos\theta\Big(\frac{1}{r}\frac{\partial^2}{\partial r\partial\theta} - \frac{1}{r^2}\frac{\partial}{\partial\theta}\Big) \\
\frac{\partial^2}{\partial y^2} &= \sin^2\theta\frac{\partial^2}{\partial r^2} + \cos^2\theta\Big(\frac{1}{r}\frac{\partial}{\partial r} + \frac{1}{r^2}\frac{\partial^2}{\partial\theta^2}\Big) \\
&\quad + 2\sin\theta\cos\theta\Big(\frac{1}{r}\frac{\partial^2}{\partial r\partial\theta} - \frac{1}{r^2}\frac{\partial}{\partial\theta}\Big) \\
\frac{\partial^2}{\partial x\partial y} &= \sin\theta\cos\theta\Big(\frac{\partial^2}{\partial r^2} - \frac{1}{r}\frac{\partial}{\partial r} - \frac{1}{r^2}\frac{\partial^2}{\partial\theta^2}\Big) \\
&\quad + (\cos^2\theta - \sin^2\theta)\Big(\frac{1}{r}\frac{\partial^2}{\partial r\partial\theta} - \frac{1}{r^2}\frac{\partial}{\partial\theta}\Big)
\end{aligned} \tag{21.163}$$

We now calculate the stress components in the polar coordinate system. Substituting Eqs. (21.156), (21.161), and (21.163) into Eq. (21.162), we have

$$
\begin{aligned}
\sigma_{rr} &= \cos^2\theta\sigma_{xx} + \sin^2\theta\sigma_{yy} + 2\sin\theta\cos\theta\sigma_{xy} \\
&= \frac{E}{1-\nu^2}\Big\{\cos^2\theta\Big[\frac{\partial u}{\partial x} + \nu\frac{\partial v}{\partial y} - z\Big(\frac{\partial^2 w}{\partial x^2} + \nu\frac{\partial^2 w}{\partial y^2}\Big) - (1+\nu)\alpha\tau\Big] \\
&\quad + \sin^2\theta\Big[\frac{\partial v}{\partial y} + \nu\frac{\partial u}{\partial x} - z\Big(\frac{\partial^2 w}{\partial y^2} + \nu\frac{\partial^2 w}{\partial x^2}\Big) - (1+\nu)\alpha\tau\Big] \\
&\quad + (1-\nu)\sin\theta\cos\theta\Big[\frac{\partial u}{\partial y} + \frac{\partial v}{\partial x} - 2z\frac{\partial^2 w}{\partial x\partial y}\Big]\Big\} \\
&= \frac{E}{1-\nu^2}\Big\{\cos^2\theta\Big(\cos\theta\frac{\partial}{\partial r} - \sin\theta\frac{1}{r}\frac{\partial}{\partial\theta}\Big)(u_r\cos\theta - u_\theta\sin\theta) \\
&\quad + \nu\cos^2\theta\Big(\sin\theta\frac{\partial}{\partial r} + \cos\theta\frac{1}{r}\frac{\partial}{\partial\theta}\Big)(u_r\sin\theta + u_\theta\cos\theta) \\
&\quad + \sin^2\theta\Big(\sin\theta\frac{\partial}{\partial r} + \cos\theta\frac{1}{r}\frac{\partial}{\partial\theta}\Big)(u_r\sin\theta + u_\theta\cos\theta) \\
&\quad + \nu\sin^2\theta\Big(\cos\theta\frac{\partial}{\partial r} - \sin\theta\frac{1}{r}\frac{\partial}{\partial\theta}\Big)(u_r\cos\theta - u_\theta\sin\theta) \\
&\quad + (1-\nu)\sin\theta\cos\theta\Big(\sin\theta\frac{\partial}{\partial r} + \cos\theta\frac{1}{r}\frac{\partial}{\partial\theta}\Big)(u_r\cos\theta - u_\theta\sin\theta) \\
&\quad + (1-\nu)\sin\theta\cos\theta\Big(\cos\theta\frac{\partial}{\partial r} - \sin\theta\frac{1}{r}\frac{\partial}{\partial\theta}\Big)(u_r\sin\theta + u_\theta\cos\theta) \\
&\quad - z\cos^2\theta\Big[\cos^2\theta\frac{\partial^2 w}{\partial r^2} + \sin^2\theta\Big(\frac{1}{r}\frac{\partial w}{\partial r} + \frac{1}{r^2}\frac{\partial^2 w}{\partial\theta^2}\Big) \\
&\quad - 2\sin\theta\cos\theta\Big(\frac{1}{r}\frac{\partial^2 w}{\partial r\partial\theta} - \frac{1}{r^2}\frac{\partial w}{\partial\theta}\Big)\Big] \\
&\quad - \nu z\cos^2\theta\Big[\sin^2\theta\frac{\partial^2 w}{\partial r^2} + \cos^2\theta\Big(\frac{1}{r}\frac{\partial w}{\partial r} + \frac{1}{r^2}\frac{\partial^2 w}{\partial\theta^2}\Big) \\
&\quad + 2\sin\theta\cos\theta\Big(\frac{1}{r}\frac{\partial^2 w}{\partial r\partial\theta} - \frac{1}{r^2}\frac{\partial w}{\partial\theta}\Big)\Big] \\
&\quad - z\sin^2\theta\Big[\sin^2\theta\frac{\partial^2 w}{\partial r^2} + \cos^2\theta\Big(\frac{1}{r}\frac{\partial w}{\partial r} + \frac{1}{r^2}\frac{\partial^2 w}{\partial\theta^2}\Big) \\
&\quad + 2\sin\theta\cos\theta\Big(\frac{1}{r}\frac{\partial^2 w}{\partial r\partial\theta} - \frac{1}{r^2}\frac{\partial w}{\partial\theta}\Big)\Big] \\
&\quad - \nu z\sin^2\theta\Big[\cos^2\theta\frac{\partial^2 w}{\partial r^2} + \sin^2\theta\Big(\frac{1}{r}\frac{\partial w}{\partial r} + \frac{1}{r^2}\frac{\partial^2 w}{\partial\theta^2}\Big) \\
&\quad - 2\sin\theta\cos\theta\Big(\frac{1}{r}\frac{\partial^2 w}{\partial r\partial\theta} - \frac{1}{r^2}\frac{\partial w}{\partial\theta}\Big)\Big] \\
&\quad - 2(1-\nu)z\sin\theta\cos\theta\Big[\sin\theta\cos\theta\Big(\frac{\partial^2 w}{\partial r^2} - \frac{1}{r}\frac{\partial w}{\partial r} - \frac{1}{r^2}\frac{\partial^2 w}{\partial\theta^2}\Big)
\end{aligned}
$$

$$+ (\cos^2\theta - \sin^2\theta)\Big(\frac{1}{r}\frac{\partial^2 w}{\partial r\partial\theta} - \frac{1}{r^2}\frac{\partial w}{\partial\theta}\Big)\Big] - (1+\nu)\alpha\tau\Big\}$$
$$= \frac{E}{1-\nu^2}\Big\{\frac{\partial u_r}{\partial r} + \nu\Big(\frac{u_r}{r} + \frac{1}{r}\frac{\partial u_\theta}{\partial\theta}\Big)$$
$$- z\Big[\frac{\partial^2 w}{\partial r^2} + \nu\Big(\frac{1}{r}\frac{\partial w}{\partial r} + \frac{1}{r^2}\frac{\partial^2 w}{\partial\theta^2}\Big)\Big] - (1+\nu)\alpha\tau\Big\} \qquad \text{(Answer)}$$

$$\sigma_{\theta\theta} = \sin^2\theta\sigma_{xx} + \cos^2\theta\sigma_{yy} - 2\sin\theta\cos\theta\sigma_{xy}$$
$$= \frac{E}{1-\nu^2}\Big\{\sin^2\theta\Big[\frac{\partial u}{\partial x} + \nu\frac{\partial v}{\partial y} - z\Big(\frac{\partial^2 w}{\partial x^2} + \nu\frac{\partial^2 w}{\partial y^2}\Big) - (1+\nu)\alpha\tau\Big]$$
$$+ \cos^2\theta\Big[\frac{\partial v}{\partial y} + \nu\frac{\partial u}{\partial x} - z\Big(\frac{\partial^2 w}{\partial y^2} + \nu\frac{\partial^2 w}{\partial x^2}\Big) - (1+\nu)\alpha\tau\Big]$$
$$- (1-\nu)\sin\theta\cos\theta\Big[\frac{\partial u}{\partial y} + \frac{\partial v}{\partial x} - 2z\frac{\partial^2 w}{\partial x\partial y}\Big]\Big\}$$
$$= \frac{E}{1-\nu^2}\Big\{\sin^2\theta\Big(\cos\theta\frac{\partial}{\partial r} - \sin\theta\frac{1}{r}\frac{\partial}{\partial\theta}\Big)(u_r\cos\theta - u_\theta\sin\theta)$$
$$+ \nu\sin^2\theta\Big(\sin\theta\frac{\partial}{\partial r} + \cos\theta\frac{1}{r}\frac{\partial}{\partial\theta}\Big)(u_r\sin\theta + u_\theta\cos\theta)$$
$$+ \cos^2\theta\Big(\sin\theta\frac{\partial}{\partial r} + \cos\theta\frac{1}{r}\frac{\partial}{\partial\theta}\Big)(u_r\sin\theta + u_\theta\cos\theta)$$
$$+ \nu\cos^2\theta\Big(\cos\theta\frac{\partial}{\partial r} - \sin\theta\frac{1}{r}\frac{\partial}{\partial\theta}\Big)(u_r\cos\theta - u_\theta\sin\theta)$$
$$- (1-\nu)\sin\theta\cos\theta\Big(\sin\theta\frac{\partial}{\partial r} + \cos\theta\frac{1}{r}\frac{\partial}{\partial\theta}\Big)(u_r\cos\theta - u_\theta\sin\theta)$$
$$- (1-\nu)\sin\theta\cos\theta\Big(\cos\theta\frac{\partial}{\partial r} - \sin\theta\frac{1}{r}\frac{\partial}{\partial\theta}\Big)(u_r\sin\theta + u_\theta\cos\theta)$$
$$- z\sin^2\theta\Big[\cos^2\theta\frac{\partial^2 w}{\partial r^2} + \sin^2\theta\Big(\frac{1}{r}\frac{\partial w}{\partial r} + \frac{1}{r^2}\frac{\partial^2 w}{\partial\theta^2}\Big)$$
$$- 2\sin\theta\cos\theta\Big(\frac{1}{r}\frac{\partial^2 w}{\partial r\partial\theta} - \frac{1}{r^2}\frac{\partial w}{\partial\theta}\Big)\Big]$$
$$- \nu z\sin^2\theta\Big[\sin^2\theta\frac{\partial^2 w}{\partial r^2} + \cos^2\theta\Big(\frac{1}{r}\frac{\partial w}{\partial r} + \frac{1}{r^2}\frac{\partial^2 w}{\partial\theta^2}\Big)$$
$$+ 2\sin\theta\cos\theta\Big(\frac{1}{r}\frac{\partial^2 w}{\partial r\partial\theta} - \frac{1}{r^2}\frac{\partial w}{\partial\theta}\Big)\Big]$$
$$- z\cos^2\theta\Big[\sin^2\theta\frac{\partial^2 w}{\partial r^2} + \cos^2\theta\Big(\frac{1}{r}\frac{\partial w}{\partial r} + \frac{1}{r^2}\frac{\partial^2 w}{\partial\theta^2}\Big)$$
$$+ 2\sin\theta\cos\theta\Big(\frac{1}{r}\frac{\partial^2 w}{\partial r\partial\theta} - \frac{1}{r^2}\frac{\partial w}{\partial\theta}\Big)\Big]$$
$$- \nu z\cos^2\theta\Big[\cos^2\theta\frac{\partial^2 w}{\partial r^2} + \sin^2\theta\Big(\frac{1}{r}\frac{\partial w}{\partial r} + \frac{1}{r^2}\frac{\partial^2 w}{\partial\theta^2}\Big)$$

$$
\begin{aligned}
&- 2\sin\theta\cos\theta\Big(\frac{1}{r}\frac{\partial^2 w}{\partial r\partial\theta} - \frac{1}{r^2}\frac{\partial w}{\partial\theta}\Big)\Big]\\
&+ 2(1-\nu)z\sin\theta\cos\theta\Big[\sin\theta\cos\theta\Big(\frac{\partial^2 w}{\partial r^2} - \frac{1}{r}\frac{\partial w}{\partial r} - \frac{1}{r^2}\frac{\partial^2 w}{\partial\theta^2}\Big)\\
&+ (\cos^2\theta - \sin^2\theta)\Big(\frac{1}{r}\frac{\partial^2 w}{\partial r\partial\theta} - \frac{1}{r^2}\frac{\partial w}{\partial\theta}\Big)\Big] - (1+\nu)\alpha\tau\Big\}\\
&= \frac{E}{1-\nu^2}\Big\{\nu\frac{\partial u_r}{\partial r} + \Big(\frac{u_r}{r} + \frac{1}{r}\frac{\partial u_\theta}{\partial\theta}\Big) - z\Big[\nu\frac{\partial^2 w}{\partial r^2} + \Big(\frac{1}{r}\frac{\partial w}{\partial r} + \frac{1}{r^2}\frac{\partial^2 w}{\partial\theta^2}\Big)\Big]\\
&- (1+\nu)\alpha\tau\Big\} \qquad \text{(Answer)}
\end{aligned}
$$

$$
\begin{aligned}
\sigma_{r\theta} &= \sin\theta\cos\theta(\sigma_{yy} - \sigma_{xx}) + (\cos^2\theta - \sin^2\theta)\sigma_{xy}\\
&= \frac{E}{1-\nu^2}\Big\{\sin\theta\cos\theta\Big[-(1-\nu)\frac{\partial u}{\partial x} + (1-\nu)\frac{\partial v}{\partial y}\\
&- (1-\nu)z\Big(\frac{\partial^2 w}{\partial y^2} - \frac{\partial^2 w}{\partial x^2}\Big)\Big]\\
&+ (\cos^2\theta - \sin^2\theta)\frac{1-\nu}{2}\Big(\frac{\partial u}{\partial y} + \frac{\partial v}{\partial x} - 2z\frac{\partial^2 w}{\partial x\partial y}\Big)\Big\}\\
&= \frac{E}{2(1+\nu)}\Big\{2\sin\theta\cos\theta\Big[-\frac{\partial u}{\partial x} + \frac{\partial v}{\partial y} - z\Big(\frac{\partial^2 w}{\partial y^2} - \frac{\partial^2 w}{\partial x^2}\Big)\Big]\\
&+ (\cos^2\theta - \sin^2\theta)\Big(\frac{\partial u}{\partial y} + \frac{\partial v}{\partial x} - 2z\frac{\partial^2 w}{\partial x\partial y}\Big)\Big\}\\
&= \frac{E}{2(1+\nu)}\Big\{-2\sin\theta\cos\theta\Big(\cos\theta\frac{\partial}{\partial r} - \sin\theta\frac{1}{r}\frac{\partial}{\partial\theta}\Big)(u_r\cos\theta - u_\theta\sin\theta)\\
&+ 2\sin\theta\cos\theta\Big(\sin\theta\frac{\partial}{\partial r} + \cos\theta\frac{1}{r}\frac{\partial}{\partial\theta}\Big)(u_r\sin\theta + u_\theta\cos\theta)\\
&+ (\cos^2\theta - \sin^2\theta)\Big(\sin\theta\frac{\partial}{\partial r} + \cos\theta\frac{1}{r}\frac{\partial}{\partial\theta}\Big)(u_r\cos\theta - u_\theta\sin\theta)\\
&+ (\cos^2\theta - \sin^2\theta)\Big(\cos\theta\frac{\partial}{\partial r} - \sin\theta\frac{1}{r}\frac{\partial}{\partial\theta}\Big)(u_r\sin\theta + u_\theta\cos\theta)\\
&- 2z\sin\theta\cos\theta\Big[\sin^2\theta\frac{\partial^2 w}{\partial r^2} + \cos^2\theta\Big(\frac{1}{r}\frac{\partial w}{\partial r} + \frac{1}{r^2}\frac{\partial^2 w}{\partial\theta^2}\Big)\\
&+ 2\sin\theta\cos\theta\Big(\frac{1}{r}\frac{\partial^2 w}{\partial r\partial\theta} - \frac{1}{r^2}\frac{\partial w}{\partial\theta}\Big)\\
&- \cos^2\theta\frac{\partial^2 w}{\partial r^2} - \sin^2\theta\Big(\frac{1}{r}\frac{\partial w}{\partial r} + \frac{1}{r^2}\frac{\partial^2 w}{\partial\theta^2}\Big)\\
&+ 2\sin\theta\cos\theta\Big(\frac{1}{r}\frac{\partial^2 w}{\partial r\partial\theta} - \frac{1}{r^2}\frac{\partial w}{\partial\theta}\Big)\Big]\\
&- 2z(\cos^2\theta - \sin^2\theta)\Big[\sin\theta\cos\theta\Big(\frac{\partial^2 w}{\partial r^2} - \frac{1}{r}\frac{\partial w}{\partial r} - \frac{1}{r^2}\frac{\partial^2 w}{\partial\theta^2}\Big)
\end{aligned}
$$

$$+ (\cos^2\theta - \sin^2\theta)\Big(\frac{1}{r}\frac{\partial^2 w}{\partial r\partial\theta} - \frac{1}{r^2}\frac{\partial w}{\partial\theta}\Big)\Big]\Big\}$$
$$= \frac{E}{2(1+\nu)}\Big[\frac{1}{r}\frac{\partial u_r}{\partial\theta} + r\frac{\partial}{\partial r}\Big(\frac{u_\theta}{r}\Big) - 2z\frac{\partial}{\partial r}\Big(\frac{1}{r}\frac{\partial w}{\partial\theta}\Big)\Big] \qquad \text{(Answer)}$$

Problem 21.17. Derive the resultant forces N_r, N_θ, and $N_{r\theta}$ for a circular plate by use of the stress function F given by Eq. (21.20).

Solution. The resultant forces constitute a second-order tensor, so thatN_r, N_θ, and $N_{r\theta}$ are transformed by the relations given by Eq. (21.162), namely

$$\begin{aligned} N_r &= \cos^2\theta N_x + \sin^2\theta N_y + 2\sin\theta\cos\theta N_{xy} \\ N_\theta &= \sin^2\theta N_x + \cos^2\theta N_y - 2\sin\theta\cos\theta N_{xy} \\ N_{r\theta} &= \sin\theta\cos\theta(N_y - N_x) + (\cos^2\theta - \sin^2\theta)N_{xy} \end{aligned} \qquad (21.164)$$

Now, the relations of the resultant forces N_x, N_y, and N_{xy} with the stress function F are given by Eq. (21.11), namely

$$N_x = \frac{\partial^2 F}{\partial y^2}, \quad N_y = \frac{\partial^2 F}{\partial x^2}, \quad N_{xy} = -\frac{\partial^2 F}{\partial x\partial y} \qquad (21.165)$$

By the substitution of Eq. (21.165) into Eq. (21.164), we have

$$\begin{aligned} N_r &= \cos^2\theta\frac{\partial^2 F}{\partial y^2} + \sin^2\theta\frac{\partial^2 F}{\partial x^2} - 2\sin\theta\cos\theta\frac{\partial^2 F}{\partial x\partial y} \\ N_\theta &= \sin^2\theta\frac{\partial^2 F}{\partial y^2} + \cos^2\theta\frac{\partial^2 F}{\partial x^2} + 2\sin\theta\cos\theta\frac{\partial^2 F}{\partial x\partial y} \\ N_{r\theta} &= \sin\theta\cos\theta\Big(\frac{\partial^2 F}{\partial x^2} - \frac{\partial^2 F}{\partial y^2}\Big) - (\cos^2\theta - \sin^2\theta)\frac{\partial^2 F}{\partial x\partial y} \end{aligned} \qquad (21.166)$$

The relations for the differential operators such as $\partial^2/\partial x^2, \partial^2/\partial y^2, \partial^2/\partial x\partial y$ are given by Eq. (21.163)

$$\begin{aligned} \frac{\partial^2}{\partial x^2} &= \cos^2\theta\frac{\partial^2}{\partial r^2} + \sin^2\theta\Big(\frac{1}{r}\frac{\partial}{\partial r} + \frac{1}{r^2}\frac{\partial^2}{\partial\theta^2}\Big) \\ &\quad -2\sin\theta\cos\theta\Big(\frac{1}{r}\frac{\partial^2}{\partial r\partial\theta} - \frac{1}{r^2}\frac{\partial}{\partial\theta}\Big) \\ \frac{\partial^2}{\partial y^2} &= \sin^2\theta\frac{\partial^2}{\partial r^2} + \cos^2\theta\Big(\frac{1}{r}\frac{\partial}{\partial r} + \frac{1}{r^2}\frac{\partial^2}{\partial\theta^2}\Big) \\ &\quad +2\sin\theta\cos\theta\Big(\frac{1}{r}\frac{\partial^2}{\partial r\partial\theta} - \frac{1}{r^2}\frac{\partial}{\partial\theta}\Big) \end{aligned}$$

$$\frac{\partial^2}{\partial x \partial y} = \sin\theta\cos\theta\Big(\frac{\partial^2}{\partial r^2} - \frac{1}{r}\frac{\partial}{\partial r} - \frac{1}{r^2}\frac{\partial^2}{\partial \theta^2}\Big) + (\cos^2\theta - \sin^2\theta)\Big(\frac{1}{r}\frac{\partial^2}{\partial r \partial \theta} - \frac{1}{r^2}\frac{\partial}{\partial \theta}\Big) \tag{21.167}$$

By the substitution of Eq. (21.167) into Eq. (21.166), we can evaluate the resultant forces N_r, N_θ, and $N_{r\theta}$

$$\begin{aligned}
N_r = {} & \cos^2\theta\Big[\sin^2\theta\frac{\partial^2 F}{\partial r^2} + \cos^2\theta\Big(\frac{1}{r}\frac{\partial F}{\partial r} + \frac{1}{r^2}\frac{\partial^2 F}{\partial \theta^2}\Big) \\
& + 2\sin\theta\cos\theta\Big(\frac{1}{r}\frac{\partial^2 F}{\partial r\partial\theta} - \frac{1}{r^2}\frac{\partial F}{\partial\theta}\Big)\Big] \\
& + \sin^2\theta\Big[\cos^2\theta\frac{\partial^2 F}{\partial r^2} + \sin^2\theta\Big(\frac{1}{r}\frac{\partial F}{\partial r} + \frac{1}{r^2}\frac{\partial^2 F}{\partial \theta^2}\Big) \\
& - 2\sin\theta\cos\theta\Big(\frac{1}{r}\frac{\partial^2 F}{\partial r\partial\theta} - \frac{1}{r^2}\frac{\partial F}{\partial\theta}\Big)\Big] \\
& - 2\sin\theta\cos\theta\Big[\sin\theta\cos\theta\Big(\frac{\partial^2 F}{\partial r^2} - \frac{1}{r}\frac{\partial F}{\partial r} - \frac{1}{r^2}\frac{\partial^2 F}{\partial \theta^2}\Big) \\
& + (\cos^2\theta - \sin^2\theta)\Big(\frac{1}{r}\frac{\partial^2 F}{\partial r\partial\theta} - \frac{1}{r^2}\frac{\partial F}{\partial\theta}\Big)\Big] \\
= {} & \frac{1}{r}\frac{\partial F}{\partial \theta} + \frac{1}{r^2}\frac{\partial^2 F}{\partial \theta^2} \qquad \text{(Answer)} \\
N_\theta = {} & \sin^2\theta\Big[\sin^2\theta\frac{\partial^2 F}{\partial r^2} + \cos^2\theta\Big(\frac{1}{r}\frac{\partial F}{\partial r} + \frac{1}{r^2}\frac{\partial^2 F}{\partial \theta^2}\Big) \\
& + 2\sin\theta\cos\theta\Big(\frac{1}{r}\frac{\partial^2 F}{\partial r\partial\theta} - \frac{1}{r^2}\frac{\partial F}{\partial\theta}\Big)\Big] \\
& + \cos^2\theta\Big[\cos^2\theta\frac{\partial^2 F}{\partial r^2} + \sin^2\theta\Big(\frac{1}{r}\frac{\partial F}{\partial r} + \frac{1}{r^2}\frac{\partial^2 F}{\partial \theta^2}\Big) \\
& - 2\sin\theta\cos\theta\Big(\frac{1}{r}\frac{\partial^2 F}{\partial r\partial\theta} - \frac{1}{r^2}\frac{\partial F}{\partial\theta}\Big)\Big] \\
& + 2\sin\theta\cos\theta\Big[\sin\theta\cos\theta\Big(\frac{\partial^2 F}{\partial r^2} - \frac{1}{r}\frac{\partial F}{\partial r} - \frac{1}{r^2}\frac{\partial^2 F}{\partial \theta^2}\Big) \\
& + (\cos^2\theta - \sin^2\theta)\Big(\frac{1}{r}\frac{\partial^2 F}{\partial r\partial\theta} - \frac{1}{r^2}\frac{\partial F}{\partial\theta}\Big)\Big] \\
= {} & \frac{1}{r^2}\frac{\partial^2 F}{\partial r^2} \qquad \text{(Answer)}
\end{aligned}$$

$$\begin{aligned}
N_{r\theta} &= \sin\theta\cos\theta\Big[(\cos^2\theta - \sin^2\theta)\frac{\partial^2 F}{\partial r^2} \\
&\quad - (\cos^2\theta - \sin^2\theta)\Big(\frac{1}{r}\frac{\partial F}{\partial r} + \frac{1}{r^2}\frac{\partial^2 F}{\partial \theta^2}\Big) \\
&\quad - 4\sin\theta\cos\theta\Big(\frac{1}{r}\frac{\partial^2 F}{\partial r\partial\theta} - \frac{1}{r^2}\frac{\partial F}{\partial \theta}\Big)\Big] \\
&\quad - (\cos^2\theta - \sin^2\theta)\Big[\sin\theta\cos\theta\frac{\partial^2 F}{\partial r^2} - \sin\theta\cos\theta\Big(\frac{1}{r}\frac{\partial F}{\partial r} + \frac{1}{r^2}\frac{\partial^2 F}{\partial \theta^2}\Big) \\
&\quad + (\cos^2\theta - \sin^2\theta)\Big(\frac{1}{r}\frac{\partial^2 F}{\partial r\partial\theta} - \frac{1}{r^2}\frac{\partial F}{\partial \theta}\Big)\Big] \\
&= -\frac{\partial}{\partial r}\Big(\frac{1}{r}\frac{\partial F}{\partial \theta}\Big) \qquad \text{(Answer)}
\end{aligned}$$

Problem 21.18. Derive the fundamental relation of thermal buckling given by Eqs. (21.24) for a circular plate.

Solution. The fundamental equation of thermal buckling for the Cartesian coordinate system is given by Eq. (21.14)

$$\nabla^2\nabla^2 w = \frac{1}{D}\Big(p - \frac{1}{1-\nu}\nabla^2 M_T + N_x\frac{\partial^2 w}{\partial x^2} + N_y\frac{\partial^2 w}{\partial y^2} + 2N_{xy}\frac{\partial^2 w}{\partial x\partial y}\Big) \tag{21.168}$$

Now, the Laplace operator ∇^2 is

$$\nabla^2 = \frac{\partial^2}{\partial x^2} + \frac{\partial^2}{\partial y^2} = \frac{\partial^2}{\partial r^2} + \frac{1}{r}\frac{\partial}{\partial r} + \frac{1}{r^2}\frac{\partial^2}{\partial \theta^2} \tag{21.169}$$

The resultant forces N_x, N_y, and N_{xy} are transformed into the form derived from Eq. (21.164)

$$\begin{aligned}
N_x &= \cos^2\theta N_r + \sin^2\theta N_\theta - 2\sin\theta\cos\theta N_{r\theta} \\
N_y &= \sin^2\theta N_r + \cos^2\theta N_\theta + 2\sin\theta\cos\theta N_{r\theta} \\
N_{xy} &= \sin\theta\cos\theta(N_r - N_\theta) + (\cos^2\theta - \sin^2\theta)N_{r\theta}
\end{aligned} \tag{21.170}$$

The relations for the differential operators $\partial^2/\partial x^2, \partial^2/\partial y^2, \partial^2/\partial x\partial y$ are given by Eq. (21.167)

$$\begin{aligned}
\frac{\partial^2}{\partial x^2} &= \cos^2\theta\frac{\partial^2}{\partial r^2} + \sin^2\theta\Big(\frac{1}{r}\frac{\partial}{\partial r} + \frac{1}{r^2}\frac{\partial^2}{\partial \theta^2}\Big) \\
&\quad - 2\sin\theta\cos\theta\Big(\frac{1}{r}\frac{\partial^2}{\partial r\partial\theta} - \frac{1}{r^2}\frac{\partial}{\partial \theta}\Big) \\
\frac{\partial^2}{\partial y^2} &= \sin^2\theta\frac{\partial^2}{\partial r^2} + \cos^2\theta\Big(\frac{1}{r}\frac{\partial}{\partial r} + \frac{1}{r^2}\frac{\partial^2}{\partial \theta^2}\Big)
\end{aligned}$$

$$+2\sin\theta\cos\theta\Big(\frac{1}{r}\frac{\partial^2}{\partial r\partial\theta}-\frac{1}{r^2}\frac{\partial}{\partial\theta}\Big)$$

$$\frac{\partial^2}{\partial x\partial y}=\sin\theta\cos\theta\Big(\frac{\partial^2}{\partial r^2}-\frac{1}{r}\frac{\partial}{\partial r}-\frac{1}{r^2}\frac{\partial^2}{\partial\theta^2}\Big)$$
$$+(\cos^2\theta-\sin^2\theta)\Big(\frac{1}{r}\frac{\partial^2}{\partial r\partial\theta}-\frac{1}{r^2}\frac{\partial}{\partial\theta}\Big) \tag{21.171}$$

Making use of Eqs. (21.170) and (21.171), we can calculate the following relation

$$\begin{aligned}
&N_x\frac{\partial^2 w}{\partial x^2}+N_y\frac{\partial^2 w}{\partial y^2}+2N_{xy}\frac{\partial^2 w}{\partial x\partial y}\\
&=[\cos^2\theta N_r+\sin^2\theta N_\theta-2\sin\theta\cos\theta N_{r\theta}]\\
&\quad\times\Big[\cos^2\theta\frac{\partial^2 w}{\partial r^2}+\sin^2\theta\Big(\frac{1}{r}\frac{\partial w}{\partial r}+\frac{1}{r^2}\frac{\partial^2 w}{\partial\theta^2}\Big)\\
&\qquad-2\sin\theta\cos\theta\Big(\frac{1}{r}\frac{\partial^2 w}{\partial r\partial\theta}-\frac{1}{r^2}\frac{\partial w}{\partial\theta}\Big)\Big]\\
&\quad+[\sin^2\theta N_r+\cos^2\theta N_\theta+2\sin\theta\cos\theta N_{r\theta}]\\
&\quad\times\Big[\sin^2\theta\frac{\partial^2 w}{\partial r^2}+\cos^2\theta\Big(\frac{1}{r}\frac{\partial w}{\partial r}+\frac{1}{r^2}\frac{\partial^2 w}{\partial\theta^2}\Big)\\
&\qquad+2\sin\theta\cos\theta\Big(\frac{1}{r}\frac{\partial^2 w}{\partial r\partial\theta}-\frac{1}{r^2}\frac{\partial w}{\partial\theta}\Big)\Big]\\
&\quad+2[\sin\theta\cos\theta(N_r-N_\theta)+(\cos^2\theta-\sin^2\theta)N_{r\theta}]\\
&\quad\times\Big[\sin\theta\cos\theta\Big(\frac{\partial^2 w}{\partial r^2}-\frac{1}{r}\frac{\partial w}{\partial r}-\frac{1}{r^2}\frac{\partial^2 w}{\partial\theta^2}\Big)\\
&\quad+(\cos^2\theta-\sin^2\theta)\Big(\frac{1}{r}\frac{\partial^2 w}{\partial r\partial\theta}-\frac{1}{r^2}\frac{\partial w}{\partial\theta}\Big)\Big]\\
&=N_r\frac{\partial^2 w}{\partial r^2}+N_\theta\Big(\frac{1}{r}\frac{\partial w}{\partial r}+\frac{1}{r^2}\frac{\partial^2 w}{\partial\theta^2}\Big)+2N_{r\theta}\frac{\partial}{\partial r}\Big(\frac{1}{r}\frac{\partial w}{\partial\theta}\Big)
\end{aligned} \tag{21.172}$$

Substituting Eq. (21.172) into Eq. (21.168), we have

$$\begin{aligned}
\nabla^2\nabla^2 w=\frac{1}{D}\Big[&p-\frac{1}{1-\nu}\nabla^2 M_T+N_r\frac{\partial^2 w}{\partial r^2}\\
&+N_\theta\Big(\frac{1}{r}\frac{\partial w}{\partial r}+\frac{1}{r^2}\frac{\partial^2 w}{\partial\theta^2}\Big)+2N_{r\theta}\frac{\partial}{\partial r}\Big(\frac{1}{r}\frac{\partial w}{\partial\theta}\Big)\Big]
\end{aligned} \qquad \text{(Answer)}$$

Part III
Thermal Stresses—Advanced Theory and Applications

Chapter 22
Heat Conduction

In this chapter heat conduction problems are presented. Employing the first law of thermodynamics, the problems in the rectangular Cartesian coordinates, cylindrical coordinates, and the spherical coordinates are solved. The method of treatments of the nonhomogeneous boundary and differential equations are given and the lumped formulation of the heat conduction problems are discussed.

22.1 Problems in Rectangular Cartesian Coordinates

The general form of the governing equation of heat conduction in solids in rectangular Cartesian coordinates for the anisotropic material is

$$\frac{\partial}{\partial x}\left(k_x\frac{\partial T}{\partial x}\right)+\frac{\partial}{\partial y}\left(k_y\frac{\partial T}{\partial y}\right)+\frac{\partial}{\partial z}\left(k_z\frac{\partial T}{\partial z}\right)=-R+\rho c\frac{\partial T}{\partial t} \tag{22.1}$$

where T is the absolute temperature and k_x, k_y, and k_z are the coefficients of thermal conduction along the coordinate axes x, y, and z, respectively. We will consider the analytical methods of solution of this partial differential equation and we will discuss some examples.

22.1.1 Steady Two-Dimensional Problems: Separation of Variables

The general form of the governing partial differential equation for a steady two-dimensional problem in x and y directions is

$$\frac{\partial}{\partial x}\left(k_x\frac{\partial T}{\partial x}\right)+\frac{\partial}{\partial y}\left(k_y\frac{\partial T}{\partial y}\right)=-R \tag{22.2}$$

where the thermal conductivities in x and y directions are assumed to be variable.

M. Reza Eslami et al., *Theory of Elasticity and Thermal Stresses*, Solid Mechanics and Its Applications 197, DOI: 10.1007/978-94-007-6356-2_22,

The solution of the partial differential equation (22.2), when k_x is a function of x and k_y is a function of y, is obtained by the method of separation of variables, which is the most common method of solution of this partial differential equation. When the boundary conditions of a problem are specified, then according to this method, the solution is sought as the product of functions of each coordinate separately. This allows the constants of integration in each separated function to be found directly from the homogeneous boundary conditions, and the non-homogeneous boundary conditions be treated by using the concept of expansion into a series.

To show the method let us assume a general form of Eq. (22.2)

$$a_1(x)\frac{\partial^2 T}{\partial x^2}+a_2(x)\frac{\partial T}{\partial x}+a_3(x)T+b_1(y)\frac{\partial^2 T}{\partial y^2}+b_2(y)\frac{\partial T}{\partial y}+b_3(y)T=0 \quad (22.3)$$

The solution of this equation may be taken in the product form as

$$T(x,y)=X(x)Y(y) \quad (22.4)$$

where $X(x)$ is a function of x alone, and $Y(y)$ is a function of y alone. Upon substitution of Eq. (22.3) into Eq. (22.4) and after dividing of the whole equation by XY, we get

$$\left[a_1(x)\frac{d^2X}{dx^2}+a_2(x)\frac{dX}{dx}+a_3(x)X\right]\frac{1}{X}=-\left[b_1(y)\frac{d^2Y}{dy^2}+b_2(y)\frac{dY}{dy}+b_3(y)Y\right]\frac{1}{Y} \quad (22.5)$$

The left-hand side of Eq. (22.5) is a function of the variable x only and the right-hand side is a function of y only. Therefore, we conclude that the only way that the above equation can hold is when both sides are equal to a constant, say, $\pm\lambda^2$. This constant is called the *separation constant*. Considering this, the equation reduces to the following ordinary differential equations:

$$a_1(x)\frac{d^2X}{dx^2}+a_2(x)\frac{dX}{dx}+[a_3(x)\mp\lambda^2]X=0 \quad (22.6)$$

$$b_1(y)\frac{d^2Y}{dy^2}+b_2(y)\frac{dY}{dy}+[b_3(y)\pm\lambda^2]Y=0 \quad (22.7)$$

These equations may be solved by the techniques of the ordinary differential equations as two independent equations, and the constants of integration then may be found using the boundary conditions.

In solving a partial differential equation by the method of separation of variables two questions may be raised: (1) Is it always possible to find the constants of integration using the given boundary conditions? (2) What sign should be considered for the separation constant? The answer to question (1) is positive, provided that the problem's geometry is classical (rectangular or circular domains). The answer to question (2) depends on the problem's geometry and the thermal boundary conditions. The nature of the solution must be compatible with the given boundary conditions. A pair of the homogeneous boundary conditions in a given direction requires harmonic solution in that direction. Therefore, the sign of the separation constant is selected in such a way that the solution in the direction of a pair of homogeneous boundary conditions leads to a harmonic solution. Referring to Eqs. (22.6) and (22.7), two boundary conditions are required in each x and y direction. If the boundary conditions in x direction, as an example, are homogeneous, the sign of the separation constant must be selected so that the solution in x direction leads to a harmonic function.

The method described may be readily extended to the three-dimensional and transient problems as they will be discussed subsequently. Before we show some examples of this method, we may study the expansion into a Fourier series, as it will be needed in the treatment of such problems.

22.1.2 Fourier Series

Consider the following series

$$g(x) = \frac{a_o}{2} + \sum_{n=1}^{\infty}[a_n \cos nx + b_n \sin nx] \tag{22.8}$$

where a_o, a_n, and b_n, $(n = 1, 2, \ldots)$ are constants. If this series converges for $-\pi \leq x \leq +\pi$, then the function $g(x)$ is a periodic function with a period 2π, since $\sin(nx)$ and $\cos(nx)$ are periodic functions of period 2π, that is

$$g(x) = g(x + 2\pi) \tag{22.9}$$

Now, we need to know whether it is possible to find a series solution of the above form for a given function $f(x)$ so that upon expansion into a series of *sine* and *cosine* functions it converges to the given function.

Consider an arbitrary function $f(x)$. Let us assume that this function may be represented by a trigonometric series

$$f(x) = \frac{a_0}{2} + \sum_{n=1}^{\infty}[a_n \cos nx + b_n \sin nx] \tag{22.10}$$

It will be assumed that the above series converges to the value of the function $f(x)$. With this assumption we may integrate both sides of Eq. (22.10) from $-\pi$ to π

$$\int_{-\pi}^{\pi} f(x)dx = \int_{-\pi}^{\pi} \frac{a_0}{2}dx + \sum_{n=1}^{\infty}\left(\int_{-\pi}^{\pi} (a_n \cos nxdx + \int_{-\pi}^{\pi} b_n \sin nxdx\right) \quad (22.11)$$

Evaluating each term gives

$$\int_{-\pi}^{\pi} \frac{a_0}{2}dx = \pi a_0$$

$$\int_{-\pi}^{\pi} a_n \cos(nx)dx = a_n \int_{-\pi}^{\pi} \cos nxdx = \frac{a_n \sin nx}{n}|_{-\pi}^{\pi} = 0$$

$$\int_{-\pi}^{\pi} b_n \sin nxdx = b_n \int_{-\pi}^{\pi} \sin nxdx = -\frac{b_n \cos nx}{n}|_{-\pi}^{\pi} = 0 \quad (22.12)$$

Therefore,

$$\int_{-\pi}^{\pi} f(x)dx = \pi a_0 \quad (22.13)$$

or

$$a_0 = \frac{1}{\pi}\int_{-\pi}^{\pi} f(x)dx \quad (22.14)$$

To calculate a_n and b_n we need to know certain definite integrals. If n and k are integer numbers, then the following relations for $n \neq k$ hold

$$\int_{-\pi}^{\pi} \cos nx \cos kxdx = 0$$

$$\int_{-\pi}^{\pi} \cos nx \sin kxdx = 0$$

$$\int_{-\pi}^{\pi} \sin nx \sin kxdx = 0 \quad (22.15)$$

and for $n = k$

$$\int_{-\pi}^{\pi} \cos^2 nxdx = \pi \quad (n \neq 0)$$

$$\int_{-\pi}^{\pi} \cos nx \sin nxdx = 0$$

$$\int_{-\pi}^{\pi} \sin^2 nxdx = \pi \quad (n \neq 0) \quad (22.16)$$

With the aid of the relations (d) and (e) we can compute the coefficients a_n and b_n of the series (22.10). To find the coefficients a_n for $n \neq 0$, multiplying both sides of

Eq. (22.10) by $\cos kx$ and integrating from $-\pi$ to π yields

$$\int_{-\pi}^{\pi} f(x)\cos kx dx = \frac{a_0}{2}\int_{-\pi}^{\pi}\cos kx dx + \sum_{n=1}^{\infty}\left(a_n\int_{-\pi}^{\pi}\cos nx\cos kx dx\right.$$
$$\left. + b_n\int_{-\pi}^{\pi}\sin nx\cos kx dx\right) \tag{22.17}$$

From relations (22.15) and (22.16), all the integrals on the right-hand side of Eq. (22.17) equal to zero, except the integral with the coefficient a_k, that is

$$\int_{-\pi}^{\pi} f(x)\cos kx dx = a_k\int_{-\pi}^{\pi}\cos^2 kx dx = a_k\pi \tag{22.18}$$

or

$$a_k = \frac{1}{\pi}\int_{-\pi}^{\pi} f(x)\cos kx dx \tag{22.19}$$

To obtain b_n, we multiply both sides of Eq. (22.10) by $\sin kx$ and integrate from $-\pi$ to π

$$\int_{-\pi}^{\pi} f(x)\sin kx dx = b_k\int_{-\pi}^{\pi}\sin^2 kx dx = b_k\pi \tag{22.20}$$

or

$$b_k = \frac{1}{\pi}\int_{-\pi}^{\pi} f(x)\sin kx dx \tag{22.21}$$

The coefficients a_0, a_k, and b_k determined by Eqs. (22.14), (22.19), and (22.21) are called *Fourier coefficients* of the function $f(x)$, and the trigonometric series (22.10) is called *Fourier series* of the function $f(x)$.

One may ask what properties must the function $f(x)$ possess in order to be possible to expand it into a Fourier series. The following theorem deals with this matter:

Theorem. If a periodic function $f(x)$ with period 2π is piecewise monotonic and bounded in the interval $[-\pi, \pi]$, then Fourier series constructed for this function converges at all points where $f(x)$ is continuous, and it converges to the average of the right- and left-hand limits of $f(x)$ at each point where $f(x)$ is discontinuous.

22.1.3 Double Fourier Series

We have discussed the expansion of a function of a single variable into *sine* and *cosine* Fourier series. In many practical problems where a function is defined in terms of two independent variables, the solutions may be sought in a form of a series of products of *sine* and *cosine* functions. In this situation, the concept of expansion of a given function of two variables into a double series will be an essential tool to handle the non-homogeneous boundary conditions involved in the solution. Expansion of a function of two independent variables into a double Fourier series is discussed in this section. The method will be applied to the solutions of problems of heat conduction, thermal stresses, and deflection of plates.

To expand a given function of two variables into a double Fourier series, we follow precisely the same procedure as for single variable functions. Consider a given function $f(x, y)$ defined in the rectangular region $-a < x < a$, $-b < y < b$. It is easily verified that for a set of two orthogonal functions, $\sin m\pi x/a$, $m = 1, 2, 3, \ldots$, and $\cos n\pi y/b$, $n = 0, 1, 2, 3, \ldots$, the product of any two functions obeys the following rules:
For $m \neq i$ or $n \neq j$

$$\begin{aligned}
&\int_{-a}^{a}\int_{-b}^{b}\left(\sin\frac{m\pi x}{a}\sin\frac{n\pi y}{b}\right)\left(\sin\frac{i\pi x}{a}\sin\frac{j\pi y}{b}\right)dxdy = 0\\
&\int_{-a}^{a}\int_{-b}^{b}\left(\sin\frac{m\pi x}{a}\cos\frac{n\pi y}{b}\right)\left(\sin\frac{i\pi x}{a}\cos\frac{j\pi y}{b}\right)dxdy = 0\\
&\int_{-a}^{a}\int_{-b}^{b}\left(\cos\frac{m\pi x}{a}\sin\frac{n\pi y}{b}\right)\left(\cos\frac{i\pi x}{a}\sin\frac{j\pi y}{b}\right)dxdy = 0\\
&\int_{-a}^{a}\int_{-b}^{b}\left(\cos\frac{m\pi x}{a}\cos\frac{n\pi y}{b}\right)\left(\cos\frac{i\pi x}{a}\cos\frac{j\pi y}{b}\right)dxdy = 0
\end{aligned} \tag{22.22}$$

The only products of such functions whose integrals do not vanish over the rectangular region are those for which $m = i$ and $n = j$

$$\begin{aligned}
&\int_{-a}^{a}\int_{-b}^{b}\left(\sin\frac{m\pi x}{a}\sin\frac{n\pi y}{b}\right)^2 dxdy = \int_{-a}^{a}\int_{-b}^{b}\left(\sin\frac{m\pi x}{a}\cos\frac{n\pi y}{b}\right)^2 dxdy\\
&= \int_{-a}^{a}\int_{-b}^{b}\left(\cos\frac{m\pi x}{a}\sin\frac{n\pi y}{b}\right)^2 dxdy = \int_{-a}^{a}\int_{-b}^{b}\left(\cos\frac{m\pi x}{a}\cos\frac{n\pi y}{b}\right)^2 dxdy\\
&= ab \quad m = n = 1, 2, 3, \ldots
\end{aligned} \tag{22.23}$$

If $n = 0$ in the second, $m = 0$ in the third, or either $m = 0$ or $n = 0$ in the last of the above integrals, the results of the integrations is doubled. If $m = n = 0$ in the last integral, its value is $4ab$.

With the above relations, and following the procedure given for Fourier expansion of a function of a single variable, we may expand a given function $f(x, y)$ into a double Fourier series as

$$
\begin{aligned}
f(x, y) = & \sum_{m=1}^{\infty}\sum_{n=1}^{\infty} A_{mn} \sin\frac{m\pi x}{a}\sin\frac{n\pi y}{b} + \sum_{m=1}^{\infty}\sum_{n=0}^{\infty} B_{mn} \sin\frac{m\pi x}{a}\cos\frac{n\pi y}{b} \\
& + \sum_{m=0}^{\infty}\sum_{n=1}^{\infty} C_{mn} \cos\frac{m\pi x}{a}\sin\frac{n\pi y}{b} + \sum_{m=0}^{\infty}\sum_{n=0}^{\infty} D_{mn} \cos\frac{m\pi x}{a}\cos\frac{n\pi y}{b}
\end{aligned}
\tag{22.24}
$$

To find the coefficients of the series, both sides of the above equation are multiplied by one of the functions $(\sin m\pi x/a)(\sin n\pi y/b)$, $(\sin m\pi x/a)(\cos n\pi y/b)$, $(\cos m\pi x/a)(\sin n\pi y/b)$, or $(\cos m\pi x/a)(\cos n\pi y/b)$, and integrated from $-a$ to a with respect to x, and from $-b$ to b with respect to y. Using the results of Eqs. (22.22) and (22.23), we find

$$
\begin{aligned}
A_{mn} &= \frac{1}{ab}\int_{-a}^{a}\int_{-b}^{b} f(x, y) \sin\frac{m\pi x}{a}\sin\frac{n\pi y}{b} dxdy \\
B_{mn} &= \frac{1}{ab}\int_{-a}^{a}\int_{-b}^{b} f(x, y) \sin\frac{m\pi x}{a}\cos\frac{n\pi y}{b} dxdy \\
C_{mn} &= \frac{1}{ab}\int_{-a}^{a}\int_{-b}^{b} f(x, y) \cos\frac{m\pi x}{a}\sin\frac{n\pi y}{b} dxdy \\
D_{mn} &= \frac{1}{ab}\int_{-a}^{a}\int_{-b}^{b} f(x, y) \cos\frac{m\pi x}{a}\cos\frac{n\pi y}{b} dxdy
\end{aligned}
\tag{22.25}
$$

The values of $B_{m0}, C_{0n}, D_{0n}, D_{m0}$ are one half, and D_{00} is one quarter of the above values.

If the function $f(x, y)$ is an odd function of both x and y, then $B_{mn} = C_{mn} = D_{mn} = 0$ and it follows that

$$
f(x, y) = \sum_{m=1}^{\infty}\sum_{n=1}^{\infty} A_{mn} \sin\frac{m\pi x}{a}\sin\frac{n\pi y}{b} \tag{22.26}
$$

where A_{mn} is obtained from the first of Eq. (22.25).

22.1.4 Bessel Functions and Fourier-Bessel Series

Bessel functions

Consider a differential equation of the form

$$
x^2\frac{d^2y}{dx^2} + x\frac{dy}{dx} + (x^2 - p^2)y = 0 \tag{22.27}
$$

where p is any real, imaginary, or complex constant. The solution of this differential equation is sought in the form of a series of products of some power of x, that is

$$y = x^r \sum_{k=0}^{\infty} a_k x^k \tag{22.28}$$

The coefficient a_0 is a nonzero constant. Equation (22.28) may be rewritten as

$$y = \sum_{k=0}^{\infty} a_k x^{r+k} \tag{22.29}$$

A solution presented by this equation is complete if the coefficients a_k are computed in such a way that the series (22.29) satisfies the differential equation (22.27). Therefore, taking the derivatives

$$y' = \sum_{k=0}^{\infty} (r+k) a_k x^{r+k-1}$$

$$y'' = \sum_{k=0}^{\infty} (r+k)(r+k-1) a_k x^{r+k-2}$$

and substituting in Eq. (22.27) yields

$$x^2 \sum_{k=0}^{\infty} (r+k)(r+k-1) a_k x^{r+k-2} + x \sum_{k=0}^{\infty} (r+k) a_k x^{r+k-1} + (x^2 - p^2) \sum_{k=0}^{\infty} a_k x^{r+k} = 0 \tag{22.30}$$

To satisfy Eq. (22.30) for all the values of x, the coefficients of x to the power r, $r+1, r+2, \ldots, r+k$ must be equal to zero, which yields a system of equations

$$[(r-1)r + r - p^2]a_0 = 0$$

$$[(r+1)r + (r+1) - p^2]a_1 = 0$$

$$[(r+2)(r+1) + (r+2) - p^2]a_2 + a_0 = 0$$

$$\cdots\cdots\cdots\cdots\cdots\cdots\cdots\cdots$$

$$[(r+k)(r+k-1) + (r+k) - p^2]a_k + a_{k-2} = 0$$

These equations can be written in the form

$$[r^2 - p^2]a_0 = 0$$

$$
\begin{aligned}
&[(r+1)^2 - p^2]a_1 = 0 \\
&[(r+2)^2 - p^2]a_2 + a_0 = 0 \\
&\dots\dots\dots\dots\dots\dots\dots\dots \\
&[(r+k)^2 - p^2]a_k + a_{k-2} = 0
\end{aligned} \tag{22.31}
$$

The last of Eq. (22.31) can be presented as

$$[(r+k-p)(r+k+p)]a_k + a_{k-2} = 0$$

Since a_0 cannot be zero, therefore, from the first of Eq. (22.31)

$$r^2 - p^2 = 0$$

The roots are $r_1 = p, r_2 = -p$. From the second of Eq. (22.31), $a_1 = 0$.

First, a solution for $r_1 = p > 0$ is considered. From the system of Eq. (22.31) all the coefficients $a_1, a_2, \dots$ are computed in succession in terms of a_0. For instance, if we put $a_0 = 1$, then

$$a_k = -\frac{a_{k-2}}{k(2p+k)} \tag{22.32}$$

which for different values of k yields

$$a_1 = 0, \quad a_3 = 0, \quad \text{and in general} \quad a_{2\nu+1} = 0$$

$$a_2 = -\frac{1}{2(2p+2)}$$

$$a_4 = \frac{1}{2 \times 4(2p+2)(2p+4)}$$

$$a_{2\nu} = (-1)^\nu \frac{1}{2 \times 4 \times 6 \times \dots \times 2\nu(2p+2)(2p+4) \times \dots \times (2p+2\nu)} \tag{22.33}$$

where ν is a natural number. Upon substitution of the coefficients a_k from Eq. (22.33) into Eq. (22.28), we receive

$$
\begin{aligned}
y_1 &= x^p \left[1 - \frac{x^2}{2(2p+2)} + \frac{x^4}{2 \times 4(2p+2)(2p+4)} \right. \\
&\left. - \frac{x^6}{2 \times 4 \times 6(2p+2)(2p+4)(2p+6)} + \dots \right]
\end{aligned} \tag{22.34}
$$

All the coefficients $a_{2\nu}$ are determined, as for every k the coefficient of a_k in Eqs. (22.31), $(r+k)^2 - p^2$, is different from zero. The function y_1 from Eq. (22.34) is a particular solution of Eq. (22.27).

Now, we will establish a condition under which all the coefficients a_k will be determined for the second root $r_2 = -p$. This occurs if for any even natural number k the following inequalities are satisfied

$$(r_2 + k)^2 - p^2 \neq 0 \tag{22.35}$$

or

$$\pm(r_2 + k) \neq p$$

But $p = r_1$, hence

$$\pm(r_2 + k) \neq r_1$$

Therefore, condition (e) is in this case equivalent to

$$\pm(r_1 - r_2) \neq k$$

where k is an even natural number. But

$$r_1 = p, \quad r_2 = -p$$

Thus

$$r_1 - r_2 = 2p$$

Therefore, if p is not equal to an integer, it is possible to write a second particular solution that is obtained from expression (22.34) by substituting $-p$ for p

$$\begin{aligned} y_2 = x^{-p} \bigg[1 - \frac{x^2}{2(-2p+2)} + \frac{x^4}{2 \times 4(-2p+2)(-2p+4)} \\ - \frac{x^6}{2 \times 4 \times 6(-2p+2)(-2p+4)(-2p+6)} + \ldots \bigg] \end{aligned} \tag{22.36}$$

It is easily verified that both series from expressions (22.34) and (22.36) converge for all the values of x. The series for y_1, in Eq. (22.34), multiplied by a certain constant, is called *Bessel function of the first kind of order* p and is designated by J_p. The series y_2 in Eq. (22.36) is then J_{-p}.

The general solution to Eq. (22.27) for p not equal to an integer is

$$y = C_1 J_p(x) + C_2 J_{-p}(x) \tag{22.37}$$

where C_1 and C_2 are arbitrary constants and $J_p(x)$ and $J_{-p}(x)$ are defined as

$$J_p(x) = \sum_{k=0}^{\infty} (-1)^k \frac{(\frac{x}{2})^{p+2k}}{k!\Gamma(p+k+1)} \tag{22.38}$$

and

$$J_{-p}(x) = \sum_{k=0}^{\infty}(-1)^k \frac{(\frac{x}{2})^{-p+2k}}{k!\Gamma(-p+k+1)} \tag{22.39}$$

The function $\Gamma(.)$ appearing in these equations is called the *Gamma function* and is defined for integer numbers as

$$\begin{aligned} \Gamma(p+1) &= p\Gamma(p) = p! \\ \Gamma(1) &= 0! = 1 \end{aligned} \tag{22.40}$$

When p is a real number, then $\Gamma(.)$ is defined as

$$\Gamma(p)\Gamma(p-1) = \frac{\pi}{\sin \pi p} \tag{22.41}$$

The solution (22.37) is valid as long as p is not an integer number. When $p = n$, a natural number, $J_p(x)$ and $J_{-p}(x)$ are not two independent solutions of Eq. (22.27) and it is easily verified that for $p = n$ the following relation exists

$$J_{-n}(x) = (-1)^n J_n(x) \tag{22.42}$$

Thus, the functions given in Eq. (22.37) are a constant multiple of the other and the solution is not complete, and therefore we must look for another independent solution to be combined with $J_n(x)$ to give the complete solution. To obtain a second linearly independent solution, the function $Y_p(x)$ is defined as

$$Y_p(x) = \frac{J_p(x)\cos p\pi - J_{-p}(x)}{\sin p\pi} \tag{22.43}$$

This combination of $J_p(x)$ and $J_{-p}(x)$ is obviously a solution of Bessel equation for p not being an integer, as $\cos \pi p$ and $\sin \pi p$ are constant numbers. Note that $Y_p(x)$ is linearly independent of $J_p(x)$. When $p = n$ is an integer, $Y_p(x)$ assumes the indeterminate form 0/0. A rather tedious mathematical manipulation indicates that the limit when $p \to n$ exists and $Y_p(x)$ in this case is a solution of Bessel equation.

We, therefore, conclude that Eqs. (22.38) and (22.43) are two linearly independent solutions of Bessel Eq. (22.27) when p is any real or imaginary number, and the general solution of Eq. (22.27) can be written as

$$y(x) = C_1 J_p(x) + C_2 Y_p(x) \tag{22.44}$$

The function $Y_p(x)$ is known as *Bessel function of the second kind of order* p.

In the differential equation (22.27) replacing x by $\pm ix$ we arrive at

$$x^2\frac{d^2y}{dx^2} + x\frac{dy}{dx} - (x^2 + p^2)y = 0 \tag{22.45}$$

This equation is called the *modified Bessel equation* and its solution is readily obtained by replacing x by ix in Eq. (22.44), thus

$$y(x) = C_1 J_p(ix) + C_2 Y_p(ix) \tag{22.46}$$

According to the definition, from Eq. (22.38) it follows that

$$\begin{aligned} J_p(ix) &= \sum_{k=0}^{\infty} \frac{(-1)^k (\frac{1}{2} ix)^{p+2k}}{k!\Gamma(p+k+1)} \\ &= i^p \sum_{k=0}^{\infty} \frac{(\frac{1}{2}x)^{p+2k}}{k!\Gamma(p+k+1)} \end{aligned}$$

Defining the *modified Bessel function of the first kind of order* p as

$$I_p(x) = \sum_{k=0}^{\infty} \frac{(\frac{1}{2}x)^{p+2k}}{k!\Gamma(p+k+1)} \tag{22.47}$$

we find that

$$J_p(ix) = i^p I_p(x) \tag{22.48}$$

When $p = n$ is a natural number

$$I_n(x) = i^{-n} J_n(ix)$$

and by interchanging n by $-n$ we receive

$$I_{-n}(x) = i^n J_{-n}(ix) = (-1)^n i^n J_n(ix) = i^{-n} J_n(ix)$$

The comparison shows that

$$I_n(x) = I_{-n}(x) \tag{22.49}$$

Also, since k is an even natural number, from Eq. (22.47) we find that

$$I_n(-x) = (-1)^n I_n(x) \tag{22.50}$$

Equation (22.50) indicates that $I_n(x)$ is odd or even, depending upon the value of n. When n is an even integer $I_n(x)$ is an even function of x, and when n is odd integer $I_n(x)$ is an odd function of x.

The second solution of Eq. (22.45), which is linearly independent of $I_p(x)$, is $I_{-p}(x)$ if p is not an integer. But, since $I_{-p}(x)$ is not in general independent of $I_p(x)$ and for integer values of p it is a constant multiple of $I_p(x)$, we introduce the function $K_p(x)$ as

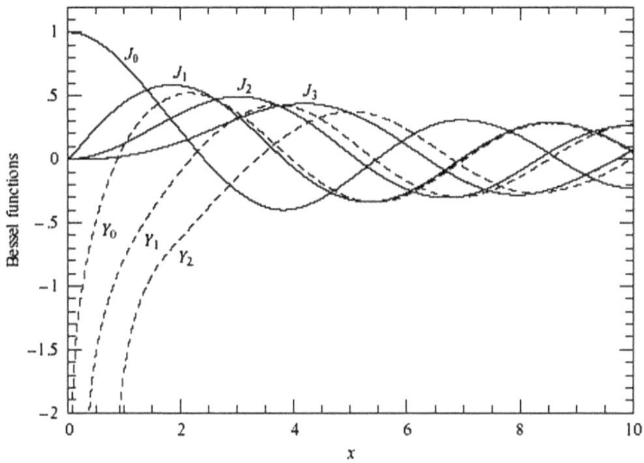

Fig. 22.1 Bessel functions of the first and second kind

$$K_p(x) = \frac{1}{2}\pi \frac{[I_{-p}(x) - I_p(x)]}{\sin p\pi} \tag{22.51}$$

The function $K_p(x)$ is called the *modified Bessel function of the second kind of order* p, and it is a solution linearly independent of $I_p(x)$. The complete solution of Eq. (22.45) is thus shown to be

$$y(x) = C_1 I_p(x) + C_2 K_p(x) \tag{22.52}$$

The graphical representations of Bessel functions and the modified Bessel functions of the first and second kinds, respectively, are presented in Figs. 22.1 and 22.2. More mathematical details are left out and the reader interested in Bessel functions may consult specialized books on the subject.

Fourier-Bessel series

Let us consider the piecewise continuous function $f(x)$ defined in the interval $0 < x < a$. Fourier-Bessel series is defined as

$$f(x) = \sum_{j=1}^{\infty} A_j J_n(\xi_j x) \tag{22.53}$$

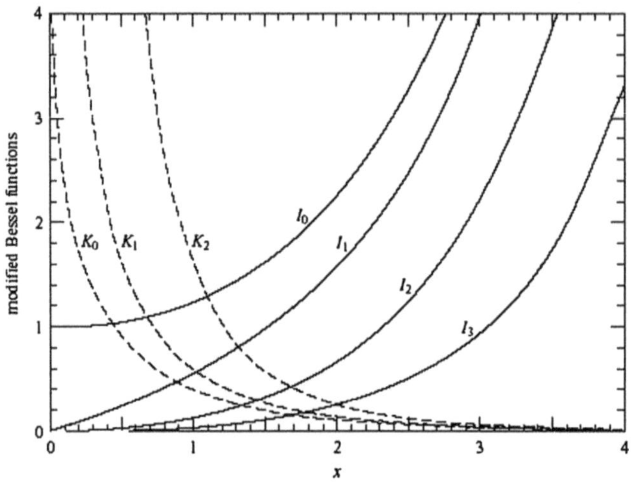

Fig. 22.2 Modified Bessel functions of the first and second kind

where

$$A_j = \frac{1}{\| J_n(\xi_j x) \|^2} \int_0^a x f(x) J_n(\xi_j x) dx \quad j = 1, 2, \ldots \tag{22.54}$$

and $\| J_n(\xi_j x) \|^2$ is the square of the norm of $J_n(\xi_j x)$ on the interval $0 < x < a$, with weight function x, and is defined by

$$\begin{aligned} \| J_n(\xi_j x) \|^2 &= \int_0^a x[J_n(\xi_j x)]^2 dx \\ &= \frac{\xi_j^2 a^2 [J_n'(\xi_j a)]^2 + (\xi_j^2 a^2 - n^2)[J_n(\xi_j a)]^2}{2\xi_j^2} \end{aligned} \tag{22.55}$$

Now the coefficient A_j in Eq. (22.53) is obtained for three special cases.
(1) if ξ_j, $(j = 1, 2, \ldots)$ are the positive roots of the equation

$$J_n(\xi a) = 0 \tag{22.56}$$

then the coefficient A_j is given by

$$A_j = \frac{2}{a^2 [J_{n+1}(\xi_j a)]^2} \int_0^a x f(x) J_n(\xi_j x) dx \quad j = 1, 2, \ldots \tag{22.57}$$

(2) if ξ_j, $(j = 1, 2, \ldots)$ are the positive roots of the equation

$$b J_n(\xi a) + \xi a J_n'(\xi a) = 0 \quad (b \geq 0, b + n > 0) \tag{22.58}$$

which can also be written in the form

$$(b+n)J_n(\xi a) - \xi a J_{n+1}(\xi a) = 0 \tag{22.59}$$

then the coefficient A_j is given as

$$A_j = \frac{2\xi_j^2}{(\xi_j^2 a^2 - n^2 + b^2)[J_n(\xi_j a)]^2} \int_0^a x f(x) J_n(\xi_j x) dx \qquad j = 1, 2, \ldots \tag{22.60}$$

(3) if $n = 0$ in Eq. (22.53) and ξ_j, $(j = 1, 2, \ldots)$ are the positive roots of the equation

$$J_0'(\xi a) = 0 \tag{22.61}$$

which can also be written in the form

$$J_1(\xi a) = 0 \tag{22.62}$$

then the coefficient A_j is given as

$$A_j = \frac{2}{a^2 [J_0(\xi_j a)]^2} \int_0^a x f(x) J_0(\xi_j x) dx \qquad j = 1, 2, \ldots \tag{22.63}$$

In the latter case, since $\xi_1 = 0$ and $J_0(0) = 1$, it is more convenient to write Eq. (22.53) in the form

$$f(x) = A_1 + \sum_{j=2}^{\infty} A_j J_0(\xi_j x) \tag{22.64}$$

where

$$\begin{aligned} A_1 &= \frac{2}{a^2} \int_0^a x f(x) dx \\ A_j &= \frac{2}{a^2 [J_0(\xi_j a)]^2} \int_0^a x f(x) J_0(\xi_j x) dx \quad j = 2, 3, \end{aligned} \tag{22.65}$$

With the general discussion presented for the method of separation of variables, Fourier expansion of a function, Bessel function, and Fourier-Bessel expansion, we may now consider the solution to some problems.

22.1.5 Nonhomogeneous Differential Equations and Boundary Conditions

In the previous section we studied a heat conduction problem in two-dimensional rectangular coordinates governed by a homogeneous differential equation, where all boundary conditions except one were homogeneous, and the application of the nonhomogeneous boundary condition resulted in obtaining the final solution to the problem. Now the question arises on how we should deal with problems in which:

(a) Two or more boundary conditions are nonhomogeneous.
(b) The governing differential equation is nonhomogeneous.

The first type of problems is easily handled by the principle of superposition. Since the governing differential equation for heat conduction in solids is linear, therefore, we may use the concept of linear superposition of the auxiliary problems. To describe the method we consider the following example.

22.2 Problems and Solutions Related to Heat Conduction

Problem 22.1. Find the steady-state temperature distribution in the rectangular plate shown in Fig. 22.3.

Solution: The heat conduction equation and associated boundary conditions are

$$\begin{aligned} &\frac{\partial^2 T}{\partial x^2} + \frac{\partial^2 T}{\partial y^2} = 0, && 0 \le x \le a, \quad 0 \le y \le b \\ &T(0, y) = f(y), && T(a, y) = T_0 \\ &K\frac{\partial T(x,0)}{\partial y} = q'', && K\frac{\partial T(x,b)}{\partial y} + h(T(x,b) - T_\infty) \end{aligned} \tag{22.66}$$

Applying the simple transform $\theta = T - T_\infty$, see Fig. 22.4, the heat conduction Eq. (22.66) simplifies to

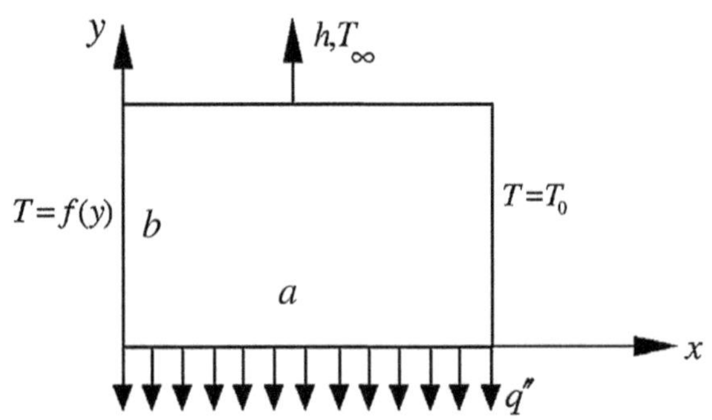

Fig. 22.3 Rectangular plate

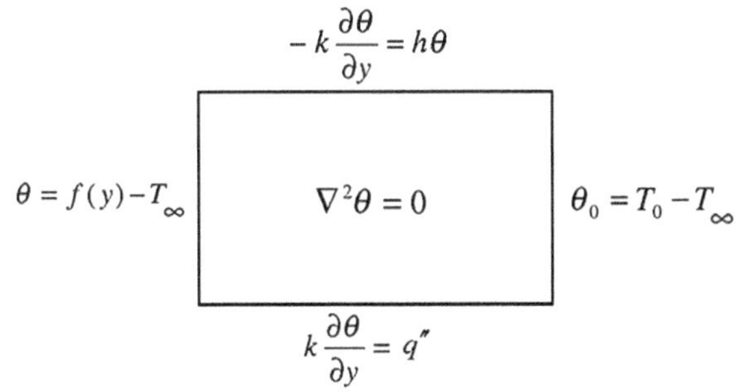

Fig. 22.4 The problem in terms of θ

$$\begin{aligned}
&\frac{\partial^2\theta}{\partial x^2} + \frac{\partial^2\theta}{\partial y^2} = 0, && 0 \le x \le a, \quad 0 \le y \le b \\
&\theta(0, y) = \theta(y), && \theta(a, y) = \theta_0 = T_0 - T_\infty \\
&K\frac{\partial\theta(x,0)}{\partial y} = q'', && K\frac{\partial\theta(x,b)}{\partial y} + h\theta(x,b) = 0
\end{aligned} \tag{22.67}$$

where $\theta(y) = f(y) - T_\infty$. To obtain a solution for boundary value problem (22.67), the solution is divided into three parts as $\theta = \theta_1 + \theta_2 + \theta_3$, as shown in Fig. 22.5. Employing this transformation and separating the equations, each equation has one non-homogeneous boundary condition. Note that in each figure the coordinate system is placed on the sides where boundary conditions are homogeneous. The results are three boundary value problems as

$$\begin{aligned}
&\frac{\partial^2\theta_1}{\partial x^2} + \frac{\partial^2\theta_1}{\partial y^2} = 0, && 0 \le x \le a, \quad 0 \le y \le b \\
&\theta_1(0, y) = \theta(y), && \theta_1(a, y) = 0, \\
&K\frac{\partial\theta_1(x,0)}{\partial y} = 0, && K\frac{\partial\theta_1(x,b)}{\partial y} + h\theta_1(x,b) = 0
\end{aligned} \tag{22.68}$$

$$\begin{aligned}
&\frac{\partial^2\theta_2}{\partial x^2} + \frac{\partial^2\theta_2}{\partial y^2} = 0, && 0 \le x \le a, \quad 0 \le y \le b \\
&\theta_2(0, y) = 0, && \theta_2(a, y) = 0, \\
&K\frac{\partial\theta_2(x,0)}{\partial y} = q'', && K\frac{\partial\theta_2(x,b)}{\partial y} + h\theta_2(x,b) = 0
\end{aligned} \tag{22.69}$$

$$\begin{aligned}
&\frac{\partial^2\theta_3}{\partial x^2} + \frac{\partial^2\theta_3}{\partial y^2} = 0, && \le x \le a, \quad 0 \le y \le b \\
&\theta_3(0, y) = 0, && \theta_3(a, y) = \theta_0, \\
&K\frac{\partial\theta_3(x,0)}{\partial y} = 0, && K\frac{\partial\theta_3(x,b)}{\partial y} + h\theta_3(x,b) = 0
\end{aligned} \tag{22.70}$$

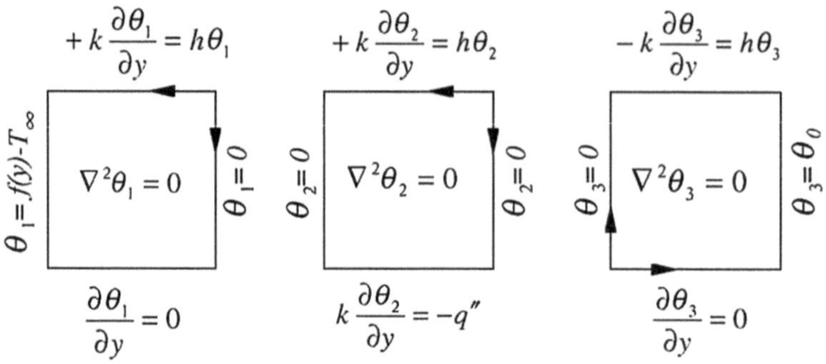

Fig. 22.5 Decomposition of the main problem into three separate problems with one nonhomogeneous boundary condition for each one

where new coordinates, $x = a - x$ and $y = b - y$ are used for Eqs. (22.69) and (22.70). Now each of the Eqs. (22.68)–(22.70) have to be solves separately.

Solution for θ_1: Following the method of separation of variables, $\theta_1(x, y) = X_1(x)Y_1(y)$. Thus, the governing equation for θ_1 transforms to

$$
\begin{aligned}
&X_1''(x) - \lambda^2 X_1(x) = 0\\
&Y_1''(y) + \lambda^2 Y_1(y) = 0, \quad Y_1'(0) = 0, \qquad Y_1'(b) + \frac{h}{K} Y_1(b) = 0
\end{aligned}
\tag{22.71}
$$

Finding the solution of the eigenvalue problem $Y_1(y)$ gives us $Y_{1n}(y) = \cos(\lambda_n y)$, where λ_n is the nth. positive real root of the equation $\lambda \tan(\lambda b) = \frac{h}{K}$. Therefore, the eigenvalues are known and the solution of θ_1 may be written in the form

$$
\theta_1 = \sum_{n=1}^{\infty} (A_{1n} \sinh(\lambda_n x) + B_{1n} \cosh(\lambda_n x)) \cos(\lambda_n y) \tag{22.72}
$$

The introduced coefficients A_{1n} and B_{1n} have to be obtained using the boundary conditions along the x-axis. Employing the first and second boundary conditions from Eq. (22.68) into the above equation results in give

$$
\begin{aligned}
&\sum_{n=1}^{\infty} (A_{1n} \sinh(\lambda_n a) + B_{1n} \cosh(\lambda_n a)) \cos(\lambda_n y) = 0\\
&\sum_{n=1}^{\infty} B_{1n} \cos(\lambda_n y) = \theta(y)
\end{aligned}
\tag{22.73}
$$

Solving the system of Eq. (22.73) for A_{1n} and B_{1n} and substituting the coefficients into Eq. (22.72) gives the final form of the solution for θ_1 as

$$\theta_1 = \sum_{n=1}^{\infty} \frac{2\lambda_n\left(\cosh(\lambda_n x) - \coth(\lambda_n a)\sinh(\lambda_n x)\right)}{b\lambda_n + \sin(\lambda_n b)\cos(\lambda_n b)} \int_0^b \theta(y)\cos(\lambda_n y)dy \quad (22.74)$$

Solving θ_2: By means of separation of variables method we may write the function $\theta_2(x, y)$ in the form $\theta_2(x, y) = X_2(x)Y_2(y)$. In this case the governing equations for X_2 and Y_2 become

$$\begin{aligned} &Y_2^{''}(y) - \mu^2 Y_2(y) = 0 \\ &X_2^{''}(x) + \mu^2 X_2(x) = 0, \quad X_2(0) = 0, \quad X_2(a) = 0 \end{aligned} \quad (22.75)$$

Solving the eigenvalue problem $X_2(x)$ gives $X_{2n}(x) = \sin\left(\frac{n\pi x}{a}\right)$. Hence the solution for θ_2 may be written in the following form

$$\theta_2 = \sum_{n-1}^{\infty} \left(\left(A_{2n}\cosh\left(\frac{n\pi y}{a}\right) + B_{2n}\sinh\left(\frac{n\pi y}{a}\right)\right)\sin\left(\frac{n\pi x}{a}\right)\right. \quad (22.76)$$

Now the constants A_{2n} and B_{2n} have to be obtained by means of the associated boundary conditions on $y = 0, b$. These boundary conditions according to Eq. (22.69) are

$$\begin{aligned} &\sum_{n=1}^{\infty} \frac{n\pi}{a} B_{2n}\sin\left(\frac{n\pi x}{a}\right) = \frac{1}{K}q^{''} \\ &\sum_{n=1}^{\infty}\left(A_{2n}\left(\frac{h}{K} + \frac{n\pi}{a}\tan\left(\frac{n\pi b}{a}\right)\right) + B_{2n}\left(\frac{h}{K}\tan\left(\frac{n\pi b}{a}\right) + \frac{n\pi}{a}\right)\right)\sin\left(\frac{n\pi x}{a}\right) = 0 \end{aligned} \quad (22.77)$$

Solving the system of Eq. (22.77) for A_{2n} and B_{2n} and substituting the results into Eq. (22.76) gives

$$\begin{aligned} \theta_2 = &\sum_{n=1}^{\infty} \frac{2aq^{''}}{n^2\pi^2 K}(1 - (-1)^n) \\ &\times \left\{\sinh\left(\frac{n\pi y}{a}\right) - \frac{\frac{n\pi}{a} + \frac{h}{K}\tanh\left(\frac{n\pi b}{a}\right)}{\frac{n\pi}{a}\tanh\left(\frac{n\pi b}{a}\right) + \frac{h}{K}}\cosh\left(\frac{n\pi y}{a}\right)\right\}\sin\left(\frac{n\pi x}{a}\right) \end{aligned} \quad (22.78)$$

Solving θ_3: Following the method of separation of variables for $\theta_3(x, y)$ we may write $\theta_3(x, y) = X_3(x)Y_3(y)$. Thus the governing equation for θ_3 reduces to

$$X_3''(x) - \lambda^2 X_3(x) = 0$$
$$Y_3''(y) + \lambda^2 Y_3(y) = 0, \quad Y_3'(0) = 0, \quad Y_3'(b) + \frac{h}{K} Y_3(b) = 0 \tag{22.79}$$

Finding the solution of eigenvalue problem $Y_3(y)$ gives $Y_{3n}(y) = \cos(\lambda_n y)$, where λ_n is the nth positive real root of the equation $\lambda \tan(\lambda b) = \dfrac{h}{K}$. Therefore, the eigenvalues are known and the solution of θ_3 may be written in the form

$$\theta_3 = \sum_{n=1}^{\infty} (A_{3n} \sinh(\lambda_n x) + B_{3n} \cosh(\lambda_n x)) \cos(\lambda_n y) \tag{22.80}$$

where the introduces coefficients A_{3n} and B_{3n} have to be obtained by means of the boundary conditions along the x-axis. Employing the first and second boundary conditions of Eq. (22.70) gives

$$\sum_{n=1}^{\infty} (A_{3n} \sinh(\lambda_n a) + B_{3n} \cosh(\lambda_n a)) \cos(\lambda_n y) = \theta_0$$
$$\sum_{n=1}^{\infty} B_{3n} \cos(\lambda_n y) = 0 \tag{22.81}$$

Solving the system of Eq. (22.81) for A_{3n} and B_{3n} and substituting the coefficients into Eq. (22.80) gives the final form of the solution for θ_3 as

$$\theta_3 = \theta_0 \sum_{n=1}^{\infty} \frac{2 \sin(\lambda_n b)}{\sinh(\lambda_n a)(\lambda_n b + \sin(\lambda_n b) \cos(\lambda_n b))} \sinh(\lambda_n x) \cos(\lambda_n y) \tag{22.82}$$

After finding θ_1, θ_2, and θ_3 the temperature distribution becomes

$$T = T_\infty + \theta_1 + \theta_2 + \theta_3 \tag{22.83}$$

Problem 22.2. Consider an infinitely long bar of square cross section floating in a fluid of constant temperature T_0. If the heat transfer coefficient between the bar and the fluid is large compared with that of the bar and ambient, find a steady-state distribution of the temperature in the cross section of the bar.

Solution: The heat conduction equation and associated boundary conditions are

$$\frac{\partial^2 T}{\partial x^2} + \frac{\partial^2 T}{\partial y^2} = 0, \quad 0 \le x \le L, \quad 0 \le y \le L$$
$$T(0, y) = T_0$$
$$T(x, 0) = T_0$$

$$
\begin{aligned}
-K\frac{\partial T(L,y)}{\partial x} &= h(T(L,y)-T_\infty)\\
-K\frac{\partial T(x,L)}{\partial y} &= h(T(x,L)-T_\infty)
\end{aligned} \tag{22.84}
$$

where K is thermal conductivity coefficient. Defining $\theta = T - T_\infty$, Eq. (22.84) transforms to

$$
\begin{aligned}
&\frac{\partial^2\theta}{\partial x^2}+\frac{\partial^2\theta}{\partial y^2}=0, \quad 0\le x\le L, \quad 0\le y\le L\\
&\theta(0,y)=\theta_0\\
&\theta(x,0)=\theta_0\\
&-K\frac{\partial\theta(L,y)}{\partial x}=h\theta(L,y)\\
&-K\frac{\partial\theta(x,L)}{\partial y}=h\theta(x,L)
\end{aligned} \tag{22.85}
$$

in which $\theta_0 = T_0 - T_\infty$. As a solution of the Eq. (22.85), lets assume $\theta = \theta_1 + \theta_2$. This transformation is adopted to divide the boundary value problem (22.85) into two partial differential equations where each of them has at least two homogeneous boundary conditions. Therefore the following governing equations are resulted for θ_2 and θ_2.

$$
\begin{aligned}
&\frac{\partial^2\theta_1}{\partial x^2}+\frac{\partial^2\theta_1}{\partial y^2}=0\\
&\theta_1(0,y)=0, \qquad \theta_1(x,0)=\theta_0\\
&-K\frac{\partial\theta_1(L,y)}{\partial x}=h\theta_1(L,y), \quad -K\frac{\partial\theta_1(x,L)}{\partial y}=h\theta_1(x,L)
\end{aligned} \tag{22.86}
$$

$$
\begin{aligned}
&\frac{\partial^2\theta_2}{\partial x^2}+\frac{\partial^2\theta_2}{\partial y^2}=0\\
&\theta_2(0,y)=\theta_0, \qquad \theta_2(x,0)=0\\
&-K\frac{\partial\theta_2(L,y)}{\partial x}=h\theta_2(L,y), \quad -K\frac{\partial\theta_2(x,L)}{\partial y}=h\theta_2(x,L)
\end{aligned}
$$

Solving θ_1: Following the method of separation of variables lets define $\theta_1 = X_1(x)Y_1(y)$. Substituting the aforementioned separation into Eq. (22.86) reveals us

$$
\begin{aligned}
&X_1''(x)+\lambda^2X_1(x)=0, \quad X_1(0)=0, \quad KX_1'(L)+hX_1(L)=0\\
&Y_1''(y)-\lambda^2Y_1(y)=0
\end{aligned} \tag{22.87}
$$

Solving the upper eigenvalue problem gives us $X_{1n}(x) = \sin(\lambda_n x)$, where λ_n is defined as the nth positive real root of the equation $\tan(\lambda L) + \frac{K}{h}\lambda = 0$. Now the eigenvalues λ_n are known and after finding Y_{1n}, the solution of θ_1 may be written in the form

$$\theta_1 = \sum_{n=1}^{\infty} (A_n \cosh(\lambda_n y) + B_n \sinh(\lambda_n y)) \sin(\lambda_n x) \tag{22.88}$$

The constants A_n and B_n have to be obtained by means of boundary conditions on $y = 0, L$. By means of series expansion (22.88) one may reach to

$$\begin{aligned}
&\sum_{n=1}^{\infty} A_n \sin(\lambda_n x) = \theta_0 \\
&\sum_{n=1}^{\infty} \left\{ A_n(\lambda_n \sinh(\lambda_n L) + \frac{h}{K}\cosh(\lambda_n L)) \right. \\
&\quad \left. + B_n(\lambda_n \cosh(\lambda_n L) + \frac{h}{K}\sinh(\lambda_n L))) \right\} \sin(\lambda_n x) = 0
\end{aligned} \tag{22.89}$$

Solving the upper system of equations gives us

$$\begin{aligned}
A_n &= \frac{2\theta_0(1 - \cos(\lambda_n L))}{L + \frac{h}{K}\cos^2(\lambda_n L)} \\
B_n &= -\frac{\frac{h}{K\lambda_n} + \tanh(\lambda_n L)}{1 + \frac{h}{K\lambda_n}\tanh(\lambda_n L)} A_n
\end{aligned} \tag{22.90}$$

Finally, substituting the Eq. (22.90) into Eq. (22.88), gives the solution of θ_1 as

$$\begin{aligned}
\theta_1 = &\sum_{n=1}^{\infty} \frac{2\theta_0(1 - \cos(\lambda_n L))}{L + \frac{h}{K}\cos^2(\lambda_n L)} \\
&\times \left(\cosh(\lambda_n y) - \frac{\frac{h}{K\lambda_n} + \tanh(\lambda_n L)}{1 + \frac{h}{K\lambda_n}\tanh(\lambda_n L)} \sinh(\lambda_n y) \right) \sin(\lambda_n x)
\end{aligned} \tag{22.91}$$

Solving θ_2: Following the method of separation of variables lets define $\theta_2 = X_2(x)Y_2(y)$. Substituting this separation into Eq. (22.87) reveals us

$$\begin{aligned}
&Y_2''(y) + \mu^2 Y_2(y) = 0, \quad Y_2(0) = 0, \quad KY_2'(L) + hY_2(L) = 0 \\
&X_2''(x) - \mu^2 X_2(x) = 0
\end{aligned} \tag{22.92}$$

Solving the upper eigenvalue problem gives us $Y_{2n}(y) = \sin(\mu_n y)$, where μ_n is defined as the nth positive real root of the equation $\tan(\mu L) + \frac{K}{h}\mu = 0$. As seen $\mu_n = \lambda_n$ and therefore in the rest of present work, we use λ_n instead of μ_n. Now the eigenvalues λ_n are known and after finding X_{2n}, the solution of θ_2 may be written in the form

$$\theta_2 = \sum_{n=1}^{\infty} (C_n \cosh(\lambda_n x) + D_n \sinh(\lambda_n x)) \sin(\lambda_n y) \tag{22.93}$$

The constants C_n and D_n have to be obtained by means of boundary conditions on $x = 0, L$. Recalling Eq. (22.87), with the simultaneous aid of Eq. (22.93) one may reach to

$$\begin{aligned} &\sum_{n=1}^{\infty} C_n \sin(\lambda_n y) = \theta_0 \\ &\sum_{n=1}^{\infty} \left\{ C_n(\lambda_n \sinh(\lambda_n L) + \frac{h}{K}\cosh(\lambda_n L)) + \right. \\ &\left. D_n(\lambda_n \cosh(\lambda_n L) + \frac{h}{K}\sinh(\lambda_n L)) \right\} \sin(\lambda_n y) = 0 \end{aligned} \tag{22.94}$$

Solving the upper system of equations gives us

$$\begin{aligned} C_n &= \frac{2\theta_0(1 - \cos(\lambda_n L))}{L + \frac{h}{K}\cos^2(\lambda_n L)} \\ D_n &= -\frac{\frac{h}{K\lambda_n} + \tanh(\lambda_n L)}{1 + \frac{h}{K\lambda_n}\tanh(\lambda_n L)} C_n \end{aligned} \tag{22.95}$$

Finally, substituting the Eq. (22.95) into Eq. (22.93), gives the solution of θ_2 as

$$\begin{aligned} \theta_2 = &\sum_{n=1}^{\infty} \frac{2\theta_0(1 - \cos(\lambda_n L))}{L + \frac{h}{K}\cos^2(\lambda_n L)} \\ &\times \left(\cosh(\lambda_n x) - \frac{\frac{h}{K\lambda_n} + \tanh(\lambda_n L)}{1 + \frac{h}{K\lambda_n}\tanh(\lambda_n L)} \cosh(\lambda_n x) \right) \sin(\lambda_n y) \end{aligned} \tag{22.96}$$

Now, the solution for $T(x, y)$ is equal to $T = T_\infty + \theta_1 + \theta_2$ which may be written also in the following closed form expression

$$
\begin{aligned}
T &= T_\infty \\
&+ 2\theta_0 \sum_{n=1}^{\infty} \left(\frac{1-\cos(\lambda_n L)}{L + \frac{h}{K}\cos^2(\lambda_n L)} \right) \\
&\times (\sin(\lambda_n y)\cosh(\lambda_n x) + \sin(\lambda_n x)\cosh(\lambda_n y)) \\
&- 2\theta_0 \sum_{n=1}^{\infty} \left(\frac{1-\cos(\lambda_n L)}{L + \frac{h}{K}\cos^2(\lambda_n L)} \right) \left(\frac{\frac{h}{K\lambda_n} + \tanh(\lambda_n L)}{1 + \frac{h}{K\lambda_n}\tanh(\lambda_n L)} \right) \\
&\times (\sin(\lambda_n x)\sinh(\lambda_n y) + \sin(\lambda_n y)\sinh(\lambda_n x))
\end{aligned} \tag{22.97}
$$

22.2.1 Lumped Formulation

In many practical cases the differential formulation of heat conduction requires the solution in complex geometries where the boundary conditions are complicated. Furthermore, the nature of the geometry is such that the detailed analysis in particular direction does not provide valuable information. In this case, we may ignore the differential formulation in that direction and by averaging the temperature distribution simplify the solution while satisfying the boundary conditions. A *lumped formulation* of a problem in a specific direction means that it is independent of space variable in that direction. For this reason, proper consideration of boundary conditions in the lumped direction must be observed to insure the correctness of the solution. Depending on the problem's geometry, we are allowed to lump in one or more space directions. The result of lumped formulation of a problem in any direction should, however, result in simplification of the solution while the boundary conditions and the required accuracy are maintained. In general, when a problem is lumped in one or more than one direction, the general form of distributed law of heat conduction is no longer valid and the governing equation is obtained by consideration of the heat balance of the lumped element. The following examples illustrate the lumped formulation for a triangular fin and a turbine blade.

Problem 22.3. Consider a portion of a gas turbine blade shown in Fig. 22.6a. An element of the blade is shown in Fig. 22.6b and an idealized configuration of the cross section for the calculation purpose is shown in Fig. 22.6c. The blade is cooled over its base and receives heat over its other surfaces in convective and radiative form as a result of the flow of hot gases. The boundary conditions are shown in Fig. 22.6a. We may assume that the tip of the blade is insulated. We are to find a temperature field within the blade.

Solution: The coordinates x and y are defined in Fig. 22.6a. For an element shown in Fig. 22.6b, the energy balance equation may be written in the form

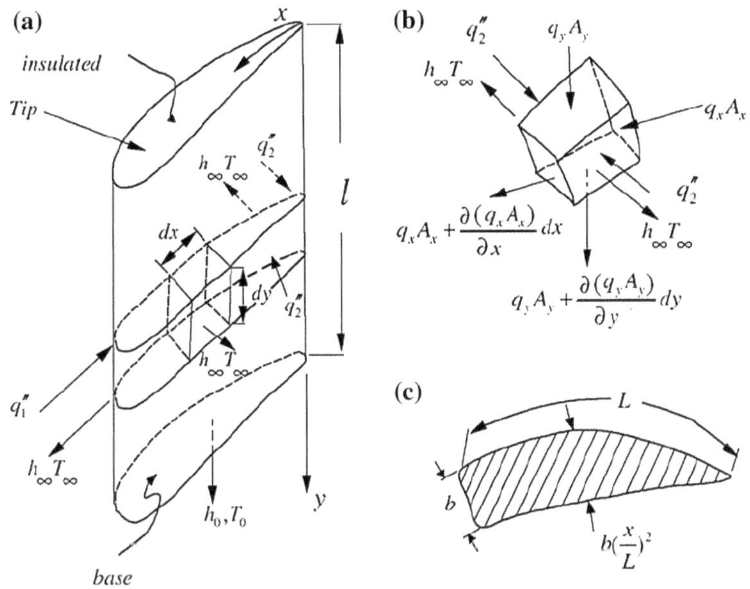

Fig. 22.6 Turbine blade

$$-\frac{\partial(q_x A_x)}{\partial x}dx - \frac{\partial(q_y A_y)}{\partial y}dy - 2h_\infty(T - T_\infty)dxdy + 2q''dxdy = 0 \quad (22.98)$$

where in the above equation, T_∞ is the reference temperature and h_∞ is the heat transfer coefficient. Noting that $A_x = b\left(\frac{x}{L}\right)^2 dy$, $A_y = b\left(\frac{y}{L}\right)^2 dx$, $q_x = -K\frac{\partial\theta}{\partial x}$ and $q_y = -K\frac{\partial\theta}{\partial y}$, where $\theta = T - T_\infty$. Therefore the above equations simplifies to

$$\frac{\partial}{\partial x^2}\left(x^2\frac{\partial\theta}{\partial x}\right) + x^2\frac{\partial^2\theta}{\partial y^2} - m^2\theta = -n \quad (22.99)$$

Here, for the sake of simplicity the constants $m^2 = \frac{2h_\infty L^2}{Kb}$ and $n = \frac{2q_2'' L^2}{Kb}$ are introduced. The associated boundary conditions for the above equation are

$$\begin{aligned}
&\theta(0, y) = finite \\
&-K\frac{\partial\theta(L, y)}{\partial x} = -q_1'' + h_\infty\theta(L, y) \\
&\frac{\partial\theta(x, 0)}{\partial y} = 0 \\
&-K\frac{\partial\theta(x, l)}{\partial y} = h_0(\theta(x, l) - \theta_0)
\end{aligned} \quad (22.100)$$

where $\theta_0 = T_0 - T_\infty$. To obtain a solution for Eq. (22.99), $\theta(x, y)$ is divided as the sum of two functions in the form of $\theta(x, y) = \psi(x, y) + \phi(x)$, where the governing equation for each of these functions and their associated boundary conditions are

$$\begin{aligned}
&\frac{d}{dx}\left(x^2\frac{d\phi}{dx}\right) - m^2\phi = -n \\
&\phi(0) = finite, \\
&-K\frac{d\phi(L)}{dx} = -q'' + h_\infty\phi(L)
\end{aligned} \tag{22.101}$$

$$\begin{aligned}
&\frac{\partial}{\partial x}\left(x^2\frac{\partial\psi}{\partial x}\right) + x^2\frac{\partial^2\psi}{\partial y^2} - m^2\psi = 0 \\
&\psi(0, y) = finite \\
&\frac{\partial(x, 0)}{\partial y} = 0 \\
&-K\frac{\partial\psi(L, y)}{\partial x} = h_\infty\psi(L, y) \\
&-K\frac{\partial\psi(x, l)}{\partial y} = h_0(\psi(x, 0) + \phi(x) - \theta_0)
\end{aligned} \tag{22.102}$$

At first, the solution of $\phi(x)$ has to be obtained. The solution of this equation according to Eq. (22.101) may be expressed as

$$\phi(x) = Ax^{\alpha-\frac{1}{2}} + Bx^{-\alpha-\frac{1}{2}} + \frac{n}{m^2} \tag{22.103}$$

where $\alpha = \sqrt{(m^2 + \frac{1}{4})}$. Constants A and B have to be calculated using the boundary conditions (22.101). As seen, for $\alpha \geq \frac{1}{2}$ and $B \neq 0$, $\lim_{x\to 0^+}\phi(x) = \infty$ which is in contradiction with $\phi(0) =$ finite, hence $B = 0$. Applying the other boundary condition gives the constant A as

$$A = \frac{q_1'' - q_2''}{L^{\alpha-\frac{3}{2}}\left(Lh_\infty + K(\alpha - \frac{1}{2})\right)}. \tag{22.104}$$

And so the solution of $\phi(x)$ is accomplished in the form

$$\phi(x) = \frac{q_1'' - q_2''}{L^{\alpha-\frac{3}{2}}(Lh_\infty + K(\alpha - \frac{1}{2}))}x^{\alpha-\frac{1}{2}} + \frac{q_2''}{h_\infty} \tag{22.105}$$

The solution of boundary value problem (22.102) may be obtained by means of separation of variables method. Assuming the solution of Eq. (22.102) in the form

$\psi(x, y) = X(x)Y(y)$, the governing equations for $X(x)$ and $Y(y)$ become

$$
\begin{aligned}
&x^2\frac{d^2X}{dx^2} + 2x\frac{dX}{dx} + (\lambda^2x^2 - m^2)X = 0 \\
&\frac{d^2Y}{dy^2} - \lambda^2 Y = 0
\end{aligned}
\tag{22.106}
$$

The exact solutions of the Eq. (22.106) may be written as

$$
\begin{aligned}
Y(y) &= C_1e^{\lambda y} + C_2e^{-\lambda y} \\
X(x) &= C_3x^{-\frac{1}{2}}J_\alpha(\lambda x) + C_4x^{-\frac{1}{2}}Y_\alpha(\lambda x)
\end{aligned}
\tag{22.107}
$$

where J_α and Y_α are the Bessel functions of the first and second kind of order α. Constants $C_i, i = 1, 2, 3, 4$ have to be obtained by means of boundary conditions (22.102). Due to property of the Bessel function of second kind $\lim_{x\to 0^+} Y_\alpha(\lambda x) = -\infty$. Therefore when $C_4 \neq 0$, $\lim_{x\to 0^+} X(x) = \infty$, which is in contradiction with the boundary condition $\psi(0, y) =$ finite. Therefore $C_4 = 0$. The other boundary condition on x results in

$$
J_\alpha(\lambda L)\left(\frac{1}{2L} - \frac{h_\infty}{K}\right) = \lambda J'_\alpha(\lambda L) \tag{22.108}
$$

Assuming λ_n as the nth positive real root of the Eq. (22.108), the eigenfunctions $X_n(x)$ become

$$
X_n(x) = x^{-\frac{1}{2}}J_\alpha(\lambda_n x) \tag{22.109}
$$

Finally the solution of function $\psi(x, y)$ may be written as the linear summation of functions $\psi_n(x, y)$ as

$$
\psi(x, y) = \sum_{n=1}^{\infty}\psi_n(x, y) = \sum_{n=1}^{\infty}\left(C_{1n}e^{\lambda_n y} + C_{2n}e^{-\lambda_n y}\right)x^{-\frac{1}{2}}J_\alpha(\lambda_n x) \tag{22.110}
$$

Considering the boundary condition $\dfrac{\partial\psi(x, 0)}{\partial y} = 0$ gives us $C_{1n} = C_{2n}$. Defining $C_n = \dfrac{1}{2}C_{1n} = \dfrac{1}{2}C_{2n}$ simplifies the Eq. (22.110) to the following form

$$
\psi(x, y) = \sum_{n=1}^{\infty} C_n\cosh(\lambda_n y)x^{-\frac{1}{2}}J_\alpha(\lambda_n x) \tag{22.111}
$$

Employing the boundary condition $-K\dfrac{\partial\psi(x, l)}{\partial y} = h_0(\psi(x, l) + \phi(x) - \theta_0)$ gives us

$$\sum_{n=1}^{\infty} C_n(K\lambda_n \cosh(l\lambda_n) + h_0 \sinh(l\lambda_n))x^{-\frac{1}{2}} J_\alpha(\lambda_n x) = h_0(\theta_0 - \phi(x)) \quad (22.112)$$

Now to find the constants C_n, the associated weight function for X_n's have to be determined. Based on Eq. (22.106), functions $X_n(x), n = 1, 2, \ldots, \infty$ are orthogonal functions with respect to each other when the weight function $w(x) = x^2$ is also taken into account. To obtain the constants C_n both sides of the Eq. (22.112) are multiplied by $X_m(x)w(x)$ and integrated over $x = [0, L]$, which reveals us

$$C_n = \frac{h_0 \int_0^L (\theta_0 - \phi(x))x^{\frac{3}{2}} J_\alpha(\lambda_n x)dx}{(K\lambda_n \sinh(l\lambda_n) + h_0 \cosh(l\lambda_n)) \int_0^L x J_\alpha^2(\lambda_n x)dx} \quad (22.113)$$

Note that, the function $\phi(x)$ is known in Eq. (22.105), and therefore constants C_n have to be evaluated through Eq. (22.113). Substituting constants C_n into Eq. (22.112) reveals the solution for function $\psi(x, y)$. The temperature distribution through the gas turbine blade is then $\theta(x, y) = \psi(x, y) + \phi(x)$.

22.3 Problems in Cylindrical Coordinates

A general form of the governing equation of heat conduction in cylindrical coordinates is

$$k\left(\frac{\partial^2 T}{\partial r^2} + \frac{1}{r}\frac{\partial T}{\partial r} + \frac{1}{r^2}\frac{\partial^2 T}{\partial \phi^2} + \frac{\partial^2 T}{\partial z^2}\right) + R = \rho c \frac{\partial T}{\partial t} \quad (22.114)$$

In the following, we will discuss one-, two-, and three-dimensional problems, both steady-state and transient.

Problem 22.4. Consider a hollow thick cylinder of inside radius a and outside radius b. The initial temperature of the cylinder is $T = 0$ at $t = 0$. At time $t > 0$, a constant heat flux q'' is radiated to the side of the cylinder. The side $z = 0$ is insulated and $z = L$ is at ambient temperature T_∞. The inside surface is kept at constant temperature T_0. Find the transient temperature distribution in the cylinder.

Solution: Assuming the axisymmetric temperature distribution, the problem is finding the temperature distribution $T(r, z, t)$ which satisfies the following heat conduction equation and associated initial and boundary conditions

$$\begin{aligned}
&\frac{\partial^2 T}{\partial r^2}+\frac{1}{r}\frac{\partial T}{\partial r}+\frac{\partial^2 T}{\partial z^2}=\frac{1}{\kappa}\frac{\partial T}{\partial t}\\
&\frac{\partial T}{\partial r}(b,z,t)=-\frac{1}{K}q'', \qquad \frac{\partial T}{\partial z}(r,0,t)=0\\
&T(r,L,t)=T_\infty, \qquad T(a,z,t)=T_0\\
&T(r,z,0)=0
\end{aligned} \tag{22.115}$$

Due to the non-homogeneity of boundary conditions, the solution of the partial differential equation (22.115) is assumed in the form

$$T=\psi(r,z,t)+\theta(r,z)+T_\infty \tag{22.116}$$

Substituting the Eq. (22.116) into the Eq. (22.115), the following set of conditions are obtained for $\theta(r,z)$

$$\begin{aligned}
&\frac{\partial^2 \theta}{\partial r^2}+\frac{1}{r}\frac{\partial \theta}{\partial r}+\frac{\partial^2 \theta}{\partial z^2}=0\\
&\theta(a,z)=\theta_0, \qquad \frac{\partial \theta}{\partial r}(b,z)=-\frac{1}{K}q''\\
&\theta(r,L)=0, \qquad \frac{\partial \theta}{\partial z}(r,0)=0
\end{aligned} \tag{22.117}$$

And the governing equation for $\psi(r,z,t)$ may be written as

$$\begin{aligned}
&\frac{\partial^2 \psi}{\partial r^2}+\frac{1}{r}\frac{\partial \psi}{\partial r}+\frac{\partial^2 \psi}{\partial z^2}=\frac{1}{\kappa}\frac{\partial \psi}{\partial t}\\
&\frac{\partial \psi}{\partial r}(b,z,t)=0, \qquad \frac{\partial \psi}{\partial z}(r,0,t)=0\\
&\psi(r,L,t)=0, \qquad \psi(a,z,t)=0,\\
&\psi(r,z,0)=-\theta(r,z)-T_\infty
\end{aligned} \tag{22.118}$$

Each of the functions $\theta(r,z)$ and $\psi(r,z,t)$ have to be obtained. According to the initial condition of the function $\psi(r,z,t)$, at first the function $\theta(r,z)$ has to be obtained.

Solving $\theta(r,z)$: By means of separation of variables method, lets separate the function θ in the form $\theta(r,z)=R(r)Z(z)$, which transform the Eq. (22.117) to

$$\frac{rR''(r)+R'(r)}{rR(r)}=-\frac{Z''(z)}{Z(z)}=\lambda^2, \quad Z'(0)=0, \quad Z(L)=0 \tag{22.119}$$

The eigenvalues and eigenfunctions of the Eq. (22.119) are obtained as

$$\lambda_n=\frac{(2n-1)\pi}{2L}, \quad Z_n(z)=\cos(\lambda_n z) \tag{22.120}$$

Solving the second order differential equation for $R(r)$ gives us

$$R_n(r) = AI_0(\lambda_n r) + BK_0(\lambda_n r) \tag{22.121}$$

Finally the solution for $\theta(r, z)$ may be written as the linear summation of functions $\theta_n(r, z) = R_n(r)Z_n(z)$ as

$$\theta(r, z) = \sum_{n=1}^{\infty} (A_n I_0(\lambda_n r) + B_n K_0(\lambda_n r)) \cos(\lambda_n z) \tag{22.122}$$

where A_n and B_n are constants have to be evaluated using the first an second boundary conditions in Eq. (22.117). Employing these boundary conditions and using the relations $I_0^{'}(x) = I_1(x)$, $K_0^{'}(x) = -K_1(x)$, the coefficients A_n and B_n are derived as

$$\begin{aligned} A_n &= \frac{2(-1)^{n+1}}{LK\lambda_n^2}\left\{\frac{\theta_0 K\lambda_n K_1(\lambda_n b) - q^{''} K_0(\lambda_n a)}{K_1(\lambda_n b)I_0(\lambda_n a) + K_0(\lambda_n a)I_1(\lambda_n b)}\right\} \\ B_n &= \frac{2(-1)^{n+1}}{LK\lambda_n^2}\left\{\frac{\theta_0 K\lambda_n I_1(\lambda_n b) + q^{''} I_0(\lambda_n a)}{K_1(\lambda_n b)I_0(\lambda_n a) + K_0(\lambda_n a)I_1(\lambda_n b)}\right\} \end{aligned} \tag{22.123}$$

Therefore, the solution for $\theta(r, z)$ is accomplished.

Solution of $\psi(r, z, t)$: Using the method of separation of variables, we seek for a solution in the form $\psi(r, z, t) = \rho(r)Y(z)\tau(t)$. Therefore Eq. (22.118) transforms to

$$\begin{aligned} \frac{r\rho^{''}(r) + \rho'(r)}{r\rho(r)} - \frac{1}{\kappa}\frac{\tau^{'}(t)}{\tau(t)} = -\frac{Y^{''}(z)}{Y(z)} = \mu^2 \\ Y^{'}(0) = 0, Y(L) = 0 \\ \rho(a) = 0, \rho^{'}(b) = 0 \end{aligned} \tag{22.124}$$

The eigenvalues and eigenfunctions associated with second order differential equation for $Y(z)$ are

$$\mu_n = \frac{(2n-1)\pi}{2L}, \quad Y_n(z) = \cos(\lambda_n z) \tag{22.125}$$

Thus, the governing equation for $\rho(r)$ and $\tau(t)$ may be described in the following form

$$\begin{aligned} & r^2\rho^{''}(r) + r\rho^{'}(r) + r^2\beta^2\rho(r) = 0, \quad \rho(a) = 0 \quad \rho^{'}(b) = 0 \\ & \frac{1}{\kappa}\frac{\tau^{'}(t)}{\tau(t)} + \mu_n^2 = -\beta^2 \end{aligned} \tag{22.126}$$

when the governing eigen-problem for $\rho(r)$ is solved, reveals us that the eigenfunctions of the corresponding equation are

$$\rho_m(r) = Y_0(\beta_m r)J_0(\beta_m a) - J_0(\beta_m r)Y_0(\beta_m a) \tag{22.127}$$

where β_m is the mth positive real root of the equation

$$Y_1(\beta b)J_0(\beta a) + J_1(\beta b)Y_0(\beta a) = 0 \tag{22.128}$$

Finally solving the governing equation for $\tau(t)$ gives us

$$\tau_{mn}(t) = e^{-(\mu_n^2+\beta_m^2)\kappa t} \tag{22.129}$$

Solution of the equation $\psi(r, z, t)$ may be written as the summation of the functions $\psi_{mn}(r, z, t)$ as below

$$\begin{aligned}\psi(r, z, t) = \sum_{n=1}^{\infty}\sum_{m=1}^{\infty} C_{mn}\left(Y_0(\mu_m r)J_0(\mu_m a) - J_0(\mu_m r)Y_0(\mu_m a)\right)\\ \times \cos(\lambda_n z)e^{-(\mu_n^2+\beta_m^2)\kappa t}\end{aligned} \tag{22.130}$$

where C_{mn}'s are constants have to be obtained by means of initial condition (22.118). For this purpose, the function T_∞ is expanded in terms of functions $Y_n(z)$ as below

$$T_\infty = \frac{2T_\infty}{L}\sum_{n=1}^{\infty}\frac{(-1)^{n+1}}{\lambda_n}\cos(\lambda_n z) \tag{22.131}$$

substituting Eqs. (22.122), (22.130) and (22.131) into the initial condition (22.118) results in

$$\begin{aligned}&\sum_{m=1}^{\infty} C_{mn}\left(Y_0(\mu_m r)J_0(\mu_m a) - J_0(\mu_m r)Y_0(\mu_m a)\right)\\ &= -A_n I_0(\lambda_n r) - B_n K_0(\lambda_n r) - \frac{2T_\infty(-1)^{n+1}}{L\lambda_n}\end{aligned} \tag{22.132}$$

The constant C_{mn} may be evaluated by mean of orthogonality properties. Considering Eq. (22.126), one can obtain that, functions $\rho_n(r)$ are orthogonal with respect to each other when the weight function $w(r) = r$ is also taken into account. Now, both sides of the Eq. (22.131) are multiplied by $w(r)\rho_m(r)$ and integrated over $[a, b]$, which gives the constants C_{mn} as below

$$C_{mn} = -\frac{\int_a^b r\left(A_n I_0(\lambda_n r) + B_n K_0(\lambda_n r) + \frac{2T_\infty(-1)^{n+1}}{L\lambda_n}\right)\rho_m(r)dr}{\int_a^b r\rho_m^2(r)dr} \tag{22.133}$$

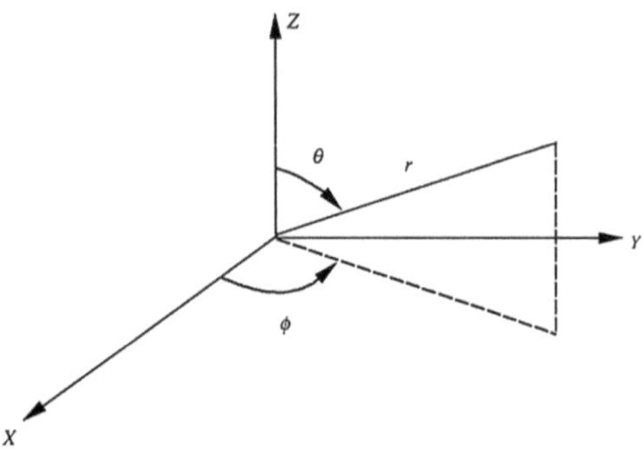

Fig. 22.7 Spherical coordinates

when coefficients C_{mn} are substituted into Eq. (22.130), the function $\psi(r, z, t)$ is known. Consequently, the temperature profile is obtained based on Eq. (22.116).

22.4 Problems in Spherical Coordinates

The governing heat conduction equation in spherical coordinates has the form

$$\frac{1}{r}\frac{\partial^2}{\partial r^2}(rT) + \frac{1}{r^2 \sin\theta}\frac{\partial}{\partial\theta}\left(\sin\theta\frac{\partial T}{\partial\theta}\right) + \frac{1}{r^2 \sin^2\theta}\frac{\partial^2 T}{\partial\phi^2} + \frac{R}{k} = \frac{\rho c}{k}\frac{\partial T}{\partial t}$$

where the variables are in accordance with Fig. 22.7. In its very general form, the analytical solution of this equation may be obtained by the use of separation of variables. The one- and two-dimensional cases are discussed in the subsections which follow.

22.4.1 Steady-State Two- and Three-Dimensional Problems

In spherical coordinates a method of separation of variables results in the Legendre differential equation. For this reason, before proceeding further, we discuss the Legendre equation and Legendre series.

The differential equation

$$(1 - x^2)\frac{d^2 y}{dx^2} - 2x\frac{dy}{dx} + n(n+1)y = 0 \tag{22.134}$$

where n is a constant, is called the *Legendre differential equation*. A general solution to this equation is obtained in a similar manner to that of Bessel equation, and it is

$$y(x) = AP_n(x) + BQ_n(x) \tag{22.135}$$

where $P_n(x)$ is the *Legendre polynomial of the first kind and degree n*, and $Q_n(x)$ is the *Legendre polynomial of the second kind and degree n*, and both polynomials have some definite relations with the variable x, as will be shown. The method at which we arrive at Eq. (22.135) is by assuming a series solution of Eq. (22.134) as

$$y(x) = \sum_{k=0}^{\infty} a_k x^k \tag{22.136}$$

which upon substitution in Eq. (22.134) results in a recurrence relation for a_k

$$a_{k+2} = -a_k \frac{(n-k)(n+k+1)}{(k+1)(k+2)} \qquad k = 0, 1, 2, \ldots \tag{22.137}$$

Specifically, we define

$$\begin{aligned}
a_0 &= a_0 \qquad a_1 = a_1 \qquad a_2 = -\frac{n(n+1)}{2!}a_0 \\
a_3 &= -\frac{(n-1)(n+2)}{3!}a_1 \qquad a_4 = \frac{n(n+1)(n-2)(n+3)}{4!}a_0 \\
a_5 &= \frac{(n-1)(n+2)(n-3)(n+4)}{5!}a_1
\end{aligned}$$

We see that all a_i are expressible in terms of a_0 and a_1 and, therefore, a solution being the sum of two independent solutions to Eq. (22.134) is obtained as

$$\begin{aligned}
y_1(x) &= a_0\left[1 - \frac{n(n+1)}{2!}x^2 + \frac{n(n+1)(n-2)(n+3)}{4!}x^4 - \ldots\ldots\ldots\right] \\
&\quad + a_1 x\left[1 - \frac{(n-1)(n+2)}{3!}x^2 + \frac{(n-1)(n+2)(n-3)(n+4)}{5!}x^4 - \ldots\ldots\ldots\right]
\end{aligned} \tag{22.138}$$

This solution is convergent as long as $|x| < 1$.

It can be easily verified that when n is a positive even integer, the first expression in Eq. (22.138) has a finite number of terms, but the second expression remains to have infinite number of terms. In this case, when n is a positive even integer, the first expression reduces to a polynomial called the *Legendre polynomial*, but the second expression remains an infinite series. When n is a positive odd integer the situation is opposite, that is, the second expression in Eq. (22.138) has a finite number of terms, called a *Legendre polynomial*, and the first expression is an infinite series. To obtain a standard form for the *Legendre polynomial* in any case, when n is even or odd, it is customary to multiply the finite sum occurring in Eq. (22.138) by one of the following factors

$$\frac{(-1)^{n/2} n!}{2^n[(\frac{n}{2})!]} \quad \text{when } n \text{ is even integer}$$

$$\frac{(-1)^{(n-1)/2}(n+1)!}{2^n(\frac{n-1}{2})!(\frac{n+1}{2})!} \quad \text{when } n \text{ is odd integer}$$

This leads to the following general relations for the series in Eq. (22.138), which we call here $P_n(x)$ and $Q_n(x)$ as indicated in Eq. (22.135):

(a) When n is a positive odd integer greater than or equal to 3, then

$$\begin{aligned} P_n(x) &= (-1)^{(n-1)/2}\frac{1\times 3\times 5\times \ldots\ldots n}{2\times 4\times 6\times \ldots\ldots (n-1)} \\ &\times \left[x - \frac{(n-1)(n+2)}{3!}x^3 + \frac{(n-1)(n+2)(n-3)(n+4)}{5!}x^5 + \ldots\ldots\right] \\ Q_n(x) &= (-1)^{(n+1)/2}\frac{2\times 4\times 6\times \ldots\ldots (n-1)}{1\times 3\times 5\times \ldots\ldots n} \\ &\times \left[1 - \frac{n(n+1)}{2!}x^2 + \frac{n(n+1)(n-2)(n+3)}{4!}x^4 + \ldots\ldots\right] \end{aligned} \tag{22.139}$$

(b) When n is a positive even integer greater than or equal to 2, then

$$\begin{aligned} P_n(x) &= (-1)^{n/2}\frac{1\times 3\times 5\times \ldots\ldots (n-1)}{2\times 4\times 6\times \ldots . n} \\ &\times \left[1 - \frac{n(n+1)}{2!}x^2 + \frac{n(n+1)(n-2)(n+3)}{4!}x^4 + \ldots .\right] \\ Q_n(x) &= (-1)^{n/2}\frac{2\times 4\times 6\times \ldots\ldots n}{1\times 3\times 5\times \ldots .(n-1)} \\ &\times \left[x - \frac{(n-1)(n+2)}{3!}x^3 + \frac{(n-1)(n+2)(n-3)(n+4)}{5!}x^5 + \ldots .\right] \end{aligned} \tag{22.140}$$

It is easily verified that the solution to the Legendre differential equation as given by Eq. (22.135) is in terms of the Legendre polynomials $P_n(x)$, which has a finite number of terms depending on the value of n, and Legendre function $Q_n(x)$, which has an infinite number of terms, as defined in Eqs. (22.139) and (22.140). The first seven Legendre polynomials $P_n(x)$ are

$$\begin{aligned} P_0(x) &= 1 \\ P_1(x) &= x \\ P_2(x) &= \frac{1}{2}(3x^2 - 1) \\ P_3(x) &= \frac{1}{2}(5x^3 - 3x) \end{aligned}$$

$$
\begin{aligned}
P_4(x) &= \frac{1}{8}(35x^4 - 30x^2 + 3) \\
P_5(x) &= \frac{1}{8}(63x^5 - 70x^3 + 15x) \\
P_6(x) &= \frac{1}{16}(231x^6 - 315x^4 + 105x^2 - 5)
\end{aligned}
\tag{22.141}
$$

The Legendre polynomial $P_n(x)$ is also expressible in the following form

$$
P_n(x) = \frac{1}{2^n n!} \frac{d^n (x^2 - 1)^n}{dx^n} \tag{22.142}
$$

which is called the *Rodrigues formula*. We also notice that the Legendre polynomials have the following properties for all values of n

$$
\begin{aligned}
P_n(1) &= 1 \\
P_n(-1) &= (-1)^n \\
P_n(-x) &= (-1)^n P_n(x)
\end{aligned}
\tag{22.143}
$$

Differentiating the Legendre differential equation (22.134) m times using Leibnitz formula we obtain the *associated Legendre differential equation* as

$$
(1 - x^2)\frac{d^2 y}{dx^2} - 2x\frac{dy}{dx} + [n(n+1) - \frac{m^2}{1 - x^2}]y = 0 \tag{22.144}
$$

The standard solution to this differential equation is

$$
y(x) = AP_n^m(x) + BQ_n^m(x) \tag{22.145}
$$

where

$$
P_n^m(x) = (1 - x^2)^{m/2} \frac{d^m P_n(x)}{dx^m} \tag{22.146}
$$

is called the *associated Legendre polynomial of the first kind of degree n and order m* and

$$
Q_n^m(x) = (1 - x^2)^{m/2} \frac{d^m Q_n(x)}{dx^m} \tag{22.147}
$$

is called the *associated Legendre polynomial of the second kind of degree n and order m*. The form of $P_n^m(x)$ for some values of m and n, when $x = \cos\theta$, is

$$
\begin{aligned}
P_1^1(x) &= (1-x^2)^{1/2} = \sin\theta \\
P_2^1(x) &= 3x(1-x^2)^{1/2} = 3\cos\theta\sin\theta \\
P_2^2(x) &= 3(1-x^2) = 3\sin^2\theta \\
P_3^1(x) &= \frac{3}{2}(5x^2-1)(1-x^2)^{1/2} = \frac{3}{2}(5\cos^2\theta-1)\sin\theta \\
P_3^2(x) &= 15x(1-x^2) = 15\cos\theta\sin^2\theta \\
P_3^3(x) &= 15(1-x^2)^{3/2} = 15\sin^3\theta \\
P_4^1(x) &= \frac{5}{2}(7x^3-3x)(1-x^2)^{1/2} = \frac{5}{2}(7\cos^3\theta-3\cos\theta)\sin\theta \\
P_4^2(x) &= \frac{15}{2}(7x^2-1)(1-x^2) = \frac{15}{2}(7\cos^2\theta-1)\sin^2\theta \\
P_4^3(x) &= 105x(1-x^2)^{3/2} = 105\cos\theta\sin^3\theta \\
P_4^4(x) &= 105x(1-x^2)^2 = 105\sin^4\theta
\end{aligned}
\tag{22.148}
$$

In discussing the Legendre differential equation it was indicated that to have a solution, the condition $|x| < 1$ must be satisfied and, furthermore, the limit of convergence for the solution in the Legendre polynomial was set to be $-1 < x < +1$. This suggests that we may take the variable in the form of $x = \cos\theta$. With such a change of variable, Eq. (22.134) reduces to

$$\frac{1}{\sin\theta}\frac{d}{d\theta}\left(\sin\theta\frac{dy}{d\theta}\right) + n(n+1)y = 0 \tag{22.149}$$

This is the Legendre differential equation in terms of the variable θ, and thus the solution follows from Eq. (22.135):

$$y = AP_n(\cos\theta) + BQ_n(\cos\theta) \tag{22.150}$$

Equation (22.149) is usually obtained when Laplace equation in spherical coordinates is solved by the method of separation of variables and the variable θ is considered as the co-latitude coordinate.

Similarly to the trigonometric functions $\cos nx$ and $\sin nx$, the Legendre polynomials $P_n(x)$ are orthogonal functions with respect to a weighting function $w(x) = 1$, over the interval -1 to $+1$, that is

$$\int_{-1}^{+1} P_m(x)P_n(x)dx = 0 \quad \text{if } m \neq n \tag{22.151}$$

and when $m = n$

$$\int_{-1}^{+1} [P_n(x)]^2 dx = \frac{2}{2n+1} \qquad n = 0, 1, 2, \ldots. \tag{22.152}$$

This property is used to expand any arbitrary function in terms of the Legendre polynomials. Suppose $f(x)$ is a continuous function and its derivative is continuous in the interval $(-1, 1)$. Then the function $f(x)$ can be expanded in a series of the form

$$f(x) = \sum_{n=0}^{\infty} a_n P_n(x) \tag{22.153}$$

The coefficients a_n are obtained by multiplying both sides of Eq. (22.153) by $P_m(x)$ and integrating over the interval $(-1, 1)$. This gives

$$\int_{-1}^{+1} f(x)P_m(x)dx = \sum_{n=0}^{\infty} a_n \int_{-1}^{+1} P_m(x)P_n(x)dx$$

Using Eqs. (22.151) and (22.152) yields

$$\int_{-1}^{+1} f(x)P_n(x)dx = a_n \int_{-1}^{+1} [P_n(x)]^2 dx = \frac{2a_n}{2n+1}$$

and thus

$$a_n = \frac{2n+1}{2} \int_{-1}^{+1} f(x)P_n(x)dx \tag{22.154}$$

The integral in Eq. (22.154) is obtained as long as we know $f(x)$. Similarly to $P_n(x)$, we notice that $P_n^m(x)$ are also orthogonal in the same interval of $(-1, 1)$ with respect to the weighting function $w(x) = 1$, that is

$$\int_{-1}^{+1} P_n^m(x)P_k^m(x)dx = 0 \quad \text{for } n \neq k \tag{22.155}$$

and

$$\int_{-1}^{+1} [P_n^m(x)]^2 dx = \frac{(n+m)!}{(n-m)!} \frac{2}{(2n+1)} \tag{22.156}$$

This property suggests that, like in the case of Legendre polynomial $P_n(x)$, we can expand a given function $F(x)$ in terms of $P_n^m(x)$ as

$$F(x) = \sum_{n=0}^{\infty} C_n P_n^m(x)$$

Multiplying both sides of the above equation by $P_k^n(x)$ and integrating from -1 to 1 and using Eqs. (22.155) and (22.156) gives

$$C_n = \frac{(n-m)!}{(n+m)!} \cdot \frac{(2n+1)}{2} \int_{-1}^{+1} F(x) P_n^m(x) dx$$

Often, while solving Laplace equation in spherical coordinates, we have to deal with a series expansion involving $P_n^m(x)$ of the following form

$$F(\theta, \phi) = \sum_{n=0}^{\infty} (a_{mn} \cos m\phi + b_{mn} \sin m\phi) P_n^m(\cos\theta)$$

In this case we need to find the coefficients a_{mn} and b_{mn} in terms of the known function $F(\theta, \phi)$. To find a_{mn} we note that

$$\int_0^{2\pi} F(\theta, \phi) \cos k\phi d\phi = \sum_{n=0}^{\infty} \sum_{m=0}^{\infty} a_{mn} \pi P_n^m(\cos\theta) \delta_{mk}$$

where δ_{mk} is the Kronecker symbol. This gives

$$\int_0^{2\pi} F(\theta, \phi) \cos m\phi d\phi = \sum_{n=0}^{\infty} a_{mn} \pi P_n^m(\cos\theta)$$

Using the orthogonality condition for $P_n^m(x)$, multiplying both sides by $\sin\theta P_s^k(\cos\theta)$, integrating from 0 to π, and interchanging k with m in the final result, yields

$$a_{mn} = \frac{(n-m)!}{(n+m)!} \frac{(2n+1)}{2\pi} \int_0^{\pi} \sin\theta P_n^m(\cos\theta) d\theta \int_0^{2\pi} F(\theta, \phi) \cos m\phi d\phi$$

In a similar manner b_{mn} is found.

In Figs. 22.8, 22.9, 22.10 and 22.11 the plots of Legendre polynomials with respect to θ and x are shown.

We now present some useful relations for the Legendre polynomials and the associated Legendre functions, as

$$\begin{aligned} P_n^m(-x) &= (-1)^{n+m} P_n^m(x) \\ P_n^m(\pm 1) &= 0 \quad \text{for } m > 0 \end{aligned} \tag{22.157}$$

The recurrence relations for $P_n(x)$ and $Q_n(x)$ are

$$\begin{aligned} P_n(x) &= \sum_{k=0}^{n/2} \frac{(-1)^k (2n-2)!}{2^n k!(n-2k)!(n-k)!} x^{n-2k} \\ Q_n(x) &= \sum_{k=0}^{\infty} \frac{2^n (n+k)!(n+2k)!}{k!(2n+2k+1)!} x^{-(n+2k+1)} \end{aligned}$$

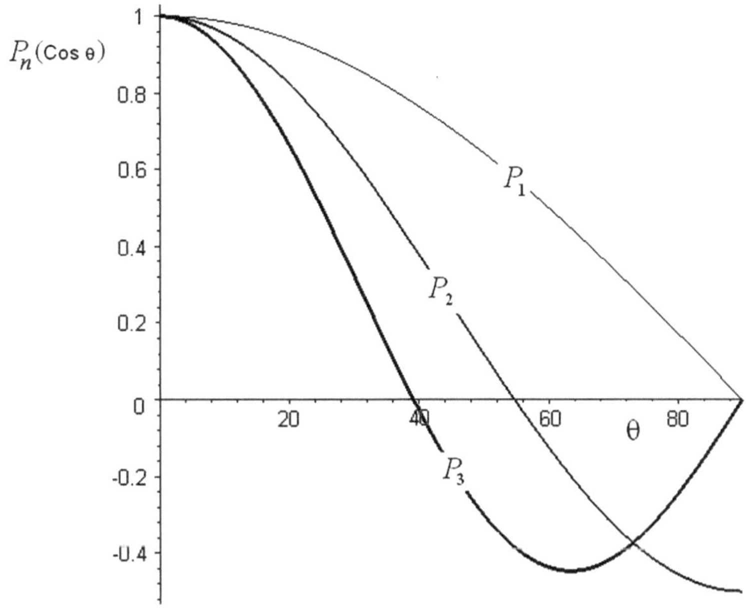

Fig. 22.8 Legendre polynomial $P_n(\cos\theta)$ versus θ

$$(2n+1)xP_n(x) = (n+1)P_{n+1}(x) + nP_{n-1}(x) \qquad n = 1, 2, 3, \ldots$$
$$P'_{n+1}(x) - P'_{n-1}(x) = (2n+1)P_n(x) \qquad n = 1, 2, 3, \ldots$$

$$(n+1)Q_{n+1}(x) - (2n+1)xQ_n(x) + nQ_{n-1}(x) = 0$$
$$xQ'_n(x) - Q'_{n-1}(x) - nQ_n(x) = 0 \tag{22.158}$$

Knowing the properties of the Legendre polynomials, we return to our discussion on a general solution of the heat conduction equation in spherical coordinates. We may consider heat conduction in a spherical, or partly spherical, solid body which is exposed to some kind of steady thermal fields. The steady-state temperature will then satisfy Laplace equation

$$\nabla^2 T = 0 \tag{22.159}$$

which in spherical coordinates is

$$r^2 \sin\theta \frac{\partial^2 T}{\partial r^2} + 2r\sin\theta \frac{\partial T}{\partial r} + \sin\theta \frac{\partial^2 T}{\partial \theta^2} + \cos\theta \frac{\partial T}{\partial \theta} + \frac{1}{\sin\theta} \frac{\partial^2 T}{\partial \phi^2} = 0 \tag{22.160}$$

Any solution of this equation is called a *spherical harmonic*. This equation will be solved by the method of separation of variables by taking

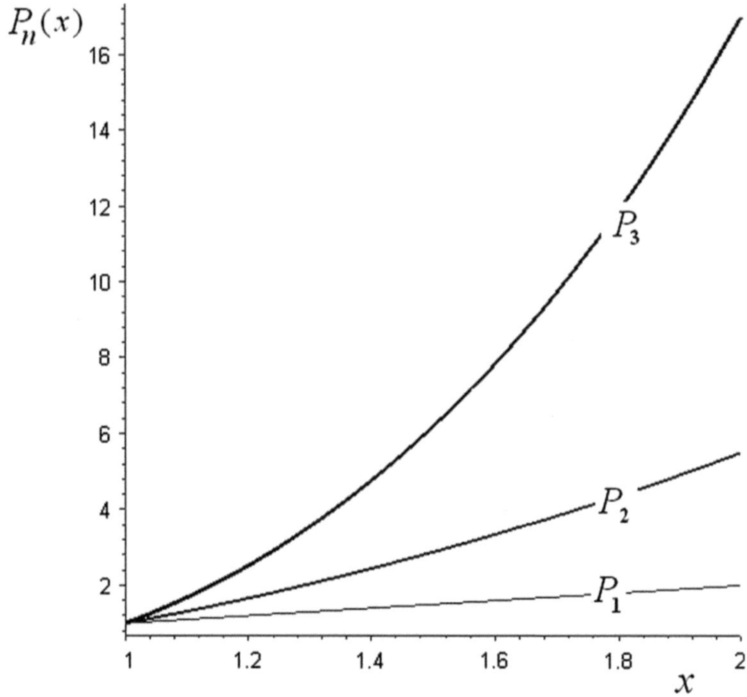

Fig. 22.9 Legendre polynomial $P_n(x)$ versus x

$$T(r, \theta, \phi) = \mathcal{R}(r)F(\theta, \phi) \tag{22.161}$$

Substituting this in Eq. (22.161) yields

$$r^2 \sin\theta \frac{d^2\mathcal{R}}{dr^2} F + 2r \sin\theta \frac{d\mathcal{R}}{dr} F + \sin\theta \mathcal{R} \frac{\partial^2 F}{\partial \theta^2} + \cos\theta \mathcal{R} \frac{\partial F}{\partial \theta} + \frac{\mathcal{R}}{\sin\theta} \frac{\partial^2 F}{\partial \phi^2} = 0$$

Dividing by $\mathcal{R}F \sin\theta$, rearranging and equating the separated function to a constant such as λ, gives

$$\begin{aligned} &\frac{r^2}{\mathcal{R}} \frac{d^2\mathcal{R}}{dr^2} + \frac{2r}{\mathcal{R}} \frac{d\mathcal{R}}{dr} \\ &= -\left(\frac{1}{F} \frac{\partial^2 F}{\partial \theta^2} + \frac{\cos\theta}{F \sin\theta} \frac{\partial F}{\partial \theta} + \frac{1}{F \sin^2\theta} \frac{\partial^2 F}{\partial \phi^2} \right) = \lambda \end{aligned} \tag{22.162}$$

It will be seen later that taking the separation constant $\lambda = n(n+1)$ is more convenient. Doing so, results in

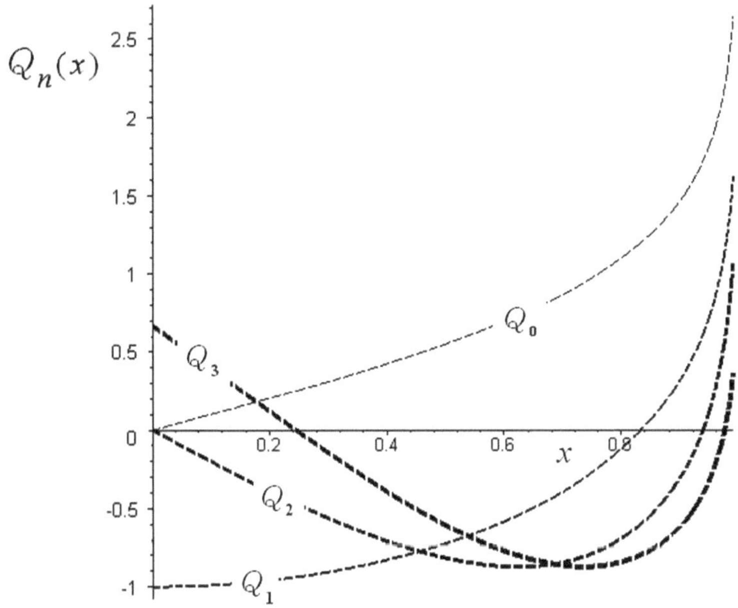

Fig. 22.10 Legendre polynomial $Q_n(x)$ versus x

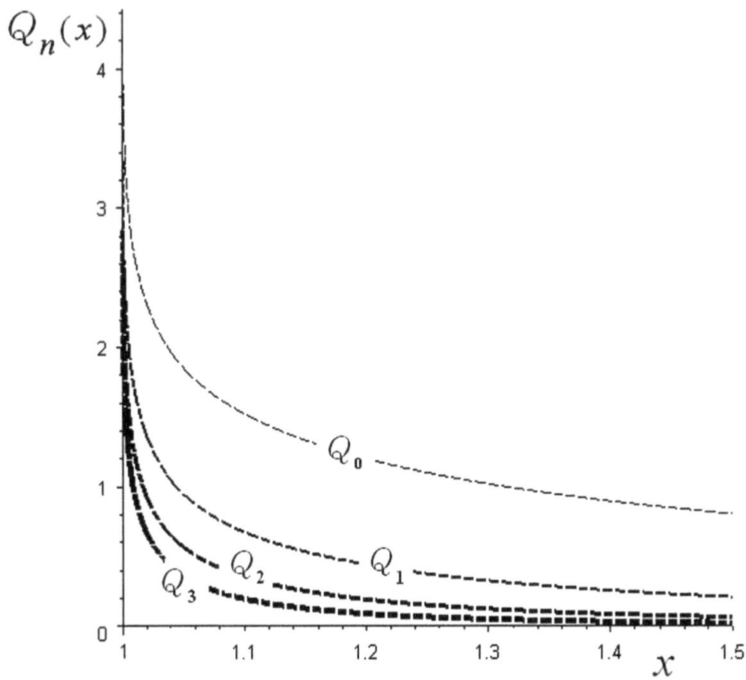

Fig. 22.11 Legendre polynomial $Q_n(x)$ versus x

$$r^2\frac{d^2\mathcal{R}}{dr^2}+2r\frac{d\mathcal{R}}{dr}-n(n+1)\mathcal{R}=0 \tag{22.163}$$

$$\frac{\partial^2 F}{\partial\theta^2}+\frac{\cos\theta}{\sin\theta}\frac{\partial F}{\partial\theta}+\frac{1}{\sin^2\theta}\frac{\partial^2 F}{\partial\phi^2}+n(n+1)F=0 \tag{22.164}$$

Equation (22.163) is the Euler differential equation and is readily solved to give

$$\mathcal{R}(r)=A_1r^n+\frac{A_2}{r^{n+1}} \tag{22.165}$$

where A_1 and A_2 are constants of integration. Equation (22.164) still has to be separated, and thus by taking

$$F(\theta,\phi)=\Theta(\theta)\Phi(\phi) \tag{22.166}$$

and substituting in Eq. (22.164) we find

$$\frac{d^2\Theta}{d\theta^2}\Phi+\frac{\cos\theta}{\sin\theta}\frac{d\Theta}{d\theta}\Phi+\frac{1}{\sin^2\theta}\Theta\frac{d^2\Phi}{d\phi^2}+n(n+1)\Theta\Phi=0$$

Dividing this equation by $\Theta\Phi/\sin^2\theta$, and calling the separation constant m^2, gives

$$\frac{\sin^2\theta}{\Theta}\frac{d^2\Theta}{d\theta^2}+\frac{\sin\theta\cos\theta}{\Theta}\frac{d\Theta}{d\theta}+n(n+1)\sin^2\theta=-\frac{1}{\Phi}\frac{d^2\Phi}{d\phi^2}=m^2$$

This yields

$$\sin^2\theta\frac{d^2\Theta}{d\theta^2}+\sin\theta\cos\theta\frac{d\Theta}{d\theta}+[n(n+1)\sin^2\theta-m^2]\Theta=0 \tag{22.167}$$

$$\frac{d^2\Phi}{d\phi^2}+m^2\Phi=0 \tag{22.168}$$

Solution to Eq. (22.168) is

$$\Phi=A_3\cos m\phi+A_4\sin m\phi \tag{22.169}$$

while the solution to Eq. (22.167) is obtained in terms of the Legendre functions. In fact, Eq. (22.167) is the associated Legendre equation which by taking $x=\cos\theta$ and knowing that

$$\frac{d\Theta}{d\theta}=\frac{d\Theta}{dx}\frac{dx}{d\theta}=-\sin\theta\frac{d\Theta}{dx}$$

$$\frac{d^2\Theta}{d\theta^2}=\frac{d}{d\theta}\left(-\sin\theta\frac{d\Theta}{dx}\right)=-\cos\theta\frac{d\Theta}{dx}+\sin^2\theta\frac{d^2\Theta}{dx^2}$$

it is modified to

$$(1-x^2)\frac{d^2\Theta}{dx^2} - 2x\frac{d\Theta}{dx} + \left[n(n+1) - \frac{m^2}{1-x^2}\right]\Theta = 0 \tag{22.170}$$

Equation (22.170) is the same as Eq. (22.144) and, therefore, its solution, after substitution of $x = \cos\theta$, is

$$\Theta = A_5 P_n^m(\cos\theta) + A_6 Q_n^m(\cos\theta) \tag{22.171}$$

Substituting Eqs. (22.165), (22.169), and (22.171) into Eqs. (22.166) and (22.161), the general solution to Laplace equation in spherical coordinates becomes

$$T(r,\theta,\phi) = \left(A_1 r^n + \frac{A_2}{r^{n+1}}\right)(A_3\cos m\phi + A_4\sin m\phi)$$
$$[A_5 P_n^m(\cos\theta) + A_6 Q_n^m(\cos\theta)] \tag{22.172}$$

There are six constants of integration, A_1 through A_6, which can be evaluated by using the thermal boundary conditions.

We may notice that if the problem is designed in such a way that the temperature distribution is independent of the meridional angle ϕ, then the temperature is only a function of r and θ and the equation in θ direction, since $m = 0$, is reduced to

$$(1-x^2)\frac{d^2\Theta}{dx^2} - 2x\frac{d\Theta}{dx} + n(n+1)\Theta = 0 \tag{22.173}$$

which is the Legendre differential equation and thus, from Eq. (22.135),

$$\Theta = A_5 P_n(\cos\theta) + A_6 Q_n(\cos\theta) \tag{22.174}$$

Equation (22.174) is a solution in θ-direction for $m = 0$.

Problem 22.5. Consider a thick sphere of inside and outside radii a and b, respectively. At $t = 0$, the sphere is at uniform temperature T_∞. The sphere is suddenly exposed to a constant heat flux q'' from one side. Find the transient temperature distribution in the sphere if the inside surface is kept at constant temperature T_0. The sphere is cooled by convection from the outer surface to the ambient at (h, T_∞).

Solution: The transient axisymmetric heat conduction equation for spherical bodies is

$$\frac{1}{r}\frac{\partial^2}{\partial r^2}(rT) + \frac{1}{r^2\sin\theta}\frac{\partial}{\partial\theta}\left(\sin\theta\frac{\partial T}{\partial\theta}\right) = \frac{\rho c}{K}\frac{\partial T}{\partial t} \tag{22.175}$$

in which, $T(r,\theta,t)$ is the temperature distribution through the sphere and $a \le r \le b$ and $0 \le \theta \le \pi$. Furthermore, ρ, c and K stand for the mass density, heat capacity and conductivity, respectively. For the problem in hand the initial and boundary conditions are as follows

$$\begin{aligned}
&T(r,\theta,0) = T_\infty \\
&T(a,\theta,t) = T_0 \\
&-K\frac{\partial T}{\partial r}(b,\theta,t) = \begin{cases} h(T(b,\theta,t) - T_\infty) & \pi/2 \le \theta \le \pi \\ h(T(b,\theta,t) - T_\infty) - q'' \cos\theta & 0 \le \theta \le \pi/2 \end{cases}
\end{aligned} \tag{22.176}$$

With the aid of simple transformation $\theta(r,\theta,t) = T(r,\theta,t) - T_\infty$, the heat conduction equation transforms to

$$\frac{1}{r}\frac{\partial^2}{\partial r^2}(r\Theta) + \frac{1}{r^2 \sin\theta}\frac{\partial}{\partial \theta}\left(\sin\theta\frac{\partial \Theta}{\partial \theta}\right) = \frac{\rho c}{K}\frac{\partial \Theta}{\partial t} \tag{22.177}$$

and the associated initial and boundary conditions are

$$\begin{aligned}
&\Theta(r,\theta,0) = 0 \\
&\Theta(a,\theta,t) = \Theta_0 \\
&-K\frac{\partial \Theta}{\partial r}(b,\theta,t) = \begin{cases} h\Theta(b,\theta,t) & \pi/2 \le \theta \le \pi \\ h\Theta(b,\theta,t) - q'' \cos\theta & 0 \le \theta \le \pi/2 \end{cases}
\end{aligned} \tag{22.178}$$

where, $\Theta_0 = T_0 - T_\infty$. Solution of the above equation is written in terms of two distinct functions in the form $\Theta(r,\theta,t) = \Theta_1(r,\theta) + \Theta_2(r,\theta,t)$. The resulted governing equation and boundary condition for Θ_1 are

$$\begin{aligned}
&\frac{1}{r}\frac{\partial^2}{\partial r^2}(r\Theta_1) + \frac{1}{r^2 \sin\theta}\frac{\partial}{\partial \theta}(\sin\theta\frac{\partial \Theta_1}{\partial \theta}) = 0 \\
&\Theta_1(a,\theta) = \Theta_0 \\
&-K\frac{\partial \Theta_1}{\partial r}(b,\theta) = \begin{cases} h\Theta_1(b,\theta) & \pi/2 \le \theta \le \pi \\ h\Theta_1(b,\theta) - q'' \cos\theta & 0 \le \theta \le \pi/2 \end{cases}
\end{aligned} \tag{22.179}$$

and for Θ_2 we have

$$\begin{aligned}
&\frac{1}{r}\frac{\partial^2}{\partial r^2}(r\Theta_2) + \frac{1}{r^2 \sin\theta}\frac{\partial}{\partial \theta}\left(\sin\theta\frac{\partial \Theta_2}{\partial \theta}\right) = \frac{\rho c}{K}\frac{\partial \Theta_2}{\partial t} \\
&\Theta_2(r,\theta,0) = -\Theta_1(r,\theta) \\
&\Theta_2(a,\theta,t) = 0 \\
&-K\frac{\partial \Theta_2}{\partial r}(b,\theta,t) = h\Theta_2(b,\theta,t)
\end{aligned} \tag{22.180}$$

Each of the above partial differential equations has to be solved, separately.
Solution of Θ_1:

To solve the system of Eq. (22.179), the separation of variables method is adopted. With the transformation $\Theta_1(r,t) = R_1(r)G_1(\theta)$ the following ordinary differential equations are resulted

$$\frac{1}{R_1(r)}(r^2 R_1^{''}(r) + 2r R_1^{'}(r)) = -\frac{1}{\sin\theta G_1(\theta)}(\cos\theta G_1^{'}(\theta) + \sin\theta G_1^{''}(\theta)) = \mu \tag{22.181}$$

where μ is constant. The differential equation associated to the function $G_1(\theta)$ simplifies to

$$G_1^{''}(\theta) + \cot\theta G_1^{'}(\theta) + \mu G_1(\theta) = 0 \tag{22.182}$$

Taking the constant μ equal to $n(n+1)$ is more convenient, since the solution of the above equation is obtained exactly as

$$G_1(\theta) = G_{1n} P_n(\cos\theta) + G_{2n} Q_n(\sin\theta) \tag{22.183}$$

in which G_{1n} and G_{2n} are constants of differential equation and P_n and Q_n are the Legendre polynomial of the first and second kind, respectively. For the problem in hand, $G_{2n} = 0$, since $G_1(\theta)$ has to be bounded at $\theta = 0$.

Therefore from Eq. (22.181) the governing equation for $R_1(r)$ is extracted as

$$r^2 R_1^{''}(r) + 2r R_1^{'}(r) - n(n+1)R_1(r) = 0 \tag{22.184}$$

The above equation is a second-order Euler equation which has the exact solution of the form

$$R_{1n}(r) = A_n r^n + B_n r^{-(1+n)} \tag{22.185}$$

The complete solution for $\Theta_1(r,\theta)$ may be written in the following form

$$\Theta_1(r,\theta) = \sum_{n=0}^{\infty} \left(A_n r^n + B_n r^{-(1+n)}\right) P_n(\cos\theta) \tag{22.186}$$

where, the constants A_n and B_n have to be obtained according to the boundary conditions (22.179). According to the first boundary condition (22.179) on $r = a$, the following equality is concluded from (22.186)

$$\Theta_0 = \sum_{n=0}^{\infty} \left(A_n a^n + B_n a^{-(1+n)}\right) P_n(\cos\theta) \tag{22.187}$$

Here we use the orthogonality conditions of the Legendre functions. According to Eq. (22.182) the eigenfunctions (22.184) are orthogonal with respect to each other when the weighting function $\sin\theta$ is also taken into account. Multiplying both sides of the Eq. (22.187) by $P_m(\cos\theta)\sin\theta$ and integrating over the domain $[0,\pi]$ gives us

$$\int_0^\pi \Theta_0 P_m(\cos\theta)\sin\theta d\theta$$
$$= \sum_{n=0}^{\infty}\left\{\left(A_n a^n + B_n a^{-(1+n)}\right)\int_0^\pi P_n(\cos\theta)P_m(\cos\theta)\sin\theta d\theta\right\} \tag{22.188}$$

Due to the orthogonality of the Legendre polynomials, among all terms of the right-hand-side of the above equation only the case $n \neq m$ results in a nonzero integral and therefore Eq. (22.188) simplifies to

$$A_m a^m + B_m a^{-(1+m)} = \frac{\int_0^\pi \Theta_0 P_m(\cos\theta)\sin\theta d\theta}{\int_0^\pi P_m^2(\cos\theta)\sin\theta d\theta} \tag{22.189}$$

The above equation is nonzero only for $m = 0$, since $P_0(\cos\theta) = 1$. Therefore, the relation between the constants A_n and B_n is extracted as

$$\begin{aligned} A_0 &= \Theta_0 - \frac{B_0}{a} \\ A_n &= -B_n a^{-(1+2n)} \quad n \geq 1 \end{aligned} \tag{22.190}$$

When the above constants are inserted into the Eq. (22.186) one may write

$$\Theta_1 = \Theta_0 + \sum_{n=0}^{\infty} B_n\left(r^{-1-n} - r^n a^{-(1+2n)}\right)P_n(\cos\theta) \tag{22.191}$$

with the aid of the above equation, the second boundary condition (22.179) is rewritten in the form

$$\sum_{n=0}^{\infty} K B_n\left(b^{-2-n}\left(n+1+\frac{bh}{K}\right) - b^{n-1}a^{-1-2n}\left(n+\frac{bh}{K}\right)\right)P_n(\cos\theta)$$
$$= \begin{cases} -h\Theta_0 & \pi/2 \leq \theta \leq \pi \\ -h\Theta_0 + q'' \cos\theta & 0 \leq \theta \leq \pi/2 \end{cases} \tag{22.192}$$

Again, both sides of the above equation are multiplied by $P_m(\cos\theta)\sin\theta$ and integrated over the domain $[0, \pi]$ which results in

$$B_n = \frac{2n+1}{2}\left\{\frac{-\int_0^\pi h\Theta_0 P_n(\cos\theta)d\theta + \int_0^{\pi/2} q'' \cos\theta P_n(\cos\theta)d\theta}{K\left(b^{-2-n}\left(n+1+\frac{bh}{K}\right) - b^{n-1}a^{-1-2n}\left(n+\frac{bh}{K}\right)\right)}\right\} \tag{22.193}$$

The constants B_n are obtained and therefore solution of Θ_1 is accomplished according to Eq. (22.191).

Solution of Θ_2:

With the aid of the separation of variables method for $\theta_2(r, \theta, t)$, we may write, $\Theta_2(r, \theta, t) = R_2(r)G_2(\theta)H(t)$. Therefore, the heat conduction Eq. (22.180) is rewritten as

$$\frac{\frac{1}{r}\frac{\partial^2}{\partial r^2}(rR_2(r)G_2(\theta)) + \frac{1}{r^2 \sin\theta}\frac{\partial}{\partial\theta}\left(\sin\theta R_2(r)\frac{\partial G_2(\theta)}{\partial\theta}\right)}{R_2(r)G_2(\theta)} = \frac{\rho c}{KH(t)}\frac{dH(t)}{dt} = -\lambda_m^2 \tag{22.194}$$

where λ_m is a set of constant parameters. The solution of the first order ordinary differential equation associated to $H(t)$ is equal to

$$H(t) = A_m e^{-\kappa\lambda_m^2 t} \tag{22.195}$$

In which $\kappa = \frac{K}{\rho c}$, A_m is the constant of integration. The right-hand-side of the Eq. (22.195) is written in the present form since $H(t)$ has to be bounded. Separating the functions $R_2(r)$ and $G_2(\theta)$ in Eq. (22.194) gives us

$$\frac{1}{R_2(r)}(r^2 R_2''(r) + 2rR_2'(r)) + \lambda_m^2 r^2 = -\frac{1}{\sin\theta G_2(\theta)}(\cos\theta G_2'(\theta) + \sin\theta G_2''(\theta)) = \xi \tag{22.196}$$

where ξ is constant. The differential equation associated to the function $G_2(\theta)$ simplifies to

$$G_2''(\theta) + \cot\theta G_2'(\theta) + \xi G_2(\theta) = 0 \tag{22.197}$$

Same as the process developed before for $G_1(\theta)$, the solution of the above equation is obtained as

$$G_2(\theta) = G_{4n} P_n(\cos\theta) \tag{22.198}$$

And finally, from Eq. (22.196) the governing differential equation for $R_2(r)$ is written in the form

$$r^2 R_2''(r) + 2rR_2'(r) + (\lambda_m^2 r^2 - n(n+1))R_2(r) = 0 \tag{22.199}$$

Exact solution of the above equation can be obtained with the aid of a proper transformation. With the definition of $S(r) = \sqrt{r}R_2(r)$ we may write

$$r^2 S''(r) + rS'(r) + \left(\lambda_m^2 r^2 - \left(n + \frac{1}{2}\right)^2\right)S(r) = 0 \tag{22.200}$$

The solution of the above equation according to the Bessel functions may be written in the form

$$S(r) = S_{1n} J_{n+\frac{1}{2}}(\lambda_m r) + S_{2n} Y_{n+\frac{1}{2}}(\lambda_m r) \tag{22.201}$$

Since the solution of each of the ordinary differential equations is obtained, recalling the separation of variables method, previously introduced, reaches us to the following solution for $\Theta_2(r, \theta, t)$

$$\begin{aligned}\Theta_2&(r, \theta, t)\\ &= \sum_{m=0}^{\infty}\sum_{n=0}^{\infty} \frac{1}{\sqrt{r}} e^{-\kappa\lambda_m^2 t} P_n(\cos\theta) \left(C_{nm} J_{n+\frac{1}{2}}(\lambda_m r) + D_{nm} Y_{n+\frac{1}{2}}(\lambda_m r) \right)\end{aligned} \tag{22.202}$$

Here constants C_{nm}, D_{nm} and parameters λ_m are still unknown.

Based on the boundary conditions (22.180), which are both homogeneous, we have the following equalities

$$\begin{aligned}& J_{n+\frac{1}{2}}(\lambda_m a) C_{nm} + Y_{n+\frac{1}{2}}(\lambda_m a) D_{nm} = 0\\ & \left(\lambda_m J'_{n+\frac{1}{2}}(\lambda_m b) + \frac{h}{K} J_{n+\frac{1}{2}}(\lambda_m b) \right) C_{nm}\\ & \quad + \left(\lambda_m Y'_{n+\frac{1}{2}}(\lambda_m b) + \frac{h}{K} Y_{n+\frac{1}{2}}(\lambda_m b) \right) D_{nm} = 0\end{aligned} \tag{22.203}$$

To obtain the parameter λ_m the determinant of coefficient matrix of the above system of equations has to set equal to zero. Consequently, the parameter λ_m is obtained as the mth positive real root of the following equation

$$\begin{aligned}& J_{n+\frac{1}{2}}(\lambda_m a) \left(\lambda_m Y'_{n+\frac{1}{2}}(\lambda_m b) + \frac{h}{K} Y_{n+\frac{1}{2}}(\lambda_m b) \right)\\ & \quad - Y_{n+\frac{1}{2}}(\lambda_m a) \left(\lambda_m J'_{n+\frac{1}{2}}(\lambda_m b) + \frac{h}{K} J_{n+\frac{1}{2}}(\lambda_m b) \right) = 0\end{aligned} \tag{22.204}$$

The solution of Θ_2 is obtained as

$$\begin{aligned}\Theta_2(r, \theta, t) = & \sum_{m=0}^{\infty}\sum_{n=0}^{\infty} C_{nm} \frac{1}{\sqrt{r}} e^{-\kappa\lambda_m^2 t} P_n(\cos\theta)\\ & \times \left(J_{n+\frac{1}{2}}(\lambda_m r) - \frac{J_{n+\frac{1}{2}}(\lambda_m a)}{Y_{n+\frac{1}{2}}(\lambda_m a)} Y_{n+\frac{1}{2}}(\lambda_m r) \right)\end{aligned} \tag{22.205}$$

Constants C_{nm} are obtained with the aid of initial condition. Accordingly, the following equation is obtained with the consideration of initial condition (22.180)

$$\sum_{m=0}^{\infty}\sum_{n=0}^{\infty} C_{nm}\frac{1}{\sqrt{r}}P_n(\cos\theta)\left(J_{n+\frac{1}{2}}(\lambda_m r)-\frac{J_{n+\frac{1}{2}}(\lambda_m a)}{Y_{n+\frac{1}{2}}(\lambda_m a)}Y_{n+\frac{1}{2}}(\lambda_m r)\right)$$
$$=\Theta_0+\sum_{n=0}^{\infty} B_n\left(r^{-1-n}-r^n a^{-(1+2n)}\right)P_n(\cos\theta) \tag{22.206}$$

In which constants B_n are previously introduced in Eq. (22.193). The function Θ_0 has to be expanded in terms of the Legendre function. To this end we may write

$$\Theta_0=\sum_{n=0}^{\infty} E_n P_n(\cos\theta) \tag{22.207}$$

which results in

$$E_n=\begin{cases}\Theta_0 & n=0\\ 0 & n>0\end{cases} \tag{22.208}$$

Substituting Eq. (22.207) into Eq. (22.206) results in the following equation for each integer number n

$$\sum_{m=0}^{\infty} C_{nm}\frac{1}{\sqrt{r}}\left(J_{n+\frac{1}{2}}(\lambda_m r)-\frac{J_{n+\frac{1}{2}}(\lambda_m a)}{Y_{n+\frac{1}{2}}(\lambda_m a)}Y_{n+\frac{1}{2}}(\lambda_m r)\right)$$
$$=E_n+B_n\left(r^{-1-n}-r^n a^{-(1+2n)}\right) \tag{22.209}$$

or

$$\sum_{m=0}^{\infty} C_{nm}R_m(r)=E_n+B_n\left(r^{-1-n}-r^n a^{-(1+2n)}\right) \tag{22.210}$$

To obtain the constants C_{nm} the orthogonality property of the eigenfunctions is implemented. According to Eq. (22.199), the eigenfunction $R_n(r)$ are orthogonal with respect to each other when the weight function r^2 is taken into account. Therefore multiplying both sides of the above equation by $r^2R_p(r)$ and integrating over the domain $[a,b]$ yields the constants C_{np} as

$$C_{np}=\frac{\int_a^b r^2R_p(r)\left(E_n+B_n\left(r^{-1-n}-r^n a^{-(1+2n)}\right)\right)}{\int_a^b r^2R_p^2(r)dr} \tag{22.211}$$

Since the constants C_{np} are known the function Θ_2 is obtained (see Eq. (22.205)) which accomplishes the solution.

Problem 22.6. Consider a thick hollow cylinder of inside radius a and outside radius b. The initial temperature of the cylinder is $T=0$ at time $t=0$. At times $t>0$ a constant heat flux q'' is radiated to one side of the cylinder. The side $z=0$ is insulated and $z = L$ is at ambient temperature. The inside surface is at constant temperature T_0. The cylinder is cooled from the outer surface to the ambient at (h, T_∞). Find the transient temperature distribution through the cylinder.

Solution: The transient three-dimensional heat conduction equation is

$$\frac{\partial^2\theta}{\partial r^2} + \frac{1}{r}\frac{\partial\theta}{\partial r} + \frac{1}{r^2}\frac{\partial^2\theta}{\partial\phi^2} + \frac{\partial^2\theta}{\partial z^2} = \frac{1}{\kappa}\frac{\partial\theta}{\partial t} \tag{22.212}$$

where, $\theta = T - T_\infty$. The associated boundary conditions for Eq. (22.212) are

$$\begin{aligned}
&K\frac{\partial\theta}{\partial r} + h\theta = q'' f(\phi) && at && r = b\\
&\theta = \theta_0 = T_0 - T_\infty && at && r = a\\
&\theta = 0 && at && z = L\\
&\frac{\partial\theta}{\partial z} = 0 && at && z = 0
\end{aligned}$$

where, $f(\phi) = \cos\phi$ for $-\pi/2 < \phi < \pi/2$ and is equal to zero else-where. To find a solution, we start by dividing a solution into two different parts. One with homogeneous boundary conditions and the other one with nonhomogeneous boundary conditions. Dependency of the solution on time is included in the first part. Therefore we have

$$\theta(r, \phi, z, t) = \theta_1(r, \phi, z, t) + \theta_2(r, \phi, z) \tag{22.213}$$

with the following governing equations for θ_1 and θ_2

$$\begin{aligned}
&\frac{\partial^2\theta_1}{\partial r^2} + \frac{1}{r}\frac{\partial\theta_1}{\partial r} + \frac{1}{r^2}\frac{\partial^2\theta_1}{\partial\phi^2} + \frac{\partial^2\theta_1}{\partial z^2} = \frac{1}{\kappa}\frac{\partial\theta_1}{\partial t}\\
&\frac{\partial^2\theta_2}{\partial r^2} + \frac{1}{r}\frac{\partial\theta_2}{\partial r} + \frac{1}{r^2}\frac{\partial^2\theta_2}{\partial\phi^2} + \frac{\partial^2\theta_2}{\partial z^2} = 0
\end{aligned} \tag{22.214}$$

The associated boundary and initial conditions for Eq. (22.214) are

$$\begin{aligned}
K\frac{\partial\theta_1}{\partial r} + h\theta_1 = 0, \quad K\frac{\partial\theta_2}{\partial r} + h\theta_2 = q'' f(\phi) && at && r = b\\
\theta_1 = 0, \quad \theta_2 = \theta_0 = T_0 - T_\infty && at && r = a\\
\theta_1 = 0, \quad \theta_2 = 0 && at && z = L\\
\frac{\partial\theta_1}{\partial z} = 0, \quad \frac{\partial\theta_2}{\partial z} = 0 && at && z = 0
\end{aligned} \tag{22.215}$$

Solution for θ_2:
The boundary conditions of θ_2 on z are homogeneous and allow us to express the solution of the function θ_2 in terms of the proper trigonometric functions. Considering

the boundary conditions on $z = 0, L$ reveals that the eigenfunctions and eigenvalues for $Z(z)$ are

$$Z_k(z) = \cos(\beta_k z), \quad \beta_k = \frac{(2k-1)\pi}{2L} \quad k = 1, 2, \ldots \tag{22.216}$$

Also due to periodicity conditions the solution for $\Phi(\phi)$ may be written in the following form

$$\Phi_n(\phi) = \cos(\alpha_n \phi), \sin(\alpha_n \phi) \qquad \alpha_n = 0, 1, 2, \ldots \tag{22.217}$$

Therefore the solution for the function θ_2 may be written in the following form

$$\begin{aligned} &\theta_2(r, \phi, z) \\ &= \sum_{k=0}^{\infty} \left\{ \frac{1}{2} R_k^0(r) + \sum_{n=0}^{\infty} \left(R_{nk}^c(r) \cos(n\phi) + R_{nk}^s(r) \sin(n\phi) \right) \right\} \cos(\beta_k z) \end{aligned} \tag{22.218}$$

Substituting the Eq. (22.218) into the Eq. (22.214) gives us

$$\begin{aligned} &\left(\frac{\partial^2}{\partial r^2} + \frac{1}{r}\frac{\partial}{\partial r} - \frac{n^2}{r^2} - \beta_k^2 \right) R_{nk}^c(r) = 0 \\ &\left(\frac{\partial^2}{\partial r^2} + \frac{1}{r}\frac{\partial}{\partial r} - \frac{n^2}{r^2} - \beta_k^2 \right) R_{nk}^s(r) = 0 \\ &\left(\frac{\partial^2}{\partial r^2} + \frac{1}{r}\frac{\partial}{\partial r} - \beta_k^2 \right) R_k^0(r) = 0 \end{aligned} \tag{22.219}$$

Equations (22.219) are the well-known modified Bessel equations and their solutions are

$$\begin{aligned} R_{nk}^c(r) &= C_1 I_n(\beta_k r) + C_2 K_n(\beta_k r) \\ R_{nk}^s(r) &= C_3 I_n(\beta_k r) + C_4 K_n(\beta_k r) \\ R_k^0(r) &= C_5 I_0(\beta_k r) + C_6 K_0(\beta_k r) \end{aligned} \tag{22.220}$$

Defining

$$R_{nk}(r, \phi) = \frac{1}{2} R_k^0(r) + \sum_{n=0}^{\infty} \left(R_{nk}^c(r) \cos(n\phi) + R_{nk}^s(r) \sin(n\phi) \right) \tag{22.221}$$

results in the solution of $\theta_2(r, \phi, z)$ in the form

$$\theta_2(r, \phi, z) = \sum_{k=1}^{\infty}\sum_{n=0}^{\infty} R_{nk}(r)\cos(\beta_k z) \tag{22.222}$$

with boundary conditions (22.215) for θ_2 on $r = a, b$, we may write the next equalities

$$\begin{aligned}
&\sum_{k=1}^{\infty}\sum_{n=0}^{\infty} R_{nk}(a)\cos(\beta_k z) = \theta_0 \\
&\sum_{k=1}^{\infty}\sum_{n=0}^{\infty}(K R_{nk}^{'}(b) + hR_{nk}(b))\cos(\beta_k z) = q^{''} f(\phi)
\end{aligned} \tag{22.223}$$

when both sides of the above equations are multiplied by $Z_l(z)$ and integrated over $z = [0, L]$, the next system of equations are revealed

$$\begin{aligned}
&\sum_{n=0}^{\infty} R_{nk}(a) = \frac{2(-1)^{k+1}}{L\beta_k}\theta_0 \\
&\sum_{n=0}^{\infty}\left(K R_{nk}^{'}(b) + h R_{nk}(b)\right) = \frac{2(-1)^{k+1}}{L\beta_k} q^{''} f(\phi)
\end{aligned} \tag{22.224}$$

which also maybe interpreted in the following form according to Eq. (22.221),

$$\begin{aligned}
&\frac{1}{2}R_k^0(a) + \sum_{n=0}^{\infty}\left(R_{nk}^c(a)\cos(n\phi) + R_{nk}^s(a)\sin(n\phi)\right) = \frac{2(-1)^{k+1}}{L\beta_k}\theta_0 \\
&\frac{1}{2}\left(K R_k^{0'}(b) + h R_k^0(b)\right) \\
&+\sum_{n=0}^{\infty}\left\{\left(K R_{nk}^{c'}(b) + h R_{nk}^{c'}(b)\right)\cos(n\phi) + \left(K R_{nk}^{s'}(b) + h R_{nk}^{s}(b)\right)\sin(n\phi)\right\} \\
&= \frac{2(-1)^{k+1}}{L\beta_k} q^{''} f(\phi)
\end{aligned} \tag{22.225}$$

The coefficients of the complete Fourier series are given in the form

$$\begin{aligned}
&\frac{1}{2}R_k^0(a) = \frac{2(-1)^{k+1}}{L\beta_k}\theta_0 \\
&R_{nk}^c(a) = 0 \\
&R_{nk}^s(a) = 0 \\
&K R_k^{0'}(b) + h R_k^0(b) = \frac{2(-1)^{k+1}}{L\pi\beta_k} q^{''}\int_{-\pi}^{\pi} f(\phi)d\phi
\end{aligned}$$

$$K R_{nk}^{c'}(b) + h R_{nk}^{c}(b) = \frac{2(-1)^{k+1}}{L\pi\beta_k} q'' \int_{-\pi}^{\pi} f(\phi)\cos\phi d\phi$$
$$K R_{nk}^{s'}(b) + h R_{nk}^{s}(b) = \frac{2(-1)^{k+1}}{L\pi\beta_k} q'' \int_{-\pi}^{\pi} f(\phi)\sin\phi d\phi \tag{22.226}$$

where according to Eq. (22.220) can be written in terms of constants $C_1, \ldots, C_6$ as

$$\begin{aligned}
&\frac{1}{2}(C_5 I_0(\beta_k a) + C_6 K_0(\beta_k a)) = \frac{2(-1)^{k+1}}{L\beta_k}\theta_0 \\
&C_1 I_n(\beta_k a) + C_2 K_n(\beta_k a) = 0 \\
&C_3 I_n(\beta_k a) + C_4 K_n(\beta_k a) = 0 \\
&\left(K\beta_k I_0'(\beta_k b) + h I_0(\beta_k b)\right) C_5 \\
&\quad + \left(K\beta_k K_0'(\beta_k b) + h K_0(\beta_k b)\right) C_6 = \frac{2(-1)^{k+1}}{L\pi\beta_k} q'' \int_{-\pi}^{\pi} f(\phi) d\phi \\
&\left(K\beta_k I_n'(\beta_k b) + h I_n(\beta_k b)\right) C_1 \\
&\quad + \left(K\beta_k K_n'(\beta_k b) + h K_n(\beta_k b)\right) C_2 = \frac{2(-1)^{k+1}}{L\pi\beta_k} q'' \int_{-\pi}^{\pi} f(\phi)\cos\phi d\phi \\
&\left(K\beta_k I_n'(\beta_k b) + h I_n(\beta_k b)\right) C_3 \\
&\quad + \left(K\beta_k K_n'(\beta_k b) + h K_n(\beta_k b)\right) C_4 = \frac{2(-1)^{k+1}}{L\pi\beta_k} q'' \int_{-\pi}^{\pi} f(\phi)\sin\phi d\phi
\end{aligned} \tag{22.227}$$

The above system contains six equations and six unknowns $C_1, \ldots, C_6$. Evaluating constants $C_i, i = 1 \ldots, 6$, the solution of θ_2 can be considered to be accomplished according to Eqs. (22.218) and (22.220).

Solution for θ_1:

The boundary conditions on z allow us to express the solution of the function θ_1 in terms of the proper trigonometric functions. Considering the boundary conditions on $z = 0, L$ reveals that the eigenfunctions and eigenvalues for $Z(z)$ are

$$Z_k(z) = \cos(\beta_k z), \quad \beta_k = \frac{(2k-1)\pi}{2L} \quad k = 1, 2, \ldots \tag{22.228}$$

Also due to periodicity conditions the solution for $\Phi(\phi)$ may be written in the following form

$$\Phi_n(\phi) = \cos(\alpha_n \phi), \sin(\alpha_n \phi) \quad \alpha_n = 0, 1, 2, \ldots \tag{22.229}$$

Therefore the solution for the function θ_1 may be written in the following form

$$\theta_1(r,\phi,z,t)=\sum_{k=1}^{\infty}\left\{\sum_{n=0}^{\infty}\left(S^c_{nk}(r,t)\cos(n\phi)+S^s_{nk}(r,t)\sin(n\phi)\right)\right\}\cos(\beta_k z) \tag{22.230}$$

Substituting Eq. (22.230) into the Eq. (22.214) gives the following partial differential equation

$$\begin{aligned}\left(\frac{\partial^2}{\partial r^2}+\frac{1}{r}\frac{\partial}{\partial r}-\frac{n^2}{r^2}-\beta_k^2\right)S^c_{nk}(r,t)&=\frac{1}{\kappa}\frac{\partial S^c_{nk}(r,t)}{\partial t}\\ \left(\frac{\partial^2}{\partial r^2}+\frac{1}{r}\frac{\partial}{\partial r}-\frac{n^2}{r^2}-\beta_k^2\right)S^s_{nk}(r,t)&=\frac{1}{\kappa}\frac{\partial S^s_{nk}(r,t)}{\partial t}\end{aligned} \tag{22.231}$$

The solution of each of the above equations can be extracted via the separation of variables method. To this end, let's write

$$\begin{aligned}S^c_{nk}(r,t)&=T^c(t)U^c(r)\\ S^s_{nk}(r,t)&=T^s(t)U^s(r)\end{aligned} \tag{22.232}$$

Substituting the above equations into the Eq. (22.231), and applying the boundary conditions (22.215) on $r=a,b$ reveals the solution of $r-$dependent functions as

$$\begin{aligned}U^c(r)&=u^c_{nm}\left(J_n(\zeta_{nm}r)Y_n(\zeta_{nm}a)-J_n(\zeta_{nm}a)Y_n(\zeta_{nm}r)\right)\\ U^c(r)&=u^s_{nm}\left(J_n(\zeta_{nm}r)Y_n(\zeta_{nm}a)-J_n(\zeta_{nm}a)Y_n(\zeta_{nm}r)\right)\end{aligned} \tag{22.233}$$

where u^c_{nm} and u^s_{nm} are constants and ζ_{nm} is the mth positive real root of the characteristic equation which is deduced as

$$\begin{aligned}&\left(\frac{n}{b}+\frac{h}{K}\right)\left[J_n(\zeta_{nm}b)Y_n(\zeta_{nm}a)-J_n(\zeta_{nm}a)Y_n(\zeta_{nm}b)\right]\\ &=\zeta_{nm}\left[J_{n+1}(\zeta_{nm}b)Y_n(\zeta_{nm}a)-J_n(\zeta_{nm}a)Y_{n+1}(\zeta_{nm}b)\right]\end{aligned} \tag{22.234}$$

And the solution of the time dependent functions; $T^c(t)$ and $T^s(t)$ can be unified as $T(t)$ as

$$T(t)=e^{-\kappa(\zeta^2_{mn}+\beta_k^2)t} \tag{22.235}$$

Substituting the Eqs. (22.232), (22.233) and (22.235) into Eq. (22.230), one arrives at

$$\begin{aligned}\theta_1(r,\phi,z,t)&=\sum_{k=1}^{\infty}\sum_{m=1}^{\infty}\sum_{n=0}^{\infty}\left(u^c_{nm}\cos n\phi+u^s_{nm}\sin n\phi\right)\\ &\{J_n(\zeta_{nm}r)Y_n(\zeta_{nm}a)-J_n(\zeta_{nm}a)Y_n(\zeta_{nm}r)\}\cos(\beta_k z)e^{-\kappa(\zeta^2_{mn}+\beta_k^2)t}\end{aligned} \tag{22.236}$$

The unknown constants have to be determined according to the initial condition. Let's assume that the initial condition through the cylinder is known by a function

$\alpha(r, \phi, z)$. Therefore, we may write

$$\theta(r, \phi, z, 0) = \alpha(r, \phi, z) \tag{22.237}$$

or equivalently

$$\theta_1(r, \phi, z, 0) = -\theta_2(r, \phi, z) + \alpha(r, \phi, z) \tag{22.238}$$

The right hand side of the above equation is known. To obtain the unknown constants on the left-hand-side, both sides of the above equation are multiplied by

$$r\left\{J_n(\zeta_{nm}r)Y_n(\zeta_{nm}a) - J_n(\zeta_{nm}a)Y_n(\zeta_{nm}r)\right\}\cos(\beta_k z) = rU_{nm}(r)Z_k(z) \tag{22.239}$$

and integrated over $[0, L] \times [a, b]$. According to the orthogonality properties of the eigenfunctions, following is concluded

$$\begin{aligned}\sum_{n=0}^{\infty}&\left(u_{nm}^c\cos(n\phi) + u_{nm}^s\sin(n\phi)\right)\\ &= \frac{\int_0^L\int_a^b rU_{nm}(r)Z_k(z)(\alpha(r, \phi, z) - \theta_2(r, \phi, z))drdz}{\int_0^L\int_a^b rU_{nm}^2(r)Z_k^2(z)drdz} = \Phi_{nm}(\phi)\end{aligned} \tag{22.240}$$

which is the form of a complete Fourier series. Finally the constants u_{nm}^c and u_{nm}^s can be extracted in the next form

$$\begin{aligned}u_{0m}^c &= \frac{1}{2\pi}\int_{\pi}^{\pi}\Phi_{nm}(\phi)d\phi\\ u_{nm}^c &= \frac{1}{\pi}\int_{\pi}^{\pi}\Phi_{nm}(\phi)\cos(n\phi)d\phi, \quad n = 1, 2, \ldots\\ u_{nm}^s &= \frac{1}{\pi}\int_{\pi}^{\pi}\Phi_{nm}(\phi)\sin(n\phi)d\phi, \quad n = 1, 2, \ldots\end{aligned} \tag{22.241}$$

Having developed the coefficients u_{nm}^c and u_{nm}^s, the solution of the boundary value problem θ_1 is completed. The temperature profile through the cylinder may be assumed as the sum of θ_1 and θ_2 developed in Eqs. (22.218) and (22.236).

Chapter 23
Thermal Stresses in Beams

Beam are one of the basic elements of structural design problems. Thermal stresses in beams are discussed in this chapter. The elementary beam theory is employed and the equations for thermal stress and thermal deflection are presented.

23.1 Thermal Stresses in Beams

In accordance with Euler-Bernoulli hypothesis, a beam deflects in such a way that its plane sections remain plane after deformation and perpendicular to the beam neutral axis. Now, consider a beam under axial and lateral loads in x-y plane, as shown in Fig. 23.1 Consider two line elements of the beam, $\overline{EF}$ and $\overline{GH}$, which are straight and along the axial direction with equal lengths before the load is applied. Element $\overline{EF}$ lies on the neutral axis, while the element $\overline{GH}$ is at a distance y from the neutral axis. The beam is assumed to be under the bending and axial loads so that it deflects in lateral direction. Considering Euler-Bernoulli hypothesis, the elongation of $\overline{EF}$ and $\overline{GH}$ elements may be written as

$$\begin{aligned} \widehat{E'F'} &= (1+\epsilon_0)\overline{EF} \\ \widehat{G'H'} &= (1+\epsilon)\overline{GH} \\ \frac{\widehat{G'H'}}{\widehat{E'F'}} &= \frac{r_y+y}{r_y} \end{aligned} \tag{23.1}$$

where ϵ and ϵ_0 are strains of $\overline{GH}$ and $\overline{EF}$ elements, respectively, and r_y is the radius of curvature of the beam axis at $y = 0$ in the xy-plane. Dividing the second of Eq. (23.1) by the first equation and using the last of Eq. (23.1) gives

$$1+\frac{y}{r_y} = \frac{(1+\epsilon)}{(1+\epsilon_0)} \tag{23.2}$$

M. Reza Eslami et al., *Theory of Elasticity and Thermal Stresses*, Solid Mechanics and Its Applications 197, DOI: 10.1007/978-94-007-6356-2_23,

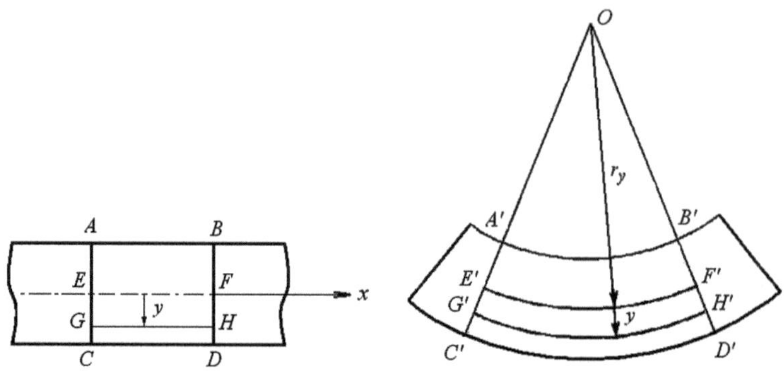

Fig. 23.1 Deflection of an element of beam

Then, using the small deformation theory, yields

$$\epsilon = \epsilon_0 + \frac{y}{r_y} + \frac{y}{r_y}\epsilon_0 \cong \epsilon_0 + \frac{y}{r_y} \tag{23.3}$$

Now, consider a beam with thermal gradients along the y and z-directions. In accordance with Euler-Bernoulli assumption, the axial displacement is a linear function of the coordinates y and z in the plane of cross section of the beam. Thus

$$u = C_1(x) + C_2(x)y + C_3(x)z \tag{23.4}$$

where $C_1,\ C_2$, and C_3 are coefficients, which are functions of x, the beam axis. Assuming thermal loading only, these coefficients may be obtained using the boundary conditions. Since the beam is in static equilibrium, the axial force and bending moments in y and z directions must vanish. These conditions in terms of the axial stress in the beam yield the following relations:

$$\int_A \sigma_{xx} dA = 0, \qquad \int_A \sigma_{xx} y dA = 0, \qquad \int_A \sigma_{xx} z dA = 0 \tag{23.5}$$

where $dA = dydz$. Equations (23.5) are sometimes called the equilibrium equations of the beam. In order to find $C_1,\ C_2$, and C_3, the axial strain is written from Eq. (23.4) as

$$\epsilon_{xx} = \frac{du}{dx} = \frac{dC_1}{dx} + \frac{dC_2}{dx}y + \frac{dC_3}{dx}z = \epsilon_0 + \frac{y}{r_y} + \frac{z}{r_z} \tag{23.6}$$

where r_y and r_z are the radii of curvatures of the beam axis in xy and xz planes, respectively, and ϵ_0 is the axial strain of the beam on the x-axis. The stress, according to Hooke's law is

$$\sigma_{xx} = E(\epsilon_{xx} - \alpha\theta) \tag{23.7}$$

where $\theta = T - T_0$. Thus

$$\sigma_{xx} = E[\epsilon_0 + \frac{y}{r_y} + \frac{z}{r_z} - \alpha\theta] \tag{23.8}$$

Substituting Eq. (23.8) in Eq. (23.5) and noting that ϵ_0, r_y, and r_z are functions of x only, they may be taken outside of the integral over the area $dy\,dz$, thus

$$\begin{aligned}
\epsilon_0 \int_A dA + \frac{1}{r_y}\int_A ydA + \frac{1}{r_z}\int_A zdA &= \int_A \alpha\theta dA \\
\epsilon_0 \int_A ydA + \frac{1}{r_y}\int_A y^2 dA + \frac{1}{r_z}\int_A yzdA &= \int_A \alpha\theta ydA \\
\epsilon_0 \int_A zdA + \frac{1}{r_y}\int_A yzdA + \frac{1}{r_z}\int_A z^2 dA &= \int_A \alpha\theta zdA
\end{aligned} \tag{23.9}$$

From the above system of equations ϵ_0, r_y, and r_z are calculated and substituted into Eq. (23.8) to obtain the axial stress.

If we select the y and z axes as the centroid axes of the cross section, then

$$\int_A ydA = \int_A zdA = 0 \tag{23.10}$$

and since, by definition, the moments of inertia and the product of inertia of the cross section are

$$\begin{aligned}
\int_A y^2 dA &= I_z \\
\int_A z^2 dA &= I_y \\
\int_A yzdA &= I_{yz}
\end{aligned} \tag{23.11}$$

then the system of Eq. (23.9) may be solved for ϵ_0, r_y, and r_z to give

$$\begin{aligned}
\epsilon_0 &= \frac{P_T}{EA} \\
r_y &= \frac{E(I_y I_z - I_{yz}^2)}{I_y M_{Tz} - I_{yz} M_{Ty}} \\
r_z &= \frac{E(I_y I_z - I_{yz}^2)}{I_z M_{Ty} - I_{yz} M_{Tz}}
\end{aligned} \tag{23.12}$$

where

$$
\begin{aligned}
P_T &= \int_A E\alpha\theta dA \\
M_{Ty} &= \int_A E\alpha\theta z dA \\
M_{Tz} &= \int_A E\alpha\theta y dA
\end{aligned}
\tag{23.13}
$$

Upon substitution of ϵ_0, r_y, and r_z into Eq. (23.8) we obtain

$$
\sigma_{xx} = -E\alpha\theta + \frac{P_T}{A} + \left(\frac{I_y M_{Tz} - I_{yz} M_{Ty}}{I_y I_z - I_{yz}^2}\right) y + \left(\frac{I_z M_{Ty} - I_{yz} M_{Tz}}{I_y I_z - I_{yz}^2}\right) z \tag{23.14}
$$

Equation (23.14) can be further simplified by taking the y and z axes in the principal directions of the cross sectional area of the beam. In this case $I_{yz} = 0$, and Eq. (23.14) simplifies to the form

$$
\sigma_{xx} = -E\alpha\theta + \frac{P_T}{A} + \frac{M_{Tz} y}{I_z} + \frac{M_{Ty} z}{I_y} \tag{23.15}
$$

Equation (23.14) gives the axial stress in a beam subjected to thermal loading when the temperature distribution is a function of y and z. To find the strains, radii of curvature and thermal stresses due to the combined mechanical and thermal loads, the thermal moments must be replaced by the total moments acting on the beam in Eqs. (23.12), (23.14) and (23.15). Also, the term P_T must be replaced by $P_T + P_M$, where P_M is the axial load due to the external forces applied on the beam and the reaction forces at the boundary. Thus, in general, when both mechanical and thermal moments are present, the relations for the axial strain, radii of curvature, and axial stress will be written in the form

$$
\begin{aligned}
\epsilon_0 &= \frac{P}{EA} \\
r_y &= \frac{E(I_y I_z - I_{yz}^2)}{I_y M_z - I_{yz} M_y} \\
r_z &= \frac{E(I_y I_z - I_{yz}^2)}{I_z M_y - I_{yz} M_z}
\end{aligned}
\tag{23.16}
$$

and

$$
\sigma_{xx} = -E\alpha\theta + \frac{P}{A} + \left(\frac{I_y M_z - I_{yz} M_y}{I_y I_z - I_{yz}^2}\right) y + \left(\frac{I_z M_y - I_{yz} M_z}{I_y I_z - I_{yz}^2}\right) z \tag{23.17}
$$

in which the total moments and the axial load are

$$
\begin{aligned}
M_y &= M_{Ty} + M_{My} \\
M_z &= M_{Tz} + M_{Mz} \\
P &= P_T + P_M
\end{aligned}
\tag{23.18}
$$

where M_T is the moment due to the thermal gradient, and M_M is the mechanical moment due to action of the external forces and the reaction forces at the boundary of the beam.

23.2 Deflection Equation of Beams

Consider a beam of arbitrary cross section as shown in Fig. 23.2. The dimensions of the beam in y and z directions are assumed to be small enough so that σ_{yy} and σ_{zz} are negligible. Denoting deflection in y and z directions by v and w, respectively, the differential equation for the beam deflection is derived as follows. From the elementary beam theory and for a small deflection, the radii of curvature are related to the deflections of the beam as

$$
\begin{aligned}
\frac{1}{r_y} &= -\frac{\dfrac{d^2v}{dx^2}}{\left[1+\left(\dfrac{dv}{dx}\right)^2\right]^{3/2}} \cong -\frac{d^2v}{dx^2} \\
\frac{1}{r_z} &= -\frac{\dfrac{d^2w}{dx^2}}{\left[1+\left(\dfrac{dw}{dx}\right)^2\right]^{3/2}} \cong -\frac{d^2w}{dx^2}
\end{aligned}
\tag{23.19}
$$

where r_y and r_z are the radii of curvature of the beam axis in xy and xz planes, respectively. Substituting the relations for radii of curvature given by Eq. (23.16) into Eq. (23.19) leads to the deflection equations of the beam

$$
\begin{aligned}
\frac{d^2v}{dx^2} &= -\frac{I_y M_z - I_{yz} M_y}{E(I_y I_z - I_{yz}^2)} \\
\frac{d^2w}{dx^2} &= -\frac{I_z M_y - I_{yz} M_z}{E(I_y I_z - I_{yz}^2)}
\end{aligned}
\tag{23.20}
$$

When both the mechanical and the thermal moments act, the total moments are

$$
\begin{aligned}
M_y &= M_{Ty} + M_{My} \\
M_z &= M_{Tz} + M_{Mz}
\end{aligned}
\tag{23.21}
$$

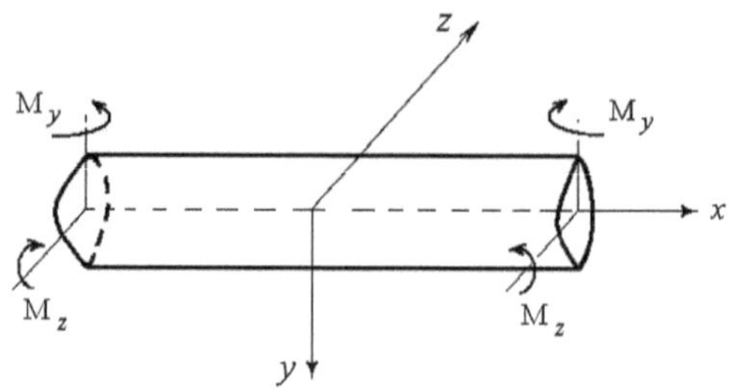

Fig. 23.2 Positive directions for the moments applied on the beam

where M_T is the moment due to the thermal gradient, and M_M is the mechanical moment due to the action of the external forces and the reaction forces at the boundary of the beam.

When the principal directions of the cross section of the beam are selected as the y and z axes, $I_{yz} = 0$ and Eq. (23.20) simplify to the form

$$\frac{d^2v}{dx^2} = -\frac{M_z}{EI_z}$$
$$\frac{d^2w}{dx^2} = -\frac{M_y}{EI_y} \tag{23.22}$$

In addition to the lateral deflections of the beam, the axial displacement of the beam may be of interest. The integration of Eq. (23.6) with respect to x, using the relations for the axial strain and radii of curvature given by Eq. (23.16), results in an expression for the axial displacement as

$$u = u_0 + \int_0^x \left\{ \frac{P}{EA} + \left(\frac{I_y M_z - I_{yz} M_y}{E(I_y I_z - I_{yz}^2)} \right) y + \left(\frac{I_z M_y - I_{yz} M_z}{E(I_y I_z - I_{yz}^2)} \right) z \right\} dx \tag{23.23}$$

where u_0 is the axial displacement at $x = 0$. Sometimes the average axial displacement for the cross section of the beam is of interest. The average of the axial displacement for the cross section is obtained from Eq. (23.23) as

$$\begin{aligned} u_{av} =& \frac{1}{A} \int_A u \, dA \\ =& \frac{1}{A} \int_A u_0 \, dA + \int_0^x \frac{P}{EA} dx + \int_0^x \left\{ \left(\frac{I_y M_z - I_{yz} M_y}{E(I_y I_z - I_{yz}^2)} \right) \left(\int_A y \, dA \right) \right. \\ & \left. + \left(\frac{I_z M_y - I_{yz} M_z}{E(I_y I_z - I_{yz}^2)} \right) \left(\int_A z \, dA \right) \right\} dx \end{aligned} \tag{23.24}$$

Since the relations for the axial strain and radii of curvature given by Eq. (23.16) are obtained when the y and z axes are the centroid axes of the cross section, Eq. (23.24) reduces to

$$u_{av} = \frac{1}{A}\int_A u_0 \, dA + \int_0^x \frac{P}{EA} dx \tag{23.25}$$

It may be found from this equation that when a cantilever beam of constant cross section and length L is exposed to a thermal gradient and axial mechanical loads, the total average displacement of the free end of the beam is

$$u_{av} = \frac{PL}{EA} \tag{23.26}$$

and when the temperature rise, θ, is uniformly distributed through the cross section of the cantilever beam, Eq. (23.26) reduces to

$$u_{av} = \alpha\theta L + \frac{P_M L}{EA} \tag{23.27}$$

which is identical to the elongation of a bar under a uniform temperature rise θ and the axial mechanical load P_M.

23.3 Boundary Conditions

Consider a beam of arbitrary cross section subjected to both mechanical and thermal loads. Take the x-axis along the axis of the beam. The following boundary conditions may exist at any end of the beam:

1- Simply supported end Let assume that the end $x = L$ of the beam is simply supported in y direction. Thus, the deflection and the moment at this end must be zero, that is

$$w|_{x=L} = 0 \qquad M_{My}|_{x=L} = \left(-EI_y \frac{d^2w}{dx^2} - M_{Ty}\right)_{x=L} = 0 \tag{23.28}$$

2- Built-in end At the built-in end the deflection and the slope of the beam must be zero. Thus, if the end $x = L$ is assumed to be built-in, it follows that

$$w|_{x=L} = 0 \qquad \frac{dw}{dx}|_{x=L} = 0 \tag{23.29}$$

3- Free end At the free end the moment and the shear force must be zero, thus, if the end $x = L$ is assumed to be free, it follows that

$$\begin{aligned} M_{My}|_{x=L} &= \left(-EI_y \frac{d^2w}{dx^2} - M_{Ty}\right)_{x=L} = 0 \\ (Q_z)_{x=L} &= \left(\frac{dM_{My}}{dx}\right)_{x=L} = -\frac{d}{dx}\left(EI_y \frac{d^2w}{dx^2}\right)_{x=L} - \left(\frac{dM_{Ty}}{dx}\right)_{x=L} = 0 \end{aligned} \tag{23.30}$$

23.4 Problems and Solutions of Beams

Problem 23.1. Consider a rectangular cross section beam ($b \times a$) of length L. The temperature distribution across the beam thickness is given below.

$$T(y,t) = T_0 + \frac{aq''}{24K}\left[\frac{24\kappa t}{a^2} + \frac{12y^2}{a^2} + \frac{12y}{a} - 1 \right.$$
$$\left. - \frac{48}{\pi^2}\sum_{n=1}^{\infty}\frac{(-1)^n}{n^2}e^{-\frac{n^2\pi^2\kappa t}{a^2}}\cos\frac{n\pi(2y+a)}{2a}\right] \tag{23.31}$$

in which κ is thermal diffusivity and K is the thermal conductivity. Calculate the transverse shear stress distribution in the beam cross section if two ends are simply supported.

Solution: The temperature distribution through the beam is given by Eq. (23.31). Next, thermal force and thermal moment through the beam can be obtained by the following equations

$$P_T = \int_A E\alpha(T - T_0)dA$$
$$M_{Tz} = \int_A yE\alpha(T - T_0)dA \tag{23.32}$$

where $dA = bdy$. When Eq. (23.31) is substituted into Eq. (23.32), thermal moment and thermal force resultants are obtained as

$$P_T = \frac{E\alpha A q'' \kappa t}{aK}$$
$$M_{Tz} = \frac{E\alpha q'' I_z}{2K}\left(1 - \frac{96}{\pi^4}\sum_{n=1,3,5,\ldots}^{\infty}\frac{1}{n^4}e^{-\frac{n^2\pi^2\kappa t}{a^2}}\right) \tag{23.33}$$

The deflected shape of the beam which is subjected to the elevated temperature is governed by the following equation

$$\frac{d^2v}{dx^2} = -\frac{M_{Tz} + M_{0z} + Q_0 x}{EI_z} \tag{23.34}$$

where M_{0z} and Q_0 are the bending moment and shear force at the left boundary of the beam. Boundary conditions for a beam which is simply supported at both ends are

$$v(0) = v(L) = \frac{d^2v(0)}{dx^2} + \frac{M_{Tz}}{EI_z} = \frac{d^2v(L)}{dx^2} + \frac{M_{Tz}}{EI_z} = 0 \tag{23.35}$$

Solving the ordinary differential Eq. (23.34) with boundary conditions (23.35) reveals that the shear force and bending moment at the edge $x = 0$ are zero and therefore total axial stress of the beam is equal to

$$\sigma_{xx} = -E\alpha(T - T_0) + \frac{P_T}{A} + \frac{M_{Tz}y}{I_z} \tag{23.36}$$

Substituting Eqs. (23.31) and (23.33) into the above equation gives the expression for axial stress of the beams as

$$\sigma_{xx} = \frac{E\alpha a q''}{2K}\left[\left(\frac{1}{12} - \frac{y^2}{a^2}\right) + \frac{4}{\pi^2}\sum_{n=1}^{\infty}\frac{(-1)^n}{n^2}e^{-\frac{n^2\pi^2\kappa t}{a^2}}\cos\frac{n\pi(2y+a)}{2a}\right.$$
$$\left. - \frac{96y}{\pi^4 a}\sum_{n=1,3,5,\ldots}^{\infty}\frac{1}{n^4}e^{-\frac{n^2\pi^2\kappa t}{a^2}}\right] \tag{23.37}$$

When the axial stress through the beam is known, the shear stress may be evaluated by

$$\sigma_{xy} = -\int_y^{a/2}\frac{\partial\sigma_{xx}}{\partial x}dy \tag{23.38}$$

As seen, no shear stress is produced through the beam since the normal stress is not a function of x component.

Problem 23.2. A beam of rectangular cross section ($b \times a$) is exposed to a constant heat flux q'' on the top surface ($y = a/2$) and the to convective heat transfer at the bottom surface ($y = -a/2$) by (h, T_∞). The heat is generated in the beam material at the rate of R per unit volume. The ends of the beam are simply supported. Obtain:

(a) Temperature distribution
(b) Thermal stresses
(c) Thermal deflection
(d) The transverse shear stress across the thickness

The thermal boundary conditions make the sides of the beam insulated so that the temperature variation is only in y-direction.

Solution:

(a): Assuming the one-dimensional heat conduction in y direction, heat conduction equation and the associated boundary conditions become

$$
\begin{aligned}
&K\frac{d^2T}{dy^2}+R=0\\
&K\frac{dT(a/2)}{dy}=q''\\
&K\frac{dT(-a/2)}{dy}=h\left(T(-a/2)-T_\infty\right)
\end{aligned}
\tag{23.39}
$$

Solving the above second order differential equation and employing the related boundary conditions gives the temperature distribution across the thickness as

$$
T(y)=-\frac{1}{2K}Ry^2+\frac{1}{K}\left(q''+\frac{1}{2}Ra\right)y+R\left(\frac{3a^2}{8K}+\frac{a}{h}\right)+q''\left(\frac{1}{h}+\frac{a}{2K}\right)+T_\infty
\tag{23.40}
$$

(**b**): Considering α and E as thermal expansion coefficient and elasticity module of the beam, the thermal force becomes

$$
\begin{aligned}
P_T&=\int_A E\alpha(T(y)-T_0)dA=bE\alpha\int_{-\frac{a}{2}}^{\frac{a}{2}}(T(y)-T_0)dy\\
&=EA\alpha\left(\frac{1}{3K}Ra^2+\frac{1}{h}\left(Ra+q''+\frac{1}{2K}q''ah\right)+T_\infty-T_0\right)
\end{aligned}
\tag{23.41}
$$

and the thermal moment is equal to

$$
\begin{aligned}
M_{Tz}&=\int_A E\alpha y(T(y)-T_0)dA=bE\alpha\int_{\frac{-a}{2}}^{\frac{a}{2}}y(T(y)-T_0)dy\\
&=\frac{1}{K}E\alpha I_z\left(q''+\frac{1}{2}Ra\right)
\end{aligned}
\tag{23.42}
$$

The deflected shape of the beam which is subjected to the elevated temperature is governed by the following equation

$$
\frac{d^2v}{dx^2}=-\frac{M_{Tz}+M_{0z}+Q_0x}{EI_z}
\tag{23.43}
$$

where M_{0z} and Q_0 are the bending moment and shear force at the left boundary of the beam. Boundary conditions for a beam which is simply supported at both ends are

$$
v(0)=v(L)=\frac{d^2v(0)}{dx^2}+\frac{M_{Tz}}{EI_z}=\frac{d^2v(L)}{dx^2}+\frac{M_{Tz}}{EI_z}=0
\tag{23.44}
$$

Solving the ordinary differential Eq. (23.43) with boundary conditions (23.44) reveals that the shear force and bending moment at the edge $x=0$ are zero and therefore total axial stress of the

$$\sigma_{xx} = -E\alpha(T - T_0) + \frac{P_T}{A} + \frac{M_{Tz}y}{I_z} \tag{23.45}$$

After simplifications the normal stress reduces to

$$\sigma_{xx} = E\alpha\frac{Ra^2}{8K}\left(\left(\frac{2y}{a}\right)^2 - \frac{1}{3}\right) \tag{23.46}$$

(**c**): The governed lateral deflection and associated mechanical boundary conditions are

$$\begin{aligned} \frac{d^2v}{dx^2} &= -\frac{M_{Tz} + M_{0z} + Q_0 x}{EI_z} \\ v(0) = v(L) &= \frac{d^2v(0)}{dx^2} + \frac{M_{Tz}}{EI_z} = \frac{d^2v(L)}{dx^2} + \frac{M_{Tz}}{EI_z} = 0 \end{aligned} \tag{23.47}$$

Here, M_{0x} and Q_0 are the mechanical moment and shear reaction at $x = 0$, respectively. Solving this second order differential equation and applying the boundary conditions gives the lateral deflection of the beam as

$$v(x) = \frac{\alpha}{2K}\left(q'' + \frac{Ra}{2}\right)(xL - x^2) \tag{23.48}$$

(**d**): When axial stress is known through the beam, the shear stress may be evaluated as

$$\sigma_{xy} = -\int_y^{a/2} \frac{\partial \sigma_{xx}}{\partial x} dy \tag{23.49}$$

As seen, the shear stress vanishes through the beam. The reason is that the normal stress is independent of the variable x.

Problem 23.3. The beam of previous section is reconsidered. The initial temperature at $t = 0$ is T_0. At this instant of time, the heat is generated at the rate of R per unit volume and unit time. The side at $z = \pm b/2$ are thermally insulated while the top and bottom surfaces at $y = \pm a/2$ are exposed to ambient at (h, T_∞). Both ends of the beam are clamped. Find:

(a) Temperature distribution
(b) Thermal stresses
(c) Thermal deflection
(d) The transverse shear stress across the thickness

Solution:

(**a**): Defining $\theta = T - T_\infty$, the one-dimensional transient heat conduction equation including heat generation becomes

$$K\frac{\partial^2\theta}{\partial y^2} + R = \rho c\frac{\partial\theta}{\partial t} \tag{23.50}$$

and the associated initial and boundary conditions are

$$\begin{aligned} -K\frac{\partial\theta(a/2,t)}{\partial y} &= h\theta(a/2,t) \\ K\frac{\partial\theta(-a/2,t)}{\partial y} &= h\theta(-a/2,t) \\ \theta(y,0) &= \theta_0 = T_0 - T_\infty \end{aligned} \tag{23.51}$$

As a solution, we let $\theta(y, t) = \psi(y, t) + \varphi(y)$. In this case the governing equation for $\varphi(y)$ and the associated boundary conditions become

$$\begin{aligned} K\frac{\partial^2\varphi}{\partial y^2} + R &= 0 \\ -K\frac{d\varphi(a/2)}{dy} &= h\varphi(a/2) \\ K\frac{d\varphi(-a/2)}{dy} &= h\varphi(-a/2) \end{aligned} \tag{23.52}$$

Solving Eq. (23.52) provides

$$\varphi(y) = \frac{R}{2K}\left(\frac{a^2}{4} + \frac{aK}{h} - y^2\right) \tag{23.53}$$

Defining $\kappa = \dfrac{K}{\rho c}$ the governed equation for $\psi(y, t)$ and its initial and boundary conditions are

$$\begin{aligned} \kappa\frac{\partial^2\psi}{\partial y^2} &= \frac{\partial\psi}{\partial t} \\ \psi(y,0) &= \theta_0 - \varphi(y) \\ -K\frac{\partial\psi(a/2,t)}{\partial y} &= h\psi(a/2,t) \\ K\frac{\partial\psi(-a/2,t)}{\partial y} &= h\psi(-a/2,t) \end{aligned} \tag{23.54}$$

Following the method of separation of variables, the solution for $\psi(y, t)$ becomes

$$\psi(y,t) = \sum_{n=0}^{\infty} B_n e^{-\lambda_n^2\kappa t}\left(\sin\lambda_n(a/2 - y) + \frac{K\lambda_n}{h}\cos\lambda_n(a/2 - y)\right) \tag{23.55}$$

where λ_n is the nth positive real root of the equation

$$\tan(a\lambda) = \frac{2\frac{h}{K}\lambda}{\lambda^2 - \frac{h^2}{K^2}}. \tag{23.56}$$

Applying the initial condition (23.54) to the Eq. (23.55) leads us to

$$\sum_{n=0}^{\infty} B_n \left(\sin \lambda_n(a/2 - y) + \frac{K\lambda_n}{h} \cos \lambda_n(a/2 - y) \right) = \theta_0 - \varphi(y) \tag{23.57}$$

The unknown constants B_n have to be obtained by means of orthogonality condition of the eigenfunctions. To this end both sides of the above equation are multiplied by

$$\sin \lambda_n(a/2 - y) + \frac{K\lambda_n}{h} \cos \lambda_n(a/2 - y) \tag{23.58}$$

and integrated over $[-a/2, a/2]$ which results in

$$B_n = \frac{\int_{-\frac{a}{2}}^{\frac{a}{2}} (\theta_0 - \varphi(y)) \left(\sin \lambda_n(a/2 - y) + \frac{K\lambda_n}{h} \cos \lambda_n(a/2 - y) \right) dy}{\int_{-\frac{a}{2}}^{\frac{a}{2}} \left(\sin \lambda_n(a/2 - y) + \frac{K\lambda_n}{h} \cos \lambda_n(a/2 - y) \right)^2 dy} \tag{23.59}$$

(b): Considering α and E as thermal expansion coefficient and elasticity modulus of the beam, respectively. The thermal force become

$$\begin{aligned} P_T &= \int_A E\alpha(T(y,t) - T_0) dA \\ &= bE\alpha \int_{-\frac{a}{2}}^{\frac{a}{2}} (\psi(y,t) + \varphi(y) + T_\infty - T_0) dy \\ &= E\alpha A \left(T_\infty - T_0 + \frac{Ra}{2K} \left(\frac{K}{h} + \frac{a}{6} \right) + \right. \\ &= \left. \sum_{n=0}^{\infty} B_n \frac{K\lambda \sin(a\lambda_n) + h - h\cos(a\lambda_n)}{ah\lambda_n} e^{-\lambda_n^2 \kappa t} \right) \end{aligned} \tag{23.60}$$

and thermal moment is equal to

$$M_{Tz} = \int_A Ey\alpha(T(y,t) - T_0) dA$$

$$
\begin{aligned}
&= bE\alpha \int_{\frac{-a}{2}}^{\frac{a}{2}} y(\psi(y,t) + \varphi(y) + T_\infty - T_0)dy \\
&= E\alpha I_z \sum_{n=0}^{\infty} \left\{ \frac{-6B_n}{a^3\lambda_n^2 h} e^{-\lambda_n^2 \kappa t} \times \right. \\
&\quad \left. ((2h + \lambda_n^2 aK)\sin(a\lambda_n) + \lambda_n(2K - ah)\cos(a\lambda_n) - 2K\lambda_n - \lambda_n ah) \right\}
\end{aligned} \tag{23.61}
$$

For beams with both ends clamped, the governing equation for deflected shape and boundary conditions are

$$
\begin{aligned}
\frac{d^2v}{dx^2} &= -\frac{M_{Tz} + M_{0z} + Q_0 x}{EI_z} \\
v(0) = v(L) &= \frac{dv(0)}{dx} = \frac{dv(L)}{dx} = 0
\end{aligned} \tag{23.62}
$$

Here, M_{0x} and Q_0 are the mechanical moment and shear reaction at $x = 0$. Solving this second order differential equation and applying the boundary conditions reveals that the produced mechanical moment at $x = 0$ is equal to $-M_{Tz}$ and the shear force vanishes at $x = 0$. Therefore, when beam is free to expand, the normal stress which is produced due to elevated temperature is

$$
\sigma_{xx} = -E\alpha(T - T_0) + \frac{P_T}{A} \tag{23.63}
$$

After some simplifications, the normal stress reduces to

$$
\begin{aligned}
\sigma_{xx} =& E\alpha \frac{Ra^2}{8K}\left(\left(\frac{2y}{a}\right)^2 - \frac{1}{3}\right) \\
&+ E\alpha \sum_{n=0}^{\infty} B_n e^{-\lambda_n^2 \kappa t} \left\{ \sin\lambda_n(y - a/2) - \frac{K\lambda_n}{h}\cos\lambda_n(a/2 - y) \right. \\
&\left. + \frac{K\lambda_n \sin(a\lambda_n) + h - h\cos(a\lambda_n)}{ah\lambda_n} \right\}
\end{aligned} \tag{23.64}
$$

(c): The governed lateral deflection equation and associated mechanical boundary conditions are

$$
\frac{d^2v}{dx^2} = -\frac{M_{Tz} + M_{0z} + Q_0 x}{EI_z} \tag{23.65}
$$

$$
v(0) = v(L) = \frac{dv(0)}{dx} = \frac{dv(L)}{dx} = 0 \tag{23.66}
$$

Here, M_{0x} and Q_0 are the mechanical moment and shear reaction at $x = 0$. Solving this second order differential equation and applying the boundary conditions gives the lateral deflection of the beam as

$$v(x) = 0 \tag{23.67}$$

which means that the beam remains flat.

(d): When axial stress is known through the beam, the shear stress may be evaluated as

$$\sigma_{xy} = -\int_y^{a/2} \frac{\partial \sigma_{xx}}{\partial x} dy \tag{23.68}$$

As seen, shear stress vanishes through the beam, because normal stress is independent of the variable x.

Problem 23.4. Obtain the general expression for the axial thermal stresses of a beam, if the cross section is circular.

Solution: The general form of axial thermal stress for an arbitrary cross section beam is

$$\sigma_{xx} = -E\alpha\theta + \frac{P}{A} + \frac{I_y M_z - I_{yz} M_y}{I_y I_z - I_{yz}^2} y + \frac{I_z M_z - I_{yz} M_z}{I_y I_z - I_{yz}^2} z \tag{23.69}$$

Here, $P = P_T + P_M$, $M_z = M_{Tz} + M_{Mz}$, and $M_y = M_{Ty} + M_{My}$. A subscript M indicates the mechanical reactions which are produced due to the external forces/moments or the reactions of the boundary conditions. By taking y and z axis as the principal directions of the circular cross section $I_{yz} = 0$ and $I_y = I_z = \frac{\pi}{4} a^4$. In the polar coordinates, $y = r\cos\phi$ and $z = r\sin\phi$, (see Fig. 23.3) therefore thermal force and moments are evaluated as

$$\begin{aligned}
P_T &= \int_A E\alpha\theta dA = \int_0^{2\pi}\int_0^a E\alpha r\theta(r,\phi) dr d\phi \\
M_{Ty} &= \int_A E\alpha\theta z dA = \int_0^{2\pi}\int_0^a E\alpha r^2 \sin\phi\,\theta(r,\phi) dr d\phi \\
M_{Tz} &= \int_A E\alpha\theta y dA = \int_0^{2\pi}\int_0^a E\alpha r^2 \cos\phi\,\theta(r,\phi) dr d\phi
\end{aligned} \tag{23.70}$$

Thus, the total axial stress produced due to the thermal loading becomes

$$\begin{aligned}
\sigma_{xx} = &- E\alpha\theta(r,\phi) + \frac{1}{\pi a^2}\int_0^{2\pi}\int_0^a E\alpha r\theta(r,\phi) dr d\phi \\
&+ \frac{4r\sin\phi}{\pi a^4}\int_0^{2\pi}\int_0^a E\alpha r^2 \sin\phi\,\theta(r,\phi) dr d\phi
\end{aligned}$$

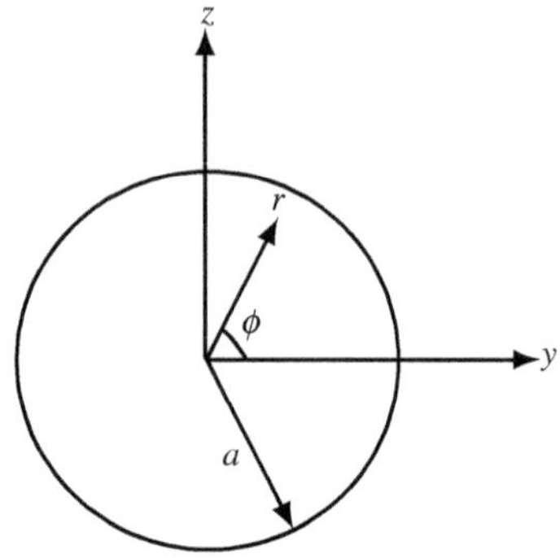

Fig. 23.3 Circular cross section beam

$$+\frac{4r\cos\phi}{\pi a^4}\int_0^{2\pi}\int_0^a E\alpha r^2\cos\phi\theta(r,\phi)drd\phi$$
$$+\frac{P_M}{\pi a^2}+\frac{4M_{Mz}r\cos\phi}{\pi a^4}+\frac{4M_{My}r\sin\phi}{\pi a^4} \quad (23.71)$$

As an especial case, considering constant thermo-elastic properties and axisymmetric temperature loading $\theta = \theta(r)$, thermal moments vanish along both y and z directions and the total axial thermal stress reduces to

$$\sigma_{xx} = -E\alpha\theta(r)+\frac{2}{a^2}\int_0^a E\alpha r\theta(r)dr$$
$$+\frac{P_M}{\pi a^2}+\frac{4M_{Mz}r\cos\phi}{\pi a^4}+\frac{4M_{My}r\sin\phi}{\pi a^4} \quad (23.72)$$

Problem 23.5. A beam of the height $2a$ and width b is considered. The initial temperature of the beam is T_0. The beam is suddenly exposed to a rate of heat flux q'' at the top surface, while the bottom surface is in equilibrium with the ambient through the convective heat transfer at (h, T_∞). The side surfaces of the beam are thermally insulated. Obtain the expression for thermal stresses.

Solution: Assuming the one-dimensional heat conduction, the heat conduction equation and associated initial and boundary conditions may be written in the form

$$\frac{\partial^2 T}{\partial u^2} = \frac{1}{\kappa}\frac{\partial T}{\partial t}$$
$$T(u,0) = T_0, \qquad 0 \le u \le 2a$$
$$K\frac{\partial T(2a,t)}{\partial t} = q''$$
$$K\frac{\partial T(0,t)}{\partial t} = h(T(0,t)-T_\infty) \quad (23.73)$$

where K is thermal conductivity and κ is defined as $\kappa = \dfrac{K}{\rho c}$. Substituting $\theta = T - T_0$, this equation transforms to

$$\begin{aligned} \frac{\partial^2\theta}{\partial u^2} &= \frac{1}{\kappa}\frac{\partial\theta}{\partial t} \\ \theta(u,0) &= \theta_0 = T_0 - T_\infty \\ K\frac{\partial\theta(2a,t)}{\partial u} &= q^{''} \\ K\frac{\partial\theta(0,t)}{\partial u} &= h\theta(0,t) \end{aligned} \tag{23.74}$$

As a solution for $\theta(u,t)$, let us define $\theta(u,t) = \theta_1(u,t) + F(u)$. The heat conduction equation is divides into two differential equations for $F(u)$ and $\theta_1(u,t)$. The ordinary differential equation for $F(u)$ and the associated boundary conditions are

$$F^{''}(u) = 0, \qquad KF^{'}(0) - hF(0) = 0, \qquad KF^{'}(2a) - q^{''} \tag{23.75}$$

Solving this second order differential equation provides

$$F(u) = \frac{1}{K}q^{''}\left(u + \frac{K}{h}\right) \tag{23.76}$$

The governing equation and the initial and boundary conditions for $\theta_1(u,t)$ are

$$\begin{aligned} \frac{\partial^2\theta_1}{\partial u^2} &= \frac{1}{\kappa}\frac{\partial\theta_1}{\partial t} \\ \theta(u,0) &= \theta_0 - F(u) \\ K\frac{\partial\theta_1(2a,t)}{\partial u} &= 0 \\ K\frac{\partial\theta_1(0,t)}{\partial u} &= h\theta(0,t) \end{aligned} \tag{23.77}$$

To solve the function $\theta_1(u,t)$, the method of separation of variables is adopted. For this purpose, the function $\theta_1(u,t)$ is assumed as $\theta_1(u,t) = U(u)T(t)$. With the aid of this transformation, Eq. (23.77) simplifies to two ordinary differential equation as

$$\begin{aligned} &U^{''}(u) + \lambda^2 U(u) = 0, \qquad U^{'}(0) - \frac{h}{K}U(0) = 0, \qquad U^{'}(2a) = 0 \\ &T^{'}(t) + \kappa\lambda^2 T(t) = 0 \end{aligned} \tag{23.78}$$

The above eigenvalue problem provides $U_n(u) = \cos\lambda_n(u - 2a)$ and $T_n(t) = e^{-\lambda_n^2\kappa t}$, where λ_n is the nth positive real root of the equation $\lambda\tan(2a\lambda) = \dfrac{h}{K}$. Finally, the general expression for $\theta_1(u, t)$ may be written in the following form

$$\theta_1(u, t) = \sum_{n=1}^{\infty} A_n e^{-\lambda_n^2 \kappa t} \cos \lambda_n (u - 2a) \tag{23.79}$$

The constants A_n has to be obtained using the initial condition $\theta_1(u, 0) = \theta_0 - F(u)$. Applying the initial condition gives

$$\theta_0 - F(u) = \sum_{n=1}^{\infty} A_n \cos \lambda_n (u - 2a) \tag{23.80}$$

Recalling the orthogonality property of the eigenfunctions, multiplying both side of this equation by $\cos \lambda_m (u - 2a)$ and integrating over the interval $[0, 2a]$ gives the constants A_n as

$$A_n = \frac{\int_0^{2a} (\theta_0 - F(u)) \cos \lambda_n (u - 2a) du}{\int_0^{2a} (\cos \lambda_n (u - 2a))^2 \, du} \tag{23.81}$$

The constant A_n is obtained after substituting $F(u) = \frac{1}{K} q'' \left(u + \frac{K}{h} \right)$. With some simplifications, we obtain

$$A_n = \frac{\lambda_n \left(\theta_0 - \frac{1}{h} q'' \right) \sin(2a\lambda_n) - \frac{1}{K} q'' (1 - \cos(2a\lambda_n))}{a\lambda_n^2 + \frac{h}{2K} \cos^2(2a\lambda_n)} \tag{23.82}$$

For simplicity, a new coordinate $y = u + a$ is introduced. The solution of $T(y, t)$ may be obtained as

$$T(y, t) = \sum_{n=1}^{\infty} A_n e^{-\lambda_n^2 \kappa t} \cos \lambda_n (y - a) + \frac{1}{K} q'' \left(y + a + \frac{K}{h} \right) + T_\infty \tag{23.83}$$

Now, to obtain the thermal stress through the beam, the boundary conditions should be known. Let us assume a beam with both edges simply supported. From the definition of thermal force we have

$$\begin{aligned} P_T &= \int_A E\alpha (T(y, t) - T_0) dA = E\alpha b \int_{-a}^{a} (T(y, t) - T_0) dy \\ &= E\alpha A \left\{ \sum_{n=1}^{\infty} \frac{1}{2a\lambda_n} A_n e^{-\lambda_n^2 \kappa t} \sin(2a\lambda_n) + \frac{a}{K} q'' + \frac{1}{h} q'' - \theta_0 \right\} \end{aligned} \tag{23.84}$$

and for thermal moment

$$M_{Tz} = \int_A E\alpha y(T(y,t) - T_0)dA = E\alpha b \int_{-a}^{a} y(T(y,t) - T_0)dy$$
$$= E\alpha I_z \left\{ \frac{1}{K} q'' + \frac{3}{2a^3} \sum_{n=1}^{\infty} \frac{1}{\lambda_n^2} (1 - a\lambda_n \sin(2a\lambda_n) - \cos(2a\lambda_n)) e^{-\lambda_n^2 \kappa t} \right\} \tag{23.85}$$

In these equations $A = 2ab$ and $I_z = \frac{2}{3} ba^3$. Now the boundary reactions have to be obtained by means of solving the curve shape equation of the beam and associated boundary conditions. For a beam with both ends simply supported, we have

$$\frac{d^2 v}{dx^2} = -\frac{M_{Tz} + M_{0z} + Q_0 x}{EI_z}$$
$$v(0) = v(L) = \frac{d^2 v(0)}{dx^2} + \frac{M_{Tz}}{EI_z} = \frac{d^2 v(L)}{dx^2} + \frac{M_{Tz}}{EI_z} = 0 \tag{23.86}$$

Here, M_{0x} and Q_0 are the mechanical moment and shear reaction at $x = 0$, respectively. Solving this second order differential equation and applying the boundary conditions gives $M_{0z} = Q_0 = 0$ and therefore total stress is calculated as

$$\sigma_{xx} = -E\alpha(T - T_0) + \frac{P_T}{A} + \frac{M_{Tz} y}{I_z} \tag{23.87}$$

Substituting expressions for P_T and M_{Tz} into the Eq. (23.87) gives the normal stress as

$$\sigma_{xx} = E\alpha \sum_{n=1}^{\infty} \left\{ \frac{q''}{K} \left(y + a + \frac{K}{h} \right) + A_n \left[\frac{\sin(2a\lambda_n)}{2a\lambda_n} - \cos \lambda_n (y - a) \right. \right.$$
$$\left. \left. + \frac{3y}{2a^3 \lambda_n^2} (1 - \cos(2a\lambda_n) - a\lambda_n \sin(2a\lambda_n)) \right] \right\} e^{-\lambda_n^2 \kappa t} \tag{23.88}$$

When the normal stress through the beam is known, the shear stress may be obtained with the following equation

$$\sigma_{xy} = -\int_y^a \frac{\partial \sigma_{xx}}{\partial x} dy \tag{23.89}$$

As seen, shear stress vanishes through the beam, because normal stress is independent of the variable x.

Problem 23.6. A beam of the height $2a$ and width b is generating heat through its body at the rate given by

$$R(z,t) = A(t) \sin(\eta z)$$

where η is constant. The beam is initially at a reference temperature T_0. Thermal boundary conditions are convective heat transfer from the top and bottom surfaces, where the side surfaces are thermally insulated. Find the thermal stress in the beam.

Solution: Assuming the one-dimensional heat conduction, the heat conduction equation and associated initial and boundary conditions may be written in the form

$$\begin{aligned}\frac{\partial^2 T}{\partial z^2} &= -\frac{1}{K}R(z,t) + \frac{1}{\kappa}\frac{\partial T}{\partial t}\\ T(z,0) &= T_0\\ -K\frac{\partial T(a,t)}{\partial t} &= h(T(a,t) - T_\infty)\\ K\frac{\partial T(-a,t)}{\partial t} &= h(T(-a,t) - T_\infty)\end{aligned} \tag{23.90}$$

where K is thermal conductivity and κ is defined as $\kappa = \dfrac{K}{\rho c}$. Substituting $\theta = T - T_0$, the upper boundary value problem transforms to

$$\begin{aligned}\frac{\partial^2 \theta}{\partial z^2} &= -\frac{1}{K}R(z,t) + \frac{1}{\kappa}\frac{\partial \theta}{\partial t}\\ \theta(z,0) &= \theta_0 = T_0 - T_\infty\\ -K\frac{\partial \theta(a,t)}{\partial u} &= h\theta(a,t)\\ K\frac{\partial \theta(-a,t)}{\partial u} &= h\theta(-a,t)\end{aligned} \tag{23.91}$$

To solve the system of boundary value problem (23.91), we should find the eigenvalues and eigenfunctions of the corresponding homogeneous equation. Considering the homogeneous part of the equation for θ and following the method of separation of variables, the second order differential equation and boundary conditions for the z domain become

$$\begin{aligned}&Z'' + \mu^2 Z = 0\\ &Z'(a) + \frac{h}{K}Z(a) = 0, \qquad Z'(-a) + \frac{h}{K}Z(-a) = 0\end{aligned} \tag{23.92}$$

where μ is eigenvalue. Solving Eq. (23.92), the eigenfunctions $Z_n(z)$ are obtained as

$$Z_n(z) = h_1 \sin \mu_n(z+a) + \mu_n \cos \mu_n(z+a) \tag{23.93}$$

where μ_n is the nth positive real root of the equation $\tan(2a\mu) = \dfrac{2\mu h_1}{\mu^2 - h_1^2}$ and h_1 is defined as $h_1 = \dfrac{h}{K}$. Now as a solution for the non-homogeneous transient equation for θ, the next series expansion is considered

$$\begin{aligned}\theta(z,t) &= \sum_{n=1}^{\infty} B_n(t) Z_n(z)\\ &= \sum_{n=1}^{\infty} B_n(t)(h_1 \sin\mu_n(z+a) + \mu_n \cos\mu_n(z+a))\end{aligned} \tag{23.94}$$

Here, the constants B_n are time dependent coefficients which have to be evaluated using the initial condition. The above solution satisfies both of the boundary conditions in Eq. (23.90), while the governing equation and initial condition are not satisfied yet. To obtain the time-dependent functions B_n, the function $\sin(\eta z)$ has to be expanded by means of the orthogonal functions $Z_n(z)$ in the form

$$\begin{aligned}\sin(\eta z) &= \sum_{n=1}^{\infty} P_n(t) Z_n(z)\\ &= \sum_{n=1}^{\infty} P_n(t)(h_1 \sin\mu_n(z+a) + \mu_n \cos\mu_n(z+a))\end{aligned} \tag{23.95}$$

The constants P_n may be easily obtained by means of the orthogonality property of the eigenfunctions. To this end, Eq. (23.95) is multiplied by $Z_m(z)$ and integrated over the interval $z[-a, a]$ which results in

$$P_n = \frac{\int_{-a}^{a} \sin(\eta z)(h_1 \sin\mu_n(z+a) + \mu_n \cos\mu_n(z+a))dz}{\int_{-a}^{a}(h_1 \sin\mu_n(z+a) + \mu_n \cos\mu_n(z+a))^2 dz} \tag{23.96}$$

After integration and proper mathematical simplifications, the constants P_n are obtained as

$$\begin{aligned}P_n =& \frac{1}{2(ah_1^2 + a\mu_n^2 + h_1)} \times \left[h_1 \frac{\sin a(\eta - 2\mu_n)}{\eta - \mu_n} - h_1 \frac{\sin a(\eta + 2\mu_n)}{\eta + \mu_n} \right.\\ &\left. -\mu_n \frac{\cos a(\eta + 2\mu_n)}{\eta + \mu_n} - \mu_n \frac{\cos a(\eta - 2\mu_n)}{\eta - \mu_n} + \frac{2h_1\mu \sin(\eta a) + 2\mu\eta \cos(\eta a)}{(\eta^2 - \mu_n^2)} \right]\end{aligned} \tag{23.97}$$

Substituting the expanded forms of $\theta(z, t)$ and $\sin(\eta z)$ into the associated partial differential equation for θ gives an ordinary differential equation for $B_n(t)$ as follows

$$B_n'(t) + \kappa\mu_n^2 B_n(t) = \frac{\kappa}{K} P_n A(t) \tag{23.98}$$

The solution of the first-order differential equation (2.6-9) is

$$B_n(t) = \frac{\kappa}{K} P_n \int_0^t A(\tau) e^{-\kappa\mu_n^2(t-\tau)} d\tau + B_n(0) e^{-\kappa\mu_n^2 t} \tag{23.99}$$

The constants $B_n(0)$ have to be evaluated by means of the initial condition $\theta(z, 0) = \theta_0$. Considering the expanded form of $\theta(z, t)$ we have

$$\theta_0 = \sum_{n=1}^{\infty} B_n(0)(h_1 \sin \mu_n(z + a) + \mu_n \cos \mu_n(z + a)) \tag{23.100}$$

The constants $B_n(0)$ are obtained by the same method that the constants P_n were calculated. After some simplifications, constants $B_n(0)$ are evaluated as

$$B_n(0) = \frac{\theta_0}{ah_1^2 + a\mu_n^2 + h_1} \left(\sin(2a\mu_n) + \frac{h_1}{\mu_n}(1 - \cos(2a\mu_n)) \right) \tag{23.101}$$

From the definition of the thermal force resultant, we have

$$\begin{aligned} P_T &= \int_A E\alpha(T(y) - T_0)dA = bE\alpha \int_{-a}^{a} (T(y) - T_0)dy \\ &= E\alpha A \sum_{n=1}^{\infty} \frac{1}{2a} B_n(t) \left(h_1 \frac{1 - \cos(2a\mu_n)}{\mu_n} + \sin(2a\mu_n) \right) - E\alpha A\theta_0 \end{aligned} \tag{23.102}$$

and the thermal moments is

$$\begin{aligned} M_{Tz} &= \int_A yE\alpha(T(y) - T_0)dA = bE\alpha \int_{-a}^{a} y(T(y) - T_0)dy \\ &= E\alpha b \sum_{n=1}^{\infty} B_n(t) \left(\left(\frac{h_1}{\mu_n^2} + a \right) \sin(2a\mu_n) + \frac{(1 - ah_1)\cos(2a\mu_n) - ah_1 - 1}{\mu_n} \right) \end{aligned} \tag{23.103}$$

To evaluate the axial thermal stress, boundary conditions of the beam should be known. Assume a beam with both ends simply-supported. In this case the boundary conditions are

$$v(0) = v(L) = \frac{d^2v(0)}{dx^2} + \frac{M_{Tz}}{EI_z} = \frac{d^2v(L)}{dx^2} + \frac{M_{Tz}}{EI_z} = 0 \tag{23.104}$$

and the governing equation for deflection shape of the beam is

$$\frac{d^2v}{dx^2} = -\frac{M_{Tz} + M_{0z} + Q_0 x}{EI_z} \tag{23.105}$$

Solving this equation and using the associated boundary conditions gives $Q_0 = M_{0z} = 0$ and thus the normal stress is evaluates as

$$\sigma_{xx} = -E\alpha(T - T_0) + \frac{P_T}{A} + \frac{M_{Tz}}{EI_z} \tag{23.106}$$

Substituting the temperature distribution in the beam from Eq. (23.94) and P_T and M_{Tz} from Eqs. (23.102) and (23.103) into the thermal stress Eq. (23.106) leads us to

$$\begin{aligned}\sigma_{xx} = &- E\alpha \sum_{n=1}^{\infty} B_n(t)\,(h_1 \sin \mu_n(z+a) + \mu_n \cos \mu_n(z+a))\\ &+ \frac{1}{2a}E\alpha \sum_{n=1}^{\infty} B_n(t)\left(h_1 \frac{1-\cos(2a\mu_n)}{\mu_n} + \sin(2a\mu_n)\right)\\ &+ \frac{3E\alpha}{2a^3} \sum_{n=1}^{\infty} B_n(t)\left(\left(\frac{h_1}{\mu_n^2} + a\right)\sin(2a\mu_n) + \frac{(1-ah_1)\cos(2a\mu_n) - ah_1 - 1}{\mu_n}\right) z\end{aligned} \tag{23.107}$$

When the axial stress through the beam is known, the shear stress may be obtained using the following equation

$$\sigma_{xz} = -\int_z^a \frac{\partial \sigma_{xx}}{\partial x} dz \tag{23.108}$$

As seen, the shear stress is zero through the beam, as the axial stress is independent of the variable x.

Chapter 24
Thick Cylinders and Spheres

Thick cylinders and spheres are components of many structural systems. Due to their capacity to withstand high pressures, radial loads, and radial temperature gradients, the problem of thermal stress calculations is an important design issue in these types of problems. This chapter presents the method to calculate thermal stresses in such structural members which are made from the homogeneous/isotropic materials. The application of Michell conditions to derive thermal stresses in a multiply-connected region, such as thick walled cylinders, is shown through the solution of problems.

24.1 Problems and Solutions of Cylinders and Spheres

Problem 24.1. Obtain the Michell conditions in cylindrical coordinates. For a hollow thick cylinder under radial temperature distribution, $T = T(r)$, check the existence of thermal stresses through the Michell conditions.

Solution: In the Cartesian coordinates, Michell conditions are obtained as

$$\begin{aligned}
&\frac{1}{E}\int_{C_s}\left(x\frac{\partial\nabla^2\Phi}{\partial s}-y\frac{\partial\nabla^2\Phi}{\partial n}\right)ds+\alpha\int_{C_s}\left(x\frac{\partial T}{\partial s}-y\frac{\partial T}{\partial n}\right)ds\\
&-\frac{1+\nu}{E}\int_{C_s}(t_y^n)ds=0\\
&\frac{1}{E}\int_{C_s}\left(y\frac{\partial\nabla^2\Phi}{\partial s}+x\frac{\partial\nabla^2\Phi}{\partial n}\right)ds+\alpha\int_{C_s}\left(y\frac{\partial T}{\partial s}+x\frac{\partial T}{\partial n}\right)ds\\
&+\frac{1+\nu}{E}\int_{C_s}(t_x^n)\,ds=0\\
&\frac{1}{E}\int_{C_s}\frac{\partial\nabla^2\Phi}{\partial n}ds+\alpha\int_{C_s}\frac{\partial T}{\partial n}ds=0
\end{aligned}\tag{24.1}$$

M. Reza Eslami et al., *Theory of Elasticity and Thermal Stresses*, Solid Mechanics and Its Applications 197, DOI: 10.1007/978-94-007-6356-2_24,

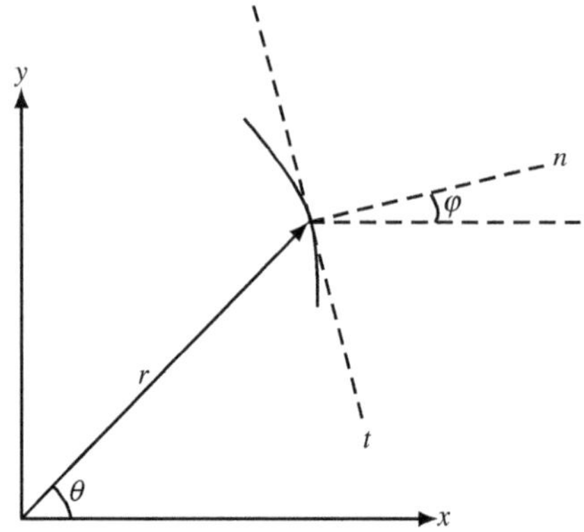

Fig. 24.1 Polar, cartesian and tangential coordinate systems

The above conditions should be transformed to the polar coordinates. The geometric relation between the three coordinates systems (x, y), (r, θ), and (t, s) is shown in Fig. 24.1

Based on the chain rule of integration, we write

$$\frac{\partial}{\partial s}=\frac{\partial}{\partial x}\frac{\partial x}{\partial s}+\frac{\partial}{\partial y}\frac{\partial y}{\partial s}=\frac{\partial}{\partial r}\frac{\partial r}{\partial x}\frac{\partial x}{\partial s}+\frac{\partial}{\partial \theta}\frac{\partial \theta}{\partial x}\frac{\partial x}{\partial s}+\frac{\partial}{\partial r}\frac{\partial r}{\partial y}\frac{\partial y}{\partial s}+\frac{\partial}{\partial \theta}\frac{\partial \theta}{\partial y}\frac{\partial y}{\partial s} \tag{24.2}$$

which simplifies to

$$-\frac{\partial}{\partial r}\cos\theta\sin\phi-\frac{1}{r}\frac{\partial}{\partial\theta}\sin\theta\sin\phi+\frac{\partial}{\partial r}\sin\theta\cos\phi+\frac{1}{r}\frac{\partial}{\partial\theta}\cos\theta\cos\phi \tag{24.3}$$

or

$$\frac{\partial}{\partial s}=\sin(\theta-\phi)\frac{\partial}{\partial r}+\frac{1}{r}\cos(\theta-\phi)\frac{\partial}{\partial\theta} \tag{24.4}$$

In a similar manner one may reach to

$$\frac{\partial}{\partial n}=\cos(\theta-\phi)\frac{\partial}{\partial r}-\frac{1}{r}\sin(\theta-\phi)\frac{\partial}{\partial\theta} \tag{24.5}$$

Components $\vec{t}^n$ of the surface traction force on a plane whose unit outer normal vector is $\vec{n}$ should be interpreted in the polar coordinates. It is simply to show that

$$\begin{aligned} t_y^n &= t_r^n\sin\theta+t_\theta^n\cos\theta \\ t_x^n &= t_r^n\cos\theta-t_\theta^n\sin\theta \end{aligned} \tag{24.6}$$

Finally, the operator ∇^2 should be transformed into the polar coordinates. By means of transformation $x=r\cos\theta$ and $y=r\sin\theta$, this operator in the polar coordinates is

$$\nabla^2 = \frac{\partial^2}{\partial x^2} + \frac{\partial^2}{\partial y^2} = \frac{\partial^2}{\partial r^2} + \frac{1}{r}\frac{\partial}{\partial r} + \frac{1}{r^2}\frac{\partial^2}{\partial \theta^2} \tag{24.7}$$

Now, when Eqs. (24.2)–(24.4) are substituted into Eq. (24.1), the following equations are obtained

$$\begin{aligned}
&\frac{1}{E}\int_{C_s}\left(x£_1\partial^2\Phi - y£_2\partial^2\Phi\right)ds + \alpha\int_{CS}(x£_1 T - y£_2 T)\,ds - \\
&\frac{1+\nu}{E}\int_{C_s}\left(t_r^n \sin\theta + t_\theta^n \cos\theta\right)ds = 0 \\
&\frac{1}{E}\int_{C_s}\left(y£_1\partial^2\Phi + x£_2\partial^2\Phi\right)ds + \alpha\int_{CS}(y£_1 T + x£_2 T)\,ds + \\
&\frac{1+\nu}{E}\int_{C_s}\left(t_r^n \cos\theta - t_\theta^n \sin\theta\right)ds = 0 \\
&\frac{1}{E}\int_{C_s}£_2\partial^2\Phi ds + \alpha\int_{C_s}£_2 T ds = 0
\end{aligned} \tag{24.8}$$

with the following definitions

$$\begin{aligned}
£_1 &= \sin(\theta-\phi)\frac{\partial}{\partial r} + \frac{1}{r}\cos(\theta-\phi)\frac{\partial}{\partial\theta} \\
£_2 &= \cos(\theta-\phi)\frac{\partial}{\partial r} - \frac{1}{r}\sin(\theta-\phi)\frac{\partial}{\partial\theta}
\end{aligned} \tag{24.9}$$

As an especial case, the Michell conditions are obtained for a hollow cylinder. In this case $\theta = \phi$ and r is constant. Then the operators of Eq. (24.9) simplifies to

$$\begin{aligned}
£_1 &= \frac{1}{r}\frac{\partial}{\partial\theta} \\
£_2 &= \frac{\partial}{\partial r}
\end{aligned} \tag{24.10}$$

and Eq. (24.8) simplify to

$$\begin{aligned}
&\frac{1}{E}\int_0^{2\pi}\left(\frac{x}{r}\frac{\partial\nabla^2\Phi}{\partial\theta} - y\frac{\partial\nabla^2\Phi}{\partial r}\right)rd\theta + \alpha\int_0^{2\pi}\left(\frac{x}{r}\frac{\partial T}{\partial\theta} - y\frac{\partial T}{\partial r}\right)rd\theta \\
&-\frac{1+\nu}{E}\int_0^{2\pi}\left(t_r^n\sin\theta + t_\theta^n\cos\theta\right)rd\theta = 0 \\
&\frac{1}{E}\int_0^{2\pi}\left(\frac{y}{r}\frac{\partial\nabla^2\Phi}{\partial\theta} + \frac{x}{r}\frac{\partial\nabla^2\Phi}{\partial\theta}\right)rd\theta + \alpha\int_0^{2\pi}\left(\frac{y}{r}\frac{\partial T}{\partial\theta} + x\frac{\partial T}{\partial r}\right)rd\theta \\
&+\frac{1+\nu}{E}\int_0^{2\pi}\left(t_r^n\cos\theta - t_\theta^n\sin\theta\right)rd\theta = 0
\end{aligned}$$

$$\frac{1}{E}\int_0^{2\pi}\frac{\partial\nabla^2\Phi}{\partial r}rd\theta+\alpha\int_0^{2\pi}\frac{\partial T}{\partial r}rd\theta=0 \tag{24.11}$$

For a hollow cylinder without traction force and subjected to thermal loads only, the Michell conditions are

$$\begin{aligned}
&\int_0^{2\pi}\left(y\frac{\partial\nabla^2\Phi}{\partial r}-\frac{x}{r}\frac{\partial\nabla^2\Phi}{\partial\theta}\right)rd\theta+E\alpha\int_0^{2\pi}\left(y\frac{\partial T}{\partial r}-\frac{x}{r}\frac{\partial T}{\partial\theta}\right)rd\theta=0\\
&\int_0^{2\pi}\left(x\frac{\partial\nabla^2\Phi}{\partial r}+\frac{y}{r}\frac{\partial\nabla^2\Phi}{\partial\theta}\right)rd\theta+E\alpha\int_0^{2\pi}\left(x\frac{\partial T}{\partial r}+\frac{y}{r}\frac{\partial T}{\partial\theta}\right)rd\theta=0\\
&\int_0^{2\pi}\frac{\partial\nabla^2\Phi}{\partial r}rd\theta+E\alpha\int_0^{2\pi}\frac{\partial T}{\partial r}rd\theta=0
\end{aligned} \tag{24.12}$$

where Φ is the Airy stress function which satisfies the equation:

$$\nabla^4\Phi+E\alpha\nabla^2T=0 \tag{24.13}$$

By means of the Fourier expansion for Φ in the form

$$\Phi\left(r,\theta\right)=f_0(r)+\sum_{n=1}^{\infty}\left(f_n(r)\cos n\theta+g_n(r)\sin n\theta\right) \tag{24.14}$$

Then we have

$$\begin{aligned}
&\frac{d^4f_0(r)}{dr^4}+\frac{2}{r}\frac{d^3f_0(r)}{dr^3}-\frac{1}{r^2}\frac{d^2f_0(r)}{dr^2}+\frac{1}{r^3}\frac{df_0(r)}{dr}+\\
&\nabla^4\left(\sum_{n=1}^{\infty}f_n(r)\cos n\theta+g_n(r)\sin n\theta\right)\\
&=E\alpha\left(\frac{d^2T}{dr^2}+\frac{1}{r}\frac{dT}{dr}\right)
\end{aligned} \tag{24.15}$$

Note that term $\frac{1}{r^2}\frac{d^2T}{d\theta^2}$ in the right-hand-side of the above equation is neglected due to the assumed radial temperature distribution. By means of the method of evaluating Fourier expansion coefficients, it is found that $f_n=g_n=0$ for $n\geq 1$ and therefore Eq. (24.15) results in

$$\frac{d^4f_0}{dr^4}+\frac{2}{r}\frac{d^3f_0}{dr^3}-\frac{1}{r^2}\frac{d^2f_0}{dr^2}+\frac{1}{r^3}\frac{df_0}{dr}=E\alpha\left(\frac{d^2T}{dr^2}+\frac{1}{r}\frac{dT}{dr}\right) \tag{24.16}$$

To obtain complete solution of this equation, the temperature distribution must be known. For the steady-state temperature distribution, without heat generation, we have

$$T(r) = T_a + \frac{T_b - T_a}{\ln \dfrac{b}{a}} \ln \frac{r}{a} \tag{24.17}$$

Here, a and b are the inner and outer radii of the cylinder, respectively. Solving Eq. (24.16) for f_0 results in

$$\Phi(r) = f_0(r) = c_1 + c_2 \ln \frac{r}{a} + c_3 \left(\frac{r}{a}\right)^2 + c_4 \left(\frac{r}{a}\right)^2 \ln \frac{r}{a} \tag{24.18}$$

where c_1, c_2, c_3 and c_4 are constants have to be obtained with the aid of boundary conditions. Substituting the Eqs. (24.17) and (24.18) into the Michell conditions (24.12), reveals that the first and second Michell conditions are automatically satisfied, while the third one results in

$$\frac{4}{a^2} c_4 + E\alpha \frac{T_b - T_a}{\ln \dfrac{b}{a}} = 0 \tag{24.19}$$

Based to the definition of Airy stress function, the components of stress field are

$$\begin{aligned} \sigma_{rr} &= \frac{1}{r}\frac{df_0}{dr} = \frac{c_2}{r^2} + \frac{2c_3}{a^2} + \frac{c_4}{a^2}\left(1 + 2\ln\frac{r}{a}\right) \\ \sigma_{\theta\theta} &= \frac{d^2 f_0}{dr^2} = -\frac{c_2}{r^2} + \frac{2c_3}{a^2} + \frac{c_4}{a^2}\left(3 + 2\ln\frac{r}{a}\right) \end{aligned} \tag{24.20}$$

Since the cylinder is only subjected to temperature loads, radial stress vanishes at inner and outer radii of the cylinder. Based to the Eq. (24.20) following equations for constants c_2, c_3 and c_4 are obtained

$$\begin{aligned} &c_2 + 2c_3 + c_4 = 0 \\ &c_2 + 2\left(\frac{b}{a}\right)^2 c_3 + \left(\frac{b}{a}\right)^2 c_4(1 + 2\ln(b/a)) = 0 \end{aligned} \tag{24.21}$$

Solving the Eqs. (24.19) and (24.21) for c_1, c_2 and c_3 and substituting the results into Eq. (24.21) yield the components of stress tensor as

$$\begin{aligned} \sigma_{rr} &= E\alpha \frac{T_b - T_a}{2\ln \dfrac{b}{a}} \left[-\ln\frac{r}{a} + \frac{b^2}{b^2 - a^2}\left(1 - \frac{a^2}{r^2}\right)\ln\frac{b}{a}\right] \\ \sigma_{\theta\theta} &= E\alpha \frac{T_b - T_a}{2\ln \dfrac{b}{a}} \left[-1 - \ln\frac{r}{a} + \frac{b^2}{b^2 - a^2}\left(1 + \frac{a^2}{r^2}\right)\ln\frac{b}{a}\right] \end{aligned} \tag{24.22}$$

Problem 24.2. Consider a temperature change distribution in a thick hollow cylinder of the form

$$\theta(r, \phi) = \sum_{n=1}^{\infty} \left(A_n r^n + B_n r^{-n}\right) \cos n\phi \tag{24.23}$$

where A_n and B_n are known constants and $\theta = T - T_\infty$, T_∞ being the reference temperature. The related stress function satisfies the equation

$$\nabla^2 \Phi = \sum_{m=1}^{\infty} \left(c_{1m} r^m + c_{2m} r^{-m}\right) \left(c_{3m} \sin m\phi + c_{4m} \cos m\phi\right) \tag{24.24}$$

Check all the three Michell conditions to identify the terms of temperature which do not satisfy the conditions. Then use the Navier equations in polar coordinates and with the use of Papkovich function obtain the stresses.

Solution: The given temperature distribution in Eq. (24.23) must satisfy the heat conduction equation in polar coordinates as

$$\nabla^2 \theta = \frac{\partial^2 \theta}{\partial r^2} + \frac{1}{r}\frac{\partial \theta}{\partial r} + \frac{1}{r^2}\frac{\partial^2 \theta}{\partial \phi^2} = 0 \tag{24.25}$$

It would be easy to verify that the given temperature distribution $\theta(r, \phi)$ satisfies this equation. Next, the given form of stress function in Eq. (24.24) should satisfy the compatibility equation which is given by

$$\nabla^4 \Phi + E_1 \alpha_1 \nabla^2 \theta = 0 \tag{24.26}$$

It is easy to show that the given form of $\nabla^2 \Phi$ satisfies this equation. Finally all three Michell conditions in terms of stress function and temperature profile in polar coordinates are given by

$$\begin{aligned}
&\int_0^{2\pi} \left(y \frac{\partial \nabla^2 \Phi}{\partial r} - \frac{x}{r}\frac{\partial \nabla^2 \Phi}{\partial \phi}\right) r d\phi = -E_1 \alpha_1 \int_0^{2\pi} \left(y \frac{\partial \theta}{\partial r} - \frac{x}{r}\frac{\partial \theta}{\partial \phi}\right) r d\phi \\
&\int_0^{2\pi} \left(x \frac{\partial \nabla^2 \Phi}{\partial r} + \frac{y}{r}\frac{\partial \nabla^2 \Phi}{\partial \phi}\right) r d\phi = -E_1 \alpha_1 \int_0^{2\pi} \left(x \frac{\partial \theta}{\partial r} + \frac{y}{r}\frac{\partial \theta}{\partial \phi}\right) r d\phi \\
&\int_0^{2\pi} \frac{\partial \nabla^2 \Phi}{\partial r} r d\phi = -E_1 \alpha_1 \int_0^{2\pi} \frac{\partial \theta}{\partial r} r d\phi \qquad at \qquad r = a
\end{aligned} \tag{24.27}$$

Recalling the following properties of the trigonometric functions

$$\int_0^{2\pi} \cos n\phi \cos m\phi d\phi = \begin{cases} 0 & n \neq m \\ 2\pi & n = m \neq 0 \end{cases}$$

$$\int_0^{2\pi} \sin n\phi \sin m\phi d\phi = \begin{cases} 0 & n \neq m \\ 2\pi & n = m \neq 0 \end{cases}$$

$$\int_0^{2\pi} \sin n\phi \cos m\phi d\phi = 0 \tag{24.28}$$

and substituting the proposed form of the temperature and Airy stress function from Eqs. (24.23) and (24.24) into the first of Michell conditions, i.e. the first of Eq. (24.28) reveals that the right hand side of this equation is equal to zero, while in the left hand side all terms are vanished except the one associated to $m = 1$. Therefore, the first Michell condition implies that

$$c_{21}c_{31} = 0 \tag{24.29}$$

Similarly, it is seen that in the second Michell condition described by the second of Eq. (24.28), all terms in both right and left sides of the equality are vanished except for $m = 1$, which results in

$$c_{21}c_{41} \neq 0 \tag{24.30}$$

The third Michell condition, i.e. the third of Eq. (24.28) is held automatically. Therefore, all three Michell conditions are satisfied for $n > 1$ and for $n = 1$ conditions are given by Eqs. (24.29) and (24.30). From these two conditions, one may concluded that $c_{21} \neq 0$ and $c_{31} = 0$. Therefore, among all infinite number of terms in temperature and stress function profile, the followings are not satisfied automatically in Michell conditions

$$\begin{aligned} \theta(r, \phi) &= \left(A_1 r + B_1 r^{-1}\right) \cos\phi \\ \nabla^2 \Phi(r, \phi) &= \left(c_{11} r + c_{21} r^{-1}\right) c_{41} \cos\phi \end{aligned} \tag{24.31}$$

The complete solution of the stress function can be decomposed as $\Phi = \Phi_h + \Phi_p$, where Φ_h is the general solution of the stress function in Eq. (24.31) with the right hand side equal to zero and is obtained with the aid of the Fourier series expansion as

$$\Phi_h = \sum_{n=0}^{\infty} \left\{ \left(D_{1n} r^n + D_{2n} r^{-n}\right) \cos n\phi + \left(D_{3n} r^n + D_{4n} r^{-n}\right) \sin n\phi \right\} \tag{24.32}$$

and Φ_p is the particular solution of this equation and is extracted as

$$\Phi_p = \left(\frac{1}{8} c_{11} c_{41} r^3 + \frac{1}{2} c_{21} c_{41} r \ln r - \frac{1}{4} c_{21} c_{41} r\right) \cos\phi \tag{24.33}$$

The boundary condition along the interior radii of the cylinder, i.e. $r = a$ are

$$\begin{aligned} \Phi &= a_1 x + a_2 y + a_3 \\ \frac{\partial \Phi}{\partial r} &= a_1 \cos\phi + a_2 \sin\phi \end{aligned} \tag{24.34}$$

and along the exterior radii, i.e. $r = b$ are

$$\Phi = \frac{\partial \Phi}{\partial r} = 0 \tag{24.35}$$

Using the boundary condition (24.34) and considering the solution of the stress function as the sum of Eqs. (24.32) and (24.33), results in $D_{1n} = D_{2n} = D_{3n} = D_{4n} = 0$ for $n > 1$. Besides, due to the temperature distribution through the cylinder, which is a cosine function, the constants D_{10}, D_{20}, D_{31}, D_{41} are ignored and therefore the stress function is obtained in the form

$$\Phi = \left(C_1 \frac{r}{a} + C_2 \frac{a}{r} + C_3 \left(\frac{r}{a}\right)^3 + C_4 \frac{r}{a} \ln \frac{r}{a} \right) \cos\phi \tag{24.36}$$

where C_i's are constants. The stress components are now obtained using the Airy stress function as

$$\begin{aligned}
\sigma_{rr} &= \frac{1}{r}\frac{\partial \Phi}{\partial r} + \frac{1}{r^2}\frac{\partial^2 \Phi}{\partial \phi^2} = \left(-\frac{2a}{r^3} C_2 + \frac{2a}{r^3} C_3 + \frac{1}{ar} C_4 \right) \cos\phi \\
\sigma_{\phi\phi} &= \frac{\partial^2 \Phi}{\partial r^2} = \left(\frac{2a}{r^3} C_2 + \frac{6r}{a^3} C_3 + \frac{1}{ar} C_4 \right) \cos\phi \\
\sigma_{r\phi} &= -\frac{\partial}{\partial r}\left(\frac{1}{r}\frac{\partial \Phi}{\partial \theta} \right) = \left(-\frac{2a}{r^3} C_2 + \frac{2r}{a^3} C_3 + \frac{1}{ar} C_4 \right) \sin\phi
\end{aligned} \tag{24.37}$$

where in the above equations, three constants C_2, C_3, and C_4 have to be determined using the given boundary conditions. It is worth mentioning that in the temperature profile of Eq. (24.31), the term $A_1 r \cos\phi = A_1 x$ is not included into the formulations as it is a linear function and does not produce any thermal stresses. The three constants are obtained when the stress function (24.36) and temperature profile are substituted into the second Michell condition and the boundary conditions (24.33) and stress free conditions at $r = a, b$ are satisfied. The results are

$$\begin{aligned}
C_2 &= -\frac{E_1 \alpha_1 B_1 a b^2}{4(a^2 + b^2)} \\
C_3 &= \frac{E_1 \alpha_1 B_1 a^3}{4(a^2 + b^2)} \\
C_4 &= -\frac{E_1 \alpha_1 B_1 a}{2}
\end{aligned} \tag{24.38}$$

The stress components are computed upon substitution of Eq. (24.38) into Eq. (24.37) to give

$$\sigma_{rr} = \frac{E_1 \alpha_1 r}{2\left(a^2 + b^2\right)} \left(1 - \frac{a^2}{r^2}\right) \left(1 - \frac{b^2}{r^2}\right) B_1 \cos\phi$$

$$\sigma_{r\phi} = \frac{E_1\alpha_1 r}{2\left(a^2+b^2\right)}\left(1-\frac{a^2}{r^2}\right)\left(1-\frac{b^2}{r^2}\right)B_1\sin\phi$$
$$\sigma_{\phi\phi} = \frac{E_1\alpha_1 r}{2\left(a^2+b^2\right)}\left(3-\frac{a^2+b^2}{r^2}-\frac{a^2b^2}{r^4}\right)B_1\cos\phi \tag{24.39}$$

Here, $E_1\alpha_1$ have to be replaced by $\frac{E\alpha}{1-\nu}$ for the case of plane stress conditions.

Problem 24.3. Consider a finite solid circular cylinder of the length $2l$ and radius r_2.

(a) Use the Papkovich function to obtain the expression for displacement components.

(b) For the given temperature distribution

$$T(r,z) = T_0 + a_0(r) + \sum_{n=1}^{\infty} a_n(r)\cos k_n z \tag{24.40}$$

where $a_n(r)$ are prescribed and $k_n = \frac{n\pi}{l}$ are constant values, obtain the associated thermal stresses.

Solution:

a: In the case of axisymmetric temperature distribution the strain-displacement relations are

$$\epsilon_r = \frac{\partial u_r}{\partial r}$$
$$\epsilon_\phi = \frac{u_r}{r}$$
$$\epsilon_z = \frac{\partial u_z}{\partial z}$$
$$\epsilon_{rz} = \frac{1}{2}\left(\frac{\partial u_r}{\partial z}+\frac{\partial u_z}{\partial r}\right) \tag{24.41}$$

where, in the above equations, u_r and u_z are displacements along r radial and aixal directions respectively. Besides, ϵ_r, ϵ_ϕ, ϵ_z and ϵ_{rz} are radial strain, hoop strain, axial strain and shear strain respectively. The equilibrium conditions in the symmetrical case of deformations are

$$\frac{\partial\sigma_r}{\partial r}+\frac{\partial\sigma_{rz}}{\partial z}+\frac{\sigma_r-\sigma_\phi}{r}=0$$
$$\frac{\partial\sigma_{rz}}{\partial r}+\frac{\partial\sigma_z}{\partial z}+\frac{1}{r}\sigma_{rz}=0 \tag{24.42}$$

where σ_r, σ_z, σ_ϕ and $\sigma_{r\phi}$ are the radial, axial, hoop and shear stresses, respectively. The stress-strain relations in symmetrical case of deformations are

$$\begin{aligned}\sigma_r &= (2\mu+\lambda)\,\epsilon_r + \lambda\left(\epsilon_\phi+\epsilon_z\right) - \beta\,(T-T_0)\\ \sigma_\phi &= (2\mu+\lambda)\,\epsilon_\phi + \lambda\,(\epsilon_r+\epsilon_z) - \beta\,(T-T_0)\\ \sigma_z &= (2\mu+\lambda)\,\epsilon_z + \lambda\left(\epsilon_\phi+\epsilon_r\right) - \beta\,(T-T_0)\\ \sigma_{rz} &= 2\mu\epsilon_{rz}\end{aligned} \tag{24.43}$$

Combining Eqs. (24.41), (24.42) and (24.43) provides the Navier equations in terms of the displacement components for axisymmetric cylindrical coordinates as

$$\begin{aligned}&(2\mu+\lambda)\left(\frac{\partial^2 u_r}{\partial r^2}+\frac{1}{r}\frac{\partial u_r}{\partial r}-\frac{1}{r^2}u_r\right)+\mu\frac{\partial^2 u_z}{\partial z^2}+\\ &(\lambda+\mu)\,\frac{\partial^2 u_z}{\partial r\partial z}=\beta\frac{\partial T}{\partial r}\\ &\mu\left(\frac{\partial^2 u_z}{\partial z^2}+\frac{1}{r}\frac{\partial u_z}{\partial z}-\frac{1}{r^2}u_z\right)+(2\mu+\lambda)\,\frac{\partial^2 u_z}{\partial z^2}+\\ &(\lambda+\mu)\left(\frac{\partial^2 u_r}{\partial r\partial z}+\frac{1}{r}\frac{\partial u_r}{\partial z}\right)=\beta\frac{\partial T}{\partial z}\end{aligned} \tag{24.44}$$

Using the Papkovich-Neuber solution, displacement field is assumed in the form

$$\vec{u}=\vec{u}^T+\vec{u}^*=\vec{\nabla}\Phi+4\,(1-\nu)\,\vec{B}-\vec{\nabla}\left(\vec{B}.\vec{r}+\vec{B}_0\right) \tag{24.45}$$

Here, B is a harmonic vector, B_0 is a harmonic scalar and Φ is the thermoelastic potential function. Components of the above equation may also be written as

$$\begin{aligned}u_r^T &= \frac{\partial\Phi}{\partial r}\\ u_z^T &= \frac{\partial\Phi}{\partial z}\\ u_r^* &= 4\,(1-\nu)\,B_r-\frac{\partial}{\partial r}\,(rB_r+zB_z+B_0)\\ u_z^* &= 4\,(1-\nu)\,B_z-\frac{\partial}{\partial z}\,(rB_r+zB_z+B_0)\end{aligned} \tag{24.46}$$

Substituting these relations into the Eq. (24.44) results into governing equations for B_r, B_z, B_0, and Φ as

$$\frac{\partial^2\Phi}{\partial r^2}+\frac{1}{r}\frac{\partial\Phi}{\partial r}+\frac{\partial^2\Phi}{\partial z^2}=\frac{1+\nu}{1-\nu}\alpha\,(T-T_0)$$

$$\frac{\partial^2 B_r}{\partial r^2} + \frac{1}{r}\frac{\partial B_r}{\partial r} + \frac{\partial^2 B_r}{\partial z^2} - \frac{B_r}{r^2} = 0$$
$$\frac{\partial^2 B_z}{\partial r^2} + \frac{1}{r}\frac{\partial B_z}{\partial r} + \frac{\partial^2 B_z}{\partial z^2} = 0$$
$$\frac{\partial^2 B_0}{\partial r^2} + \frac{1}{r}\frac{\partial B_0}{\partial r} + \frac{\partial^2 B_0}{\partial z^2} = 0 \tag{24.47}$$

b: The temperature distribution given by Eq. (24.40) may be written in the next form

$$\Theta\,(r, z) = T\,(r, z) - T_0 = a_0\,(r) + \sum_{n=1}^{\infty} a_n\,(r) \cos k_n z \tag{24.48}$$

the above equation is the cosine Fourier series expansion for the function $\Theta\,(r, z)$. Therefore the constants a_n may be considered to be known by the following equations.

$$a_0(r) = \frac{1}{l}\int_0^l \Theta\,(r, z)\,dz$$
$$a_n(r) = \frac{2}{l}\int_0^l \Theta\,(r, z) \cos k_n z dz \tag{24.49}$$

Now, each of the functions Φ, B_r, B_z and B_0 have to be solved. Based to the first of Eq. (24.47), the function $\Phi(r, z)$ may be written as

$$\Phi\,(r, z) = \frac{1+\nu}{1-\nu}\alpha \sum_{n=0}^{\infty} b_n\,(r) \cos k_n z \tag{24.50}$$

Substituting the expressions for Φ from Eq. (24.50) and T from (24.48) into the first of relations in Eq. (24.47) reaches us to

$$r\frac{d}{dr}\left(rb_n'(r)\right) - k_n^2 r^2 b_n(r) = r^2 a_n(r), \qquad n = 0, 1, 2, ... \tag{24.51}$$

The particular solution of the above equations may be written as

$$b_0\,(r) = \int_0^r \frac{1}{r}\left(\int_0^r r a_0\,(r)\,dr\right)dr$$
$$b_n\,(r) = I_0\,(k_n r)\int_0^r r K_0\,(k_n r)\,a_n\,(r)\,dr -$$
$$K_0\,(k_n r)\int_0^r r I_0\,(k_n r)\,a_n\,(r)\,dr, \quad n = 1, 2, ... \tag{24.52}$$

And displacement components, u_r^T and u_z^T can be obtained by mean of the first of Eq. (24.46) as

$$u_z^T = -\frac{1+\nu}{1-\nu}\alpha \sum_{n=1}^{\infty} k_n b_n(r) \sin k_n z \tag{24.53}$$

$$u_r^T = \frac{1+\nu}{1-\nu}\alpha \frac{db_0(r)}{dr} + \frac{1+\nu}{1-\nu}\alpha \sum_{n=1}^{\infty} \frac{db_n(r)}{dr} \cos k_n z$$

From Eqs. (24.41) and (24.43) the associated stress components are

$$\begin{aligned}
\sigma_r^T &= 2\mu \left\{ \frac{1+\nu}{1-\nu}\alpha \left(\frac{d^2 b_0(r)}{dr^2} - a_0(r) \right) \right. \\
&\quad \left. + \frac{1+\nu}{1-\nu}\alpha \sum_{n=1}^{\infty} \left(\frac{d^2 b_n(r)}{dr^2} - a_n(r) \right) \cos k_n z \right\} \\
\sigma_\phi^T &= 2\mu \left\{ \frac{1+\nu}{1-\nu}\alpha \left(\frac{1}{r}\frac{db_0(r)}{dr} - a_0(r) \right) \right. \\
&\quad \left. + \frac{1+\nu}{1-\nu}\alpha \sum_{n=1}^{\infty} \left(\frac{1}{r}\frac{db_n(r)}{dr} - a_n(r) \right) \cos k_n z \right\} \\
\sigma_z^T &= -2\mu \frac{1+\nu}{1-\nu}\alpha \sum_{n=1}^{\infty} \left(k_n^2 b_n(r) + a_n(r) \right) \cos k_n z \\
\sigma_{rz}^T &= -2\mu \frac{1+\nu}{1-\nu}\alpha \sum_{n=1}^{\infty} k_n \frac{db_n(r)}{dr} \sin k_n z
\end{aligned} \tag{24.54}$$

When the temperature distribution is symmetric with respect to the plane $z = 0$, the solution of u_r^* and u_z^* must satisfy the conditions $u_r^*(-z) = u_r^*(z)$ and $u_z^*(-z) = u_z^*(z)$. We split the solution into two parts; $u_r^* = u_r^I + u_r^{II}$ and $u_z^* = u_z^I + u_z^{II}$. The first part with superscript I will be constructed for arbitrary boundary condition on $r = r_2$. The related stresses to this solution must be in the form of complete orthogonal series on domain $[-l, l]$ so B_r and B_0 will be chosen in the form

$$\begin{aligned}
B_r(r) &= g_0(r) + \sum_{n=1}^{\infty} g_n(r) \cos k_n z \\
B_0(r) &= h_0(r) + \sum_{n=1}^{\infty} h_n(r) \cos k_n z
\end{aligned} \tag{24.55}$$

and also in this case B_z can be set equal to zero. Substitution of Eq. (24.55) into the second and fourth relations of Eq. (24.47) results in

$$g_0'' + \frac{1}{r} g_0' - \frac{1}{r^2} g_0 = 0$$

$$h_0'' + \frac{1}{r}h_0' = 0$$
$$g_n'' + \frac{1}{r}g_n' - \left(k_n^2 + \frac{1}{r^2}\right)g_n = 0$$
$$h_n'' + \frac{1}{r}h_n' - k_n^2 h_n = 0 \tag{24.56}$$

and the solutions of these equations are

$$g_0(r) = \alpha_0 r + \frac{\alpha_0'}{r}$$
$$g_n(r) = a_n I_1(k_n r) + \alpha_n' k_1(k_n r)$$
$$h_0(r) = \beta_0 r + \beta_0' \ln r$$
$$h_n(r) = \beta_n I_0(k_n r) + \beta_n' k_0(k_n r) \tag{24.57}$$

Here, $\alpha_0, \alpha_0', \beta_0, \beta_0', \alpha_n, \beta_n, \alpha_n', \beta_n'$ are arbitrary constants. Since the cylinder is solid, the above-mentioned functions have to be finite at $r = 0$ which results in

$$\alpha_0' = 0, \beta_0', \alpha_n' = 0, \beta_n' = 0 \tag{24.58}$$

With the simultaneous aids of Eqs. (24.46) and (24.55), we have

$$u_z^I = \sum_{n=1}^{\infty} \{\alpha_n k_n r I_1(k_n r) + \beta_n k_n I_0(k_n r)\} \sin k_n z$$
$$u_r^I = 2(1 - 2\nu)\alpha_0 r +$$
$$\sum_{n=1}^{\infty} \{\alpha_n [4(1-\nu) I_1(k_n r) - k_n r I_0(k_n r)] - \beta_n k_n I_1(k_n r)\} \cos k_n z \tag{24.59}$$

When constructing the second part u_r^{II} and u_z^{II} we can assume the functions B_r, B_0 and B_z in the form

$$B_r(r) = 0$$
$$B_z(r) = p_0(z) + \sum_{j=1}^{\infty} p_j(z) J_0(\lambda_j r)$$
$$B_0(r) = q_0(z) + \sum_{j=1}^{\infty} q_j(z) J_0(\lambda_j r) \tag{24.60}$$

This form is chosen according to the second and fourth of the Eq. (24.47). Substituting the Eq. (24.60) into the second and fourth of Eq. (24.47) reveals that λ_j's are roots of the equations $J_1\left(\lambda_j r_2\right) = 0$. Simultaneously, the governing equations for the eigenfunctions are

$$\begin{aligned}
&p_0^{''}(z) = 0 \\
&q_0^{''}(z) = 0 \\
&p_j^{''} - \lambda_j^2 p_j = 0 \\
&q_j^{''} - \lambda_j^2 q_j = 0
\end{aligned} \tag{24.61}$$

Solving these equations and considering that p_j and q_j are odd and even functions, respectively, the solutions of these functions are

$$\begin{aligned}
&p_0(z) = \gamma_0 z \\
&q_0(z) = \delta_0 \\
&p_j(z) = \gamma_j \sinh \lambda_j z \\
&q_j(z) = \delta_j \cosh \lambda_j z
\end{aligned} \tag{24.62}$$

Finally, u_r^{II} and u_z^{II} are given as

$$\begin{aligned}
u_r^{II} &= \sum_{j=1}^{\infty} \left\{\gamma_j \lambda_j z \sinh \lambda_j z + \delta_j \lambda_j \cosh \lambda_j z\right\} J_1\left(\lambda_j z\right) \\
u_z^{II} &= 2(1-2\nu)\gamma_0 z + \\
&\sum_{j=1}^{\infty} \left\{\gamma_j \left[(3-4\nu)\sinh \lambda_j z - \lambda_j z \cosh \lambda_j z\right] - \delta_j \lambda_j \sinh \lambda_j z\right\} J_0\left(\lambda_j z\right)
\end{aligned} \tag{24.63}$$

The complete solution of the functions u_t^* and u_z^* is equal to the summation of Eqs. (24.53) and (24.63), that are

$$\begin{aligned}
u_r^* &= 2(1-2\nu)\alpha_0 r \\
&\quad + \sum_{n=1}^{\infty} \left\{\alpha_n \left[4(1-\nu) I_1(k_n r) - k_n r I_0(k_n r)\right] - \beta_n k_n I_1(k_n r)\right\} \cos k_n z \\
&\quad + \sum_{j=1}^{\infty} \left\{\gamma_j \lambda_j z \sinh \lambda_j z + \delta_j \lambda_j \cosh \lambda_j z\right\} J_1\left(\lambda_j r\right) \\
u_z^* &= 2(1-2\nu)\gamma_0 z + \sum_{n=1}^{\infty} \left\{\alpha_n k_n r I_1(k_n r) + \beta_n k_n I_0(k_n r)\right\} \sin k_n z
\end{aligned}$$

$$+\sum_{j=1}^{\infty}\left\{\gamma_j\left[(3-4\nu)\sinh\lambda_j z-\lambda_j z\cosh\lambda_j z\right]-\delta_j\lambda_j\sinh\lambda_j z\right\}J_0\left(\lambda_j r\right) \tag{24.64}$$

Using Eqs. (24.41), (24.43), and (24.64) the related stress components are found to be

$$\begin{aligned}
\frac{\sigma_r^*}{2\mu} &= 2\alpha_0+2\nu\gamma_0 \\
&\quad-\sum_{j=1}^{\infty}\frac{1}{r}\left(\gamma_j\lambda_j z\sinh\lambda_j z+\delta_j\lambda_j\cosh\lambda_j z\right)J_1\left(\lambda_j r\right) \\
&\quad+\sum_{n=1}^{\infty}\left\{\alpha_n\left[-4(1-\nu)\frac{1}{r}I_1(k_n r)-k_n^2 r I_1(k_n r)+(3-2\nu)k_n I_0(k_n r)\right]\right\} \\
&\quad\times\cos k_n z+\sum_{n=1}^{\infty}\left\{\beta_n k_n\left[\frac{1}{r}I_1(k_n r)-k_n I_0(k_n r)\right]\right\}\cos k_n z \\
&\quad+\sum_{j=1}^{\infty}\left\{\gamma_j\left(\lambda_j^2 z\sinh\lambda_j z+2\nu\lambda_j\cosh\lambda_j z\right)+\delta_j\lambda_j^2\cosh\lambda_j z\right\}J_0\left(\lambda_j r\right) \\
\frac{\sigma_\phi^*}{2\mu} &= 2\alpha_0+2\nu\gamma_0 \\
&\quad+\sum_{n=1}^{\infty}\left\{\alpha_n\left[4(1-\nu)\frac{1}{r}I_1(k_n r)-(1-2\nu)k_n I_0(k_n r)\right]\right. \\
&\qquad\left.-\beta_n k_n\frac{1}{r}I_1(k_n r)\right\}\cos k_n z \\
&\quad+\sum_{j=1}^{\infty}\frac{1}{r}\left\{\gamma_j\lambda_j z\sinh\lambda_j z+\delta_j\lambda_j\cosh\lambda_j z\right\}J_1\left(\lambda_j r\right) \\
&\quad+\sum_{j=1}^{\infty}2\nu\left\{\gamma_j\lambda_j\cosh\lambda_j z\right\}J_0\left(\lambda_j r\right) \\
\frac{\sigma_z^*}{2\mu} &= 2(1-\nu)\gamma_0+4\nu\alpha_0 \\
&\quad+\sum_{n=1}^{\infty}\left\{\alpha_n\left[k_n^2 r I_1(k_n r)+2\nu k_n I_0(k_n r)\right]+\beta_n k_n^2 I_0(k_n r)\right\}\cos k_n z \\
&\quad+\sum_{j=1}^{\infty}\left\{\gamma_j\left[2(1-\nu)\lambda_j\cosh\lambda_j z-\lambda_j^2 z\sinh\lambda_j z\right]-\delta_j\lambda_j^2\cosh\lambda_j z\right\} \\
&\quad\times J_0\left(\lambda_j r\right)
\end{aligned}$$

$$\frac{\sigma_{rz}^*}{2\mu} = \sum_{n=1}^{\infty} \left\{ \alpha_n k_n \left[-2(1-\nu) I_1(k_n r) + k_n r I_0(k_n r) \right] + \beta_n k_n^2 I_1(k_n r) \right\} \sin k_n z$$
$$+ \sum_{j=1}^{\infty} \left\{ \gamma_j \lambda_j \left[\lambda_j z \cosh \lambda_j z - (1-2\nu) \sinh \lambda_j z \right] + \delta_j \lambda_j^2 \sinh \lambda_j z \right\}$$
$$\times J_1(\lambda_j z) \tag{24.65}$$

when Eqs. (24.54) and (24.65) are added together, the complete solution of thermal stresses is obtained. It should be emphasized that for the case when cylinder is subjected to thermal loadings only, shear and radial stresses have to be vanished at $r = r_2$. Also axial and shear stresses have to be omitted at $z = \mp l/2$. In such a case the series expansion have to be truncated into a finite number to enable us to calculate the constants α_n, β_n, γ_j and δ_j.

Problem 24.4. Find the thermal stresses in a thick hollow sphere of inside and outside radii a and b, respectively, subjected to the general temperature variation of the form $T = T(r, \phi)$ and with the stress-free boundary conditions.

Solution: In spherical coordinates and in the case of axisymmetric heat conduction, the governing heat conduction equation is

$$\frac{1}{r^2}\frac{\partial}{\partial r}\left(r^2\theta\right) + \frac{1}{\sin\varphi}\frac{\partial}{\partial\varphi}(\sin\varphi\theta) = 0 \tag{24.66}$$

where $\theta = T(r, \varphi) - T_0$. Using the method of separation of variables

$$\theta(r, \varphi) = R(r)\phi(\varphi) \tag{24.67}$$

Substituting Eq. (24.67) into the heat conduction equation (24.66) gives

$$\left(r^2 R'(r)\right)' - \mu^2 R(r) = \left(\phi'(\varphi)\sin\varphi\right)' + \mu^2\phi(\varphi)\sin\varphi = 0 \tag{24.68}$$

where, μ is constant. The above equations have to solved as an eigenvalue problem. The eigenvalues and eigenfunctions are

$$\begin{aligned} \mu_n^2 &= n(n+1) \\ \phi_n(\varphi) &= P_n(\cos\varphi) \end{aligned} \tag{24.69}$$

Substituting the eigenvalues μ_n form Eq. (24.69) into the Eq. (24.68) and solving for $R_n(r)$ gives us Solving for $R(r)$ gives

$$R_n(r) = \alpha_n r^n + \beta_n r^{-n-1} \tag{24.70}$$

Having developed the functions R_n and $\phi(\varphi)$, the temperature profile becomes

$$T(r,\varphi) = T_0 + \sum_{n=0}^{\infty} \left(\alpha_n r^n + \beta_n r^{-n-1} \right) P_n(\cos\varphi) \tag{24.71}$$

The components of the displacement vector, as a sum of the general u^* and particular u^T solutions, have the form

$$\begin{aligned} u_r &= u_r^T + u_r^* \\ u_\varphi &= u_\varphi^* + u_\varphi^T \end{aligned} \tag{24.72}$$

where, each of these components can be interpreted in terms of Papkovich-Neuber functions, displacement field is assumed in the form

$$\begin{aligned} u_r^T &= \frac{\partial \psi}{\partial r} \\ u_r^* &= 4(1-\nu) B_r - \frac{\partial}{\partial r}(rB_r + B_0) \\ u_\varphi^T &= \frac{1}{r}\frac{\partial \psi}{\partial \varphi} \\ u_\varphi^* &= 4(1-\nu) B_\varphi - \frac{1}{r}\frac{\partial}{\partial \varphi}(rB_r + B_0) \end{aligned} \tag{24.73}$$

Here, B_r and B_φ are the elements of a harmonic vector, B_0 is a harmonic scalar and ψ is the thermoelastic potential function.

In spherical coordinates, the strain - displacement relations are

$$\begin{aligned} \epsilon_r &= \frac{\partial u_r}{\partial r} \\ \epsilon_\varphi &= \frac{1}{r}\left(u_r + \frac{\partial u_\varphi}{\partial \varphi} \right) \\ \epsilon_\theta &= \frac{1}{r}\left(u_r + u_\varphi \cot\varphi \right) \\ \epsilon_{r\varphi} &= \frac{1}{2}\left(\frac{\partial u_\varphi}{\partial r} + \frac{1}{r}\frac{\partial u_r}{\partial \varphi} - \frac{u_\varphi}{r} \right) \end{aligned} \tag{24.74}$$

and the equilibrium equations are

$$\begin{aligned} &\frac{\partial \sigma_r}{\partial r} + \frac{1}{r}\left(\frac{\partial \sigma_{r\varphi}}{\partial \varphi} + 2\sigma_r - \sigma_\varphi - \sigma_\theta + \sigma_{r\varphi}\cot\varphi \right) = 0 \\ &\frac{\partial \sigma_{r\varphi}}{\partial r} + \frac{1}{r}\left(\frac{\partial \sigma_\varphi}{\partial \varphi} + \left(\sigma_\varphi - \sigma_\theta \right)\cot\varphi + 3\sigma_{r\varphi} \right) = 0 \end{aligned} \tag{24.75}$$

The linear thermoelastic stress-strain relations are

$$\begin{aligned}\sigma_r &= (2\mu+\lambda)\,\epsilon_r + \lambda\left(\epsilon_\theta+\epsilon_\varphi\right) - \beta\theta \\ \sigma_\theta &= (2\mu+\lambda)\,\epsilon_\theta + \lambda\left(\epsilon_r+\epsilon_\varphi\right) - \beta\theta \\ \sigma_\varphi &= (2\mu+\lambda)\,\epsilon_\varphi + \lambda\left(\epsilon_\theta+\epsilon_r\right) - \beta\theta \\ \sigma_{r\varphi} &= 2\mu\epsilon_{r\varphi}\end{aligned} \tag{24.76}$$

where, μ and λ are Lame constants. Combining Eqs. (24.73)–(24.76) yields the equilibrium equations in terms of the Papkovich-Neuber functions, that are

$$\begin{aligned}&\frac{1}{r^2}\frac{\partial}{\partial r}\left(r^2\frac{\partial\psi}{\partial r}\right) + \frac{1}{r^2}\frac{1}{\sin\varphi}\frac{\partial}{\partial\varphi}\left(\sin\varphi\frac{\partial\psi}{\partial\varphi}\right) = \frac{1+\nu}{1-\nu}\alpha\,(T-T_0) \\ &\frac{\partial e_B}{\partial r} - \frac{1}{r\sin\varphi}\frac{\partial}{\partial\varphi}\left(\sin\varphi w_B\right) = 0 \\ &\frac{\partial e_B}{\partial\varphi} + \frac{\partial}{\partial r}\left(r w_B\right) = 0\end{aligned} \tag{24.77}$$

where we have set

$$\begin{aligned}e_B &= \frac{1}{r}\left(\frac{1}{r}\frac{\partial}{\partial r}\left(r^2 B_r\right) + \frac{1}{\sin\varphi}\frac{\partial}{\partial\varphi}\left(\sin\varphi B_\varphi\right)\right) \\ w_B &= \frac{1}{r}\left(\frac{\partial}{\partial\varphi}\left(r B_\varphi\right) - \frac{\partial B_r}{\partial\varphi}\right)\end{aligned} \tag{24.78}$$

Solution for ψ: Solution of the function $\psi(r,\phi)$ has to be obtained by means of the first of Eq. (24.77). Since the temperature profile is written in terms of the Legendre polynomials, we seek the particular solution of ψ in the form

$$\psi\,(r,\varphi) = \sum_{n=0}^{\infty} b_n\,(r)\,P_n\,(\cos\varphi) \tag{24.79}$$

Substituting Eqs. (24.71) and (24.79) into the first of Eq. (24.77) results in

$$\frac{d}{dr}\left(r^2 b_n'(r)\right) - n\,(n+1)\,b_n(r) = \frac{1+\nu}{1-\nu}\alpha r^2 R_n(r) \tag{24.80}$$

which has the exact solution for $b_n(r)$'s as

$$b_n\,(r) = \frac{1+\nu}{1-\nu}\frac{\alpha}{2n+1}\left(r^n\int_{r_{in}}^{r}\xi^{1-n}R_n\,(\xi)\,d\xi - r^{-n-1}\int_{r_{in}}^{r}\xi^{n+2}R_n\,(\xi)\,d\xi\right) \tag{24.81}$$

with the substitution of Eq. (24.70) into the above equation, closed-form expressions for $b_n(r)$ are deduced

$$b_n(r) = \frac{1+\nu}{1-\nu}\alpha\left(\frac{\alpha_n}{2(2n+3)}r^{n+2} - \frac{\beta_n}{2(2n-1)}r^{-n+1}\right) \tag{24.82}$$

Having developed the function $\psi(r, \phi)$, the particular solution for displacement components can be evaluated based on Eq. (24.73). Substitution of the displacements into the stress-displacement relations gives us the particular stress components in the form

$$\begin{aligned}
\frac{\sigma_r^T}{2\mu} &= \sum_{n=0}^{\infty}\left\{-\frac{2}{r}b_n^{'}(r) + \frac{n(n+1)}{r^2}b_n(r)\right\}P_n(\cos\varphi)\\
\frac{\sigma_\varphi^T}{2\mu} &= \sum_{n=0}^{\infty}\left\{-\frac{1}{r}\left(rb_n^{'}(r)\right)^{'}P_n(\cos\varphi) + b_n(r)\frac{dP_n(\cos\varphi)}{d\varphi}\cot\varphi\right\}\\
\frac{\sigma_\theta^T}{2\mu} &= \sum_{n=0}^{\infty}\left\{-\frac{1}{r}\left(rb_n^{'}(r)\right)^{'}P_n(\cos\varphi) - n(n+1)b_n(r)P_n(\cos\varphi)\right.\\
&\quad\left. - b_n(r)\frac{dP_n(\cos\varphi)}{d\varphi}\cot\varphi\right\}\\
\frac{\sigma_{r\varphi}^T}{2\mu} &= \sum_{n=1}^{\infty}\left\{-\frac{1}{r^2}b_n(r) + \frac{1}{r}b_n^{'}(r)\right\}\frac{dP_n(\cos\varphi)}{d\varphi}
\end{aligned} \tag{24.83}$$

with the substitution of Eq. (24.83) into the Eq. (24.82), above equations simplify to

$$\begin{aligned}
\frac{\sigma_r^T}{2\mu} &= \frac{\alpha(1+\nu)}{2(1-\nu)}\sum_{n=0}^{\infty}\left\{\frac{n^2-n-4}{2n+3}\alpha_n r^n - \frac{n^2+3n-2}{2n-1}\beta_n r^{-n-1}\right\}P_n(\cos\varphi)\\
\frac{\sigma_\varphi^T}{2\mu} &= \frac{\alpha(1+\nu)}{2(1-\nu)}\sum_{n=0}^{\infty}\left\{\left(\frac{(n+2)^2}{2n+3}\alpha_n r^n - \frac{(n-1)^2}{2n-1}\beta_n r^{-n-1}\right)P_n(\cos\varphi)\right\}\\
&\quad + \frac{\alpha(1+\nu)}{2(1-\nu)}\sum_{n=0}^{\infty}\left\{\left(\frac{\alpha_n r^n}{2n+3} + \frac{\beta_n r^{-n-1}}{2n-1}\right)\frac{dP_n(\cos\varphi)}{d\varphi}\cot\varphi\right\}\\
\frac{\sigma_\theta^T}{2\mu} &= -\frac{\alpha(1+\nu)}{2(1-\nu)}\sum_{n=0}^{\infty}\left\{\left(\frac{3n+4}{2n+3}\alpha_n r^n + \frac{3n-1}{2n-1}\beta_n r^{-n-1}\right)P_n(\cos\varphi)\right\}\\
&\quad - \frac{\alpha(1+\nu)}{2(1-\nu)}\sum_{n=0}^{\infty}\left\{\left(\frac{\alpha_n r^n}{2n+3} - \frac{\beta_n r^{-n-1}}{2n-1}\right)\frac{dP_n(\cos\varphi)}{d\varphi}\cot\varphi\right\}\\
\frac{\sigma_{r\varphi}^T}{2\mu} &= \frac{\alpha(1+\nu)}{2(1-\nu)}\sum_{n=1}^{\infty}\left\{\frac{n+1}{2n+3}\alpha_n r^n + \frac{n}{2n-1}\beta_n r^{-n-1}\right\}\frac{dP_n(\cos\varphi)}{d\varphi}
\end{aligned} \tag{24.84}$$

When the sphere is solid $\beta_n = 0$, and when spherical coordinates is used for a half-space $\alpha_n = 0$.

Now we turn to determination of u_r^* and u_φ^*. To find the general solution for e_B and w_B, these function are written in terms of Legendre polynomials such as

$$
\begin{aligned}
e_B &= \sum_{n=-\infty}^{\infty} e_{Bn} = \sum_{n=-\infty}^{\infty} C_n^{'} r^n P_n(\cos\varphi) \\
w_B &= \sum_{n=-\infty}^{\infty} w_{Bn} = \sum_{n=-\infty}^{\infty} D_n^{'} r^n \frac{dP_n(\cos\varphi)}{d\varphi}
\end{aligned}
\tag{24.85}
$$

Substitution of these equations into the third of Eq. (24.77). results in the relation between constants $C_n^{'}$ and $D_n^{'}$ as

$$
C_n^{'} + (n+1) D_n^{'} = 0 \tag{24.86}
$$

Recalling the above equation and substituting Eq. (24.85) into Eq. (24.77) gives

$$
\begin{aligned}
&\frac{1}{r}\frac{\partial}{\partial r}\left(r^2 B_{rn}\right) + \frac{1}{\sin\varphi}\frac{\partial}{\partial\varphi}\left(\sin\varphi B_{\varphi n}\right) = -(n+1) D_n^{'} r^{n+1} P_n(\cos\varphi) \\
&\frac{\partial\left(rB_{\varphi n}\right)}{\partial r} - \frac{\partial B_{rn}}{\partial\varphi} = D_n^{'} r^{n+1} \frac{dP_n(\cos\varphi)}{d\varphi}
\end{aligned}
\tag{24.87}
$$

A particular solution of the above equations is easy to obtain in the form

$$
\begin{aligned}
B_{rn} &= -(n+1) A_n r^{n+1} P_n(\cos\varphi) \\
B_{\varphi n} &= A_n r^{n+1} \frac{dP_n(\cos\varphi)}{d\varphi}
\end{aligned}
\tag{24.88}
$$

where in the above equation, the new constant $A_n = \dfrac{D_n^{'}}{2n+3}$ is introduced. It should be emphasized that there is no need to obtain the particular solutions for B_{rn} and $B_{\varphi n}$ since the boundary conditions will be applied next when governing equations associated to B_0 are solved. Thus the general solution for B_r and B_φ are not necessary. In the next we present the function B_0 as the sum of the terms B_{0n} as

$$
B_0 = \sum_{n=-\infty}^{\infty} B_{0n} = \sum_{n=-\infty}^{\infty} -B_n r^n P_n(\cos\varphi) \tag{24.89}
$$

Substituting the later expression and Eq. (24.88) into Eq. (24.73) and using the relation $P_n(\cos\varphi) = P_{-n-1}(\cos\varphi)$ one may reach to

$$
\begin{aligned}
u_r^* =& \sum_{n=0}^{\infty} \left\{ (n+1)(n-2+4\nu) A_n r^{n+1} + n B_n r^{n-1} \right\} P_n(\cos\varphi) \\
&+ \sum_{n=0}^{\infty} \left\{ (n+3-4\nu) C_n r^{-n} - (n+1) D_n r^{-n-2} \right\} P_n(\cos\varphi) \\
u_\varphi^* =& \sum_{n=0}^{\infty} \left\{ (n+5-4\nu) A_n r^{n+1} + B_n r^{n-1} \right. \\
&\left. + (4-n-4\nu) C_n r^{-n} + D_n r^{-n-2} \right\} \frac{dP_n(\cos\varphi)}{d\varphi}
\end{aligned}
\tag{24.90}
$$

where in the above equation the new constants $C_n = A_{-n-1}$ and $D_n = B_{-n-1}$ are introduced. Stress components associated to displacements u_r^* and u_φ^* are

$$
\begin{aligned}
\frac{\sigma_r^*}{2\mu} =& \sum_{n-0}^{\infty} \left\{ (n+1)\left(n^2-n-2-2\nu\right) A_n r^n + n(n-1) B_n r^{n-2} \right\} \\
&+ \sum_{n=0}^{\infty} \left\{ -n\left(n^2+3n-2\nu\right) C_n r^{-n-1} + (n+1)(n+2) D_n r^{-n-3} \right\} \\
\frac{\sigma_\varphi^*}{2\mu} =& -\sum_{n=0}^{\infty} \left\{ (n+1)\left(n^2+4n+2+2\nu\right) A_n r^n + n^2 B_n r^{n-2} \right\} P_n(\cos\varphi) \\
&- \sum_{n=0}^{\infty} \left\{ -n\left(n^2-2n-1+2\nu\right) C_n r^{-n-1} + (n+1)^2 D_n r^{-n-3} \right\} P_n(\cos\varphi) \\
&- \sum_{n=0}^{\infty} \left\{ (n+5-4\nu) A_n r^n + B_n r^{n-2} \right. \\
&\left. + (4-n-4\nu) C_n r^{-n-1} + D_n r^{-n-3} \right\} \frac{dP_n(\cos\varphi)}{d\varphi} \cot\varphi \\
\frac{\sigma_\theta^*}{2\mu} =& \sum_{n=0}^{\infty} \left\{ (n+1)(n-2-2\nu-4n\nu) A_n r^n + n B_n r^{n-2} \right\} P_n(\cos\varphi) \\
&+ \sum_{n=0}^{\infty} \left\{ n(n+3-2\nu-4n\nu) C_n r^{-n-1} - (n+1) D_n r^{-n-3} \right\} P_n(\cos\varphi) \\
&+ \sum_{n=0}^{\infty} \left\{ (n+5-4\nu) A_n r^n + B_n r^{n-2} \right. \\
&\left. + (4-n-4\nu) C_n r^{-n-1} + D_n r^{-n-3} \right\} \frac{dP_n(\cos\varphi)}{d\varphi} \cot\varphi
\end{aligned}
$$

$$\frac{\sigma^*_{r\varphi}}{2\mu} = \sum_{n=1}^{\infty} \left\{ \left(n^2 + 2n - 1 + 2\nu \right) A_n r^n + (n-1) B_n r^{n-2} \right\} \frac{dP_n(\cos\varphi)}{d\varphi}$$
$$+ \sum_{n=1}^{\infty} \left\{ \left(n^2 - 2 + 2\nu \right) C_n r^{-n-1} - (n+2) D_n r^{-n-3} \right\} \frac{dP_n(\cos\varphi)}{d\varphi} \tag{24.91}$$

The constants A_n, B_n, C_n, and D_n are determined using the given boundary conditions. Since the sphere is subjected to thermal loads only, the stress-free conditions at inner and outer radii should be satisfied. In this case we may write

$$\begin{aligned} r = a, b: &\quad \sigma_r = \sigma^*_r + \sigma^T_r = 0 \\ r = a, b: &\quad \sigma_{r\varphi} = \sigma^*_{r\varphi} + \sigma^T_{r\varphi} = 0 \end{aligned} \tag{24.92}$$

We will consider the cases $n = 0$, $n = 1$ and $n \geq 2$, separately. For the case $n = 0$, $\frac{dP_0}{d\varphi} = 0$, and using Eqs. (24.84) and (24.91), $\sigma_{r\varphi} = 0$ and other stresses are function of r only. From the four constants of integration, there remains only A_0 and D_0. These constants are determined by the two algebraic equations

$$\begin{aligned} 2(1+\nu) A_0 - \frac{2}{a^3} D_0 &= \frac{-2}{a} b_0'(a) \\ 2(1+\nu) A_0 - \frac{2}{b^3} D_0 &= \frac{-2}{b} b_0'(b) \end{aligned} \tag{24.93}$$

Solving these equations for A_0 and D_0 gives us

$$\begin{aligned} A_0 &= \frac{1}{(1+\nu)\left(b^3 - a^3\right)} \left[b^2 b_0'(b) - a^2 b_0'(a) \right] \\ D_0 &= \frac{a^2 b^2}{b^3 - a^3} \left[b b_0'(a) - a b_0'(b) \right] \end{aligned} \tag{24.94}$$

In the case of steady state temperature distribution, from Eq. (24.82)

$$b_0(r) = \frac{1+\nu}{1-\nu} \alpha \left(\frac{\alpha_0}{6} r^2 + \frac{\beta_0}{2} r \right) \tag{24.95}$$

which yields the stresses to be

$$\begin{aligned} \sigma_r &= -4\mu \left(\frac{1}{r} b_0' + (1+\nu) A_0 - \frac{1}{r^3} D_0 \right) \\ \sigma_\theta = \sigma_\varphi &= -2\mu \left(b_0'' + \frac{1}{r} b_0' + 2(1+\nu) A_0 + \frac{2}{r^3} D_0 \right) \end{aligned} \tag{24.96}$$

In the case when $n = 1$, the coefficient B_1 in the associated equation vanishes (see the first of Eq. (24.91)). For the determination of the remaining constants A_1, C_1, and D_1 we have

$$
\begin{aligned}
&-4(1+\nu)aA_1 - \frac{2(2-\nu)}{a^2}C_1 + \frac{6}{a^4}D_1 = \frac{2}{a}\left(b_1^{'}(a) - \frac{1}{a}b_1(a)\right) = \lambda_1 \\
&2(1+\nu)aA_1 - \frac{1-2\nu}{a^2}C_1 - \frac{3}{a^4}D_1 = \frac{1}{a}\left(\frac{1}{a}b_1(a) - b_1^{'}(a)\right) = \lambda_2 \\
&-4(1+\nu)bA_1 - \frac{2(2-\nu)}{b^2}C_1 + \frac{6}{b^4}D_1 = \frac{2}{b}\left(b_1^{'}(b) - \frac{1}{b}b_1(b)\right) = \lambda_3 \\
&2(1+\nu)bA_1 - \frac{1-2\nu}{b^2}C_1 - \frac{3}{b^4}D_1 = \frac{1}{b}\left(\frac{1}{b}b_1(b) - b_1^{'}(b)\right) = \lambda_4
\end{aligned}
\tag{24.97}
$$

The existence of a solution for the system of equations (24.97) requires that the determinant of the following matrix vanishes.

$$
\begin{bmatrix}
-4(1+\nu)a & -\dfrac{2(2-\nu)}{a^2} & \dfrac{6}{a^4} & -\lambda_1 \\
2(1+\nu)a & -\dfrac{1-2\nu)}{a^2} & -\dfrac{3}{a^4} & -\lambda_2 \\
-4(1+\nu)b & -\dfrac{2(2-\nu)}{b^2} & \dfrac{6}{b^4} & -\lambda_3 \\
2(1+\nu)b & -\dfrac{1-2\nu)}{b^2} & -\dfrac{3}{b^4} & -\lambda_4
\end{bmatrix}
\tag{24.98}
$$

This determinant has the value

$$
\Delta = \frac{36}{a^6 b^6}\left(b^5 - a^5\right)\left(1-\nu^2\right)\left[a^2(\lambda_1 + 2\lambda_2) - b^2(\lambda_3 + 2\lambda_4)\right] \tag{24.99}
$$

and the solution is

$$
\begin{aligned}
&C_1 = 0 \\
&A_1 = \frac{b^2\left[bb_1^{'}(b) - b_1(b)\right] - a^2\left[ab_1^{'}(a) - b_1(a)\right]}{2(1+\nu)\left(a^5 - b^5\right)} \\
&D_1 = \frac{a^4 b^4}{3\left(a^5 - b^5\right)}\left\{\frac{a}{b}\left[b_1^{'}(b) - \frac{1}{b}b_1(b)\right] - \frac{b}{a}\left[b_1^{'}(a) - \frac{1}{a}b_1(a)\right]\right\}
\end{aligned}
\tag{24.100}
$$

For $n \geq 2$, similar procedure is followed. The solution of the problem can be considered to be accomplished.

Chapter 25
Piping Systems

Piping systems are essential components in many industries such as refineries, power plants, and chemical plants, where their prime purpose is the transport of fluid from one piece of equipment to another. Normally, the content fluid of the pipe is hot, and since the piping system is initially designed at reference temperature, the temperature change causes thermal expansion. If the ends of the piping system are restricted, which is usually the case, forces and moments are produced through the pipe system and at the supports of the pipes causing thermal stresses in the system. The art of piping flexibility analysis is to give enough flexibility to the piping system so that the resulting stresses at all points of the system remain under a safe limit. Usually, this flexibility is designed with a loop in the system or flexible joints at the ends. Therefore, the design procedure of a piping system is to consider the isometric of the piping system at the reference temperature. Then, by means of analytical or numerical methods the restrained forces and moments at the support of the piping system are calculated and, finally, by sketching the free body diagram of each piping member using the appropriate codes and standards, the stresses are computed and compared with the safe limit. If the calculated stresses are above the allowable limit, a loop for the piping system at a proper location may be considered and the calculation procedure is repeated.

25.1 Thermal Expansion of Piping Systems

There are many methods for the calculation of the restrained forces and moments of a piping system under thermal expansion. In this chapter we discuss an analytical technique based on the elastic center method. The stiffness approach or the finite element method may also be employed for pipeline analysis. The stiffness approach is essentially derived from the structural analysis under mechanical loads. Since thermal loads behave similarly to mechanical loads, they may be included in the stiffness method of analysis of structures. In the following sections the elastic center method is described.

M. Reza Eslami et al., *Theory of Elasticity and Thermal Stresses*, Solid Mechanics and Its Applications 197, DOI: 10.1007/978-94-007-6356-2_25,

25.1.1 Definition of the Elastic Center

Consider a piping system in two-dimensions with clamped ends A and B. The isometric lines of the piping system are shown in Fig. 25.1. The piping system is assumed to be under a uniform reference temperature T_0. If we ignore the weight of the pipes, the reaction forces and moments at ends A and B at the reference temperature T_0 are zero. Now, a hot fluid is passed through the piping system and the temperature of the piping system is raised to T. It is again assumed that the temperature T is constant through the length of the piping system. If the end B is considered free, due to thermal expansion of the piping system, it travels to point B', as seen from Fig. 25.1 To bring point B' to B, considering the clamped condition, forces F_x, F_y, and a bending moment M_B must be applied at point B, see Fig. 25.2. The application of these forces and moment at point B produce opposite and equal reaction forces F_x and F_y at A, as well as a bending moment M_A which in general is not equal to M_B. The pipeline analysis entails computation of the reaction forces and moments at the ends of the piping system, based upon free-body diagrams of each pipe element. The stress analysis of the pipe is then carried out in conjunction with an acceptable engineering code.

To obtain the reaction forces F_x and F_y and the reaction moments M_A and M_B, the method of elastic center may be used. According to this method, the elastic center of the piping system is found and a coordinate system is fixed to it. In the

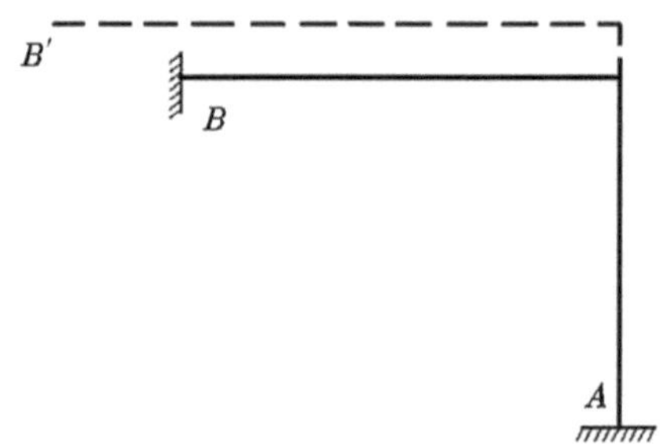

Fig. 25.1 Pipe at reference temperature and at raised temperature, without constraint at B

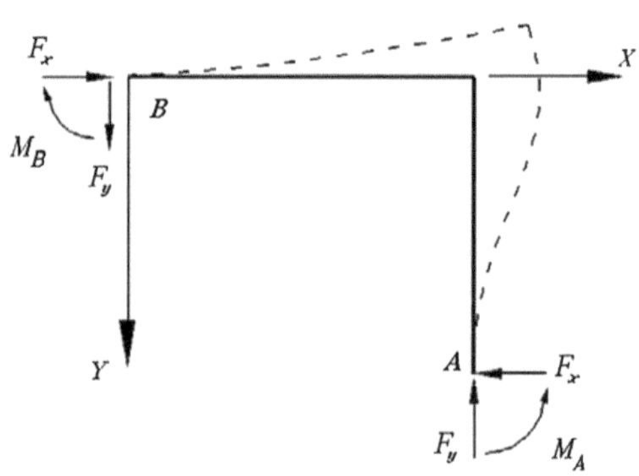

Fig. 25.2 Pipe at elevated temperature

coordinate system fixed to the elastic center, the bending moments do not appear and, therefore, the problem is reduced to two equations for two unknown forces F_x and F_y. To find the elastic center and the associated coordinate system, the Maxwell reciprocity theorem is used. A general treatment of this theorem is given in Sect. 15 of Chap. 2 of Ref. [3]. The general reciprocity theorem is reduced to a simple law for a piping system in two-dimensions.

The Maxwell Reciprocity Theorem

The work done by the loads of the first state on the corresponding deformations of the second state is equal to the work done by the loads of the second state on the corresponding deformations of the first state.

To apply the reciprocity theorem to a piping system in two-dimensions, consider Figs. 25.3a and b with the first and second states as defined below:

The first state:

- The load is the bending moment M_B acting at point B.
- The deformations are the horizontal and vertical displacements δ_{xo} and δ_{yo}, respectively, and the rotation ϕ_o at an arbitrary point O (which is elastically connected to the piping system).

The second state:

- The loads are F_x and F_y applied at point O.
- The deformations are δ_{xB}, δ_{yB}, and the rotation ϕ_B, produced at point B.

According to Maxwell theorem

$$M_B\phi_B = F_x\delta_{xo} + F_y\delta_{yo} \tag{25.1}$$

Now, if the point O is selected so that the application of forces F_x and F_y at O do not produce any rotation in the piping system (and, consequently, no bending moment

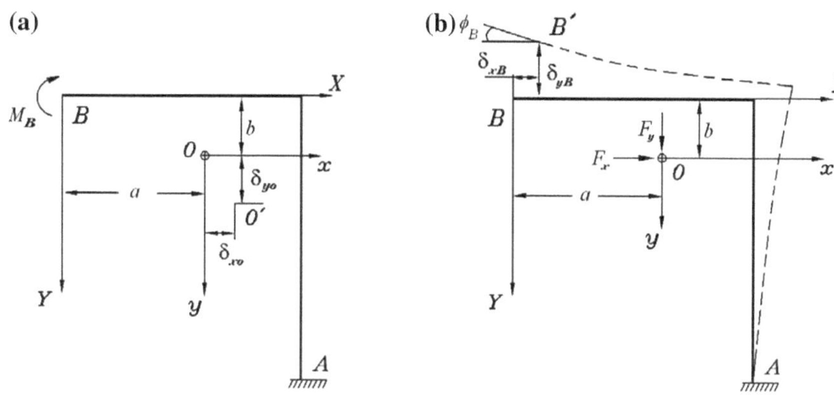

Fig. 25.3 (**a**) The first and (**b**) second states

is produced), then $\phi_B = 0$. From Eq. (25.1), since $\phi_B = 0$, the left-hand side is zero and thus

$$F_x \delta_{xo} + F_y \delta_{yo} = 0 \tag{25.2}$$

Since F_x and F_y are, in general, different from zero, it follows that

$$\begin{aligned} \delta_{xo} &= 0 \\ \delta_{yo} &= 0 \end{aligned} \tag{25.3}$$

Since δ_{xo} and δ_{yo} are the displacements produced by M_B acting at point B, Eq. (25.3) provide a method of calculation of the location of the point O. This point is called the *elastic center*, and it is a fictitious point in the piping system. Its property is that if the forces F_x and F_y act at that point, no rotation and thus the moments appear at any point on the piping system. But, since F_x and F_y are needed at points other than O, moments M_A and M_B will appear in the system.

Now consider a two-dimensional piping system with fixed ends A and B under elevated temperature, as shown in Fig. 25.4. The coordinate axes are fixed to the end B. The position of the elastic center is shown at point O with the coordinates a and b from the end B. We let the end B remain free to move. Due to the action of the bending moment M_B at point B, this point is displaced to point B', as shown in Fig. 25.5, with horizontal and vertical displacements u_B and v_B and the end line rotation ϕ_B. The location of the elastic center O is found using Eq. (25.3). From the elementary theory of strength of materials, the slope of a pipe element ds due to the constant bending moment M_B is

$$d\phi = \frac{M_B}{EI}\, ds$$

or

$$\phi_B = M_B \int_L \frac{ds}{EI} \tag{25.4}$$

where the integration is carried over the length of the pipe L. The horizontal displacement of the end B due to the rotation of an element ds at the vertical distance y

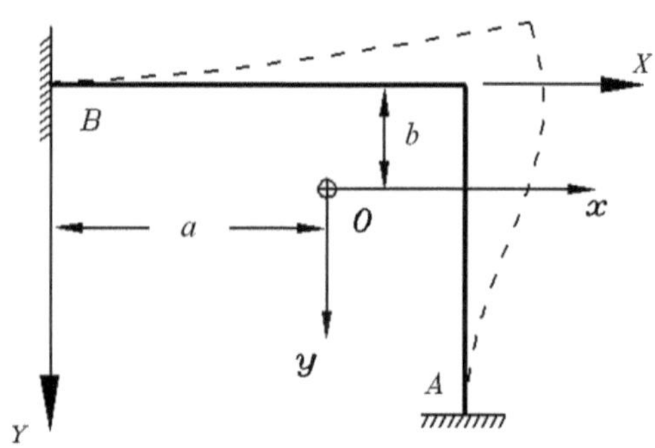

Fig. 25.4 Position of the elastic center

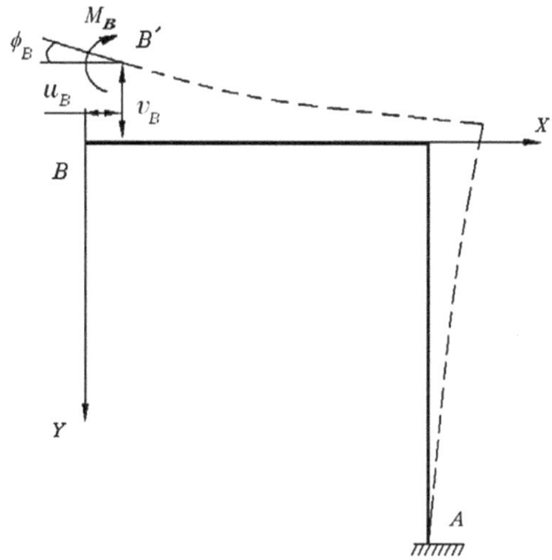

Fig. 25.5 Displacement of the end point B due to the bending moment M_B

from B is

$$u_B = \int_L y\,d\phi = M_B \int_L \frac{yds}{EI} \tag{25.5}$$

Similarly, the vertical displacement of B due to the rotation of an element ds at the horizontal distance x from B is

$$v_B = \int_L x\,d\phi = M_B \int_L \frac{xds}{EI} \tag{25.6}$$

The displacements u_B and v_B and the rotation ϕ_B at the end B cause the point O to be displaced. We may assume that a rigid bar connects points O and B. The total horizontal and vertical displacements of point O due to the displacements u_B and v_B and the rotation ϕ_B at point B are, see Fig. 25.3b

$$\begin{aligned} \delta_{xo} &= u_B - b\phi_B \\ \delta_{yo} &= a\phi_B - v_B \end{aligned} \tag{25.7}$$

Therefore, when the material and the pipe properties remain constant through the length of the piping system, from Eq. (25.3) and upon the substitution from Eq. (25.4), the coordinates a and b of the elastic center O are

$$\begin{aligned} a &= \frac{v_B}{\phi_B} = \frac{\int_L xds}{\int_L ds} = \bar{x} \\ b &= \frac{u_B}{\phi_B} = \frac{\int_L yds}{\int_L ds} = \bar{y} \end{aligned} \tag{25.8}$$

It is noticed that the elastic center O of a two-dimensional piping system coincides with the center of mass of the isometric of the piping system.

25.2 Piping Systems in Two-Dimensions

Consider a piping system in two-dimensions, as shown in Fig. 25.6. A hot fluid is passed through the piping system and its temperature is raised from T_0 to T. It is assumed that T_0 and T are uniform along the piping system. The global coordinate axes are fixed at one of the ends of the piping system. The rule is that the considered fixed end is arbitrarily made free, while the other end is fixed, and the global coordinate axes are chosen in opposite direction of the thermal expansion of the free end. Let, arbitrarily, the end B be free, while the end A remain fixed. The global coordinate system (x, y) is fixed to the point B and the axes are pointed in opposite directions to their thermal expansion. The location of the elastic center O is calculated and the global coordinate system is transferred to the point O.

The deformation of the piping system in the coordinates xy fixed to the elastic center O, and in terms of the thermal deflections, are

$$\begin{aligned} \Delta_{xx} + \Delta_{yx} &= \Delta_x \\ \Delta_{xy} + \Delta_{yy} &= \Delta_y \end{aligned} \tag{25.9}$$

where

Δ_{xx} = deflection of the piping system due to force F_x in x direction.
Δ_{yx} = deflection of the piping system due to force F_y in x direction.
Δ_x = thermal expansion of the piping system in x direction due to the temperature change $(T - T_0)$.
Symbols Δ_{xy}, Δ_{yy}, and Δ_y are similarly defined.

To obtain the thermal deformation, Castigliano theorem may be used. According to Castigliano theorem, the deflection of a piping system in a specific direction is

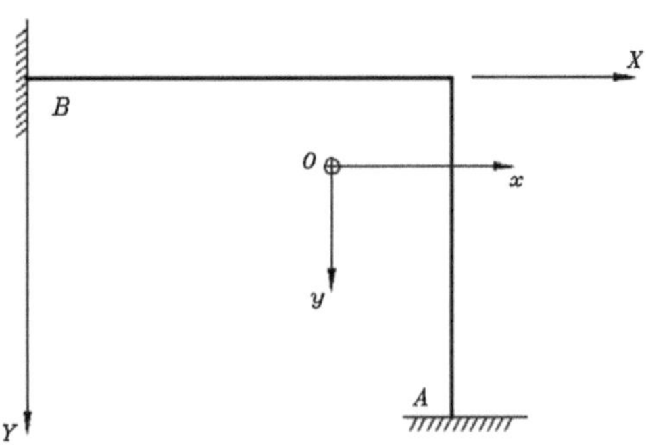

Fig. 25.6 A piping system in two-dimensions

the partial derivative of the strain energy with respect to the force in that specific direction.

From the elementary theory of strength of materials, the strain energy of a pipe segment of length L under the axial force P and bending moment M is

$$U = \int_0^L \frac{P^2 ds}{2AE} + \int_0^L \frac{M^2 ds}{2EI} \tag{25.10}$$

where E is the modulus of elasticity, A is the cross sectional area of the pipe, and I is the moment of inertia of the pipe cross section. Neglecting the axial strain energy as small compared to the bending strain energy, we have

$$U = \int_0^L \frac{M^2 ds}{2EI} \tag{25.11}$$

Consider an element ds of the piping system in the local coordinate system xy fixed to the elastic center. To obtain Δ_{xx}, we apply the force F_x to point O, Fig. 25.7. Since the deflection in x direction is required, an auxiliary force F_{xa} is applied at O, where the deflection is

$$\Delta_{xx} = \frac{\partial U}{\partial F_{xa}}|_{(F_{xa}=0)} \tag{25.12}$$

From Fig. 25.7, the moment about the element ds is

$$M = (F_x + F_{xa})y$$

where the moment in the clockwise direction is considered positive. Thus

$$\frac{\partial U}{\partial F_{xa}} = \int \frac{M}{EI} \frac{\partial M}{\partial F_{xa}} ds$$

Therefore, substituting into Eq. (25.12) and using $M = (F_x + F_{xa})y$, yields

$$\Delta_{xx} = \frac{F_x I_{xx}}{EI} \tag{25.13}$$

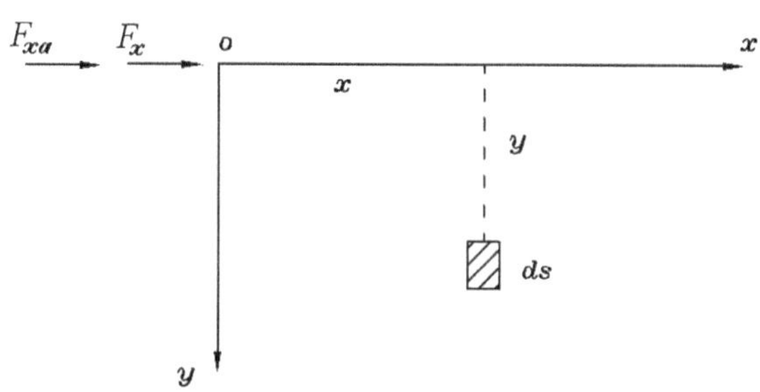

Fig. 25.7 Calculation of Δ_{xx}

Fig. 25.8 Calculation of Δ_{yx}

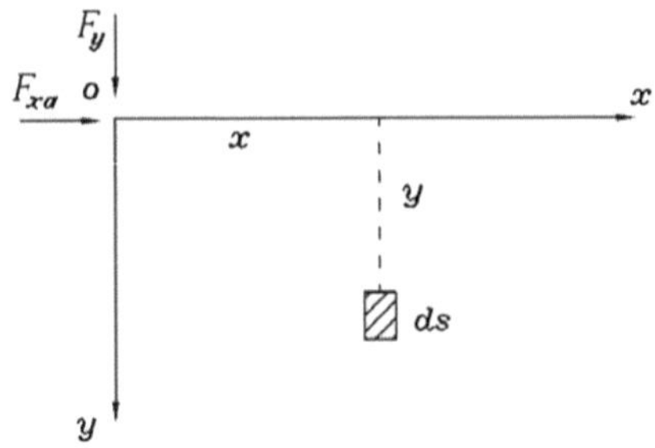

where $I_{xx} = \int y^2 ds$ is the line moment of inertia of the piping system about the x-axis.

Deformation Δ_{yx} is calculated by applying the force F_y at point O along the y direction and the auxiliary force F_{xa} along the x-axis, as shown in Fig. 25.8. The deformation Δ_{yx} is

$$\Delta_{yx} = \frac{\partial U}{\partial F_{xa}}|_{(F_{xa}=0)} = \int \frac{M}{EI}\frac{\partial M}{\partial F_{xa}} ds \tag{25.14}$$

The bending moment of forces acting at O about the element ds is

$$M = F_{xa}y - F_y x$$

Substituting into Eq. (25.14) gives

$$\Delta_{yx} = -\frac{F_y I_{xy}}{EI} \tag{25.15}$$

where $I_{xy} = \int xyds$ is the line product of inertia of the piping system about the xy-axes.

To calculate Δ_{xy}, we apply the force F_x at point O and the auxiliary force F_{ya} at O along the y-axis, as shown in Fig. 25.9. The deformation Δ_{xy} is

Fig. 25.9 Calculation of Δ_{xy}

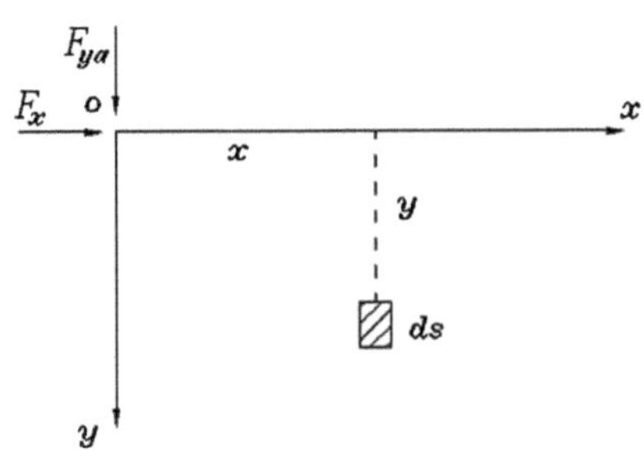

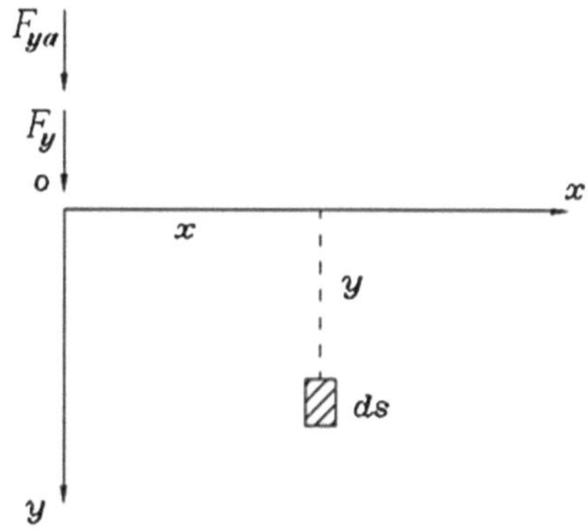

Fig. 25.10 Calculation of Δ_{yy}

$$\Delta_{xy} = \frac{\partial U}{\partial F_{ya}}|_{(F_{ya}=0)} = \int \frac{M}{EI}\frac{\partial M}{\partial F_{ya}}\,ds \tag{25.16}$$

The bending moment about *ds* is

$$M = F_x y - F_{ya} x$$

and the deformation Δ_{xy} is

$$\Delta_{xy} = -\frac{F_x I_{xy}}{EI} \tag{25.17}$$

Finally, the deformation Δ_{yy} is obtained by applying the force F_y and the auxiliary force F_{ya} at point O along the y-axis, as shown in Fig. 25.10. The deformation is

$$\Delta_{yy} = \frac{\partial U}{\partial F_{ya}}|_{(F_{ya}=0)} = \int \frac{M}{EI}\frac{\partial M}{\partial F_{ya}}\,ds \tag{25.18}$$

The bending moment of forces about the element *ds* is

$$M = -(F_y + F_{ya})x$$

Substituting into Eq. (25.18) gives

$$\Delta_{yy} = \frac{F_y I_{yy}}{EI} \tag{25.19}$$

where $I_{yy} = \int x^2 ds$ is the line moment of inertia of the piping system about the y-axis. Substituting Eqs. (25.13), (25.15), (25.17), and (25.19) into Eq. (25.9) yields the system of equilibrium equations of the piping system in terms of the forces in the coordinate system fixed to the elastic center as

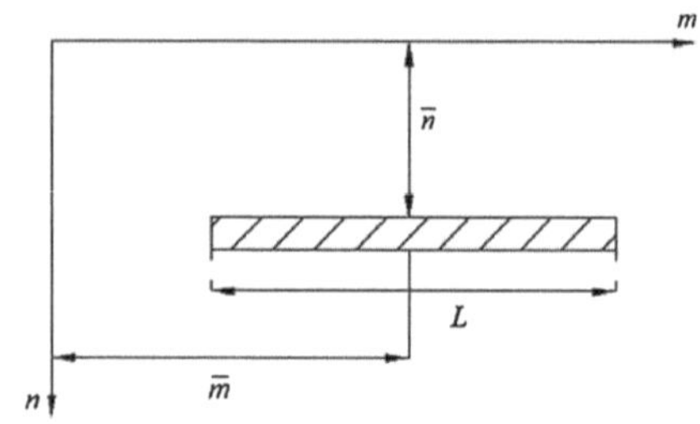

Fig. 25.11 Moment of inertia of a straight pipe

$$
\begin{aligned}
F_x I_{xx} - F_y I_{xy} &= EI\Delta_x \\
-F_x I_{xy} + F_y I_{yy} &= EI\Delta_y
\end{aligned}
\tag{25.20}
$$

This is a system of two equations for two unknowns F_x and F_y. Once the configuration of the piping system in xy-coordinates is known, the line moments of inertia are calculated. Given the pipe material and the pipe cross section properties, the constants E and I are known. The values of Δ_x and Δ_y are thermal expansions of the piping system in x and y directions, which are calculated knowing the coefficient of thermal expansion, temperature change, and the projection of the piping system between the end points. These known values are substituted into Eq. (25.20) and the system of equations is solved for F_x and F_y. The calculated forces are in the coordinate system fixed to the elastic center. The calculated forces with their moments are transferred to the end where the global coordinate is fixed. The forces F_x and F_y with their moments are transferred to the other end, changing their directions. The following example illustrates the technique discussed in this section.

As a note from statics, the line moments of inertia of a pipe of length L in $m - n$ plane, see Fig. 25.11, is

$$
\begin{aligned}
I_{mm} &= L\bar{n}^2 \\
I_{nn} &= \frac{L^3}{12} + L\bar{m}^2 \\
I_{mn} &= L\bar{m}\bar{n}
\end{aligned}
\tag{25.21}
$$

Equation (25.21) show that I_{mm} and I_{nn} are always positive, but I_{mn} may be positive or negative, depending on the coordinates of its elastic center. The algebraic sign of $\bar{m}$ and $\bar{n}$ must be considered in calculating I_{mn}.

25.3 Piping Systems in Three-Dimensions

The equilibrium equation of a three-dimensional piping system under thermal expansion in the coordinate system fixed to the elastic center can be established by means of the equilibrium of thermal deflection. In each direction, the total deflection of the

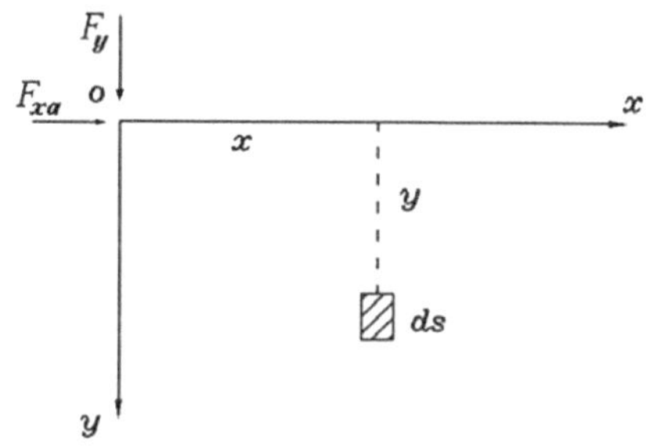

Fig. 25.12 Applied forces at elastic center in x-y plane

piping system due to the reaction forces must be equal to the thermal expansion. For a piping system in three-dimensions the deflection equations are

$$\begin{aligned} \Delta_{xx} + \Delta_{yx} + \Delta_{zx} &= \Delta_x \\ \Delta_{yx} + \Delta_{yy} + \Delta_{yz} &= \Delta_y \\ \Delta_{zx} + \Delta_{zy} + \Delta_{zz} &= \Delta_z \end{aligned} \tag{25.22}$$

where Δ_{ij} is the deflection of the piping system in j-direction due to a force in i-direction and Δ_x, Δ_y, and Δ_z are the total thermal expansions of the piping system in x, y, and z direction, respectively. To evaluate the deflection Δ_{ij}, Castigliano theorem may be used (Fig. 25.12).

The strain energy of a pipe segment of length L under the action of an axial force P, bending moment M_b, and torsional moment M_t, is

$$U = \int_0^L \frac{P^2 ds}{2AE} + \int_0^L \frac{M_b^2 ds}{2EI} + \int_0^L \frac{M_t^2 ds}{2GJ} \tag{25.23}$$

where I and J are the moment of inertia and the polar moment of inertia of the pipe cross section, respectively. In the evaluation of Δ_{ij}, the strain energy of the axial force may be ignored since it is small compared to the other terms. Since $G = E/2(1+\nu)$ and for the pipe cross section $J = 2I$, therefore, $GJ = EI/(1+\nu)$. The strain energy of a pipe under the bending moment M_b and the torque M_t is thus

$$U = \int_0^L \frac{M_b^2 ds}{2EI} + \int_0^L \frac{M_t^2 (1+\nu) ds}{2EI} \tag{25.24}$$

which suggests that the torsional strain energy of a pipe may be calculated similarly to that of the bending strain energy, but with an equivalent length of $(1+\nu)ds$. Now, consider a pipe element ds in x-y projection of the piping system, as shown in Fig. 25.12.

To obtain Δ_{yx}, for instance, the real force F_y and the auxiliary force F_{xa} are applied at the elastic center O and the strain energy of the pipe element ds under the action of F_y and F_{xa} is calculated. From Castigliano theorem the deformation is

$$\Delta_{yx} = \frac{\partial U}{\partial F_{xa}}|_{F_{xa}=0} = \int \frac{M_b}{EI}\frac{\partial M_b}{\partial F_{xa}} ds \qquad \text{for in-plane pipe member} \tag{25.25}$$

$$\Delta_{yx} = \frac{\partial U}{\partial F_{xa}}|_{F_{xa}=0} = (1+\nu)\int \frac{M_t}{EI}\frac{\partial M_t}{\partial F_{xa}} ds \qquad \text{for out-of-plane pipe member} \tag{25.26}$$

Forces F_y and F_{xa} produce bending moment $M_b = F_{xa}y - F_y x$ about element ds, if ds lies in x-y plane. Otherwise, that is, if ds is an element of out-of-plane pipe, forces F_y and F_{xa} produce a torque $M_t = F_{xa}y - F_y x$. Upon substitution of the moment M_b, or the torque M_t, in Eq. (25.25), or (25.26), and integrating over the length of the pipe, we obtain

$$\Delta_{yx} = -F_y \frac{I_{xy}}{EI} \tag{25.27}$$

where $I_{xy} = \int xyds$ is the product of inertia of the isometric line of the pipe system in x-y plane. In a similar manner, the remaining deformations in the x-y plane, Δ_{xx}, Δ_{yy}, and Δ_{xy}, may be calculated. Projection of the piping system in the y-z plane provides four deformations Δ'_{yy} (not to be confused with Δ_{yy} in the x-y plane), Δ_{yz}, Δ_{zy}, and Δ_{zz}, and projection in the x-z plane produce another set of four displacements Δ'_{xx}, Δ_{xz}, Δ_{zx}, and Δ'_{zz}, which in general, are

$$\Delta_{ii} = F_i \frac{I_{ii}}{EI} \qquad i \text{ not summed} \tag{25.28}$$

and

$$\Delta_{ji} = -F_j \frac{I_{ij}}{EI} \qquad i \neq j \quad j \text{ not summed} \tag{25.29}$$

Upon substitution into Eq. (25.22), the equilibrium equations of the piping systems in three-dimensions under thermal expansions are obtained as

$$\begin{aligned} (I_{xx} + I'_{xx})F_x - I_{xy}F_y - I_{xz}F_z &= EI\,\Delta_x \\ -I_{xy}F_x + (I_{yy} + I'_{yy})F_y - I_{yz}F_z &= EI\,\Delta_y \\ -I_{xz}F_x - I_{yz}F_y + (I_{zz} + I'_{zz})F_z &= EI\,\Delta_z \end{aligned} \tag{25.30}$$

Here, I_{xx} is the moment of inertia in the x-y plane and I'_{xx} is the moment of inertia in the x-z plane. Similarly, I_{yy} and I'_{yy} are the line moments of inertia in the x-y and y-z planes, and I_{zz} and I'_{zz} are the line moments of inertia in the y-z and x-z planes, respectively. In these equations the forces F_x, F_y, and F_z are unknowns. The quantity I is the moment of inertia of the pipe's cross section, E is the modulus of elasticity at the design temperature, Δ_x, Δ_y, and Δ_z are the total thermal expansions of the piping system in x, y, and z directions, respectively, which are obtained knowing the coefficient of thermal expansion of the pipe's material.

25.4 Problems and Solutions of Piping Systems

Problem 25.1. Consider a three dimensional pipeline configuration, where instead of fixed ends, the ends are supported by flexible joints. The flexible joints are assumed to work only along the pipeline axes. Modify the three-dimensional equilibrium equations of the pipeline system to include this type of boundary conditions.

Solution: Consider two flexible joints at the ends of the piping system, where their stiffness are denoted by $(k_{x_1}, k_{y_1}, k_{z_1})$ and $(k_{x_2}, k_{y_2}, k_{z_2})$.

The equilibrium equations of a three dimensional piping system under thermal expansion with flexible joints may be established by means of the equilibrium of thermal deflection. In each direction, the total deflection of the piping system due to the reaction forces must be equal to the thermal expansion minus the total deflection of each flexible joint as

$$\begin{aligned}
\Delta_{xT} - \Delta_{kx_1} - \Delta_{kx_2} &= \Delta_{xx} + \Delta_{yx} + \Delta_{zx} \\
\Delta_{yT} - \Delta_{ky_1} - \Delta_{ky_2} &= \Delta_{xy} + \Delta_{yy} + \Delta_{zy} \\
\Delta_{zT} - \Delta_{kz_1} - \Delta_{kz_2} &= \Delta_{xz} + \Delta_{yz} + \Delta_{zz}
\end{aligned} \tag{25.31}$$

where Δ_{ij} is the deflection of piping system in the $j-$direction due to a force in $i-$direction, and Δ_{xT}, Δ_{yT}, and Δ_{zT} are total thermal expansions of the piping system in the x, y, and z directions, respectively.

To evaluate the deflection Δ_{ij}, the Castigliano theorem may be used. For this purpose, total strain energy of the pipe, including the strain energies of the axial load P, bending moment M_b, and torsional moment M_t should be calculated. For a pipe with length L, one may obtain

$$U = \int_0^L \frac{P^2}{2AE} ds + \int_0^L \frac{M_b^2}{2EI} ds + \int_0^L \frac{M_t^2}{2JG} ds \tag{25.32}$$

For a pipe with circular cross section the polar moment of inertia J is twice of moment of inertia I, i.e. $J = 2I$. Also, the shear modulus G may be replaced in terms of the elasticity modulus E and Poisson's ratio ν, as $G = \dfrac{E}{2(1+\nu)}$. The strain energy of the axial load is negligible in comparison with the bending and torsional strain energies and may be neglected. Therefore, the following is obtained as the simplified form of the total strain energy

$$U = \int_0^L \frac{M_b^2}{2EI} ds + \int_0^L \frac{M_t^2(1+\nu)}{2EI} ds \tag{25.33}$$

Now, to obtain Δ_{yx}, for instance, the force F_y and an imaginary force F_{xa} are considered to be applied on the elastic center O, as shown in Fig. 25.13.

Fig. 25.13 Pipe element ds in x-y plane

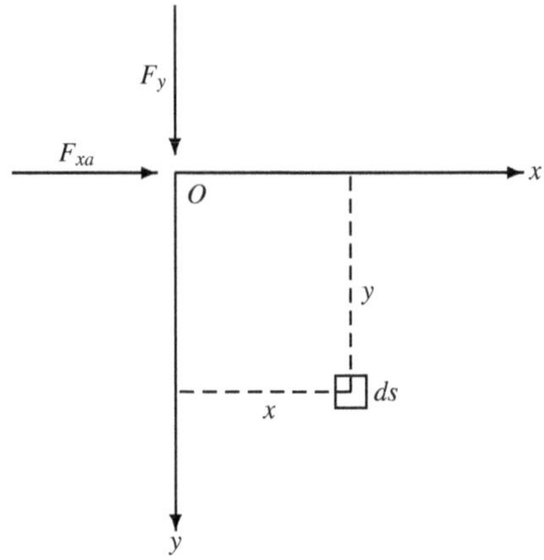

Based on the Castigliano theorem, the deformation for in-plane pipe members is

$$\Delta_{yx} = \frac{\partial U}{\partial F_{xa}} |_{F_{xa}=0} = \int \frac{M_b}{EI} \frac{\partial M_b}{\partial F_{xa}} ds \tag{25.34}$$

and for the out of plane pipe members is

$$\Delta_{yx} = \frac{\partial U}{\partial F_{xa}} |_{F_{xa}=0} = (1+\nu) \int \frac{M_t}{EI} \frac{\partial M_t}{\partial F_{xa}} ds \tag{25.35}$$

where the produced bending moment due to the action of forces F_{xa} and F_y is equal to $M_b = F_{xa}y - F_y x$, when the element lies in the x–y plane. If the element is out of x–y plane, a torsional moment is produced which is equal to $M_t = F_{xa}y - F_y x$. Now, substituting the bending moment, or torsional moment, into Eq. (25.34) or (25.35) gives the deflection of the pipe as

$$\Delta_{yx} = -F_y \frac{I_{xy}}{EI} \tag{25.36}$$

Note that in the case of torsional moment, the length of the pipe is assumed to be $(1+\nu)ds$ instead of ds. Here, I_{xy} is the product of inertia of the isometric line of the pipe in the x–y plane, that is $I_{xy} = \int xyds$. Using the same progress, other deflections in the x–y plane may be obtained. When the pipe is projected in the y–z plane, four deformations are produced which are Δ'_{yy}, Δ_{yz}, Δ_{zy}, and Δ_{zz}. Also, projection in the $x-z$ plane produces Δ'_{xx}, Δ_{xz}, Δ_{zx}, and Δ'_{zz}. In general, all of the deformations may be calculated as

$$\Delta_{ii} = F_i \frac{I_{ii}}{EI} \qquad \Delta_{ji} = -F_j \frac{I_{ij}}{EI} \tag{25.37}$$

Also, $\Delta_{k_{im}}$, $i = x, y, z$, $m = 1, 2$, is the deflection of the flexible joint in the i–direction, which is equal to

$$\Delta_{k_{im}} = \frac{F_i}{k_{im}} \tag{25.38}$$

Eqs. (25.37) and (25.38) have to be substituted into Eq. (25.31) to give the modified three dimensional equations of the pipeline systems as

$$\begin{aligned}
&\left(I_{xx} + I'_{xx} + \frac{EI}{k_{x_1}} + \frac{EI}{k_{x_2}}\right) F_x - I_{xy}F_y - I_{xz}F_z = EI\Delta_{xT} \\
-I_{yz}F_x + &\left(I_{yy} + I'_{yy} + \frac{EI}{k_{y_1}} + \frac{EI}{k_{y_2}}\right) F_y - I_{yz}F_z = EI\Delta_{yT} \\
-I_{zx}F_x - I_{zy}F_y + &\left(I_{zz} + I'_{zz} + \frac{EI}{k_{z_1}} + \frac{EI}{k_{z_2}}\right) F_z = EI\Delta_{zT}
\end{aligned} \tag{25.39}$$

Here, I_{xx} and I'_{xx} are the line moment of inertia in the x–y and x–z planes, respectively. Similarly, I_{yy} and I'_{yy} are the line moments of inertia in the x–y and y–z planes, respectively. The inertia moments in the y–z and x–z planes are denoted by I_{zz} and I'_{zz}. Thc system of Eq. (25.39) should be solved for the unknowns F_x, F_y, and F_z.

Problem 25.2. Consider the pipeline isometric, as shown in Fig. 25.14, under temperature change $\Delta T = 400°$ C. The pipe is No. 8, schedule 30 with the modulus of elasticity $E = 20600\,\text{kN/cm}^2$. The geometric properties of the pipe are

$$\begin{aligned}
D &= 21.908\,\text{cm} \\
I &= 2638.9\,\text{cm}^4 \\
t &= 0.688\,\text{cm}
\end{aligned}$$

The radius of elbows is $R = 150\,\text{cm}$. Find the pipeline reaction forces and moments at end points A and H.

Solution: The mean radii of the pipe is evaluated as

$$r = \frac{D-t}{2} = 10.61\,\text{cm} \tag{25.40}$$

The von-Karman bending rigidity factor K is obtained as

$$K = \frac{12\lambda^2 + 10}{10\lambda^2 + 1} = 1 + \frac{9r^4}{12t^2R^2 + r^4} = 1.812 \tag{25.41}$$

Now, the piping system should be projected in each plane (Fig. 25.14).
The $x-y$ plane: The length of the elements when are projected in the $x-y$ plane, are

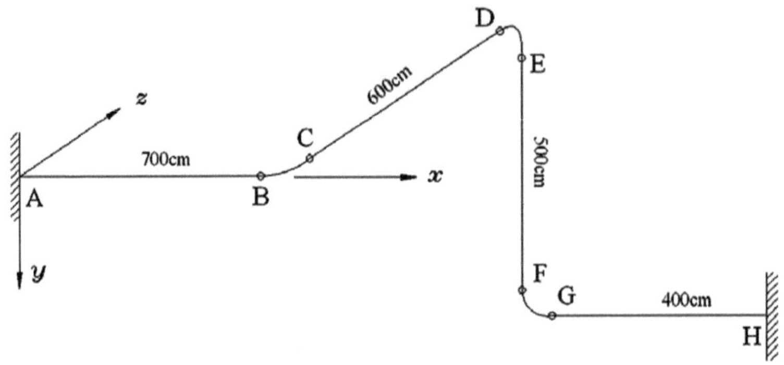

Fig. 25.14 Pipeline isometric

$$\begin{aligned}
AB &= 700\,\text{cm}\\
BC &= 1.15 \times 235.62 = 271\,\text{cm}\\
CD &= 1.3 \times 600 = 780\,\text{cm}\\
DE &= 271\,\text{cm}\\
EF &= 500\,\text{cm}\\
FG &= 1.812 \times 235.62 = 427\,\text{cm}\\
GH &= 400\,\text{cm}
\end{aligned} \tag{25.42}$$

The inertia moment I_{elbow} and the position of the elastic center $E.C._{elbow}$ of the elbow are found as

$$I_{elbow} = R^3\left(\frac{\pi}{4} - \frac{2}{\pi}\right) = 0.1488R^3, \qquad E.C._{elbow} = \frac{2}{\pi}R \tag{25.43}$$

The elastic center of the projection in x–y plane is evaluated by

$$\begin{aligned}
\overline{x} &= \frac{\sum_{i=1}^{7}\overline{x}_i L_i}{\sum_{i=1}^{7} L_i}\\
\overline{y} &= \frac{\sum_{i=1}^{7}\overline{y}_i L_i}{\sum_{i=1}^{7} L_i}
\end{aligned} \tag{25.44}$$

where L_i is the length of the ith pipe and $\overline{x}_i$ and $\overline{y}_i$ indicate the position of center of each pipe with respect to a prescribed point. These quantities are evaluated as

$$\sum_{i=1}^{7} L_i = 700 + 2 \times 271 + 780 + 500 + 427 + 400$$

$$\sum_{i=1}^{7} \bar{x}_i L_i = -700 \times 500 - 271 \times (1 - 0.6366) \times 150 + 780 \times 0$$
$$+ 271 \times 0 + 500 \times 0 + 427 \times (1 - 0.6366) \times 150 + 400 \times 350$$
$$\sum_{i=1}^{7} \bar{y}_i L_i = 700 \times 0 + 271 \times 0 + 780 \times 0 + 271 \times (1 - 0.6366) \times 150$$
$$+ 500 \times 400 + 427 \times (650 + 0.6366 \times 150) + 400 \times 800 \quad (25.45)$$

where $\bar{x}_i$ and $\bar{y}_i$ are measured from points F and A, respectively. After substitution of Eq. (25.45) into (25.44), one may reach to

$$\bar{x} = -60.166\,\text{cm}, \qquad \bar{y} = 254.73\,\text{cm} \quad (25.46)$$

The local coordinate axes are transferred to the elastic center and the line inertia moments of projection of the system are calculated as

$$\begin{aligned} I_{xx} = {} & 700 \times 254.73^2 + 271 \times (254.73)^2 + 1.15 \times 0.1488 \times 150^3 \\ & + 780 \times 254.73^2 + 1.15 \times 0.1488 \times 150^3 \\ & + 271 \times (254.73 - 150 \times (1 - 0.6366))^2 + 500 \times (400 - 254.73)^2 \\ & + \frac{1}{12} \times 500^3 + 1.812 \times 0.1488 \times 150^3 \\ & + 427 \times (650 + 0.6366 \times 150 - 254.73)^2 \\ & + 400 \times (800 - 254.73)^2 = 369283775.966\,\text{cm}^3 \\ I_{yy} = {} & 700 \times (500 - 60.166)^2 + \frac{1}{12} \times 700^3 \\ & + 1.15 \times 0.1488 \times 150^3 + 271 \times (-60.166 + 150 \times 0.3634)^2 \\ & + 780 \times (-60.166)^2 + 1.15 \times 0.1488 \times 150^3 \\ & + 271 \times (-60.166)^2 + 500 \times (-60.166)^2 + 1.812 \times 0.1488 \times 150^3 \\ & + 427 \times (60.166 + 150 \times 0.3634)^2 + \frac{1}{12} \times 400^3 \\ & + 400 \times (350 + 60.166)^2 = 233071384.763\,\text{cm}^3 \end{aligned}$$

$$\begin{aligned} I_{xy} = {} & 700 \times (500 - 60.166) \times 254.73 + 780 \times (-60.166) \times 254.73 \\ & + 500 \times 60.166 \times (400 - 254.73) + 1.812 \times (-\frac{1}{2} \times 150^3 \\ & + 150^2 \times (210.66 - 395.27) + 397.27 \times 210.166 \times 150 \times \frac{\pi}{2}) \\ & + 400 \times (350 + 60.166) \times (800 - 254.73) = 185148262.708\,\text{cm}^3 \end{aligned} \quad (25.47)$$

The piping system should be projected on other planes. For the sake of simplicity the calculations are omitted.
The $x - z$ plane: Similar to the process developed for x–y plane, the length of each pipe in this plane is

$$\begin{aligned}
AB &= 700\,\text{cm}\\
BC &= 1.812 \times 235.62 = 427\,\text{cm}\\
CD &= 600\,\text{cm}\\
DE &= 271\,\text{cm}\\
EF &= 1.3 \times 500 = 650\,\text{cm}\\
FG &= 1.15 \times 235.62 = 271\,\text{cm}\\
GH &= 400\,\text{cm}
\end{aligned} \tag{25.48}$$

The position of the elastic center

$$\overline{x} = -65.84\,\text{cm}, \qquad \overline{y} = 515.61\,\text{cm} \tag{25.49}$$

where $\overline{x}$ and $\overline{z}$ are measured from points F and A, respectively. And the magnitudes of line inertia moments are

$$\begin{aligned}
I'_{xx} &= 524206944.335\,\text{cm}^3\\
I_{zz} &= 247670711.248\,\text{cm}^3\\
I_{xz} &= 233095278.126\,\text{cm}^3
\end{aligned} \tag{25.50}$$

The y–z plane: The length of each pipe when is projected in this plane is

$$\begin{aligned}
AB &= 1.3 \times 700 = 910\,\text{cm}\\
BC &= 1.15 \times 235.62 = 271\,\text{cm}\\
CD &= 600\,\text{cm},\\
DE &= 427\,\text{cm}\\
EF &= 500\,\text{cm}\\
FG &= 1.15 \times 235.62 = 271\,\text{cm}\\
GH &= 1.3 \times 400 = 520\,\text{cm}
\end{aligned} \tag{25.51}$$

The position of elastic center is then obtained as

$$\overline{y} = 240.44\,\text{cm}, \qquad \overline{z} = 383.367\,\text{cm} \tag{25.52}$$

where $\overline{y}$ and $\overline{z}$ are measured from points A and F, respectively.

The magnitudes of line inertia moments are

$$I'_{yy} = 559408274.77\,\text{cm}^3$$
$$I'_{zz} = 374761821.725\,\text{cm}^3$$
$$I_{yz} = 239520234.191\,\text{cm}^3 \tag{25.53}$$

Now, the thermal expansions of the pipe should be evaluated. At 400°C, the the thermal expansion coefficient of the pipe is equal to $4.9153\,\text{cm}/10\,\text{m}$. Therefore, total thermal deformations in each direction are

$$\Delta_{xT} = 4.9153 \times 1.4 = 6.8814\,\text{cm}$$
$$\Delta_{yT} = 4.9153 \times 0.8 = 3.9322\,\text{cm}$$
$$\Delta_{zT} = 4.9153 \times 0.9 = 4.4238\,\text{cm} \tag{25.54}$$

The governing equilibrium equations for the forces at the boundaries are

$$(I_{xx} + I'_{xx})F_x - I_{xy}F_y - I_{xz}F_z = EI\Delta_{xT}$$
$$-I_{yz}F_x + (I_{yy} + I'_{yy})F_y - I_{yz}F_z = EI\Delta_{yT}$$
$$-I_{zx}F_x - I_{zy}F_y + (I_{zz} + I'_{zz})F_z = EI\Delta_{zT} \tag{25.55}$$

Substituting the magnitudes of total thermal deflections and moments of inertia into the system of Eq. (25.55) and solving for the unknowns F_x, F_y and F_z, yields

$$F_x = 0.8085\,\text{KN}$$
$$F_y = 0.7376\,\text{KN}$$
$$F_x = 0.9084\,\text{KN} \tag{25.56}$$

The moments acting on the boundaries due to the subjected thermal loading are

$$M_{zA} = 376.613\,\text{KN.cm}$$
$$M_{yA} = -295.44\,\text{KN.cm}$$
$$M_{zA} = 162.647\,\text{KN.cm}$$
$$M_{zB} = -9.1986\,\text{KN.cm}$$
$$M_{yB} = 248.62\,\text{KN.cm}$$
$$M_{zB} = 225.52\,\text{KN.cm} \tag{25.57}$$

Problem 25.3. Reconsider the problem of the previous section, where both ends are supported by axial flexible joints with axial spring constant $k_a = 600\,\text{N/m}$. Calculate the pipeline reaction forces and moments. Discuss the results.

Solution: The displacements of axial springs must be incorporated into the deflection Eq. (25.55) of problem (2) in x-direction. The effect of elastic boundaries have been reported in Eq. (25.39) of problem (1). In this equation, the force-equilibrium conditions for a piping system with all edges flexible are obtained. For two flexible joints in the x-direction, these equations simplify to

$$\left(I_{xx} + I'_{xx} + \frac{2EI}{k_a}\right) F_x - I_{xy}F_y - I_{xz}F_z = EI\Delta_{xT}$$
$$-I_{yz}F_x + (I_{yy} + I'_{yy})F_y - I_{yz}F_z = EI\Delta_{yT}$$
$$-I_{zx}F_x - I_{zy}F_y + (I_{zz} + I'_{zz})F_z = EI\Delta_{zT} \tag{25.58}$$

Substituting $k_a = 0.006\,\text{KN/cm}$ and the magnitudes of line moment of inertia from Eqs. (25.47), (25.50) and (25.53) of problem (2) into the Eq. (25.31) of problem (1) and solving the resulted system for unknowns F_x, F_y, and F_z gives us

$$F_x = 0.0304\,\text{KN}$$
$$F_y = 0.4391\,\text{KN}$$
$$F_z = 0.5282\,\text{KN} \tag{25.59}$$

The moments acting on the boundaries due to the subjected thermal loading are then evaluated by

$$M_{zA} = 339.094\,\text{KN.cm}$$
$$M_{yA} = -398.572\,\text{KN.cm}$$
$$M_{zA} = 99.8515\,\text{KN.cm}$$
$$M_{zB} = -251.355\,\text{KN.cm}$$
$$M_{yB} = 313.637\,\text{KN.cm}$$
$$M_{zB} = 127.25\,\text{KN.cm} \tag{25.60}$$

Proper design of flexible joints at the end connections of a piping system will reduce the induced forces. In this system, springs reduce the reaction forces in x-direction, significantly. Solving the system of equations shows that the reactions forces in other directions are also reduced.

Problem 25.4. Reconsider the problem 25.2 and find the reaction forces and moments when point I (the middle of pipe CD) is fixed. Discuss the results.

Solution: The piping system is divided into two systems, AI and IH.
Piping system AI: This system is lied in x–z plane. Length of each element in this plane are

$$AB = 700\,\text{cm}$$
$$BC = 427\,\text{cm}$$
$$CI = 300\,\text{cm} \tag{25.61}$$

The moments of inertia for the piping system are easily obtained as

$$
\begin{aligned}
I_{xx} &= 22436877.516\,\text{cm}^3 \\
I_{zz} &= 108120315.685\,\text{cm}^3 \\
I_{xz} &= 29197502.746\,\text{cm}^3
\end{aligned} \tag{25.62}
$$

and thermal expansions due to the evaluated temperature are

$$
\begin{aligned}
\Delta_x &= 4.9153 \times 0.85 = 4.178\,\text{cm} \\
\Delta_z &= 4.9153 \times 0.45 = 2.212\,\text{cm}
\end{aligned} \tag{25.63}
$$

For the problem in hand, elasticity modulus and moment of inertia are

$$
E = 20600\,\text{KN/cm}^2 \qquad I = 2638.9\,\text{cm}^4 \tag{25.64}
$$

Therefore, the forces acting on the boundaries are obtained when the following system of equations is solved

$$
\begin{aligned}
I_{xx}F_x - I_{xz}F_z &= EI\,\Delta_{xT} \\
-I_{xz}F_x + I_{zz}F_z &= EI\,\Delta_{zT}
\end{aligned} \tag{25.65}
$$

Solving the upper system of equations gives the boundary forces as

$$
\begin{aligned}
F_x &= 17.8387\,\text{KN} \\
F_z &= 5.9294\,\text{KN}
\end{aligned} \tag{25.66}
$$

The moments on points A and I are evaluated as

$$
\begin{aligned}
M_{zA} &= 2072.92\,\text{KN.m} \\
M_{zI} &= 5060.367\,\text{KN.m}
\end{aligned} \tag{25.67}
$$

Piping system *IH*:

The x–y plane:

In a similar manner discussed for piping system AI, we have

$$
\begin{aligned}
GH &= 400\,\text{cm} \\
EF &= 500\,\text{cm} \\
DI &= 300\,\text{cm} \\
FG &= 427\,\text{cm} \\
DI &= 271\,\text{cm}
\end{aligned} \tag{25.68}
$$

The moments of inertia for the piping system projected in the x–y plane are calculated as

$$
\begin{aligned}
I_{xx} &= 219933678.88\,\mathrm{cm}^3 \\
I_{yy} &= 48763445.612\,\mathrm{cm}^3 \\
I_{xy} &= 51780878.72\,\mathrm{cm}^3
\end{aligned}
\tag{25.69}
$$

The x–z plane:

The effective lengths of the piping system projected in the x–z plane are

$$
\begin{aligned}
ID &= 300\,\mathrm{cm} \\
FG &= 271\,\mathrm{cm} \\
DE &= 271\,\mathrm{cm} \\
EF &= 650\,\mathrm{cm} \\
GH &= 400\,\mathrm{cm}
\end{aligned}
\tag{25.70}
$$

The moments of inertia for the piping system projected in the x–z plane are calculated as

$$
\begin{aligned}
I'_{xx} &= 25408381.97\,\mathrm{cm}^3 \\
I_{zz} &= 43632719\,\mathrm{cm}^3 \\
I_{xz} &= -30071031.25\,\mathrm{cm}^3
\end{aligned}
\tag{25.71}
$$

The y–z plane:

The effective lengths of the piping system projected in the y–z plane are

$$
\begin{aligned}
ID &= 300\,\mathrm{cm} \\
DE &= 427\,\mathrm{cm} \\
EF &= 500\,\mathrm{cm} \\
GH &= 520\,\mathrm{cm}
\end{aligned}
\tag{25.72}
$$

The moments of inertia for the piping system projected in the $y - z$ plane are calculated as

$$
\begin{aligned}
I'_{yy} &= 25647804.852\,\mathrm{cm}^3 \\
I'_{zz} &= 22584345.511\,\mathrm{cm}^3 \\
I'_{yz} &= 41787016.45\,\mathrm{cm}^3
\end{aligned}
\tag{25.73}
$$

The thermal expansions due to the elevated temperature are

$$\begin{aligned}\Delta_{xT} &= 4.9153 \times 0.55 = 2.7034\,\text{cm}\\ \Delta_{yT} &= 4.9153 \times 0.8 = 3.9324\,\text{cm}\\ \Delta_{zT} &= 4.9153 \times 0.45 = 2.212\,\text{cm}\end{aligned} \tag{25.74}$$

Material properties of the pipe are

$$E = 20600\,\text{KN/cm}^2 \qquad I = 2638.9\,\text{cm}^4 \tag{25.75}$$

Therefore, the forces acting on the boundaries are obtained when the following system of equations is solved

$$\begin{aligned}(I_{xx} + I'_{xx})F_x - I_{xy}F_y - I_{xz}F_z &= EI\,\Delta_{xT}\\ -I_{yz}F_x + (I_{yy} + I'_{yy})F_y - I_{yz}F_z &= EI\,\Delta_{yT}\\ -I_{zx}F_x - I_{zy}F_y + (I_{zz} + I'_{zz})F_z &= EI\,\Delta_{zT}\end{aligned} \tag{25.76}$$

Solving the above equations give the boundary forces as

$$\begin{aligned}F_x &= 2.7466\,\text{KN}\\ F_y &= 5.9052\,\text{KN}\\ F_z &= 1.9965\,\text{KN}\end{aligned} \tag{25.77}$$

The moments on points I and H are evaluated as

$$\begin{aligned}M_{zI} &= -693.63\,\text{KN.cm}\\ M_{yI} &= 420.55\,\text{KN.cm}\\ M_{xI} &= 149.352\,\text{KN.cm}\\ M_{zH} &= -1744.183\,\text{KN.cm}\\ M_{yH} &= 782.634\,\text{KN.cm}\\ M_{xH} &= 433.37\,\text{KN.cm}\end{aligned} \tag{25.78}$$

As seen, the fixed point I divides the piping system into two parts causing the length of each of the piping system to become shorter. The results show that for shorter piping elements, the forces and moments become larger.

The reason is that a piping system with shorter pipe length has smaller lines of moments of inertia. While the right hand side of the equilibrium equations are the same (similar elevated temperature and pipe's moments of inertia of the cross section), reducing the numerical values of the lines moments of inertia results into larger end forces. Let us examine the case mathematically. The mathematical nature of a linear system of equations is the same as a linear single equation

$$AF = B \tag{25.79}$$

The force F is thus

$$F = \frac{B}{A} \tag{25.80}$$

Now, the value of F becomes large if B is large or A is small. In case of piping system, the value of B is related to the temperature change, modulus of elasticity of the pipe material, and the cross section moment of inertia of the run pipe. None of these values may be reduced for a piping system which is designed to transport the piping content. Thus, to reduce the force F, the value of A must be increased. This may be done by increasing the piping length, designing proper loops and large radius elbows, and placing flexible joints at the end points of a piping system. Proper cold spring of the end points of the piping system is also an efficient method to reduce the induces forces.

The example of this section is intended to show that fixing of a point in a piping system cause to increase the end forces, and thus the resulting moments, in a pipeline.

Chapter 26
Coupled Thermoelasticity

When a structure is under the thermal shock load, the governing equation of thermoelasticity and the first law of thermodynamics are coupled. This thermal shock may be applied to the surface of a body or may be caused through the body heating. When the period of applied thermal shock is considerably smaller than the time period of the first natural frequency of the structure, then coupled equations may be justified to be employed to obtain the stress and deformation of the structure. In this chapter the governing equations for the classical coupled thermoelasticity theory are given. Some basic problems, such as solid sphere and one-dimensional rod, are considered and their behavior under thermal shock loads are discussed.

26.1 Governing Equations, Classical Theory

Returning to displacement formulations, and introducing the displacement vector $\mathbf{U} = u\mathbf{i} + v\mathbf{j} + w\mathbf{k}$, the vectorial form of governing equations of the classical coupled thermoelasticity are

$$k\nabla^2 T - \rho c\dot{T} - \alpha T_0(3\lambda + 2\mu)\ \text{div}\ \dot{\mathbf{U}} = -R \tag{26.1}$$

and

$$\mu\nabla^2\mathbf{U} + (\lambda + \mu)\ \text{grad div}\mathbf{U} - (3\lambda + 2\mu)\alpha\ \text{grad}\ T = \rho\ddot{\mathbf{U}} \tag{26.2}$$

The displacement vector can now be written as the sum of an irrotational and a potential part as given in the form

$$\mathbf{U} = \text{grad}\ \psi + \text{curl}\ \Omega \tag{26.3}$$

where ψ is a scalar potential, and Ω is a vector potential. We may substitute Eq. (26.3) into Eqs. (26.2) and (26.1) to arrive at

M. Reza Eslami et al., *Theory of Elasticity and Thermal Stresses*, Solid Mechanics and Its Applications 197, DOI: 10.1007/978-94-007-6356-2_26,

$$\nabla^2\psi - \frac{1}{c_1^2}\ddot{\psi} = \frac{(3\lambda + 2\mu)}{\lambda + 2\mu}\alpha(T - T_0)$$
$$\nabla^2\Omega_i - \frac{1}{c_2^2}\ddot{\Omega}_i = 0 \qquad i = 1, 2, 3$$
$$k\nabla^2 T - \rho c\dot{T} - \alpha T_0(3\lambda + 2\mu)\nabla^2\dot{\psi} = -R \tag{26.4}$$

where c_1 and c_2 are the speed of propagation of the elastic longitudinal wave and the speed of the shear wave, respectively.

Elimination of T between the first and the last of Eq. (26.4) results in a single equation for ψ, namely

$$\left(\nabla^2 - \frac{1}{\kappa}\frac{\partial}{\partial t}\right)\left(\nabla^2 - \frac{1}{c_1^2}\frac{\partial^2}{\partial t^2}\right)\psi - \frac{\beta^2 T_0}{(\lambda + 2\mu)k}\nabla^2\dot{\psi} = -\frac{m_1 R}{k} \tag{26.5}$$

with

$$m_1 = \frac{\beta}{\lambda + 2\mu} \qquad \beta = (3\lambda + 2\mu)\alpha \qquad \kappa = \frac{k}{\rho c} \tag{26.6}$$

and the equation for the components of vector $\mathbf{\Omega}$ remains as

$$\nabla^2\Omega_i - \frac{1}{c_2^2}\ddot{\Omega}_i = 0 \qquad i = 1, 2, 3 \tag{26.7}$$

For *quasi-steady* problems, when the variation of temperature with respect to time is slow and the inertia effects are neglected, the system of equations reduces to

$$\nabla^2 T - \frac{1}{\kappa}\dot{T} - \frac{\alpha T_0(3\lambda + 2\mu)}{k}\nabla^2\dot{\psi} = -\frac{R}{k}$$
$$\nabla^2\psi = \frac{\alpha(3\lambda + 2\mu)}{\lambda + 2\mu}(T - T_0) \tag{26.8}$$

The function ψ can be eliminated from Eq. (26.8) and the equation for heat conduction takes the form

$$\nabla^2 T - m\dot{T} = -\frac{R}{k} \tag{26.9}$$

where

$$m = \frac{1}{\kappa} + \frac{\alpha^2 T_0(3\lambda + 2\mu)^2}{k(\lambda + 2\mu)} \tag{26.10}$$

which is the uncoupled heat conduction equation in a solid body.

26.2 Problems and Solutions of Coupled Thermoelasticity

Problem 26.1. Solve Eq. (26.5) for a solid spherical domain with the radial thermal flow when $R = 0$

Solution: When the heat generation, R, thought the elastic medium is neglected, Eq. (26.5) reduces to

$$\left(\nabla^2 - \frac{1}{\kappa}\frac{\partial}{\partial t}\right)\left(\nabla^2 - \frac{1}{c_1^2}\frac{\partial^2}{\partial t^2}\right)\psi - \frac{\alpha^2(3\lambda+2\mu)^2 T_0}{(\lambda+2\mu)k}\nabla^2\dot{\psi} = 0 \tag{26.11}$$

The Laplace operator in spherical coordinates is defined as

$$\nabla^2 f = \frac{1}{r^2}\frac{\partial}{\partial r}\left(r^2\frac{\partial f}{\partial r}\right) + \frac{1}{r^2\sin\phi}\frac{\partial}{\partial\phi}\left(\sin\phi\frac{\partial f}{\partial\phi}\right) + \frac{1}{r^2\sin^2\phi}\frac{\partial^2 f}{\partial\theta^2} \tag{26.12}$$

Assuming thermal heat flow is symmetrically distributed only, Eq. (26.12) reduces to

$$\nabla^2 f = \frac{1}{r^2}\frac{\partial}{\partial r}\left(r^2\frac{\partial f}{\partial r}\right) \tag{26.13}$$

Now, as a solution of Eq. (26.11), the following separated solution is assumed for $\psi(r,t)$

$$\psi(r,t) = R(r)e^{pt} \tag{26.14}$$

Here, p is a constant. Substituting Eq. (26.14) into Eq. (26.11) gives us

$$\nabla^4 R(r) - \left(\frac{p^2}{c_1^2} + \frac{p}{\kappa} + \frac{p\alpha^2(2\lambda+3\mu)^2 T_0}{(\lambda+2\mu)k}\right)\nabla^2 R(r) + \frac{p^3}{\kappa c_1^2}R(r) = 0 \tag{26.15}$$

Introducing the constants $C = \dfrac{p\alpha^2(2\lambda+3\mu)^2 T_0}{(\lambda+2\mu)\rho c}$ and $\eta = \dfrac{\kappa p}{c_1^2}$ transforms the Eq. (26.15) to

$$\nabla^4 R(r) - \frac{p^2}{c_1^2}\left(1 + \frac{1+C}{\eta}\right)\nabla^2 R(r) + \frac{p^4}{\eta c_1^4}R(r) = 0 \tag{26.16}$$

which also may be written as

$$\left(\nabla^2 + \delta_1^2\right)\left(\nabla^2 + \delta_2^2\right)R(r) = 0 \tag{26.17}$$

where

$$\delta_{1,2}^2 = -\frac{p^2}{2c_1^2}\left\{\left(1+\frac{1+C}{\eta}\right) \pm \left[1+\frac{2(C-1)}{\eta}+\left(\frac{1+C}{\eta}\right)^2\right]^{1/2}\right\} \tag{26.18}$$

The solution of Eq. (26.17) is divided as the summation of two distinct functions $R_1(r)$ and $R_2(r)$, where each of these functions satisfy the following equations

$$\begin{aligned}(\nabla^2+\delta_1^2)R_1(r) &= 0\\ (\nabla^2+\delta_2^2)R_2(r) &= 0\end{aligned} \tag{26.19}$$

Now, each of these equations have to be solved distinctly. Considering Eq. (26.13), the governing equation for $R_1(r)$ may also be written as

$$R_1''(r) + \frac{2}{r}R_1'(r) + \delta_1^2 R_1(r) = 0 \tag{26.20}$$

To find an exact solution for this equation, the transformation $R_1(r) = \frac{1}{r}G_1(r)$ is adopted. Equation (26.20) in terms of function $G_1(r)$ simplifies to

$$G_1''(r) + \delta_1^2 G_1(r) = 0 \tag{26.21}$$

The exact solution of this equation is given as

$$G_1(r) = C_1 \sinh(\delta_1 r) + C_2 \cosh(\delta_1 r) \tag{26.22}$$

In similar manner, the solution of the function $R_2(r)$ is obtained. Now the exact solution of Eq. (26.17) has the following form

$$R(r) = \frac{1}{r}\left(C_1 \sinh(\delta_1 r) + C_2 \cosh(\delta_1 r) + C_3 \sinh(\delta_2 r) + C_4 \cosh(\delta_3 r)\right) \tag{26.23}$$

For a solid sphere, the solution at $r = 0$ has to be finite. Therefore, the constants C_2 and C_4 are equal to zero and the final solution is obtained as

$$R(r) = \frac{1}{r}\left(C_1 \sinh(\delta_1 r) + C_3 \sinh(\delta_2 r)\right) \tag{26.24}$$

The constants C_3 and C_1 have to be determined by means of the prescribed boundary conditions at the outer surface of the sphere.

Problem 26.2. Consider a rod of length L thermally insulated along its length. The initial temperature at $x = L$ is suddenly raised by $T(L,t) = T_0 e^{-t/t_0}$ while the temperature at side $x = 0$ is kept at T_0. Find the temperature and displacement distribution for the fixed boundary conditions at $x = 0$ and the free end $x = L$. The body is initially at rest and at the uniform temperature T_0.

Solution: Since the beam is thermally insulated along the length, the temperature distribution varies only as a function of length coordinate and problem is considered as a one-dimensional classical coupled thermoelasticity. The governing equations for this case, in the absence of heat generation, are

$$\begin{aligned}(\lambda + 2\mu)\frac{\partial^2 u}{\partial x^2} - \rho\frac{\partial u^2}{\partial t^2} - (3\lambda + 2\mu)\alpha\frac{\partial T}{\partial x} &= 0\\ K\frac{\partial^2 T}{\partial x^2} - \rho c\frac{\partial T}{\partial t} - (3\lambda + 2\mu)\alpha T_0\frac{\partial^2 u}{\partial x \partial t} &= 0\end{aligned} \tag{26.25}$$

where λ and μ are the Lamé constants and ρ, K, and c indicate the mass density, thermal conductivity, and the specific heat capacity. The initial conditions of the beam are

$$\begin{aligned}u(x,0) &= 0\\ \frac{\partial u}{\partial t}(x,0) &= 0\\ T(x,0) &= T_0\end{aligned} \tag{26.26}$$

and the boundary conditions are

$$\begin{aligned}u(0,t) &= 0,\\ \sigma_x(L,t) &= 0\\ T(0,t) &= T_0\\ T(L,t) &= T_0 e^{-t/t_0}\end{aligned} \tag{26.27}$$

in which σ_x is the axial stress through the beam and is defined by

$$\sigma_x = (\lambda + 2\mu)\frac{\partial u}{\partial x} - (3\lambda + 2\mu)\alpha(T - T_0) \tag{26.28}$$

The system of Eqs. (26.25)–(26.27) may be transferred into the nondimensional form. To this end, the following dimensionless quantities are introduced

$$\begin{aligned}\overline{x} &= \frac{c_1}{\kappa}x, & \overline{t} &= \frac{c_1^2}{\kappa}t\\ \overline{u} &= \frac{c_1}{\kappa}u, & c_1 &= \sqrt{\frac{\lambda + 2\mu}{\rho}}\end{aligned}$$

$$\overline{T} = \frac{T - T_0}{T_0}, \qquad \eta = \frac{(3\lambda + 2\mu)\alpha}{\rho c}$$
$$\gamma = \frac{3\lambda + 2\mu}{\lambda + 2\mu}\alpha T_0 \tag{26.29}$$

where $\kappa = \dfrac{K}{\rho c}$. With the introduction of Eq. (26.29), the governing Eq. (26.25) reduce to

$$\frac{\partial^2 \overline{T}}{\partial \overline{x}^2} - \frac{\partial \overline{T}}{\partial \overline{t}} = \eta \frac{\partial^2 \overline{u}}{\partial \overline{x} \partial \overline{t}}$$
$$\frac{\partial^2 \overline{u}}{\partial \overline{x}^2} - \frac{\partial^2 \overline{u}}{\partial \overline{t}^2} = \gamma \frac{\partial \overline{T}}{\partial \overline{x}} \tag{26.30}$$

and the transformed initial condition are

$$\overline{u}(\overline{x}, 0) = 0$$
$$\frac{\partial \overline{u}}{\partial \overline{t}}(\overline{x}, 0) = 0$$
$$\overline{T}(\overline{x}, 0) = 0 \tag{26.31}$$

and the boundary conditions are

$$\overline{u}(0, \overline{t}) = 0, \qquad \frac{\partial \overline{u}}{\partial \overline{x}}(\overline{L}, \overline{t}) = \gamma(e^{-\overline{t}/\overline{t_0}} - 1)$$
$$\overline{T}(0, \overline{t}) = 0, \qquad \overline{T}(\overline{L}, \overline{t}) = e^{-\overline{t}/\overline{t_0}} - 1 \tag{26.32}$$

where $\overline{L} = \dfrac{c_1}{\kappa} L$ and $\overline{t_0} = \dfrac{c_1^2}{\kappa} t_0$.

The time-domain solution of the problem is accomplished via the Laplace transform. The Laplace operator for the function $f(\overline{t})$ is defined by

$$f^*(s) = \pounds[f(\overline{t})] = \int_0^\infty e^{-s\overline{t}} f(\overline{t}) d\overline{t} \tag{26.33}$$

Applying the Laplace transform to Eq. (26.30) with the consideration of initial conditions (26.31) and defining the operator $D = d/d\overline{x}$ reaches us to the following system of homogeneous equations

$$(D^2 - s)\overline{T}^*(\overline{x}, s) - \eta s D\overline{u}^*(\overline{x}, s) = 0$$
$$\gamma D\overline{T}^*(\overline{x}, s) - (D^2 - s^2)\overline{u}^*(\overline{x}, s) = 0 \tag{26.34}$$

The above equations has nonzero solution if and if only the determinant of the coefficient matrix is set equal to zero, which results in

$$D^4 - s(s+1+\gamma\eta)D^2 + s^3 = 0 \tag{26.35}$$

Accordingly, the characteristic equation associated to the (26.32) takes the form

$$k^4 - s(s+1+\gamma\eta)k^2 + s^3 = 0 \tag{26.36}$$

This equation has four roots, i.e. $\pm k_1$ and $\pm k_2$, which are defined by

$$k_{1,2} = \frac{s(s+1+\gamma\eta) \pm \sqrt{s(s+1+\gamma\eta)^2 - 4s^3}}{2} \tag{26.37}$$

Consequently, the solutions for $T^*(\overline{x}, s)$ and $u^*(\overline{x}, s)$ take the following form

$$\begin{aligned} T^*(\overline{x}, s) &= A_1 e^{-k_1\overline{x}} + A_2 e^{k_1\overline{x}} + A_3 e^{-k_2\overline{x}} + A_4 e^{k_2\overline{x}} \\ u^*(\overline{x}, s) &= B_1 e^{-k_1\overline{x}} + B_2 e^{k_1\overline{x}} + B_3 e^{-k_2\overline{x}} + B_4 e^{k_2\overline{x}} \end{aligned} \tag{26.38}$$

It should be pointed out that the constants B_i and A_i are related to each other according to Eq. (26.34). According to the second of Eq. (26.34), one may deduce

$$\begin{aligned} A_1 &= -\frac{k_1^2 - s^2}{\gamma k_1} B_1 \\ A_2 &= \frac{k_1^2 - s^2}{\gamma k_1} B_2 \\ A_3 &= -\frac{k_2^2 - s^2}{\gamma k_2} B_3 \\ A_4 &= \frac{k_2^2 - s^2}{\gamma k_2} B_4 \end{aligned} \tag{26.39}$$

Therefore, the system of Eq. (26.38), take the form

$$\begin{aligned} T^*(\overline{x}, s) = &- \frac{k_1^2 - s^2}{\gamma k_1} B_1 e^{-k_1\overline{x}} + \frac{k_1^2 - s^2}{\gamma k_1} B_2 e^{k_1\overline{x}} \\ &- \frac{k_2^2 - s^2}{\gamma k_2} B_3 e^{-k_2\overline{x}} + \frac{k_2^2 - s^2}{\gamma k_2} B_4 e^{k_2\overline{x}} \\ u^*(\overline{x}, s) = &\, B_1 e^{-k_1\overline{x}} + B_2 e^{k_1\overline{x}} + B_3 e^{-k_2\overline{x}} + B_4 e^{k_2\overline{x}} \end{aligned} \tag{26.40}$$

The constants B_i are determined using the given boundary conditions. Applying the Laplace transform to the boundary conditions (26.32), gives

$$\begin{aligned}
\overline{u}^*(0, s) &= 0\\
\frac{\partial \overline{u}^*}{\partial \overline{x}}(\overline{L}, s) &= \frac{-\gamma t_0}{s(1 + t_0 s)}\\
\overline{T}^*(0, s) &= 0\\
\overline{T}^*(\overline{L}, s) &= \frac{-t_0}{s(1 + t_0 s)}
\end{aligned} \tag{26.41}$$

Using the boundary conditions (26.41) and substituting into Eq. (26.38) yield the following system of non-homogeneous equations for B_i's

$$\begin{bmatrix}
1 & 1 & 1 & 1\\
-k_1 e^{-k_1 \overline{L}} & k_1 e^{k_1 \overline{L}} & -k_2 e^{-k_2 \overline{L}} & k_2 e^{k_2 \overline{L}}\\
-\dfrac{k_1^2 - s^2}{\gamma k_1} & \dfrac{k_1^2 - s^2}{\gamma k_1} & -\dfrac{k_2^2 - s^2}{\gamma k_2} & \dfrac{k_2^2 - s^2}{\gamma k_2}\\
-\dfrac{k_1^2 - s^2}{\gamma k_1} e^{-k_1 \overline{L}} & \dfrac{k_1^2 - s^2}{\gamma k_1} e^{k_1 \overline{L}} & -\dfrac{k_2^2 - s^2}{\gamma k_2} e^{-k_2 \overline{L}} & \dfrac{k_2^2 - s^2}{\gamma k_2} e^{k_2 \overline{L}}
\end{bmatrix}
\begin{Bmatrix} B_1\\ B_2\\ B_3\\ B_4 \end{Bmatrix}
= \begin{Bmatrix} 0\\ \dfrac{-\gamma t_0}{s(1 + t_0 s)}\\ 0\\ \dfrac{-t_0}{s(1 + t_0 s)} \end{Bmatrix} \tag{26.42}$$

The constants B_i are obtained from the above system of equations in terms of the Laplace parameter s. Therefore, the functions $\overline{u}^*(\overline{x}, s)$ and $\overline{T}^*(\overline{x}, s)$ are known in the Laplace domain. Definition of Laplace inverse should be implemented herein to extract the functions $\overline{u}(\overline{x}, \overline{t})$ and $\overline{T}(\overline{x}, \overline{t})$. Since the closed-form solution for the inverse Laplace transform for these functions may not be simple to obtain, to trace the displacement and temperature profiles in time domain, numerical Laplace inverse method may be used. Such methods are given by Durbin[1]. For instance, according to Durbin method, function $f(\overline{t})$ can be obtained from Eq. (26.33) as follows

$$\begin{aligned}
f(\overline{t}) \sim \frac{e^{aN}}{N} \Bigg\{ &\frac{1}{2} \Re(f^*(a)) + \sum_{n=1}^{\infty} \Re\left(f^*(a + in\frac{\pi}{N})\right) \cos\left(n\frac{\pi}{N}\overline{t}\right)\\
&- \sum_{n=0}^{\infty} \Im\left(f^*(a + in\frac{\pi}{N})\right) \sin\left(n\frac{\pi}{N}\overline{t}\right) \Bigg\}
\end{aligned} \tag{26.43}$$

[1] F. Durbin, Numerical Inversion of Laplace Transforms: An E±cient Improvement to Dubner and Abate's Method. Computer J. **17**, 371–376 (1974)

where $\Re(x)$ and $\Im(x)$ stand for the real and imaginary parts of the number x. The numbers a and N should be chosen according to the method conditions.

Problem 26.3. Consider a rod of length L. The heat described by

$$Q(x,t) = Q_1(t)\cos(x/L) \tag{26.44}$$

is generated along the rod. Both ends of the beam at $x = 0, L$ are fixed. The temperature distribution and displacement distribution along the rod are to be obtained.

Solution: The governing equations for one-dimensional coupled thermoelasticity are

$$\begin{aligned} k\frac{\partial^2 T}{\partial x^2} - \rho c\frac{\partial T}{\partial t} - (3\lambda + 2\mu)\alpha T_0\frac{\partial^2 u}{\partial x \partial t} + Q = 0 \\ (\lambda + 2\mu)\frac{\partial^2 u}{\partial x^2} - \rho\frac{\partial^2 u}{\partial t^2} - (3\lambda + 2\mu)\alpha\frac{\partial T}{\partial x} = 0 \end{aligned} \tag{26.45}$$

When both ends of the rod are fixed, the following assumed displacement function satisfies the boundary conditions

$$u(x,t) = \frac{\rho c L}{\alpha(3\lambda + 2\mu)}F(\tau)\sin(x/L) \tag{26.46}$$

where $\tau = \dfrac{kt}{\rho c L^2}$. As seen the assumed form in Eq. (26.46) satisfies the fixed conditions on $x = 0, L$. Also, the following form is assumed for the axial temperature distribution through the rod

$$T = T_0 G(\tau)\cos(x/L) \tag{26.47}$$

When Eqs. (26.46) and (26.47) are substituted into Eq. (26.45), two coupled ordinary differential equations are obtained as

$$\begin{aligned} G(\tau) + \frac{dG(\tau)}{d\tau} + \frac{dF(\tau)}{d\tau} = \frac{Q_1(\tau)L^2}{T_0 k} \\ F(\tau) + K^2\frac{d^2F(\tau)}{d\tau^2} - \delta G(\tau) = 0 \end{aligned} \tag{26.48}$$

where we have set

$$\begin{aligned} \delta &= \frac{(3\lambda + 2\mu)^2\alpha^2 T_0}{(\lambda + 2\mu)\rho c} \\ K &= \frac{k}{L\rho c}\sqrt{\frac{\rho}{\lambda + 2\mu}} \end{aligned} \tag{26.49}$$

The initial conditions for a rod which is initially at rest in terms of functions $F(\tau)$ and $G(\tau)$ are

$$F(0) = \frac{dF}{d\tau}(0) = G(0) = 0 \tag{26.50}$$

The complementary solution, which is denoted by a subscript c, may be written as

$$\begin{aligned} F_c(\tau) &= C_1 e^{m_1\tau} + C_2 e^{m_2\tau} + C_3 e^{m_3\tau} \\ G_c(\tau) &= D_1 e^{m_1\tau} + D_2 e^{m_2\tau} + D_3 e^{m_3\tau} \end{aligned} \tag{26.51}$$

When Eq. (26.51) are substituted into Eq. (26.48), it is found that the constants C_i and D_i are the nontrivial roots of

$$\begin{aligned} &m_j C_j + (1 + m_j) D_j = 0 \\ &(1 + m_j^2 K^2) C_j - \delta D_j = 0, \qquad j = 1, 2, 3 \end{aligned} \tag{26.52}$$

The above system of equations has non-trivial solution only when the determinant of the coefficient matrix has set equal to zero. Therefore, the m_j's are the roots of the following determinantal equation

$$(1 + m_j)(1 + m_j^2 K^2) + m_j \delta = 0, \qquad j = 1, 2, 3 \tag{26.53}$$

The constants C_j and D_j are related to each other as

$$\frac{D_j}{C_j} = \frac{1 + m_j^2 K^2}{\delta} = r_j, \qquad j = 1, 2, 3 \tag{26.54}$$

Note that the determinant of Eq. (26.53) is equal to

$$\left(1 + \delta - \frac{K^2}{3}\right)^3 \frac{K^6}{27} + \left(\frac{\delta}{2} - 1 - \frac{K^2}{9}\right)^2 \frac{K^8}{9} = 0 \tag{26.55}$$

which is positive when K is small compared to unity. Therefore, Eq. (26.53) have one real root and two conjugate complex roots. Hence, the roots of Eq. (26.53) may be written as

$$m_{1,2} = -p \pm iq, \qquad m_3 = -n \tag{26.56}$$

where p, q and r satisfy the following equalities

$$\begin{aligned} 2p + n &= 1 \\ K^2(p^2 + q^2 + 2np) &= 1 + \delta \\ K^2 n(p^2 + q^2) &= 1 \end{aligned} \tag{26.57}$$

In term of p, q, and n the ratios r_j from Eq. (26.54) become

$$
\begin{aligned}
r_{1,2} &= \frac{1}{\delta}\left(1+(p^2-q^2)K^2 \pm 2ipqK^2\right)\\
r_3 &= \frac{1+n^2K^2}{\delta}
\end{aligned}
\tag{26.58}
$$

Therefore, Eq. (26.51) may be rewritten in terms of constants p, q and n in the form

$$
\begin{aligned}
F_c &= e^{-p\tau}\left\{A\cos q\tau - B\sin q\tau\right\} + Ce^{-n\tau}\\
G_c &= \frac{1}{\delta}e^{-p\tau}\left\{\left[(1+K^2p^2-K^2q^2)A + 2pqK^2B\right]\cos q\tau\right.\\
&\quad \left. -\left[(1+K^2p^2-K^2q^2)B - 2pqK^2A\right]\sin q\tau\right\} + \frac{1}{\delta}(1+n^2K^2)Ce^{-n\tau}
\end{aligned}
\tag{26.59}
$$

The above functions are the general solution of the Eq. (26.49). The particular solution of these equations, also has to be found for a complete solution. As an example, lets assume $Q_1(\tau) = Q_0(1-e^{-\tau/\tau_0})$. Recalling Eq. (26.49), it is easy to obtain the particular solutions for F and G as

$$
\begin{aligned}
F_p(\tau) &= \frac{Q_0L^2\delta}{T_0k}\left(1+\frac{\tau_0^3 e^{-\tau/\tau_0}}{D}\right)\\
G_p(\tau) &= \frac{Q_0L^2}{T_0k}\left(1+\frac{\tau_0(K^2+\tau_0)e^{-\tau/\tau_0}}{D}\right)
\end{aligned}
\tag{26.60}
$$

in which the newly introduced parameters are defined by

$$
\begin{aligned}
D &= (K^2+\tau_0^2)(1-\tau_0) + \delta\tau_0^2\\
\tau_0 &= \frac{kt_0}{\rho c L^2}
\end{aligned}
\tag{26.61}
$$

Now the solutions of each of functions F and G are known as the sum of associated particular and complementary components, i.e.

$$
\begin{aligned}
F(\tau) &= F_p(\tau) + F_c(\tau)\\
G(\tau) &= G_p(\tau) + G_c(\tau)
\end{aligned}
\tag{26.62}
$$

The constants A, B, and C which are appeared in Eq. (26.59) have to be obtained by means of the initial conditions given by Eq. (26.50). This constants are easy to obtain and the final form of displacement and temperature profiles are can then be obtained as

$$
\begin{aligned}
u(\tau,x) = &\sin(x/L)\left(\frac{(3\lambda+2\mu)Q_0L^3\alpha}{(\lambda+2\mu)kD\left[(p-n)^2+q^2\right]}\right)\\
&\times\ \left\{\left(\tau_0^3e^{-\tau/\tau_0}+D(1-e^{-nr})\right)\left((p-n)^2+q^2\right)\right.\\
&+\ e^{-p\tau}\cos q\tau\left\{\left[K^2(1-\tau_0)+\tau_0^2(1+\delta)\right](2p-n)n+\tau_0(1-2p\tau_0)\right\}\\
&+\ \frac{1}{q}e^{-p\tau}\sin q\tau\left\{\left[K^2(1-\tau_0)+\tau_0^2(1+\delta)\right](p^2-q^2-np)n\right.\\
&\left.+\ \tau_0(p-n)-\tau_0^2(p^2-q^2-n^2)\right\}-e^{-nr}\left\{\left[K^2(1-\tau_0)\right.\right.\\
&\left.+\ \tau_0^2(1+\delta)\right](2p-n)n+\tau_0(1-2p\tau_0)\\
&\left.\left.+\ \tau_0^3\left((p-n)^2+q^2\right)\right\}\right\}
\end{aligned}
\tag{26.63}
$$

$$
\begin{aligned}
T(\tau,x) = &\cos(x/L)\left(\frac{Q_0L^2}{kD\left[(p-n)^2+q^2\right]}\right)\\
&\times\ \left\{\left\{\tau_0(\tau_0^2+K^2)e^{-\tau/\tau_0}+D\left[1-(1+n^2K^2)e^{-nr}\right]\right\}[(p-n)^2+q^2]\right.\\
&+\ e^{-p\tau}\cos q\tau\left\{\left[K^2(1-\tau_0)+\tau_0^2(1+\delta)\right](2p-n+1)n\right.\\
&\left.+\ 2K^2np\tau_0(1-n\tau_0)+\tau_0\left(1-2\tau_0p-\frac{1}{n}\right)\right\}\\
&+\ \frac{1}{q}e^{-p\tau}\sin q\tau\left\{\left[K^2(1-\tau_0)+\tau_0^2(1+\delta)\right]\right.\\
&\times\ \left[\frac{1}{nK^2}+n(p^2-q^2-p-np)\right]+2nq^2K^2\tau_0(1-\tau_0n)\\
&\left.+\ \tau_0(p-n)\left(1+\frac{1}{n}\right)-\tau_0^2\left(p^2-q^2-n^2-n+\frac{1}{n^2K^2}\right)\right\}\\
&-\ e^{-nr}\left\{\left[K^2(1-\tau_0)+\tau_0^2(1+\delta)\right](2p-n)n+\tau_0-2p\tau_0^2\right.\\
&\left.\left.+\ \tau_0^3[(p-n)^2+q^2]\right\}(1+n^2K^2)\right\}
\end{aligned}
\tag{26.64}
$$

Part IV
Numerical Methods

Chapter 27
The Method of Characteristics

The purpose of this chapter is to develop the method of characteristics for the solution of dynamic problems in thermoelasticity. The theoretical analysis of dynamic stresses due to impact loadings has generally been performed by the Laplace transform method. Due to inversion difficulties, the Laplace transform method is usually limited to simple wave problems. The need for the numerical methods to the solution of dynamic problems is dictated by the well-known difficulty of obtaining the exact solutions. Among the various numerical methods, the method of characteristics has the advantages of giving a simple description of the wave fronts and it can give numerical solutions readily to problems with any types of input functions. In mathematics, the method of characteristics is a method of numerical integration of a system of partial differential equations of hyperbolic type. The method is to reduce the hyperbolic partial differential equations to a family of ordinary differential equations, each of which is valid along a different family of characteristic lines (called the characteristics). These equations (called the characteristic equations) are more suitable for numerical analysis because the use of these equations makes it possible to obtain the solutions by a step-by-step integration procedure.

27.1 Basic Equations for Plane Thermoelastic Waves

For the shake of motivation, we confine our attention to the solution of one-dimensional transient, uncoupled dynamic thermal stresses in plane thermoelastic media subjected to sudden temperature change. The equations which govern the propagation of plane waves in thermoelastic media is given by

(1) Equation of motion

$$\rho\frac{\partial U}{\partial t} = \frac{\partial \sigma_{xx}}{\partial x} \quad ; \quad U = \frac{\partial u}{\partial t} \tag{27.1}$$

DOI: 10.1007/978-94-007-6356-2_27,

(2) Constitutive equations

$$\frac{\partial \sigma_{xx}}{\partial t} = (\lambda + 2\mu)\frac{\partial U}{\partial x} - \beta\frac{\partial T}{\partial t} \tag{27.2}$$

where x is the Cartesian coordinate and t is the time; u is the displacement and $U = \partial u/\partial t$ is the particle velocity in the direction x of the wave propagation; σ_{xx} is the normal stress; T is the temperature to be determined independently of the mechanical state of the thermoelastic media; ρ is the density; $\beta = \alpha(3\lambda + 2\mu)$, where α is the coefficient of thermal expansion; λ and μ are Lamé constants.

Eliminating σ_{xx} from Eqs. (27.1) and (27.2), we obtain the governing equation in terms of the displacement u

$$\frac{\partial^2 u}{\partial x^2} = \frac{\rho}{\lambda + 2\mu}\left(\frac{\partial^2 u}{\partial t^2} + \frac{\beta}{\rho}\frac{\partial T}{\partial x}\right) \tag{27.3}$$

Thus, the dynamic thermoelasticity theory results in the displacement field governed by a hyperbolic second-order partial differential equation, which predicts the finite propagation velocity for mechanical disturbances.

27.2 Characteristics and Characteristic Equations

Equation (27.3) is more convenient for the application of the Laplace transform method. For the method of characteristics, we use a system of two linear first-order partial differential equations (27.1) and (27.2) with U and σ_{xx} as two dependent variables, because the expression for the boundary conditions are simple when the dependent variables are U and σ_{xx}. In the $(x-t)$ plane, certain curves may exist, along which these variables are continuous, but their first partial derivatives may be discontinuous. These curves will be called the characteristics, physical characteristics, or waves and the differential equations governing the propagation of discontinuities (waves) along characteristics will be called the characteristic equations. The characteristics and the characteristic equations may be derived by the directional derivative method.

The total differentials of two dependent variables are written as

$$dU = \frac{\partial U}{\partial t}dt + \frac{\partial U}{\partial x}dx \tag{27.4}$$

$$d\sigma_{xx} = \frac{\partial \sigma_{xx}}{\partial t}dt + \frac{\partial \sigma_{xx}}{\partial x}dx \tag{27.5}$$

Discontinuity in the derivatives of U and σ_{xx} implies that four derivatives $\partial U/\partial t$, $\partial U/\partial x$, $\partial \sigma_{xx}/\partial t$, $\partial \sigma_{xx}/\partial x$ are indeterminate along the characteristics.

If these derivatives are considered as unknown variables, they are related by the four equations (27.1), (27.2) and (27.4), (27.5) as follows:

$$\mathbf{AX} = \mathbf{B} \ : \ \sum_{j=1}^{4} a_{ij} x_j = b_i, \quad (i = 1\text{–}4) \tag{27.6}$$

where the matrix A and vectors $\mathbf{X}$ and $\mathbf{B}$ denote

$$A = [a_{ij}] = \begin{bmatrix} \rho & 0 & 0 & -1 \\ 0 & -(\lambda + 2\mu) & 1 & 0 \\ dt & dx & 0 & 0 \\ 0 & 0 & dt & dx \end{bmatrix}$$

$$\mathbf{X} = [x_i] = \begin{bmatrix} \partial U/\partial t \\ \partial U/\partial x \\ \partial \sigma_{xx}/\partial t \\ \partial \sigma_{xx}/\partial x \end{bmatrix}, \quad \mathbf{B} = [b_i] = \begin{bmatrix} 0 \\ -\beta \partial T/\partial t \\ dU \\ d\sigma_{xx} \end{bmatrix}$$

Solving the linear equations (27.6) for $\partial U/\partial t$, we obtain

$$\frac{\partial U}{\partial t} = \frac{|A_1|}{|A|} \tag{27.7}$$

where the matrix A_1 is obtained by exchanging the first column of the matrix A by the column vector $\mathbf{B}$. Therefore, when both the numerator and denominator become zero, the derivative $\partial U/\partial t$ becomes indeterminate. The vanishing of the denominator of Eq. (27.7) yields two characteristics I_i, $(i = 1, 2)$,

$$I_i \ : \ \frac{dx}{dt} = (c_L, -c_L) = (V_1, V_2) = V_i, \quad (i = 1, 2) \tag{27.8}$$

where c_L is the dilatation wave speed defined by

$$c_L = \sqrt{\frac{\lambda + 2\mu}{\rho}} \tag{27.9}$$

For $I_i : dx/dt = V_i$, $(i = 1, 2)$, the vanishing of the numerator of Eq. (27.7) yields two characteristic equations along two characteristics I_1 and I_2. These characteristic equations are

$$d\sigma_{xx} - \rho V_i dU = -\beta \frac{\partial T}{\partial t} dt, \quad (i = 1, 2) \tag{27.10}$$

Solving the system of Eq. (27.6) for $\partial U/\partial x$, $\partial \sigma_{xx}/\partial t$, or $\partial \sigma_{xx}/\partial x$ gives the same result as Eq. (27.10). The I_1 characteristic represents propagation of the discontinuity in the derivatives at velocity c_L, traveling to the right in the $(x-t)$ plane (direct characteristics). The I_2 characteristic gives the propagation towards the left (return characteristics). For homogeneous materials, the velocity c_L is constant throughout the medium and the characteristics are straight lines of equal slope. Finally, the values of U and σ_{xx} may be found by solving Eq. (27.10) subjected to appropriate initial and boundary conditions. The boundary inputs may be in the form of specified time functions of any one of two variables.

27.3 Derivation of Difference Equations and Numerical Procedure

A numerical procedure involving stepwise integration of characteristic equations along characteristics is employed to solve problems. The $(x-t)$ plane is subdivided into a grid system by the two characteristics (27.8) as shown in Fig. 27.1. A step of integration on length Δx is connected with a step of integration on time Δt by linear correlation $\Delta x = c_L \Delta t$. If we assume a linear variation of variables between these closely spaced mesh points, the integration of the characteristic equations (27.10) between points A and D along I_1 characteristic, and points C and D along I_2 characteristic yield their finite-difference equivalents.

$$\int_i^D d\sigma_{xx} - \rho V_i \int_i^D dU = -\beta \int_i^D \frac{\partial T}{\partial t} dt, \quad (i = 1, 2) \tag{27.11}$$

Therefore, we obtain

$$\sigma_{xxD} - \rho V_i U_D = \sigma_{xxi} - \rho V_i U_i - \beta (T_D - T_E), \quad (i = 1, 2) \tag{27.12}$$

In Eq. (27.12), the quantities with subscripts i take the values at the points $A(i = 1)$ and $C(i = 2)$, respectively. Therefore, the two unknowns at typical point D can be calculated successively from these two finite-difference equations (27.12) if all the quantities at neighboring points A, C and E are known from the previous calculations. Along the boundary point where one of these variables is prescribed, the analysis is the same except that the characteristic equation along the characteristic extending outside of the region should be replaced by the prescribed boundary condition.

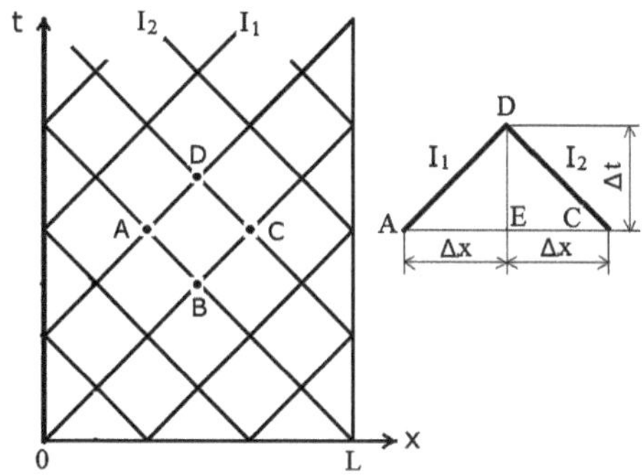

Fig. 27.1 Characteristics network for numerical procedure

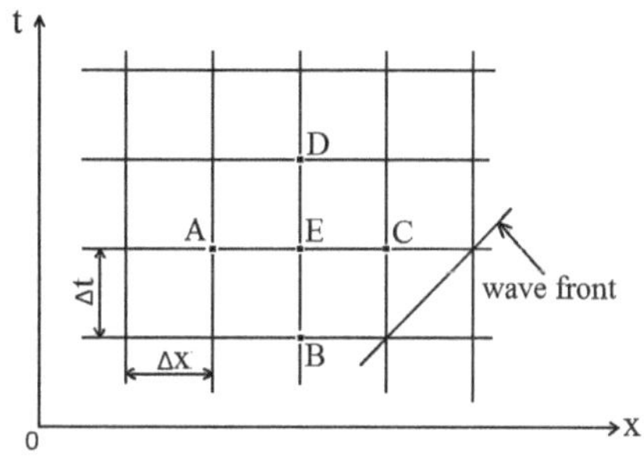

Fig. 27.2 Finite-difference and wave front diagram

27.4 Finite-Difference Solution for Temperature

The one-dimensional uncoupled, unsteady heat conduction equation is given by

$$\frac{\partial T}{\partial t} = \frac{k}{\rho c_v} \frac{\partial^2 T}{\partial x^2} \tag{27.13}$$

where k is the thermal conductivity; c_v is the specific heat. In the method of characteristics, the temperature can be obtained by writing Eq. (27.13) in the explicit finite-difference form in the $(x-t)$ characteristic plane. A typical finite-difference mesh and a wave front diagram in the $(x-t)$ plane is shown in Fig. 27.2. We can use approximate values of the partial derivatives at a typical point E by the central difference expressions:

$$\left(\frac{\partial T}{\partial t}\right)_E = \frac{T_D - T_B}{2\Delta t}, \quad \left(\frac{\partial^2 T}{\partial x^2}\right)_E = \frac{T_C - 2T_E + T_A}{(\Delta x)^2} \tag{27.14}$$

The "leap-frog" method replaces the value at point E by the arithmetic mean of the values at points B and D.

$$T_E = \frac{T_D + T_B}{2} \tag{27.15}$$

Applying Eqs. (27.14) and (27.15) into Eq. (27.13), we obtain

$$T_D = \frac{1-2p}{1+2p} T_B + \frac{2p}{1+2p}(T_C + T_A) \tag{27.16}$$

where

$$p = \frac{\Delta t}{(\Delta x)^2}, \quad q = \frac{\Delta t}{\Delta x} \tag{27.17}$$

The leap frog equation (27.16) is stable if $\Delta x \geq c_L \Delta t$. Therefore the temperature at a typical mesh point D can be calculated if the temperature at three neighboring point A, C and B are known from the previous calculations.

27.5 Problems and Solutions Related to The Method of Characteristics

Problem 27.1. The equations governing the propagation of cylindrical and spherical waves are given by

$$\rho \frac{\partial U}{\partial t} = \frac{\partial \sigma_{rr}}{\partial r} + N \frac{(\sigma_{rr} - \sigma_{\theta\theta})}{r} \quad ; \quad U = \frac{\partial u}{\partial t} \tag{27.18}$$

$$\frac{\partial \sigma_{rr}}{\partial t} = (\lambda + 2\mu) \frac{\partial U}{\partial r} + N\lambda \frac{U}{r} - \beta \frac{\partial T}{\partial t} \tag{27.19}$$

$$\frac{\partial \sigma_{\theta\theta}}{\partial t} = \lambda \frac{\partial U}{\partial r} + N \left\{ \lambda + 2\mu - (N-1)\mu \right\} \frac{U}{r} - \beta \frac{\partial T}{\partial t} \tag{27.20}$$

where N is a constant, with values of 1 and 2, corresponding to the cylindrical and spherical waves, respectively: r is the radial coordinate. Derive the characteristics and characteristic equations for cylindrical and spherical waves.

Solution. Equations (27.18), (27.19) and (27.20) constitute three linear first-order partial differential equations with U, σ_{rr} and $\sigma_{\theta\theta}$ as three dependent variables. The total differentials of three dependent variables U, σ_{rr} and $\sigma_{\theta\theta}$ are written as

$$dU = \frac{\partial U}{\partial t} dt + \frac{\partial U}{\partial r} dr \tag{27.21}$$

$$d\sigma_{rr} = \frac{\partial \sigma_{rr}}{\partial t} dt + \frac{\partial \sigma_{rr}}{\partial r} dr \tag{27.22}$$

$$d\sigma_{\theta\theta} = \frac{\partial \sigma_{\theta\theta}}{\partial t} dt + \frac{\partial \sigma_{\theta\theta}}{\partial r} dr \tag{27.23}$$

If six derivatives $\partial U/\partial t$, $\partial U/\partial r$, $\cdots \partial \sigma_{\theta\theta}/\partial t$, $\partial \sigma_{\theta\theta}/\partial r$ are considered as six unknown variables, they are related by the six equations Eqs. (27.18)–(27.20) and Eqs. (27.21)–(27.23) as follows:

$$\mathbf{AX} = \mathbf{B} \ : \ \sum_{j=1}^{6} a_{ij} x_j = b_i, \quad (i = 1\text{–}6) \tag{27.24}$$

where the matrix A and vectors $\mathbf{X}$ and $\mathbf{B}$ denote

$$A = [a_{ij}] = \begin{bmatrix} \rho & 0 & 0 & -1 & 0 & 0 \\ 0 & -(\lambda + 2\mu) & 1 & 0 & 0 & 0 \\ 0 & -\lambda & 0 & 0 & 1 & 0 \\ dt & dr & 0 & 0 & 0 & 0 \\ 0 & 0 & dt & dr & 0 & 0 \\ 0 & 0 & 0 & 0 & dt & dr \end{bmatrix}$$

$$\mathbf{X} = [x_i] = \begin{bmatrix} \partial U/\partial t \\ \partial U/\partial r \\ \partial \sigma_{rr}/\partial t \\ \partial \sigma_{rr}/\partial r \\ \partial \sigma_{\theta\theta}/\partial t \\ \partial \sigma_{\theta\theta}/\partial r \end{bmatrix}$$

$$\mathbf{B} = [b_i] = \begin{bmatrix} N(\sigma_{rr} - \sigma_{\theta\theta})/r \\ N\lambda U/r - \beta \partial T/\partial t \\ N\{\lambda + 2\mu - (N-1)\mu\}U/r - \beta \partial T/\partial t \\ dU \\ d\sigma_{rr} \\ d\sigma_{\theta\theta} \end{bmatrix}$$

Solving these linear equations (27.24) for the k-th derivative x_k, we obtain

$$x_k = \frac{|A_k|}{|A|}, \quad (k = 1\text{–}6) \tag{27.25}$$

where the matrix A_k is obtained by exchanging the k-th column of the matrix A by the column vector $\mathbf{B}$.

The vanishing of the denominator of Eq. (27.25) yields three characteristics $I_i, \ (i = 1\text{–}3)$

$$I_i \ : \ \frac{dr}{dt} = (c_L, -c_L, 0) = (V_1, V_2, V_3) = V_i, \quad (i = 1\text{–}3) \tag{27.26}$$

where c_L is the dilatation wave speed defined by Eq. (27.9).

For $I_j : dr/dt = V_j, (j = 1, 2)$, the vanishing of the numerator of Eq. (27.25) yields two characteristic equations along two characteristics I_1 and I_2. These characteristic equations are

$$l_1^{(j)} d\sigma_{rr} + l_2^{(j)} dU = l_3^{(j)} \frac{U}{r} dt + l_4^{(j)} \frac{\sigma_{rr} - \sigma_{\theta\theta}}{r} dt - \beta \frac{\partial T}{\partial t} dt, \quad (j = 1, 2) \quad (27.27)$$

where

$$l_1^{(j)} = 1, \ l_2^{(j)} = -\rho V_j, \ l_3^{(j)} = N\lambda, \ l_4^{(j)} = -N V_j \quad (27.28)$$

From Eq. (27.25), the characteristic equation along the characteristic $I_3 : dr = 0$ cannot be obtained, since $(dr)^2$ appears in the numerator and denominator as a common factor. Eliminating $\partial U/\partial r$ from Eqs. (27.19) and (27.20), we obtain the characteristic equation along $I_3 : dr/dt = V_3 = 0$

$$l_1^{(3)} d\sigma_{rr} + l_2^{(3)} d\sigma_{\theta\theta} = l_3^{(3)} \frac{U}{r} dt + \frac{2\mu\beta}{\lambda} dT \quad (27.29)$$

where

$$\begin{aligned} l_1^{(3)} &= 1, \ l_2^{(3)} = -\frac{\lambda + 2\mu}{\lambda} \\ l_3^{(3)} &= N \left[\lambda - \frac{\lambda + 2\mu}{\lambda} \left\{ \lambda + 2\mu - (N - 1)\mu \right\} \right] \end{aligned} \quad (27.30)$$

The characteristic equation (27.29) is merely a restatement of Eqs. (27.19) and (27.20), which gives the static relation between the differentials of velocity and stresses at any constant r. The I_3 is a degenerate dynamic wave expressing conditions along lines with $r =$ constant.

Problem 27.2. Derive the difference equations for cylindrical and spherical waves.

Solution. The $(r-t)$ plane is subdivided into a grid system by the three characteristics (27.26) as shown in Fig. 27.3. The integration of the characteristic equations (27.27) between points A and D along I_1 characteristic, and points C and D along I_2 yield their finite-difference equivalents.

$$\begin{aligned} &\left(l_1^{(j)} - \frac{l_4^{(j)} \Delta t}{2r_D} \right) \sigma_{rrD} + \frac{l_4^{(j)} \Delta t}{2r_D} \sigma_{\theta\theta D} + \left(l_2^{(j)} - \frac{l_3^{(j)} \Delta t}{2r_D} \right) U_D \\ &= \left(l_1^{(j)} + \frac{l_4^{(j)} \Delta t}{2r_j} \right) \sigma_{rrj} - \frac{l_4^{(j)} \Delta t}{2r_j} \sigma_{\theta\theta j} + \left(l_2^{(j)} + \frac{l_3^{(j)} \Delta t}{2r_j} \right) U_j \\ &\quad - \beta(T_D - T_E), \quad (j = 1, 2) \end{aligned} \quad (27.31)$$

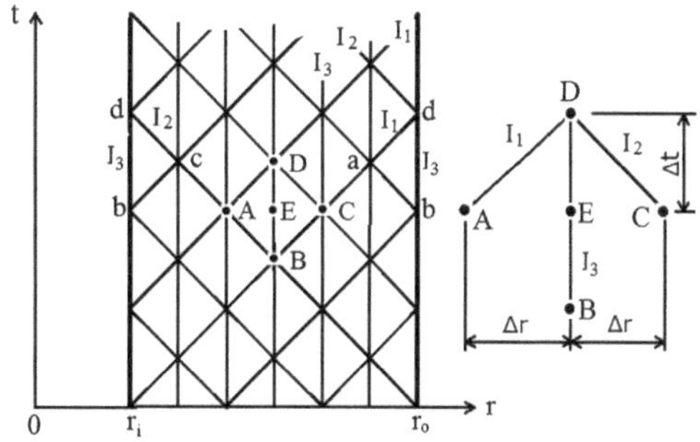

Fig. 27.3 Characteristics network for cylindrical and spherical waves

Along I_3 from points B to D, Eq. (27.29) is also expressed in finite-difference algebraic form

$$l_1^{(3)}\sigma_{rrD} + l_2^{(3)}\sigma_{\theta\theta D} - \frac{l_3^{(3)}\Delta t}{r_D}U_D = l_1^{(3)}\sigma_{rrB} + l_2^{(3)}\sigma_{\theta\theta B} + \frac{l_3^{(3)}\Delta t}{r_B}U_B + \frac{2\mu\beta}{\lambda}(T_D - T_B) \tag{27.32}$$

Therefore, the three unknowns at typical point D can be calculated successively from these three finite-difference equations (27.31) and (27.32) if all the quantities at neighboring points A, C and B are known from the previous calculations.

Problem 27.3. Based on the generalized theory of thermoelasticity proposed by Lord and Shulman, the equations that govern the propagation of one-dimensional thermal and thermal stress waves in linear elastic, isotropic and homogeneous materials under plane strain are given by one set of generalized equations

$$\rho\frac{\partial U}{\partial t} = \frac{\partial \sigma_{xx}}{\partial x} \quad : \quad U = \frac{\partial u}{\partial t} \tag{27.33}$$

$$\frac{\partial \sigma_{xx}}{\partial t} = (\lambda + 2\mu)\frac{\partial W}{\partial x} + \beta\frac{\partial T}{\partial t} \tag{27.34}$$

$$\tau\frac{\partial q_x}{\partial t} + k\frac{\partial T}{\partial x} = -q_x \tag{27.35}$$

$$\frac{\partial q_x}{\partial x} + \rho c_v\frac{\partial T}{\partial t} + T_0\beta\frac{\partial U}{\partial x} = 0 \tag{27.36}$$

where q_x is the heat flux; T is the temperature change from the absolute reference temperature T_0. In Eq. (27.35), the classical Fourier's law is modified by adding a thermal relaxation time τ to eliminate the paradox of infinite thermal speed of the classical theory of thermoelasticity. In Eq. (27.36), a coupling between thermal and mechanical fields is taken into account. Derive the characteristics and the characteristic equations.

Solution. Equations (27.33)–(27.36) constitute a system of four linear first-order partial differential equations with U, σ_{xx}, q_x and T as four dependent variables. Equations (27.33)–(27.36) and total differentials of these four variables may be considered as eight linear equations with eight derivatives $\partial U/\partial t$, $\partial U/\partial x$, $\cdots \partial T/\partial t$, $\partial T/\partial x$. They are expressed in the matrix form as

$$\mathbf{AX} = \mathbf{B} \quad : \quad \sum_{j=1}^{8} a_{ij} x_j = b_i, \quad (i = 1\text{–}8) \tag{27.37}$$

By Cramer's formula of linear equations, the k-th solution of Eq. (27.37) is

$$x_k = \frac{|A_k|}{|A|} \tag{27.38}$$

The vanishing of the denominator of Eq. (27.38) yields

$$\left(\frac{dx}{dt}\right)^4 - \left(\frac{\lambda + 2\mu}{\rho} + \frac{k}{\rho c_v \tau} + \frac{T_0 \beta^2}{\rho^2 c_v}\right)\left(\frac{dx}{dt}\right)^2 + \frac{k}{\rho^2 c_v \tau} = 0 \tag{27.39}$$

If we denote the four solutions of Eq. (27.39) as $\pm c_1, \pm c_2$, we obtain

$$\left(\frac{dx}{dt}\right)^2 = \begin{Bmatrix} c_1^2 \\ c_2^2 \end{Bmatrix} = \frac{1}{2}\left[\left(c_L^2 + \delta c_L^2 + \frac{\kappa}{\tau}\right) \pm \sqrt{(c_L^2 + \delta c_L^2 + \frac{\kappa}{\tau})^2 - 4\frac{\kappa c_L^2}{\tau}}\right] \tag{27.40}$$

where κ is the thermal diffusivity and δ is the thermomechanical coupling parameter, respectively defined by

$$\kappa = \frac{k}{\rho c_v}, \quad \delta = \frac{\beta^2 T_0}{\rho c_v (\lambda + 2\mu)} \tag{27.41}$$

Therefore, the characteristics are found to be composed of four families of characteristic lines I_j, $(j = 1\text{–}4)$

$$\frac{dx}{dt} = (c_1, -c_1, c_2, -c_2) = (V_1, V_2, V_3, V_4) = V_j, \quad (j = 1\text{–}4) \tag{27.42}$$

The corresponding characteristic equations along characteristics are obtained by calculating the determinant of matrix A_k be zero ($|A_k| = 0$). Therefore, the

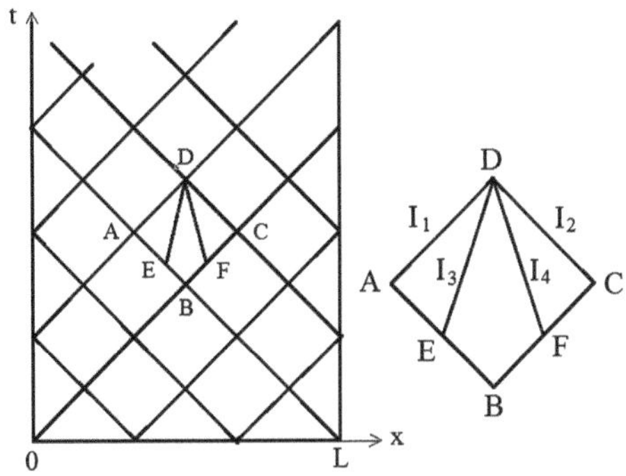

Fig. 27.4 Characteristics network for numerical procedure

characteristic equations along characteristics $I_j : dx/dt = V_j, (j = 1\text{--}4)$ are given as follows:

$$l_1^{(j)} d\sigma_{xx} + l_2^{(j)} dU + l_3^{(j)} dq_x + l_4^{(j)} dT = l_5^{(j)} q_x dt \tag{27.43}$$

where

$$l_1^{(j)} = 1, \quad l_2^{(j)} = -\rho V_j, \quad l_3^{(j)} = \frac{\beta \tau V_j}{(k - \rho c_v \tau V_j^2)}$$
$$l_4^{(j)} = l_3^{(j)} \times \frac{k}{\tau V_j}, \quad l_5^{(j)} = -l_3^{(j)} \times \frac{1}{\tau} \tag{27.44}$$

Finally, the values of U, σ_{xx}, q_x, and T may be found by solving Eq. (27.43) subjected to the appropriate initial and boundary conditions.

A characteristic network constructed by two families of lines $dx/dt = \pm c_1$ is shown in Fig. 27.4. At a typical mesh point D, we draw two characteristic lines ED, FD with slopes $\pm c_2$. The location and values of variables at points E, F may be found from the locations and values of points A, B and C, B, respectively. Then, the values at point D are obtained by integration of the characteristic equations (27.43) along the characteristics (27.42).

Chapter 28
Finite Element of Coupled Thermoelasticity

Due to the mathematical complexities encountered in analytical treatment of the coupled thermoelasticity problems, the finite element method is often preferred. The finite element method itself is based on two entirely different approaches, the variational approach based on the Ritz method, and the weighted residual methods. The variational approach, which for elastic continuum is based on the extremum of the total potential and kinetic energies has deficiencies in handling the coupled thermoelasticity problems due to the controversial functional relation of the first law of thermodynamics. On the other hand, the weighted residual method based on the Galerkin technique, which is directly applied to the governing equations, is quite efficient and has a very high rate of convergence.

28.1 Galerkin Finite Element

The general governing equations of the classical coupled thermoelasticity are the equation of motion and the first law of thermodynamics as

$$\sigma_{ij,j} + X_i = \rho \ddot{u}_i \qquad \text{in V} \tag{28.1}$$

$$q_{i,i} + \rho c \dot{\theta} + \beta T_0 \dot{\epsilon}_{ii} = R \qquad \text{in V} \tag{28.2}$$

These equations must be simultaneously solved for the displacement components u_i and temperature change θ. The thermal boundary conditions are satisfied by either of the equations

$$\theta = \theta_s \qquad \text{on } A \qquad \text{for } t > t_0 \tag{28.3}$$

$$\theta_{,n} + a\theta = b \qquad \text{on } A \qquad \text{for } t > t_0 \tag{28.4}$$

where $\theta_{,n}$ is the gradient of temperature change along the normal to the surface boundary A, and a and b are either constants or given functions of temperature on

M. Reza Eslami et al., *Theory of Elasticity and Thermal Stresses*, Solid Mechanics and Its Applications 197, DOI: 10.1007/978-94-007-6356-2_28,

the boundary. The first condition is related to the specified temperature and the second condition describes the convection and radiation on the boundary.

The mechanical boundary conditions are specified through the traction vector on the boundary. The traction components are related to the stress tensor through the Cauchy's formula given by

$$t_i^n = \sigma_{ij} n_j \qquad \text{on } A \qquad \text{for } t > t_0 \tag{28.5}$$

where t_i^n is the prescribed traction component on the boundary surface whose outer unit normal vector is $\vec{n}$. For displacement formulations, using the constitutive laws of linear thermoelasticity along with the strain-displacement relations, the traction components can be related to the displacements as

$$t_i^n = \mu(u_{i,j} + u_{j,i})n_j + \lambda u_{k,k} n_i - (3\lambda + 2\mu)\alpha\theta n_i \tag{28.6}$$

where $\theta = T - T_0$ is the temperature change above the reference temperature T_0. It is further possible to have kinematical boundary conditions where the displacements are specified on the boundary as

$$u_i = \bar{u}_i(s) \qquad \text{on } A \qquad \text{for } t > t_0 \tag{28.7}$$

The system of coupled Eqs. (28.1) and (28.2) does not have a general analytical solution. A finite element formulation may be developed based on the Galerkin method. The finite element model of the problem is obtained by discretizing the solution domain into a number of arbitrary elements. In each base element (e), the components of displacement and temperature change are approximated by the shape functions

$$u_i^{(e)}(x_1, x_2, x_3, t) = U_{mi}(t) N_m(x_1, x_2, x_3) \tag{28.8}$$

$$\theta^{(e)}(x_1, x_2, x_3, t) = \theta_m(t) N_m(x_1, x_2, x_3) \qquad m = 1, 2, ..., r \tag{28.9}$$

where r is the total number of nodal points in the base element (e). The summation convention is used for the dummy index m. This is a Kantrovitch type of approximation, where the time and space functions are separated into distinct functions. Here $U_{mi}(t)$ is the component of displacement at each nodal point, and $\theta_m(t)$ is the temperature change at each nodal point, all being functions of time. The shape function $N_m(x_1, x_2, x_3)$ is function of space variables.

Substituting Eqs. (28.8) and (28.9) into Eq. (28.1) and applying the weighted residual integral with respect to the weighting functions $N_m(x_1, x_2, x_3)$, the formal Galerkin approximation reduces to

$$\int_{V(e)} (\sigma_{ij,j} + X_i - \rho \ddot{u}_i) N_l dV = 0 \qquad l = 1, 2, ..., r \tag{28.10}$$

Applying the weak formulation to the first term, yields

$$\int_{V(e)} (\sigma_{ij,j}) N_l dv = \int_{A(e)} \sigma_{ij} n_j N_l dA - \int_{V(e)} \frac{\partial N_l}{\partial x_j} \sigma_{ij} dV \tag{28.11}$$

where n_j is the component of the unit outer normal vector to the boundary. Substituting Eq. (28.11) in Eq. (28.10) gives

$$\int_{A(e)} \sigma_{ij} n_j N_l dA - \int_{V(e)} \frac{\partial N_l}{\partial x_j} \sigma_{ij} dV + \int_{V(e)} X_i N_l dV - \int_{V(e)} \rho \ddot{u}_i N_l dV = 0 \tag{28.12}$$

According to Cauchy's formula, the traction force components acting on the boundary are related to the stress tensor as

$$t_i = \sigma_{ij} n_j \tag{28.13}$$

Thus, the first term of Eq. (28.12) is

$$\int_{A(e)} \sigma_{ij} n_j N_l dA = \int_{A(e)} t_i N_l dA \tag{28.14}$$

From Hooke's law, the stress tensor is related to the strain tensor, or the displacement components, and temperature change θ as

$$\sigma_{ij} = G(u_{i,j} + u_{j,i}) + \lambda u_{k,k} \delta_{ij} - \beta \theta \delta_{ij} \tag{28.15}$$

Substituting for σ_{ij} in the second term of Eq. (28.12) yields

$$\int_{V(e)} \frac{\partial N_l}{\partial x_j} \sigma_{ij} dV = \int_{V(e)} \frac{\partial N_l}{\partial x_j} [G(u_{i,j} + u_{j,i}) + \lambda u_{k,k} \delta_{ij} - \beta \theta \delta_{ij}] dV \tag{28.16}$$

Substituting this expression in Eq. (28.12) gives

$$\begin{aligned} &\int_{V(e)} \rho \ddot{u}_i N_l dV + \int_{V(e)} \frac{\partial N_l}{\partial x_j} [G(u_{i,j} + u_{j,i}) + \lambda u_{k,k} \delta_{ij}] dV \\ &- \int_{V(e)} \beta \theta \frac{\partial N_l}{\partial x_i} dV = \int_{V(e)} X_i N_l dV + \int_{A(e)} t_i N_l dA \end{aligned} \tag{28.17}$$

Now, the base element (e) with r nodal points is considered and the displacement components and temperature change in the element (e) are approximated by Eqs. (28.8) and (28.9). Using these approximation, Eq. (28.17) becomes

$$\left(\int_{V(e)} \rho N_l N_m dV \right) \ddot{U}_{mi} + \left(\int_{V(e)} G \frac{\partial N_l}{\partial x_j} \frac{\partial N_m}{\partial x_j} d \right) U_{mi}$$

$$
\begin{aligned}
&+\left(\int_{V(e)} G\frac{\partial N_l}{\partial x_j}\frac{\partial N_m}{\partial x_i}dV\right)U_{mj}+\left(\int_{V(e)} \lambda\frac{\partial N_l}{\partial x_i}\frac{\partial N_m}{\partial x_j}dV\right)U_{mj}\\
&-\left(\int_{V(e)} \beta\frac{\partial N_l}{\partial x_i}N_m dV\right)\theta_m=\int_{V(e)} X_i N_l dV+\int_{A(e)} t_i N_l dA\\
&\qquad\qquad l,m=1,2,...,r \quad i,j=1,2,3
\end{aligned}
\tag{28.18}
$$

Equation (28.18) is the finite element approximation of the equation of motion.

The Galerkin approximation of the energy equation given by Eq. (28.1) becomes

$$
\int_{V(e)} (q_{i,i}+\rho c\,\frac{\partial\theta}{\partial t}+T_0\beta\dot{u}_{i,i}-R)N_l dV=0 \quad l=1,2,\cdots,r \tag{28.19}
$$

The weak formulation of the heat flux gradient $q_{i,i}$ gives

$$
\begin{aligned}
&\int_{V(e)} q_{i,i}N_l dV=\int_{V(e)}\left(\frac{\partial q_x}{\partial x}+\frac{\partial q_y}{\partial y}+\frac{\partial q_z}{\partial z}\right)N_l dV=\int_{A(e)}(\vec{q}\cdot\vec{n})N_l dA\\
&-\int_{V(e)} q_i\,\frac{\partial N_l}{\partial x_i}dV
\end{aligned}
\tag{28.20}
$$

where $A(e)$ is the boundary surface of the base element (e). Substituting Eq. (28.20) in Eq. (28.19) and rearranging the terms gives

$$
\begin{aligned}
&\int_{V(e)} \rho c\,\frac{\partial\theta}{\partial t}\,N_l dV-\int_{V(e)} q_i\,\frac{\partial N_l}{\partial x_i}dV+\int_{V(e)} T_0\beta\dot{u}_{i,i}N_l dV\\
&=\int_{V(e)} RN_l dV-\int_{A(e)}(\vec{q}\cdot\vec{n})N_l dA \quad l=1,2,\cdots,r
\end{aligned}
\tag{28.21}
$$

Substituting for the displacement components u_i and temperature change θ their approximate values in the base element (e) from Eqs. (28.8) and (28.9) give

$$
\begin{aligned}
&\left(\int_{V(e)} k\,\frac{\partial N_m}{\partial x_i}\frac{\partial N_l}{\partial x_i}dV\right)\theta_m+\left(\int_{V(e)} T_0\beta\,\frac{\partial N_m}{\partial x_i}\,N_l dV\right)\dot{U}_{mi}\\
&+\left(\int_{V(e)} \rho c N_m N_l dV\right)\dot{\theta}_m=\int_{V(e)} RN_l dV-\int_{A(e)}(\vec{q}\cdot\vec{n})N_l dA
\end{aligned}
\tag{28.22}
$$

Equation (28.22) is the finite element approximation of the coupled energy equation.

Equations (28.18) and (28.22) are assembled into a matrix form resulting in the general finite element coupled equation given by

$$
[M]\{\ddot{\Delta}\}+[C]\{\dot{\Delta}\}+[K]\{\Delta\}=\{F\} \tag{28.23}
$$

where $[M]$, $[C]$ and $[K]$ are the mass, damping, and the stiffness matrices, respectively. Matrix $\{\Delta\}^T = < U_i, \theta >$ is the matrix of unknowns and $\{F\}$ is the known mechanical and thermal force matrix.

For a two-dimensional problem, l and m take the values $1, 2, ...r$. In this case, Eq. (28.18) reduces into two equations in x and y-directions, as

$$\begin{aligned}&\left(\int_{V(e)} \rho N_l N_m dV\right) \ddot{U}_m + \left[\int_{V(e)} (2G+\lambda) \frac{\partial N_l}{\partial x} \frac{\partial N_m}{\partial x} dV\right.\\&\left. + \int_{V(e)} G \frac{\partial N_l}{\partial y} \frac{\partial N_m}{\partial y} dV\right] U_m + \left[\int_{V(e)} G \frac{\partial N_l}{\partial y} \frac{\partial N_m}{\partial x} dV\right.\\&\left. + \int_{V(e)} \lambda \frac{\partial N_l}{\partial x} \frac{\partial N_m}{\partial y} dV\right] V_m - \left[\int_{V(e)} \beta N_m \frac{\partial N_l}{\partial x} dV\right] \theta_m\\&= \int_{V(e)} X N_l dV + \int_{A(e)} t_x N_l dA \end{aligned} \tag{28.24}$$

$$\begin{aligned}&\left(\int_{V(e)} \rho N_l N_m dV\right) \ddot{V}_m + \left[(2G+\lambda) \int_{V(e)} \frac{\partial N_l}{\partial y} \frac{\partial N_m}{\partial y} dV\right.\\&\left. + \int_{V(e)} G \frac{\partial N_l}{\partial x} \frac{\partial N_m}{\partial x} dV\right] V_m + \left[\int_{V(e)} G \frac{\partial N_l}{\partial x} \frac{\partial N_m}{\partial y} dV\right.\\&\left. + \int_{V(e)} \lambda \frac{\partial N_l}{\partial y} \frac{\partial N_m}{\partial x} dV\right] U_m - \left[\int_{V(e)} \beta N_m \frac{\partial N_l}{\partial y} dV\right] \theta_m\\&= \int_{V(e)} Y N_l dV + \int_{A(e)} t_y N_l dA \end{aligned} \tag{28.25}$$

The energy Eq. (28.22) for a two-dimensional problem becomes

$$\begin{aligned}&\left(\int_{V(e)} T_0 \beta \frac{\partial N_m}{\partial x} N_l dV\right) \dot{U}_m + \left(\int_{V(e)} T_0 \beta \frac{\partial N_m}{\partial y} N_l dV\right) \dot{V}_m\\&+ \left(\int_{V(e)} \rho c N_m N_l dV\right) \dot{\theta}_m + \left(\int_{V(e)} k \frac{\partial N_m}{\partial x} \frac{\partial N_l}{\partial x} dV\right.\\&\left. + \int_{V(e)} k \frac{\partial N_m}{\partial y} \frac{\partial N_l}{\partial y} dV\right) \theta_m = \int_{V(e)} R N_l dV - \int_{V(e)} (\vec{q} \cdot \vec{n}) N_l dA \end{aligned} \tag{28.26}$$

The elements of the mass, damping, stiffness, and the force matrices of the base element (e) are

$$[M]^{(e)} = \begin{bmatrix} [\int_{V(e)} \rho N_l N_m dV] & 0 & 0 \\ 0 & [\int_{V(e)} \rho N_l N_m dV] & 0 \\ 0 & 0 & 0 \end{bmatrix} \tag{28.27}$$

The damping matrix is

$$[C]^{(e)} = \begin{bmatrix} 0 & 0 & 0 \\ 0 & 0 & 0 \\ \int_{v(e)} T_0\beta \frac{\partial N_m}{\partial x} N_l dV & \int_{v(e)} T_0\beta \frac{\partial N_m}{\partial y} N_l dV & \int_{v(e)} \rho c N_m N_l dV \end{bmatrix} \tag{28.28}$$

and the stiffness matrix is

$$[k]^{(e)} = \begin{bmatrix} k_{11} & k_{12} & k_{13} \\ k_{21} & k_{22} & k_{23} \\ k_{31} & k_{32} & k_{33} \end{bmatrix} \tag{28.29}$$

where

$$\begin{aligned}
[k_{11}^{lm}] &= \left[\int_{V(e)} (2G+\lambda) \frac{\partial N_l}{\partial x} \frac{\partial N_m}{\partial x} dV + \int_{V(e)} G \frac{\partial N_l}{\partial y} \frac{\partial N_m}{\partial y} dV \right] \\
[k_{12}^{lm}] &= \left[\int_{V(e)} G \frac{\partial N_l}{\partial y} \frac{\partial N_m}{\partial x} dV + \int_{V(e)} \lambda \frac{\partial N_l}{\partial x} \frac{\partial N_m}{\partial y} dV \right] \\
[k_{13}^{lm}] &= -\left[\int_{V(e)} \beta N_m \frac{\partial N_l}{\partial x} dV \right] \\
[k_{21}^{lm}] &= \left[\int_{V(e)} G \frac{\partial N_l}{\partial x} \frac{\partial N_m}{\partial y} dV + \int_{V(e)} \lambda \frac{\partial N_l}{\partial y} \frac{\partial N_m}{\partial x} dV \right] \\
[k_{22}^{lm}] &= \left[\int_{V(e)} (2G+\lambda) \frac{\partial N_l}{\partial y} \frac{\partial N_m}{\partial y} dV + \int_{V(e)} G \frac{\partial N_l}{\partial x} \frac{\partial N_m}{\partial x} dV \right] \\
[k_{23}^{lm}] &= -\left[\int_{V(e)} \beta N_m \frac{\partial N_l}{\partial y} dV \right] \\
[k_{31}^{lm}] &= [k_{32}]^{ml} = 0 \\
[k_{33}^{lm}] &= \left[\int_{V(e)} k \frac{\partial N_m}{\partial x} \frac{\partial N_l}{\partial x} dV + \int_{V(e)} k \frac{\partial N_m}{\partial y} \frac{\partial N_l}{\partial y} dV \right]
\end{aligned} \tag{28.30}$$

The force matrix is

$$\{f\}_l^{(e)} = \begin{Bmatrix} \{\int_{V(e)} X N_l dV + \int_{A(e)} t_x N_l dA\} \\ \{\int_{V(e)} Y N_l dV + \int_{A(e)} t_y N_l dA\} \\ \{\int_{V(e)} R N_l dV - \int_{V(e)} (\vec{q} \cdot \vec{n}) N_l dA\} \end{Bmatrix} \tag{28.31}$$

and the unknown matrix is

$$\{\delta\}^{(e)} = \begin{Bmatrix} \{U\} \\ \{V\} \\ \{\theta\} \end{Bmatrix} \tag{28.32}$$

The initial and the general form of the thermal boundary conditions are one, or the combinations, of the following:

$$
\begin{aligned}
&\theta(x, y, z, 0) = 0(x, y, z) \quad \text{at } t = 0 \\
&\theta(x, y, z, t) = \theta_s \quad \text{on } A_1 \quad \text{and } t > 0 \\
&q_x l + q_y m + q_z n = -q^{''} \quad \text{on } A_2 \text{ and } t > 0 \\
&q_x l + q_y m + q_z n = h(\theta + T_0 - T_\infty) \quad \text{on } A_3 \text{ and } t > 0 \\
&q_x l + q_y m + q_z n = \sigma\epsilon(\theta + T_0)^4 - \alpha_{ab} q_r \quad \text{on } A_4 \text{ and } t > 0
\end{aligned} \tag{28.33}
$$

where $T_0(x, y, z)$ is the known initial temperature, θ_s is the known specified temperature change on a part of the boundary surface A_1, $q^{''}$ is the known heat flux on the boundary A_2, h and T_∞ are the convection coefficient and ambient temperature specified on a part of the boundary surface A_3, respectively, σ is the Stefan-Boltzmann constant, ϵ is the radiation coefficient of the boundary surface, α_{ab} is the boundary surface absorption coefficient, and q_r is the rate of thermal flux reaching the boundary surface per unit area all specified on boundary surface A_4. The cosine directors of the unit outer normal vector to the boundary in x, y, and z-directions are shown by l, m, and n, respectively. According to the boundary conditions given by Eq. (28.33), the last surface integral of the energy Eq. (28.22) may be decomposed into four integrals over A_1 through A_4 as

$$
\begin{aligned}
\int_{A(e)} (\vec{q} \cdot \vec{n}) N_l dA &= \int_{A_2} q^{''} N_l dA - \int_{A_3} h(\theta + T_0 - T_\infty) N_l dA \\
&- \int_{A_4} (\sigma\epsilon(\theta + T_0)^4 - \alpha_{ab} q_r) N_l dA \qquad l = 1, 2, \cdots r
\end{aligned} \tag{28.34}
$$

Note that the signs of the integrals in Eq. (28.34) depend upon the direction of the heat input. The positive sign is defined when the heat is given to the body, and is negative when the heat is removed from the body. That is, q'' is defined positive in Eq. (28.34), since we have assumed that the heat flux is given to the body. On the other hand, we have assumed negative convection on the surface area A_3, which means the heat is removed from A_3 boundary by convection. Similarly, the boundary A_4 is assumed to radiate to the ambient, as the sign of this integral is considered negative.

28.2 One-Dimensional Problem

In order to discuss the method in more detail a one-dimensional problem is considered. The equation of motion in terms of displacement is

$$
(\lambda + 2G)\frac{\partial^2 u}{\partial x^2} - \beta\frac{\partial \theta}{\partial x} = \rho\frac{\partial^2 u}{\partial t^2} \tag{28.35}
$$

and the first law of thermodynamics reduces to

$$k\frac{\partial^2\theta}{\partial x^2} - \rho c\frac{\partial\theta}{\partial t} - \beta T_0\frac{\partial^2 u}{\partial x\partial t} = 0 \tag{28.36}$$

Taking a line element of length L, the approximating function for axial displacement for the base element (e) is assumed to be linear in x as

$$u(x,t)^{(e)} = N_i U_i + N_j U_j =< N >^{(e)} \ \{U\}^{(e)} \tag{28.37}$$

where the piecewise linear shape function $< N >$ is $N_i = (L-\eta)/L$, $N_j = \eta/L$, and $\eta = x - x_i$. Similarly the temperature change is approximated by

$$\theta(x,t)^{(e)} = N_i\theta_i + N_j\theta_j =< N >^{(e)} \ \{\theta\}^{(e)} \tag{28.38}$$

Employing the formal Galerkin method and applying the weak form to the first and second terms of Eq. (28.35) and first term of Eq. (28.36) results in the following system of equations

$$\begin{aligned}&\left(\int_0^L \rho N_l N_m d\eta\right)\ddot{U}_m + \left(\int_0^L (2G+\lambda)\frac{\partial N_l}{\partial\eta}\frac{\partial N_m}{\partial\eta}d\eta\right)U_m \\ &-\left(\int_0^L \beta N_m\frac{\partial N_l}{\partial\eta}d\eta\right)\theta_m = \mathbf{t}_x N_l\Big|_i^j + \int_0^L X N_l d\eta\end{aligned} \tag{28.39}$$

$$\begin{aligned}&\left(\int_0^L T_0\beta\frac{\partial N_m}{\partial\eta}\ N_l d\eta\right)\dot{U}_m + \left(\int_0^L \rho c N_m N_l d\eta\right)\dot{\theta}_m \\ &+\left(\int_0^L k\ \frac{\partial N_m}{\partial\eta}\ \frac{\partial N_l}{\partial\eta}d\eta\right)\theta_m = -\ (\vec{q}\cdot\vec{n})N_l|_i^j + \int_0^L RN_l d\eta\end{aligned} \tag{28.40}$$

This system of equations may be written in matrix form as

$$[M]\{\ddot{\Delta}\} + [C]\{\dot{\Delta}\} + [K]\{\Delta\} = \{F\} \tag{28.41}$$

where the mass, damping, stiffness and force matrices for first order element are 4×4 matrices and are defined as

$$[M]^{(e)} = \int_0^L \begin{bmatrix} \rho N_i N_i & 0 & \rho N_i N_j & 0 \\ 0 & 0 & 0 & 0 \\ \rho N_j N_i & 0 & \rho N_j N_j & 0 \\ 0 & 0 & 0 & 0 \end{bmatrix} d\eta \tag{28.42}$$

$$[C]^{(e)} = \int_0^L \begin{bmatrix} 0 & 0 & 0 & 0 \\ T_0\beta\, N_i \dfrac{\partial N_i}{\partial \eta} & \rho c N_i N_i & T_0\beta\, N_i \dfrac{\partial N_j}{\partial \eta} & \rho c N_i N_j \\ 0 & 0 & 0 & 0 \\ T_0\beta\, N_j \dfrac{\partial N_i}{\partial \eta} & \rho c N_j N_i & T_0\beta\, N_j \dfrac{\partial N_j}{\partial \eta} & \rho c N_j N_j \end{bmatrix} d\eta \tag{28.43}$$

$$[K]^{(e)} = \int_0^L \begin{bmatrix} (2G+\lambda)\left(\dfrac{\partial N_i}{\partial \eta}\right)^2 & -\beta N_i \dfrac{\partial N_i}{\partial \eta} & (2G+\lambda)\dfrac{\partial N_i}{\partial \eta}\dfrac{\partial N_j}{\partial \eta} & -\beta N_j \dfrac{\partial N_i}{\partial \eta} \\ 0 & k\left(\dfrac{\partial N_i}{\partial \eta}\right)^2 & 0 & k\,\dfrac{\partial N_j}{\partial \eta}\dfrac{\partial N_i}{\partial \eta} \\ (2G+\lambda)\dfrac{\partial N_j}{\partial \eta}\dfrac{\partial N_i}{\partial \eta} & -\beta N_i \dfrac{\partial N_j}{\partial \eta} & (2G+\lambda)\left(\dfrac{\partial N_j}{\partial \eta}\right)^2 & -\beta N_j \dfrac{\partial N_j}{\partial \eta} \\ 0 & k\,\dfrac{\partial N_i}{\partial \eta}\dfrac{\partial N_j}{\partial \eta} & 0 & k\left(\dfrac{\partial N_j}{\partial \eta}\right)^2 \end{bmatrix} d\eta \tag{28.44}$$

$$\{F\}^{(e)} = \begin{Bmatrix} \mathbf{t}_x N_i|_0^L + \int_0^L X N_i d\eta \\ -\, q_x N_i|_0^L + \int_0^L R N_i d\eta \\ \mathbf{t}_x N_j\big|_L + \int_0^L X N_j d\eta \\ -\, q_x N_j\big|_0^L + \int_0^L R N_j d\eta \end{Bmatrix} \tag{28.45}$$

Upon substitution of the shape functions in the foregoing equations, the submatrices for the base element (e) are

$$[M]^{(e)} = \begin{bmatrix} \frac{\rho L}{3} & 0 & \frac{\rho L}{6} & 0 \\ 0 & 0 & 0 & 0 \\ \frac{\rho L}{6} & 0 & \frac{\rho L}{3} & 0 \\ 0 & 0 & 0 & 0 \end{bmatrix} \tag{28.46}$$

$$[C]^{(e)} = \begin{bmatrix} 0 & 0 & 0 & 0 \\ -\frac{T_0\beta}{2} & \frac{\rho c L}{3} & \frac{T_0\beta}{2} & \frac{\rho c L}{6} \\ 0 & 0 & 0 & 0 \\ -\frac{T_0\beta}{2} & \frac{\rho c L}{6} & \frac{T_0\beta}{2} & \frac{\rho c L}{3} \end{bmatrix} \tag{28.47}$$

$$[K]^{(e)} = \begin{bmatrix} \frac{(2G+\lambda)}{L} & \frac{\beta}{2} & -\frac{(2G+\lambda)}{L} & \frac{\beta}{2} \\ 0 & \frac{k}{L} & 0 & -\frac{k}{L} \\ -\frac{(2G+\lambda)}{L} & -\frac{\beta}{2} & \frac{(2G+\lambda)}{L} & -\frac{\beta}{2} \\ 0 & -\frac{k}{L} & 0 & \frac{k}{L} \end{bmatrix} \tag{28.48}$$

$$\{F\}^{(e)} = \begin{Bmatrix} -\, t_x|_0 + \frac{XL}{2} \\ q_x|_0 + \frac{RL}{2} \\ t_x|_L + \frac{XL}{2} \\ -\, q_x|_L + \frac{RL}{2} \end{Bmatrix} \tag{28.49}$$

and the matrix of unknown nodal value is

$$\{\Delta\}^{(e)} = \begin{Bmatrix} U_i \\ \theta_i \\ U_j \\ \theta_j \end{Bmatrix} \tag{28.50}$$

28.3 Problems and Solutions Related to Coupled Thermoelasticity

Problem 28.1. Find the dynamic response of a layer made of functionally graded materials based on the Lord-Shulman (LS) theory. The power law form function is assumed for the material properties distribution.

Solution: The Functionally Graded Materials (FGMs) are high-performance heat resistant materials able to withstand ultra high temperatures and extremely large thermal gradients used in the aerospace industries. The FGMs are microscopically inhomogeneous in which the mechanical properties vary smoothly and continuously from one surface to the other. Typically, these materials are made from a mixture of ceramic and metal.

Consider a ceramic-metal FG layer with thickness of L and assume that the properties of FG layer obey a power law function as

$$P = \left(\frac{x}{L}\right)^n (P_m - P_c) + P_c \tag{28.51}$$

where x is the position from the ceramic rich side of the layer, P is the effective property of FGM, n is the power law index that governs the distribution of the constituent materials through the thickness of the layer, and P_m and P_c are the properties of metal and ceramic, respectively. Meanwhile, the subscripts m and c indicate the metal and ceramic features, respectively.

For the LS theory, in the absence of body forces and heat supply, when the derivative of the relaxation time with respect to the position variable is neglected, the governing equations of an FG layer in terms of displacement and temperature, are as follow

$$(\lambda+2\mu)\frac{\partial^2 u}{\partial x^2}+\frac{\partial(\lambda+2\mu)}{\partial x}\frac{\partial u}{\partial x}-\frac{\partial[\beta(T-T_0)]}{\partial x}=\rho\frac{\partial^2 u}{\partial t^2} \tag{28.52}$$

$$\frac{\partial}{\partial x}\left(k\frac{\partial T}{\partial x}\right)-\rho c\frac{\partial T}{\partial t}-\rho c t_0\frac{\partial^2 T}{\partial t^2}-\beta T_0\left(t_0\frac{\partial^2}{\partial t^2}+\frac{\partial}{\partial t}\right)\frac{\partial u}{\partial x}=0 \tag{28.53}$$

where t_0 is the relaxation time proposed by Lord and Shulman. The preceding equations may be introduced in dimensionless form for convenience. The nondimensional parameters are defined as

$$\begin{aligned}
\bar{x}&=\frac{x c_m\sqrt{\rho_m(\lambda_m+2\mu_m)}}{k_m}\ ;\ \bar{T}=\frac{T-T_0}{T_d}\\
\bar{t}&=\frac{t(\lambda_m+2\mu_m)c_m}{k_m}\ ;\ \bar{t}_0=\frac{t_0(\lambda_m+2\mu_m)c_m}{k_m}\\
\bar{q}_x&=\frac{q_x}{c_m T_d\sqrt{\rho_m(\lambda_m+2\mu_m)}}\ ;\ \bar{\sigma}_{xx}=\frac{\sigma_{xx}}{\beta_m T_d}\\
\bar{u}&=\frac{(\lambda_m+2\mu_m)^{3/2}\rho_m^{1/2}\ c_m\ u}{k_m\beta_m T_d}
\end{aligned} \tag{28.54}$$

where the subscript m denotes the metal properties and term T_d is a characteristic temperature used for normalizing the temperature. Using the dimensionless parameters, the governing Eqs. (28.52) and (28.53) appear in the form

$$\left[\frac{(\lambda+2\mu)}{(\lambda_m+2\mu_m)}\frac{\partial^2}{\partial\bar{x}^2}+\frac{1}{(\lambda_m+2\mu_m)}\frac{\partial(\lambda+2\mu)}{\partial\bar{x}}\frac{\partial}{\partial\bar{x}}-\frac{\rho}{\rho_m}\frac{\partial^2}{\partial\bar{t}^2}\right]\bar{u}-\frac{1}{\beta_m}\left(\frac{\partial\beta}{\partial\bar{x}}+\beta\frac{\partial}{\partial\bar{x}}\right)\bar{T}=0 \tag{28.55}$$

$$\left[\frac{k}{k_m}\frac{\partial^2}{\partial\bar{x}^2}+\frac{1}{k_m}\frac{\partial k}{\partial\bar{x}}\frac{\partial}{\partial\bar{x}}-\frac{\rho c}{\rho_m c_m}(\frac{\partial}{\partial\bar{t}}+\bar{t}_0\frac{\partial^2}{\partial\bar{t}^2})\right]\bar{T}-\frac{\beta_m T_0}{\rho_m c_m(\lambda_m+2\mu_m)}\beta\left(\bar{t}_0\frac{\partial^2}{\partial\bar{t}^2}+\frac{\partial}{\partial\bar{t}}\right)\frac{\partial\bar{u}}{\partial\bar{x}}=0 \tag{28.56}$$

Also, the dimensionless stress-displacement-temperature relation and the heat conduction equation for the functionally graded layer based on the LS theory are

$$\bar{\sigma}_{xx}=\frac{(\lambda+2\mu)}{(\lambda_m+2\mu_m)}\frac{\partial\bar{u}}{\partial\bar{x}}-\frac{\beta}{\beta_m}\bar{T} \tag{28.57}$$

$$\bar{q}_x+\bar{t}_0\frac{\partial\bar{q}_x}{\partial\bar{t}}=-\frac{k}{k_m}\frac{\partial\bar{T}}{\partial\bar{x}} \tag{28.58}$$

The layer is occupied in the region $0\le\bar{x}\le 1$ and the corresponding dimensionless boundary conditions are

$$\bar{q}_x = 1 - (1 + 100\bar{t})e^{-100\bar{t}} \quad ; \quad \bar{\sigma}_{xx} = 0 \quad at \quad \bar{x} = 0$$
$$\bar{T} = 0 \quad ; \quad \bar{u} = 0 \quad at \quad \bar{x} = 1 \tag{28.59}$$

To solve the coupled system of equations, the transfinite element method may be employed. To this end, the equations may be transformed to the space domain using the Laplace transformation. Assume that the layer is initially at rest and the initial displacement, velocity, temperature, and temperature rate are zero. Applying the Laplace transformation to Eqs. (28.55) and (28.56) give

$$\begin{aligned} &\left[\frac{(\lambda + 2\mu)}{(\lambda_m + 2\mu_m)}\frac{\partial^2}{\partial \bar{x}^2} + \frac{1}{(\lambda_m + 2\mu_m)}\frac{\partial(\lambda + 2\mu)}{\partial \bar{x}}\frac{\partial}{\partial \bar{x}} - \frac{\rho}{\rho_m}s^2\right]\bar{u}^* \\ &-\frac{1}{\beta_m}\left(\frac{\partial \beta}{\partial \bar{x}} + \beta\frac{\partial}{\partial \bar{x}}\right)\bar{T}^* = 0 \end{aligned} \tag{28.60}$$

$$\begin{aligned} &\left[\frac{k}{k_m}\frac{\partial^2}{\partial \bar{x}^2} + \frac{1}{k_m}\frac{\partial k}{\partial \bar{x}}\frac{\partial}{\partial \bar{x}} - \frac{\rho c}{\rho_m c_m}s(1 + \bar{t}_0 s)\right]\bar{T}^* \\ &-\frac{\beta_m T_0}{\rho_m c_m(\lambda_m + 2\mu_m)}\beta s(1 + \bar{t}_0 s)\frac{\partial \bar{u}^*}{\partial \bar{x}} = 0 \end{aligned} \tag{28.61}$$

To find the solution of the equations using the transfinite element method, the geometry of the layer may be divided into a number of discretized elements through the thickness of the layer. In the base element, the Kantorovich approximation for the displacement u and temperature T with identical shape functions is assumed as

$$\bar{u}^{*(e)} = \sum_{i=1}^{\ell} N_i \bar{U}_i^* \qquad \bar{T}^{*(e)} = \sum_{i=1}^{\ell} N_i \bar{T}_i^* \tag{28.62}$$

where N_i is the shape function and terms $\bar{U}_i^*$ and $\bar{T}_i^*$ are the unknown nodal values of displacement and temperature, respectively. Substituting Eq. (28.62) into Eqs. (28.60) and (28.61) and then employing the Galerkin finite element method, the following system of equations, applying the weak form to the terms of second order of derivatives of the space variable, is obtained:

$$\begin{bmatrix} [K_{11}] & [K_{12}] \\ [K_{21}] & [K_{22}] \end{bmatrix} \begin{Bmatrix} \bar{U}^* \\ \bar{T}^* \end{Bmatrix} = \begin{Bmatrix} F^* \\ Q^* \end{Bmatrix} \tag{28.63}$$

The submatrices $[K_{11}]$, $[K_{12}]$, $[K_{21}]$, $[K_{22}]$, F^* and Q^* are

$$[K_{11}^{ij}] = \int_{\bar{x}_f}^{\bar{x}_e} \left\{ \frac{(\lambda + 2\mu)}{(\lambda_m + 2\mu_m)} \frac{\partial N_i}{\partial \bar{x}} \frac{\partial N_j}{\partial \bar{x}} + \frac{\rho s^2}{\rho_m} N_i N_j \right\} d\bar{x} \tag{28.64}$$

$$[K_{12}^{ij}] = \frac{1}{\beta_m} \int_{\bar{x}_f}^{\bar{x}_e} \left\{ \beta N_j \frac{\partial N_j}{\partial \bar{x}} + \frac{\partial \beta}{\partial \bar{x}} N_i N_j \right\} d\bar{x} \tag{28.65}$$

$$[K_{21}^{ij}] = \frac{s\beta_m T_0}{\rho_m c_m(\lambda_m + 2\mu_m)} \int_{\bar{x}_f}^{\bar{x}_e} \beta(1+\bar{t}_0 s)N_i \frac{\partial N_j}{\partial \bar{x}} d\bar{x} \tag{28.66}$$

$$[K_{22}^{ij}] = \int_{\bar{x}_f}^{\bar{x}_e} \left\{ \frac{k}{k_m}\frac{\partial N_i}{\partial \bar{x}}\frac{\partial N_j}{\partial \bar{x}} + \frac{\rho c}{\rho_m c_m} s(1+\bar{t}_0 s)N_i N_j \right\} d\bar{x} \tag{28.67}$$

$$\{F^*\} = \begin{Bmatrix} \frac{\beta_c}{\beta_m}\bar{T}_1^* \\ 0 \\ . \\ . \\ 0 \end{Bmatrix}; \{Q^*\} = \begin{Bmatrix} (1+\bar{t}_0 s)\bar{q}_x^* \\ 0 \\ . \\ . \\ 0 \end{Bmatrix} \tag{28.68}$$

In these equations $\bar{x}_f$ and $\bar{x}_e$ are the first and last nodes of the solution domain, respectively.

System of Eq. (28.63) is solved in the space domain. To transform the results from the Laplace transform domain into the real time domain, the numerical inverse Laplace transform technique is used.

Consider an FG layer composed of aluminum and alumina as metal and ceramic constituents, respectively. The reference temperature is assumed to be $T_0 = 298K$. The relaxation time of aluminum and alumina are assumed to be $\bar{t}_{0m} = 0.64$, $\bar{t}_{0c} = 1.5625$, respectively. The linear Lagrangian polynomials are used for the shape functions in the base element. Figures 28.1 and 28.2 show the temperature and stress waves propagation and reflection from the boundaries of the layer for $n = 1$. In Fig. 28.1, the times $\bar{t} = 0.2, 0.4, 0.6, 0.8, 1.7$ show the temperature wave propagation through the thickness of the layer, while the reflection of the temperature wave occurred at time $\bar{t} = 1.9$. Figure 28.2 shows that the maximum of stress occurs at the

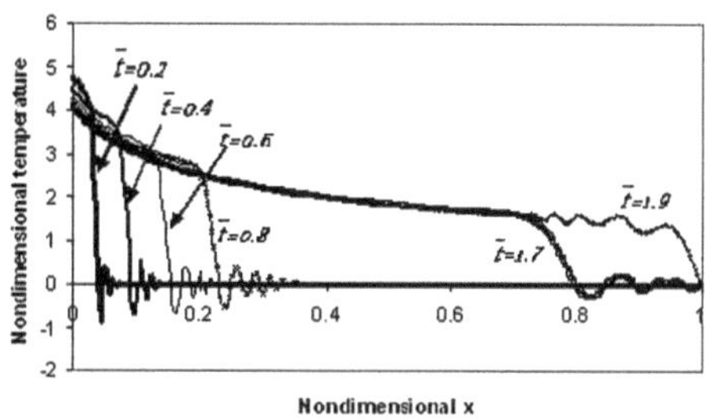

Fig. 28.1 Distribution of the nondimensional temperature through the thickness of the layer for $n = 1$

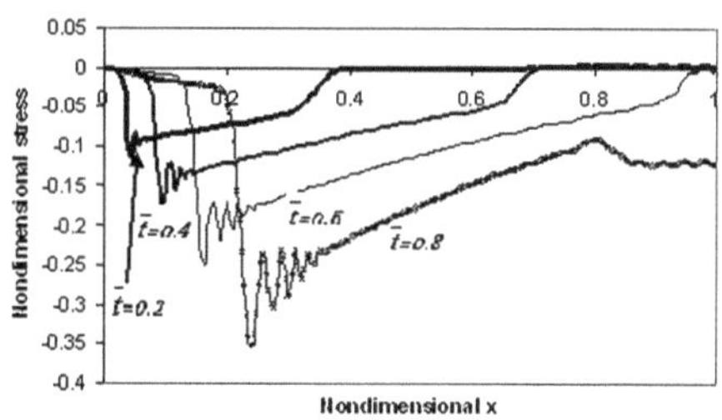

Fig. 28.2 Distribution of the nondimensional stress through the thickness of the layer for $n = 1$

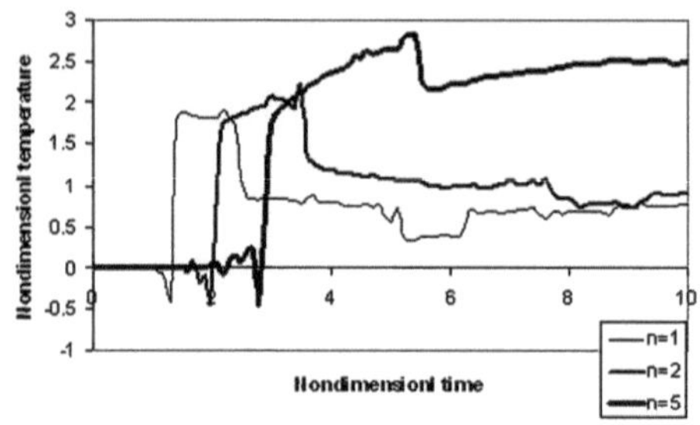

Fig. 28.3 Variations of the nondimensional temperature at the middle point of the thickness of the layer for different values of power index

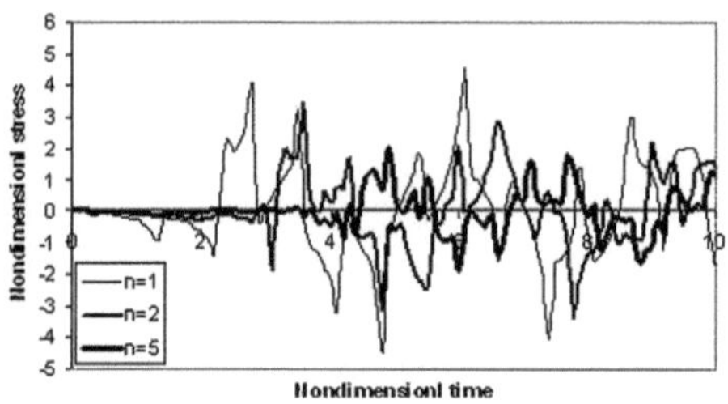

Fig. 28.4 Variations of the nondimensional stress at the middle point of the thickness of the layer for different values of power index

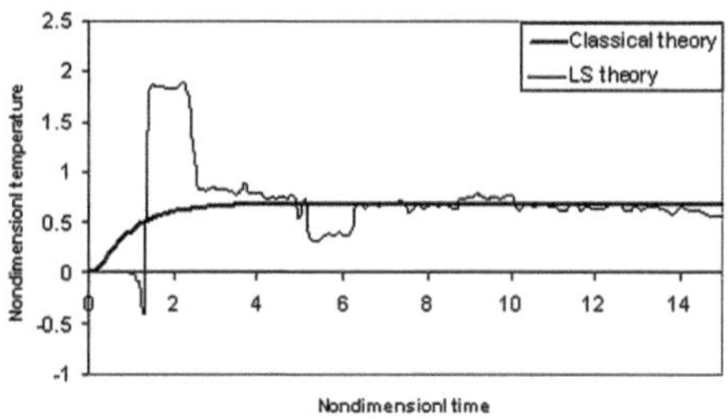

Fig. 28.5 Variations of the nondimensional temperature at the middle point of the thickness of the layer for the classical and LS theories

temperature wave front. In Figs. 28.1 and 28.2 it is seen that a conversion between the mechanical and thermal energies occurs at the temperature wave front. It may be found from the figures that the propagation velocity of waves varies through the thickness of the layer.

The effect of power law index, n, on variation of the temperature and stress at a point located at the middle point of the layer thickness is shown in Figs. 28.3 and 28.4. It is seen from Fig. 28.3 that when n increases the speed of temperature wave decreases. In Fig. 28.4 it is shown that the amplitude of stress variation is decreased with the increase of n.

The relaxation time effect on variation of the temperature and stress at the middle point of the thickness of the layer is investigated and are shown in Figs. 28.5 and 28.6. The value of $n = 1$ is considered for the power law index. It is seen that for the classical theory of thermoelasticity, the case when $t_0 = 0$, smaller values for amplitude of temperature and resulting stress variations are obtained. Since with the increase of relaxation time the propagation velocity of temperature wave decreases, these maximum values of variations occurs at the later times.

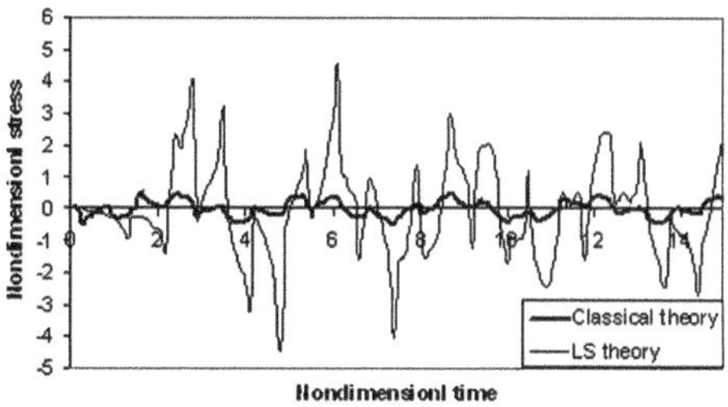

Fig. 28.6 Variations of the nondimensional stress at the middle point of the thickness of the layer for the classical and LS theories

Problem 28.2. Consider a thick-walled sphere of inside and outside radii r_{in} and r_{out}, respectively. The sphere is under symmetric thermal shock load on its inside surface. Find the distribution of the temperature and stresses along the radial direction.

Solution: For symmetric thermal shock loading condition

$$\sigma_{\theta\theta} = \sigma_{\phi\phi} \qquad \epsilon_{\theta\theta} = \epsilon_{\phi\phi} \tag{28.69}$$

the equation of motion is

$$\frac{\partial \sigma_{rr}}{\partial r} + 2\frac{\sigma_{rr} - \sigma_{\theta\theta}}{r} = \rho \ddot{u} \tag{28.70}$$

and the strain-displacement relations are

$$\epsilon_{rr} = \frac{\partial u}{\partial r} \qquad \epsilon_{\theta\theta} = \frac{u}{r} = \epsilon_{\phi\phi} \tag{28.71}$$

From Hooke's law

$$\begin{aligned} \sigma_{rr} &= \frac{E}{(1+\nu)(1-2\nu)}[(1-\nu)\epsilon_{rr} + 2\nu\epsilon_{\theta\theta} - (1+\nu)\alpha(T-T_0)] \\ \sigma_{\theta\theta} &= \frac{E}{(1+\nu)(1-2\nu)}[\epsilon_{\theta\theta} + \nu\epsilon_{rr} - (1+\nu)\alpha(T-T_0)] \end{aligned} \tag{28.72}$$

The first law of thermodynamics for coupled condition is

$$\frac{\partial^2 T}{\partial r^2} + \frac{2}{r}\frac{\partial T}{\partial r} - \frac{\rho c}{k}\dot{T} = \gamma_1(\dot{\epsilon}_{rr} + 2\dot{\epsilon}_{\theta\theta}) - R \tag{28.73}$$

where $\gamma_1 = (3\lambda + 2\mu)\alpha T_0/k$. Elimination of the stresses from Eqs. (28.70), (28.71), and (28.72) results in the equation of motion in terms of the displacement

$$\frac{\partial^2 u}{\partial r^2} + \frac{2}{r}\frac{\partial u}{\partial r} - \frac{2u}{r^2} - \alpha\frac{(1+\nu)}{(1-\nu)}\frac{\partial T}{\partial r} = \frac{(1+\nu)(1-2\nu)\rho}{(1-\nu)E}\ddot{u} \tag{28.74}$$

The energy equation, after substitution for strains, becomes

$$\frac{\partial^2 T}{\partial r^2} + \frac{2}{r}\frac{\partial T}{\partial r} - \frac{\rho c}{k}\dot{T} = \frac{\gamma_1}{r^2}\frac{\partial r}{\partial}(r^2\dot{u}) - R \tag{28.75}$$

The boundary conditions are in general given as

$$\sigma_{rr} \times n_r = \mathbf{t}_r \qquad -k\frac{\partial T}{\partial r} = q_r \tag{28.76}$$

where n_r is the unit vector in radial direction. In terms of displacement, the boundary conditions at inside and outside radii are:

At $r = r_{in}$

$$\frac{E(1-\nu)}{(1+\nu)(1-2\nu)}\left[\frac{\partial u}{\partial r} + \frac{\nu}{(1-\nu)}\frac{2u}{r} - \frac{(1+\nu)}{(1-\nu)}\alpha(T-T_0)\right] = -P_a(t)$$
$$T(t) = T_0\left\{2 + \left[(\frac{\rho\, c\, c_1^2\, t}{k})^2 - \frac{\rho\, c\, c_1^2\, t}{k} - 1\right]e^{-\frac{\rho\, c\, c_1^2\, t}{k}}\right\} \tag{28.77}$$

At $r = r_{out}$

$$u = 0$$
$$k\frac{\partial T}{\partial r} = -h_o[T - T_0] \tag{28.78}$$

where h_o is the convection coefficients at inside and outside surfaces of the sphere, respectively, $T_i(t)$ is the inside surface temperature which is assumed to vary in time and is applied as a thermal shock to the inside surface, T_0 is the constant outside ambient temperature, $P_a(t)$ is the applied pressure shock at the inside surface which may be considered zero.

The governing equations are changed into dimensionless form through the following formulas

$$\bar{T} = \frac{(T-T_0)}{T_0}$$
$$\bar{u} = \left(\frac{1-\nu}{1+\nu}\right)\frac{\rho\, c\, c_1\, u}{k\, \alpha\, T_0}$$
$$\bar{\sigma}_{rr} = \frac{(1-2\nu)\sigma_{rr}}{E\alpha T_0}$$
$$\bar{r} = \frac{\rho\, c\, c_1\, r}{k}$$

$$\bar{t} = \frac{\rho\, c\, c_1^2\, t}{k}$$
$$c_1 = \sqrt{\frac{(1-\nu)E}{(1+\nu)(1-2\nu)\rho}} \qquad (28.79)$$

Using these quantities, and in the absence of heat generation, the governing equations are expressed in dimensionless form (bar is dropped for convenience)

$$\frac{\partial r}{\partial}\left(\frac{\partial u}{\partial r} + \frac{2u}{r}\right) - \frac{\partial T}{\partial r} = \ddot{u} \qquad (28.80)$$
$$\frac{\partial r}{\partial}\left(\frac{\partial T}{\partial r}\right) + \frac{2}{r}\frac{\partial T}{\partial r} - \dot{T} = C\frac{1}{r^2}\frac{\partial r}{\partial}(r^2\dot{u}) \qquad (28.81)$$

The boundary conditions are

$$\frac{\partial u}{\partial r} + \gamma_2\frac{u}{r} - T = -\frac{1-2\nu}{E\alpha T_0}P_a(t)$$
$$T = 1 + (t^2 - t - 1)\, e^{-t} \qquad \text{at } r = a \qquad (28.82)$$

and

$$u = 0$$
$$\frac{\partial T}{\partial r} = -\eta_o T \qquad \text{at } r = b \qquad (28.83)$$

where a and b are the dimensionless inner and outer radii of the sphere, respectively, and the parameters used in these equations are defined as

$$C = \frac{T_0(1+\nu)E\alpha^2}{\rho c(1-\nu)(1-2\nu)};\ \gamma_2 = \frac{2\nu}{(1-\nu)};\ \eta_o = \frac{h_o}{\rho c c_1} \qquad (28.84)$$

Due to the radial symmetry of loading conditions the variations of the dependent functions are along the radius of the sphere. Thus, the radius of the sphere is divided into a number of line elements (NE) with nodes i and j for the base element (e), as shown in Fig. 28.7.

The displacement of element (e) is described by the linear shape function

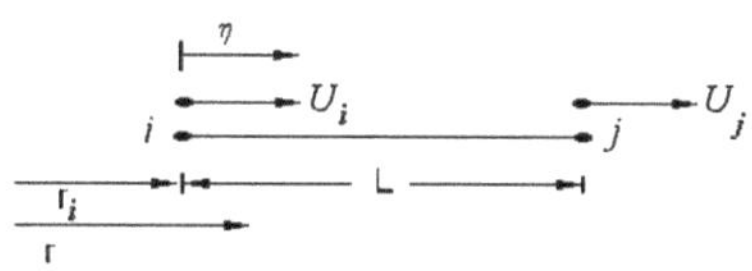

Fig. 28.7 The element (e) along the radius

$$u(r,t) = \alpha_1(t) + \alpha_2(t)r \tag{28.85}$$

in terms of nodal unknown variables, the unknown coefficients $\alpha_1(t)$ and $\alpha_2(t)$ are found from Eq. (28.85) as

$$\begin{aligned} U_i(t) &= \alpha_1 + \alpha_2 r_i \\ U_j(t) &= \alpha_1 + \alpha_2 r_j \end{aligned} \tag{28.86}$$

Solving for α_1 and α_2 and substituting in Eq. (28.85) yields

$$u(\eta,t) = \frac{L-\eta}{L}\, U_i + \frac{\eta}{L} U_j \tag{28.87}$$

where η is the variable in local coordinates, $\eta = r - r_i$, and L is the element length $L = r_j - r_i$. Defining the linear shape functions as

$$N_i = \frac{L-\eta}{L} \qquad\qquad N_j = \frac{\eta}{L} \tag{28.88}$$

the displacement u is allowed to vary linearly in the base element (e) as

$$u^{(e)}(\eta,t) = N_i U_i + N_j U_j =< N_i \quad N_j > \begin{Bmatrix} U_i \\ U_j \end{Bmatrix} =< N >^{(e)} \; \{U\}^{(e)} \tag{28.89}$$

Similarly, the temperature variation in the element (e) is assumed to vary linearly

$$T^{(e)}(\eta,t) = N_i T_i + N_j T_j =< N >^{(e)} \; \{T\}^{(e)} \tag{28.90}$$

Using Eqs. (28.89) and (28.90) and applying the formal Galerkin method to the governing Eq. (28.80) and (28.81) for the base element (e) yields

$$\int_{r_i}^{r_j} \left[\frac{\partial r}{\partial} \left(\frac{\partial u}{\partial r} + \frac{2u}{r} \right) - \frac{\partial T}{\partial r} - \frac{\partial^2 u}{\partial t^2} \right] r^2 N_m dr = 0 \tag{28.91}$$

$$\int_{r_i}^{r_j} \left[\frac{\partial r}{\partial} \left(\frac{\partial T}{\partial r} \right) + \frac{2}{r}\frac{\partial T}{\partial r} - \frac{\partial T}{\partial t} - C \frac{1}{r^2}\frac{\partial r}{\partial} \left(r^2 \frac{\partial u}{\partial t} \right) \right] r^2 N_m dr = 0$$

$$m = i, j \tag{28.92}$$

Considering the change of variable $\eta = r - r_i$ and applying the weak formulation to the terms of second order derivatives gives

$$\begin{aligned} &\int_0^L \frac{\partial[(\eta+r_i)^2 N_m]}{\partial \eta} \frac{\partial u}{\partial \eta} d\eta - \int_0^L (\eta+r_i)^2 N_m \left(\frac{2}{(\eta+r_i)} \frac{\partial u}{\partial \eta} - \frac{2u}{(\eta+r_i)^2} \right) d\eta \\ &+ \int_0^L (\eta+r_i)^2 N_m \frac{\partial T}{\partial \eta} d\eta + \int_0^L (\eta+r_i)^2 N_m \ddot{u} d\eta = (\eta+r_i)^2 N_m \frac{\partial u}{\partial \eta} |_0^L \end{aligned} \tag{28.93}$$

$$\int_0^L \frac{\partial[(\eta+r_i)^2 N_m]}{\partial\eta}\frac{\partial T}{\partial\eta}d\eta - \int_0^L 2(\eta+r_i)N_m\frac{\partial T}{\partial\eta}d\eta + \int_0^L (\eta+r_i)^2 N_m \dot{T} d\eta$$
$$+ C\int_0^L N_m \frac{\partial\eta}{\partial[}(\eta+r_i)^2\dot{u}]d\eta = (\eta+r_i)^2 N_m \frac{\partial T}{\partial\eta}|_0^L$$
$$m = i, j \tag{28.94}$$

Substituting the shape functions for u and T from Eqs. (28.89) and (28.90) yields

$$\int_0^L \left(\frac{d[(\eta+r_i)^2 N_m]}{d\eta} < \frac{dN}{d\eta} > -2N_m[(\eta+r_i) < \frac{dN}{d\eta} > - < N >]\right) d\eta\{U\}$$
$$+ \int_0^L (\eta+r_i)^2 N_m < \frac{dN}{d\eta} > d\eta\{T\} + \int_0^L (\eta+r_i)^2 N_m < N > d\eta\{\ddot{U}\}$$
$$= (\eta+r_i)^2 N_m \frac{\partial u}{\partial\eta}|_0^L \tag{28.95}$$

$$\int_0^L \left(\frac{d[(\eta+r_i)^2 N_m]}{d\eta} < \frac{dN}{d\eta} > -2(\eta+r_i)N_m < \frac{dN}{d\eta} >\right) d\eta\{T\}$$
$$+ \int_0^L (\eta+r_i)^2 N_m < N > d\eta\{\dot{T}\} + C\int_0^L N_m < \frac{d[(\eta+r_i)^2 N]}{d\eta} > d\eta\{\dot{U}\}$$
$$= (\eta+r_i)^2 N_m \frac{dT}{d\eta}|_0^L$$
$$m = i, j \tag{28.96}$$

The terms on the right-hand side of Eqs. (28.95) and (28.96) are derived through the weak formulation and coincide with the natural boundary conditions. They cancel out each other between any two adjacent elements except the first node of the first element and the last node of the last element which coincide with the given boundary conditions on inside and outside surfaces of the sphere. These boundary conditions are

$$-a^2\frac{\partial u}{\partial\eta}|_1 = 2a\gamma_2 U_1 - a^2 T_1 + a^2\frac{1-2\nu}{E\alpha T_0}P_a(t)$$
$$U_M = 0$$
$$T_1 = 1 + (t^2 - t - 1)\, e^{-t}$$
$$b^2\frac{\partial T}{\partial\eta}|_M = -b^2\eta_o T_M \tag{28.97}$$

where the index 1 denotes the first note of the first element of the solution domain at $r = a$ and the index M denotes the last node of the last element of the solution domain at $r = b$. It is to be noted that due to the assumed boundary conditions at $r = a$ and $r = b$ terms $-a^2\frac{\partial T}{\partial\eta}|_1$ and $b^2\frac{\partial u}{\partial\eta}|_M$ vanish. Equations (28.95) and (28.96) are solved for the nodal unknown of the element (e) and are finally arranged in the form of the following matrix equations

$$[M]\{\ddot{\Delta}\} + [C]\{\dot{\Delta}\} + [K]\{\Delta\} = \{F\} \tag{28.98}$$

The definitions of the mass, damping, stiffness, and force matrices of Eq. (28.98) for the base element (e) are

Mass matrix

$$[M]^{(e)} == \begin{bmatrix} m_{11} & 0 & m_{13} & 0 \\ 0 & 0 & 0 & 0 \\ m_{31} & 0 & m_{33} & 0 \\ 0 & 0 & 0 & 0 \end{bmatrix} \tag{28.99}$$

where the components of the mass matrix are

$$\begin{aligned} m_{11} &= \int_0^L (\eta + r_i)^2 N_i N_i d\eta \\ m_{13} &= \int_0^L (\eta + r_i)^2 N_i N_j d\eta \\ m_{31} &= \int_0^L (\eta + r_i)^2 N_j N_i d\eta \\ m_{33} &= \int_0^L (\eta + r_i)^2 N_j N_j d\eta \end{aligned} \tag{28.100}$$

Damping matrix

$$[C]^{(e)} = \begin{bmatrix} 0 & 0 & 0 & 0 \\ C_{21} & C_{22} & C_{23} & C_{24} \\ 0 & 0 & 0 & 0 \\ C_{41} & C_{42} & C_{43} & C_{44} \end{bmatrix} \tag{28.101}$$

where the components of the damping matrix are

$$\begin{aligned} C_{21} &= C \int_0^L N_i \frac{d[(\eta + r_i)^2 N_i]}{d\eta} d\eta \\ C_{23} &= C \int_0^L N_i \frac{d[(\eta + r_i)^2 N_j]}{d\eta} d\eta \\ C_{41} &= C \int_0^L N_j \frac{d[(\eta + r_i)^2 N_i]}{d\eta} d\eta \\ C_{43} &= C \int_0^L N_j \frac{d[(\eta + r_i)^2 N_j]}{d\eta} d\eta \\ C_{22} &= \int_0^L (\eta + r_i)^2 N_i N_i d\eta \end{aligned}$$

$$
\begin{aligned}
C_{24} &= \int_0^L (\eta + r_i)^2 N_i N_j d\eta \\
C_{42} &= \int_0^L (\eta + r_i)^2 N_j N_i d\eta \\
C_{44} &= \int_0^L (\eta + r_i)^2 N_j N_j d\eta
\end{aligned} \tag{28.102}
$$

Stiffness matrix

$$
[K]^{(e)} = \begin{bmatrix} K_{11} & K_{12} & K_{13} & K_{14} \\ 0 & K_{22} & 0 & K_{24} \\ K_{31} & K_{32} & K_{33} & K_{34} \\ 0 & K_{42} & 0 & K_{44} \end{bmatrix} \tag{28.103}
$$

where the components of the stiffness matrix are

$$
\begin{aligned}
K_{11} &= \int_0^L \left(\frac{d[(\eta + r_i)^2 N_i]}{d\eta} \frac{dN_i}{d\eta} - 2(\eta + r_i) N_i \frac{dN_i}{d\eta} + 2N_i N_i) \right) d\eta \\
K_{13} &= \int_0^L \left(\frac{d[(\eta + r_i)^2 N_i]}{d\eta} \frac{dN_j}{d\eta} - 2(\eta + r_i) N_i \frac{dN_j}{d\eta} + 2N_i N_j) \right) d\eta \\
K_{31} &= \int_0^L \left(\frac{d[(\eta + r_i)^2 N_j]}{d\eta} \frac{dN_i}{d\eta} - 2(\eta + r_i) N_j \frac{dN_i}{d\eta} + 2N_j N_i) \right) d\eta \\
K_{33} &= \int_0^L \left(\frac{d[(\eta + r_i)^2 N_j]}{d\eta} \frac{dN_j}{d\eta} - 2(\eta + r_i) N_j \frac{dN_j}{d\eta} + 2N_j N_j) \right) d\eta \\
K_{12} &= \int_0^L (\eta + r_i)^2 N_i \frac{dN_i}{d\eta} d\eta \\
K_{14} &= \int_0^L (\eta + r_i)^2 N_i \frac{dN_j}{d\eta} d\eta \\
K_{32} &= \int_0^L (\eta + r_i)^2 N_j \frac{dN_i}{d\eta} d\eta \\
K_{34} &= \int_0^L (\eta + r_i)^2 N_j \frac{dN_j}{d\eta} d\eta \\
K_{22} &= \int_0^L \left(\frac{d[(\eta + r_i)^2 N_i]}{d\eta} \frac{dN_i}{d\eta} - 2(\eta + r_i) N_i \frac{dN_i}{d\eta} \right) d\eta \\
K_{24} &= \int_0^L \left(\frac{d[(\eta + r_i)^2 N_i]}{d\eta} \frac{dN_j}{d\eta} - 2(\eta + r_i) N_i \frac{dN_j}{d\eta} \right) d\eta \\
K_{42} &= \int_0^L \left(\frac{d[(\eta + r_i)^2 N_j]}{d\eta} \frac{dN_i}{d\eta} - 2(\eta + r_i) N_j \frac{dN_i}{d\eta} \right) d\eta \\
K_{44} &= \int_0^L \left(\frac{d[(\eta + r_i)^2 N_j]}{d\eta} \frac{dN_j}{d\eta} - 2(\eta + r_i) N_j \frac{dN_j}{d\eta} \right) d\eta
\end{aligned} \tag{28.104}
$$

Also, the matrix $\{F\}$ for this special case is related to the boundary conditions. With the inside temperature and pressure shocks and outside surface insulated, as given by Eq. (28.97), the final assembled form of this matrix becomes

$$\{F\} = \begin{Bmatrix} 2a\gamma_2 U_1 - a^2 T_1 + a^2 \frac{1-2\nu}{E\alpha T_0} P_a(t) \\ 1 + (t^2 - t - 1)\, e^{-t} \\ 0 \\ 0 \\ . \\ . \\ 0 \\ 0 \\ -b^2 \eta_o T_M \end{Bmatrix} \tag{28.105}$$

Moreover, the matrix of unknown nodal values is

$$\{\Delta\}^{(e)} = \begin{Bmatrix} U_i \\ T_i \\ U_j \\ T_j \end{Bmatrix} \tag{28.106}$$

Using the linear shape functions (28.89) and (28.90) for $\{U\}$ and $\{T\}$ the component of the matrices are simplified as

Components of mass matrix

$$\begin{aligned} m_{11} &= \frac{L(L^2 + 5r_i L + 10r_i^2)}{30} \\ m_{13} = m_{31} &= \frac{L(3L^2 + 10r_i L + 10ri^2)}{60} \\ m_{33} &= \frac{L(6L^2 + 15r_i L + 10r_i^2)}{30} \end{aligned} \tag{28.107}$$

Components of damping matrix

$$\begin{aligned} C_{21} &= C\frac{(L^2 + 4r_i L - 6r_i^2)}{12} \\ C_{23} &= C\frac{(3L^2 + 8r_i L + 6r_i^2)}{12} \\ C_{41} &= -C\frac{(L^2 + 4r_i L + 6r_i^2)}{12} \\ C_{43} &= C\frac{(9L^2 + 16r_i L + 6r_i^2)}{12} \end{aligned}$$

$$
\begin{aligned}
C_{22} &= L\frac{(L^2 + 5r_iL + 10r_i^2)}{30} \\
C_{24} &= C_{42} = L\frac{(3L^2 + 10r_iL + 10r_i^2)}{60} \\
C_{44} &= L\frac{(6L^2 + 15r_iL + 10r_i^2)}{30}
\end{aligned} \tag{28.108}
$$

Components of stiffness matrix

$$
\begin{aligned}
K_{11} &= K_{33} = \frac{L^2 + r_iL + r_i^2}{L} \\
K_{13} &= K_{31} = -\frac{r_i(L + r_i)}{L} \\
K_{12} &= -K_{14} = -\frac{1}{12}L^2 - \frac{1}{3}r_iL - \frac{1}{2}r_i^2 \\
K_{32} &= -K_{34} = -\frac{1}{4}L^2 - \frac{2}{3}r_iL - \frac{1}{2}r_i^2 \\
K_{22} &= K_{44} = -K_{24} = -K_{42} = \frac{L^2 + 3r_iL + 3r_i^2}{3L}
\end{aligned} \tag{28.109}
$$

The element matrices given by Eqs. (28.102)–(28.104) are generated within a loop to construct the general matrices of Eq. (28.98), where after assembly of all the elements in the solution domain, they are solved using one of the numerical techniques of either time marching or modal analysis methods.

As a numerical example, a thick sphere is considered with the following properties: $E = 70 \times 10^9 \mathrm{N/m^2}$, $\nu = 0.3$, $\rho = 2707\,\mathrm{Kgr/m^3}$, $k = 204\,\mathrm{W/m\text{-}K}$, $\alpha = 23 \times 10^{-6}\,1/\mathrm{K}$, $c = 903\,\mathrm{J/Kgr\text{-}K}$, $T_0 = 298\,\mathrm{K}$. The pressure at the inner surface of the sphere is assumed to be zero (traction free condition) and the outer surface of the sphere is insulated (with $h_o = 0$). The plot of internal thermal shock is shown in Fig. 28.8. The distribution of temperature, radial displacement, radial stress, and hoop stress at different times are plotted in Figs. 28.9 28.10, 28.11, and 28.12. Figure 28.13 show the variations of radial and hoop stresses versus radius at different times. Figure 28.14 shows the time variation of the radial and hoop stresses and temperature at mid-point of the thickness of the sphere.

Problem 28.3. Employ the transfinite element method, using the Laplace transform, to solve the coupled equations for an axisymmetrically loaded disk in the transformed domain. By the inverse numerical method obtain the distribution of the temperature and stresses in the real time domain. Consider elements with various orders to investigate the effects of the number of nodes in an element.

Solution: In the absence of the heat source and body forces and for isotropic materials, the nondimensionalized form of the generalized coupled thermoelastic equations

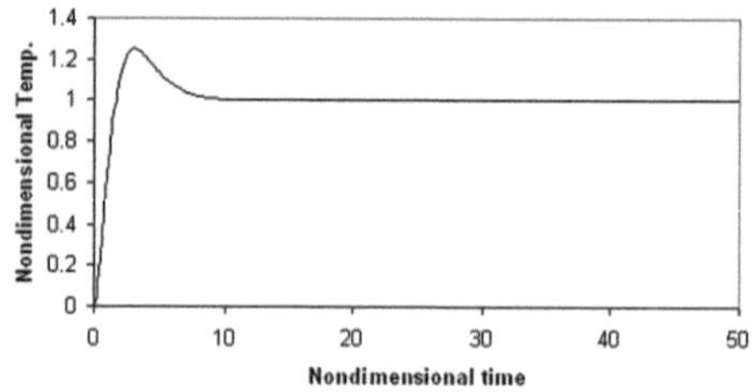

Fig. 28.8 The temperature shock applied to the inner surface of the sphere

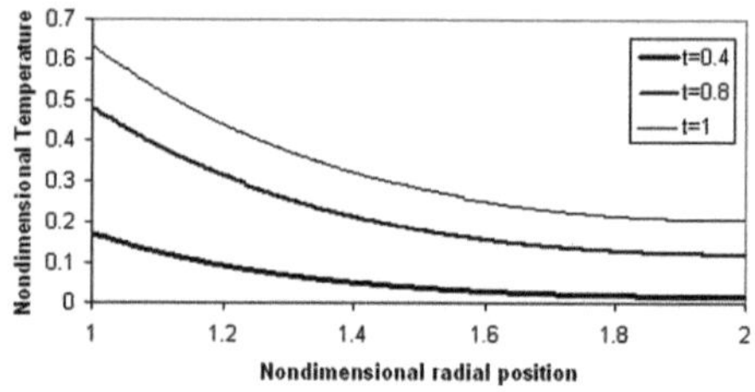

Fig. 28.9 The temperature distribution at different times

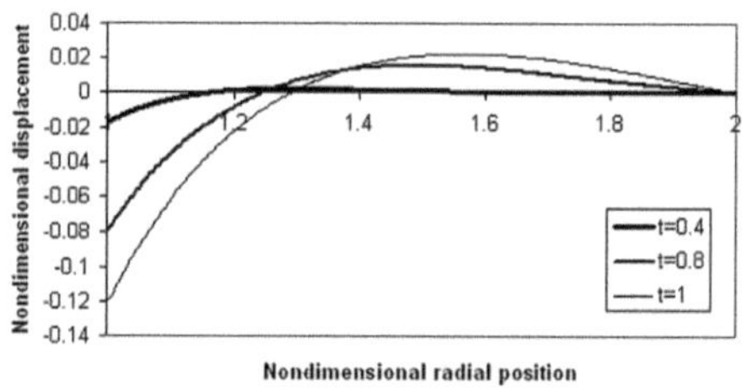

Fig. 28.10 The displacement distribution at different times

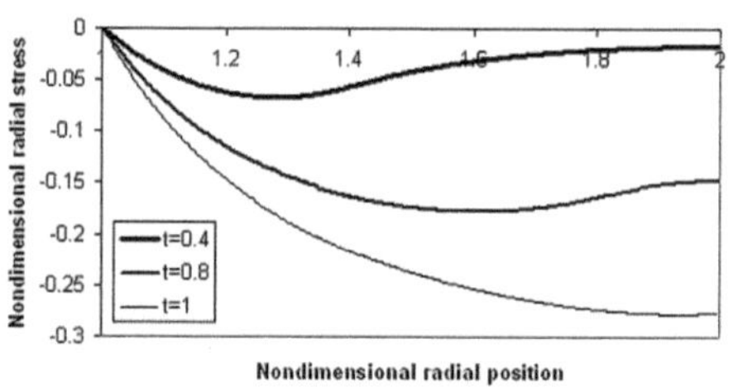

Fig. 28.11 The radial stress distribution at different times

of the axisymmetrically loaded circular disk based on the Lord-Shulman theory in terms of the displacement and temperature may be written as

$$\left\{\frac{\partial^2}{\partial r^2}+\frac{1}{r}\frac{\partial}{\partial r}-\frac{1}{r^2}-\frac{\partial^2}{\partial t^2}\right\}u-\frac{\partial T}{\partial r}=0 \tag{28.110}$$

$$\left\{\frac{\partial^2}{\partial r^2}+\frac{1}{r}\frac{\partial}{\partial r}-\frac{\partial}{\partial t}\left(1+t_0\frac{\partial}{\partial t}\right)\right\}T$$
$$-C\left\{t_0\left[\frac{\partial^3}{\partial r\partial t^2}+\frac{1}{r}\frac{\partial^2}{\partial t^2}\right]+\frac{\partial^2}{\partial r\partial t}+\frac{1}{r}\frac{\partial}{\partial t}\right\}u=0 \tag{28.111}$$

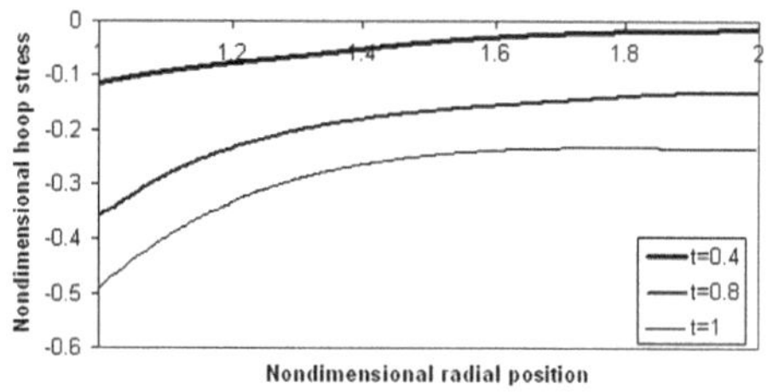

Fig. 28.12 The hoop stress distribution at different times

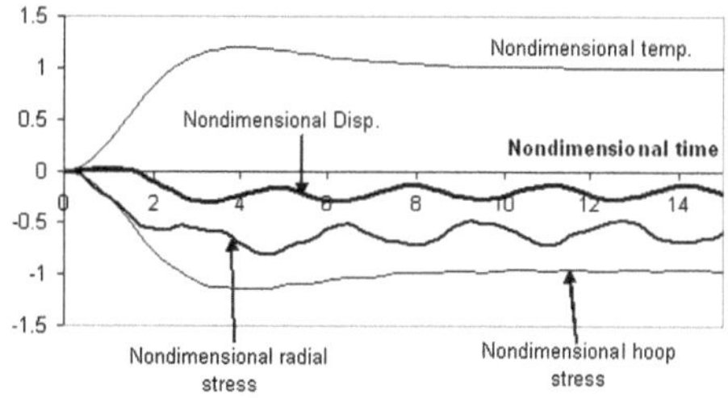

Fig. 28.13 Time history of temperature, displacement and stresses at mid-radius point of the thickness of the sphere

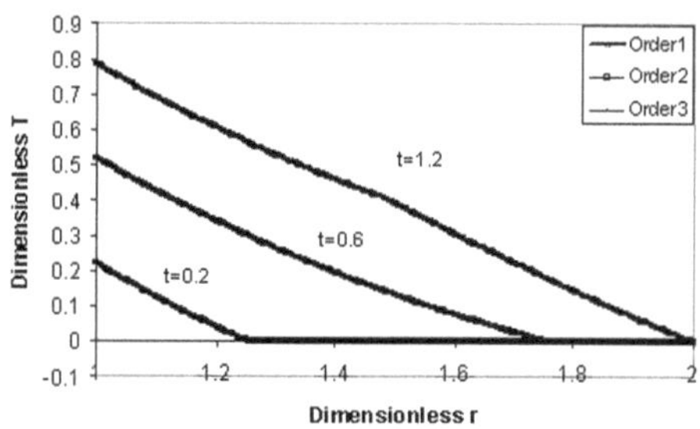

Fig. 28.14 Distribution of the dimensionless temperature along the radius of the disk at three values of the time for three types of elements

Here, $C = T_0\bar{\beta}^2/\left[\rho c_e\left(\bar{\lambda}+2\mu\right)\right]$ is the coupling coefficient. For the plane stress condition $\bar{\lambda} = \frac{2\mu}{\lambda+2\mu}\lambda$ and $\bar{\beta} = \frac{2\mu}{\lambda+2\mu}\beta$. In the preceding equations $\rho,\ u,\ T_0,\ T,\ \bar{\beta},\ c_e$, and t_0 are the density, radial displacement, reference temperature, temperature change, stress-temperature moduli, thermal conductivity, specific heat and the relaxation time (proposed by Lord and Shulman), respectively, while λ and μ are Lamé constants. The dimensionless thermal and mechanical boundary conditions are

$$
\begin{aligned}
&q_{in} = -\frac{\partial T}{\partial r}; \qquad u = 0 && at\ \ r = a \\
T = 0;\qquad &\sigma_{rr} = \frac{\partial u}{\partial r} + \frac{\bar{\lambda}}{\bar{\lambda}+2\mu}\frac{u}{r} - T = 0 && at\ \ r = b
\end{aligned}
\tag{28.112}
$$

where σ_{rr} and a and b are the radial stress and dimensionless inner and outer radii, respectively.

In order to derive the transfinite element formulation, the Laplace transformation is used to transform the equations into the Laplace transform domain. Applying the

Galerkin finite element method to the governing Eqs. (28.110) and (28.111) for the base element (e), yields

$$\int_0^L \left\{ -\left[\left\{ \frac{1}{(\eta + r_i)} \frac{\partial}{\partial \eta} - \frac{1}{(\eta + r_i)^2} - s^2 \right\} u - \frac{\partial T}{\partial \eta} \right] N_m(\eta + r_i) + \frac{\partial (N_m(\eta + r_i))}{\partial \eta} \frac{\partial u}{\partial \eta} \right\} d\eta = N_m(\eta + r_i) \frac{\partial u}{\partial \eta} \bigg|_0^L \tag{28.113}$$

$$\int_0^L \left\{ -\left[\left\{ \frac{1}{(\eta + r_i)} \frac{\partial}{\partial \eta} - s(1 + t_0 s) \right\} T - C \left(t_0 s^2 + s \right) \left(\frac{\partial}{\partial \eta} + \frac{1}{\eta + r_i} \right) u \right] \times N_m(\eta + r_i) + \frac{\partial (N_m(\eta + r_i))}{\partial \eta} \frac{\partial T}{\partial \eta} \right\} d\eta = N_m(\eta + r_i) \frac{\partial T}{\partial \eta} \bigg|_0^L \tag{28.114}$$

where $u = \sum_{m=1}^{n} N_m U_m$ and $T = \sum_{m=1}^{n} N_m T_m$. In the preceding equations, s, N_m, $\eta = r - r_i$, r_i, L, U_m, and T_m are the Laplace parameter, shape function, local coordinates, the radius of the i-th node of the base element, the length of element in the radial direction, nodal displacement, and the nodal temperature respectively. The terms on the right-hand sides of Eqs. (28.113) and (28.114) cancel each other between any two adjacent elements, except the nodes located on the boundaries of the solution domain. These boundary conditions are

$$-a \frac{\partial T}{\partial \eta} \bigg|_1 = a q_{in} \quad ; \quad U_1 = 0$$

$$T_M = 0 \quad ; \quad b \frac{\partial u}{\partial \eta} \bigg|_M = -\frac{\bar{\lambda}}{\bar{\lambda} + 2\mu} U_M + b T_M \tag{28.115}$$

The subscript 1 and M are referred to the first and last nodes of the solution domain, respectively.

To investigate the accuracy of the method, a numerical example is considered. The material of the disk is assumed to be aluminum. The dimensionless inside and outside radii are $a = 1$ and $b = 2$. The dimensionless input heat flux is defined as the Heaviside unit step function. Since the applied boundary conditions are assumed to be axisymmetric, the radius of the disk is divided into 100 elements. Three types of shape functions, linear, second order, and third order polynomials are used for the finite element model of the problem. Results for each of these orders are plotted and are compared.

Figures 28.15 and 28.16 show the wave propagation of the temperature and radial displacement along the radial direction. The numerical values of the coupling parameter and the dimensionless relaxation time are assumed to be 0.01 and 0.64, respectively. The waves propagation are shown at several times. Two wave fronts for elastic and temperature waves are detected from the figures, as expected from the LS model. It is seen from the figures that the results of the three types of shape functions

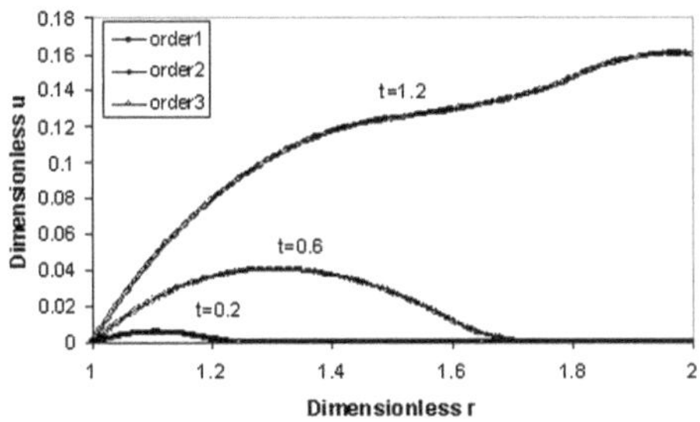

Fig. 28.15 Distribution of the dimensionless displacement along the radius of the disk at three values of the time for three types of elements

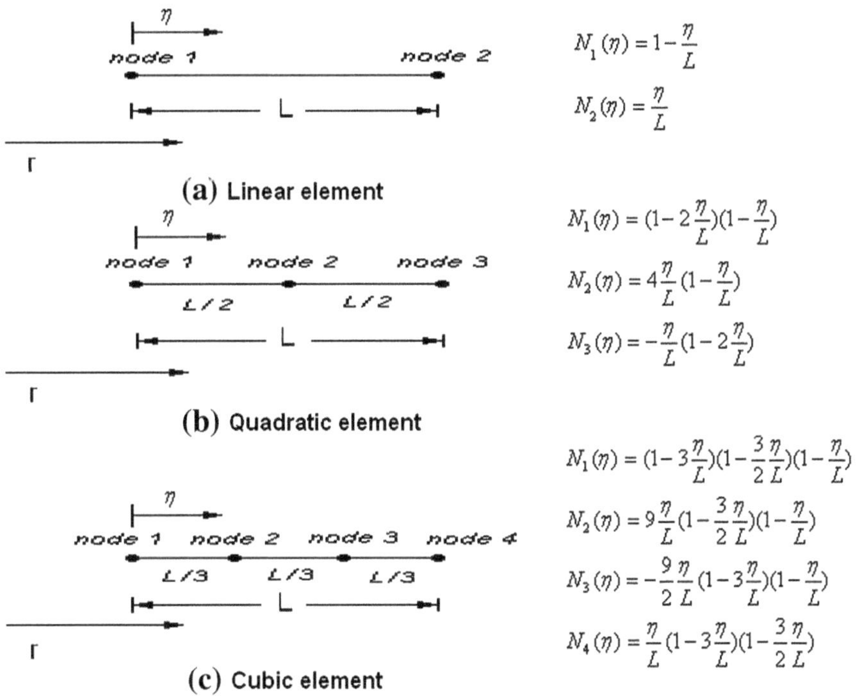

Fig. 28.16 Elements with linear, second, and third order shape functions

for the assumed number of elements coincide. For smaller number of elements, the difference between the results obtained for different shape functions increase noticeably. For the assumed number of elements, the curves for radial displacement and temperature distribution are checked against the known data in the literature, where very close agreement is observed. Figure 28.15 clearly shows the temperature wave front (the second sound effect), which is propagating along the radius of the disk.

Chapter 29
Boundary Element, Coupled Thermoelasticity

In this chapter, considering the Lord and Shulman's theory, a Laplace-transform boundary element method is developed for the dynamic problems in coupled thermoelasticity with relaxation time involving a finite two dimensional domain. The boundary element formulation is presented and a single heat excitation is used to drive the boundary element formulations. Aspect of numerical implementation are discussed. It is shown that the distributions of temperature, displacement, and stress show jumps at their wave fronts. The thermo-mechanical waves propagation in a finite domain and the influence of relaxation time on them are presented. The results of this section are compared with the classical coupled theory (CCT) and the Green-Lindsay theory (GL). It is verified that the LS theory of the generalized thermoelasticity results into significant differences in the patterns and wave fronts of temperature, displacement, and stress compared to the CCT and GL theories, although the material and geometrical properties of the solution domain are identical. The details of these differences are given in the result section.

29.1 Governing Equations

A homogeneous isotropic thermoelastic solid is considered. In the absence of body forces and heat flux, the governing equations for the dynamic coupled generalized thermoelasticity in the time domain based on the Lord and Shulman theory are written as [1]

$$(\lambda + \mu)u_{j,ij} + \mu u_{i,jj} - \gamma T_0 T_{,i} - \rho \ddot{u}_i = 0 \tag{29.1}$$

$$kT_{,jj} - \rho c_e \dot{T} - \rho c_e t_0 \ddot{T} - \gamma(t_0 \ddot{u}_{j,j} + \dot{u}_{j,j}) = 0 \tag{29.2}$$

[1] Tehrani, P., Eslami, M.R., Boundary Element Analysis of Finite Domain Under Thermal and Mechanical Shock with the Lord-Shulman Theory, Proceedings of I.Mech.E, Journal of Strain, Analysis, **38**(1), 53–64 (2003).

M. Reza Eslami et al., *Theory of Elasticity and Thermal Stresses*, Solid Mechanics and Its Applications 197, DOI: 10.1007/978-94-007-6356-2_29,

Throughout this section, the summation convention on repeated indices is used. A dot indicates time differentiation and the subscript i after a comma is partial differentiation with respect to $x_i (i = 1, 2)$. In these equations $\lambda, \mu, u_i, \rho, T, T_0, k, \gamma, c_e$, and t_0 are the Lame's constants, the components of displacement vector, mass density, absolute temperature, reference temperature, coefficient of thermal conductivity, stress-temperature modulus, specific heat at constant strain, and relaxation times proposed by Lord and Shulman, respectively. When t_0 vanish, Eq. (29.2) reduces to the classical coupled theory. In the Lord and Shulman's theory, Fourier's law of heat conduction is modified by introducing the relaxation time t_0. It is convenient to introduce the usual dimensionless variables as

$$\hat{x} = \frac{x}{\alpha}; \quad \hat{t} = \frac{tC_s}{\alpha} \quad \hat{t_0} = \frac{t_0 C_s}{\alpha};$$
$$\hat{\sigma}_{ij} = \frac{\sigma_{ij}}{\gamma T_0}; \quad \hat{u}_i = \frac{(\lambda + 2\mu)u_i}{\alpha \cdot \gamma \cdot T_0}; \quad T = \frac{T - T_0}{T_0} \tag{29.3}$$

where $\alpha = k/\rho c_e C_s$ is the dimensionless characteristic length and $C_s = \sqrt{(\lambda + 2\mu)/\rho}$ is the velocity of the longitudinal wave. Equations (29.1) and (29.2) take the form (dropping the hat for convenience)

$$\frac{\mu}{\lambda + 2\mu} u_{i,jj} + \frac{\lambda + \mu}{\lambda + 2\mu} u_{j,ij} - T_{,i} - \ddot{u}_i = 0 \tag{29.4}$$

$$T_{,jj} - \dot{T} - t_0 \ddot{T} - \frac{T_0 \gamma^2}{\rho c_e (\lambda + 2\mu)} (\dot{u}_{j,j} + t_0 \ddot{u}_{j,j}) = 0 \tag{29.5}$$

At the above dimensionless equations, the stress wave speed is one and the speed of the temperature wave may be computed as

$$C_t = \sqrt{\frac{1}{t_0}} \tag{29.6}$$

Transferring Eqs. (29.4) and (29.5) to Laplace domain yield

$$\frac{\mu}{\lambda + 2\mu} \tilde{u}_{i,jj} + \frac{\lambda + \mu}{\lambda + 2\mu} \tilde{u}_{j,ij} - \tilde{T}_{,i} - s^2 \tilde{u}_i = 0 \tag{29.7}$$

$$\tilde{T}_{,jj} - s\tilde{T} - t_0 s^2 \tilde{T} - \frac{T_0 \gamma^2}{\rho c_e (\lambda + 2\mu)} (s\tilde{u}_{j,j} + t_0 s^2 \tilde{u}_{j,j}) = 0 \tag{29.8}$$

Equations (29.7) and (29.8) are rewritten in matrix form as

$$L_{ij} U_j = 0 \tag{29.9}$$

For two-dimensional domain the operator L_{ij} reduces to

$$
L_{ij} = \begin{bmatrix} \frac{\mu}{\lambda+2\mu}\Delta + \frac{\lambda+\mu}{\lambda+2\mu}D_1^2 - s^2 & \frac{\lambda+\mu}{\lambda+2\mu}D_1D_2 & -D_1 \\ \frac{\lambda+\mu}{\lambda+2\mu}D_1D_2 & \frac{\mu}{\lambda+2\mu}\Delta + \frac{\lambda+\mu}{\lambda+2\mu}D_2^2 - s^2 & -D_2 \\ -\frac{T_0\gamma^2}{\rho c_e(\lambda+2\mu)}s(1+t_0s)D_1 & -\frac{T_0\gamma^2}{\rho c_e(\lambda+2\mu)}s(1+t_0s)D_2 & \Delta - s(1+st_0) \end{bmatrix}
$$

$$
U_i = < \tilde{u} \quad \tilde{v} \quad \tilde{T} >
$$

where $D_i = \frac{\partial}{\partial x_i}$ (i = 1, 2) and Δ denotes the Laplacian. The effect of the relaxation time t_0 is apparent in Eq. (29.9). The boundary conditions are assumed to be as follow

$$
\begin{aligned}
u_i &= \bar{u}_i && \text{on} \quad \Gamma_u \\
\tau_i &= \bar{\tau}_i = \sigma_{ij}n_j && \text{on} \quad \Gamma_\tau \\
T &= \bar{T} && \text{on} \quad \Gamma_T \\
q &= \bar{q}_n = q_i n_i && \text{on} \quad \Gamma_q
\end{aligned}
\tag{29.10}
$$

where $\bar{u}_i$, $\bar{\tau}_i$, $\bar{T}$, and $\bar{q}_n$ are the given values of the displacement, traction, temperature, and heat flux vector on the boundary.

29.2 Boundary Integral Equation

In order to derive the boundary integral equation, we start with the following weak formulation of the differential equation (29.9) for the fundamental solution tensor V_{ik}^*

$$
\int_\Omega (L_{ij}U_j)V_{ik}^* d\Omega = 0 \tag{29.11}
$$

After integrating by parts over the domain and taking a limiting procedure by taking the internal source point to the boundary point, we may obtain the following boundary integral equation

$$
\begin{aligned}
C_{kj}U_k(y,s) = &\int_\Gamma \tilde{\tau}_\alpha(x,s)V_{\alpha j}^*(x,y,s) - U_\alpha(x,s)\Sigma_{\alpha j}^*(x,y,s)d\Gamma(x) \\
&+ \int_\Gamma \tilde{T}_{,n}(x,s)V_{3j,n}^*(x,y,s) - \tilde{T}(x,s)V_{3j,n}^*(x,y,s)d\Gamma(x)
\end{aligned}
\tag{29.12}
$$

where $U_\alpha = \tilde{u}_\alpha(\alpha = 1, 2)$ and $U_3 = \tilde{T}$ and C_{kj} denotes the shape coefficient tensor. The kernel $\Sigma_{\alpha j}^*$ in Eq. (29.12) is defined by

$$\Sigma_{\alpha j}^{*}=\left[\left(\frac{\lambda}{\lambda+2\mu}V_{kj,k}^{*}+\frac{T_{0}\gamma^{2}(s+t_{0}s^{2})}{\rho C_{e}(\lambda+2\mu)}V_{3j}^{*}\right)\delta_{\alpha\beta}+\frac{\mu}{\lambda+2\mu}(V_{\alpha j,\beta}^{*}+V_{\beta j,\alpha}^{*})\right]n_{\beta} \tag{29.13}$$

where n_β is the component of the normal vector to the boundary. Here, the fundamental solution tensor V_{jk} must satisfy the differential equation

$$l_{ij}V_{jk}^{*}=-\delta_{ik}\delta(x-y) \tag{29.14}$$

where l_{ij} is the adjoint operator of L_{ij} in Eq. (29.9) and is given by

$$l_{ij}=\begin{bmatrix} \frac{\mu}{\lambda+2\mu}\Delta+\frac{\lambda+\mu}{\lambda+2\mu}D_1^2-s^2 & \frac{\lambda+\mu}{\lambda+2\mu}D_1D_2 & \frac{T_0\gamma^2}{\rho c_e(\lambda+2\mu)}(s+t_0s)D_1 \\ \frac{\lambda+\mu}{\lambda+2\mu}D_1D_2 & \frac{\mu}{\lambda+2\mu}\Delta+\frac{\lambda+\mu}{\lambda+2\mu}D_2^2+s^2 & \frac{T_0\gamma^2}{\rho c_e(\lambda+2\mu)}(s+t_0s)D_2 \\ D_1 & D_2 & \Delta-s(1+t_0s) \end{bmatrix}$$

29.3 Fundamental Solution

In order to construct the fundamental solution, we put the fundamental solution tensor V_{ij}^* of Eq. (29.14) in the following potential representation using the transposed co-factor operator μ_{ij} of l_{ij} and scalar function Φ^*

$$V_{ij}^{*}(x,y,s)=\mu_{ij}\Phi^{*}(x,y,s) \tag{29.15}$$

After substitution of Eq. (29.15) into (29.14), the following differential equations is obtained

$$\Lambda\Phi^{*}=-\delta(x-y) \tag{29.16}$$

where the operation Λ is

$$\Lambda=det(l_{ij})=\frac{\mu}{\lambda+2\mu}(\Delta-h_1^2)(\Delta-h_2^2)(\Delta-h_3^2) \tag{29.17}$$

and h_i^2 are the solution of

$$\begin{aligned} h_1^2&=\frac{\lambda+2\mu}{\mu}s^2\\ h_2^2+h_3^2&=s^2+s(1+t_0s)+\frac{T_0\gamma^2}{\rho C_e(\lambda+2\mu)}s(1+t_0s)\\ h_2^2h_3^2&=s^3(1+t_0s) \end{aligned} \tag{29.18}$$

where h_1 = longitudinal wave velocity, h_2 = shear wave velocity, and h_3 = rotational wave velocity. Note that h_2 and h_3 are functions of the relaxation times t_0.

The solution for Φ^* from Eq. (29.16) with the help of Eq. (29.17) is thus

$$\Phi^* = \frac{\lambda + 2\mu}{2\pi\mu}\left[\frac{K_0(h_1 r)}{(h_2^2 - h_1^2)(h_3^2 - h_1^2)} + \frac{K_0(h_2 r)}{(h_3^2 - h_2^2)(h_1^2 - h_2^2)} + \frac{K_0(h_3 r)}{(h_1^2 - h_3^2)(h_2^2 - h_3^2)}\right] \tag{29.19}$$

The fundamental solution tensor V_{ij}^* for two dimensional domain is found as

$$\begin{aligned}
V_{\alpha\beta}^* &= \sum_{k=1}^{3}(\psi_k(r)\delta_{\alpha\beta} - \kappa_k r_{,\alpha} r_{,\beta}) \quad (\alpha, \beta = 1, 2)\\
V_{3\alpha}^* &= \sum_{k=1}^{3} \acute{\xi}_k(r) r_{,\alpha}\\
V_{\alpha 3}^* &= \sum_{k=1}^{3} \xi_k(r) r_{,\alpha}\\
V_{33}^* &= \sum_{k=1}^{3} \zeta_k(r)
\end{aligned} \tag{29.20}$$

where

$$\begin{aligned}
\psi_k(r) &= \frac{W_k}{2\pi}\left[\left(h_k^2 - m_2\right)\left(h_k^2 - m_1\right) + \left(\frac{\lambda+\mu}{\mu}\right)\left(h_k^2 - m_1 - m_3 C\frac{\lambda+2\mu}{\lambda+\mu}\right)\right) h_k^2\right] K_0(h_k r)\\
&\quad + \frac{W_k(\lambda+\mu)}{2\pi\mu}\left[h_k^2 - m_1 - m_3 C\frac{\lambda+2\mu}{\lambda+\mu}\right)\right]\frac{h_k}{r} K_1(h_k r)\\
\kappa_k(r) &= \frac{W_k(\lambda+\mu)}{2\pi\mu}\left[h_k^2 - m_1 - m_3 C\frac{\lambda+2\mu}{\lambda+\mu}\right)\right] K_2(h_k r)\\
\acute{\xi}_k(r) &= \frac{W_k}{2\pi} m_4\left(h_k^2 - m_2\right) h_k K_1(h_k r)\\
\xi_k(r) &= \frac{W_k}{2\pi} C m_3\left(h_k^2 - m_2\right) h_k K_1(h_k r)\\
\zeta_k(r) &= \frac{W_k}{2\pi}\left(h_k^2 - m_2\right)\left(h_k^2 - s^2\right) K_0(h_k r)
\end{aligned} \tag{29.21}$$

and

$$\begin{aligned}
&r = \|x - y\|; \quad m_1 = s(1 + t_0 s); \quad m_2 = \frac{\lambda+2\mu}{\mu} s^2\\
&m_3 = s(1 + t_0 s); \quad C = \frac{T_0\gamma^2}{\rho c_e(\lambda+2\mu)}\\
&W_i = \frac{-1}{(h_i^2 - h_j^2)(h_k^2 - h_i^2)} \quad (i = 1, 2, 3 \;\; j = 2, 3, 1 \;\; k = 3, 2, 1)
\end{aligned} \tag{29.22}$$

Here, K_0, K_1, and K_2 are the modified Bessel function of second kind and zero, first, and second orders, respectively.

In order to solve numerically the boundary element integral equation (29.12), the standard procedure is applied. When transformed numerical solutions are specified, transient solutions can be obtained using an appropriate numerical inversion technique. In this section, the method presented by Durbin[2], which combines the Fourier cosine and sine transform to reduce numerical error, is adopted for numerical inversion. This formulation yields time-domain functional values $F(t_n)$ as

$$F(t_n) = \frac{2e^{n\beta\Delta t}}{t_N}\left[-\frac{1}{2}Re\tilde{F}(s_0) + Re\left\{\sum_{k=0}^{N-1}(A_k + iB_k)W^{nk}\right\}\right]$$
$$\text{for} \quad n = 0, 1, \ldots, N-1 \quad \text{and} \quad t_n = n\Delta t \tag{29.23}$$

where $\tilde{F}$ is the function in the Laplace-domain, and

$$\begin{aligned} A_k &= \sum_{l=0}^{L} Re\tilde{F}(s_{k+lN}), \\ B_k &= \sum_{l=0}^{L} Im\tilde{F}(s_{k+lN}), \\ W &= e^{2\pi i/N}, \\ s_m &= \beta + 2\pi i m/t_N, \end{aligned} \tag{29.24}$$

with the real constant $\beta = 6/t_n$, based upon experience and the recommendation of Durbin. Different values of $5 \leq \beta t_n \leq 10$, according to Durbin's recommendation, were considered in the analysis and verification of the method. The best results were achieved by choosing $(\beta t_n) = 6$. This value was used for all examples given in this section. Notice that from Eqs. (29.23) and (29.24), the determination of $F(t_n)$ for $n = 0, 1, \ldots, N-1$ depends upon the values of $\tilde{F}(s_m)$ for $m = 0, 1, \ldots, M-1$, where $M = N(L+1)$. Consequently, the boundary integral equations

$$\begin{aligned} C_{kj}U_k(y, s_m) &= \int_\Gamma \tilde{\tau}_\alpha(x, s_m)V^*_{\alpha j}(x, y, s_m) - U_\alpha(x, s_m)\Sigma^*_{\alpha j}(x, y, s_m)d\Gamma(x) \\ &+ \int_\Gamma \tilde{T}_{,n}(x, s_m)V^*_{3j,n}(x, y, s_m) - \tilde{T}(x, s_m)V^*_{3j,n}(x, y, s_m)d\Gamma(x) \end{aligned} \tag{29.25}$$

must be solved independently at each of the M discrete values s_m of the transform parameters. In the current implementation, $L = 0$ is considered to minimize the computational effort.

[2] F. Durbin, Numerical Inversion of Laplace Transforms: An Efficient Improvement to Dubner and Abate's Method. Computer J. **17**, 371–376 (1974).

29.4 Numerical Formulation

The boundary of the solution domain Γ is divided into N elements, such that Eq. (29.12) becomes

$$C_{kj}U_k(y,s) = \sum_{n=1}^{N}\int_{\Gamma_n}\left[\tilde{\tau}_{\alpha}(x,s)V^*_{\alpha j}(x,y,s) - U_{\alpha}(x,s)\Sigma^*_{\alpha j}(x,y,s)\right]d\Gamma_n(x)$$
$$+\sum_{n=1}^{N}\int_{\Gamma_n}\left[\tilde{T}_{,n}(x,s)V^*_{3j,n}(x,y,s) - \tilde{T}(x,s)V^*_{3j,n}(x,y,s)\right]d\Gamma_n(x) \tag{29.26}$$

where $\Gamma = \sum_{n=1}^{N}\Gamma_n$. On each element the boundary parameter x (with components x_j), the unknown displacement and temperature fields $U_j(x,s)$, and the traction and heat flux fields $\tilde{\tau}_{\alpha}(x,s)$, $\tilde{T}_{,n}(x,s)$ are approximated using the interpolation functions in the form

$$x = \sum_{\theta=1}^{m} M^{\theta}x^{\theta}$$
$$U = \sum_{\theta=1}^{m} N^{\theta}U^{\theta}$$
$$\tilde{\tau} = \sum_{\theta=1}^{m} N^{\theta}\tilde{\tau}^{\theta} \tag{29.27}$$

where M^{θ} and N^{θ} are called shape functions and are polynomials of degree $m-1$. The property of these shape functions is such that they are equal to 1 at node θ, and 0 at all other nodes. Here, x^{θ}, U^{θ} and $\tilde{\tau}^{\theta}$ are the values of the functions at node θ.

The different choices of M^{θ} and N^{θ} lead to different boundary element formulations. If $M^{\theta} = N^{\theta}$, the formulation is referred to as isoparametric, and if M^{θ} is of higher order polynomial than N^{θ} then the formulation is referred to as superparametric. Conversely, if M^{θ} is of lower order polynomial than N^{θ} then the formulation is referred to as subparametric.

For the present two-dimensional formulation, the isoparametric element is considered. The shape functions M^{θ} are defined in terms of the non-dimensional coordinates $\xi(-1 \le \xi \le 1)$ and can be derived from the Lagrange polynomials defined, for degree $(m-1)$, as

$$M^{\theta}(\xi) = \prod_{i=0;i\neq\theta}^{m}\frac{\xi-\xi_i}{\xi_{\theta}-\xi_i} \tag{29.28}$$

It is seen that $M^\theta(\xi)$ is given by the product of m linear factors. The Lagrangian shape functions $M^\theta(\xi)$ can be shown to have the following properties:
At node β

$$M^\theta(\xi_\beta) = \delta_{\alpha\beta}, \qquad \xi_0 < \xi_\beta < \xi_m; \tag{29.29}$$

Also

$$\sum_{\theta=1}^{m} M^\theta(\xi) = 1 \quad \text{and} \quad \sum_{\theta=1}^{m} \frac{dM^\theta(\xi)}{d\xi} = 0 \tag{29.30}$$

For constant and quadratic elements, the shape functions $M^\theta(\xi)$ are listed as:
For constant element $m = 1$ and $M(\xi) = 1$
For quadratic element considering $l_1 = 1/2(1-\xi), l_2 = 1/2(1+\xi)$

$$M^1(\xi) = l_1(2l_1 - 1), \quad M^2(\xi) = 4l_1 l_2, \quad M^3(\xi) = l_2(2l_2 - 1) \tag{29.31}$$

A discretized boundary element formulation can be obtained by substituting the expressions (29.27) and (29.28), with $M^\theta = N^\theta$, into the integral equation (29.26) to obtain

$$C_{kj} U_k(y, s) + \sum_{n=1}^{N} \sum_{\theta=1}^{m} P_{jk}^{n\theta} U_k^{n\theta} = \sum_{n=1}^{N} \sum_{\theta=1}^{m} Q_{jk}^{n\theta} \tau_k^{n\theta} \qquad k, j = 1, 2, 3 \tag{29.32}$$

The coefficients $P_{jk}^{n\theta}$ and $Q_{jk}^{n\theta}$ are defined in terms of the integrals over Γ_n, where $d\Gamma_n(\xi)$ becomes $J^n(\xi)d\xi$. That is

$$P_{jk}^{n\theta} = \int_{-1}^{1} M^\theta(\xi) \tilde{\tau}_{jk}(y, x(\xi), s) J^n(\xi) d\xi$$
$$Q_{jk}^{n\theta} = \int_{-1}^{1} M^\theta(\xi) U_{jk}(y, x(\xi), s) J^n(\xi) d\xi \tag{29.33}$$

In general, the Jacobian of a transformation $J(\xi)$ is given as

$$J(\xi) = \left[\left[\frac{dx_1}{d\xi} \right]^2 + \left[\frac{dx_2}{d\xi} \right]^2 \right]^{1/2} \tag{29.34}$$

Therefore

$$J^n(\xi) = \left[\left[\sum_{\theta=1}^{m} \frac{dM^\theta(\xi)}{d\xi} x_1^{n\theta} \right]^2 + \left[\sum_{\theta=1}^{m} \frac{dM^\theta(\xi)}{d\xi} x_2^{n\theta} \right]^2 \right]^{1/2}. \tag{29.35}$$

29.5 Problems and Solutions Related to Coupled Thermoelasticity

Problem 29.1. Consider a rectangular two-dimensional plate of length $l \times l$, as shown in Fig. 29.1. The plate is exposed to a thermal shock load at its edge at $x = 0$. Let us first consider the uncoupled condition, where the temperature equation is separately solved for the temperature distribution and is independent of the stress field.

Solution. Since the data presented in Figs. 29.2, 29.3 and 29.4 are related to the axis of symmetry of the plate, thus $q_y = k\frac{\partial T}{\partial y}$ is zero and temperature distribution is symmetric about the x-axis. Consequently, the distribution of temperature along the axis of symmetry of a two-dimensional domain coincides with that of the one-dimensional solution of the half-space. For the coupled solution, however, the influence of the stress field in temperature distribution of a two-dimensional domain results into lower curve of temperature compared to the half-space solution. The displacement curves of the two-dimensional domain along the axis of symmetry of the domain almost coincide with those of half-space. The reason is that due to symmetry, $v = 0$ along the x-axis and the only non-zero displacement is u. This condition coincides with the one-dimensional half-space solution. The stress plot of two-dimensional domain along the axis of symmetry, shown in Fig. 29.4, is different from those of the half-space. The main reason for the difference is the existence of non-zero ϵ_y in the two-dimensional domain. The effect of ϵ_y causes the peak compressive stress to occur at shorter time. The tensile stress produced by the application of thermal shock is larger in a finite domain compared to the half-space. This is shown in Fig. 29.4 for the range of time where the wave reflection is not produced yet.

Problem 29.2. Now, the influence of the finite domain in wave reflection may be studied. Consider a coated surface subjected to a laser beam. When the length is large compared to thickness, as it is in the case of a coated surface, the problem may be modelled as shown in Fig. 29.5. A number of 40 constant elements uniformly distributed along the boundary are considered, as shown in Fig. 29.5.

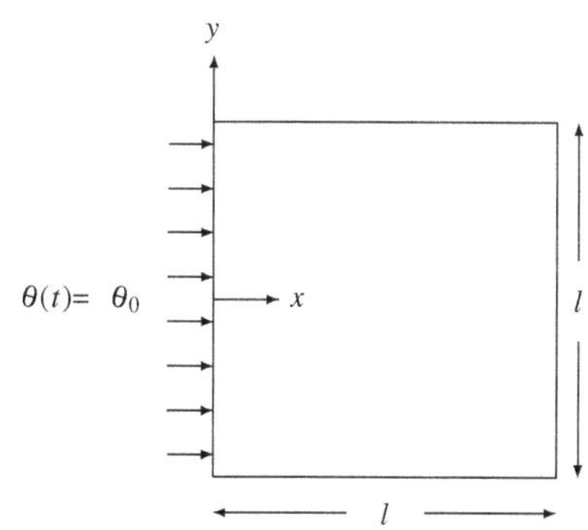

Fig. 29.1 A square plate subjected to thermal loading

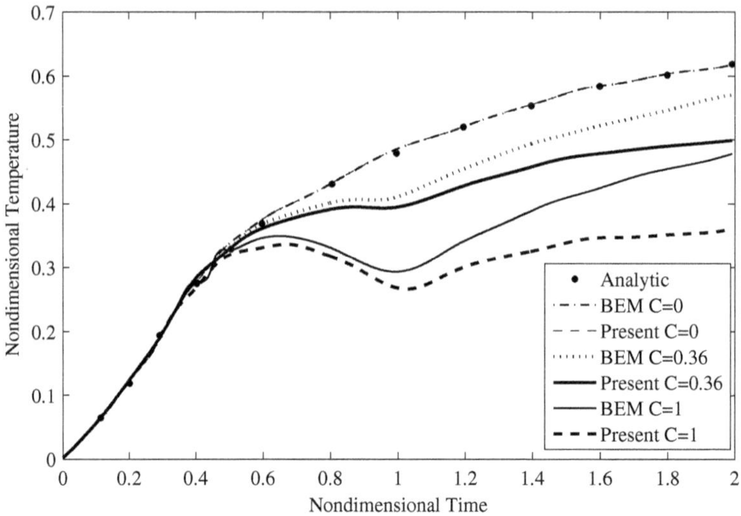

Fig. 29.2 Comparison of the dimensionless temperature at $x = 1$. Curve identified by 1 is related to $C = 0$, curve identified by 2 is related to $C = 0.36$, and curve identified by 3 is related to $C = 1$

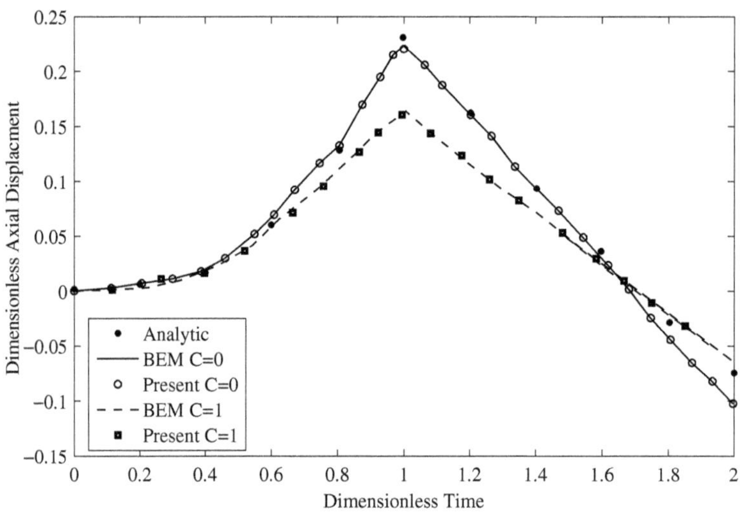

Fig. 29.3 Comparison of the dimensionless axial displacement at $x = 1$

Solution. To compare the results with the solutions available for the half-space, the square plate ($l = 2$ nondimensional) of Fig. 29.5 is considered to experience a step function heating in one side and to be insulated at the other sides. The u-displacement at $x = l$ and v-displacements at $y = \pm l/2$ are assumed to be zero. The thermoelastic wave propagation, reflection, and the effect of relaxation time due to the Lord and

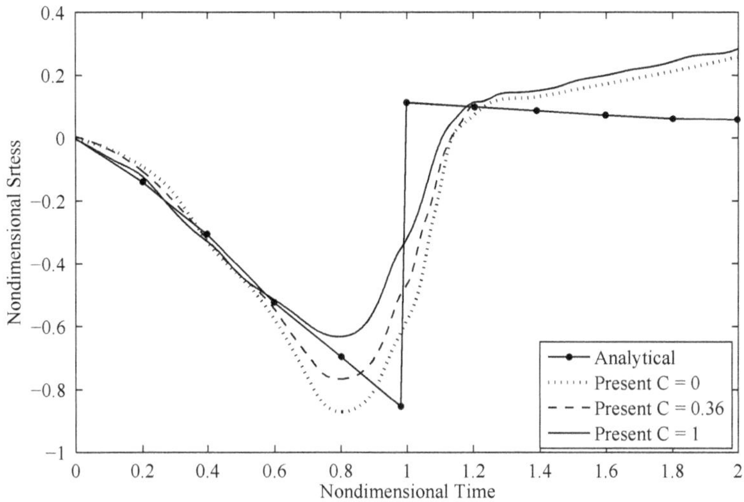

Fig. 29.4 Comparison of the dimensionless axial stress at $x = 1$

Fig. 29.5 Model of a layer subjected to thermal loading

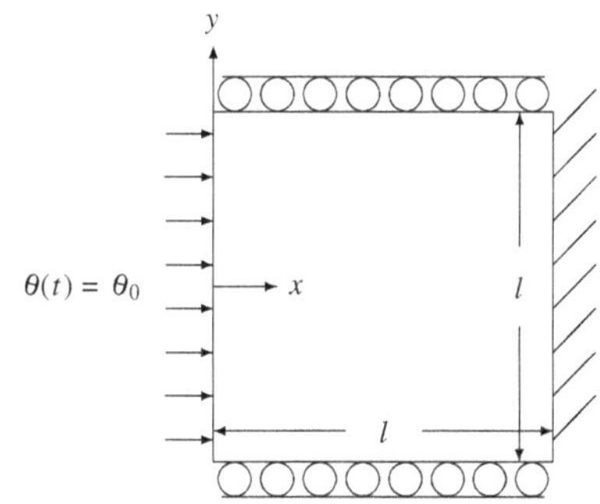

Shulman model are investigated along the axis of symmetry of the plate in unit dimensionless length and are compared with the half-space results reported by Chen and Lin[3]. We Consider a unit step function temperature rise at edge $x = 0$. Figure 29.6 shows the dimensionless temperature distribution versus dimensionless time for the LS theory. The temperature distribution shows oscillations after the temperature wave front, due to the temperature wave reflection from the boundaries of the plate. The distribution of the axial displacement based the LS model is shown in Fig. 29.7. In the LS model, the temperature variation has no effect on the displacement distribution.

Problem 29.3. Now, the discussion proceeds on the coupled thermoelastic response of a square plate under thermal and mechanical shocks based on classical and Lord and Shulman model. Consider a finite two-dimensional square domain of $l \times l$. The

[3] H. Chen, H. Lin, Study of Transient Coupled Thermoelastic Problems with Relaxation Time, Trans. ASME. J. Appl. Mech. **62**, 208–215 (1995).

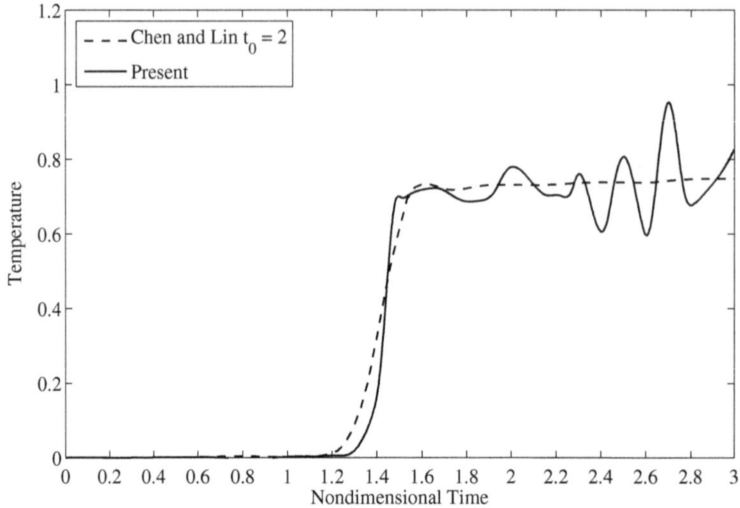

Fig. 29.6 Comparison of the dimensionless temperature at middle of the plate for L.S. theory

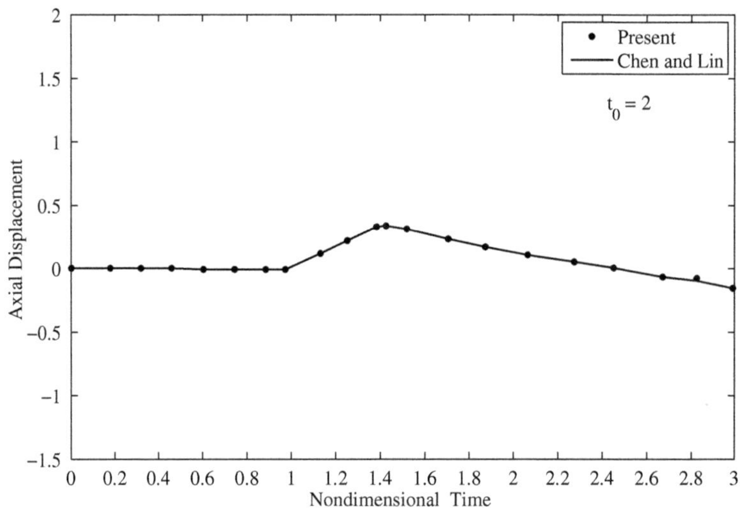

Fig. 29.7 Comparison of the dimensionless axial displacement at middle of the plate for L.S. theory

plate is fixed at $x = l$ and free at other edges (Fig. 29.8). The plate is initially stress free and at reference temperature $T_0 = 0$. At $t > 0$, the edge $x = 0$ experiences thermal and/or mechanical shocks given by (Fig. 29.9)

Fig. 29.8 Model of A plate subjected to thermal loading

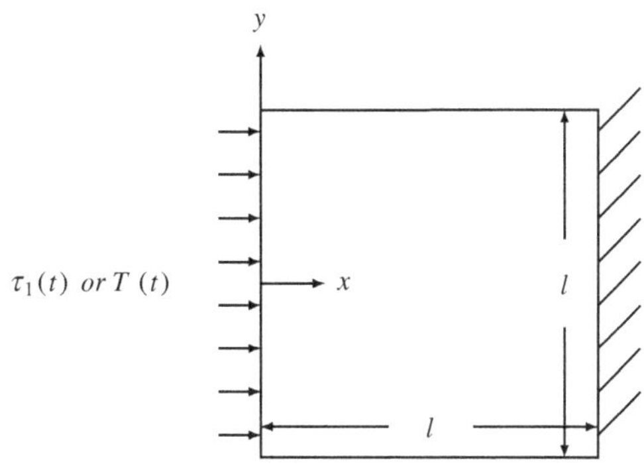

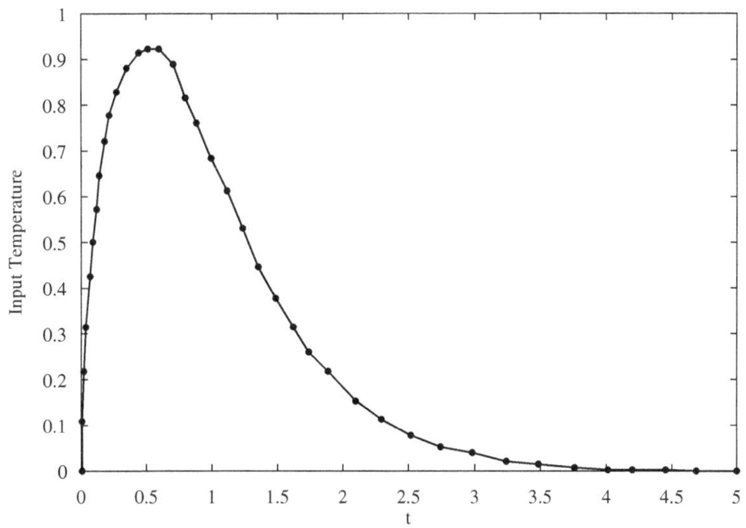

Fig. 29.9 Pattern of thermal loading

$$T(t) = 5t\ \exp^{-2t} \tag{29.36}$$

$$\tau_1 = 5t\ \exp^{-2t} \tag{29.37}$$

Solution. When thermal shock is applied to the edge $x = 0$, $\tau_i = 0$ on $x = 0$, where τ_i are the traction components. Temperature shock is in the form of heat input, and the traction shock is applied in the positive x-direction.

Comparison of the constant and quadratic elements:

The temperature shock of Eq. (29.36) is applied to the edge $x = 0$. We consider the uncoupled equations ($C = 0$) and study the accuracy of the boundary element solution by assuming the constant and quadratic elements. The boundary of the plate is divided into 40 and 20 number of elements for the constant and quadratic elements,

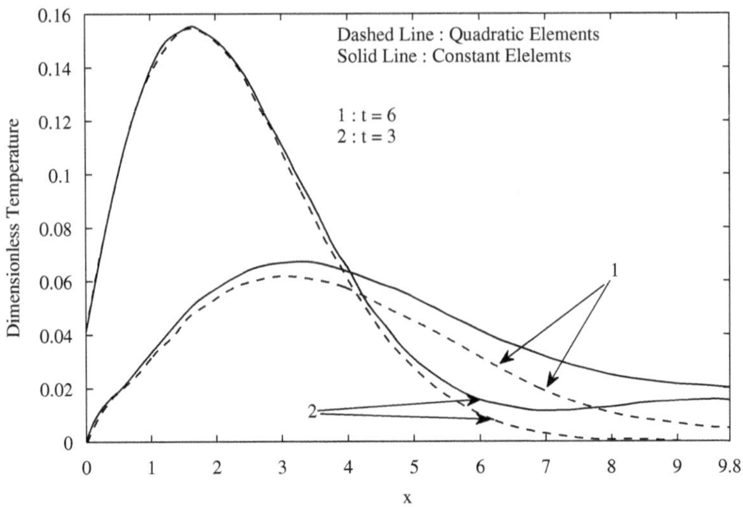

Fig. 29.10 Comparison of the dimensionless temperature at middle of the plate for temperature loading

respectively. Figures 29.10, 29.11 and 29.12 show comparison of the temperature T, axial displacement u, and the axial stress σ_{xx} along the x-axis at dimensionless time $t = 3$ and $t = 6$. The figures show that the constant elements over estimates the temperature distribution especially for the longer x-values and under estimate the axial displacement and the stress distribution. It is clearly shown that the constant elements approach is not capable to represent the sharp variation of stress near the free edge at $x = 0$, and its error in temperature distribution increases with the increase of the distance from the free edge $x = 0$, where the temperature shock is applied.

To show briefly the effects of LS model, in the following figures the effect of thermal shock alone, traction shock alone, and the combination of two shocks applied simultaneously are studied. All curves are plotted along the x-axis of symmetry of the plate. The boundary shown in Fig. 29.8 is divided uniformly into 20 quadratic elements.

A—Thermal loading: Consider the plate under thermal shock alone on its edge at $x = 0$. Figures 29.13, 29.14 and 29.15 show the axial temperature, displacement, and the axial stress along the x-axis for dimensionless times $t = 3$ and 6. The curves are plotted for two different conditions. When $t_0 = 0.64$, the speed of propagation of the thermal wave is $C_t = 1.25$ and the speed of propagation of the stress wave is $C_s = 1$ ($C_t > C_s$). On the other hand, for $t_0 = 1.5625$, $C_t = 0.8$ and $C_s = 1$ thus $C_s > C_t$. The speed of temperature wave is C_t and when $t_0 = 0.64$, the speed of temperature wave is faster than the stress wave ($C_t > C_s$). At the stress wave front, a negative thermal gradient is appeared in temperature curve. At this point the thermal energy is converted to the mechanical energy resulting into negative temperature gradient. This is shown in Fig. 29.13. On the other hand, when $t_0 = 1.5625$ $C_s > C_t$ and the

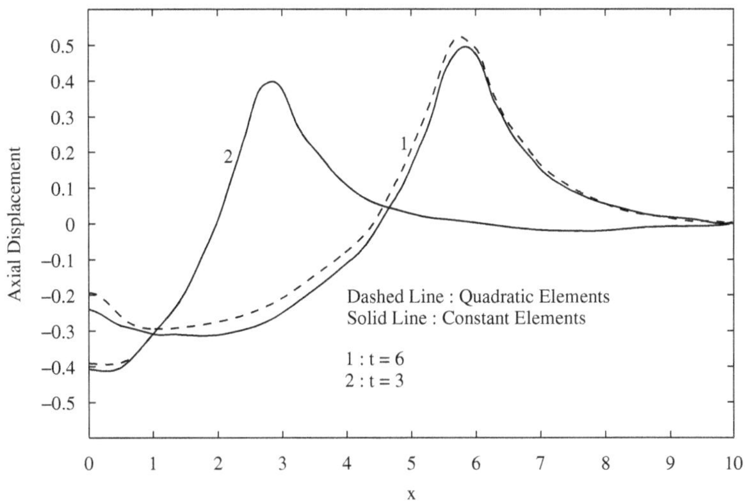

Fig. 29.11 Comparison of the dimensionless axial displacement at middle of the plate for temperature loading

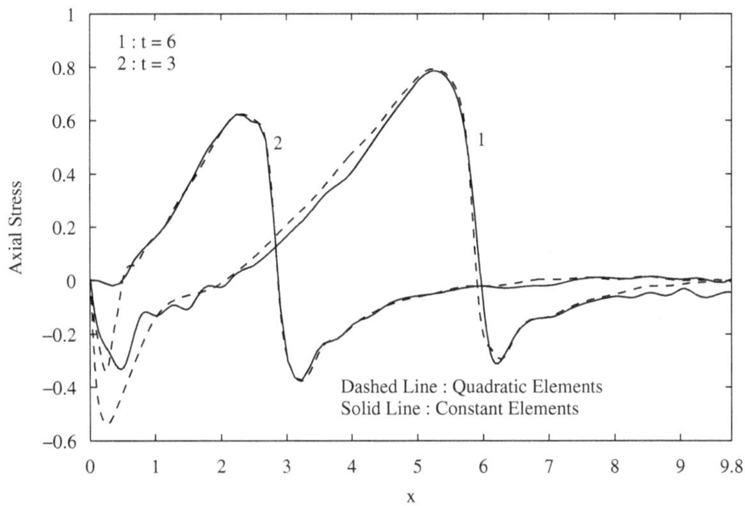

Fig. 29.12 Comparison of the dimensionless axial stress at middle of the plate for temperature loading

speed of stress wave is faster than the temperature wave. At the stress wave front, due to the induced compressive stress, positive temperature gradient is observed due to the conversion of mechanical energy into the thermal energy. This is again seen from Fig. 29.13. The peak of the temperature curve for $C_t < C_s$ is greater compared to the condition when $C_s > C_t$. The reason is for larger t_0 (smaller C_t) time available

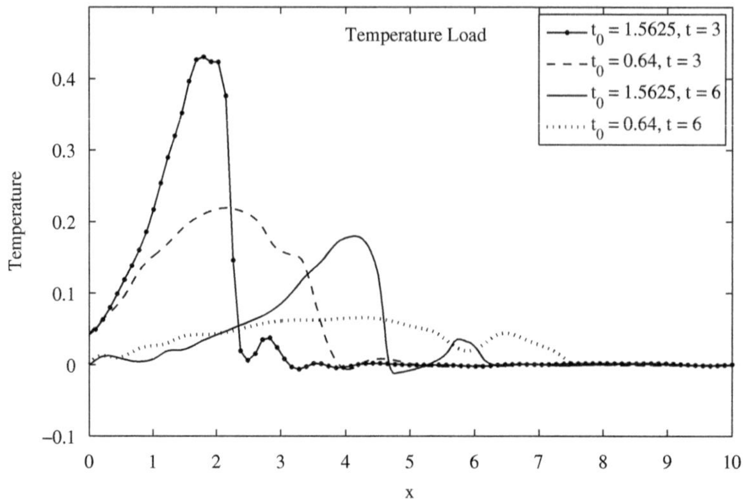

Fig. 29.13 Comparison of the dimensionless temperature at middle of the plate for temperature loading

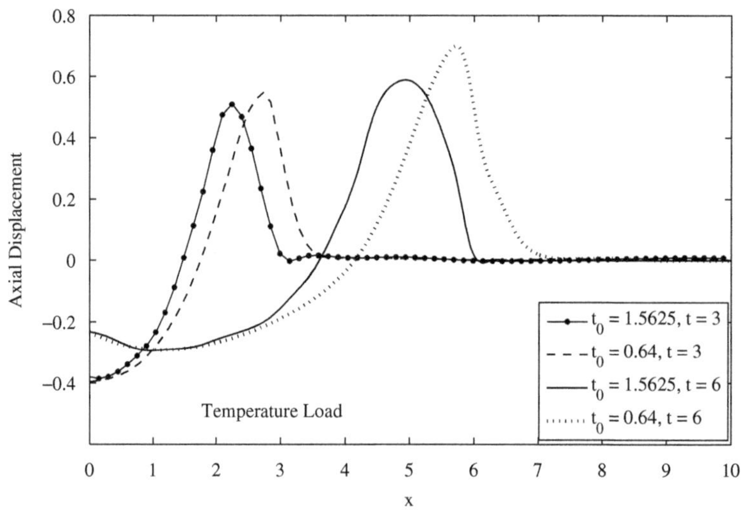

Fig. 29.14 Comparison of the dimensionless axial displacement at middle of the plate for temperature loading

for the exchange of thermal energy of pulse with domain is larger and the peak of temperature wave is thus higher.

Figure 29.14 is a plot of the axial displacement of the middle of the plate at times $t = 3$ and 6 (dimensionless time). In comparison with Fig. 29.13, it seems that when $C_t > C_s$ (curve corresponding to $t_0 = 0.64$) larger displacement is taken place. The

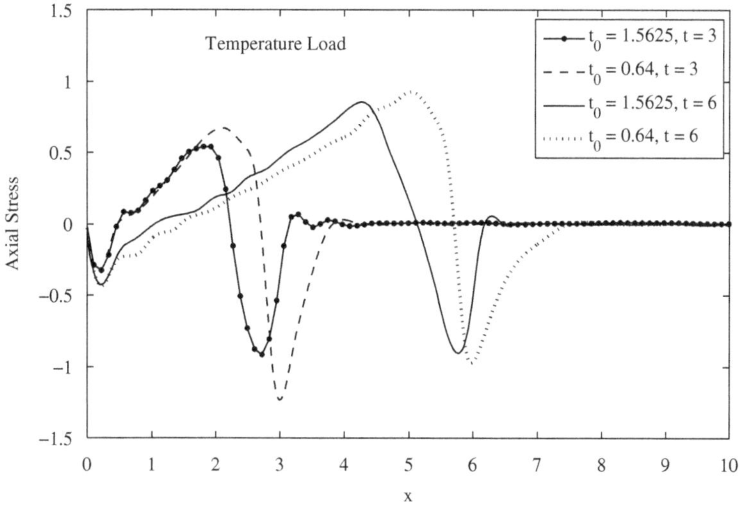

Fig. 29.15 Comparison of the dimensionless axial stress at middle of the plate for temperature loading

plot of axial stress along the axis of symmetry of the plate is shown in Fig. 29.15. The comparison of this figure with the plot of the temperature, Fig. 29.13, shows that the stress wave front coincide with the temperature wave front indicating that the conversion of the mechanical energy into thermal energy takes place at these locations. When $C_t > C_s$, larger compressive axial displacement $(+u)$ is produced resulting into larger compressive axial stress.

Now, the results may be compared with the solution based on the Green-Lindsay (GL) model[4]. The same geometrical and material properties are assumed for the solution domains. The coupling parameter is the same and $t_1 = t_2 = t_0$, where t_1 and t_2 are the relaxation times associated with the Green-Lindsay model. The temperature distribution between the LS and GL model are about the same for pure thermal loadings, but the temperature distribution across the plate for LS theory is much higher than the CCT. The temperature wave front for the LS and GL models are about the same, but quite different with the CCT theory, as the CCT theory principally do not predict any temperature wave. The pattern of stress distribution for the same type of thermal loading for the LS model is quite different with that of the GL model. The stress wave front for LS model is much smaller than the GL model.

B—Mechanical loading: Consider the plate under compressive traction load of Eq. (29.37) on its boundary at $x = 0$. The axial displacement, temperature, and axial stress are shown in Figs. 29.16, 29.17 and 29.18. Figure 29.16 is the plot of temperature along the x-axis. The scale of temperature along the temperature axis shows the temperature rise due to a mechanical shock applied at the boundary $x = 0$ and is small, as expected. The mechanical energy is transferred into heat energy

[4] P. Tehrani, M.R. Eslami, Boundary Element Analysis of Green and Lindsay Theory under Thermal and Mechanical Shocks in a Finite Domain. J. Thermal Stresses **23**(8), 773–792 (2000).

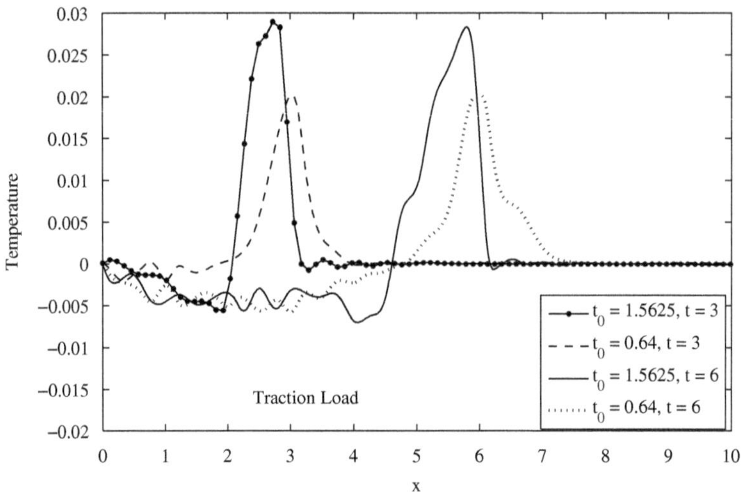

Fig. 29.16 Comparison of the dimensionless temperature at middle of the plate for traction loading

through the expression $C(\dot{u}_{i,i} + t_0\ddot{u}_{i,i})$ in Eq. (29.2). For larger value of t_0, more mechanical energy is changed into thermal energy and thus for $C_s > C_t$ the plot of temperature shows higher values. Figure 29.17 shows the plot of axial displacement along the x-axis. The differences between the two cases of $t_0 = 0.64$ and $t_0 = 1.5625$

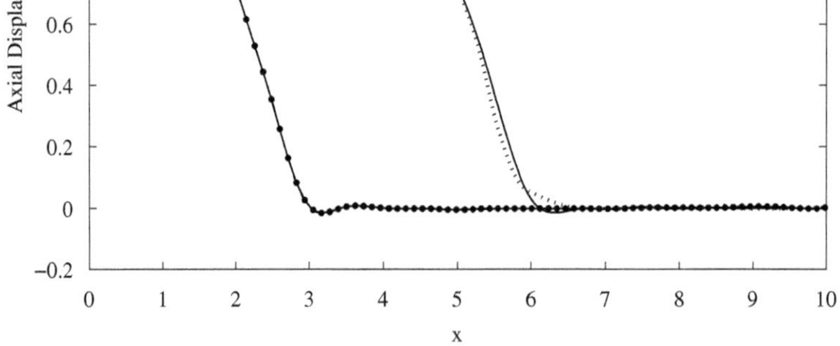

Fig. 29.17 Comparison of the dimensionless axial displacement at middle of the plate for traction loading

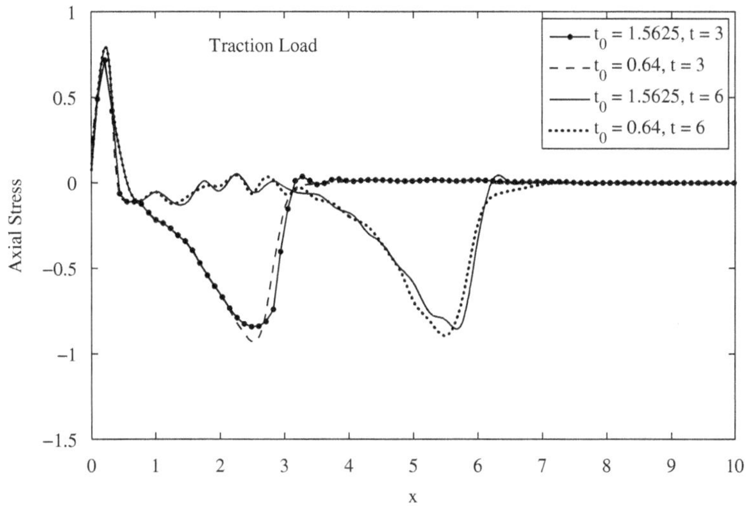

Fig. 29.18 Comparison of the dimensionless axial stress at middle of the plate for traction loading

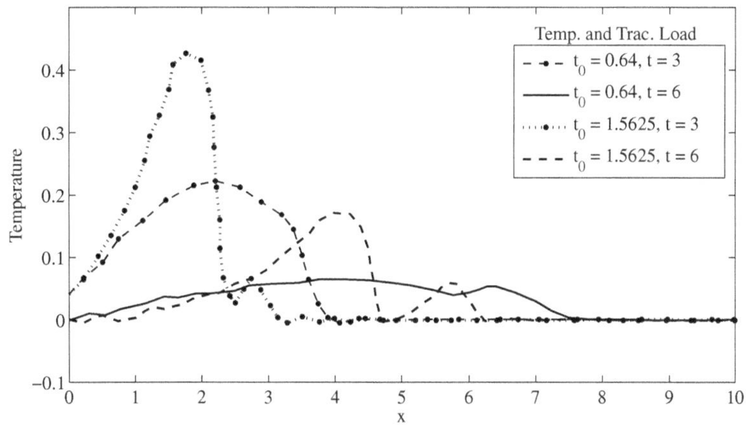

Fig. 29.19 Comparison of the dimensionless temperature at middle of the plate for temperature and traction loading

are negligible. Figure 29.18 is the plot of axial stress, and shows the stress wave fronts at $t = 3$ and $t = 6$ (dimensionless time). The difference between the two cases are negligible. The curves related to $t = 6$ clearly show the after shock vibrations (with smaller amplitudes).

Now, the results may be compared with the solution obtained using the GL model. The geometrical and material properties of the solution domains are considered similar. Also, the relaxation times associated with the GL and LS models are identical and the coupling parameter is the same. The pattern of the temperature distributions

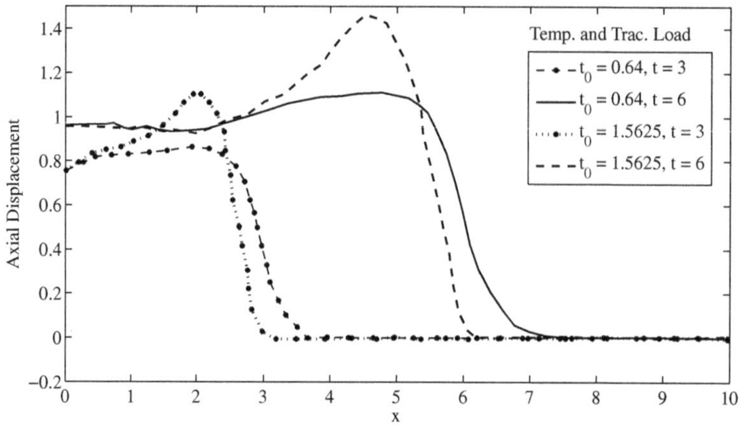

Fig. 29.20 Comparison of the dimensionless axial displacement at middle of the plate for temperature and traction loading

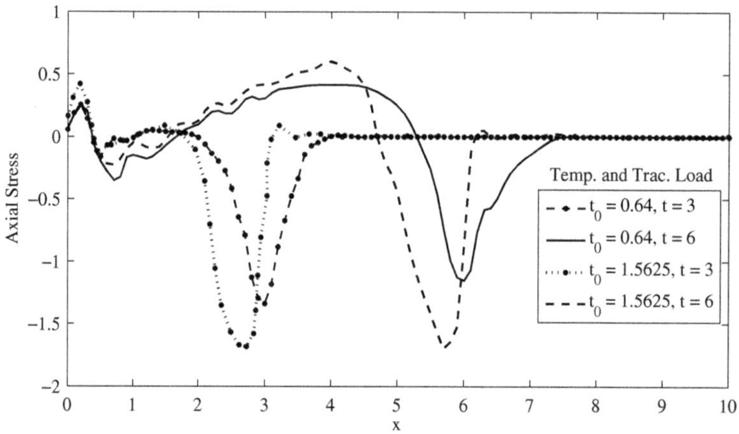

Fig. 29.21 Comparison of the dimensionless axial stress at middle of the plate for temperature and traction loading

between the LS and GL models for pure mechanical loading are quite different. The temperature wave front based on the LS model is much larger of that of the GL model. The stress wave front are about the same for both LS and GL models.

C—Combined mechanical and thermal loading: The thermal and mechanical shocks given by Eqs. (29.36) and (29.37) are applied simultaneously at edge $x = 0$. The plot of the axial displacement, temperature, and the axial stress are shown in Figs. 29.19, 29.20 and 29.21. In this loading condition the response of the plate due to the thermal and mechanical shock loadings are the superimposed results of the cases A and B above, as expected.

Dear Reader,
While studying this book, you might have been wondering what the Latin sentence at the top of *Preface* means.
Here is the translation:

We did what we could;
let those who are able, do it better.

Errata to Three Books

The Mathematical Theory of Elasticity, Second Edition

R. B. Hetnarski and J. Ignaczak

1. Page xxii, delete first 7 lines, that is, $J - L\{f(t)\}$
2. Page 60, Eq. (2.3.36), $(p - a)^n$ should read: $(p - a)$
3. Page 169, line 1 and 6, a unique solution $\mathbf{u}$ should read: a solution $\mathbf{u}$
4. Page 174, line 14, $\hat{W}(\mathbf{S})^{\perp}$ should read: $\hat{W}(\mathbf{S}^{\perp})$
5. Page 224, Eq. (4.2.62), $\tilde{\mathbf{B}} = \hat{\nabla}\mathbf{b} + \rho\ddot{T}\mathbf{1}$ should read: $\tilde{\mathbf{B}} = \hat{\nabla}\mathbf{b} + \alpha\rho\ddot{T}\mathbf{1}$
6. Page 229, line 3, $[\tilde{\mathbf{u}}_0, \dot{\mathbf{u}}_0]$ should read: $[\tilde{\mathbf{u}}_0, \dot{\tilde{\mathbf{u}}}_0]$
7. Page 250, line 5 and 9 from bottom, $\Box_1^2$ should read: $\Box_0^2$
8. Page 251, line 4 from bottom, $\Box_1^2$ should read: $\Box_0^2$
9. Page 252, line 2, $\Box_1^2$ should read: $\Box_0^2$
10. Page 266, Eq. (5.1.65), ∂B should read: B
11. Page 773, Eq. (A.2.21), $\varepsilon^n(\nabla^2)$ should read: $\varepsilon^n(\nabla^2)^n$
12. Page 773, Eq. (A.2.22), $\dfrac{\varepsilon^n(\nabla^2)}{(n!)^2}\dfrac{1}{R}$ should read: $\dfrac{\varepsilon^n(\nabla^2)^n}{(n!)^2}$
13. Page 774, Eq. (A.2.25), $\dfrac{\varepsilon^n(\nabla^2)}{(n!)^2}\dfrac{1}{R}$ should read: $\dfrac{\varepsilon^n}{(n!)^2}$
14. Page 787, line 7 under "F", Frederick the Great, 5 should read: Frederick the Great, 6

Thermal Stresses, Second Edition

N. Noda, R. B. Hetnarski and Y. Tanigawa

1. Page 10, line 4, (1.5) should read: (1.20)
2. Page 10, line 8, (1.6) should read: (1.21)

M. Reza Eslami et al., *Theory of Elasticity and Thermal Stresses,* Solid Mechanics and Its Applications 197, DOI: 10.1007/978-94-007-6356-2,

3. Page 50, line 5, in Example 2.1 should read: as shown in Fig. 2.7
4. Page 78, lines 1 and 2, Calculate the thermal stress and curvature produced in a beam if the elongation at $y = 0$ is restrained to zero, as in Example 2.1. should read: Calculate the thermal stress and curvature produced in two parallel beams clamped to rigid plates in Fig. 2.7, if the elongation at $y = 0$ is restrained to zero.
5. Page 154, footnote, Kellog should read: Kellogg
6. Page 300, Eq. (a), $\times\,[(T_i - T_b)(-1)^n b + (T_b - T_a)a]$ should read: $\times[(T_i - T_b)(-1)^n b + (T_i - T_a)a]$
7. Page 414, Eq. (9.95), $\left(\frac{\partial^2 w}{\partial x^2} + \frac{\partial^2 w}{\partial y^2}\right) w - \frac{N_0}{D} w = F$

   should read: $\left(\frac{\partial^2 w}{\partial x^2} + \frac{\partial^2 w}{\partial y^2}\right) - \frac{N_0}{D} = F$
8. Page 420, Eq. (9.119), $N_{r\theta} = (1 - \nu)D\frac{\partial}{\partial r}\left(\frac{1}{r}\frac{\partial w}{\partial \theta}\right)$

   should read: $M_{r\theta} = (1 - \nu)D\frac{\partial}{\partial r}\left(\frac{1}{r}\frac{\partial w}{\partial \theta}\right)$

Thermal Stresses—Advanced Theory and Applications

R. B. Hetnarski and M. R. Eslami

1. Page XXV, line 6 from bottom, Eight should read: Eighth
2. Page XXIX, line 5, Warszaawa should read: Warszawa
3. Page 3, line 8, 0 and Eq. (1.2-3) should read: 0, Eq. (1.2-3)
4. Page 11, line 1, The difference of the should read: The difference of squares of the
5. Page 32, line 12 from bottom, strain deformation should read: strain
6. Page 37, Eq. (1.13-12), third term $-a \oint_{L_i} \frac{\partial T}{\partial n} ds$ should read: $-a \oint_{C_i} \frac{\partial T}{\partial n} ds$
7. Page 38, Eq. (1.13-17), second line, remove $\square$.
8. Page 133, Example 2, line 4, Assume a rate of internal heat generation per unit volume R ... should read: Assume a rate of internal heat generation per unit volume per unit time R ...
9. Page 135, line 10, Yoshinbo should read: Yoshinobu
10. Page 163, under Eq. (4.2-68), change Eq. (4.2-53) to Eq. (4.2-68)
11. Page 164, two lines under Eq. (e), Thus, A_{mn} is obtained from the first of Eq. (4.2-17) should read: A_{mn} is obtained from the first of Eq. (4.2-27)
12. Page 171, second of Eqs. (b), $\frac{\partial\theta(0,y)}{\partial y}$ should read: $\frac{\partial\theta(0,t)}{\partial y}$
13. Page 176, Example 15, Consider the hollow cylinder of Example 12 should read: Consider the hollow cylinder of Example 14
14. Page 233, Eq. (5.8-1), $R = q_0 e^{-\mu a/2} \cosh \mu z$ should read: $R = q_0 e^{-\mu a/2} \cos \mu z$

15. Page 253, 3 lines from bottom, Thick hollow cylinders or spheres under axisymmetric ... should read: Thick hollow cylinders under axisymmetric ....
16. Page 254, line 1, The reason is that Michell conditions are automatically satisfied for symmetric loadings should read: The reason is that it is known the Michell conditions are not automatically satisfied for symmetric thermal loading in thick cylinders and thermal stresses must be calculated for this type of temperature distribution
17. Page 257, Eq. (6.2-23), $\sigma_{zz} = r(\sigma_{rr} - \sigma_{\phi\phi}) + E\alpha(\bar{\theta} - \theta)$ should read: $\sigma_{zz} = \nu(\sigma_{rr} - \sigma_{\phi\phi}) + E\alpha(\bar{\theta} - \theta)$
18. Page 265, Eq. (6.6-1), part of the equation $+\sum_{n=0}^{\infty} F_n(r)\cos(n\phi)$ should read: $+\sum_{n=1}^{\infty} F_n(r)\cos(n\phi)$
19. Page 272, line under Eq. (6.7-10), From Eqs. (6.7-6), (6.7-9) and (6.7-10) should read: From Eqs. (6.7-7), (6.7-9) and (6.7-10)
20. Page 280, under Eq. (6.8-8), change Eq. (6.7-7) to (6.8-7)
21. Page 285, line 4, general from should read: general form
22. Page 291, Eq. (6.9-12), after Eq. (6.9-12) description is needed. This description is: "where p_1, p_2, q_1, are q_2 are the proper values of p and q in Eqs. (6.9-10). The values of p and q in the expression for e_7 to e_10 are replaced with p_1 and q_1".
23. Page 293, Eq. (6.9-19) $\frac{e_7}{8}$ should read: $\frac{e_7}{e_8}$
24. Page 299, line 5 under Eq. (6.10-20), ... are obtained from Eq. (6.10-21) should read: .... are obtained from Eq. (6.10-20)
25. Page 305, line under Eq. (6.11-17), Equation (6.11-19) should read: Equation (6.11-17)
26. Page 361, Eq. (8.5-2), the first parenthesis "$\left(\frac{\partial^2}{\partial \bar{x}^2} - \frac{\partial^2}{\partial t^2}\right)$" should be changed to "$\left(\frac{\partial^2}{\partial \bar{x}^2} - \frac{\partial^2}{\partial \bar{t}^2}\right)$".
27. Page 391, Caption of Fig. 8.9-2, change $x = 1$ to $X = 1$.
28. Page 416, two lines below Eq. (9.2-18), change Eq. (9.2-1) to (9.2-2).
29. Page 438, first line from top, change Eqs. (9.4-34) to (9.4-36) to (9.4-31) to (9.4-36).
30. Page 448 and 449, in the captions of Figs. 9.5-3 to 9.5-6, change "disk" to "sphere".
31. Page 471, three lines below Eq. (9.8-25), change "The solution of Eq. (9.8-22)" to "The solution of Eq. (9.8-25)".
32. Page 472, Figures (9.8-2) and (9.8-3), the reference number [4] must be changed to [51].
33. Page 478, eight line from top, change "… a single heat excitation is used to drive …" should be changed to "… a single heat excitation is used to derive …".

Bibliography

1. R.B. Hetnarski, J. Ignaczak, *The Mathematical Theory of Elasticity*, 2nd edn. (CRC Press, Boca Raton, 2011)
2. N. Noda, R.B. Hetnarski, Y. Tanigawa, *Thermal Stresses*, 2nd edn. (Taylor & Francis, New York, 2003)
3. R. B. Hetnarski, M. R. Eslami, *Thermal Stresses—Advanced Theory and Applications* (Springer, New York, 2009)
4. B.A. Boley, J.H. Weiner, *Theory of Thermal Stresses* (Wiley, New York, 1960)
5. D. Burgreen, *Elements of Thermal Stress Analysis* (C. P. Press, Jamaica, 1971)
6. R.S. Cukić, *Solution of Thermoelasticity Problems by Integral Transforms (in Serbian)* (Naucna Knjiga, Beograd, 1979)
7. B.E. Gatewood, *Thermal Stresses* (McGraw Hill, New York, 1957)
8. A.E. Green, W. Zerna, *Theoretical Elasticity* (Oxford University Press, Oxford, 1968)
9. M.E. Gurtin, *The Linear Theory of Elasticity, Encyclopedia of Physics*, Chief editor S. Flügge, vol. VIa/2, ed. by C. Truesdell (Springer, Berlin, 1972), pp. 1–295
10. J. Ignaczak, M. Ostoja-Starzewski, *Thermoelasticity with Finite Wave Speeds* (Oxford University Press, Oxford, 2010)
11. D.J. Johns, *Thermal Stress Analysis* (Pergamon Press, London, 1965)
12. W. Krzyś, M. Życzkowski, *Elasticity and Plasticity—Collection of Problems and Examples (in Polish)* (PWN-Polish Scientific Publishers, Warszawa, 1962)
13. A. Lanin, I. Fedik, *Thermal Stress Resistance of Materials (translation from Russian)* (Springer, New York, 2008)
14. R. W. Little, *Elasticity* (Prentice-Hall, Englewood Cliffs, 1973)
15. E. Melan, H. Parkus, *Wärmespannungen infolge stationärer Temperaturfelder* (Springer, Wien, 1953)
16. W. Nowacki, *Thermoelasticity*, 2nd edn. (Pergamon Press and PWN-Polish Scientific Publishers, Warszawa, 1986)
17. W. Nowacki, *Dynamic Problems of Thermoelasticity (translation from Polish)*, ed. by P.H. Francis, R.B. Hetnarski (Noordhoff Int. Publ., Leyden, 1975)
18. J.L. Nowiński, *Theory of Thermoelasticity with Applications* (Sijthoff and Noordhoff, Alphen aan den Rijn, 1978)
19. H. Parkus, *Instationäre Wärmespannungen* (Springer, Wien, 1959)
20. B. Straughan, *Heat Waves* (Springer, New York, 2011)
21. Y. Takeuti, N. Noda, *Analyses of Thermal Stresses (in Japanese)* (Nisshin, Tokyo, 1981)
22. R.B. Hetnarski (ed.), *Thermal Stresses I* (Elsevier, Amsterdam, 1986)
23. R.B. Hetnarski (ed.), *Thermal Stresses II* (Elsevier, Amsterdam, 1987)
24. R.B. Hetnarski (ed.), *Thermal Stresses III* (Elsevier, Amsterdam, 1989)
25. R.B. Hetnarski (ed.), *Thermal Stresses IV* (Elsevier, Amsterdam, 1996)
26. R.B. Hetnarski (ed.), *Thermal Stresses V* (Lastran Corp., Rochester, 1999)

M. Reza Eslami et al., *Theory of Elasticity and Thermal Stresses*, Solid Mechanics and Its Applications 197, DOI: 10.1007/978-94-007-6356-2,

Index

M. Reza Eslami et al., *Theory of Elasticity and Thermal Stresses,* Solid Mechanics and Its Applications 197, DOI: 10.1007/978-94-007-6356-2,

MIX
Papier aus verantwortungsvollen Quellen
Paper from responsible sources
FSC® C105338

If you have any concerns about our products, you can contact us on
ProductSafety@springernature.com

In case Publisher is established outside the EU, the EU authorized representative is:
Springer Nature Customer Service Center GmbH
Europaplatz 3, 69115 Heidelberg, Germany

Printed by Libri Plureos GmbH
in Hamburg, Germany